T0213376

METHODOLOGICAL AND TECHNOLOGICAL ISSUES IN TECHNOLOGY TRANSFER

Effective global response to climate change requires the development and transfer of environmentally sound technologies between and within countries, both for adapting to climate change as well as for mitigating the effects of greenhouse gas emissions. This Special Report of the Intergovernmental Panel on Climate Change (IPCC) provides a state-of-the-art overview of how to achieve and enhance this transfer. 185 eminent experts from around the world provide accurate, unbiased, policy-relevant information on technology transfer, such as capacity building, the promotion of an enabling environment, and mechanisms for technology transfer from developed to developing countries. The transfer of both hardware as well as software (knowledge) is discussed, and the importance of the involvement of various stakeholders is emphasised. The report was written in response to the request of the Subsidiary Body on Scientific and Technological Advice (SBSTA) of the United Nations Framework Convention on Climate Change.

This IPCC Special Report is the most comprehensive assessment available on technology transfer, and provides information to serve industry, policy-makers, environmental organisations, and researchers in global change, technology, engineering and economics.

Bert Metz is Co-Chair of IPCC Working Group III and Head of the Global and European Environmental Assessment Division of the National Institute of Public Health and Environment, Bilthoven, The Netherlands.

Ogunlade R. Davidson is Co-Chair of IPCC Working Group III and Dean of the Faculty of Engineering and Professor of Mechanical Engineering at the University of Sierra Leone.

Jan-Willem Martens is a member of the IPCC Working Group III Technical Support Unit and works as an energy economist at the Netherlands Energy Research Foundation, Petten/Amsterdam, The Netherlands.

Sascha N. M. van Rooijen is a member of the IPCC Working Group III Technical Support Unit and works as an environmental economist at the Netherlands Energy Research Foundation, Petten/Amsterdam, The Netherlands.

Laura Van Wie McGrory was a member of the IPCC Working Group II Technical Support Unit and currently works as a scientific engineering associate at the Lawrence Berkeley National Laboratory in Washington, DC, USA.

Methodological and Technological Issues in Technology Transfer

Edited by

Bert Metz

Ogunlade R. Davidson

Jan-Willem Martens

Sascha N.M. van Rooijen

Laura Van Wie McGrory

A Special Report of IPCC Working Group III

Published for the Intergovernmental Panel on Climate Change

CAMBRIDGE
UNIVERSITY PRESS

CAMBRIDGE
UNIVERSITY PRESS

32 Avenue of the Americas, New York NY 10013-2473, USA

Cambridge University Press is part of the University of Cambridge.

It furthers the University's mission by disseminating knowledge in the pursuit of education, learning and research at the highest international levels of excellence.

www.cambridge.org
Information on this title: www.cambridge.org/9780521804943

First published 2000

A catalogue record for this publication is available from the British Library

ISBN 978-0-521-80082-2 Hardback
ISBN 978-0-521-80494-3 Paperback

This report is dedicated to

John Turkson, Ghana
Lead Author, Chapter 4 and 5

Dr John Turkson died at the age of 46 years on a plane crash while he was en route to establish a CDM Pilot project in Uganda. He was a Senior Energy Economist at the UNEP Collaborating Centre for Energy and Environment at RISØ National Laboratory in Denmark. Before joining RISØ, he was a lecturer at the University of Science and Technology, Kumasi, Ghana. John was one of the few well-known energy economists on the African continent who published extensively in international journals, conference proceedings and books in the international fora on energy economics and climate change. He has initiated several regional projects in energy and climate change in Africa because of his belief that the climate change debate provides an opportunity for transforming African economies to more sustainable development paths.

His keen sense of duty earned him respect of colleagues he had worked with and helped him to build a network of eminent energy specialists and economists not only from Ghana, but from all over Africa and beyond. He was married to Gifty who no doubt had to bear the intensity with which John normally took his work. He will be missed, but his contribution to IPCC will always be remembered by his colleagues and friends.

Contents

Foreword

The Intergovernmental Panel on Climate Change (IPCC) was jointly established by the World Meteorological Organization (WMO) and the United Nations Environment Programme (UNEP) to asses available information on the science, impacts and the economics of climate change and of mitigation options to address it. It provides also, on request, scientific/technical/socio-economic advice to the Conference of the Parties (COP) to the United Nations Framework Convention on Climate Change (UNFCCC). Since its inception the IPCC has produced a series of Assessment Reports, Special Reports, Technical Papers, methodologies and other products which have become standard works of reference, widely used by policymakers, scientists and other experts.

This Special Report has been prepared by IPCC Working Group III in response to a request by the Subsidiary Body for Scientific and Technological Advice (SBSTA) to the UNFCC. Innovation and enhanced efforts to transfer environmentally sound technology to limit greenhouse gas emissions and to adapt to climate change will be required to meet the objective of the Climate Convention and to reduce vulnerability to climate change impacts. The report addresses the technology transfer problem in the context of climate change while emphasizing the sustainable development perspective. Technology transfer is defined as the broad set of processes covering the flows of know-how, experience and equipment and is the result of many day-to-day decisions of the different stakeholders involved. A number of social, economic, political, legal, and technological factors influence the flow and quality of technology transfer. Essential elements of successful transfers include consumer and business awareness, access to information, availability of a wide range of technical, business, management and regulatory skills locally, and sound economic policy and regulatory frameworks. Technology transfers that meet local needs and priorities are more likely to be successful. But there is no pre-set answer to enhancing technology transfer. Interactions and barriers vary according to sector, type of technology and country, and recent trends in international financial flows that drive technology transfer are altering the relative capacities and roles of different stakeholders. Policy actions therefore need to be tailored to the specific context and interests. The report elaborates on what governments can do to facilitate and enhance the transfer of Environmentally Sound Technologies, but it also aims at reaching decision makers in the private sector, lending institutions, multilateral agencies, non-governmental organizations, and the interested public.

As usual in the IPCC, success in producing this report has depended first and foremost on the enthusiasm and cooperation of scientists and other experts worldwide. These individuals have devoted enormous time and effort to producing this report and we are extremely

We would like to express our sincere thanks to:
- Robert T. Watson, the Chairman of the IPCC;
- The Co-chairs of Working Group III Bert Metz and Ogunlade Davidson;
- The Section Coordinators Kilaparti Ramakrishna, Jayant Sathaye, Youba Sokona, William Chandler, Stephen O. Andersen and Ajay Mathur;
- The staff of the Working Group III and II Technical Support Units, including Rob Swart, Ms Sascha van Rooijen, Jan-Willem Martens, Ms Laura VanWie McGrory, Ms Flo Ormond and Marlies Kamp;
- N. Sundararaman, Secretary of the IPCC, Renate Christ, Deputy Secretary of the IPCC and the staff of the IPCC Secretariat Rudie Bourgeois, Chantal Ettori and Annie Courtin.

G.O.P. Obasi

Secretary-General
World Meteorological
Organisation

K. Töpfer

Executive Director
United Nations Enviroment Programme
and Director-General
United Nations Office
in Nairobi

Preface

The Intergovernmental Panel on Climate Change (IPCC) was established jointly by the World Meteorological Organisation (WMO) and the United Nations Environment Programme (UNEP) to assess periodically the science, impacts and socioeconomics of climate change and of adaptation and mitigation options. The IPCC provides, on request, scientific and technical advice to the Conference of Parties (CoP) to the United Nations Framework Convention on Climate Change (UNFCCC) and its subsidiary bodies. The CoP at its first session in Berlin 1995 requested the IPCC to include in its assessments an elaboration of the terms under which transfer of environmentally sound technologies and know-how could take place.

As a further elaboration of the COP-1 request, IPCC was requested by the Subsidiary Body for Scientific and Technological Advice (SBSTA) to prepare a Technical Paper on methodological and technological aspects of technology transfer (see FCCC/SBSTA/ 1996/8, Annex III). The objective of the paper would be to synthesise information from the Second Assessment Report on experiences with:
  (i)   types of transfer, technology evaluation, and options;
  (ii)   sectors targeted;
  (iii) role of participants (for example governments, private sector, IGOs, NGOs);
  (iv)   approaches to promote co-operation;
  (v)   issues related to capacity building.

According to IPCC procedures, Technical Papers should be based on material already present in the IPCC reports. However, the Second Assessment Report did not contain sufficient information to prepare a Technical Paper that would address the questions raised. Therefore, the IPCC decided at its Twelfth Plenary Session in Mexico City to prepare a Special Report on Methodological and Technological Issues in Technology Transfer.

In order to provide structure in the wide variety of subjects, the writing team chose to divide the Report in three sections:

Section I provides a framework for analysis of the complex and multi-facetted nature of the technology transfer process, emphasising the sustainable development perspective. It examines broad trends of technology transfer in recent years, explores the international political context, discusses policy tools for overcoming key barriers and creating enabling environments and provides an overview of financing and partnerships.

Section II provides a sectoral perspective on the transfer of adaptation and mitigation technologies. Every chapter discusses the prevalent climate mitigation and adaptation technologies, the magnitude of current and future transfers, technology transfer issues within and between countries and the lessons learned in that particular sector.

Section III includes a wide variety of case studies to illustrate the issues discussed in section I and II and demonstrates the distinctive problems and special opportunities that stakeholders are likely to encounter in dealing with technology transfer.

In accordance with the wide scope of technology transfer, the team of authors put together to prepare the report represented a multitude of disciplines and a broad geographical distribution. The writing team consisted of 8 Section Coordinators, 24 Coordinating Lead Authors, 120 Lead Authors and 53 Contributing Authors. In accordance with the revised IPCC Procedures, 20 Review Editors were appointed to oversee the review process.

Over 180 Expert and Government Reviewers submitted valuable suggestions for improvement during the review process. All the comments have been afforded appropriate consideration by the writing team and genuine scientific controversies have been reflected adequately in the text of the report as confirmed in the Review Editors report. The revised document was submitted to the Working Group III Plenary in Kathmandu, Nepal, that took place from 8 to 10 March, 2000. There, the Summary for Policymakers was approved in detail and the underlying report accepted. The IPCC Plenary finally accepted the report and the Summary for Policy makers during its Sixteenth Session that took place in Montreal, Canada, from 1-8 May 2000.

We wish to commend all Section Coordinators, Coordinating Lead Authors, Lead Authors, Contributing Authors and all Review Editors for all the effort they put into the compilation of this Report and deeply appreciate the commitment they have shown.

It is with profound sadness and regret that we have to convey the message that three of our dear colleagues and team members passed away during the writing process of this Report: Katsuo Seiki (August 1998), David Hall (August 1999) and John Turkson (January 2000). They were highly appreciated members of the team, John Turkson as Lead Author of Chapter 4 and 5 and David Hall as review editor of Chapter 12. Katsuo Seiki was envisaged CLA of then Chapter 18 and showed as a Vice-Chair of IPCC much interest in the issue of technology transfer. We will remember their excellent work and enjoyable personalities.

We are grateful to:
  • The Tata Energy Research Institute in New Delhi, India, and in particular Dr. Pachauri, the Director and vice chair of IPCC for hosting the first Lead Authors meeting;
  • The United Kingdom Climate Impact Programme of the Environmental Change Unit at the University of Oxford, United Kingdom, for hosting the second Lead Authors meeting with the support of the United Kingdom Department of Environment, Transport and the Regions;
  • The Department of Hydrology and Meteorology of the

Government of Nepal, for hosting the Fifth Plenary of the IPCC Working Group III from 8 to 10 March, 2000, where the Summary for Policymakers was approved line by line and the underlying Report accepted.

We would finally like to express our gratitude to the three successive Report Co-ordinators at the Technical Support Units: Laura van Wie-McGrory (TSU WG II), and Sascha van Rooijen and Jan-Willem Martens (TSU WG III) for their never ending dedication to get the report in its current shape. We thank Flo Ormond of the Technical Support Unit of Working Group II and Marlies Kamp of the Technical Support Unit of Working Group III for their invaluable support throughout the preparation of the Report. Also other members of the Technical Support Units of Working Group II and III have provided much appreciated assistance, including Rob Swart, Anita Meier, Jiahua Pan, Remko Ybema and Dave Dokken. Dr. N. Sundararaman, Secretary of the IPCC, and the staff of the IPCC Secretariat in Geneva ensured the essential services of providing government liaison and travel of experts from the developing and transitional economy countries as well as making the arrangements with the Government of Nepal. We are also grateful to Renate Christ, Deputy Secretary of the IPCC, for her substantive inputs on various occasions during the preparation of the Report.

We would like to encourage the readers, which include policymakers, scientists, managers, professionals and academics, to evaluate the contents of this work, adjust it to their own conditions and ensure a rapid and widespread replication of its lessons across the world. We sincerely hope that this Report will thus contribute to the widespread use of environmentally sound technologies and assist in achieving the objectives of the Climate Convention.

Ogunlade Davidson, Co-chair of Working Group III
Bert Metz, Co-chair of Working Group III

SUMMARY FOR POLICYMAKERS

METHODOLOGICAL AND TECHNOLOGICAL ISSUES IN TECHNOLOGY TRANSFER

*A Special Report of Working Group III
of the Intergovernmental Panel on Climate Change*

Based on a draft prepared by:

*Stephen O. Andersen (USA), William Chandler (USA), Renate Christ (Austria), Ogunlade Davidson (Sierra Leone),
Sukumar Devotta (India), Michael Grubb (UK), Joyeeta Gupta (The Netherlands), Thomas C. Heller (USA)
Maithili Iyer (India), Daniel M. Kammen (USA), Richard J.T. Klein (The Netherlands/Germany), Dina Kruger (USA),
Ritu Kumar (India), Mark Levine (USA), Lin Erda (China), Patricia Iturregui (Peru), Merylyn McKenzie Hedger (UK),
Anthony McMichael (UK), Mark Mansley (UK), Jan-Willem Martens (The Netherlands), Eric Martinot (USA),
Ajay Mathur (India), Bert Metz (The Netherlands), John Millhone (USA), Jose Roberto Moreira (Brazil),
Tongroj Onchan (Thailand), Mark Radka (USA), Kilaparti Ramakrishna (India), N.H. Ravindranath (India),
Sascha van Rooijen (The Netherlands), Jayant Sathaye (USA), Youba Sokona (Mali), Sergio C. Trindade (Brazil),
David Wallace (UK), Ernst Worrell (The Netherlands)*

1. Introduction

Background

Article 4.5 of the United Nations Framework Convention on Climate Change (UNFCCC) states that developed country Parties and other developed Parties included in Annex II "shall take all practicable steps to promote, facilitate and finance, as appropriate, the transfer of, or access to, environmentally sound technologies and know-how to other Parties, particularly developing country Parties, to enable them to implement the provisions of the Convention." The Subsidiary Body for Scientific and Technological Advice (SBSTA) identified at its first session a list of areas in which it could draw upon the assistance of the IPCC. This Special Report was prepared in response to this request. It addresses the technology transfer problem in the context of all relevant UNFCCC provisions, including decisions of the Conference of Parties (CoP), and Chapter 34 in Agenda 21. It attempts to respond to recent development in the UNFCCC debate on technology transfer, by providing available scientific and technical information to enable Parties to address issues and questions identified in Decision 4/CP.4 adopted by CoP4.

The role of technology transfer in addressing climate change

Achieving the ultimate objective of the UNFCCC, as formulated in Article 2[1], will require technological innovation and the rapid and widespread transfer and implementation of technologies, including know-how for mitigation of greenhouse gas emissions. Transfer of technology for adaptation to climate change is also an important element of reducing vulnerability to climate change.

This technological innovation must occur fast enough and continue over a period of time to allow greenhouse gas concentrations to stabilise and reduce vulnerability to climate change. Technology for mitigating and adapting to climate change should be environmentally sound technology (EST) and should support sustainable development.

Sustainable development on a global scale will require radical technological and related changes in both developed and developing countries. Economic development is most rapid in developing countries, but it will not be sustainable if these countries follow the historic greenhouse gas emission trends of developed countries. Development with modern knowledge offers many opportunities to avoid past unsustainable practices and move more rapidly towards better technologies, techniques and asso-

ciated institutions. The literature indicates that to achieve this developing countries require assistance with developing human capacity (knowledge, techniques and management skills), developing appropriate institutions and networks, and with acquiring and adapting specific hardware. Technology transfer, in particular from developed countries to developing countries, must therefore operate on a broad front covering these software and hardware challenges, and ideally within a framework of helping to find new sustainable paths for economies as a whole. There is, however, no simple definition of a "sustainable development agenda" for developing countries. Sustainable development is a context driven concept and each society may define it differently, based on Agenda 21. Technologies that may be suitable in each of such contexts may differ considerably. This makes it important to ensure that transferred technologies meet local needs and priorities, thus increasing the likelihood that they will be successful, and that there is an appropriate enabling environment for promoting environmentally sound technologies (ESTs).

The Report analyses the special challenges of transferring ESTs to address climate change in the context of sustainable development. The literature provides ample evidence of the many problems in current processes of technology transfer which makes it very unlikely to meet this challenge without additional actions for the transfer of mitigation and adaptation technologies.

What do we mean by technology transfer?

The Report defines the term "technology transfer" as a broad set of processes covering the flows of know-how, experience and equipment for mitigating and adapting to climate change amongst different stakeholders such as governments, private sector entities, financial institutions, NGOs and research/education institutions. Therefore, the treatment of technology transfer in this Report is much broader than that in the UNFCCC or of any particular Article of that Convention. The broad and inclusive term "transfer" encompasses diffusion of technologies and technology cooperation across and within countries. It covers technology transfer processes between developed countries, developing countries and countries with economies in transition, amongst developed countries, amongst developing countries and amongst countries with economies in transition. It comprises the process of learning to understand, utilise and replicate[2] the technology, including the capacity to choose it and adapt it to local conditions and integrate it with indigenous technologies.

The Report generally makes a distinction between developed and developing countries. Although economies in transition are included as developed countries under the UNFCCC, they may have characteristics in common with both developed and developing countries.

---

[1] "The ultimate objective of this Convention and any related legal instruments that the Conference of Parties may adopt is to achieve, in accordance with the relevant provisions of the Convention, stabilisation of greenhouse gas concentrations in the atmosphere at such a level that would prevent dangerous interference with the climate system. Such a level should be achieved within a timeframe sufficient to allow ecosystems to adapt naturally to climate change, to ensure that food production is not threatened and to enable economic development to proceed in a sustainable manner."

[2] The final stage of the five basic stages of technology transfer (assessment, agreement, implementation, evaluation and adjustment, replication) as defined in the Report as a combination of actions that lead to the deployment of a given technology, once transferred, to meet a new demand elsewhere.

Trends of technology transfer

It is difficult to quantify how much climate-relevant hardware is successfully transferred annually. When software elements such as education, training and other capacity building activities are included, the task of quantification is further complicated. Financial flows, often used as proxies, allow only a limited comparison of technology transfer trends over time. The 1990s have seen broad changes in the types and magnitudes of the international financial flows that drive technology transfer.

Official Development Assistance (ODA) experienced a downward trend in the period of 1993 to 1997, both in absolute terms and as a percentage of funding for projects with significant impact on technology flows to developing countries. However, in 1998 there was an increase in ODA funding. ODA is still important for those parts of the world and sectors where private sector flows are comparatively low, like agriculture, forestry, human health and coastal zone management. Moreover, it can support the creation of enabling conditions, which may leverage larger flows of private finance into ESTs in the context of overall sustainable development goals in the recipient countries.

Levels of foreign direct investment (FDI), commercial lending, and equity investment all increased greatly in recent years. These are the dominant means by which the private sector makes technology-based investments in developing countries and economies in transition, often in the industry, energy supply and transportation sectors. However, private sector investment in the form of FDI in developing countries has favoured East and South East Asia, and Latin America.

These trends are altering the relative capacities and roles of different stakeholders. The importance of the private sector has increased substantially. However, there is a definite role for governments both in providing an enabling environment for the technology transfer process as well as participating directly in it. Many NGOs support technology transfer activities.

Stakeholders, pathways, stages and barriers

Technology transfer results from actions taken by various stakeholders. Key stakeholders include developers; owners; suppliers, buyers, recipients and users of technology (such as private firms, state enterprises, and individual consumers); financiers and donors; governments; international institutions; NGOs and community groups. Some technology is transferred directly between government agencies or wholly within vertically integrated firms, but increasingly technology flows depend also on the coordination of multiple organisations such as networks of information service providers, business consultants and financial firms. Although stakeholders play different roles, there is a need for partnerships among stakeholders to create successful transfers. Governments can facilitate such partnerships.

There is a large number of pathways through which stakeholders can interact to transfer technologies. They vary depending on sectors, country circumstances and type of technology. Pathways may be different for "close to market" technologies and for technology innovations still in the development phase. Common pathways include government assistance programmes, direct purchases, licensing, foreign direct investment, joint ventures, cooperative research arrangements and co-production agreements, education and training, and government direct investment.

While technology transfer processes can be complex and intertwined, certain stages can be identified. These may include the identification of needs, choice of technology, assessment of conditions of transfer, agreement and implementation[2]. Evaluation and adjustment to local conditions, and replication[2] are other important stages.

Barriers to the transfer of ESTs may arise at each stage of the process. These vary according to the specific context, for example from sector to sector, and can manifest themselves differently in developed countries, developing countries and countries with economies in transition. These barriers range from lack of information; insufficient human capabilities; political and economic barriers such as lack of capital, high transaction costs, lack of full cost pricing, and trade and policy barriers; lack of understanding of local needs; business limitations, such as risk aversion in financial institutions; and institutional limitations such as insufficient legal protection, and inadequate environmental codes and standards.[3]

There is no pre-set answer to enhancing technology transfer. The identification, analysis and prioritisation of barriers should be country based. It is important to tailor action to the specific barriers, interests and influences of different stakeholders in order to develop effective policy tools.

2.   Increase the Flow; Improve the Quality

Government actions can transform the conditions under which technology transfer takes place. The spread of proven ESTs that would diffuse through commercial transactions may be limited because of the barriers listed above.

The three major dimensions of making technology transfer more effective are capacity building, an enabling environment and mechanisms for technology transfer, all of which are discussed in more detail in the subsections below.

Building capacity

Capacity building is required at all stages in the process of technology transfer. Social structures and personal values evolve with a society's physical infrastructure, institutions, and the technologies embodied within them. New technological trajectories for an economy therefore imply new social challenges. This requires a capacity of people and organisations to continuously

---

[3.] See Technical Summary and Chapters 3, 4 and 5 of the main report.

adapt to new circumstances and to acquire new skills. This applies both for mitigation and adaptation technologies. Comparatively little consideration has been given in a systematic way to what capacity building is required for adaptation to climate change.

Human capacity

Adequate human capacity is essential at every stage of every transfer process. The transfer of many ESTs demands a wide range of technical, business, management and regulatory skills. The availability of these skills locally can enhance the flow of international capital, helping to promote technology transfer.

Developed country governments, in particular, can ensure that training and capacity building programmes they sponsor consider the full range of information, financial, legal, and business consulting and engineering services that technology transfer requires, as well as the local conditions under which these may be provided. This requires cooperation with local governments, institutions and stakeholders, commercial organisations and consumers/ users.

Developing country governments can build local capacities to gear them for technology transfer. Training and human resource development have been popular development assistance activities. Future approaches can be more effective by better stressing the integration of a total package of technology transfer, focusing less exclusively on developing technical skills and more on creating improved and accessible competence in associated services, organisational know-how, and regulatory management.

Organisational capacity

It is important to recognise the need for participatory approaches and to strengthen the networks in which diverse organisations contribute to technology transfer. In technology intensive economies, technology increasingly flows through private networks of information and assessment services, management consultants, financial firms, lawyers and accountants, and technical specialist groups. Local government agencies, consumer groups, industry associations and NGOs may ensure that technology meets local needs and demand. This organisational infrastructure can help reduce but will not eliminate risks arising from deficiencies in legal systems. Although many actions that facilitate the growth of such networks are already underway, initiatives of particular importance to EST transfer include:

- Expansion of opportunities to develop firms for management consulting, accounting, energy service, law, investment and product rating, trade, publishing and provision for communication, access to and transfer of information, such as Internet services;
- Encouragement of industry associations, professional associations and user/consumer organisations;
- Participatory approaches to enable private actors, public agencies, NGOs and grassroots organisations to engage at all levels of environmental policy-making and project formulation;

- Where appropriate, decentralisation of governmental decision-making and authority, in relation to technology transfer, to effectively meet community needs.

Information assessment and monitoring capacity

Information access and assessment are essential to technology transfer. However, focussing too narrowly on information barriers while ignoring the later stages of the transfer process can be less productive. The roles of governments and private actors in technology assessment are changing. Private information networks are proliferating through specialised consulting and evaluation services and over the Internet. Increasing FDI also demonstrates that many ESTs can diffuse rapidly without direct government action. Governments in developing countries, developed countries, and countries with economies in transition may wish to consider:

- Developing improved indicators and collecting data on availability, quality and flows of ESTs to improve monitoring of implementation;
- Developing technology performance benchmarks for ESTs to indicate the potential for technological improvements;
- Improving information systems and linking them to international or regional networks, through well-defined clearing houses (such as energy efficiency and renewable energy centres), information speciality firms, trade publications, electronic media, or NGOs and community groups.

Enabling environment and extra effort to enhance technology transfer

Governments, through inter alia sound economic policy and regulatory frameworks, transparency and political stability, can create an enabling environment for private and public sector technology transfers. Although many ESTs are in common use and could be diffused through commercial channels, their spread is hampered by risks such as those arising from weak legal protection and inadequate regulation in developed countries, developing countries and countries with economies in transition. But many technologies that can mitigate emissions or contribute to adaptation to climate change are not as yet commercially viable. Beyond an enabling environment, it will take extra efforts to develop and enhance the transfer of those potentially viable ESTs. The following actions could increase the flow of ESTs and improve its quality.

All governments may therefore wish to consider:

- Enacting measures, including well-enforced regulations, taxes, codes, standards and removal of subsidies, to internalise the externalities to capture the environmental and social costs, and assist the replication of ESTs;
- Reforming legal systems. Uncertain, slow and expensive enforcement of contracts by national courts or international arbitration and insecure property rights can discourage investment. Reforming administrative law to

reduce regulatory risk and ensuring that public regulation is accessible to stakeholders and subject to independent review;

- Protecting intellectual property rights and licenses in such a way that innovation is fostered, while avoiding misapplication, which may impede diffusion of ESTs;
- Encouraging financial reforms, competitive and open national capital markets, and international capital flows that support foreign direct investment. Governments can expand financial lending for ESTs through regulation that allows the design of specialised credit instruments, capital pools, and energy service companies;
- Simplifying and making transparent programme and project approval procedures and public procurement requirements;
- Promoting competitive and open markets for ESTs;
- Stimulating national markets for ESTs to facilitate economy of scale and other cost reducing practices;
- Encouraging multinational companies to show leadership and use the same standards for environmental performance wherever they operate;
- Creating awareness about products, processes and services that use ESTs through means such as eco-labelling, product standards, industry codes, and community education;
- Using legislation, enhancing transparency, and increasing participation by civil society to reduce corruption in conformity with international conventions.

Governments of developed countries and countries with economies in transitions may wish to consider:

- Stimulating fair competition in EST markets by discouraging restrictive business practices;
- Reforming export credit, political risk insurance and other subsidies for the export of products or production processes to encourage foreign direct investment in ESTs;
- Developing environmental guidelines for export credit agencies to avoid a bias against, and promote the transfer of ESTs, and discourage the transfer of obsolete technologies;
- Reducing the use, as trade policy measures applied to ESTs, of tied aid;
- Developing modalities and/or policies to improve the transfer of ESTs that are in the public domain;
- Increasing public funding for R&D in cleaner technologies to reflect the high rate of social return, and wherever possible, enhancing the flows of ESTs arising from their publicly funded R&D programmes by entering into cooperation with developing countries in R&D partnerships and international research institutions;
- Increasing flows of national and multilateral assistance, including funding, especially in programmes targeted to environmental technologies, including patent licensing of ESTs where appropriate. Attention should also be paid to supporting pathways for transfer of ESTs among developing countries.

Governments of developing countries may wish to consider:

- Ensuring assessment of local technology needs and social impact of technologies so that transfer of and investment in ESTs meet local demands;
- Expanding R&D programmes, aiming at the development of ESTs particularly appropriate in developing countries and adjustment to local conditions; promoting complementary policies for ESTs;
- Improving pathways for technology transfer among developing countries through information regarding the performance of ESTs in developing countries, joint R&D, demonstration programmes, and opening markets for ESTs;
- Developing physical and communications infrastructure to support private investments in ESTs and the operations of intermediary organisations providing information services;
- Improving the identification of specific barriers, needs and steps towards introduction of ESTs by consulting with priority stakeholders;
- Continuing to improve macro-economic stability to facilitate ESTs to be transferred.

Mechanisms for technology transfer
National Systems of Innovation
The literature shows that National Systems of Innovation (NSIs) which integrate the elements of capacity building, access to information and an enabling environment into comprehensive approaches to EST transfer add up to more than the individual components and support the creation of an innovation culture. Subsystems and the quality of interconnections within them can successfully influence technology transfer. The concept of NSIs can be enhanced through partnerships with international consortia. Partnerships would be system oriented, encompass all stages of the transfer process, and ensure the participation of private and public stakeholders, including business, legal, financial and other service providers from developed and developing countries.

NSI activities may include:
- Targeted capacity building, information access, and training for public and private stakeholders and support for project preparation;
- Strengthening scientific and technical educational institutions in the context of technology needs;
- Collection and assessment of specific technical, commercial, financial and legal information;
- Identification and development of solutions to technical, financial, legal, policy and other barriers to wide deployment of ESTs;
- Technology assessment, promotion of prototypes, demonstration projects and extension services through linkages between manufacturers, producers and end users;
- Innovative financial mechanisms such as public/private sector partnerships and specialised credit facilities;
- Local and regional partnerships between different stakeholders for the transfer, evaluation and adjustment to local conditions of ESTs;

- Market intermediary organisations such as Energy Service Companies.

Official Development Assistance (ODA)

Official Development Assistance (ODA) is still significant for developing countries and successful transfers of ESTs. ODA can also assist the improvement of policy frameworks and take on long-term capacity building. There is increasing recognition that ODA can best be focused on mobilising and multiplying additional financial resources.

Global Environment Facility

The Global Environment Facility, an operating entity of the UNFCCC Financial Mechanism, is a key multilateral institution for transfers of ESTs. Compared to the magnitude of the technology transfer challenge, these efforts are of modest scale, even when added to the contributions from bilateral development assistance. The GEF currently targets incremental, one-time investments in mitigation projects that test and demonstrate a variety of financing and institutional models for promoting technology diffusion, thus contributing to a host country's ability to understand, absorb and diffuse technologies. GEF also supports capacity building projects for adaptation consistent with limitations currently imposed by Convention guidance. Continued effectiveness of GEF project funding for technology transfer may depend on factors such as:

- Sustainability of market development and policy impacts achieved through GEF projects;
- Duplication of successful technology transfer models;
- Enhanced links with multilateral-bank and other financing of ESTs;
- Funding for development and licensing of ESTs;
- Coordination with other activities that support national systems of innovation and international technology partnerships;
- Attention to technology transfer among developing countries.

Multilateral Development Banks

Governments may use their leverage to direct the activities of multilateral development banks (MDBs) through their respective Boards and Councils in order to:

- Strengthen MDB programmes to account for the environmental consequences of their lending;
- Develop programmatic approaches to lending that remove institutional barriers and create enabling environments for private technology transfers;
- Encourage MDBs to participate in NSI partnerships.

The Kyoto Protocol Mechanisms and the UNFCCC

The analysis of the literature on the Kyoto Protocol Mechanisms, based on the preliminary stage of development of the rules for these, suggests that if they are implemented, the Mechanisms may have potential to affect the transfer of ESTs.

The extent to which Article 4.5 of the UNFCCC has been implemented is being reviewed by the UNFCCC. Given this evolving process, the IPCC has not been able to assess this matter.

3. Sectoral Actions

The key actions for the transfer of mitigation and adaptation technologies vary across sectors. Governments, private actors and community organisations are all involved in technology transfer in each sector, although their roles and the extent of their involvement differ within and across sectors. It is important to note the special characteristics of adaptation technologies. Adaptation in anticipation of future climate change is faced with uncertainty about location, rate and magnitude of climate change impacts. Adaptation technologies often address site-specific issues and their benefits are primarily local, which could hamper large scale replication. On the other hand, they could reduce vulnerability not only to anticipated impacts of climate change but also to contemporary hazards associated with climate variability.

Central lessons learned through the sectoral studies are: *(1)* networking among stakeholders is essential for effective technology transfer, and *(2)* most effective technology transfers focus on products and techniques with multiple benefits. Actions that have been effective in technology transfer in the sectors evaluated in the Report, are:

Buildings

World-wide, the mix of relevant ESTs will vary, depending upon the climate; rural-urban distribution, and historical context. The effective actions for the transfer of ESTs may include, *(1)* government financing for incentives for the construction of more energy efficient and environmentally-friendly homes, *(2)* building codes and guidelines, and equipment standards developed in consultation with industry to minimise adverse impacts on manufacturers; *(3)* energy and environmental performance labels on consumer products; *(4)* government programmes for more energy efficient and environmentally-friendly buildings, office appliances and other equipment, *(5)* demand-side management programmes to promote energy-efficient lighting and equipment, and *(6)* R&D to develop products in the building sector that meet community priorities.

Transport

Technological options - improved technology design and maintenance, alternative or improved fuels, vehicle use change, and modal shifts - as well as non-technical options, transport demand reduction, and improved management systems can reduce GHG emissions significantly. There are also non-transport options such as urban planning and transport demand substitution, such as telematics and improved telecommunications. Resource availability, technical know-how, and institutional capacity are among the factors that affect the cost and transfer of these options.

Government policies can promote cooperative technology agreements among companies of different countries, joint R&D, joint information networks, improved technical and management skills, and specialized training programmes. Adoption of appropriate standards and regulations can stimulate and facilitate technology transfer within and among countries. Partnership between government and the private sector and among countries can also help promote technology transfer within and among countries.

Industry

New processes, efficient energy and resource use, substitution of materials, changes in design and manufacture of products resulting in less material use, and increased recycling, can substantially reduce GHG emissions. Environmental legislation, regulation and voluntary agreements between government and industry can stimulate the development of efficient technologies and can lead to increased use of ESTs. Public technology assessment capabilities are important to provide information and capabilities to successfully transfer ESTs. Wel-defined clearinghouses can be useful in disseminating information to improve energy efficiency, especially with respect to small and medium-sized enterprises that often do not have the resources to assess technologies. Long term support for capacity building is essential, stressing the need for the cooperation of equipment and software suppliers and users. Experience has shown that investment in developing local capability to undertake adjustment to indigenous conditions is crucial to the success of industrial EST transfer.

Energy Supply

In general, the private sector plays a strong role in the transfer of energy supply technologies based in oil and gas sources and technology transfer mechanisms have been established for some time. Restructuring of the electricity sector world-wide is rapidly changing the direction of investments in the power sector with growing participation of the private sector. At the same time, the transfer of energy supply technologies for some other conventional and renewable sources, which often depend on the government to preserve or increase their presence in the market, is restricted due to institutional and socio-economic barriers. Nevertheless, the role of the government and multilateral banks are important in every sector to foster and ensure conditions for international financing, establishing appropriate regulatory frameworks and create conditions to couple new energy investments, environmentally sound projects and sustainable development. Enabling actions by governments to promote energy options, including renewable resources, that are assisting to mitigate climate change, can be crucial to mobilise private capital for ESTs and raise increased attention to energy efficiency.

Agriculture

Development of appropriate information bases on inter alia improved crop species and varieties, irrigation facilities, different tillage and crop management systems, and livestock manure treatment, including biogas recovery systems, can facilitate and promote the transfer of adaptation and mitigation technologies within and across countries and integration with indigenous solutions. Governments can create incentives for the transfer of ESTs by improving national agricultural information systems to disseminate information on ESTs, and expanding credit and savings schemes to assist farmers to manage the increased variability in their environment. The existing Consultative Group on International Agricultural Research (CGIAR) system may be one possible model for an R&D network among countries to build such an information base. Capacities to deal with climate change technologies and national agricultural research systems including those that investigate carbon storage, and early warning systems, are important elements. Efforts by developed countries and multilateral agencies can be improved to enhance this R&D system.

Forestry

Government, community, and international organisations, including conservation organisations, have dominated technology transfer in the forestry sector. More recently, private establishments have been making inroads. Transfer of practices such as sustainable forest management (including reduced-impact logging, certification techniques and silvicultural practices), recycling, bioenergy technologies and agroforestry can contribute to the mitigation of carbon dioxide emissions. Establishing clear property rights, participatory forest management, use of financial incentives and disincentives, optimal use of regulations, and strengthening of monitoring and evaluating institutions are government actions that can promote their transfer.

Waste Management

Mitigation technologies are available and can be readily deployed. Roles of governments, private sector, and other organisations are changing. National governments can act as facilitators of municipal, private sector, and community-based initiatives. The private sector plays an increasing role, because meeting future waste management needs depends on expanded private investment. The involvement of community organisations is also increasing as the link between community support and project sustainability has become clear. It is important that projects emphasise the deployment of locally-appropriate technologies, and minimise the development of conventional large, integrated waste management systems in situations where lower cost, simpler alternatives can be used without compromising public health and environmental standards.

Human Health

An effective health system can help to address the adverse health impacts of climate change. Transfer of existing health technologies within and across countries can assist in achieving this objective. Raising public awareness of likely health impacts, close monitoring of health outcomes and training of health professionals are suitable actions. Thus, in terms of technology transfer there is a need to ensure that technologies are available at national and local levels for coping with any changes in the burden of disease that might be associated with climate change.

Coastal Adaptation

Technology transfer should focus on proven technologies for coastal adaptation, including indigenous solutions. Wetland restoration and preservation are examples of such proven adaptation technologies. Effective transfers of adaptation technologies are part of integrated coastal-management plans or programmes, that utilises local expertise. Because coastal management is predominantly a public activity, technology transfer in coastal zones is driven by governments. Fragmented organisational and institutional relationships, and lack of access to financial means are major barriers to the transfer of coastal adaptation technologies. Coastal adaptation programmes, based on strong partnership between existing institutions, can provide an effective response.

TECHNICAL SUMMARY

Authors:

Stephen O. Andersen (USA), Earle N. Buckley (USA), William Chandler (USA), Renate Christ (Austria),
Ogunlade Davidson (Sierra Leone), Sukumar Devotta (India), Michael Grubb (UK), Joyeeta Gupta (The Netherlands),
Thomas C. Heller (USA), Maithili Iyer (India), Daniel M. Kammen (USA),
Richard J.T. Klein (The Netherlands/Germany), Dina Kruger (USA), Ritu Kumar (India), Mark Levine (USA),
Lin Erda (China), Patricia Iterregui (Peru), Merylyn McKenzie Hedger (UK), Anthony McMichael (UK),
Mark Mansley (UK), Jan-Willem Martens (The Netherlands), Eric Martinot (USA), Ajay Mathur (India),
Bert Metz (The Netherlands), John Millhone (USA), Jose Roberto Moreira (Brazil), Tongroj Onchan (Thailand),
Mark Radka (USA), Kilaparti Ramakrishna (India), N.H. Ravindranath (India), Jayant Sathaye (USA),
Youba Sokona (Mali), Sergio C. Trindade (Brazil), David Wallace (UK), Ernst Worrell (The Netherlands)

CONTENTS

EXECUTIVE SUMMARY

This introductory Chapter sets out the landscape for the discussion, throughout the Special Report, of the multitude of facets of managing technological change in support of the Climate Change Convention and its Protocols. The framework proposed for decision-making by government policymakers, and other relevant stakeholders, emphasises the sustainable development perspective, while exploring the national and international political settings, trends in finance and trade, the organisational and institutional context, and the meanings of technology transfer and of the innovation system. A model of the latter and of pathways in technology transfer is presented to help understand the nature, motivations, barriers to the process, and possible options to promote sustainable development in the face of the climate change challenge.

1. Background

Article 4.5 of the United Nations Framework Convention on Climate Change (UNFCCC) states that developed country Parties and other developed Parties included in Annex II to take "all practicable steps to promote, facilitate and finance, as appropriate, the transfer of, or access to, environmentally sound technologies and know-how to other Parties, particularly developing country Parties", and to "support the development and enhancement of endogenous capacities and technologies of developing country Parties", and calls on other Parties and organisations to assist in facilitating the transfer of such technologies.

The Subsidiary Body for Scientific Technological Advice (SBSTA) identified at its first session a list of areas in which it could draw upon the assistance of the IPCC. "Development and assessment of methodological and technological aspects of transfer of technology" was included in this list as an important element of the Third Assessment Report and an issue that may be appropriate for an interim or special report.

This Special Report was prepared in response to this request. It addresses the "technology transfer" problem in the context of all relevant UNFCCC provisions, including decisions of the Conference of Parties (CoP), and Chapter 34 in Agenda 21. It attempts to respond to recent developments in the UNFCCC debate on technology transfer, by providing available scientific and technical information to enable Parties to address issues and questions identified in Decision 4/CP.4 adopted by COP-4 (see Box TS 1). The focus of the Report is on the technology transfer process rather than on the assessment of technologies, which have been addressed in earlier IPCC Technical Papers and Reports.

1.1 The role of technology transfer in addressing climate change

Global economic growth is currently leading to increased consumption of raw materials, loss of natural habitats, energy use and production of waste. Achieving the ultimate objective of the UNFCCC, as formulated in Article 2[1], will require technological innovation and the rapid and widespread transfer and implementation of technologies and know-how for mitigation of greenhouse gas emissions. Transfer of technology for adaptation to climate change is also an important element of reducing vulnerability to climate change.

---

[1]."The ultimate objective of this Convention and any related legal instruments that the Conference of Parties may adopt to achieve, in accordance with the relevant provisions of the Convention, stabilisation of greenhouse gas concentrations in the atmosphere at such a level that would prevent dangerous interference with the climate system. Such a level should be achieved within a timeframe sufficient to allow ecosystems to adopt naturally to climate change, to ensure that food production is not threatened and to enable economic development to proceed in a sustainable manner"

Technology transfer has successfully contributed to the solution of a variety of local and global environmental problems. Case studies included in the Report document this experience and provide valuable lessons for climate protection. These case studies include, to varying degrees, essential elements of successful technology transfer including consumer and business awareness, access to information, capacity building, investment financing, relaxation of trade barriers, and a strong regulatory framework.

This technological innovation must occur fast enough and continue over a period of time to allow greenhouse gas concentrations to stabilise and reduce vulnerability to climate change. Technology for mitigating and adapting to climate change should be environmentally sound technology and should support sustainable development. Sustainable development on a global scale will require radical technological and related changes in both developed and developing countries. Economic development is most rapid in developing countries, but it will not be sustainable if these countries simply follow the historic greenhouse gas emission trends of developed countries. Development with modern knowledge offers many opportunities to avoid past unsustainable practices and move more rapidly towards better technologies, techniques and associated institutions. The literature indicates that to achieve this, developing countries require assistance with developing human capacity (knowledge, techniques and management skills), developing appropriate institutions and networks, and with acquiring and adapting specific hardware. Technology transfer, in particular from developed to developing countries, must therefore operate on a broad front covering these software and hardware challenges, and ideally within a framework of helping to find new sustainable paths for economies as a whole.

There is, however, no simple definition of a "sustainable development agenda" for developing countries. Sustainable development is a context driven concept and each society may define it differently, based on Agenda 21. Technologies that may be suitable in each of such contexts may differ considerably. This makes it important to ensure that transferred technologies meet local needs and priorities, thus increasing the likelihood that they will be successful, and that there is an appropriate enabling environment for promoting Environmentally Sound Technologies (ESTs).

The Report analyses the special challenges of transferring ESTs to address climate change in the context of sustainable development. The literature provides ample evidence of the many problems in current processes of technology transfer which makes it very unlikely to meet this challenge without additional actions for the transfer of mitigation and adaptation technologies.

1.2 What do we mean by technology transfer?

The Report defines the term "technology transfer" as a broad set of processes covering the flows of know-how, experience and equipment for mitigating and adapting to climate change amongst

Box TS1   QUESTIONS INCLUDED IN ANNEX TO DECISION 4/CP.4 OF THE CONFERENCE OF THE PARTIES TO THE UNFCCC THAT ARE
                     TO BE CONSIDERED IN THE CONSULTATIVE PROCESS SET UP BY THIS DECISION.

1. How should Parties promote the removal of barriers to technology transfer? Which barriers are a priority and what practical steps should be taken?

2. How should Annex II Parties promote the transfer of publicly-owned technologies?

3. What additional bilateral and multilateral efforts to promote technology cooperation to facilitate technology transfer should be initiated? What should be the priority?

4. Are existing multilateral mechanisms sufficient? Are new mechanisms needed for technology transfer? If so, what are appropriate mechanisms for the transfer of technologies among Parties in pursuance of article 4.5 of the Convention?

5. What should be the objective of collaboration with relevant multilateral institutions to promote technology transfer and what practical steps should be taken?

6. What additional guidance should be given to the financial mechanism?

7. What sort of information is needed and how can this best be done?

8. How could access to emerging technologies be facilitated?

9. What role is the private sector playing in technology transfer? What additional role can the private sector play? What barriers prevent their greater participation?

10. What technical advice on technology transfer is needed?

11. What areas should be the focus of capacity building and how should it be undertaken, e.g. what kind of activities, programmes and institutional arrangements?

12. How, to whom and in what format should developing country Parties make their requests for assistance to assess required technologies?

13. What technical, legal and economic information is needed? What practical steps should be taken to promote and enhance access to such information by national and regional centres?

14. What type of process is needed to develop a consensus on practical next steps to improve existing technology centres and networks in order to accelerate the diffusion of clean technologies in non-Annex I Party markets. What type of arrangement is needed to monitor progress?

15. What measures, programmes and

activities can best help to create an appropriate enabling environment for private sector investment?

16. How should the Convention oversee the exchange of information among Parties and other interested organisations or innovative technology cooperation approaches, and the assessment and synthesis of such information?

17. How should information be compiled and synthesised on innovative technology cooperation approaches? When should recommendations on such approaches be forwarded to the Conference of Parties?

18. How and when should information on projects and programmes of technology cooperation which Parties believe can serve as models for improving the diffusion and implementation of clean technologies internationally under the Convention be provided to the secretariat?

19. Can specific technology transfer goals be set?

20. Can we develop indicators and accounting systems to track progress on technology transfer?

21. Are particular institutional arrangements needed to monitor progress?

---

different stakeholders such as governments, private sector entities, financial institutions, NGOs and research/education institutions. Therefore, the treatment of technology transfer in this Report is much broader than that in the UNFCCC or of any particular Article of that Convention. The broad and inclusive term "transfer" encompasses diffusion of technologies and technology cooperation across and within countries. It covers technology transfer processes between developed countries, developing countries and countries with economies in transition, amongst developed countries, amongst developing countries and amongst countries with economies in transition. It comprises the process of learning to understand, utilise and replicate the technology, including the capacity to choose and adapt to local conditions and integrate it with indigenous technologies.

The Report generally makes a distinction between developed and developing countries. Although economies in transition are included as developed countries in the UNFCCC, they may have characteristics in common with both developed and developing countries.

1.3   *Stakeholders, pathways and stages*

Technology transfer results from actions taken by various stakeholders. Key stakeholders include developers, owners, suppliers, buyers, recipients and users of technology such as private firms, state enterprises, and individual consumers, financiers and donors, governments, international institutions, NGOs and community groups. Some technology is transferred directly between government agencies or wholly within vertically integrated firms, but increasingly technology flows depend also on the co-ordination of multiple organisations such as networks of information service providers, business consultants and financial firms. Although stakeholders play different roles there is a need for partnerships among stakeholders to create successful transfers. Governments can facilitate such partnerships. The rate of technology transfer is affected both by motivations that induce more rapid adoption of new techniques and by barriers that impede such transfers. Both types of factors can be influenced by policy (see Table TS 1).

The theme of technology transfer is highly interdisciplinary and has been approached from a variety of perspectives, including business, law, finance, microeconomics, international trade, international political economy, environment, geography, anthropology, education, communication, and labour studies. Although there are numerous frameworks and models put forth to cover different aspects of technology transfer, there are no corresponding overarching theories. However, the literature reveals a large number of pathways through which stakeholders can interact to transfer technologies. They vary depending on sectors, country circumstances and type of technology. Pathways may be different for "close to market" technologies and for technology innovations still in the development phase. The role of stakeholders is dependent on the pathway followed. Common pathways include government assistance programmes, direct purchases, trade, licensing, foreign direct investment, joint ventures, cooperative research arrangements and co-production agreements, education and training, and government direct investment.

While technology transfer processes can be complex and intertwined certain stages can be identified. These may include the identification of needs, choice of technology, assessment of conditions of transfer, agreement and implementation. Evaluation and adjustment to local conditions, and replication are other important stages. In order to evaluate whether technology transfer can be considered effective, different criteria can be applied. The criteria can be grouped into four categories, namely, *(i)* greenhouse gas (GHG) and environmentally related; *(ii)* economic and socially related; *(iii)* administrative, institutional and politically related; and *(iv)* process-related.

Table TS1	Principal stakeholders and their decisions or policies in technology transfer	
STAKEHOLDERS	**MOTIVATIONS**	**DECISIONS OR POLICIES THAT INFLUENCE TECHNOLOGY TRANSFER**
Governments • national/federal • regional/provincial • local/municipal	Development goals Environmental goals Competitive advantage Energy security	Tax policies (including investment tax policy) Import/export policies Innovation policies Education and capacity-building policies Regulations and institutional development Direct credit provision
Private-sector business • transnational • national • local/microenterprise (including producers, users, distributors, and financiers of technology)	Profits Market share Return on investment	Technology R&D/commercialisation decisions Marketing decisions Capital investment decisions Skills/capabilities development policies Structure for acquiring outside information Decision to transfer technology Choice of technology transfer pathway Lending/credit policies (producers, financiers) Technology selection (distributors, users)
Donors • multilateral banks • GEF • bilateral aid agencies	Development goals Environmental goals Return on investment	Project selection and design criteria Investment decisions Technical assistance design and delivery Procurement requirements Conditional reform requirements
International institutions • WTO • UNCSD • OECD	Development goals Environmental goals Policy formulation International dialogue	Policy and technology focus Selection of participants in forums Choice of modes of information dissemination
Research/extension • research centres/labs • universities • extension services	Basic knowledge Applied research Teaching Knowledge transfer Perceived credibility	Research agenda Technology R&D/commercialisation decisions Decision to transfer technology Choice of pathway to transfer technology
Media/public groups • TV, radio, newspaper • Schools • Community groups • NGOs	Information distribution Education Collective decisions Collective welfare	Acceptance of advertising Promotion of selected technologies Educational curricula Lobbying for technology-related policies
Individual consumers • urban/core • rural/periphery	Welfare Utility Expense minimisation	Purchase decisions Decision to learn more about a technology Selection of learning/information channels Ratings of information credibility by source

1.4 Trends in technology transfer

Little is known about how much climate-relevant hardware is successfully "transferred" annually. When software elements such as education, training and other capacity building activities are included, the task of quantification is further complicated. Financial flows, often used as proxies, allow only a limited comparison of technology transfer trends over time.

The 1990s have seen broad changes in the types and magnitudes of the international financial flows that drive technology transfer, at least that occurring between countries (see Figure TS 1). Official Development Assistance (ODA) experienced a downward trend in the period of 1993 to 1997, both in absolute terms and as a percentage of funding for projects with a significant impact on technology flows to developing countries. However, in 1998 there was an increase in ODA funding. ODA has become relatively less important to many developing countries given the dramatic increase in opportunities for obtaining private sector financing for technology acquisition.

Sources and amounts of development finance, some portion of which goes for technology transfer, vary widely from region to region. Countries in Sub-Saharan Africa received in 1997 an average of some US$27 per capita of foreign aid and US$3 per capita of foreign direct investment. By contrast, countries in Latin America and the Caribbean received US$13 per capita of aid and US$62 per capita of foreign direct investment. Recent initiatives to spur development progress in Africa aim to respond to these disparities.

Levels of foreign direct investment (FDI), commercial lending, and equity investment all increased dramatically during the 1990s. As a result, by 1997 private flows supplied more than three fourths of the total net resource flows from OECD member countries to developing countries compared to one third in 1990. Probable causes for this shift, and what it means for governments and the private sector, are described in detail in the Report. FDI, loans, and equity are the dominant means by which the private sector makes technology-based investments in developing countries and economies in transition, often in the industry, energy supply and transportation sectors. Private sector investment in the form of FDI in developing countries has favoured East and South East Asia, and Latin America.

Total private flows to developing countries peaked in the first half of 1997 and then fell significantly in the wake of the global financial crisis that started in Asia during the middle of that year. Most of the decline was due to reduced bank lending by the private sector, although this remained robust to Latin America. Foreign direct investment in developing countries is estimated to have increased slightly during 1998 and 1999.

Overall, FDI still represents a relatively small share of total investment in developing countries, both in absolute values and as a share of all developing country inflows. FDI exceeded 10 per cent of gross fixed capital formation in only eight countries, and

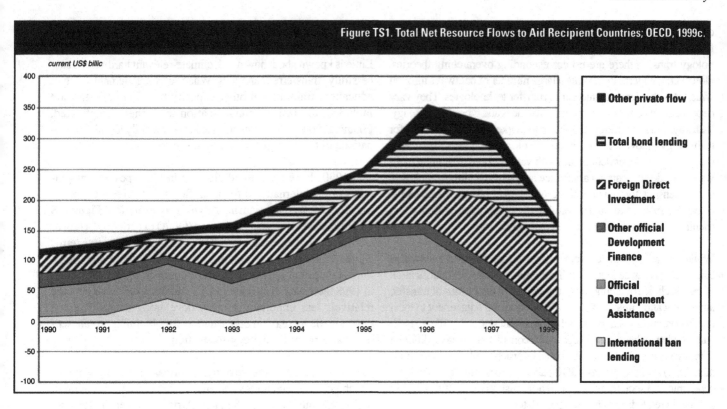

Figure TS1. Total Net Resource Flows to Aid Recipient Countries; OECD, 1999c.

current US$ billic

Legend:
- Other private flow
- Total bond lending
- Foreign Direct Investment
- Other official Development Finance
- Official Development Assistance
- International ban lending

in most it is much less than seven per cent of the total. Despite the small size of inflows, FDI is still important for many of these economies. Foreign direct investors are often manufacturers that occupy a dominant position in the supply chain and that play a major part in the industrial sectors in which they operate. In the best circumstances they bring to the host country, and the firms with whom they work, state-of-the-art technologies and high standards for environment, health and safety, and quality assurance. International financial statistics, however, indicate only the quantity, and not the quality, of FDI. Table TS 2 shows the increase in net private capital flows (of which FDI is a component) to low and middle income countries by country group and region during the period 1990-1996. Notable is the low share received by countries in Africa.

Grants by NGOs to developing countries have stayed fairly constant during the last decade, ranging between US$5 billion and US$6 billion per year since 1990. Despite their relatively modest amounts, many of these are directed at least developed countries and at capacity building efforts.

The general increase in the importance of private sector investment in developing countries does not reduce the role of ODA. First and as noted above, private sector investment has been very selective (see Table TS 2 and also Figures TS 2 and TS 3 in the Industry section). While almost all countries have benefited to some degree, a handful of countries (East Asia and Latin America) have received most of the attention. ODA is still critical for the poorest countries, particularly when it is aimed at developing basic capacities to acquire, adapt, and use foreign technologies. Second, ODA is still important for those sectors where private sector flows are comparatively low, like agriculture, forestry, human health and coastal zone management. Moreover it can be essential for certain activities including the leveraging of funds for capacity building activities and supporting the creation of enabling conditions which may leverage larger flows of private sector finance into ESTs. Third, private investment, most notably foreign portfolio equity investment and commercial lending, is volatile. Many developing countries have found to their distress that private investment can quickly dry up if investors perceive more attractive—or less risky—opportunities elsewhere.

Table TS2	Net private capital flows to low and middle income countries by country group (US$ billion) (Source: World Bank.)						
COUNTRY GROUP	1990	1991	1992	1993	1994	1995	1996
All countries	44.4	56.6	90.6	157.1	161.3	184.2	243.8
Sub-Saharan Africa	0.3	0.8	-0.3	-0.5	5.2	9.1	11.8
East Asia and the Pacific	19.3	20.8	36.9	62.4	71.0	84.1	108.7
South Asia	2.2	1.9	2.9	6.0	8.5	5.2	10.7
Europe and Central Asia	9.5	7.9	21.8	25.6	17.2	30.1	31.2
Latin America and Caribbean	12.5	22.9	28.7	59.8	53.6	54.3	74.3
Middle East and North Africa	0.6	2.2	0.5	3.9	5.8	1.4	76.9
Incomegroup							
Low income countries Excluding China and India	1.4	3.0	2.4	5.8	6.3	5.5	7.1
China and India	10.0	9.1	23.0	44.2	50.8	47.9	60.0
Middle income countries	32.0	44.0	64.8	107.1	104.2	130.7	176.7

Measuring Technology Transfer of EST

Because of the limited comparison of trends in technology transfer on the basis of financial flows, better indicators and data to quantify the level and flows of climate-relevant EST are needed to give governments better information on which to base their policy. In addition, technology performance benchmarks for different sectors could be compiled to give an indication about the real degree of implementation of EST and the potential for technological improvements. It would be useful to have simple and agreed upon criteria for measuring the transfer of ESTs.

1.5 Barriers to the transfer of Environmentally Sound Technologies

The spread of proven ESTs that would diffuse through commercial transactions may be limited because of existing barriers. Barriers to the transfer of ESTs arise at each stage of the process. These vary according to the specific context, for example from sector to sector, and can manifest themselves differently in developed countries, developing countries and countries with economies in transition. The Report provides an extensive overview of the most important barriers in developed, developing and transition economies that could impede the transfer of ESTs to mitigate and adapt to climate change (see Table TS 3 and TS 4 for further reference). Governments can promote technology transfer by reducing the barriers that are associated with each of these elements of an enabling environment.

- Lack of full-cost pricing, which internalises environmental and social costs;
- Poor macroeconomic conditions, which could include underdeveloped financial sector, high import duties, high or uncertain inflation or interest rates, uncertain stability of tax and tariff policies, investment risk;
- Low private sector involvement because of lack of access to capital, in particular inadequate financial strength of smaller firms;
- Lack of financial institutions or systems to ensure initial investments for the utilisation and extended use of transferred technologies;
- Low, often subsidised conventional energy prices resulting in negative incentives to adopt energy saving measures and renewable energy technologies;
- Lack of markets for ESTs because of lack of confidence in economic, commercial or technical viability, lack of manufacturers, lack of consumer awareness and acceptance of technologies;
- Lack of supporting legal institutions and frameworks, including codes and standards for the evaluation and implementation of environmentally sound technologies;
- Lack of understanding of the role of developed and developing countries and international institutions in the failures and successes of past technology cooperation.;
- General lack of support for an open and transparent international banking and trading system;
- Institutional corruption in both developed and developing countries;
- Reluctance to identify and make available ESTs that are in the public domain;
- Insufficient human and institutional capabilities;
- Inadequate vision about and understanding of local needs and demands;
- Inability to assess, select, import, develop and adapt appropriate technologies;
- Lack of data, information, knowledge and awareness, especially on "emerging" technologies;
- Lack of confidence in unproven technologies;
- Risk aversion and business practices that favour large projects in financial institutions including MDBs;
- Lack of science, engineering and technical knowledge available to private industry;
- Insufficient R&D because of lack of investments in R&D and inadequate science and educational infrastructure;
- Inadequate resources for project implementation;
- High transaction costs;
- Lack of access to relevant and credible information on potential partners to allow for the timely formation of effective relationships which could enhance the spread of ESTs.

There is no pre-set answer to enhancing technology transfer. The identification, analysis and prioritisation of barriers should be country based. It is important to tailor action to the specific barriers, interests and influences of different stakeholders in order to develop effective policy tools.

2. Increase the flow, improve the quality

The challenge of successfully transferring ESTs should be seen in the context of sustainable development. Sustainable development needs not restrict growth but can stimulate the emergence of a vibrant. industrial economy, a process in which technology transfer is likely to play a major role. Sustainable industrialisation is especially a challenge for developing countries, because their low initial level of development provides them with an opportunity to follow a technological trajectory which can be cleaner and more efficient than the path OECD countries have followed.

To enhance the sustainability of the development process, government actions can transform the conditions under which technology transfer takes place. The spread of proven ESTs that would diffuse through commercial transactions may be limited because of the barriers listed above. Governments can play important roles in facilitating the private transfer of ESTs by encouraging private sector trade and investment of environmentally sound technologies. Capacity building programmes and enabling environments that reduce the risks and restrictions associated with the transfer of ESTs will increase the flow of technologies close to the commercial margin. The key issue is thus to make the markets work by " opening the channels". For technologies that will not yet diffuse commercially, it is important to go further than improving market performance by enacting poli-

Table TS3	Policy Tools for Creating an Enabling Environment for Technology Transfer
POLICY TOOL	**BARRIERS ADDRESSED**
NATIONAL SYSTEMS OF INNOVATION AND TECHNOLOGY INFRASTRUCTURE (4.3)	
• Build firms' capabilities for innovation • Develop scientific and technical educational institutions • Facilitate technological innovation by modifying the form or operation of technology networks, including finance, marketing, organisation, training, and relationships between customers and suppliers	• Lack of technology development and adaptation centres • Lack of educational and skills development institutions • Lack of science, engineering and technical knowledge available to private industry • Lack of research and test facilities • Lack of information relevant for strategic planning and market development • Lack of forums for joint industry-government planning and collaboration
SOCIAL INFRASTRUCTURE AND RECOGNITION THROUGH PARTICIPATORY APPROACHES (4.4)	
• Increase the capacity of social organisations and NGOs to facilitate appropriate technology selection • Create new private-sector-focused social organisations and NGOs with the technical skills to support replication of technology transfers • Devise mechanisms and adopt processes to harness the networks, skills and knowledge of NGO movements	• Technology selection inappropriate to development priorities • Historical legacy of technology transfer in development • Problems of scaling cultural and language gaps and fostering long-term relationships
HUMAN AND INSTITUTIONAL CAPACITIES (4.5)	
• Build capacities of firms, non-governmental organisations, regulatory agencies, financial institutions, and consumers	• Inability to assess, select, import, develop and adapt appropriate technologies • Lack of information • Lack of management experience • Problems of scaling cultural and language gaps and fostering long-term relationships • Limited impact of technology because no long term capacity built to maintain innovation • Lack of joint venture capabilities for learning and integrating
MACROECONOMIC POLICY FRAMEWORKS (4.6)	
• Provide direct financial support like grants, subsidies, provision of equipment or services, loans and loan guarantees. • Provide indirect financial support, like investment tax credits • Raise energy tariffs to cover full long-run economic costs • Alter trade and foreign investment policies like trade agreements, tariffs, currency regulations, and joint venture regulations • Alter financial sector regulation (See also chapter 5 for further discussion of policy tools for financing technology transfer)	• Lack of access to capital • Lack of available long-term capital • Subsidised or average-cost (rather than marginal-cost) prices for energy • High import duties • High or uncertain inflation or interest rates • Uncertain stability of tax and tariff policies • Investment risk • Excessive banking regulation or inadequate banking supervision • Incentives for banks that are distorted against risk taking • Banks that are poorly capitalised • Risk of expropriation
SUSTAINABLE MARKETS FOR ENVIRONMENTALLY SOUND TECHNOLOGIES (4.7)	
• Conduct market transformation programmes that focus on both technology supply and demand simultaneous. • Develop capacity for technology adaptation by small- and medium-scale enterprises (SMEs) • Conduct consumer education and outreach campaigns • Targeted purchasing and demonstrations by public sector	• High transaction costs • Inadequate strength of smaller firms • Uncertainty of markets for technologies prevents manufacturers from producing them • Lack of consumer awareness and acceptance of technologies • Lack of confidence in the economic, commercial, or technical viability of a technology
NATIONAL LEGAL INSTITUTIONS (4.8)	
• Strengthen national frameworks for intellectual property protection • Strengthen administrative and law processes to assure transparency, participation in regulatory policy-making, and independent review • Strengthen legal institutions to reduce risks	• Lack of intellectual property protection • Contract risk, property risk, and regulatory risk • Corruption
CODES, STANDARDS, AND CERTIFICATION (4.9)	
• Develop codes and standards and the institutional framework to enforce them. • Develop certification procedures, and institutions, including test and measurement facilities	• High user discount rates do not necessarily result in most efficient technologies • Lack of information about technology or producer quality and characteristics • Lack of government agency capability to regulate or promote technologies • Lack of technical standards and institutions for supporting the standards
EQUITY CONSIDERATIONS (4.10)	
• Devise analytical tools and provide training for social impact assessment. • Require social impact assessments before technology is selected • Create compensatory mechanisms for "losers"	• Social impacts not adequately considered • Some stakeholders may be made worse off by technology transfer
RIGHTS TO PRODUCTIVE RESOURCES (4.11)	
• Investigate impacts of technology on property rights, test through participatory approaches, devise compensatory mechanisms for losers.	• Inadequately protected resource rights
RESEARCH AND TECHNOLOGY DEVELOPMENT (4.12)	
• Develop science and educational infrastructure by building public research laboratories, providing targeted research grants, and strengthening technical education system • Directly invest in research and development	• Insufficient investment in R&D • Inadequate science and educational infrastructure

cies that lower costs and stimulate demand in order to realise social and environmental benefits not adequately produced by private conduct. The international community could assist these extra efforts of individual countries by increasing available means for non-market transfers and creating new or improving existing mechanisms for technology transfer.

The Report clearly points out that there is no pre-set answer to enhancing technology transfer. It is important to tailor action to the specific barriers, interests and influences of different stakeholders in order to develop effective policy tools. As has been stated clearly in Agenda 21. policy tools are most effective if they are considered in the context of sustainable development. Agenda 21 provided some of the earliest recommendations for public policies to promote technology transfer for environmental benefits. These recommendations reflect not only the need for hardware, but also for building associated local capacities and for providing market intermediation. Strategies outlined in Agenda 21 include: (a) information networks and clearinghouses that disseminate information and provide advice and training; (b) government policies creating favourable conditions for both public-sector and private-sector transfers; (c) institutional support and training for assessing, developing, and managing new technologies; (d) collaborative networks of technology research and demonstration centres; (e) international programmes for cooperation and assistance in R&D and capacity building; (f) technology-assessment capabilities among international organisations; and (g) long-term collaborative arrangements between private businesses for foreign direct investment and joint ventures.

The three major dimensions of making technology transfer more effective are capacity building, an enabling environment and mechanisms for technology transfer, all of which are discussed in more detail in the subsections below.

2.1 Building capacity

Capacity building is required at all stages in the process of technology transfer. It is a slow and complex process to which long-term commitments must be made for resources and to which the host country must also be committed if results are to bear fruit. For enhancing the transfer of ESTs, the focus should be on building human, organisational and information assessment capacity. Moreover, integrating human skills, organisational develop-

Table TS4 Policy Tools for Financing and Participation (Source: Mansley et al., 1997a, b)	
POLICY TOOL	**BARRIERS ADDRESSED**
PUBLIC-SECTOR FINANCE AND INVESTMENT (5.2)	
• Provide direct finance • Provide official development assistance • Provide multilateral development bank finance	• Lack of confidence in "unproven" technology • Lack of access to capital • User acceptability • Costs of developing new public infrastructure • Lack of policy harmonisation • Uncertain future energy prices
PRIVATE-SECTOR FINANCE AND INVESTMENT (5.3)	
• Support, through a variety of policy tools, private-sector financing mechanisms such as microcredit, leasing, venture capital, project finance, "green" finance, and a range of other private-sector financing initiatives. • Reduce perceived risks through consistent policies and transparent regulatory frameworks (see Ch. 4)	• Lack of access to capital • High transaction costs • High front-end capital costs • High user discount rates • Inadequate financial strength of smaller firms • Lack of information • Lack of confidence in "unproven" technology
PRIVATE-FIRM INVESTMENTS (5.4)	
• Create incentives for firms to make environmentally sound investments, such as energy taxes, investment tax credits, and emissions fees (see Ch. 4) • Engage firms in public-private partnerships, as discussed in 5.6, particularly to overcome managerial misincentives	• Managerial misincentives • Competing purchase decision criteria • Sunk investments in existing equipment and infrastructure • Energy-supply-side financing bias • Lowest-cost equipment favoured
PUBLIC-PRIVATE PARTNERSHIPS (5.5)	
• Enter into voluntary agreements with the private-sector • Develop technical partnership programmes • Conduct informational initiatives • Provide tax incentives, guarantees, and trade finance • Promote new financial initiatives	Barriers are similar to those for public-sector and private-sector financing and investment, plus the following: • Uncertain markets for technologies • Import duties • Utility acceptance of renewable energy technologies • Permit risk • Environmental externalities • Shortage of trained personnel
TECHNOLOGY INTERMEDIARIES (5.6)	
• Create information networks, advisory centres, specialist libraries, databases, liaison services • Create and support technology intermediaries like energy service companies and national-level institutions	• High transactions costs • Lack of available information about technology costs and benefits among potential purchasers and partners • Missing connections between potential partners and credible information about partners • Disaggregated opportunities that do not provide sufficient benefits for individual firms to capture on their own • Lack of capacity to contract and conduct technology transfers

ment and information networks is the key to effective technology transfer. Social structures and personal values evolve with a society's physical infrastructure, institutions, and the technologies embodied within them. New technological trajectories for an economy therefore imply new social challenges. This requires a capacity of people and organisations to continuously adapt to new circumstances and to acquire new skills. Governments can support the establishment of dynamic flexible learning mechanisms not only at the national level, but also at a level of sub-national regions. This applies both for mitigation and adaptation technologies. Comparatively little consideration has been given in a systematic way to what capacity building is required for adaptation to climate change.

Human capacity
There are many failures of technology transfer that result from an absence of human and institutional capacity. This makes adequate human capacity essential at every stage of every transfer process. The transfer of many ESTs demands a wide range of technical, business and regulatory skills. Capacity is needed to assess, select, import, develop and adapt appropriate technologies. Accumulated technology transfer experience indicates that developing countries enterprises are not always able to effectively exploit the diversity of available technological options and ser-

vices. Policies aimed at ensuring the availability of these skills locally can enhance private international investments through which much technology is diffused.

Many ways of developing capabilities for the assessment, agreement, and implementation stages of technology transfer are suggested by development experience: (a) formal training of employees, (b) technological gatekeeping, by keeping informed of technical literature, forming links with other enterprises, professional and trade organisations, and research institutions; (c) learning by doing-operational experience such as through twinning arrangements with other firms.

Implementing such activities involves different responsibilities for developed and developing countries. The donor agency understanding or concepts of capacity building have frequently been that they can be developed or strengthened along the lines of management and organisational models of the donor countries. However, the experiences and results of over three decades of capacity and institutional building in sub-Saharan Africa, for instance, suggest that these assumptions or concepts do not necessarily hold. Developed country governments could therefore pay more attention to ensure that training and capacity building programmes they sponsor are in full co-operation with all relevant stakeholders, including local governments, institutions and stakeholders, NGOs, community organisations, commercial organisations and consumers and take into account local conditions under which these may be provided.

Developing country governments can build local capacities to gear them for technology transfer. Training and human resource development have been popular development assistance activities. Future approaches can be more effective by better stressing the integration of a total package of technology transfer, focusing less exclusively on developing technical skills and more on creating improved and accessible competence in associated services, organisational know-how, and regulatory management. The engineering and management skills required in acquiring the capacity to optimise and innovate are essential. Various kinds of high quality training are needed to embody in personnel of the receiving firm the skills, knowledge and expertise applicable to particular products and processes.

Organisational capacity

The historical legacy of development efforts, i.e. the failure of top-down, technology focused development, has provoked a reassessment of the appropriate roles of the community, government and private sector in development and technology transfer. It is now widely recognised that involvement of community institutions is an essential part of successful sustainable development and is therefore an important factor for the successful transfer of climate change mitigation and adaptation technologies. The involvement of local government agencies, consumer groups, industry associations and NGOs can help to ensure that the ESTs being adopted within their particular country/region are consistent with their sustainable development goals.

Besides the involvement of such community institutions, lessons from technology intensive economies teach that technology increasingly flows through private networks of information and assessment services, management consultants, financial firms, lawyers and accountants, and technical specialist groups. These new insights make it important for governments to strengthen the networks in which these diverse organisations can contribute to technology transfer. Although many actions that facilitate the growth of such networks are already underway, initiatives of particular importance to EST transfer include:

- Expansion of opportunities to develop firms for management consulting, accounting, energy service, law, investment and product rating, trade, publishing and provision of communication, access to and transfer of information, such as Internet services;
- Encouragement of industry associations, professional associations and user/consumer organisations;
- Participatory approaches to enable private actors, public agencies, NGO's and grassroots organisations to engage at all levels of environmental policy-making and project formulation;
- Where appropriate, decentralisation of governmental decision-making and authority, in relation to technology transfer, to effectively meet community needs.

Information assessment and monitoring capacity

Information access and assessment are essential to technology transfer. However, focussing too narrowly on information barriers while ignoring the later stages of the transfer process can be counter-productive. Technology information programmes should be demand driven and results oriented to the degree possible. Information is most useful when it supports an actual technology choice, investment decision, etc.

Because of its public good characteristics, technology infrastructure required to generate new knowledge and information may lack direct economic value to one firm, and thus individual firms may lack adequate incentives to build technology infrastructure on their own. This points to an important role for governments to create the necessary information assessment and monitoring capacity. However, the roles of governments and private actors in technology assessment are changing. Private information networks are proliferating through specialised consulting and evaluation services and over the Internet. Increasing FDI also demonstrates that many ESTs can diffuse rapidly without direct government action.

Governments in developing countries, developed countries and countries with economies in transition may wish to consider:
- Developing improved indicators and collecting data on availability, quality and flows of ESTs to improve monitoring of implementation;
- Developing technology performance benchmarks for ESTs to indicate the potential for technological improvements;
- Improving information systems and linking them to international or regional networks, through well-defined clearing houses (such as energy efficiency and renewable energy centres), information speciality firms, trade publications, electronic media or NGOs and community groups. In order to overcome information barriers, technology information centres have been widely advocated, but existing experience and literature does not provide enough certainty about exactly what is required for key stakeholders at critical stages.

2.2   Enabling environment and extra efforts to enhance technology transfer

Successful, sustainable technology transfer requires a multifacetted enabling environment. An enabling environment for technology transfer includes macroeconomic conditions, the involvement of social organisations, national institutions for technology innovation, human and institutional capacities for selecting and managing technologies, the underpinnings of sustainable markets for environmentally sound technologies, national legal institutions that reduce risk and protect intellectual property rights, codes and standards, research and technology development, and the means for addressing equity issues and respecting existing property rights.

Governments can aid in establishing an environment which promotes markets by considering what is the set of institutions which underlie markets and what are appropriate public interventions to shape those institutions. They define the property rights, contract enforcement mechanisms, and many of the rules for transactions that are necessary for markets to work well. Policies that build or facilitate markets can have a strong influence on the characteristics of those markets - for example, the relative sales share of domestic vs. foreign products, the segments of consumers participating in the market, the ability of domestic producers to participate in the market, the technologies available, and how regulations govern market behaviour.

Creating such an enabling environment is especially relevant for those ESTs that are already in common use and that could be diffused through commercial channels, but whose spread is hampered by risks such as those arising from distortive incentives, deficiencies in legal systems and inadequate regulation.

Apart from close to market technologies, many technologies that can mitigate emissions or contribute to adaptation to climate change still lie beyond the commercial frontier. Improving the enabling environment will not be sufficient to stimulate the transfer of such ESTs. Many ESTs are still in the early stages of their development and have a comparatively short track record. Market actors will not accept the extra risks or costs involved in utilising these ESTs. Governments should therefore consider extra efforts to increase the demand of these ESTs by stimulating their development, lowering their costs and reducing the associated commercial risks. Such extra efforts may include increasing the means for such non-market EST transfers and creating new and improving existing mechanisms for technology transfer.

These extra efforts will be discussed integrated with the actions to improve an enabling environment in the following sections. In order to improve the enabling environment of the transfer of ESTs, both market and non-market transfers, governments may wish to consider a number of actions. These actions have been classified according to actions for all governments, actions specific to governments of developed countries and actions which are in general more relevant for governments of developing countries.

2.2.1 Actions for all governments

Macroeconomic conditions

Macroeconomic conditions can favour the flourishing of private sector development include such factors as low inflation, stable and realistic exchange rates, deregulation, free movement of capital, promotion of competitive markets, open trade policies and transparent foreign investment policies.

An especially important barrier for the transfer of ESTs, is the existence of externalities in the economy. The lack of internalising environmental and social costs and the resultant underpriced energy deter investments in clean alternatives, for instance in the buildings, industrial, and energy sectors. Different measures to "internalise" the environmental costs of fossil fuel use and reduce the unfair commercial risk for ESTs are now being tried to improve the competitiveness of the cleaner energy sources. Such measures could include regulations, taxes, codes, standards and removal of subsidies to internalise the full environmental and social costs.

EST market transformation

A market transformation approach promotes technology transfer by catalysing sustainable markets for specific technologies, and thus harnessing the power of market-based incentives to accomplish environmental goals. The three central characteristics of a market are: *(i)* the number, nature, and capabilities of participants, *(ii)* the characteristics of the products and services, and *(iii)* the rules governing transactions. Properly functioning markets generally require the availability of information, acceptable levels of risk, appropriate skills, a system of property definitions, quality and contractual norms or standards, oversight and intermediation bodies, decision-making autonomy for buyers and sellers, and stable political and legal regimes. Government actions to encourage market transformation of ESTs include:

- Foster competitive international markets by opening their domestic markets in industry, energy and agriculture;
- Simplification and transparency of programme and project approval procedures and public procurement requirements;
- Establishment of requirements for environmental impact assessment and environmental reporting;
- Establishment of incentives for corporations to pioneer new ways of working with the community;
- Promote so-called "green-labelling" programmes to employ trademark or related principles (e.g. a not-for-profit organisation allows a vendor to use an environmental seal of approval if certain requirements are satisfied).
- Encouragement of social and technological learning between private firms and public agencies at regional (sub-national) levels;
- Conducting of programmes that focus on the demand-side of technology transfer;
- Development of capacity for technology adjustment by small- and medium-scale enterprises (SMEs);
- Conducting of consumer education and outreach campaigns;
- Influencing of WTO standards to facilitate the trade products and goods produced from the use of ESTs (PPM – products and process methods);
- Targeted purchasing and demonstrations by public sector.

Availability of and access to financing

Further efforts are needed to engage banks and other lending institutions into financing environmentally sound technologies and projects. Such efforts may consist of stimulating innovative financial mechanisms, public-private partnerships and the involvement of intermediaries.

There is a wide variety of types of traditional private sector debt and equity finance available depending on the scale and type of the project. The most flexible finance debt is secured loan and leasing. The transfer of environmentally sound technologies to developing countries will also involve increased use of innovation to structure existing financial products to new markets and to develop new ones as appropriate. Just as supporting scientific and technical innovation is seen as an appropriate use of public funds, so can financial innovation. A number of worthwhile initiatives have been undertaken to date (such as micro-credit, project finance, green finance and also the use of strategic investors) and there is scope to replicate and extend these as well as develop new concepts. Different financing arrangements are often required at both the production and acquisition stages.

Public private partnerships are increasingly seen as an effective way in which the public sector can achieve public policy objectives by working with the private sector. For the public sector they have the potential of harnessing the efficiency of the private sector, as well as overcoming budget restrictions and leveraging limited public funds. For the private sector, they aim to help overcome some of the internal and external barriers which prevent appropriate technology transfer from taking place, and to create interesting business opportunities.

Technology intermediaries, such as national-level technology transfer agencies, electric utilities, and energy service companies (ESCOs) have gained widespread acceptance to stimulate innovative financing schemes. They can effectively address financial, capability as well as other institutional market barriers. An ESCO is a firm that offers energy services to customers with performance guarantees. Typical performance contracting arrangements provide customers with feasible means of improving their competitiveness by reducing energy consumption costs. Additionally, companies' cash flows are enhanced, which adds value to their financial value. Governments and other public-sector entities can develop technology intermediaries through direct support and other interventions.

Legal systems
Uncertain, slow and expensive enforcement of contracts by national courts or international arbitration and insecure property rights can discourage investment. Three broad types of legal risk are likely to influence decisions to invest in advanced environmental technologies by foreign and domestic actors: Contract risk, Property risk and regulatory risk.

Contract risk refers to the likelihood and costs of enforcing legal obligations with suppliers, partners, distributors, managers, labour forces, construction organisations or licensors. Property risk refers both to more familiar risks associated with interference in asset ownership and to less visible, but also to essential questions of corporate governance including shareholder rights and competition laws that determine how decision making within the firm is divided and whether firms will be able to operate in competitive markets. Regulatory risk arises from the behaviour of public administrations, which influence economic returns through licensing, tariff setting, taxation, and foreign exchange and trade controls.

To reduce contract, property and regulatory risk, governments can strengthen national legal institutions for intellectual property protection; strengthen administrative and law processes to assure transparency, participation in regulatory policy-making, and independent review; and strengthen legal institutions to reduce risks and corruptionand to ensure that public regulation is accessible to stakeholders and subject to review by independent authorities.

Intellectual property rights
For harnessing the bulk of international investment, intellectual property rights (IPR) regimes are an important consideration. Overall the literature is diverse concerning the relationship between IPRs and technology transfer. Strong IPR regimes, generally lead to increased innovation and "vertical" technology transfer and increased foreign investment, although it should be kept in mind that it is not the only factor affecting investment decisions. Strong IPR regimes could, however, depending on the holder of the patents, slow down the dissemination of certain technologies, the so called horizontal technology transfer Where this is the case, countries may address this concern by taking appropriate measures. For example, the risk of unlicensed patents may be reduced by charging increasing annual maintenance fees. If the fee becomes high enough by 5 to 10 years after patent issuance, the owner might let an uncommercialised patent lapse. Another option is so called compulsory licensing as specified under the international Trade Related Aspects of Intellectual Property (TRIP) agreement and in decisions contained in Agenda 21, provided that correct procedures are followed (generally, they require the user first to seek a license through regular venues, to pay reasonable compensation for the license and to practice the invention on a limited non-exclusive basis). IPR regimes could be extended more widely to support innovation and dissemination of ESTs. Industry standards, including management standards developed through the International Standards Organisation (ISO) and sector standards for some industries, could also play an important role in fostering global dissemination of ESTs.

Multinational companies' leadership in using the same standards for environmental performance wherever they operate.
Two key ways by which private sector firms can stimulate the more rapid adoption of ESTs are leadership and participation. Leadership involves the senior executives of the company making a clear commitment to addressing environmental issues, with the consequence that all in the business are aware of these issues and sees them as an important aspect of their work (e.g. relevant towards promotion or bonuses). An example of leadership has been the recent clear announcement of multinational companies to initiate ambitious GHG reduction programmes including innovative mechanisms such as internal trading systems. Voluntary agreements can be a useful ways of obtaining high level commitment. Participation involves engaging employees (and others) in the environmental challenge, and encouraging responses and

initiatives from them. It should be seen as complementary (not an alternative) to leadership. A number of companies have successfully used participatory approaches to address environmental problems.

Consumer awareness, and product standards, industry codes and certification
Governments can work with the private sector and NGOs to establish codes, standards and labels. This provides a framework which can work to the benefit of industry and consumers. This route can help build markets for dispersed, small-scale technologies where technologies are diverse, vendors are many, and consumers face high risks in evaluating and selecting technologies and suppliers. Codes and standards also provide a means for representing the interests of end-users who are absent from purchase or construction decisions. Standards also reduce risk for consumers with regard to the equipment they are purchasing.

With regard to energy use, information programmes have proven successful in assisting energy consumers to understand and employ technologies and practices to use energy more efficiently. These programmes aim to increase consumers' awareness, acceptance, and use of particular technologies or utility energy conservation programmes. Examples of information programmes include educational brochures, hotlines, videos, design-assistance programmes, audits, energy use feedback programmes and labelling programmes.

For industry, energy audit programmes are a more targeted type of information transaction than simple advertising. Industrial customers that have undergone audits have reduced their electricity use by an average of 2 to 8%, with the higher savings rates achieved when utilities followed up their initial recommendations with strong marketing, repeated follow-up visits, and financial incentives to implement the recommended measures

An example of the effective role codes and standards can play is the International Performance Measurement and Verification Protocol (IPMVP) used for energy efficiency. Energy efficiency investments in the buildings, industrial, energy sectors have been constrained due to inconsistencies and uncertainties in their performance (i.e., actual energy savings achieved), financing for efficiency investments has been limited and inflexible. The existence of monitoring and verification protocols can help to reduce these inconsistencies and uncertainties.

Nowadays, Multilateral Development Banks such as the World Bank are using the IPMVP as the technical basis for large scale energy efficiency financing. Use of the IPMVP results in higher and more persistent levels of energy efficiency savings and in a standardised approach to contract development, implementation and monitoring. This uniform approach cuts transactions costs, allows project pooling and facilitates project financing. As a result of the rapidly increasing application of the IPMVP, there is increased energy efficiency project financing, with improved project performance, and increased availability of lower cost financing for energy efficiency projects.

2.2.2   *Actions for developed countries and countries with economies in transition*

Creation of global EST markets
Developed countries can play special roles in stimulating global transfer of ESTs. In particular they can:
*   Stimulate fair competition in EST markets by discouraging restrictive business practices such as product dumping to strive out competitors; overly restrictive conditions on the use of patents and refusal of licencing;
*   Reform export credit, political risk insurance and other subsidies for the export of products or production processes to encourage direct foreign investment in ESTs and discourage export that lower environmental quality in developing countries;
*   Reduce the use of export controls, export cartels, licensing restrictions;
*   Reduce or eliminate tied aid as trade policy measure;
*   Ensure requirements of host countries are adequately reflected in project design.

Strong and open R&D structure
Governments play an important role in providing funding for public R&D programmes as part of their industrial policies or science and technology development strategy. To promote the development of ESTs that lack short-term commercial viability, government funding and public R&D programmes are vital, and appropriate, reflecting the high rate of social return. For example, governments have been investing for three decades in R&D for environmentally sound energy technologies in the energy sector. These programmes are implemented either by government institutions or in joint partnership with the private sector. Over the past decade, about 40% of annual national R&D spending within a number of OECD countries was publicly funded. To increase the rate of development of EST, governments should therefore increase public funding for R&D in cleaner technologies.

The issue of publicly owned technology transfer and the role that national regulatory frameworks play in creating demand and market for ESTs was addressed both in the Rio Summit of 1992 and the UNGASS of 1997. Major findings of the UNCTAD study are that governments are funding public R&D programmes as part of their industrial policy aimed at improving their industrial competitiveness. Strong emphasis is placed on the commercialisation of those technologies developed from public R&D. In many cases, co-financing with the private sector also plays an important role. Many governments either transfer or license the patents of the publicly funded technologies to the private sector as part of their industrial policy and then the transferred patents follow the rules of the privately owned technologies and behave just like the other ordinary private IPRs. As a result, instead of focusing on technologies held by governments, the focus should rather be on public R&D programmes for ESTs and exploring the feasibility of enhancing the transfer of such technologies.

Developed countries could enhance flows of technology transfer arising from their public funded R&D programmes by encour-

BOX TS 2 — NATIONAL TECHNOLOGY INTERMEDIARIES IN INDIA

National-level government agencies acting as intermediaries can also be important in creating incentives and facilitating a market for cleaner technologies. The Energy Management Centre (EMC), an autonomous agency, under the Ministry of Power, Government of India, is an example of a technology intermediary for energy efficiency. EMC has been carrying out a number of initiatives to promote energy conservation and efficiency in India. To begin with, EMC set up and trained 25 agencies (public, private, NGOs), to provide specialised energy auditing and management to consumers in India. Each of these agencies are carrying out an average of 10-12 energy audits annually, and the feedback from the industry is that there is an urgent need for many more such professional agencies to be able to serve the consumers in the country. EMC also carried out a number of studies in the area of technologies for energy efficiency, issues relating to standards and labelling, as well as implementing a nation-wide energy conservation awareness project. EMC annually organises, through industry associations, about 20-25 training programmes and workshops for wider dissemination of information on ener-

gy conservation in the country. To date, it is reported that over 5,000 professionals have been provided training in different aspects of energy efficiency. Regular feedback carried out indicated that the participants have actually implemented energy efficiency projects in their organisations. EMC was the executing agency for international cooperation projects with Germany, the European Union, and the Department of Energy (USA), among others.

The initiatives of the Indian Government implemented through the EMC have resulted in a significant rise in the exposure and awareness on energy conservation technologies. It is reported that there are proposals to introduce standards for appliance and energy consuming devices and these would be mandatory. Penalties for non-compliance would be enforced once the law is passed by the Indian parliament. Under a collaborative programme with the EU, EMC has set up an Information Service on Energy Efficiency (ISEE), jointly with a national industry association. The database established is expected to contain information on technologies, guide books, manuals, best

practice programmes, a list of manufacturers, etc. and is expected to fill the gap in information for energy consumers.

The Technology Information, Forecasting and Assessment Council (TIFAC) in India was established as an autonomous organisation of the Indian Department of Science and Technology, and has been particularly successful in making the public-private sector linkages, providing information on patent issues, and supporting start-up ventures. Each of these activities provides important examples for other similar, knowledge-based technology transfer policy offices.

The Ministry of Non-conventional Energy Sources (MNES) in the nodal ministry responsible for providing the overall thrust and direction for increased adoption and installation of renewable energy devices in the country. MNES implements the programmes through the state governments and through state energy nodal agencies. MNES has separate programmes for biogas, solar thermal, solar PV, biomass gasifier, and for new technologies.

---

aging or requiring where appropriate the recipients of such support to transfer the technology as soon as practical; by entering into co-operation with developing countries on R&D partnerships and international research institutions.

National and multilateral assistance flows
There is substantial scope for increased flows of ODA aimed at addressing genuine development objectives. Within this context there is potential to support activities which assist the transfer of ESTs, also among developing countries, but the overall development objectives must remain paramount.

Trade assistance programmes supportive of ESTs
ODA is not the only form of public sector finance from developed countries to support technology transfer. Substantial assistance is provided in the form of export credit, political risk insurance and other subsidies for the export of products or production processes. Little of this assistance is provided specifically for the transfer of ESTs and substantial volumes may transfer environmental inferior technology or support environmentally damaging projects. Developed countries can explore ways, such as developing environmental guidelines for export credit agencies, to refocus

their trade assistance activities to avoid a bias against and promote the transfer of ESTs, and discourage the transfer of obsolete technologies.

Coordination within donor governments
One major issue regarding the effective transfer of environmentally sound technologies through public finance is the fact that foreign aid expenditure tends to be institutionally divorced from other powerful agencies that also have a huge influence on technology choice and investment patterns, such as trade, industry and science ministries. It is not uncommon to find governments pursuing seemingly contradictory international financial policies, and one of the strongest recommendations in this area is for greater institutional coherence within donor governments to promote transfer of ESTs.

Intellectual Property Rights and promotion of ESTs
In general, developing countries and their companies tend to have fewer resources to purchase licenses and fear that strong IPR regimes would impede their access to patented technologies. To meet this concern, bilateral and multilateral financial assistance could provide funds for licensing of relevant technologies. This

could be one way for the international regime to support technology transfer in cases where market-driven, grants, equity investment, and joint venture solutions are inadequate or infeasible.

2.2.3 Developing countries actions

Assessment of local technology needs and meeting of local demands
Participatory development is now widely recognised as a way of achieving effective technology transfer at all levels of development endeavour. This has grown from a perceived need to move from donor driven technology transfer to national needs driven approaches. It can facilitate market transformation through the involvement of firms and consumers. Governments are the most direct and influential actors for promoting a favourable environment for participation among the private sector, public sector organisations, NGOs and grassroots organisations at regional and local levels. Practical experience with participation but the step towards mainstreaming in government and development agencies still has to be made.

Meeting local demands also includes examining what the social impacts of technology transfer will be and how negative impacts can be reduced. There is a particular need for developing guidelines for ensuring that technology transfer projects do not disempower or negatively influence weaker social groups in a society. Such guidelines could draw from guidelines on integrating gender issues in technology development.

Participatory development can thus achieve:
- Better choices and identification of possibilities and opportunities in local systems
- Better commitment to projects which improves implementation and sustainability
- Opportunities to negotiate conflicts
- Empowerment- which raises awareness about the need for stakeholders to achieve solutions themselves.
- Access to additional resources for the project raised from the target project beneficiaries through payments, and time.

Property right issues and ownership
The experience in agriculture, forestry, and use of other natural resources has shown that the successful introduction of new technologies often depends on a recognition of the existing forms of ownership, or on taking steps to create an improved property rights regime. With an understanding of existing – legal and actual – forms of ownership, technologies or modified resource uses can be adapted to fit this existing system. If property issues are taken into account, those introducing new technologies or proposing modifications in land or resource use can be more assured of the support of the target populations or groups.

Appropriate R&D programmes for adjustment of ESTs to own local conditions
Developing countries' R&D efforts are often adaptive, following externally developed technology, suggesting the need for additional resources to develop indigenous innovative capacity. R&D and the process of innovation are closely linked, but innovation has been found to fail at the level of capability - the ability to focus specific sets of resources in a particular way (e.g., financial management, marketing, understanding user needs, etc.) rather than because of inadequate resources or hardware. The process of dissemination depends closely on influencing key opinion formers, at government, industry, firm or community level.

Improved pathways for South-South transfer
Most technology transfer to date has passed along a North-South axis, and given financial constraints in many developing countries and CEITs this situation is likely to continue. However, creative means of using developed country bilateral aid, multilateral programmes and increased access to world capital markets may provide opportunities to increase South-South transfers. Enhancing South-South transfers is important, because developing countries may encounter challenges that are unlikely to be found in developed countries, but for which solutions exist in other developing countries. Initiatives to improve the pathways for South-South transfer could include:
- Sharing of information regarding the performance of ESTs in developing countries;
- Joint R&D and demonstration programmes;
- Opening markets for ESTs from other developing countries.

2.3 Mechanisms for technology transfer

2.3.1 National systems of innovation

In recent years it has become clear that technology intermediation is needed to reduce barriers to technology transfer associated with information, management, technology, and financing. Research on technology innovation also highlights the role of intermediaries in the innovation process. They operate between users and suppliers of technology and help to create the links within networks and systems through bridging between institutions, encouraging interaction within the system and assisting with undertaking search, evaluation and dissemination tasks. They ensure that technological know-how is broadly dispersed within the system and can provide a compensating mechanism for weaknesses or "holes" in the system. The high value of technology intermediation is illustrated by many of the case studies presented throughout the Report (see for instance Box TS 2).

Examples of technology intermediaries include specialised government agencies, energy-service companies, non-governmental organisations, university liaison departments, regional technology centres, research and technology organisations, electric power utilities, and cross-national networks. Non-governmental organisations in particular are playing a greater role in technology intermediation; for example, there are many cases where technology intermediation by non-governmental organisations played a key role in the success of particular technology transfer efforts for renewable energy.

The key lesson which can be learnt from the literature on technology intermediaries, is the importance of mechanisms in which actions are integrated to make them more effective. Technology transfers are influenced greatly by what have been called national systems of innovation--the institutional and organisational structures which support technological development and innovation. Governments can build or strengthen scientific and technical educational institutions and modify the form or operation of technology networks—the interrelated organisations generating, diffusing, and utilising technologies.

National systems of innovation (NSI) integrate the elements of capacity building, access to information and an enabling environment into a mechanism for EST transfer that adds up to more than the individual components. Subsystems and the quality of interconnections within them can successfully influence technology transfer. NSIs can be enhanced through partnerships sponsored by international consortia. Partnerships would be system oriented, encompass all stages of the transfer process, and ensure the participation of private and public stakeholders, including business, legal, financial and other service providers from developed and developing countries.

NSI activities may include:
- Targeted capacity building, information access and training for public and private stakeholders and support for project preparation;
- Strengthening scientific and technical educational institutions in the context of technology needs;
- Collection and assessment of specific technical, commercial, financial and legal information;
- Identification and development of solutions to technical, financial, legal, policy and other barriers to wide deployment of ESTs;
- Technology assessment, promotion of prototypes, demonstration projects and extension services through linkages between manufacturers, producers and end users;
- Innovative financial mechanisms such as public/private sector partnerships and specialised credit facilities;
- Local and regional partnerships between different stakeholders for the transfer, evaluation and adjustment to local conditions of ESTs (see for instance the TCAPP example in Box TS 3);
- Market intermediary organisations, such as Energy Service Companies.

NSI offer a new solution to the challenge of technology transfer and offer a cost effective, flexible way of enhancing technology transfer. Governments and multinational organisations may wish to consider developing programmes to support NSI activities, either individually or jointly.

Comprehensive approaches to technology transfer which incorporate many of the elements listed above are beginning to emerge on both a bilateral and multilateral basis. One such activity being conducted on a bilateral basis by the United States is the Technology Cooperation Agreement Pilot Project, known as TCAPP. A similar multilateral approach is being pursued by the Climate Technology Initiative (CTI), an activity undertaken by 23 developed country Parties and the European Commission in support of the UNFCCC. Both approaches utilise a bottom-up, collaborative process under which all relevant stakeholders are engaged to jointly determine the technology selection/practices and implementation path consistent with that country's/region's sustainable development goals for one or more sectors. These approaches require a mutual commitment to explore actions which focus on achieving climate and development benefits to the host country, with active participation of relevant government, non-government, community, and industry groups. Since the goal is to translate sectoral needs into functioning projects, consideration is given to how the projects might be best structured to encourage private sector investment, including the need for regulatory or other change within the country to foster an enabling environment.

2.3.2 ODA

Donor governments have often pursued multiple interests in development assistance, including economic and geo-political goals – e.g. contracts for their domestic firms, support for friendly political regimes – which have not always been consistent with the sustainable development objectives of host countries. This has resulted in development aid, and particular tied bilateral aid, having a very mixed record. It is increasingly recognised that Official Development Assistance (ODA) can be better used to address the fundamental determinants of development, which include a sound policy environment, strong investment in human capital, well functioning institutions and governance systems and environmental sustainability rather than as a leading source for investment in environmentally sound and cleaner technologies. This realisation has arisen not least from the mixed experience with development aid programmes.

2.3.3 GEF

The Global Environment Facility, the operating entity of the UNFCCC Financial Mechanism, is a key multilateral institution for transfers of ESTs. The GEF aims to promote energy efficiency and renewable energy technologies by reducing barriers, implementation costs, and long-term technology costs. A significant aim of these programmes is to catalyse sustainable markets and enable the private sector to transfer technologies. GEF projects are testing and demonstrating a variety of financing and institutional models for promoting technology diffusion and several GEF projects are designed to directly mobilise private-sector finance. Capacity building is a central feature of most GEF projects and is resulting in indirect impacts on host countries' abilities to understand, absorb and diffuse technologies. Projects build the human resources and institutional capacities that are widely recognised as important conditions for technology adoption and diffusion. Next tot the programmes on renewable energy and energy efficiency, additional operational programmes for energy-efficient transport and carbon sequestration are now being developed.

From 1991 to 1998 the GEF approved grants totalling $610 million for 61 energy efficiency and renewable energy projects in 38 countries. The total cost of these projects is $4.8 billion, as the GEF has leveraged financing through loans and other resources from governments, other donor agencies, the private sector, and the three GEF project-implementing agencies (UN Development Programme, UN Environment Programme and World Bank Group). An additional $180 million in grants for enabling activities and short-term response measures have been approved for climate change. Compared to the magnitude of the technology transfer challenge, these efforts are of modest scale even when added to the contributions from bilateral development assistance For instance, ODA in 1997 totalled $40-50 billion and private-sector transfers from FDI reached $240-250 billion. Achieved energy savings and renewable-energy capacity installed through GEF-supported projects are small but not insignificant relative to world markets. This is especially significant for renewable energy technologies such as wind, solar thermal, solar PV home systems and geothermal.

Installed capacity or direct energy savings is only part of the GEF impact. GEF projects have attracted considerable attention among policy-makers and industry in host countries and among the international community. Through policy changes, stakeholder dialogues, and project design activities and studies, GEF projects have provided an important stimulus for technology transfer beyond direct project impacts.

Continued effectiveness of GEF project funding for technology transfer may depend on factors such as:

- Sustainability of market development and policy impacts achieved through GEF projects;
- Duplication of successful technology transfer models;
- Enhanced links with multilateral-bank and other financing of ESTs;
- Funding for development and licensing of ESTs;
- Coordination with other activities that support national systems of innovation and international technology partnerships;
- Attention to technology transfer among developing countries.

2.3.4 Multilateral Development Banks

The Multilateral Development Banks (MDB) have seen technology transfer as part of their mission to encourage development. More recently they started to focus on the challenges of the environment and the specific problems involved in transferring environmental technology. In response many have started to develop a range of initiatives and activities.

In particular, development banks have become aware of the role they can play in helping to mobilise private capital to help meet

---

BOX TS 3   TECHNOLOGY COOPERATION AGREEMENT PILOT PROJECT (TCAPP)

In 1997, the U.S. Government launched the Technology Cooperation Agreement Pilot Project (TCAPP) to provide a model for a collaborative approach to foster technology cooperation for climate change mitigation technologies. Under TCAPP, the Governments of Brazil, China, Kazakhstan, Mexico, and the Philippines are currently working with the private sector and bilateral and international donor organisations to attract private investment in clean energy technologies in their countries. Many other donor initiatives have also adopted similar collaborative approaches between country officials, businesses, and donors in fostering private investment. However, TCAPP is one of the few initiatives that has engaged climate change officials in this collaborative process to lead to actions that address both development needs and climate change goals.

TCAPP has two basic phases of activities. In the first phase, the participating countries have developed technology cooperation frameworks that define their climate change technology cooperation priorities and the actions necessary to attract private investment in these priorities. These actions include efforts aimed at capturing immediate investment opportunities (e.g., issuance of investment solicitations, investment financing, business matchmaking and capacity building, etc.) and longer-term efforts to remove market barriers. In the second phase, TCAPP assists the country teams in securing the private sector, in-country, and donor participation and support necessary to successfully implement these actions. This second phase of activities includes two major types of activities:

1   Attracting direct private investment in immediate market opportunities. This includes helping the countries develop and issue investment solicitations for large-scale opportunities and business matchmaking and financing activities. TCAPP has established an international business network to help guide the design and implementation of these activities.

2   Securing support for actions to address market barriers. These actions range from business capacity building to policy reform. This includes development of domestic implementation plans for these actions and securing necessary donor support to assist with implementation of these plans. TCAPP assists the countries in preparing implementation plans and donor proposals and in matching country needs with donor programmes.

(Sources: NREL, 1998, UNFCCC, 1999)

the needs of sustainable development and the environment, and of the potential to use financial innovation to encourage environmental projects and initiatives. While much of the earlier work they did was sporadic, the private sector arms of the Multilateral Development Banks are now seeking to identify ways they can work with international private capital to help address environmental and developmental needs.

The World Bank, including its affiliate the International Finance Corporation, has developed a number of initiatives with the potential to support environmental technology transfer. These include financing a number of environmental lending programmes at domestic financial institutions, meant for local industries. An important new initiative is the Carbon Investment Fund. This is a vehicle which will provide additional finance for CO_2 mitigating projects in return for carbon offsets, i.e. the right to transfer the credit for the CO_2 saved to the investor. It is expected to have a substantial private sector component.

The European Bank for Reconstruction and Development (EBRD), is the only one to have sustainable development incorporated in its charter. It is also quite active in working with the private sector . Thus it is not surprising that it has shown a fair amount of leadership in helping to encourage technology transfer. Examples of its initiatives include working with intermediary banks to educate them on environmental issues and the potential of clean technology, and activities to support the development of Energy Service Companies.

Other MDBs, such as the regional development banks (African Development Bank, AfDB, Asian Development Bank, ADB, Inter-American Development Bank, IDB) all play an important role in regional investment, and most have given some attention to issues of sustainable development, in varied ways. The European Investment Bank is also a huge investor, internationally particularly to the Asia-Pacific-Caribbean countries under the Lomé Convention; its investment programmes however have not generally been coordinated with the EU's goals on sustainable development and climate change.

Governments can use their leverage to direct the activities of multilateral development banks through their respective Boards and Councils in order to:

- Strengthen MDB programmes to account for the environmental consequences of their lending;
- Develop programmatic approaches to lending that remove institutional barriers and create enabling environments for private technology transfers;
- Encourage MDBs to participate in NSI partnership initiatives;
- Pay special attention to South-South technology transfer.

2.3.5   *Kyoto Protocol mechanisms and the UNFCCC*

The analysis of the literature on the Kyoto Protocol Mechanisms, based on the preliminary stage of development of the rules for these, suggests that if they are implemented, the Mechanisms may

have potential to affect the transfer of ESTs. The extent to which Article 4.5 of the UNFCCC has been implemented is being reviewed by the UNFCCC. Given this evolving process, the IPCC has not been able to assess this matter.

The Clean Development Mechanism (CDM) and Joint Implementation (JI) can provide financial incentives for ESTs and influence technology choice. As voluntary mechanisms they require cooperation among developed and between developed and developing country Parties as well as between governments, private sector entities and community organisations.

Although much about the design and governance of the CDM remains to be resolved, some notions on the CDM are starting to emerge from a variety of new literature that has been published since Kyoto:

- The CDM can be a means to build trust and strengthen capacity. It could strengthen working relations and understanding among partners, private sector, non-governmental organisations, and governments at various levels to enhance technology cooperation. Project based crediting could lead to tangible investments and development of local capacity to maintain the performance of these investments. These projects could incrementally assist developing countries to achieve multiple sustainable development objectives (economic development, improvement of local environmental quality, minimise risk to human health of local pollutants, and reduce greenhouse gases). Careful project screening and selection, including host community decision-makers, will assist multiple benefits for all participants.
- There is a need to design simple, unambiguous rules that ensure environmental performance in the context of sustainable development while also favouring investment. The multilateral oversight and governance provisions of the mechanism, and the project basis of transactions, will raise the transaction costs of investment in CDM projects as compared to mitigation reduction through other more conventional means (e.g. local options or even within other Annex I countries). This increases the complexity of the transaction and shaves a portion of the economic benefit that might otherwise be attained. Critical questions under this heading have to do with how to determine the additionality of proposed projects as well as the baseline against which to assess performance and establish transferable "credits."

One important distinction is between "bilateral" or "portfolio" approaches. The bilateral approach is closest to joint implementation programmes where the host country negotiates directly with the investor about the terms of the contract. The portfolio approach would allow host (developing) countries to advance bundles of possible projects that fit with their own sustainable development objectives. Some authors argue that both models will be needed, depending on the type of project involved.

## 3.	Technology Transfer: A Sectoral Analysis

Domestic actions, and those taken in co-operation with other countries, will require an increased penetration of environmentally sound technologies, many of which are particularly important in their application to each sector. What is the potential for the penetration of mitigation and adaptation technologies? What barriers exist to the increased penetration of such technologies? Can these be overcome through the implementation of a mix of judicious policies, programmes and other measures? What can we learn from past experience in promoting these, or similar, technologies? Is it better to intervene at the R&D stage or during the end-use of fuels and technology? The chapters in this section address these questions using examples specific to each sector. Technology transfer activities may be evaluated at three levels – macro or national, sector-specific and project-specific. Many of the options explored are at the latter two levels.

Greenhouse gas emissions from some sectors described are larger than those from other sectors, and the importance of each greenhouse gas varies across sectors and countries as well. Methane, for instance, is a much bigger contributor to emissions from agricultural activity than, for instance, from the industry sector. Table TS 5 shows the carbon emissions from energy use in 1995. Emissions from electricity generation are allocated to the respective consuming sector. Carbon emissions from the industrial sector clearly constitute the largest share, while those derived from agricultural energy use have the smallest share. In terms of growth rates of carbon emissions, however, the fastest growing sector is transport and buildings. With rapid urbanisation promoting increased use of fossil fuels for habitation and mobility, the two sectors are likely to continue to grow faster than others in the future. Carbon emissions from fossil fuels used to generate electricity amounted to 1,762 Mt C of the total for all sectors in Table TS5.

Table TS5	Carbon emissions from fossil fuel combustion in Mt C (Price *et al.*, 1998)		
SECTOR	CARBON EMISSIONS AND(%SHARE) 1995	AVARAGE ANNUAL GROWTH RATE (%)	
		(1971-90)	(1990-95)
Industry	2370 (43%)	1.7	0.4
Buildings			
• Residential	1172 (21%)	1.8	1.0
• Commercial	584 (10%)	2.2	1.0
Transport	1227 (22%)	2.6	2.4
Agriculture	223 (4%)	3.8	0.8
All Sectors	5577 (100%)	2.0	1.0
Electricity Generation*	1762 (32%)	2.3	1.7

NOTE: EMISSIONS FROM ENERGY USE ONLY; DOES NOT INCLUDE FEEDSTOCKS OR CARBON DIOXIDE FROM CALCINATION IN CEMENT PRODUCTION. BIOMASS = NO EMISSIONS.

* INCLUDES EMISSIONS ONLY FROM FUELS USED FOR ELECTRICITY GENERATION. OTHER ENERGY PRODUCTION AND TRANSFORMATION ACTIVITIES DISCUSSED IN CHAPTER 10 ARE NOT INCLUDED.

Technology transfer includes steps within and across countries by actors who are engaged in promoting the use of a particular technology along one or more pathways. The market penetration of a technology proceeds from research, development, and demonstration (RD&D) to adoption, adjustment, replication and development. At a project-specific level, the elements of the pathway are different, and may proceed from project formulation, feasibility studies, loan appraisals, implementation, monitoring and evaluation and verification of carbon benefits.

The pathways that differ from sector to sector usually include many actors, starting with laboratories for RD&D, manufacturers, financiers and project developers, and eventually the customer whose production capacity or welfare is hopefully enhanced through their use. This presumption needs to be carefully established through an assessment of technology needs of the customer. A poor needs assessment can result in ineffective technology transfer that could have been avoided had the assessment fully captured the social, and other attributes of the technology. The actors may make specific types of arrangements – joint ventures, public-private parnerships, licensing, etc., that are mutually beneficial. These arrangements will define the particular pathway chosen for technology transfer.

The spread of a technology may occur through transfer within a country and then transfer to other countries, both may occur simultaneously, or transfer across countries may precede that within a country. Generally, the spread of a technology is more likely to proceed along the first option rather than the other two, since the transfer of technologies to markets within a country is likely to be less expensive given the proximity to the market, and lower barriers to the penetration of that technology in the indigenous markets. Transfer of technology from one country to another will generally face trade and other barriers both in the initiating and recipient country, which may dissuade manufacturers and suppliers from implementing such transfer.

Many market barriers prevent the adoption of cost-effective mitigation options in developing countries. Market barriers can be divided in more common barriers which are more or less relevant for all sectors (see the above section on "Barriers to the transfer of ESTs" and tables TS3 and TS4) and barriers specific for each sector. For example, the presence of subsidies for electricity and fuels are highly relevant for ESTs in the energy sector, but also affects the transfers of ESTs in the transport sector (through subsidised fuel costs), the building and industry sector (the viability of energy efficient technologies), waste sector (electricity generation from waste) and even in agriculture and forestry (it affects the demand for biomass fuels such as agricultural waste and wood). On the other hand, barriers like the risks of drought, fire and pests are very sector-specific and mostly affect the forestry and agriculture sectors.

What conditions and policies are necessary to overcome these barriers and successfully put in place technologies for mitigation and adaptation? There is no pre-set answer to enhancing technology transfer. The combination of barriers and actors in each country creates a unique set of conditions, requiring "custom" implementation strategies. Each of the sector specific chapters discusses the barriers that are particularly important to a sector, such as fuel and electricity price subsidies, weak institutional and legal frameworks, lack of trained personnel, etc. Each chapter also provides examples and case studies to highlight the barriers, and policies, programmes and measures that were used, or could be developed, to overcome them.

Adaptation technologies

The general dynamics of technology transfer apply to the transfer of climate mitigation and adaptation technologies. Nevertheless, it is important to note the special characteristics of adaptation technologies that distinguish them from mitigation technologies.

Many impacts of climate change will impinge on collective goods and systems, such as food and water security, biodiversity and human health and safety. These impacts could affect commercial interests indirectly, but usually the strongest and most direct incentives to adapt are with the public sector. The use and transfer of many adaptation technologies world-wide has occurred because of societal interventions, not as a result of market forces. Examples of such interventions include direct governmental expenditures, regulations and policies and public choices.

Apart from the government being a dominant stakeholder in technology transfer for adaptation, four more characteristics often distinguish adaptation from mitigation to climate change. Each of these characteristics also represents a barrier to adaptation and associated technology transfer:

- Uncertainty concerning the role of greenhouse gases in causing climate change has been reduced, but uncertainty about the location, rate and magnitude of impacts is still considerable, which could hamper effective anticipatory adaptation.
- Adaptation technologies will often address site-specific issues, and will therefore have to be designed and implemented keeping local considerations in mind. This could hamper large-scale technology replication.
- As opposed to benefits of mitigation, which are global (reduced atmospheric greenhouse-gas concentrations), benefits of adaptation are primarily local. For this reason, adaptation projects thus far have attracted limited interest from the Global Environment Facility (GEF) and other donors.
- The implementation of mitigation technologies can contribute to the development of a country's energy-consuming sectors, while adaptation technologies are primarily aimed at preventing or reducing impacts on these and other sectors. As such, adaptation is often not considered a development objective.

In spite of adaptation often not being considered a development objective, governments have a number of clear incentives and opportunities to start planning for adaptation. For example, many adaptation technologies do not only reduce vulnerability to anticipated impacts of climate change but also to contemporary hazards associated with climate variability. It could be considered "no-regret" adaptation or "climate safe development", having utility both now and in the future, even if climate change were not to occur. In addition, adaptation options need to be designed keeping site-specific natural and socio-cultural circumstances in mind. Strengthening technological, institutional, legal and economic capacities as well as raising awareness are important for effective adaptation and technology

transfer, for no adaptation option will be successful when it is implemented in an environment that is not ready, willing or able to receive the option.

3.1   *Residential, Commercial, and Institutional Buildings Sector*

In the residential, commercial, and institutional building sector energy is used to heat and cool buildings, provide lighting, and services ranging from cooking to computers. The emissions from the building sector include those from the direct use of fossil fuels in buildings and emissions from the fuels used to provide electricity and heat to buildings.

Buildings are long lasting and community development patterns have even longer lives. The incremental costs of the best technologies are slight at time of construction, compared with the cost of replacing energy-wasteful buildings and equipment. The technologies themselves are varied and powerful. The IPCC Technical Paper I found an existing technical potential to meet the sector's global energy needs through 2050 with no increase in energy use from the 1990 level.

Buildings vary greatly in their size, shape, function, equipment, climate, and ownership, all of which affects the mix of technologies needed to improve their performance. In some countries, the energy used in housing is "free" or subsidised to the occupants. When they don't have to pay the full cost of the energy they use, they have less incentive to use it wisely. Where a large portion of the occupants would have great difficulty paying the full costs immediately, there is strong political pressure to continue the subsidies.

In the near term, the most successful technology transfer programmes won't be driven by their environmentally sound benefits alone, but because they also meet other human needs and wants. Examples include new energy-efficient buildings that are more comfortable and have lower energy costs, as well as lower GHG emissions; and the Kenya cookstove that is cheaper and healthier to use, and also lowers GHG emissions. The most successful technology programmes focus on new products and techniques that have multiple benefits.

The most successful mechanisms for technology transfer include government mandated energy and environmental standards for new buildings and equipment; information, education and labelling programmes; and government-supported research, development, and demonstration (RD&D) programmes. Governments also have a key role in creating a market environment for successful private sector-driven technology transfer through decisions on full cost pricing, financing, taxes, regulations, and customs and duties. Local governments can encourage successful community programmes by proactively identifying community-level needs and by encouraging and responding to community initiatives.

3.2   *Transport*

Transport-related GHG emissions are the second-fastest growing sector emissions world-wide as shown in Table TS5, but the transport sector is the least flexible to changes because of its almost dependence on petroleum-based fuels, current entrenched travel lifestyles, and lack of political will. Further, transportation is growing in all regions of the world, 3-4% annually in developed countries and higher in developing countries and air transportation is growing even more rapidly, about 5% annually world-wide. Controlling the associated GHG emissions pose serious challenges because a departure from current lifestyles and use patterns are required in addition to changes in travel movements.

However, a number of efforts such as performance gains, safety and energy intensity improvements have lead to the development of many technical and non-technical options that reduce GHG emissions. Most of these options are technically feasible but not all are economically feasible. The cost-effectiveness of most options varies among users. Resource availability, technical know-how, institutional capacity and local market are among the different factors that affect the cost of these options.

Improvements in vehicle technology such as improved engine design or vehicle body with the aim of reducing energy intensity can result in reduction of carbon emissions. Similarly, low cost actions such as proper maintenance and overall servicing of vehicles will lead to both reduction in fuel use and carbon emissions. The use of improved gasoline and diesel, and alternative transport fuels such as compressed natural gas (CNG), liquefied petroleum gas (LPG), methanol and ethanol can also lead to reductions in GHG emissions. Electric vehicles are penetrating in niche markets and hydrogen powered vehicles with the potential of much reduced GHG emissions could be feasible in the future. The energy intensity of aircrafts can be improved with engine modifications and new designs.

Wider use of public transport such as more comfortable and safer buses and non-motorised systems such as bicycles, rickshaws, and push-carts are examples of options with the supporting infrastructure such as dedicated lanes and improved signalling can result in significant environmental benefits including reduction of GHG emissions. Changes in transport infrastructure and systems to reduce travel trips improve modal choices, and increased freight volume per trip can result in reduction of GHG emissions. Also, the use of some non-transport options such as improved urban planning, and transport substitution using modern telecommunications options can lead to reduce trips and thus GHG reductions.

These options could be transferred within and across countries and regions of the world through different mechanisms. Options such as the manufacture and improvements of vehicle and aircraft, and developments in fuel technologies can be transferred through market oriented paths, while those related to transport infrastructure can be through non-market oriented paths such as bilateral and multilateral institutions. However, recently, private sector involvement in the sector is growing resulting in minimising the difference between these paths.

However, there are significant technological, economic and institutional barriers that can prevent transferring these options within and across countries and regions. Lack of suitable local firms to supply components and services required by large firms limit technology transfer within countries. Unavailability of technical and business information affects penetration within and across countries. Local firms suffer from limited access to capital, a barrier that also affects many countries especially those requiring transport infrastructure. A non-supportive environment for technology transfer for both technology supplier and recipient negatively affects the transfer of transport technologies and is amongst others reasons due to the lack of political will to take the necessary actions such as instituting standards with complimentary compliance regimes.

Various government policies and actions can facilitate the transfer of these options in addition to overcoming some of these barriers. They can also provide non-climate related benefits such as reduction in local air pollution and road congestion. Transfer of technological experiences between countries can stimulate promotion of low cost options such as proper vehicle and aircraft maintenance; enforceable regulatory systems for inspection and testing of vehicles and aircrafts; improved traffic management; and improvements in the quality of drivers. Promotion of policies aimed at reduction of transport intensities for passenger systems such new public transport modes, local market re-organisation, comfortable walkways, wider use of telecommunications, and new freight systems such as dematerialisation, regionalisation of production networks, and use of new logistics systems can lead to reductions in GHG emissions. Use of environmentally friendly transport infrastructure aimed at reduction in travel distance, affecting modal choice, and use of dedicated lanes are useful, but can be expensive, time-consuming and may require behavioural and lifestyle changes.

Encouraging cooperative technology programmes between countries and enterprises can result in the transfer of many of these options. These programmes may include joint R&D, design and manufacture, and information networks on management and specific technical skills. Encouraging sub-contracting among firms and enterprises will promote the transfer of technical and managerial skills. However, governments of technology recipients need to build further local capacities for information development and exchange, technology assessment and selection, negotiation abilities, and support infrastructure, as these will create the enabling environment for effective technology transfer and development.

The need for commitment from governments is crucial in stimulating technology transfer, both for technology outflows and technology inflows. Instituting policies that promote transport technology outflows such as special incentives for ESTs and build capacities to receive technology inflows through an improved business environment will be important in increasing the flow of ESTs in the transport sector.

3.3 *Industry*

The industrial sector is extremely diverse and involves a wide range of activities. Aggregate energy use and emissions depend on the structure of industry, and the energy and carbon intensity of each of the activities. The structure of the industry may depend on the development of the economy, as well as factors like resource availability and historical context. In 1995, industry accounted for 41% of global energy use and up to 43% of global CO_2 emissions. Besides CO_2 industry also emits various other GHGs. Although the efficiency of industrial processes has increased greatly during the past decades, the potentials for energy efficiency improvement in all processes remain large. Fundamentally new process schemes, energy and resource efficiency, substitution of materials, changes in design and manufacture of products resulting in less material use and increased recycling, can lead to substantial reduction in GHG emissions.

Technology transfer can be most effective in OECD countries if technologies comparable to that of efficient industrial facilities are adopted during stock turnover. For countries with economies in transition, technology transfer options are intimately tied to the economic redevelopment choices and the form that industrial restructuring takes. In developing countries, large potentials for adoption of energy efficient technologies exist as the role of industry is expanding in the economy. Hence, in industry, GHG emission reduction is often the result of investments in modern equipment, stressing the attention to sound and environmentally benign investment policies.

Barriers for technology transfer in the industrial sector include corporate decision-making rules, lack of information, limited capital availability, shortage of trained personnel (especially in small and medium sized enterprises), low energy prices, and the "invisibility" of energy savings. Developing countries suffer from all barriers that inhibit technology transfer in industrialised countries plus a multitude of other problems.

Global investments in industrial technology (i.e. hardware and software) are dominated by the private sector, although in some countries government-owned enterprises are still important. Foreign capital investment in the industrial sector is increasing (see Figure TS2). Foreign Direct Investment (FDI) is a large part of the total foreign investments. FDI is increasing, although concentrated on a small number of rapidly industrialising countries (see Figure TS3). Despite the concentration, the development in these countries may impact regional industrial development patterns, as seen in Southeast Asia. Private investment is increasing in other developing regions but still limited. Public funding (in industrialised and developing countries) for technology development and transfer, although still important, is decreasing. Funding for science and technology development is important to support industrial development, especially in developing countries. Public funding in the industrial sector, although small and declining in comparison to private funding, remains important for longer term development.

An effective process for technology transfer requires active networks among users, producers and developers of technology. The variety of stakeholders makes it necessary to have comprehensive industrial policies for technology transfer and cooperation, both from a technology donor and technology recipient perspective. Such a framework may include environmental, energy, (international) trade, taxation and, patent legislation, as well as a variety of well-aimed incentives. Changing economic and technical developments stress the need for innovative and flexible approaches. Instead of regulation specifying the means, policies such as voluntary agreements with well-defined goals for energy efficiency improvements, can be more successful. There is a strong need to develop the public and private capacities to assess and select technologies, in particular for state owned and small and medium-sized industries. Stakeholders (policymakers, private investors, financing institutions) in developing countries have even more difficult access to technology information, stressing the need for a clearinghouse for information on climate change abatement technology, well integrated in the policy framework. To be successful, long-term support for capacity building is essential to meet the need for public support and cooperation of technology suppliers and users.

Adjustment of technology to local conditions is essential, but practices vary widely. Countries that spend on average more on technology adjustment seem to be more successful in technology transfer. Successful technology transfer includes the development of technological capabilities of the user or host. This will accelerate the technology transfer process through assimilation, adjustment and development of new technologies.

3.4 *Energy Supply*

Major objectives of the current energy supply sector are economic development and international competitiveness. Climate change objectives, particularly the reduction of CO_2 emissions, do not play a significant role. Nevertheless, significant opportunities exist to reduce CO_2 emissions (see Table TS5) as well as other GHG emissions.

Technology transfer in the energy sector is mainly driven by the private-sector in oil, natural gas and electricity supply. On the other hand, coal, nuclear and renewable sources are often dependent on government to preserve or increase their presence in the market. Technology transfer, as presently understood, is a relatively new process since historically it was used as an euphemism for large-scale power projects financed by multilateral banks or for limited knowledge transfer from the international oil and gas companies to national industries. The oil crises in the 1970s changed the contractual terms in the oil and gas sector when powerful national oil companies were able to negotiate technology transfer on more favourable terms. In the early 1990s, the process of market globalisation and the availability of private capital on a global scale triggered investments and hence technology transfer opportunities in the electric sector also. The private sector is now playing a bigger role in electric power generation in concert with a new set of regulations and standards.

In the energy supply sector, technology transfer comes with investment. One of the keys to the transfer of technology is to promote investment through an appropriate economic and institutional framework. For some, the high initial costs of clean energy supply technologies is a major barrier for their transfer. In general, economic and institutional barriers rather than technology availability are more apt to be the cause of failure to transfer technology. In all energy sectors, the role of government in facilitating technology transfer is critical. Annex II countries could develop more effective policies to stimulate and finance private investments in clean energy sources in developing countries and CEITs. Developing countries and CEIT governments should have policies in place for liberalising the energy supply market, fostering and ensuring conditions to allow international financing, promoting infrastructure development, eliminating unnecessary regulatory and trade barriers, educating and training local workforce, protecting intellectual property rights and strengthening local R&D and environmental management regimes

Barriers
General barriers for technology transfer in the energy sector are:
- Lack of incentive for the major utilities: electric utilities sell electricity through a regulatory review process that allows the utility to recover all operating expenses, including taxes and a fair return for its pendent investments. This means that independent of the quality of the service they will be reimbursed by the operational costs;
- Lack of human qualifications in developing countries. Without investment in capacity building the existing electric service is of lower quality compared with industrialised countries, information gathering is not a priority and new technologies, which may be less costly or more environmentally friendly, are not even taken into account;
- Although many countries are revising their trade policies in order to liberalise markets, substantial tariff barriers for imports of foreign technologies including energy supply equipment remain in many cases. This limits exposure to energy efficiency improvement pressures from foreign competition on domestic suppliers and prevents early introduction of sustainable energy innovations from abroad;
- Political interference and corruption caused by powerful construction groups which prefers the installation of more generation capacity based on conventional technologies;
- Institutional and administrative difficulties to develop technology transfer contracts, which can be a necessity to qualify regional construction companies as potential partners of the entrepreneurship. There is a need for greater regional co-operation among developing countries both in R&D work and in the international commercial contracting network;
- Poor access to information. It is one thing to recognise that the information and technology desired are available, but it is quite another issue to gain access to them. That will require that developing countries strengthen their linkages with the rest of the world by investing in the infrastructure needed to receive and transfer information;
- A major requirement for successful agreement in technology transfer is the guarantee of intellectual property rights (IPR). In the energy sector, a well-developed mechanism exists for sharing IPR. It is the production-sharing contract. Under this agreement, private firms contract with local parties, usually state-owned companies or governments, to share technology with them in return for a share of the products produced. This practice has proven very successful in the international oil and gas sector and could be a model for other energy supply areas;
- Needs of the developing countries are quite different to these of the developed countries.

Policies
It is expected that markets will respond to whatever regulatory policies are adopted to promote the reduction of GHG emissions. This will stimulate technology transfer as investments are made in response to the price signals when uncertainties on policies to secure environmental goals are settled. A number of generic strategies can promote and/or facilitate the implementation of some of the emission reduction options in the energy sector. These strategies could include (but are certainly not limited to):
- Creating political and regulatory framework to allow full cost pricing and recognise the indirect benefits of renewables such as the creation of more local jobs, improvement of the environment, balance of trade, etc.
- Providing information and general training to government officials -The advantages and local/international opportunities provided by the measures and technologies must be presented to top officials in a manner that is both forceful and forthright; top officials embracing the concept;
- Providing specific information and training to local actors - Training of a very specific and practical nature must be provided to personnel at the local level;
- Encouraging the implementation and enforcement of energy and environmental standards - Standards require the capability to enforce them by recruiting and training enforcement personnel and supplying them with the necessary tools and high-level administrative support needed for credible implementation of sanctions. Setting up these regulations and support structures is an area where international assistance may be valuable for some countries;

Governments play a special role in the creation of markets for ESTs. This can be done by governments through procurement of new technologies for public missions like energy production and distribution, and provision of incentives for its development, including grants, low interest loans, import duty exemption, income tax exemption, tradable permits and competitively determined subsidies. Grants are a very common incentive used to stimulate adoption of a new technology in developed countries (e.g. The Clean Coal Program in US) An example of the use of income tax exemption is the Green Funds scheme in The Netherlands. Also in the Netherlands a pilot a trading mechanism for Green Certificates (including a spot and forward market) has been established for renewably generated electricity, which makes it feasible to reach quantified targets for renewable energy introduction efficiently. Competitively determined subsidies have been employed such as the Non-Fossil-Fuel Obligation (NFFO) in UK, and the electricity feed law (EFL) in other parts of Europe.

Fossil Fuel

Technology transfer in the fossil fuel sector is mature, and well-established mechanisms are in place. Technology is readily available from a wide variety of sources, such as the oil, gas, and coal industries, engineering contractors, equipment vendors, etc. Barriers to technology transfer are primarily economic and institutional. Specific measures and policies could include:

- Creation and/or enlargement of the market for clean energy technologies like:
  - switching to lower carbon fuels such as the substitution of natural gas for coal
  - using of combined heating (cooling) and power generation where feasible
  - using sequestration of CO_2 where feasible
  - using of high efficiency in electric power generation
- Enhancing regional co-operation in development and transportation of natural gas and electricity across national boundaries.
- Increasing efficiency by promoting dissemination of best practices.

Nuclear

Technology transfer in the nuclear power sector for water-cooled and water moderated reactors is also mature and has well-established mechanisms. Because of the unique safety, ecological and proliferation risks involved, successful transfer of nuclear technology requires major government involvement and careful consideration of the costs and benefits involved. The large capital costs, lack of public acceptance, availability of cheap domestic fossil fuel and the resolution of safety and waste disposal provide significant barriers to the use of nuclear energy. In many cases, nuclear proliferation issues are also a major problem to be addressed by governments and other international institutions. Developing infrastructure and trained personnel is required to insure the highest possible level of nuclear safety.

Renewables

In the renewable sector technology transfer has been constrained by the lack of investment and high costs. Investment has been generally limited to niche or protected markets, because of technical, institutional and economic barriers. Governments need to provide incentives for investment and to remove policies that hinder the application of renewable energy as described in the general policy measures above. They also need to promote the development of improved and more cost effective renewable technologies amongst others by:

- Developing human and institutional capacities;
- Fostering joint research and technology development;
- Promoting assessment of the potential of renewables;
- Involving local communities, mainly in small-size energy supply projects.

3.5   *Agriculture*

Agriculture is a world-wide critical strategic resource expected to double its production in 30 years to feed the world. Yet, agriculture is most directly affected by climate change through increased variability as well as temperature and moisture changes. Agriculture's adaptation to climate change will require new genetic stocks, improved irrigation efficiency, improved nutrient use efficiency and improved risk management and production management techniques.

Agriculture can contribute modestly to mitigation through carbon sequestration in soils. Emissions from manure can be turned into methane fuel. Methane from ruminants can be reduced through straw ammoniation and increased feed efficiency, and methane from rice paddies can be mitigated. Better nutrient management can reduce emissions of nitrous oxide. Agricultural soils can be managed to increase soil organic matter through improved tillage practices, and tillage agriculture can be concentrated on better soils, allowing marginal soils to be converted to grasslands or forests. Each of these mitigation opportunities, however, requires farmers to change existing practices, creating the need for technology transfer.

Adaptation to uncertainty such as climate change requires assembling a diverse portfolio of technologies and keeping the flexibility to transfer and adopt needed technology. However, small farms and related businesses are risk-averse and transfer of technology is also discouraged through lack of information; financial and human capital, transportation; temporary tenure; and unreliable equipment and supplies. These hindrances can rarely be surmounted unless it is evident that the transferred technology profitably solves a clearly identified problem.

Even though adaptation and mitigation options are clear, integrated options need consideration in technology transfer. These will meet the following criteria: based on development needs: operating at a desired capacity and adapting the technologies to local conditions. For example, technology transfer of fertiliser use, as a main source of GHG, is focused upon and must therefore be balanced by productivity needs and by abatement of GHG emissions.

The effectiveness of technology transfer in the agricultural sector in the context of climate change response strategies would depend to a great extent on the suitability of transferred technologies to the socio-economic and cultural context of the recipients, and on considering development, equity, and sustainability issues. This is particularly relevant when applied to North-South technology transfers in this sector.

Governments can create incentives for technology transfer through regulation and improving institutions, in particular if those incentives are directly influencing farmers. The following actions would make sense in this context:

- Reducing and eliminating subsidies and market distortions that mask climate variability and climate change signals
- Improving or developing national agricultural information systems to produce to disseminate information on available state-of-the-art technologies and assist users in identifying their needs;
- Expansion of credit and savings schemes, to assist rural people to manage the increased variability in their environment;
- Encouraging free movement of knowledge and trained workers in order to expose farmers to innovative practices and taking action to make new patented technologies available to small farmers.

The worry about the absence of protection for intellectual property might be the key barrier to more private sector involvement in Technology Transfer. So it is important to adopt stricter IPRs to encourage greater private investment in agricultural R&D, and greater involvement in technology transfer to increase agricultural research funding. Many (particularly developed) countries have adopted stricter intellectual property rights (IPR) regimes for agro-chemicals, agricultural machinery and biological innovations. A rationale for adopting stricter IPRs is to increase private appropriability of research benefits and to encourage greater private sector investment in agricultural R&D and greater involvement in technology transfer. Although evidence from the United States suggests that increased plant variety protection has stimulated private R&D and adoption of improved crop varieties, the issue of IPRs for genetic resources remains controversial. Particular areas of controversy are farmer and research exemptions to IPR protection, and whether and to what extent IPRs should be extended to developing countries. Recent theoretical literature suggests that there may be limits to how far IPRs should be extended internationally.

The success of a response to an actual climate demonstrates that necessary technology can be developed, transferred and adopted. A new rice variety was developed in Sierra Leone to exploit seasonal rain and require less pesticide. Once success of the variety became apparent, farmers themselves transferred it to others. This transfer demonstrates the success of a policy that responds to present needs, concentrates researchers, devises cheap technology, and promptly benefits farmers.

Some technologies will not be so easily transferred. Irrigation, a pre-eminent adaptation to climate, costs millions and requires communities to adopt unfamiliar crops and methods. Nations must deal with scarce water and environmental impacts, marshal capital to construct the dams and canals and assist in the marketing of new crops. Banks must extend credit to farmers. Research and training must be turned to irrigation design, new crops, water use efficiency and prevention of salinity. Only an integrated national effort that extends to the farm level succeeds.

Centuries of experience, much of it governmental, have demonstrated the value of new plants and useful genes of established ones. Breeding, testing and demonstrating in the diverse locales and climates where farmers must cope with drought, pests, and different lengths of season have had high payoff and are essential for adaptation to climate change.

Education lies at the heart of technological transfer. The public role is pre-eminent and must be supported. Policies and programmes that rely on practical demonstration have proven the most effective. Private business can inform about technology in advertisements or demonstrate it at fairs. Governments have a role in monitoring claims and educating broadly. For example the "training and visit" transfer requires both training technicians and getting them into the field to educate farmers.

Although the transfer of new varieties proceeds quickly and easily, transferring systems of management requires persistence. For example, the United States established the Conservation Technology Information Center in 1982 to encourage conservation tillage. Great progress has been made, but after 17 years, adoption is still ongoing.

Uncertainties cloud the outlook for the transfer of agricultural technology. Because people transfer technology most readily to solve evident problems, the uncertainty of climate change hinders transfer. When, for example, climatologists assessing the climate for the next few years cannot agree whether it will be wetter or drier, no farmer is likely to invest in irrigation or drainage.

The main flow of technology transfer is from developed to developing countries dealing with climate change, which was emphasised by UNFCCC and The Kyoto Protocol. Some cases of existing agricultural technology transfer among developing countries, such as the Consultative Group on International Agricultural Research (CGIAR) and other multilateral systems, can be most helpful in assisting countries meet climate change if they are strengthened. International organisations and relevant developed countries can make great contributions by encouraging and supporting the technology transfer among developing countries.

Efforts to transfer technologies that address the following needs are important in addressing climate change:

- Increase crop output per litre of irrigation water drawn;
- Increase demand for appropriate technology with incentives and awareness raising and insures the provision of reliable supplies and equipment that meet local situations and needs;
- Provide crops suited to warmer temperatures

3.6 Forestry Sector

Technology transfer in the forest sector provides a significant opportunity to help mitigate climate change and adapt to potential changes in the climate. Apart from reducing GHG emissions or enhancing the carbon sinks, technology transfer strategies in the forest sector have the potential to provide tangible socio-economic and local and global environmental benefits, contributing to sustainable development. However, existing financial and institutional mechanisms are inadequate and thus new policies, measures and institutions are required to promote technology transfer in the forest sector.

The forest sector includes a wide variety of environmentally sound technologies and practices such as genetically superior planting material, improved silvicultural practices, sustainable harvest and management practices, protected area management systems, substitution of fossil fuels with bioenergy, incorporation of indigenous knowledge in forest management, efficient processing and use of forest products, and monitoring of area and vegetation status of forests, particularly under afforestation and reforestation activities.

Forestry mitigation technologies often involve minimal technology transfer and are strongly linked to other environmental factors such as biodiversity. They also possess features which are unique to the forestry sector and need careful consideration in the planning of technology transfer. Such features include: long gestation periods, linkage to subsistence economy, vulnerability to natural calamities, variation with climate and location, and state control of forests.

In Annex-I countries, significant research and development (R & D) and technology transfer occurs in the private sector, for instance through increased participation of timber companies and industries (paper mills). In non-Annex-I countries a substantial part of technology transfer in the forest sector is driven by forest departments, local communities and NGOs. This is particularly true for forest conservation practices, agroforestry systems, and systems for harvesting of non-timber and other subsistence products. Currently, there is a marginal role for private sector or industry participation in technology transfer.

The existing institutional mechanisms are dominated by state forest departments, government ministries and multilateral institutions. . These institutions have several limitations to promote climate mitigation technologies, namely, limited resources and an absence of policies and institutions to process, evaluate and approve mitigation projects and activities.

Emerging technology transfer mechanisms
In the face of limitations of existing institutions, a number of climate mitigation related mechanisms have emerged in the forest sector for technology transfer, including the concept of joint implementation, GEF, Activities Implemented Jointly (AIJ), and Clean Development Mechanism (if approved under the Kyoto protocol).

Barriers
The generic barriers to the current and emerging pathways of technology transfer mechanisms are: limited financial resources, inadequate information on the costs and potential benefits, limited technical capacity, absence of policies and institutions to process, evaluate and clear mitigation projects, uncertain additionality of the mitigation and its sustainability and long gestation periods for several options. In addition, the forest sector faces land use regulation, unclear property rights, and other macro-economic policies which favour conversion of forest lands to other land uses such as agriculture and cattle ranching. Lack of methods and capability for monitoring and verification need to be overcome in order to gain the credibility needed to capture the potential benefits of forest sector response options, particularly in reducing deforestation.

Technology transfer within countries
Technology transfer within countries is crucial in the forest sector. The sources of technology and management practices for in-country technology transfer are the forest departments of local and national governments; research institutions and university laboratories; paper, pulp, timber logging and plantation industries; and indigenous communities. Th transfer can be stimulated through funding, regulations and awareness raising programmes.

Technology transfer within Annex-1 countries
For technology transfer within Annex-I countries, the policies in timber importing countries are critical to ensure environmentally sound technology transfer within timber exporting countries. Governments can play a role in timber certification or labelling, levying of import or customs duty on non-certified timber, provision of Sustainable Forest Management (SFM) principles, investment in forestry R & D, preference for SFM-certified wood for public construction, recycling of paper through financial incentives and promotion of public awareness. NGOs can create awareness in consumers to promote the use of certified timber and recycled paper. The private sector in Annex-I countries could play a role in the development of criteria and indicators for SFM, development of certification schemes and their adaptation to the national/regional level.

Technology transfer within non Annex-1 countries
Within non-Annex-I countries, the government's role includes regulations on timber extraction companies, enforcement of forest conservation, removal of subsidies to deforestation, financial incentives for adopting SFM and creating technical capacity for monitoring of forest areas and promotion of research on causes of deforestation and their impacts. The important emerging role for the private sector includes linkages between industry and farmers for future technology transfer. The role of industry is crucial in facilitating technology transfer to a large number of small and dispersed farmers. Technology transfer could be facilitated as part of a package from industry to farmers including credit, technology and marketing arrangements. NGOs could create public awareness regarding forest conservation, SFM practices and recycling, and ensure compliance with legislation and policies by the government departments, industries and timber logging.

Technology transfer between Annex-I and non-Annex-I countries
For technology transfer between Annex-I and non-Annex-I countries, the government roles could include promoting the multi-

lateral and bilateral agencies to *(1)* support funding of forest-sector mitigation projects and programmes through grants and low interest loans for SFM practices, industrial plantations, protected areas, and forest restoration programmes, *(2)* setting up forest monitoring and verification programmes in non-Annex-I countries, and *(3)* providing funding for institution and human capacity building. In the private sector, timber companies could import technologies and transfer them to farmers, co-operatives and forest departments. NGOs and dedicated international institutions could contribute to international verification and monitoring programmes on sustainable logging practices.

Technology transfer among the Annex-I and among non-Annex-I countries:
Technology transfer among the Annex-I countries is largely private sector driven. Mechanisms such as timber certification and financial incentives for sustainably logged timber could facilitate adoption of sustainable practices. So far the technology transfer among the non-Annex-I countries has been marginal. But it is important as there are ecological and socio-economic similarities among these mostly tropical countries. Currently the level of technology transfer among non-Annex 1 countries is increasing.

Adaptation
Countries where forest vegetation is likely to be adversely affected by climate change may have to set up institutions to assess the impacts of projected climate change in their region and to develop adaptation strategies as a first step towards developing strategies for increasing forest resilience. Any adverse impact of climate change will affect forest ecosystems and the local communities and economies, which depend on them. Thus, the major responsibility for developing and adopting technologies to minimise the adverse impacts and to increase forest resilience, will rest with the government. Multilateral institutions such as FAO, CGIAR institutions and World Bank will have to play a critical role in developing, transferring and implementing adaptation technologies, as the adaptation technologies may be similar for a given forest type, within the tropical or temperate forest regions.

The existing institutional mechanisms may be inadequate, and in many countries there are no significant incentives for the local governments and local communities to protect and manage forests as carbon sinks. Adoption of mechanisms to promote technology transfer in the forestry sector requires strengthening of existing institutions as well as the creation of new institutions. The majority of the new institutions are likely to be set-up in non-Annex-I countries. It is very important to establish internationally acceptable monitoring and verification procedures and institutions. The existing and emerging financial mechanisms may also have to be strengthened and reoriented to promote forestry mitigation projects. The role of the private sector is increasing in the forest sector technology transfer. Governments will have to create policy environments to facilitate private sector participation in technology transfer programmes, as in many countries forests are still largely controlled by the state forest departments.

3.7 Solid Waste Management and Wastewater Treatment

Methane is generated from solid waste and wastewater through anaerobic decomposition. Together, solid waste and wastewater disposal and treatment represent about 20 percent of human-induced methane emissions. Emissions are expected to grow in the future, with the largest increases coming from developing countries. Methane emissions can be reduced in many ways, including reducing waste generation (source reduction), diverting waste away from disposal sites (i.e., through composting, recycling, or incineration), recovering methane generated from the waste, or ensuring that waste does not decompose in an anaerobic environment. In general, any technique or technology that reduces methane generation or converts methane into carbon dioxide through combustion will reduce greenhouse gases. The most effective mitigation approaches are those that either reduce overall methane generation (because methane collection efficiencies rarely approach 100%) or ensure that the combusted methane is substituted for fossil-based energy.

Extensive technology transfer aimed at improving waste management is underway both within and between countries, although most activities have been, and will likely continue to be, domestic in nature. In many regions, large investments are still required to provide adequate waste management services. In the past, the climate-related impacts of waste management choices were not routinely considered. Mitigation technologies can be readily deployed in this sector, however, and provide benefits beyond the reduction of greenhouse gas emissions, such as reduced landfill space requirements or additional energy generation through methane recovery.

Governments play a predominant role for technology transfer in the waste management sector , with several levels of government (from the national to the municipal level) participating. Key government priorities include the establishing of appropriate policy and regulatory frameworks, supporting the expansion of private sector participation, participating in technical assistance and capacity building activities, particularly with community groups, and in some cases providing incentives to catalyse desirable actions.

Historically, the private sector (including both domestic and multinational firms, as well as more informal local enterprises) and community-based organisations have been somewhat limited participants in government-driven technology transfer. The private sector has an increasingly important role, however, because meeting future waste management needs depends on expanded private investment. Private sector driven pathways are already used routinely for some types of investments (such as methane recovery at landfills), and efforts are underway to expand private sector participation across the full range of waste management services and technologies. The involvement of community organisations is also increasing as the link between community support and project sustainability has become clear. Soliciting local input and providing local training are two ways of ensuring sustainability. In many areas, locally developed and implemented projects are also being used to quickly address serious local concerns.

Technology transfer of waste management systems between countries will be confronted with many barriers, including limited financing, especially for South-South financing, limited institutional capabilities, jurisdictional complexity and lack of clear regulatory and investment frameworks and an overemphasis on projects at the expense of capacity building activities.

Mitigation projects can be successfully integrated into larger waste management efforts provided they are able to meet the needs and priorities of end-users, decision-makers, and financial supporters. However, mitigation projects may confront additional barriers, including:
- Lack of familiarity with the potential to reduce methane generation or capture the methane emissions associated with waste management;
- Unwillingness or inability to commit additional human or financial resources to investigating and addressing the climactic implications of the waste management project; and
- Additional institutional complexity when new groups, representing issues such as energy generation or by-product marketing, are incorporated into the project.

Key issues for technology transfer in solid waste management and waste water treatment
The review of the waste management sector in the Report reveals several key findings. This sector can contribute to greenhouse gas mitigation in ways that are economically viable and meet many social priorities. Already, extensive technology transfer is underway, and it will continue due to the continuing need to provide and improve waste management services for the world's population. In this sector it is important that projects emphasise the deployment of locally appropriate technologies, and minimise the development of conventional large, integrated waste management systems (with their attendant financial, institutional and technical requirements) in situations where lower cost, simpler alternative waste management technologies can be used.

Policy and Regulatory Development
Bilateral or multilateral regulatory or policy development assistance can be very useful to countries seeking to develop an appropriate framework for waste management. Given the importance of the regulatory/policy framework for international private sector investment, private firms should be encouraged to participate in such bilateral activities. When considering technology transfer between countries aimed at improving regulatory or policy frameworks, it is important to recognise differences between host and donor countries, and to ensure that the proper solutions are developed.

Innovative Financing Approaches
The private sector can participate in technology transfer in the waste management sector by serving as the developer of bilateral

or multilateral funded projects or through direct investment. Private sector opportunities for direct investment are emerging, however, and many countries are increasingly seeking private participation in waste projects.

Technology transfer aimed at assisting government agencies, particularly at the municipal level, to privatise or otherwise encourage private participation may be useful. Especially in developing countries, however, the structure and function of the existing waste management system is likely to be very different from the norm in developed countries. If these differences are not recognised and addressed, attempts to emulate developed country models can fail.

Capacity Building
Many bilateral technical assistance and capacity building activities are already underway in the waste management sector. The Report discusses several types of technology transfer for capacity building within countries, and many of the same approaches can be used between countries. Some areas of capacity building may be particularly appropriate for bilateral activities, including:
- Training to facilitate public participation;
- Training in financial management;
- Training in alternative technologies;
- Training in formulating business plans.

Existing and future capacity building activities in the waste management area can be readily expanded to facilitate the transfer of mitigation technologies. Activities could include increased emphasis on the climate impacts of various technologies as part of basic technical assistance programmes, developing specific decision tools and information for assessing climate impacts, and working with government counterparts on how to encourage mitigation technologies using a menu of voluntary, regulatory and incentive-based programmes. Expanded involvement of local and regional government officials from both donor and host countries would be beneficial, given their role in project development and implementation. Given the still emerging understanding of these issues, technical assistance activities in all directions (North-North, North-South, South-North, and South-South) are likely to be useful.

Incentives
The Kyoto Protocol will likely provide important incentives for expanded activities between governments to develop mitigation projects. Provisions related to both the Clean Development Mechanism and emissions trading could encourage expanded bilateral investments in projects in exchange for emission reduction units.

3.8   Human Health

There are opportunities for technology transfer in relation to the potential health impacts at several levels. Minimisation of the health impacts of impending, or unavoided future, climate change can entail:

- Reduction of the exposure of local populations to climate change and its environmental and social consequences;
- Reduction of the vulnerability of local populations to those exposures;
- Improved care and treatment for those whose health is adversely affected.

With respect to the first of those three modalities, the health of communities and populations depends fundamentally upon social and natural environmental conditions. Therefore governments should be fully aware of the potential public health impacts when assessing options for adaptation to climate change.

With respect to the second modality, population vulnerability can be reduced generically by improving the social and technical circumstances and by reducing socio-economic inequalities. Vulnerability can be reduced more specifically by adaptations directed at specific exposures, or susceptibility to them. For example, improved housing design and domestic temperature regulation can reduce vulnerability to thermal stress, while vaccinations and control of "vector" insects reduce the risks of various infectious diseases. The health gains from these specific adaptations will largely be confined to the population taking action.

With respect to the third issue, national public health infrastructure is of fundamental importance. The poorest countries, where impacts are likely to be greatest, are also least able to maintain a public health infrastructure. This area should therefore be a priority for technology transfer.

Monitoring and Surveillance
The most elementary form of adaptation is good health monitoring and surveillance systems. Within the health sector, only basic measures of population health status (e.g., life expectancy) can be measured simply and uniformly around the world. However, disease (morbidity) surveillance varies widely depending on the locality, the country and the disease. Many less developed countries have inadequate surveillance systems. Weather and climate forecasts should be used, where appropriate, in preventing deaths and injuries, and in disease prevention and control.

Control of infectious disease
The last decade has seen the resurgence of several major infectious diseases which were previously controlled, such as malaria. International efforts to control malaria - a known climate-sensitive disease - are failing because of drug resistance and the socio-economic, geographical and ecological conditions in much of today's world. The recent resurgence of malaria in areas where it had previously been eradicated (Azerbaijan, Tajikistan) or under control (Iraq, Turkey) reflects declines in malaria prevention and mosquito control programmes due to conflict and economic crises. In Ethiopia, indoor spraying campaigns with DDT were effective at reducing both malaria morbidity and mortality, but over the last 20 years there has been a programmatic breakdown because of civil war and the forced movement of people. WHO has recently launched a new initiative, the "Roll Back Malaria" programme, to be implemented in close cooperation with the World Bank and other international agencies.

The coverage of existing vaccination programmes aimed at elimination of diseases such as yellow fever should be expanded. Other strategies are important to combat diseases like malaria. For example, periodic checks should be carried out on parasite sensitivity to the commonly used antimalarial drugs. The use of insecticide-impregnated bed nets has been successful in reducing malaria transmission in endemic areas.

The incidence of certain water-borne and vector-borne infections can be reduced by several environmental measures. A broader approach to preventing water- and food-borne diseases would consider the interactions between climate, vegetation, agricultural practices and human activity. Strategies to control climate-sensitive disease require intersectoral collaboration between the health, forestry, environment, and conservation sectors.

Improve coverage of health system
There are many existing health inequalities between rich and poor; these inequalities are widest in developing country cities. Strategies to reduce poverty and improve access to health services in urban environments will serve to reduce vulnerability to climate change

Role of the UN system
Public policies to reduce socio-economic and environmental vulnerability will enhance population capacity to adapt to climate change. Conventional health sectors are not equipped nor empowered for this multidimensional task. International cooperation is required, for data collection, analysis, and policy-making to reduce health impacts.

The specialised agencies of the United Nations are primarily mandated to support government planning and management in Member States, by strengthening sectors such as industry, health and agriculture. Implementation of the following three-tier strategy endorsed by the Interagency Committee for the Climate Agenda (IACCA) is entrusted to WHO:

- Assistance to Member States to promote vulnerability assessment, adaptation strategies, and the adoption of technologies to promote health and reduce GHG emissions;
- The exchange and provision of information on the health impacts of climate change and of mitigation strategies, as well as effective approaches to adaptation;
- The promotion of research on the above topics.

3.9   Coastal Adaptation

Technology transfer for coastal adaptation to climate change has characteristics that are in many ways different from explanations of transfer in other chapters of Section II of the Report. It focuses on technologies for a geographic area rather than on a specific resource, such as forestry and agriculture, or an infrastructure issue, such as transportation and building construction. Populations and infrastructure investments are continuously growing in these regions because of the large number of valuable goods and services provided by coastal systems. Yet, the coastal zone is perhaps the most vulnerable region to climate-related hazards, such as storm surges and erosion. Extensive research has shown that climate change will increase the hazard potential.

The potential impacts of climate change by itself may not always be the largest threat to natural coastal systems. Coastal ecological systems and societies are already under severe pressure from uncontrolled socio-economic growth leading to overexploitation of natural resources, pollution, decreasing fresh-water availability, and urbanisation. These non-climate stresses decrease the resilience of coastal systems to cope with natural climate variability and increase their vulnerability to climate change. However, governments often do not consider adaptation a development objective, and there may be a great disparity between government expenditures to improve social organisation and economic production and investments of public resources in coastal adaptation.

In many places, technology has been instrumental in reducing society's vulnerability to anticipated impacts of climate change in three basic ways:
- *Protect* - reduce the risk of the event by decreasing its probability of occurrence;
- *Retreat* - reduce the risk of the event by limiting its potential effects;
- *Accommodate* - increase society's ability to cope with the effects of the event.

Effective adaptation to climate change needs to consider the numerous non-climate stresses in coastal zones and be consistent with existing policy criteria and development objectives. Adaptation in coastal zones must strike a balance between current pressures resulting from climate variability and unsustainable development, and anticipated impacts of climate change and associated sea-level rise. Adaptation technologies are best implemented as part of a broader, integrated coastal-management framework that recognises immediate and longer-term sectoral needs. Win-win situations could be established when coastal-adaptation technologies also provide benefits unrelated to climate change such as improving recreatinal opportunities enhancing biodiversity and reducing vulnerability to today's hazards.

Existing coastal technologies that have been used to deal with climate variability in coastal zones and implementation of integrated coastal zone management can also be applied to accomplish each of the four main steps to adapt to climate change: *(i)* information development and awareness raising, *(ii)* planning and design, *(iii)* implementation, and *(iv)* monitoring and evaluation. Important data gathering technologies to describe coastal characteristics and processes include in-situ and airborne and satellite-based remote sensing systems. Planning and design tools include Geographical Information Systems and models. A range of opportunities exists for the application of both hard and soft technologies to complement economic, legal and institutional options for implementation of the three main coastal adaptation strategies: protect, retreat and accommodate.

A large number of effective protection technologies are available, such as dikes, seawalls, and beach nourishment. These also include traditional, indigenous, non-western technologies, such as coconut-leaf walls, coconut-fibre stone units. Technologies which are incorporated within a managed retreat strategy include rolling easements, set-back zones, and moveable structures. An accommodation strategy would employ technologies such as early warning systems for hazards, rain/waste-water management, and desalination. A number of technologies which have emerged to exchange knowledge and information to support integrated coastal zone management may also be applied to evaluate the effectiveness of coastal adaptation strategies. Despite this vast array of available technologies, many of the world's vulnerable coastal countries currently do not have access to adaptation technologies appropriate for their circumstances, nor to the knowledge or resources that are required to develop or implement these. Therefore, extra efforts should focus on promoting, adjusting, and transferring existing technologies, rather than on the development of new technologies.

The predominant nature and goal of coastal adaptation require a form of technology transfer that differs from many explanations of transfer that describe the process as a company-to-company transaction. Coastal-adaptation technologies - with few exceptions - are not developed and owned by business and industry. Economic considerations are a major force in driving technology transfer for coastal adaptation, but objectives are less focused on commercial terms. Rather, considerations of public well-being are essential, such as the reduction of loss of property and lives, and the protection of essential coastal habitats. Therefore, the strongest and most direct incentives to adapt to climate change in coastal zones are with the public sector, and government interests predominantly drive pathways of technology transfer in coastal zones. Furthermore, many coastal-technology transactions involve the exchange of information and knowledge that exist in the public domain. Knowledge transactions have characteristics that make them quite different from trade transactions.

Many barriers to effective transfer of coastal adaptation technologies are site-specific and require site-specific solutions. Four major general barriers exist: *(i)* lack of data, information and knowledge to identify adaptation needs and appropriate technologies, *(ii)* lack of local capacity and consequent dependence of customers on suppliers of technology for operation, maintenance and duplication, *(iii)* disconnected organisational and institutional relationships between relevant actors and *(iv)* access to financial means. Overcoming these barriers does not require setting up new bilateral and multilateral institutions or mechanisms. Instead, existing activities and institutions need to be refocused to improve the efficiency and effectiveness of coastal technology transfer. In addition, regional collaboration and a redirection of funds to support appropriate coastal adaptation to climate change are required.

Coastal adaptation to climate change is a transnational issue—it cannot be addressed within the borders of one country, no matter how effective and creative the decision-makers, innovative the academic relationships and dynamic the private sector are in advancing the deployment of appropriate technologies.

First, all stakeholders must recognise that successful coastal adaptation depends on many local factors, and adaptation technologies cannot be simply transferred to other vulnerable areas. The effectiveness of a particular technology depends on local circumstances, including the biophysical setting and economic, institutional, legal and socio-cultural conditions. Technologies must be adjusted, oriented and made appropriate for local conditions in the host country, possibly in the context of integrated coastal zone management. Local expertise is essential to identify and design appropriate coastal-adaptation technologies, as well as to implement, operate and maintain these. Therefore, the importance of global networks to improve and accelerate coastal-adaptation technology transfer should not be underestimated. Such networks provide access to up-to-date information and real-time tracking of global trends, accelerate the formation of joint ventures and permit direct participation in strategic locations around the world. The process for building these networks must include not only personal links but also institutional and functional linkages.

Ever since humans have lived near the sea they have developed and applied technologies to reduce their vulnerability to coastal hazards. The same technologies can be applied to adapt to anticipated impacts of climate change. However, access to these technologies in vulnerable areas can be a problem without effective technology transfer. Improving and facilitating the process of technology transfer are key challenges to reduce coastal vulnerability worldwide.

Section I

Framework for Analysis:

Technology Transfer to address Climate Change

Section Coordinators:
BERT METZ (THE NETHERLANDS), KILAPARTI RAMAKRISHNA (INDIA)

1

Managing Technological Change in Support of the Climate Change Convention: A Framework for Decision-making

Coordinating Lead Author:
SERGIO C. TRINDADE (BRAZIL)

Lead Authors:
Toufiq Siddiqi (USA), Eric Martinot (USA)

Contributing Authors:
Richard J.T. Klein (The Netherlands/Germany), Mary-Renee Dempsey-Clifford (Ireland)

Review Editors:
Li Liyan (China), Roberto Schaeffer (Brazil)

CONTENTS

EXECUTIVE SUMMARY

This introductory Chapter sets out the landscape for the discussion, throughout the Special Report, of the multitude of facets of managing technological change in support of the Climate Change Convention and its Protocols. The framework proposed for decision-making by government policymakers, and other relevant stakeholders, emphasises the sustainable development perspective, while exploring the national and international political settings, trends in finance and trade, the organisational and institutional context, and the meanings of technology transfer and of the innovation system. A model of the latter and of pathways in technology transfer is presented to help understand the nature, motivations, barriers to the process, and possible options to promote sustainable development in the face of the climate change challenge.

1.1   Introduction

This Special Report has been prepared, in response to a request (FCCC/SBSTA/1995/3) by the Subsidiary Body for Scientific and Technological Advice (SBSTA), with the following objectives in mind:

- To explain the "technology transfer" problem in the context of global climate change, the United Nations Framework Convention on Climate Change (UNFCCC, also referred to as FCCC), and the subsequent Conferences of the Parties (CoP).

- To provide an overview of the concepts of "technology transfer" in current use, and the related concepts of "technology diffusion" and "technology cooperation" via an analytical framework that includes relevant stakeholders, processes and procedures and the crucial matter of sustainable development choices.

- To list key options that are available to governments and other stakeholders, such as the private sector, to enhance "technology transfer", including the Kyoto mechanisms where relevant, and to build and improve the capacity to make informed choices for the management of climate-related technologies.

The Report is divided into three sections. Section I covers the analytical framework and database of transferring technology in support of the Climate Convention and sustainable development. Section II takes a sectoral view of the main mitigation and adaptation issues in technology transfer related to climate change. Section III illustrates the points made in the preceding sections with well-considered case studies. Since the case studies are a very important part of this Report, a listing of these is provided at the conclusion of this Chapter (Annex 1-1).

The Report is addressed primarily to the policymakers of governments that are parties to the Convention, but it also aims at reaching other stakeholders in the private sector, lending institutions, multilateral agencies, non-governmental organisations, and the interested public.

The broad objective of this introductory Chapter is to set out the landscape for the discussion of the many facets of managing technological change in support of the Climate Change Convention and its Protocols. A list of references is included at the end of the chapter to assist readers who wish to explore further some of the concepts introduced here.

1.2   Basic Concepts

For the sake of brevity, the term "Technology Transfer" as used throughout the Report includes the related concepts of "Technology Cooperation" and "Technology Diffusion". Using this terminology, the Report describes what is needed in con-

nection with technology transfer to address climate change; examines the adequacy of current initiatives in technology transfer in support of the Convention, and places the discussion in the context of the sustainable development agenda and the problem of development choices.

There is, however, no simple definition of a "sustainable development agenda" for developing countries. Sustainable development is a context-driven concept and each society may define it differently. Technologies that may be suitable in each of such contexts may differ considerably. This makes it important to ensure that transferred technologies meet local needs and priorities, thus increasing the likelihood that they will be effective. And although local needs and priorities may be challenging to ascertain by the consensus of the relevant stakeholders, transferring technologies that are to meet such needs without causing a huge economic burden to the people is even more complicated.

A special emphasis in the Report is on the direct role of governments in technology transfer and their role in creating favourable conditions for the development of a market for climate change-related goods, technologies and services (IPCC, 1996), and for taking actions. The Report stresses that current efforts and established processes of technology transfer will not be sufficient to meet this challenge, and concludes that additional actions are needed for the transfer of mitigation and adaptation technologies. Especially for technologies that will not yet diffuse commercially, it is important to go further than improving market performance. Extra efforts to enact policies that lower costs and stimulate demand will be needed to realise environmental benefits that are not adequately produced otherwise. Integrating human skills, organisational development and information networks is a key to effective technology transfer.

The Report looks at technology transfer as the result of many day-to-day decisions, and identifies the stakeholders who participate in the decision-making process regarding strategy, investment, international trade, market opportunities, etc. It also identifies social, economic, political, legal, and technological factors that influence technology transfer. Cultural preferences, consumers' awareness, social values, lifestyles, corruption, competition, etc. are reflected in technology transfer as well. Although the report contains a comprehensive list of publications dealing with the conceptual aspects of technology transfer, it emphasises actual case studies, listed in Annex 1-1, to illustrate the practical aspects of managing technological change to address global climate change concerns. The key words here are management, technology, climate and change for sustainable development.

The Report elaborates on what governments can do to facilitate technology transfer in support of the Climate Change Convention, taking into consideration their stage of development and specific sectors. The Report addresses effectiveness criteria as being subjective and dependent on perspectives and indicators on greenhouse gas abatement, economic sustainability, social acceptability, and environmental sustainability. Annex 1-2 articulates this matter further.

It is important at the outset to distinguish between technologies for mitigation and adaptation to climate change. In simple words, mitigation focuses on slowing climate change, whereas adaptation deals with the effects of climate change. The general question the Report attempts to address is how can "technology transfer", or rather the management of technological change, encourage development that is climate friendly (mitigation) and climate responsive (adaptation), taking into account the need to adjust to the effects of climate change.

Throughout this Report reference is made to Environmentally Sound (or Sounder) Technologies (ESTs), that is: "technologies which protect the environment, are less polluting, use all resources in a more sustainable manner, recycle more of their wastes and products, handle residual wastes in a more acceptable manner than the technologies for which they were substitutes, and are compatible with nationally determined socio-economic, cultural and environmental priorities. The term encompasses hard and soft technologies (United Nations, 1993)." "Examples of soft technologies include capacity building, information networks, training and research, while examples of hard technologies include equipment and products to control, reduce or prevent anthropogenic emissions of greenhouse gases in the energy, transportation, forestry, agriculture, industry and waste management sectors, to enhance removals by sinks and to facilitate adaptation (van Berkel and Arkesteijn, 1998)."

In this Report ESTs imply both mitigation and adaptation technologies. It is worth noting that technologies, which address climate change, that is, which are climate friendly and climate responsive, are not necessarily always environmentally sound. For instance, technologies for large hydroelectric plants are climate friendly but could affect the environment where they are deployed. Conversely, the catalytic processing of exhaust gases from automobile tail pipes may be environmentally sound but may not be climate friendly. However, in this Report it is assumed that both mitigation and adaptation technologies will be applied in such a way that they are environmentally sound. A finer qualification on equating climate-responsive with adaptation is the possibility that adaptation technologies may take advantage of climate change, for instance to increase agricultural productivity in specific instances.

1.3   Background

The beginning of the industrial revolution is usually used as the time frame from which increases in the emissions and the accumulation in the atmosphere of greenhouse gases (GHGs) are measured. The introduction of large numbers of new technologies was the principal characteristic of the industrial revolution. It is thus not surprising that a frequently expressed view in the worldwide deliberations on global climate change has been "If the introduction of new technologies created the problem, other new technologies will help us in solving it". In most cases, the adequate technologies already exist, but not necessarily in the locations where they could best be used to mitigate increases in the emissions of GHGs or adapt to their impacts on the environment.

Sustaining development globally will require radical technological and related changes in both developed[1] and developing countries. Economic development is most rapid in developing countries, but it will not be sustainable if these countries simply follow the historic polluting trends of industrialised countries (Munasinghe and Swart, 2000). Rapid development with modern knowledge offers many opportunities to avoid bad past practices and move more rapidly towards better technologies, techniques and associated institutions. But to achieve this developing countries will require assistance with developing human capacity (knowledge, techniques and management skills), developing appropriate institutions and networks, and with acquiring and adapting specific hardware. Technology transfer must therefore operate on a broad front covering these "software" and "hardware" challenges, and ideally within a framework of helping to find new sustainable paths for economies as a whole. A key element is choice. Hence the development of societal and organisational structures that enable well-informed choices of technologies which promote climate stability, adaptation to the effects of climate change and sustainable development is essential. To a large extent, the state of the environment today is the result of the technological choices of yesterday. Similarly, the state of the environment in the 21st century will be determined largely by the technologies we choose today (Trindade, 1991).

The bulk of technology transfers occur within the countries that generate them. The transfer of technologies from the countries and companies that developed them to other countries and entities that could put them to good use in reducing GHGs has been an important theme in international deliberations leading to the UNFCCC. The Convention, signed at the United Nations Conference on Environment and Development (UNCED) in 1992, came into force in 1994, and requires (Article 4.1.c) the parties to UNFCCC, "to promote and cooperate in the development, application, diffusion, including transfer, of technologies, practices, and processes that control, reduce, or prevent anthropogenic emissions of greenhouse gases" (UN, 1992).

Article 4.5 of the UNFCCC further states:

"The developed country Parties and other developed Parties included in Annex II shall take all practical steps to promote, facilitate and finance, as appropriate, the transfer of, or access to, environmentally sound technologies and know-how to other Parties, particularly developing country Parties, to enable them to implement the provisions of the Convention. In this process, the developed country Parties shall support the development and enhancement of endogenous capacities and technologies of developing

---

[1] If not further specified, developed (or industrialised) countries in this Report include countries with economies in transition (CEIT).

country Parties. Other Parties and organisations in a position to do so may also assist in facilitating the transfer of such technologies".

Another important outcome of UNCED was Agenda 21 (UN, 1993), a vision for the 21st Century based on the concept of sustainable development. Chapter 34 of the Agenda, on the "Transfer of environmentally sound technology, cooperation, and capacity building" calls for access to scientific and technical information, promotion of technology transfer projects, promotion of indigenous and public domain technologies, capacity building, intellectual property rights, and long-term technological partnerships between suppliers and recipients of technology. It points out that "Technology cooperation involves joint efforts by enterprises and governments, both suppliers of technology and its recipients. Therefore, such cooperation entails an interactive process involving government, the private sector, and research and development facilities to ensure the best possible results from transfer of technology". It also recommends the utilisation of existing technological information and promoting research partnerships and assessment networks and the development of new ones.

The importance of "technology transfer" was also recognised in the Kyoto Protocol to the UNFCCC, where Article 10c asks all Parties to "Cooperate in the promotion of effective modalities for the development, application and diffusion of, and take all possible steps to promote, facilitate and finance, as appropriate, the transfer of, or access to, environmentally sound technologies, know-how, practices and processes pertinent to climate change, in particular to developing countries, including the formulation of policies and programmes for the effective transfer of environmentally sound technologies that are publicly owned or in the public domain and the creation of an enabling environment for the private sector, to promote and enhance the transfer of, and access to, environmentally sound technologies". The fourth Conference of the Parties (CoP-4) meeting in Buenos Aires, in November 1998, further discussed the development and transfer of technologies, where the SBSTA made a set of specific recommendations, with a special emphasis on capacity building and consultative processes. Decision 4/CoP4 of CoP4 included a set of questions that are to be considered in these consultative processes (see Box 1.1). The CoP4 decision on technology transfer is fully consistent with Agenda 21 and the Kyoto Protocol, and added a fresh dimension to the UNFCCC. The criteria for effective technology transfer are presented in Annex 1-2, and constitute a useful checklist for policymakers and other relevant stakeholders.

BOX 1.1 QUESTIONS INCLUDED IN ANNEX TO DECISION 4/CP.4 OF THE CONFERENCE OF THE PARTIES TO THE UNFCCC THAT ARE TO BE CONSIDERED IN THE CONSULTATIVE PROCESS SET UP BY THIS DECISION.

1. How should Parties promote the removal of barriers to technology transfer? Which barriers are a priority and what practical steps should be taken?

2. How should Annex II Parties promote the transfer of publicly-owned technologies?

3. What additional bilateral and multilateral efforts to promote technology cooperation to facilitate technology transfer should be initiated? What should be the priority?

4. Are existing multilateral mechanisms sufficient? Are new mechanisms needed for technology transfer? If so, what are appropriate mechanisms for the transfer of technologies among Parties in pursuance of article 4.5 of the Convention?

5. What should be the objective of collaboration with relevant multilateral institutions to promote technology transfer and what practical steps should be taken?

6. What additional guidance should be given to the financial mechanism?

7. What sort of information is needed and how can this best be done?

8. How could access to emerging technologies be facilitated?

9. What role is the private sector playing in technology transfer? What additional role can the private sector play? What barriers prevent their greater participation?

10. What technical advice on technology transfer is needed?

11. What areas should be the focus of capacity building and how should it be undertaken, e.g. what kind of activities, programmes and institutional arrangements?

12. How, to whom and in what format should developing country Parties make their requests for assistance to assess required technologies?

13. What technical, legal and economic information is needed? What practical steps should be taken to promote and enhance access to such information by national and regional centres?

14. What type of process is needed to develop a consensus on practical next steps to improve existing technology centres and networks in order to accelerate the diffusion of clean technologies in non-Annex I Party markets. What type of arrangement is needed to monitor progress?

15. What measures, programmes and activities can best help to create an appropriate enabling environment for private sector investment?

16. How should the Convention oversee the exchange of information among Parties and other interested organisations or innovative technology cooperation approaches, and the assessment and synthesis of such information?

17. How should information be compiled and synthesised on innovative technology cooperation approaches? When should recommendations on such approaches be forwarded to the Conference of Parties?

18. How and when should information on projects and programmes of technology cooperation which Parties believe can serve as models for improving the diffusion and implementation of clean technologies internationally under the Convention be provided to the secretariat?

19. Can specific technology transfer goals be set?

20. Can we develop indicators and accounting systems to track progress on technology transfer?

21. Are particular institutional arrangements needed to monitor progress?

It is widely believed that the provision of a favourable environment must be based on equity concerns (Pachauri and Bhandari, 1994), and on participatory decision-making to improve the chances that such an enabling environment will be sustainable (Munasinghe, 2000). Others emphasise that a great deal of technology transfer will be the consequence of commercial transactions, not through aid or grants, and that the emphasis should be on developing a financial and legal framework to promote investment and trade on climate mitigation and adaptation. In specific situations, ways have to be found which establish the adequate balance between these two positions. Actually, the two different positions may not be mutually exclusive and could occur simultaneously.

Countries and organisations are affected differently by developments such as climate change. While all countries need to cooperate in addressing global climate change, their capacity to respond varies greatly. Furthermore, small developing countries and least developed countries need special attention. Industrial countries, by and large, have the knowledge and the financial resources necessary to better respond to these challenges, but most of the countries in the world are small developing countries, which have limited capacity to take initiatives and require assistance in moving towards a sustainable development path. Some organisations fear the additional costs of economic restructuring, whereas others, who are ready for it, see the challenge providing many opportunities for them (Trindade, 1994).

The idea that a country should actively seek to transfer technology to another country is a relatively new one, seen in practice only during the second half of the 20th century (Siddiqi, 1990). For most of history, countries have sought to protect knowledge of technologies, since knowledge is power — military power as well as economic power. Most technologies are improved incrementally over time, but occasionally there are opportunities for leapfrogging, especially for developing countries. It should be noted that the incremental improvement of technologies is accompanied by a continuous process of social and organisational change. The growth of multilateral organisations, such as the United Nations system, as well as of transnational corporations, and developments in communications and in intellectual property legislation were major factors in influencing technology cooperation and change in recent history.

1.4 The Many Meanings of Technology Transfer and Innovation

Differing views of technology see it as a commodity, as knowledge, or as an socio-economic process (Rosenberg, 1982). The classical economic view of technology as a commodity holds that technology can be reproduced without cost and transmitted from one agent to another. In this view, technology transfer is as simple as making a photocopy of design documents or obtaining a working artifact. But this view of

technology has been supplanted by the view of technology as knowledge (Kranzberg, 1986). This knowledge is brought about both through research and innovation (moving ideas from invention to new products, processes and services in practical use), and through a complex and often costly process involving learning from others. A useful discussion of technology in relation to climate change is provided in Rip and Kemp (1998).

Many have concluded that technology transfer is most fundamentally this complex process of learning (Kranzberg 1986). "It is not unreasonable to say that a transfer is not achieved until the transferee understands and can utilise the technology" (Chen, 1996). A test of this criterion would be the ability of the transferee to choose and adapt the technology to the local socio-economic environment and raw materials, and to sell to someone else the original technology with improvements. In the absence of such capacity to choose, transfer of inadequate, unsustainable, unsafe, or "bad", but perhaps cheaper technology and equipment can take place. The transfer during the 1980s of technology for the production of inefficient refrigerators and the international trade of used cars are cases in point.

The understanding of technology transfer is not helped by the tendency of public discourse to lump "science" together with "technology". Each may benefit from the other, but do not necessarily follow one from the other. Some countries have good scientific traditions, but are relatively weak in developing new technologies. Other countries may be world leaders in technology without having many winners of Nobel Prizes. Often technology development leads to scientific advance, which in turn may lead to new technology. Furthermore, while it may be appropriate to think of science as "Public knowledge" (Ziman, 1968), modern technology is very often "Private (or proprietary) knowledge". Whereas scientific knowledge is available freely to all that are scientifically literate, proprietary knowledge is not. As a form of knowledge, technology moves from one organisation to another in a variety of ways. "The complex ways in which knowledge moves from individual to individual and organisation to organisation raises the first problem in effective technology transfer" assert Dodgson and Bessant (1996). Effectiveness criteria for technology transfer are spelled out in Annex 1-2 and summarised in chapter 6. It may be worth repeating that technology is not simply a patent or a piece of equipment, but knowledge, processes and practices.

Also important is an understanding of the complex networks through which those involved in technology transfer can interact: "Networks are critical for market creation [because of] their contribution to learning, specifically to the generation of a broad social pool of knowledge related to the capital good in question" (Teubal *et al*, 1991). Archibugi and Michie (1997) also emphasise the learning process: "policy to support technology should address the diversity of learning mechanisms and the conditions which enhance the learning capabilities of firms". The understanding of technology transfer is different for different stakeholders: governments and end-users need to understand the costs and benefits of a technology; innovators need to understand how to adapt it; and firms need to understand how to market it and how it meets user needs. These differences are further elaborated in the typology of stakeholders presented in Annex 1-3.

An increasing number of people feel uncomfortable with the term "technology transfer". For some, the criticism is that there is an implicit view of technology as an object, and its transfer is a one-time transaction that maintains the dependency of the recipient (Heaton *et al.*, 1994). They advocate a new mindset and terminology --- technology cooperation --- to replace the notion that technology can be transferred full-blown from one economic and cultural context to another. Martinot *et al.* (1997) also prefer the concept of technology cooperation to technology transfer. Grubler and Nakicenovic (1991) are amongst those who prefer the concept of technology diffusion to that of technology transfer. For most of the people in this group, diffusion represents a process of technological change brought about by dispersed and uncoordinated decisions over time. Still others (Robinson, 1991) see technology transfer as a two-way learning process that might more appropriately be called "Technology Communication".

While recognising these different views and meanings, the term "technology transfer" is generally used throughout this Report, in conformity with the title of the Report. In the Report the term "technology transfer" is defined as the broad set of processes covering the flows of knowledge, experience and equipment amongst different stakeholders such as governments, private sector entities, financial institutions, NGOs and research/educational institutions. The broad and inclusive term "transfer" encompasses diffusion of technologies and technology cooperation across and within countries. It comprises the process of learning to understand, utilise and replicate the technology, including the capacity to choose it and adapt it to local conditions.

Technology transfer is not just from the North to the South. Most of the technology transfer today at the international level actually takes place between one industrialised country and another. Increasingly, however, technology transfer is also expected to take place from the South to the North or between South and South. The Report touches on these aspects in several Chapters.

1.5 Stakeholders, Decisions and Policies

Technology transfer is embodied in the actions taken by individuals and organisations. The investment and trade decisions made by firms, acquisition of knowledge and skills by individuals through formal education and on-the-job experience, purchase of patent rights and licenses, assimilating the published results of public or private research, development and demonstration (RD&D) activity, and migration of skilled personnel with knowledge of particular technologies, all represent different forms of technology transfer. Technology transfer can also be influenced by government aid and financing programs, and by

multilateral bank lending. Governments can implement policies that promote R&D programmes that address global climate change concerns in sectors such as energy, forestry, and transportation. The role of governments is especially important for those climate-related technologies, which are not immediately viable and profitable. An overview of how environmental policies influence technology development and diffusion in OECD countries describes public-private partnerships for developing environmental technologies and policies aiming at diffusing them (OECD, 1999, see also section 5.6 for a more elaborate treatment of public-private partnerships).

The rate of technology transfer is affected both by motivations that induce more rapid adoption of new techniques and by barriers that impede such transfers. Both types of factors can be influenced by policy. Table 1.1 shows a typology of different types of stakeholders, their motivations, and the kinds of decisions or policies they can adopt that relate to technology transfer. The taxonomy of stakeholders is further described in Annex 1-3, which could be a useful reference for policymakers to consult. Motivations of the various stakeholders can differ markedly:

- Transnational or multinational corporations are major sources of technology. They seek international sales, market share, and cheaper production costs through equipment transfers and foreign direct investment. Corporations are primarily concerned about profits, acceptable risks and ensuring protection of intellectual property.

- Recipient-country firms are also motivated to transfer technology to minimise costs, just as with transnational corporations. But other motivations may be quite different from those of supplier firms, such as: (a) technical capabilities, quality, or cost reductions that they cannot achieve on their own; (b) the higher perceived status of "international level" technologies; (c) access to managerial and marketing expertise and sources of capital; (d) access to export markets; and (e) access to new distribution networks.

- Recipient governments may seek to increase capabilities for domestic technology-development and promote foreign investment in their country. At the local level, communities and community organisations need to be reached by information networks, get organised and participate in decision-making processes to improve local living standards and the quality of the environment via appropriate technologies.

- Provider or donor governments may set up policies to encourage technology transfer and fund transfers of research and expertise via Official Development Assistance (ODA) to support development and political goals, but more often are interested in policies that expand foreign markets for their national firms and increase exports.

- Multilateral agencies with development goals, such as the World Bank, the United Nations Development Programme (UNDP), Regional Development Banks and Regional Organisations, pursue technology transfer to support development and as an instrument for achieving desired economic and policy reforms.

- Multilateral agencies with environmental goals, such as the Global Environment Facility (GEF), have the transfer of ESTs as an explicit objective, and explore new and effective means to accomplish these objectives, by catalysing sustainable markets and enabling private sector involvement in the transfer of these technologies (see Box 5.2 for further reference on the GEF).

- Non-governmental organisations have been at the forefront of concerns about technology choice and the "appropriateness" of technologies transferred through development assistance and commercial channels, the social and cultural impacts of such transfers, and the needs for technology adaptation to suit local conditions and minimise unwanted impacts.

The decisions and policies shown in Table 1.1 represent another point of departure for thinking about barriers to technology transfer and interventions for overcoming such barriers. Each of these decisions and policies will face a set of barriers that will limit their realisation. We see throughout this Report how interventions affect the decisions and policies presented in Table 1.1. (For further reference on the link between relevant policy tools with barriers see Table 4.1 and 5.1 which also indicate the relevant sectors).

The relevant aspects of the capacity for technology choice – a true measure of the capacity to make independent decisions on sustainable development – are a separate issue, and are taken up in Chapters 3–5, while the criteria for the effectiveness of technology transfer are elaborated upon in Annex 1-2 and briefly in Chapter 6.

1.6 The Innovation System and Pathways: A Framework for Analysis

The theme of technology transfer is highly interdisciplinary and has been approached from a variety of perspectives, including: business, law, finance, microeconomics, international trade, international political economy, the environment, geography, anthropology, education, communication, and labour studies. Although there are numerous frameworks and models put forth to cover different aspects of technology transfer, there are no corresponding overarching theories (Martinot *et al.*, 1997; Reddy and Zhao, 1990). This Report employs the technology transfer/innovation system as a framework for analysis. In this system (see Figure 1.1), technology is transferred as knowledge, money (investment) and goods (trade) flow among different stakeholders: governments, private-sector entities, financial institutions, non-governmental organisations, and research/teaching institu-

Table 1.1	Principal stakeholders and their decisions or policies in technology transfer	
STAKEHOLDERS	**MOTIVATIONS**	**DECISIONS OR POLICIES THAT INFLUENCE TECHNOLOGY TRANSFER**
Governments • national/federal • regional/provincial • local/municipal	Development goals Environmental goals Competitive advantage Energy security	Tax policies (including investment tax policy) Import/export policies Innovation policies Education and capacity-building policies Regulations and institutional development Direct credit provision
Private-sector business • transnational • national • local/microenterprise (including producers, users, distributors, and financiers of technology)	Profits Market share Return on investment	Technology R&D/commercialisation decisions Marketing decisions Capital investment decisions Skills/capabilities development policies Structure for acquiring outside information Decision to transfer technology Choice of technology transfer pathway Lending/credit policies (producers, financiers) Technology selection (distributors, users)
Donors • multilateral banks • GEF • bilateral aid agencies	Development goals Environmental goals Return on investment	Project selection and design criteria Investment decisions Technical assistance design and delivery Procurement requirements Conditional reform requirements
International institutions • WTO • UNCSD • OECD	Development goals Environmental goals Policy formulation International dialogue	Policy and technology focus Selection of participants in forums Choice of modes of information dissemination
Research/extension • research centres/labs • universities • extension services	Basic knowledge Applied research Teaching Knowledge transfer Perceived credibility	Research agenda Technology R&D/commercialisation decisions Decision to transfer technology Choice of pathway to transfer technology
Media/public groups • TV, radio, newspaper • Schools • Community groups • NGOs	Information distribution Education Collective decisions Collective welfare	Acceptance of advertising Promotion of selected technologies Educational curricula Lobbying for technology-related policies
Individual consumers • urban/core • rural/periphery	Welfare Utility Expense minimisation	Purchase decisions Decision to learn more about a technology Selection of learning/information channels Ratings of information credibility by source

tions. Innovation performance depends on the way in which the different stakeholders of the "innovation system" - businesses, universities and other research bodies - interact with one another at the local, national and international levels. It also depends on the innovation promotion policies of governments.

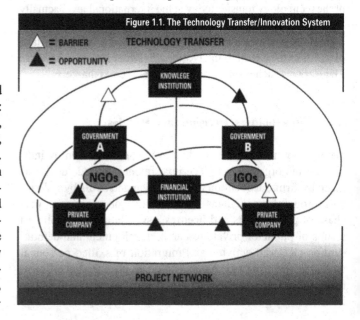

Figure 1.1. The Technology Transfer/Innovation System

There are a large number of pathways through which the various stakeholders can interact to transfer technologies. The most common include:

- direct purchases
- licensing
- franchising
- foreign direct investment
- sale of turn-key plants
- joint ventures
- subcontracting
- cooperative research arrangements and co-production agreements
- export of products and capital goods
- exchange of scientific and technical personnel
- science and technology conferences, trade shows and exhibits
- education and training (of nationals and foreigners)
- commercial visits
- open literature (journals, magazines, books, and articles)
- government assistance programmes.

Other pathways require little or no interaction among the principal stakeholders, because they involve the acquisition of technology without the consent of the provider. Among these are:

1. industrial espionage
2. end-user or third country diversions
3. reverse engineering.

Each pathway represents different types of flows of knowledge, money, goods and services among different sets of stakeholders (see also Annex 1-3). Each pathway has very different implications for the learning that occurs and ultimately the degree of technology-as-knowledge transfer that takes place beyond simple hardware transfers.

The following framework classifies pathways into three primary types:

1. *Government-driven pathways* are technology transfers initiated by government to fulfil specific policy objectives;

2. *Private-sector-driven pathways* primarily involve transfers between commercially oriented private-sector entities, and have become the dominant mode of technology transfer;

3. *Community-driven pathways* are those technology transfers involving community organisations with a high degree of collective decision-making.

Some observers have suggested that along any pathway, technology transfer follows five "stages": assessment (including identification of needs), agreement, implementation, evaluation and adjustment, and replication (Figure 1.2). The stakeholders involved and the specific decisions and actions taken at each stage differ greatly depending upon the pathway. By analysing the interests and influences of different stakeholders at each stage it is possible to determine how various barriers to technology transfer might be overcome.

Much has been written about why transnational corporations choose one pathway over another. Some of the key issues are summarised in Table 1.2. While wholly owned subsidiaries have been the dominant modes of foreign direct investment (except where joint ventures have been specifically targeted by national policies, such as in India), international joint ventures have been growing in number since the 1980s. Various theories have been advanced to explain this change. Datta (1988) suggests that: (a) host governments are increasingly requiring foreign investors to form joint ventures, (b) transnational corporations have realised that the knowledge of complex and often volatile local business environments by local partners can be a significant asset, and (c) there is a growing trend to internationalise business to reduce costs. Contractor (1991) sees transaction costs as determinants of a firm's choice of the mode of transfer. Kogut (1988) explains joint ventures in terms of transaction costs, strategic behaviour, and transfer of organisational knowledge and learning. Kogut further suggests that some forms of tacit knowledge can only be transferred through a joint venture, because the knowledge is organisationally embedded and not conducive to licensing or other forms of transfer. On the other hand, increasing global competition may deter technology transfer, especially at the cutting edge.

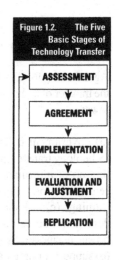

Figure 1.2. The Five Basic Stages of Technology Transfer

ASSESSMENT → AGREEMENT → IMPLEMENTATION → EVALUATION AND ADJUSTMENT → REPLICATION

Table 1.2	Key issues and factors affecting choice of technology transfer pathways
PATHWAY	**KEY ISSUES AND FACTORS AFFECTING CHOICE OF PATHWAY**
Direct sales	Import duties
	Advertising
	Product compatibility
	Standards and certification
	After-sales service and training
	Distributor capabilities
	Degree of system integration required before use by final user
	Insurance and product liabilities
Turnkey contracts	Domestic technological capabilities
	International competitive bidding
	Import duties
	Buyer training
	Corruption
Wholly owned subsidiaries	Acceptable financial risks
	Foreign investment policies of government
	Expected size of domestic market
	Export duties
	Repatriation of profits
Joint ventures	Acceptable financial risks
	Ensuring protection of intellectual property
	Expected size of domestic market
	Product adaptation
	Partner identification, appraisal, and negotiations
	Foreign investment policies of government
	Export duties
	Repatriation of profits
Licensing agreements	Intellectual property protection
	Future domestic market and strategic interests of MNC
	Acceptable financial risk
Multilateral development lending	Need for and viability of carrying out structural economic reforms
	Guarantees and credit worthiness of government and borrowers
	Economic and financial rates of return from investments
	Procurement procedures
Development aid and other grant financing (like GEF)	Donor country political agenda
	Multilateral agency priorities
	Recipient country capacity to make informed choices
	Range of stakeholders' involvement in recipient country
Twinning, conferences, symposia, and other person-to-person pathways	Ability to attend conferences, symposia
	Availability of counterpart resources
	Access to information and communication means
	Intellectual property protection

1.7 United Nations Strategies for Facilitating Technology Transfer

Agenda 21 (UN, 1993) outlines several strategies for promoting the transfer of technologies, some of which are summarised as follows:

(a) government policies that create conditions favourable for technology transfer in both the public and private sectors; (b) setting up of information networks and clearing houses that disseminate information on technologies, and provide advice and training; (c) collaboration between technology research and demonstration centres; (d) collaboration agreements between private businesses for direct foreign investment and joint ventures; (e) support for national and international organisations to undertake training in the assessment, development, and management of new technologies; and (f) international programmes for cooperation in capacity building related to technology research and development.

The United Nations Commission for Sustainable Development (UNCSD) has set up an Ad Hoc Working Group on Technology Transfer and Cooperation. The group has identified inadequate financial resources, shortage of suitably trained manpower and of appropriate institutions as major difficulties in technology transfer (UNECOSOC, 1994). The Working Group suggested several mechanisms for implementing many of the actions suggested in the preceding paragraph. The Group recommended ways to facilitate the transfer of technologies in the public sector, and also recognised the crucial role of the private sector in the transfer of technology. Subsequent Reports of UNECOSOC (1995, 1996) have elaborated on this theme, and proposed activities by governments that would contribute to the dissemination of information, capacity building and institutional development, financial mechanisms, and partnership arrangements.

1.8 Technology Transfer Related to Global Climate Change

An introductory and illustrative reference is made here on technology transfer and cooperation in addressing the challenges posed by global climate change. Some technology transfer processes are driven by the FCCC process, others are not. The important role of technology transfer in this context has been recognised in several Reports (*e.g.* IPCC, 1996) of the Intergovernmental Panel on Climate Change (IPCC), as well as in Agenda 21 and in the deliberations of the Parties to the UNFCCC, and by others (*e.g.* Grubb, 1990; MacDonald, 1992; Guertin *et al.*, 1993; Rath and Herbert-Copley, 1993; UNCTAD, 1997).

The coming into effect of the UNFCCC has assisted in helping identify the technology needs of developing countries in cooperation with various stakeholders, including investors, through technology assessment panels. An initiative of the International Energy Agency (IEA) and the government of Japan attempts to combine voluntary action by governments with incentives for private dissemination of technological information. The latter has evolved the Greenhouse Gas Technology-Information Exchange (GREENTIE), which has identified specific initiatives for certain countries (Forsyth, 1998).

While technology transfer is a common feature these days of all sectors of human activity, there are some features that are unique to the area of climate change. One salient feature is that of scale – both in terms of geography and the number of technologies. Essentially all countries of the world could be involved in the process, and the number of technologies could easily run into the thousands. Another unique feature of technology transfer in the context of global climate change is the number of persons that might benefit from the success of these efforts, since the whole world is expected to be the beneficiary.

Another aspect of technology development and transfer related to global climate change that is sometimes different from technology in many other sectors is that the payback period for the research and development expenses may be too long to be of interest to the private sector. In such cases, special incentives may have to be provided through government policies or the research and development may have to be undertaken by research organisations in the public sector. A great deal of such effort in the area of renewable energy has been undertaken in governmental laboratories in the industrialised countries, and the intellectual property associated with it thus lies in the public sector. This might facilitate the transfer of environmentally sound technologies dealing with, for example, renewable energy.

Mitigation and adaptation technologies

The focus of this Report is on the processes of technology transfer, and ways to promote cooperation in technologies dealing with global climate change, both for mitigation and adaptation. The mitigation technologies can be classified either by the specific greenhouse gas, or by sector of human activity resulting in the emissions and accumulation. Three sectors, energy (including transportation, industry and buildings) agriculture and forestry are key in determining the level of emissions affecting climate change. There are six gases under the Kyoto Protocol, but the dominant ones in climate forcing are carbon dioxide (CO_2), methane (CH_4) and nitrous oxide (N_2O). It is however important to keep in mind that a climate change effect, that is, temperature rise and its impacts, is determined more by the cumulative emissions over time than by current emissions alone. In this context, land-use change can play an important role in climate change as well. Prominent processes include measures to increase energy efficiency in general, development of renewable energy, improving the emission levels of the transport sector, mitigating emissions from the agricultural sector, building markets for environmentally sounder technologies, and facilitating investment and trade (IPCC, 1996).

While the scope of mitigation technologies is large, that of technologies for adapting to climate change may be even larger. Such technologies would include everything from building design and construction to changes in agricultural practices and to approaches to coastal zone management. The process of adaptation relates to many impacts of climate change that will impinge on collective goods and systems, such as food and water security, biodiversity and human health and safety.

Adaptation has played only a marginal part in the reports produced by IPCC so far, which is also reflected on the attention given to adaptation technology development and its transfer. One of the reasons was the existence of two distinct schools of thought about climate change, both of which have chosen not to encourage adaptation research and planning (Kates, 1997). The "preventionist" school argues that the ongoing increase of atmospheric greenhouse-gas concentrations could be catastrophic, and that drastic action is required to reduce emissions. The "adaptationist" school, on the other hand, sees the need to focus on neither adaptation nor mitigation and argue that both natural and human systems will adapt naturally to changing circumstances (Kates, 1997). The increasing awareness of climate change not as a theoretical phenomenon but as a genuine threat has led to the emergence of a third school of thought labelled the "realist" school (Klein and MacIver, 1999). The realist understand that a process must be set in motion to consider adaptation as a crucial and realistic response option along with mitigation (Parry *et al.*, 1998; Pielke, 1998).

The impacts of climate change may be positive in some cases. For example, in specific locations, and for some plant species, a rise in carbon dioxide concentration may result in higher production. Adaptation may also provide new opportunities for uses of land, and result in reduced heating needs in temperate countries. On the other hand, climate change is expected to have a negative impact in many other cases, for instance in agriculture (see chapter 11), forest lands (see chapter 12), human health (see chapter 14) and coastal zones (see chapter 15). These impacts could affect commercial interests indirectly, but usually the strongest and most direct incentives to adapt are found within the public sector. The use and transfer of many adaptation technologies worldwide has occurred because of societal interventions, not as a result of market forces. Examples of such interventions include direct governmental expenditures, regulations and policies, and public choices.

Apart from the government being a dominant stakeholder in technology transfer for adaptation, four more characteristics often distinguish adaptation from mitigation to climate change. Each of these characteristics also represents a barrier to adaptation and associated technology transfer:

- Uncertainty concerning the role of greenhouse gases in causing climate change has been reduced, but uncertainty about the location, rate and magnitude of impacts is still considerable, which could hamper effective anticipatory adaptation.
- Adaptation technologies will often address site-specific issues and will therefore have to be designed and implemented keeping local considerations in mind. This could hamper large-scale technology replication.
- As opposed to benefits of mitigation, which are global (reduced atmospheric greenhouse-gas concentrations),

benefits of adaptation are primarily local. For this reason, adaptation projects thus far have attracted limited interest from the Global Environment Facility (GEF) and other donors.
- The implementation of mitigation technologies can contribute to the development of a country's energy-consuming sectors, while adaptation technologies are primarily aimed at preventing or reducing impacts on these and other sectors. As such, adaptation is often not considered a development objective.

In spite of adaptation often not being considered a development objective, governments have a number of clear incentives and opportunities to start planning for adaptation. For example, many adaptation technologies do not only reduce vulnerability to anticipated impacts of climate change but also to contemporary hazards associated with climate variability. It could be considered "no-regret" adaptation or "climate safe development", having utility both now and in the future, even if climate change were not to occur. In addition, adaptation options need to be designed keeping site-specific natural and socio-cultural circumstances in mind. Strengthening technological, institutional, legal and economic capacities, as well as raising awareness among key stakeholders are important for effective adaptation and technology transfer, because no adaptation option will be successful when it is implemented in an environment that is not ready, willing or able to receive the option.

Kyoto Mechanisms

A specific measure which can affect the promotion of climate-related technologies is the Clean Development Mechanism (CDM) defined in Article 12 of the Kyoto Protocol (UNDP, 1998). As further discussed in the Report, it offers, inter alia, the possibility of obtaining credit from certified emission reductions occurring during the period 2000 to 2008 AD during the years 2008 to 2012 (Art.12, para. 10). This provides a strong incentive to embark on CDM projects in developing countries soon after the beginning of the new millennium. Tradable permits and Joint Implementation (JI) are other flexible mechanisms which might facilitate technology transfer (Carraro, 1999; Grubb *et al.*, 1999; Jepma and van der Gaast, 1999; Oberthur *et al.*, 1999). The CoP4 meeting in Buenos Aires, in November 1998, further discussed the development and transfer of technologies, where the SBSTA made a set of specific recommendations, with a special emphasis on capacity building and consultative processes (see also sections 3.4 for further discussion of CoP4 decisions and flexible mechanisms).

1.9 The Changing Roles of Key Stakeholders: Governments, Private Sector, Communities and NGOs

While acknowledging the crucial role that governments have to play in the formulation and implementation of policies that facilitate technology transfer, for example, via ODA and other measures, there is increasing recognition of the dominant role of the private sector in actually bringing about the successful transfer of technologies, especially in the case of mitigation. Agenda 21 and FCCC, and the statements of the CoP make frequent references to this role. Von Moltke (1992) and others have advocated letting the markets determine the choice of technology and of transfer modes, once suitable macroeconomic conditions and policies have been put in place by governments. Heaton *et al.* (1994) recommend the diffusion of environmental technology through transnational commercial networks, the development of business charters for environmental technology cooperation, and the creation of environmental technology investment corporations financed on a long-term basis through private sources. The CDM provides considerable scope for creative partnerships involving private parties fostering technology transfer for implementing the FCCC.

The private sector has always played an important role in the development of energy efficient technologies, which reduce the emissions of GHGs. It is again becoming increasingly active in developing renewable energy technologies, after a loss of interest following the sharp decline in oil prices during the 1980s. Several hundred companies are now involved in the manufacture of wind turbines, photovoltaic systems and component devices (Anderson, 1997). They are benefiting from work that is frequently sponsored directly by governments and by multilateral organisations, such as the Global Environment Facility and its Implementing Agencies - the World Bank, UNDP and UNEP.

There are an increasing number of companies that are already involved not only in the actual transfer of specific technologies, e.g. wind turbines and solar photovoltaics (Browne, 1997), but also in undertaking broad analyses dealing with technology transfer for their own use or the use of governments (Touche Ross, 1991) and transnational organisations. Illustrations of this recent trend are the International Climate Change Partnership (ICCP), formed in 1991, and one of the largest international industry coalitions focused exclusively on climate change issues, whose members include major transnational corporations and NGOs, and the Pew Center for Climate Change established in 1998, gathering some 22 companies in its Business Environmental Leadership Council.

At the community level and especially in adaptation situations, individual citizens, small-scale enterprises and NGOs are the main stakeholders, concerned with agriculture, forestry, coastal management, etc.

References

Anderson, D., 1997: Renewable Energy Technology and Policy for Development. In *Annual Review of Energy and Environment, Vol. 22.* R. H. Socolow, D. Anderson, J. Harte, (eds.), Annual Reviews Inc., Palo Alto, CA, pp. 187-215.

Archibugi, D., J. Michie (eds.), 1997: *Technology, Globalisation and Economic Performance*. Cambridge University Press, Cambridge, UK.

Asian Development Bank, 1994: *National Response Strategy for Global Climate Change: People's Republic of China,* T. A. Siddiqi, D. G. Streets, Wu Zongxin, and He Jiankun (eds.). Office of Environment, ADB, and Beijing: State Science and Technology Commission of China, Manila and Beijing.

Barnes, D. F., and W. M. Floor, 1996: Rural Energy in Developing Countries: A Challenge for Economic Development. *In Annual Review of Energy and Environment, Vol. 21,* R. H. Socolow, D. Anderson, and J. Harte (eds.), . Annual Reviews, Inc., Palo Alto, pp. 497-530.

Browne, J., 1997: *Climate Change: The New Agenda*. British Petroleum, London.

Carraro, C., (ed.), 1999: *International Environmental Agreements on Climate Change*. Kluwer, Dordrecht, The Netherlands.

Chen, M., 1996: *Managing International Technology Transfer*. International Thomson Business Press, London.

Contractor, F. J., 1991: Interfirm Technology Transfers and the Theory of Multinational Enterprise. In *The International Transfer of Technology: A Book of Readings*. R. D. Robinson, (ed.), Ballinger, Cambridge, MA, pp. 57-78.

Datta, D. K., 1988: International Joint Ventures: A Framework for Analysis. *Journal of General Management* 14(2 - Winter), 78-91.

Dixon, R.K., J. A. Sathaye, S. P. Meyers, O. R. Masera, A. A. Makarov et al., 1996: Greenhouse Gas Mitigation Strategies: Preliminary Results from the U. S. Country Studies Programme. *Ambio*, **25**, 26-32.

Dodgson, M., and J. Bessant, 1996: *Effective Innovation Policy: A New Approach*. International Thomson Business Press, London.

Forsyth, T., 1998: Climate Change Debate. *Environment*, **40**(9), 17-20 and 39-43.

Global Environment Facility, 1997: *Quarterly Operational Report*. GEF, Washington, DC.

Grubb, M., 1990: *Energy Policies and the Greenhouse Effect, Vol. 1: Policy Appraisal*. Brookfield. VT.

Grubb, M., C. Vrolijk, and D. Brack, 1999: *The Kyoto Protocol: A Guide and Assessment*. Royal Institute of International Affairs, London.

Grubler, A., 1998: *Technology and Global Change*. Cambridge University Press, Cambridge, UK.

Grubler, A., and N. Nakicenovic, 1991: Long Waves, Technology Diffusion, and Substitution. *Review*, **14**, 313-342.

Guertin, D.L., J.E. Gray, and H.C. Bailey (eds.), 1993: *Energy Technology Co-operation for Sustainable Economic Development*. University Press of America, Lanham, MD.

Heaton, G. R., R.D. Banks, and D. W. Ditz, 1994: *Missing Links: Technology and Environmental Improvement in the Industrializing World*. World Resources Institute, Washington, DC.

ICF, 1999: *An Analysis of Measures to Promote Technologies to Mitigate Emissions of GHGs*. Technology Issues Table of the Canadian National Climate Change Process.

Intergovernmental Panel on Climate Change, 1995: *IPCC Second Assessment Report*. Reports of IPCC Working Group I (*Science*), Working Group II (*Impacts, Adaptation, and Mitigation*), and Working Group III (*Economic and Social Dimensions*), and a *Synthesis of Scientific - Technical Information Relevant to Interpreting Article 2 of UNFCCC*, including *Summaries for Policymakers*, are available. World Meteorological Organization and the United Nations Environment Programme, Geneva.

Intergovernmental Panel on Climate Change, 1996: *Technologies, Policies, and Measures for Mitigating Climate Change*. R. T. Watson, M. C. Zinyowera, R. H. Moss, (eds.), IPCC Technical Paper 1. World Meteorological Organization and the United Nations Environment Programme, Geneva.

International Energy Agency, 1996: *International Energy Technology Collaboration: Benefits and Achievements*. O.E.C.D., Paris.

International Energy Agency, 1997. *Renewable Energy Policies of IEA Countries*. O.E.C.D., Paris

International Petroleum Industry Environmental Conservation Association, and the United Nations Environment Programme, 1995: Technological Cooperation and Capacity Building: The Oil Industry Experience. Contribution to Agenda 21, IPIECA, London.

Jepma, C. J., and W. van der Gaast, 1999: *On the Compatability of Flexible Instruments*. Kluwer, Dordrecht, The Netherlands.

Johansson, T. B., H. Kelly, A.K.N. Reddy, and R.H. Williams, (eds.), 1993: *Renewable Energy: Sources for Fuels and Electricity*. Island Press, Washington, D.C.

Joint Implementation Quarterly, 1998: Planned and Ongoing AIJ Pilot projects. *JIQ*, **4** (April 1998), 14.

Kates, R.W., 1997: Climate change 1995—impacts, adaptations, and mitigation. *Environment*, **39**(9), 29-33.

Klein, R.J.T., and D.C. MacIver, 1999: Adaptation to climate variability and change: methodological issues. *Mitigation and Adaptation Strategies for Global Change*, **4** (3-4), 189-198.

Kogut B., 1988: Joint ventures: Theoretical and empirical perspectives. *Strategic Management Journal*, **9**, 319-332.

Kranzberg, M., 1986: The Technical Elements in International Technology Transfer: Historical Perspectives. In *The Political Economy of International Technology Transfer*. J. R. McIntyre, D.S. Papp, (eds.), Quorum Books, New York, pp.31-46.

Levine, M. D., A. Gadgil, S. Myers, J. Sathaye, J. Stafurik, and T. Wilbanks, 1991: *Energy Efficiency, Developing Nations, and Eastern Europe*. Lawrence Berkeley Laboratory, Berkeley.

MacDonald, G.J.F., 1992: *Climate Change: A Challenge to the Means of Technology Transfer*. Institute of Global Cooperation and Conflict, University of California, San Diego, CA.

Martinot, E., J. E. Sinton, and B. M. Haddad, 1997: International Technology Transfer for Climate Change Mitigation and the Cases of Russia and China. In *Annual Review of Energy and Environment, Vol. 22.* R. H. Socolow, D. Anderson, J. Harte, (eds.), Annual Reviews Inc., Palo Alto, CA, pp. 357 – 401.

Munasinghe, M. and R. Swart (eds). 2000. *Cilmate Change and its Linkages with Development, Equity and Sustainability, Proceedings of the IPCC Expert Meeting*, Colombo, Sri Lanka, April 1999. IPCC, Geneva.

Munasinghe, M 2000. *Development, Equity and Sustainability in the Context of Climate Change*, IPCC Guidance Paper, IPCC, Geneva.

Nakicenovic, N., W. D. Nordhaus, R. Richels, and F. L. Toth, (eds.), 1994: *Integrative assessment of Mitigation, Impacts, and Adaptation to Climate Change*. International Institute for Applied Systems Analysis, Laxenburg, Austria.

Nordic Council of Ministers, 1997: *Criteria and Perspectives for Joint Implementation*. Nordic Council of Ministers, Copenhagen.

Oberthur, S., H.E. Ott, and F. Yamin (1999). *The Kyoto Protocol: International Climate Change Policy for the New Millenium*. Cambridge University Press, Cambridge.

Organisation for Economic Co-operation and Development, 1991: *Managing Technological Change in Less Advanced Developing Countries in the 1990s*. OECD, Paris:.

Organisation for Economic Co-operation and Development, 1999: *Technology and Environment: Towards Policy Integration*. (DSTI/STP(99)/FINAL. OECD, Paris:.

Pachauri, R. K., and P. Bhandari, editors, 1994: *Climate Change in Asia and Brazil: The Role of Technology Transfer*. Tata Energy Research Institute, New Delhi.

Parry, M., N. Arnell, M. Hulme, R. Nicholls, and M. Livermore, 1998: Adapting to the inevitable. *Nature*, **395**, 741.

Pielke, R.A. Jr., 1998: Rethinking the role of adaptation in climate policy. *Global Environmental Change*, **8**(2), 159-170.

Philips, M., 1991: *The Least-Cost Energy Path for developing countries: Energy-Efficiency Investments for the Multilateral Development Banks*. International Institute for Energy Conservation, Washington, DC.

Rath, A., and B. Herbert-Copley, 1993: *Green Technologies for Development: Transfer, Trade, and Co-operation.* International Development Research Centre, Ottawa.

Reddy, A. K.N., 1991: Barriers to Improvements in Energy Efficiency. *Energy Policy*, **19**, 953-961.

Reddy, M. N., and L. M. Zhao, 1990: International Technology Transfer: A Review. *Resources Policy*, **19**, 258-307.

Riemer, P. W. F., A. Y. Smith, and K. V. Thambimuthu, (eds.), 1998: *Greenhouse Gas Mitigation: Technologies for Activities Implemented Jointly.*

Rip, A., and R. Kemp, 1998: Technological Change. In *Human Choice and Climate Change, Vol. 2: Resources and Technology.* S. Rayner, E.L. Malone , {eds.), International Energy Agency , Batelle Press, Cheltenham, UK.

Robinson, R. D., 1991: *The International Communication of Technology: A Book of Readings.* Taylor and Francis, New York.

Rosenberg, N., 1982: *Inside the Black Box: Technology and Economics.* Cambridge University Press, New York.

Sagafi-nejad, T., 1991: International Technology Transfer Literature: Advances in theory, empirical research, and policy. In *The International Communication of Technology: A Book of Readings.* Taylor and Francis, New York.

Schipper, L., S. Myers, R. Howarth, and R. Steiner, 1993: *Energy Efficiency and Human Activity: Past Trends, Future Prospects.* Cambridge University Press, Cambridge.

Shukla, P. R., editor, 1997: *Energy Strategies and Greenhouse Gas Mitigation.* Allied Publishers, New Delhi.

Siddiqi, T. A., 1990: Factors Affecting the Transfer of High Technology to the Developing Countries In *Technology Transfer to the Developing Countries.* M. Chatterjee, (ed.), The MacMillan Press Ltd., London.

Socolow, R., C. Andrews, F. Berkhout, and V. Thomas, (eds.), 1994: *Industrial Ecology and Global Change.* Cambridge University Press, Cambridge, UK.

Teubal, M., T. Yinnon, and E. Zuscovitch. 1991: Networks and Market Creation. *Research Policy* **20**, 381-392.

Touche Ross Management Consultants, 1991: *Global Climate Change: The Role of Technology Transfer.* Report prepared for the UK Department of Trade and Industry and the Overseas Development Administration, Touche Ross, London.

Trindade, S.C., 1991: Environpeace. *Chemtech.*, pp. 710-711.

Trindade, S.C., 1994: Transfer of clean(er) technologies to developing countries. In *Industrial metabolism: Restructuring for sustainable development.* R.U. Ayres, U.E. Simonis, (eds.), United Nations University Press, New York, pp. 319-336.

United Nations, 1992: *Framework Convention on Climate Change.* UNEP/WMO Information Unit on Climate Change, Geneva.

United Nations, 1993: *Agenda 21: Programme of Action for Sustainable Development.* United Nations, New York.

United Nations Conference on Trade and Development, 1995a: *Technological Capacity Building and Technology Partnership; Field Findings, Country Experience, and Programmes.* UNCTAD, New York & Geneva.

United Nations Conference on Trade and Development, 1995b: *Negotiations on an International Code of Conduct on the Transfer of Technology.* Report by the Secretary - General of UNCTAD (TD/CODE TOT/60) UNCTAD, New York.

United Nations Conference on Trade and Development, 1997: *Promoting the Transfer and Use of Environmentally Sound Technologies: A Review of Policies.* UNCTAD, New York.

United Nations Development Programme, 1997: *Energy After Rio: Prospects and Challenges.* Report authored by K. Amulya, N. Reddy, R. H. Williams, and T. B. Johansson, with other contributors, Energy and Atmosphere Programme, Sustainable Energy and Environment Division, UNDP, New York.

United Nations Development Programme, 1998: *Clean Development Mechanism: Issues and Options.* Edited by J. Goldemberg. UNDP, New York.

United Nations Economic and Social Council, Commission on Sustainable Development, 1994: *Education, Science, Transfer of Environmentally Sound Technologies, Cooperation, and Capacity Building.* Report of the Inter-Sessional Ad Hoc Open-Ended Working Group on Technology Transfer and Cooperation. (E/CN.17/1994/11), 2nd Session, UNECOSOC, New York.

United Nations Economic and Social Council, Commission on Sustainable Development, 1995: *Transfer of Environmentally Sound Technologies, Cooperation, and Capacity Building*: Report of the Secretary-General (E/CN.17/1995/17), 3rd Session, UNECOSOC, New York.

United Nations Economic and Social Council, Commission on Sustainable Development, 1996: *Transfer of Environmentally Sound Technologies, Cooperation, and Capacity Building*: Report of the Secretary -General, (E/CN.17/1996/13), 4th Session, UNECOSOC, New York.

United Nations Framework Convention on Climate Change, 1997: *Kyoto Protocol to UNFCCC.* UNFCCC Secretariat, Bonn.

Van Berkel, R., and E. Arkesteijn, 1998: *Transfer of Environmentally Sound Technologies under the Climate Convention: survey of experiences, needs and opportunities among Non-Annex II Countries*: http://www.ivambv.uva.nl/climate/survey/

Von Moltke, K., 1992: International Trade, technology transfer, and climate change. In *Confronting Climate Change: Risks, Implications, and Responses.* I. Mintzer, (ed.), Cambridge University Press, Cambridge.

Von Weiszacker, E., A. B. Lovins, and L. Hunter Lovins, 1998: *Factor Four: Doubling Wealth, Halving Resource Use.* Earthscan, London.

World Bank, 1993: *Energy Efficiency and Conservation in the Developing World: The World Bank's Role.* World Bank, Washington, DC.

World Energy Council, 1994: *New Renewable Energy Resources: A Guide to the Future.* Kogan, London.

Ziman, J. M., 1968: *Public Knowledge: An Essay Concerning the Social Dimension of Science.*

Annex 1-1: List of Case Studies in Chapter 16

Case Study 1	Research, Development and Commercialisation of Cookstoves (KCJ)
Case Study 2	Public Promotion of Private Investment in Efficient Lighting
Case Study 3	Inner Mongolian Household Wind Electric Systems
Case Study 4	Hydrocarbon Refrigerator "Ecofrig" in India
Case Study 5	The Commercial Dissemination of Photovoltaic Systems in Kenya
Case Study 6	Coal Power Plants in China
Case Study 7	Butane Gas Stove in Senegal
Case Study 8	The Brazilian Fuel Alcohol Programme
Case Study 9	Bamboo Fiber Reinforced Cement Board for Carbon Sequestration
Case Study 10	Demand Side Management (DSM) in Ukraine
Case Study 11	Mitigating Transport Sector GHG Emissions: Options for Uganda
Case Study 12	USIJI as a Technology Transfer Process
Case Study 13	Technology Cooperation in Indonesia for Natural Gas Production
Case Study 14	Rural Electrification using Photovoltaics in Ladakh, India
Case Study 15	Blast Furnace Hot Stove Heat Recovery Technology for Chinese Steel Industries
Case Study 16	Coastal Zone Management for Cyprus: Transnational Technology Transfer and Diffusion
Case Study 17	CFC (ODS) Solvent Phase-out in Mexican Electronic Industries
Case Study 18	Swedish Government Programme for Biomass Boiler Conversion in the Baltic States
Case Study 19	Dissemination of Biogas Digester Technology
Case Study 20	Caribbean Planning for Adaptation to Global Climate Change (CPACC): Design and Establishment of Sea-Level/Climate Monitoring Network
Case Study 21	Concrete Armoring for the Coast - Government to Private Sector Technology Transfer
Case Study 22	World Bank/GEF India Alternative Energy Project
Case Study 23	CFC Free Refrigerators in Thailand
Case Study 24	Financing Microhydro Energy Dissemination in Peru
Case Study 25	Tree Growers Cooperatives: A Participatory Approach to Reclaim Degraded Lands
Case Study 26	Carbon Sequestration Benefits of Reduced-Impact Logging
Case Study 27	Technology Information Assessment and Dissemination in India
Case Study 28	Medicinal Plants vs. Pharmaceuticals for Tropical Rural Health Care
Case Study 29	ROK-5 Mangrove Rice Variety in Sierra Leone
Case Study 30	Use of Indigenous Technologies in the South Pacific

Annex 1-2: Criteria for Effective Technology Transfer

For technology transfer to be considered effective, several pre-conditions must first be met. The following criteria and categorisation emanate from box 2 of the IPCC Technical Paper I (IPCC, 1996), and from the preparatory process of this special report. The sectoral chapters and the case studies have by and large adopted these criteria as indicated in section 6.2, in their respective analyses of the material presented. The criteria can be grouped into four categories, namely, (1) GHG and environmentally-related; (ii) economic and socially-related; (iii) administrative, institutional and politically-related; and (iv) process-related:

1. GHG and Other Environmental Criteria

GHG reduction potential

Tons of carbon equivalent; percent of IS92a baseline and range (IS92c-e).

Other environmental considerations
Percentage change in emissions of other gases/particulates; biodiversity, soil conservation, watershed management, indoor air quality, etc.

2. Economic and Social Criteria

Cost effectiveness
- average and marginal costs must be estimated and compared to alternative options
- benefits of technology transfer must exceed its costs

From the buyer's perspective, especially for enterprises in the private sector with incentives to maximise profits or increase shareholder value, the benefits of a technology must exceed its acquisition costs. The perceived risk of the technology may play an important part in the benefit-cost calculation. For example, companies usually employ risk weighting to adjust the discount rate that they use for calculating costs and benefits associated with specific projects.

- price and conditions must provide incentives to seller

Suppliers of technology will only be willing to sell their technology if they perceive that the price received for it exceeds the costs of supply. For sellers of proprietary technology, the lack of patent protection may make this condition difficult to meet. In these situations, the suppliers of technology may believe that there is a risk of their technology being copied without payment, resulting in an inability of the supplier to recoup the costs incurred in research, development, commercialisation and profit. Suppliers must perceive the conditions associated with the terms of transfer to provide them with sufficient incentive to sell the technology.

Adequate financing
There must be adequate financing available to ensure the transfer of technology. The financing can be in the form of commercial bank loans, capital provided through the equity markets, or any one of a number of new and innovative financing schemes. In addition, financing could be provided by public sector organisations such as countries' ODA, the GEF, or the IFC.

Once the terms of transfer have been agreed upon and financing has been identified, the success of the transfer can be measured using a series of performance indicators. Factors that could be evaluated include:

Project-level considerations
Capital and operating costs, opportunity costs, incremental costs must be estimated and compared to alternative options.

Macroeconomic considerations
GDP change, jobs created or lost, effects on inflation or interest rates, implications for long-term development, foreign exchange and trade, other economic benefits or drawbacks.

Equity considerations
Differential impacts on countries, income groups or future generations.

3. Administrative, Institutional and Political Criteria

Information about technology
Buyers require accurate, balanced, and comprehensive information about the technologies they are considering acquiring. Information costs money. The costs to generate useful and useable information about technologies can be costly but, once developed and in the public domain, helps to reduce the buyer's costs. While difficult to measure its effectiveness, the wider dissemination of information about technologies can assist in the more rapid transfer of technology. Dissemination about the various pathways for transfer could also help reduce the costs of negotiating terms of transfer.

Access to technology
Buyers must have access to technology. There are various impediments that sometimes limit the buyers' access. In some instances, the obstacles may involve patent restrictions, whereas in others it could be the high level of technical know-how and costs that limit the diffusion of the technology. While in others market restrictions can be important barriers.

Administrative burden
Institutional capabilities to undertake necessary information collection, monitoring, enforcement, permitting, etc.

Political considerations
Capacity to pass through political and bureaucratic processes and sustain political support; consistency with other public policies.

Replicability
Adaptability to different geographical and socio-economic-cultural settings.

4. Process-related Criteria

Market penetration
Rate of indigenisation; geographic extent of penetration and impacts on other technologies and ancillary benefits

Long term institutional capacity building
- flexibility and capacity to adapt technology to changing circumstances and to sell back to original provider with improvements
- capacity of local staff and long-term financing
- improvements in training and management practices

Monitoring and evaluation of continuous delivery of services provided by technology and adequate financial performance
- continuous delivery of services provided by technology
- comparison of actual and intended benefits
- performance of technology
- quality of benefits
- satisfaction of beneficiaries
- distribution of benefits (equity)
- maintenance and service of equipment
- adequate financial performance
- payback period
- financial rate of return
- net present value

Leakages that reduce the impact of programme or measure

Annex 1-3: A Stakeholder Typology

Stakeholders Description

• **Sources and developers of technology**

Individuals/organisations who undertake original research to develop technology. Typical developers include scientific research organisations, R&D departments within private firms, and government-sponsored research entities. Technology can be developed in either the public or private sector.

• **Owners and suppliers of technology**

These usually include private firms, state-owned enterprises, and government agencies. Technology developed in the public sector often is "spun off" to the private sector, since the private sector is seen as better able to exploit the market potential of the technology. In some countries, however, public sector organisations now compete with suppliers of technology that are based in the private sector.

• **Buyers of technology**

The primary stakeholders in the technology transfer process. Buyers of technology usually are private firms, but can also include state-owned enterprises, government agencies, and individual entrepreneurs.

• **Financiers of technology transfer**

Those who lend to technology buyers, or invest in them to enable the buyers to acquire the technology from the suppliers. Organisations involved include commercial banks, international financial institutions (e.g., the International Finance Corporation), and individual or institutional investors.

• **Information providers**

Include organisations such as UN agencies that have no commercial interest at stake and whose objective is to facilitate match-making between the buyer's needs and the suppliers by providing objective, unbiased information. This information could include technology options, sources of technology, case studies where technologies have been used, data and data processing information, and methods for evaluating different options.

• **Market intermediaries**

Include consultants, NGOs, media, consumer groups, and trade associations. Market intermediaries usually can have significant influence on the buyer's decision by providing information about technologies. Depending on the interest of the intermediary, this information may promote certain technologies at the expense of others.

• **Governments**

The government of the buyer's country sets the rules for transactions through regulation, incentives, and frameworks governing imports of technology/foreign capital. Where the government perceives that the private costs of technology may not reflect the true costs to society (e.g., a technology may have environmental externalities), the government may be involved in expanding or limiting the range of technologies under consideration. The government of countries whose companies sell technologies may set policies to promote technology transfer in support of the climate stabilisation, via ODA and other measures.

2

Trends in Technology Transfer:
Financial Resource Flows

Coordinating Lead Author:
MARK RADKA (USA)

Lead Authors:
Jacqueline Aloisi de Larderel (France), J.P. Abeeku Brew-Hammond (Ghana),
Xu Huaqing (China)

Contributing Authors:
Julia Benn (Finland), Woodrow W. Clark, Jr. (USA),
Andrew Dearing (United Kingdom), Kevin Fay (USA), Doug McKay (UK),
Paul Metz (The Netherlands), K.P. Nyati (India), Luiz Pinguelli Rosa (Brazil)

Review Editors:
Prodipto Ghosh (India), Ramon Pichs-Madruga (Cuba)

CONTENTS

EXECUTIVE SUMMARY

The 1990s have seen broad changes in the types and magnitudes of the international financial flows that drive technology transfer, at least that occurring between countries. Official Development Assistance (ODA) from donor governments has become relatively less important to many developing countries given the dramatic increase in opportunities for obtaining private sector financing for technology acquisition. The same is true of Official Assistance provided to Countries with Economies in Transition (CEITs). Levels of foreign direct investment, commercial lending, and equity investment all increased dramatically during the 1990s, to the point where ODA became less than one quarter of the total foreign finance available to developing countries by mid-decade. During this time many developing countries instituted policy changes that made them more attractive to private investors, often with the assistance of bilateral and multilateral donors; many governments now see their main role in the transfer of technology as facilitating the role of the private sector.

However, the general increase in the importance of private sector investment in developing countries masks two points. First, private sector investment has been very selective. While almost all countries have benefited to some degree, a handful of countries have received most of the attention. ODA is still critical for the poorest countries, particularly when it is aimed at developing capacities to acquire, adapt, and use foreign technologies. Second, private investment, most notably foreign portfolio equity investment and commercial lending, is often fickle. Many developing countries have found to their distress that private investment can quickly dry up if investors perceive more attractive—or less risky—opportunities elsewhere.

A notable characteristic of technology transfer is the difficulty with which it is actually measured and little is actually known about how much climate-relevant technology is successfully transferred each year. It is reasonable to assume that there exists a relationship between international financial flows and international technology transfer, but the results of existing efforts of data collection and statistical interpretation do not make this relationship clear. Greater attention paid to quantifying types of technology transferred and interpretation of trends would give governments better information on which to base policies.

2.1   Introduction

In decision 4/CP.4 Parties to the UNFCCC raised several questions (see Box 1.1 in Chapter 1) pertaining to the amount of technology being transferred that is consistent with Convention objectives, namely:

- *Can specific technology transfer goals be set?*
- *Can we develop conditions and accounting systems to track progress on technology transfer?* and
- *Are particular institutional arrangements needed to monitor progress?*

This chapter examines broad trends in technology transfer relevant to these questions. It looks at resource flows related in part to international technology transfer efforts or initiatives, and is best seen as setting the context for the remaining chapters in the Report. Knowing something about levels of different technology transfer efforts and how they have been changing, both in absolute magnitude and relative to one another, is important to any discussion about what approaches might be more effective in meeting the challenges of global climate change.

2.2   Technology Transfer Quantified

Measuring technology transfer is difficult given the diverse and complex ways in which transfers occur. It would be so even if the task were confined to quantifying the transfer of hardware. When "software" elements such as education and training are included in a broad definition of technology transfer (as is done in Chapter 1 and then used throughout this Report), the task is further complicated; it becomes almost impossible if some judgement about effectiveness is also demanded. "How much effective technology transfer related to climate change mitigation or adaptation takes place between countries each year?" is an interesting question, but one that cannot be answered with any confidence. Data are simply not collected and analysed in a manner that informs policymakers interested in the issue. In commenting on the effect that changing patterns of international finance have on the environment, one observer has noted that "aggregate data on international financial flows obscure important variations across all relevant variables thereby failing to provide much guidance for strategic planning related to particular environmental issues or geographic areas of interest" (World Resources Institute, 1998).

Despite these difficulties it is possible to draw some general conclusions about global technology flows based on information in the literature. Various organisations collect, analyse, and present data on international trade, development assistance, foreign direct and portfolio investment, foreign lending, and related topics. One problem is that none of these corresponds directly to the government, private sector, or community dominated pathways for technology transfer used roughly as the organising scheme for analysis elsewhere in this Report, although different stakeholders generally favour different delivery mechanisms.

Used with some caution and an understanding of their limitations, however, such data can give a crude indication about levels of international technology transfer, and how these levels are changing over time.

2.2.1   International Financial Flows

Several types of international financial flows support technology transfer (WBCSD, 1998). In practice, the transfer of a particular technology may involve several of them operating simultaneously or in a coordinated sequence, particularly for large, costly projects. Among the types of financial flows are:

Official Development Assistance (ODA) and Official Aid (OA). These include grants and interest free or subsidised loans to developing countries (ODA) and countries with economies in transition (OA), primarily from member countries of the Organisation for Economic Cooperation and Development (OECD). ODA/OA includes both bilateral aid and that provided by governments indirectly through multilateral organisations.

Loans at market rates. These include loans from international institutions, including the multilateral development banks (MDBs), and commercial banks. As noted above, some of the grant portions of ODA and OA are also channelled through MDBs to subsidise their loan interest rates, blurring somewhat the lines between these categories.

Foreign Direct Investment (FDI) involves direct investment in physical plant and equipment in one country by business interests from a foreign country.

Commercial sales refer to the sale (and corresponding purchase), on commercial terms, of equipment and knowledge.

Foreign Portfolio Equity Investment (FPEI) and *Venture Capital*. These involve purchase of stock or shares of foreign companies through investment funds or directly. Venture capital is characterised by being longer term and higher risk, with a greater degree of management control exerted by the investor.

Other financial flows, include Export Credit Agencies and activities supported by non-governmental organisations active in technology transfer efforts, educational and training efforts not captured in the other indicators, and related transfers.

None of these indicators of investment flows provides a direct measure of technology transfer, nor, as noted above, do they have a unique relationship with governments, the private sector, or communities, but all capture levels of technology transfer to some extent. Many also have the virtue that they have been measured for some years, allowing an analysis of their changes over time. Trends in each indicator, as well as its strengths and weaknesses as an approximate measure for technology transfer, are discussed in the following section; Table 2.1 shows the relationship between the three broad categories of technology transfer path-

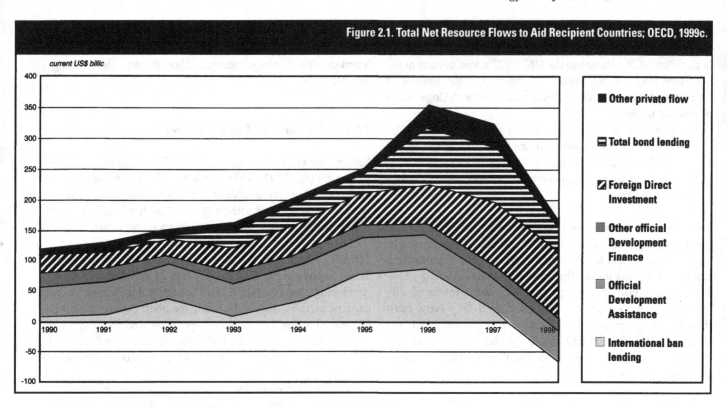

Figure 2.1. Total Net Resource Flows to Aid Recipient Countries; OECD, 1999c.

way and various types of international financial flows, while Figure 2.1 shows how magnitudes of some of these have changed in recent years.

2.2.2   *Official Development Assistance and Official Aid*

Issues
ODA/OA remains the main conduit for government supported technology transfer efforts, particularly those aimed at the poorest developing countries. By definition, support to a developing country counted as Official Development Assistance must be both developmental and concessional. ODA statistics are collected from members of the OECD's Development Assistance Committee (DAC), which consists of 21 OECD member governments (of 29 total) and the Commission of the European Communities (see also section 5.2.2 on ODA).

DAC statistics describe aid flows (grants and loans) to recipients by major category of expenditure. In addition to ODA/OA, data are collected on other official flows, private market transactions, and assistance from non-governmental organisations to each recipient country and recipient countries combined. The OECD also reports on development assistance provided by other groups (OECD, 1998a).

ODA involves much more than technology transfer and looking at overall ODA trends tells little about actual financial flows for this purpose, except to set an upper limit on official government transfers. The OECD's Creditor Reporting System (CRS) is an information system comprising more detailed data on the components of ODA and other lending by the official sector. Data relate to individual grant and loan commitments and specify their purpose, tying status, and, for loans, terms of repayment. The DAC Secretariat converts the amounts of the projects into US dollars using annual average exchange rates. Approximately 15,000 transactions are recorded annually. The CRS does not report details on aid flows from NGOs, non-DAC donors, FDI, unguaranteed bank lending, or portfolio investment (Felcke, 1997).

DAC/CRS statistics on the purpose of aid cover three dimensions: the sector of destination, the nature or form of the aid, and the policy objectives of the aid. Three of 26 principal sectors (broken down into some 200 sub-sectors) have a plausibly stronger relationship to climate change: energy generation and supply, transport, and industry. In principle, the CRS can be used to identify how much ODA (and to a lesser degree OA) is directed to these sectors. Non-sector specific aid to the environment is contained within the "multi-sector/cross-cutting" category.

The policy objectives of aid efforts are tracked across most sectors and forms of aid using a "marker system" introduced sever-

Table 2.1	The International Financing of Technology Transfer					
Technology Transfer Pathway	RELATIVE IMPORTANCE OF TYPE OF FINANCIAL FLOW TO TECHNOLOGY TRANSFER PATHWAY					
	OFFICIAL DEVELOPMENT ASSISTANCE	LOANS	COMMERCIAL SALES	FOREIGN DIRECT INVESTMENT	FOREIGN PORTFOLIO EQUITY INVESTMENT	NGO AND OTHER FLOWS
Government	+++	++	+	+	+	+
Private Sector	-	+++	+++	+++	++	-
Community	++	-	-	-	+	+++

KEY: +++PRIMARY COMPONENT OF PATHWAY   ++SECONDARY COMPONENT OF PATHWAY   +MINOR COMPONENT OF PATHWAY

al years ago to support changing DAC policy objectives. The marker system facilitates monitoring aid in relation to cross-cutting themes (that is, issues that can be addressed through aid activities across all economic sectors), such as environment. Donor governments are supposed to report an environment-oriented policy objective for aid if the activity 1) is intended to produce an improvement in the physical or biological environment of the recipient country, area, or target group concerned or 2) "includes specific action to integrate environmental concerns with a range of development objectives through institution building and/or capacity development" (OECD, 1997b). The marker system is not used by all donor governments. In the sectors of transport, energy and industry, however, the data are reasonably complete since 1996.

The DAC has recently begun a pilot study with the secretariats of the conventions on biological diversity, desertification, and climate change to see whether the marker system can be strengthened to allow tracking of developed country financial assistance required under those conventions. The goal is to avoid creating new and possibly overlapping reporting requirements for donor governments. If implemented fully the expanded marker system would track ODA directed to sectors relevant to the objectives of the UNFCCC (see section 3.4), including buildings, transport, industry, energy, agriculture, forestry, waste management and coastal adaptation. Results of the pilot study will be available in March 2000 from the DAC Working Party on Statistics, and will be presented to SBSTA 12 in June 2000 (UNFCCC, 1999).

Apart from operating various reporting systems, the DAC helps donor governments share information about each other's aid programmes to promote coordination and avoid overlap. In November 1996, the OECD Working Party on Development Assistance and Environment gathered basic information about donor strategies, policies and programmes involving what it called "cleaner technology" (OECD, 1997a). In their responses to a questionnaire, about half the donors identified cleaner technology as a major focus of their technology cooperation activities. The other half of the donors responding reported minor or negligible emphasis on cleaner technology approaches.

Support for overall capacity development to use cleaner technology was the most important goal of activities directed towards cleaner technology. Demonstration projects, training, and education were the main instruments used. This highlights the importance of ODA/OA in leveraging funds for technology transfer, particularly in developing capacity to make good technology choices and use improved technologies.

Virtually all donors saw the private sector in developing countries as a major target for cleaner technology activities, with the establishment of joint ventures between enterprises in their countries and developing countries a primary goal. This blurring of the formerly distinct role of government in ODA is one characteristic change in recent years; many governments now view their primary role in technology transfer as facilitating the role of the private sector in transferring hardware and skills, particularly those from the donating country. The creation of networks between business associations in industrialised and developing countries and the development of new types of public-private partnerships were identified as areas of growing interest for donors.

In the majority of donor countries aid organisations are not the only agencies supporting the transfer of environmental technology. Environment and trade departments, export promotion agencies, and research and development agencies are also involved in ODA efforts related to cleaner technologies, pointing to the need for coordination within both donor and recipient governments.

Ten donors responding to the OECD survey reported using export promotion programmes to support the transfer of environmental technologies, although the promotion of environmental technology is not always an explicit goal of such schemes. Risk funds and seed funds, as well as financing for small-scale investment, were identified as special financing instruments to support cleaner technology. Financing for patent right acquisition is generally not provided.

ODA Trends

The OECD (1999c) reported that in 1998 ODA reversed a five year downward trend, rising to US$51.5 billion from US$47.6 billion in 1997, an almost nine per cent increase in real terms (Figure 2.1). Official aid to economies in transition amounted to some US$5.4 billion in 1998, a slight decrease from 1997. The technical cooperation component of ODA in 1997 was US$13 billion, with some US$1.3 billion spent on students and trainees.

The OECD's summary of 1998 ODA (OECD, 1999c) notes that the recovery in aid in 1998 was in part due to the timing of contributions to multilateral agencies and short-term measures to deal with the Asian crisis, but also reflected some members' commitments to maintain or increase aid flows. Fourteen of the 21 DAC Members reported a rise in ODA in real terms, but the percentage of GNP spent on ODA averaged 0.23 per cent for DAC member countries; only Norway and Sweden maintained ODA above the United Nations target of 0.7 per cent of GNP (OECD, 1999c).

Looking at trends, DAC countries' ODA to transport, energy, and industry sectors in 1993 to 1997 amounted to US$47 billion. In comparison to their total bilateral ODA, aid to transport (US$24 billion) was nine per cent, aid to energy generation and supply (US$20 billion) eight per cent, and aid to industry (US$3.5 billion) one per cent.

Aid to transport can be further broken down to road, rail, water and air transport. Half of all activities were to support road transport and one-fourth to support rail transport. Water and air transport received 15 per cent of aid to transport each. On the basis of the "aid to environment" marker data (which are reasonably complete since 1996), over 60 per cent of aid activities in the transport sector addressed environmental concerns as a significant objective.

Regarding aid to energy projects, CRS data allow a distinction between that directed to renewable and non-renewable sources. Roughly one-third of aid to energy in 1993 to 1998 was allocated to "renewable sources", one-third to "non-renewable sources" and one-third to electrical transmission and distribution. Hydropower is by far the largest component (75 per cent) in the renewable category, but aid activities in the field of geothermal, solar and wind power have been increasing during the last few years. Projects involving coal-fired power plants make up the largest component in the non-renewable energy generation category. Over 70 per cent of all aid activities in the energy sector, and 95 per cent involving coal-fired power plants, were marked against aid to environment as a significant objective.

Aid to industry has decreased during the 1990s. In particular, there has been a substantial drop in aid to fertiliser, chemicals and cement manufacturing, and to the basic metal and transport equipment industries. Industrial development and SME development are at present the largest sub-sectors. Approximately 40 per cent of aid to industry are reported as having environment as a significant objective.

Finally, there has been an increase in aid flows to general environmental protection. This may be partly due to the revision of the DAC sector classification in 1996, which clarified the reporting of environment-related aid activities. The total for 1993 to 1997 was close to US$5 billion. Slightly over 20 per cent of this amount was reported to be for flood prevention/control, nine per cent for biosphere protection (air pollution control, ozone layer preservation and marine pollution control) and eight per cent for biodiversity conservation. The remainder was reported as aid to environmental policy/administrative management and unspecified environmental protection activities.

The sectoral data can be broken down by recipient, giving some idea about aid targeting by donor countries. Bilateral aid to energy, for example, has been highly concentrated on Asia, the region with the largest growth in CO_2 emissions. The ten largest Asian recipients - India, China, Indonesia, Pakistan, Philippines, Thailand, Viet Nam, Malaysia, Sri Lanka and Bangladesh - have received 63 per cent of DAC Members' bilateral aid to energy since 1980. The extent to which aid has influenced the path of energy development depends on the relative importance of ODA in overall energy sector finance. But as noted above, aid can contribute in ways that are not quantifiable, transfer of technology through aid projects also means transfer of knowledge, which in turn can influence developing countries' internal investments in energy infrastructure. The DAC's pilot study with the FCCC Secretariat referred to above is examining this relationship, as well as the targeting of aid in other climate change related sectors.

The OECD notes that the general decrease in ODA has occurred during a period in which there have been widespread improvements in the economic and budgetary situations of DAC Member countries. The G8 Summit leaders meeting in May 1998 reconfirmed the commitment to mobilising resources for develop-

ment, including a goal (among others) of promoting environmental sustainability in developing countries. The organisation notes that "... without a renewed commitment to invest adequate and well targeted resources, progress cannot be expected and achievement of the internationally agreed goals will be jeopardised" (OECD, 1998a).

2.2.3   *Foreign Lending*

Issues
Foreign loans for technology acquisition are provided by both governments (partly through multilateral lending institutions) and private sector financial institutions, with loan recipients falling into both categories as well[1]. Loans come with an obligation to repay interest and principal. If loan principal and interest are not repaid on schedule the credit worthiness of the borrower can be downgraded, making access to new capital difficult. Unless hedged, loans also create foreign currency exposure, with its attendant risks (WBCSD, 1998). Historically, much of MDB lending has gone for large projects, including those in the energy sector or with some other link to climate change (more on MDB in section 5.2.4).

Lending Trends
Lending by multilateral development banks to climate change relevant sectors has declined during the past decade. Loans for industrial development, for instance, decreased from US$8 billion in 1990 to US$4 billion in 1994 (UNIDO, 1997) due to the reduced emphasis on project lending by these banks and better access by developing countries to international capital markets. After a steep rise in the 1990s, the financial crisis in East and Southeast Asia brought about a sharp decrease in private external capital flows to most developing countries in that region. Net private foreign bank lending turned negative for the group of countries most affected by the crisis (ICC, 1998), and total global international bank lending to developing countries went from US$86 billion in 1996 to US$12 billion in 1997 to negative US$65 billion in 1998 (OECD, 1999c). Lending from multilateral development banks, however, did not change significantly during the same period. Bond lending to the same countries, which had increased rapidly in the 1990s, fell as well, dropping from US$83 billion in 1997 to US$37 billion a year later (OECD, 1999c).

World Bank Group lending for energy investments is about one-fifth to one-sixth of the Group's total commitments, or an average of $3 billion per year during 1995 to 1999 (World Bank, 1999). Over roughly the same period, World Bank lending for renewable energy (excluding large-scale hydropower and tradi-

---

[1] Loans to developing countries made by multilateral lending institutions are technically counted as ODA when the loan is disbursed. When the principal and interest are repaid, a negative entry is made in the ODA statistics. If the loan is cancelled, the funds provided are then considered to have been a grant. Because they share certain characteristics with commercial loans, however, MDB loans are discussed here.

tional geothermal) was $275 million in loans, credits and grants, or less than 10 per cent of the sector total. The trend, however, is more encouraging. During a period where lending for energy projects generally decreased, the Bank's commitments for energy efficiency and renewable energy increased. The number of projects that can be considered climate-friendly is projected to increase from 80 during the period 1996 to 1998 ($1.9 billion) to 160 during the period 1999 to 2001 ($2.3 billion), as shown in Table 2.2 (World Bank, 1998b and 1999).

Table 2.2	World Bank Group Environmental Strategies Record: Energy Sector			
CATEGORY	ACTUAL LENDING THREE-YEAR TOTAL 1996-1998		PIPELINE THREE-YEAR TOTAL 1999-2001	
	NUMBER OF PROJECTS	FUNDING - ALL WBG SOURCES ($ MILLION)	NUMBER OF PROJECTS	FUNDING - ALL WBG SOURCES ($ MILLION)
End-use efficiency	13	571	11	280
District heating	4	261	7	471
Supply side efficiency and T&D loss reduction	7	410	10	536
Environmentally innovative energy projects	6	201	6	100
Renewable energy (on and off grid)	11	274	23	902
Oil and gas (environmentally innovative only)	4	174	4	86
Power sector reform	31	N/A	94	N/A
Coal sector reform	4	N/A	5	N/A
Total	80	1,891	160	2,375

NOTE: WBG REFERS TO IBRD, IDA AND IFC OPERATIONS. REFORM ELEMENTS ARE INCLUDED IN MOST PROJECTS.
SOURCE: ADAPTED FROM WORLD BANK, 1999.

The World Bank Group's lending commitments in the renewable energy and energy efficiency areas have risen concomitantly with financing provided through the Global Environment Facility (GEF, 1998b) (see Box 2.1 and also section 5.2.4 and Box 5.2). The financing trends provide strong evidence that GEF funds have not displaced or substituted for World Bank financing, but may instead have helped redirect it. New renewable and energy efficiency projects are still very few in number and to a large extent can be traced to programmes supported by donor governments in specific countries, mostly in South and East Asia. The need to bring these types of projects into the mainstream is recognised in the World Bank's new energy strategy.

Apart from the IBRD and regional development banks, other multilateral development-finance institutions lend for projects in areas related to climate change issues. The OPEC Fund for International Development, an example of a South-South lending-development-finance institution, has committed loans and grants totalling US$5.1 billion since it was founded in 1976 (OPEC, 1999). Almost 50 per cent of loans have gone for projects in the energy, industry, and transportation sectors. The Fund also provides grants to support technical assistance, research, and similar activities, and has a focus on the least developed and other low-income countries.

2.2.4 Foreign Direct Investment

Issues

Given the decline in official development assistance, observers have focussed on the importance of foreign direct investment in technology transfer. Along with commercial lending, FDI is the primary investment vehicle for longer term, private-sector technology transfer. While FDI flows involve financial inputs, they also involve the capitalisation of technology, knowledge, skills, and other resources that represent a stock of assets for production (ICC, 1998; see also section 5.3 and 5.4, which provide a qualitative analysis of private sector finance and private firm investment decisions).

Most analysts agree that FDI, like other private flows, promotes technology change. The question is whether FDI is linked to the transfer of *environmentally* sound technologies and if so how to strengthen the link. UNCTAD (1998) notes that there is a "... close relationship between FDI and intangible technology flows, as well as ... [a] strong proprietary asset base of FDI." Goldenman (1999) notes that FDI flowing into Central and Eastern Europe manufacturing facilities has gone for upgrading technologies and remedying past environmental damage. Investors have brought new equipment and skills, "introduced more efficient methods of production, and demonstrated the link between good environmental practices and profitability". FDI also generates technological spill-over to national businesses through imitation, employment turnover, and the higher qual-

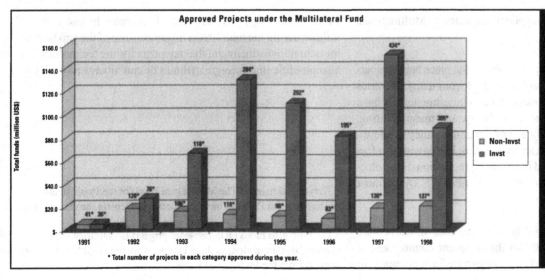

Figure 2.2 Number of Approved Projects and Commitments under the Montreal Protocol Multilateral Fund; (Multilateral Fund Secretariat, 1998.)

Approved Projects under the Multilateral Fund

* Total number of projects in each category approved during the year.

Box 2.1:	DEDICATED FUNDS PROMOTING TECHNOLOGY TRANSFER

Two multilateral funding mechanisms promote the transfer of technologies to developing countries in order to address global environmental problems: the Global Environment Facility and the Montreal Protocol Multilateral Fund[2] (see also section 5.2.4 and Box 5.2 for GEF and section 3.3.3 for the Montreal Protocol).

The Global Environment Facility (GEF) has since 1992 promoted technology transfer of energy efficiency and renewable energy technologies to developing countries[3]. The early focus was on government-driven efforts, but the aim of current GEF programmes is to catalyse sustainable markets and enable the private sector to transfer technologies. An initial three-year pilot phase focused on cost-effectiveness of greenhouse-gas emissions reductions. Following the pilot phase, the GEF in 1996 adopted specific operational programmes for promoting energy efficiency and renewable energy technologies by reducing barriers, implementation costs, and long-term technology costs (GEF 1996, 1997). Additional programmes for energy-efficient transport and carbon sequestration are being developed.

From 1991 to 1999 the GEF approved $700 million for 80 energy efficiency and renewable energy projects in 40 countries (GEF 1999). The total cost of these projects is $5.0 billion, as the GEF has leveraged financing through loans and other resources from governments, other donor agencies, the private sector, and the three GEF project-implementing agencies. An additional $200 million in grant financing has been provided for enabling activities and short-term response measures; the total number of recipient countries is 120.

GEF projects are testing and demonstrating a variety of financing and institutional models for promoting technology transfer (GEF, 1999 and GEF, 1998a). Fourteen projects diffuse photovoltaic technologies in rural areas through local community organisations, financial intermediaries, local photovoltaic dealers/entrepreneurs, and rural energy-service concessions. Several projects assist public and private project developers to install grid-based wind, biomass and geothermal technologies. For energy-efficiency technologies, projects promote technology diffusion through energy-service companies, utility-based demand-side management, private-sector sales of efficient lighting products, regulatory frameworks for municipal heating markets, and development and marketing of more efficient refrigerators and industrial boilers through foreign technology transfer.

The Multilateral Fund helps developing countries eliminate their use of ozone-depleting substances. Since the Fund was established in 1991, over $775 million has been approved for 2,565 projects in developing countries that are Parties to the Protocol. Some 85 per cent of the total (over US$660 million) has been used for investment projects, in which ODS-based technologies have been replaced by more environmentally benign alternatives, or is directly related to investment project preparation (Figure 2.2).

The Fund has also supported a variety of non-investment projects, including the preparation of country programmes, demonstration projects involving ODS-alternative technologies or best practices to reduce ODS emissions, strengthening of national institutions, technical assistance, and training. Total funds approved for the 825 projects in the non-investment category have amounted to US$113 million since the Fund's inception.

A UNEP survey of Implementing Agencies conducted in 1997 (UNEP, 1997) determined that of 863 investment projects approved through mid-1997, only 10 per cent had any sort of formal technology transfer agreement. Although all funded projects involved technology transfer in some form, much of the technology transferred was well established and in the public domain. Most projects did not require a private contractual agreement to proceed because they either involved unpatented technology or the patent had expired (Multilateral Fund Secretariat, 1998)

---

ity standards demanded of supplier companies by Multinational Corporations (MNCs).

Although most payments for technology take place between parent companies and foreign affiliates, FDI from many countries (the U.S., for example) is concentrated in investments that have climate impacts, such as energy, industry, and manufacturing. Data collected by the U.S. Bureau of Economic Analysis show a sharp rise in the 1990s in the amount of U.S. FDI invested in foreign electric utilities, caused in part by deregulation of the electricity industry in many developing countries (U.S. Department of Commerce, 1998).

An increase in FDI does not always generate an immediate increase in technology flows to the recipient country as the relationship between FDI and technology transfer is complex. In China, technology payments did not increase in line with FDI inflows during the mid-1990s, in part because of the gap between the actual investment and the payments for the technology. It is also possible that foreign affiliates do not always pay fully for

---

[2.] The official name is The Multilateral Fund for the Implementation of the Montreal Protocol on Substances that Deplete the Ozone Layer.

[3.] The GEF also funds ODS phaseout projects in countries with economies in transition, as these countries are not eligible to receive support from the Multilateral Fund.

the technology they receive or that perhaps they are not always permitted to do so (UNCTAD, 1998). In developing countries, royalty payments for manufacturing technology generally reach their peak only three to four years after the initial investment has taken place. In the case of royalties for patents that can be absorbed rapidly in new products and processes, as is often the case for patent-related transactions among developed-country companies, the time gap between the initial investments and payments received for technology is smaller (UNCTAD, 1998).

FDI data do not reflect the actual size of investments by foreign affiliates as they include only funds directly tying a parent company and its foreign affiliates. Given the many external sources of finance available, the quantity of funds used in direct investment projects raised outside multinational relationships is quite significant. UNIDO (1997) has estimated that in 1995 foreign capital accounted for only six per cent of total investment in developing countries.

As a source of funding for technology transfer, FDI is highly selective. The World Business Council for Sustainable Development has identified a number of "more intractable" impediments to FDI (WBCSD, 1998). In its view, FDI is less likely to go to countries that have small potential markets, have few skilled or well trained workers, are subject to endemic corruption or are vulnerable to social and civil disruption, or have very limited stocks of natural resources of commercial interest. That many of these conditions apply to the poorest developing countries makes it difficult for them to attract private investment, further emphasising the continued importance of ODA for establishing favourable conditions for technology transfer to least developed countries (LDCs).

Just as some donor governments have emphasised the transfer of environmentally sound technologies in their ODA programmes, some developing country recipients of FDI have attempted to channel private sector investment into ESTs. In 1998 China issued foreign investment guidelines that favour investment in some 200 industries, mostly through preferential ownership rules and tax regimes (Goldenman, 1999). Their purpose is mostly to encourage the entry of better technologies. Gentry (1999) has noted that more "environmental policy leverage" exists over FDI than over other forms of private investment, but that FDI often goes into resource extraction, infrastructure, or manufacturing operations, all with environmental (and climate change) implications. Reviewing studies on FDI and the environment, Zarsky (1999) determined that the picture was mixed. Some studies have found foreign ownership associated with increased energy efficiency and reduced emissions per unit of output resulting from the introduction of advanced technologies and better management. Other studies have found no such links. Zarsky's conclusion is that it is unwise to make overarching conclusions about FDI-environment linkages "on average".

FDI Trends
The global FDI stock, a measure of the investment underlying international production, increased fourfold between 1982 and 1994; by 1996 it was valued at $3.2 trillion. The rate of growth of FDI during the period 1986 to 1995 was more than twice that of gross fixed capital formation, indicating an increasing internationalisation of national production systems. The worldwide assets of foreign affiliates, valued at $8.4 trillion in 1994, have in recent years also increased more rapidly than world gross fixed capital formation, indicating that international production is becoming a more significant element in the world economy (UNCTAD, 1998). Despite its rapid growth, however, FDI still represents a relatively small share of total investment in developing countries. UNCTAD (1996) reports that FDI exceeded 10 per cent of gross fixed capital formation in only eight countries, and in most it is much less than seven per cent of the total, the balance primarily coming from domestic sources.

During 1995 to 1996, the share of developing countries in global inflows of FDI was 34 per cent. Although this is not much higher than the developing-country share during the investment boom at the beginning of the 1980s, qualitatively it reflects a wide variety of location-specific advantages enjoyed by developing countries over and above natural resources. The composition of the top developing-country recipients has also changed dramatically between these two investment booms, with oil producing countries now featuring far less prominently among the top recipients (UNCTAD, 1998). Also significant in the Asian context are the rapid increase in inward FDI to China and the ASEAN countries coming from newly industrialising economies in Asia. This South-South private investment, one quarter of the total in 1992 (UNCTAD, 1996), highlights the error of focusing only on technology transfer from OECD members to developing countries.

During the 1990s, every developing region saw an increase in FDI inflows; even the 48 least developed countries experienced an increase in inflows of 56 per cent in 1996, to $1.6 billion, with Cambodia the largest recipient in this group of countries. FDI to aid-recipient countries, however, dropped in 1998 to US$100 billion (from US$242 billion in 1997) in the wake of the Asian financial crisis (OECD, 1999c). Despite the small size of inflows (both in absolute values and as a share of all developing-country inflows), FDI is still important for many of these economies. Table 2.3 shows the increase in net private capital flows (of which FDI is a component) to low and middle income countries by country group and region during the period 1990 to 1996. Notable is the low share of FDI in countries in Africa, the poorer Asian countries, and the Pacific Islands.

A survey of foreign investors conducted by UNCTAD suggests that the ongoing globalisation of production will continue into the next century. Mergers and acquisitions, joint ventures and other equity and non-equity types of inter-company agreements are expected to go hand in hand with growth in FDI. Corporate restructuring in developed countries, aimed at improving efficiency and modernisation, is expected to continue, giving rise to efficiency-seeking investment.

COUNTRY GROUP	1990	1991	1992	1993	1994	1995	1996
All countries	44.4	56.6	90.6	157.1	161.3	184.2	243.8
Sub-Saharan Africa	0.3	0.8	-0.3	-0.5	5.2	9.1	11.8
East Asia and the Pacific	19.3	20.8	36.9	62.4	71.0	84.1	108.7
South Asia	2.2	1.9	2.9	6.0	8.5	5.2	10.7
Europe and Central Asia	9.5	7.9	21.8	25.6	17.2	30.1	31.2
Latin America and Caribbean	12.5	22.9	28.7	59.8	53.6	54.3	74.3
Middle East and North Africa	0.6	2.2	0.5	3.9	5.8	1.4	76.9
Incomegroup							
Low income countries Excluding China and India	1.4	3.0	2.4	5.8	6.3	5.5	7.1
China and India	10.0	9.1	23.0	44.2	50.8	47.9	60.0
Middle income countries	32.0	44.0	64.8	107.1	104.2	130.7	176.7

Table 2.3   Net private capital flows to low and middle income countries by country group (US$ billion) (Source: World Bank.)

2.2.5   Commercial Sales

Issues

The OECD has estimated that three quarters of international technology transfer arises from trade. Certainly much private sector technology transfer takes place on a commercial basis, with companies investing in new, upgraded, or expanded facilities purchasing the requisite technology from either domestic or foreign suppliers. Although funds for the purchase of equipment and training in its use can come from many of the sources described earlier, they may also come from retained earnings, commercial loans from banks in the same country, and other non-international sources. Such funding is particularly relevant to small and medium-sized companies that do not have access to international finance. Related private sector flows of an international commercial nature result from payments for intellectual property rights, related specialised services (such as consulting services), and payments arising from strategic partnerships between unaffiliated companies (UNCTAD, 1998; WBCSD, 1998).

Trends

It is difficult to assemble a comprehensive picture about international trade and technology transfer. Most government export promotion programmes are not transparent, and it is difficult to gather information about the projects they support (and often make possible). The World Bank assembles data on high technology exports for its World Development Index, but admits that the methodology "rests on a somewhat unrealistic assumption" (World Bank, 1998a). Not surprisingly, the data show that the OECD and "Asian Tiger" countries dominate the export of high technology goods, both in absolute amounts and as a percentage of total exports. Data on importing countries is not provided. UNCTAD (1998) notes, however, that imports of capital goods have consistently represented between 20 per cent and 40 per cent of gross domestic investment in Asian newly industrialising economies, although the percentage relevant to climate change cannot be determined.

Better data exist about payments for foreign royalty and license fees. As reported by the IMF in the *Balance of Payments Statistics Yearbook*, (World Bank, 1998c), developing countries and CEITs made payments in excess of US$7.3 billion and received in turn US$1.2 billion in royalty and license fees in 1997. This catego-

ry covers patents, copyrights, trademarks, industrial processes, and franchises, and for the use, through licensing agreements, of products made from prototypes, such as manuscripts and films. Some of these, but not all, are presumed to be related to the transfer of technology.

2.2.6   Foreign Portfolio Equity Investment and Venture Capital

Issues

Foreign portfolio equity investment (FPEI) is distinguished from foreign direct investment by the investor's degree of management control (UNCTAD, 1998). Portfolio investors typically purchase shares in a company, often through a fund, without any intention of influencing management decisions. The investment horizon of portfolio investors is typically shorter, and the type of investor is more often a financial institution, institutional investor, or insurance company interested in the financial returns. Venture capital is a special type of long-term, high-risk finance with returns usually taken in the form of capital gains and fees rather than dividends (see section 5.3.3). Venture capital is often focused on early-stage investment in technology-based companies. FPEI and venture capital primarily involve the private sector, although institutions such as the International Finance Corporation, Commonwealth Development Corporation, and U.S.-backed Overseas Private Investment Corporation have promoted venture capital funds in developing countries in an effort to "improve the access of small and medium-sized companies to equity finance and management expertise" (UNCTAD, 1998). FPEI's links to technology transfer are that it makes money available to companies in developing countries, which they in turn can use to purchase equipment and training.

FPEI and Venture Capital Trends

According to UNCTAD (1998), the capitalisation of stock markets in developing countries grew more than tenfold during the period 1986 to 1995, from $171 billion to more than $1.9 trillion (compared to US$13.2 trillion for the industrial economies). Much of the increase came from FPEI flowing into emerging markets in Latin America and Asia, although as with FDI the inward flows were concentrated on relatively few countries. The major factors behind the increased inflows were the liberalisation of stock markets in upper middle-income and large low-income developing countries, the globalisation of financial markets, and the concentration of financial resources in the hands of institutional investors, such as mutual and pension funds and insurance companies.

Overall, FPEI flows to developing countries have fluctuated more widely than FDI flows during the period 1986 to 1995 (UNCTAD, 1998), consistent with the shorter investment horizon of equity investors. After a steep rise, FPEI declined in 1998 because of the financial crisis in Asia, and because high returns in the U.S. market offered a more attractive alternative for many investors. Equity investment in developing countries in that year was US$12 billion, down from $38 billion in 1997 (OECD, 1999c). Foreign portfolio equity investment has provided large sums of money for the expansion of industry in a select group of developing countries, but its volatility suggests that it should be treated with caution when looked to for consistent, stable funding of technology transfer.

Venture capital is more common in the United States than in other OECD member countries. OECD governments invest an estimate $3 billion of risk finance per year in small, technology-based companies, while the private sector venture capital supply was estimated at US$100 billion in 1997 (OECD 1997c). It is not clear how much of this is directed at developing countries or climate-relevant technologies.

Box 2.2: **IMPACT OF THE ASIAN FINANCIAL CRISIS ON INTERNATIONAL FINANCIAL FLOWS TO ASIA**

Economic difficulties experienced by a number of Asian countries since mid-1997 initially caused great concern that they might spread throughout the world's financial system, leading in the worst scenarios to a global depression. In early 1998, the International Chamber of Commerce and the UNCTAD conducted a study of multinational corporations to determine what impacts the Asian financial crisis had had on their investment plans in that region. The survey covered 500 companies, including the world's largest MNCs in terms of foreign assets; the largest MNCs with origins in developing countries; and additional businesses with significant operations in Asia. Banking and finance companies were not surveyed. A total of 198 companies (40 per cent) responded.

What emerged from the survey is that foreign companies investing in Asia if anything saw the crisis as an opportunity to increase their investment in the region; their confidence in the region remained unchanged despite the turmoil. ICC and UNCTAD noted "... most MNCs typically take a long-term view with respect to their locational decisions and, given the region's past performance and solid fundamentals in a number of respects, this leads them to remain bullish about FDI in Asia. ... Indeed, in the longer-term, the attractiveness of the region is likely to be enhanced as Asian countries overcome their problems and reassert their economic dynamism on the basis of improved fundamental strengths" (ICC, 1998; OECD, 1998a).

2.2.7   Other Investment and Assistance Flows

Export credit agencies

Governments often facilitate private technology sales by making available export credits, providing various types of risk guarantees, and playing a market intermediation role through trade missions and export promotion programmes. The focus of such efforts, however, is often promoting exports rather than transferring technology to the recipient country in the broad sense used in this Report. Some observers have questioned whether export promotion programmes serve the goal of facilitating transfer of climate-friendly technology, or whether they are actually counterproductive in this regard.

Export Credit Agencies (ECAs) use a number of financial instruments to give national industries advantages: export credits, project financing, risk guarantees and insurance (see also section 5.2.3 and box 5.1). These financial supports are provided at terms that are more favourable than those provided by purely commercial banks or investment houses. Although ECAs have traditionally focused on export credits, they are increasingly providing services that support foreign direct investment (long-term project financing and investment guarantees or insurance).

Trends

On average ECA flows in the mid-1990s reached an average of $110 to $120 billion annually (World Bank, 1998b). This represents more than twice the average flows of official development assistance reaching developing countries during this period ($50 billion annually on average), and about three times the annual average for concessional financing provided by multilateral development banks, about $40 billion (World Bank, 1998b). Furthermore, over 80 per cent of ECA financing to developing countries is provided by the world's seven major industrialised economies (the G-7, or Canada, France, Germany, Italy, Japan, the United Kingdom and the United States). In parallel fashion, the destinations of ECA financing is concentrated in just 10 developing countries, and among these the top two (Indonesia and China) account for more than half of the total (Maurer and Bhandari, 2000).

A study based on information from a commercial database indicates that fully two-thirds of private financing going to developing countries between 1994 and the first quarter of 1999 ($217 billion) was concentrated in investments and exports that can be considered to be carbon or energy-intensive, such as upstream and downstream oil and gas development, fossil-fuelled power generation, heavy infrastructure, energy-intensive manufacturing, and aircraft (Maurer and Bhandari, 2000). Of this, ECAs accounted for 20 per cent, or $44 billion, of the direct financing or guarantees. More significantly, however, ECAs played some financing role in just under half of the financing ($103 billion) for these energy-intensive investments and exports. ECAs, therefore, have a significant leveraging effect and their participation in a project or trade deal draws in significant private capital (Maurer and Bhandari, 2000).

NGO flows

Many non-governmental organisations, foundations, and industry associations support technology transfer activities to developing countries, particularly through efforts that aim to develop capacity or demonstrate technologies (see also section 4.4). These range from small, not-for-profit development oriented organisations to industry consortia that have as members large utilities and technology providers.

Grants by NGOs to developing countries have stayed fairly constant during the last decade, ranging between US$5 billion and US$6 billion per year since 1990 (OECD, 1999c). Not all of these support technology transfer efforts, however, but they may be significant in the least developed countries. Little data on the distribution of grants by NGOs appears to be available. NGOs are important in that they often help community based organisations in developing countries obtain support through foundation grants and ODA.

2.3   Technology Research and Development

In the intersection of technology R&D and climate change technology transfer two questions arise. The first is how much of global R&D is conducted in areas relevant to climate change. The second is how much R&D is conducted across national borders. (see section 4.3 and 4.12 for policy related issues of R&D).

2.3.1   Energy-related Research and Development

Energy-related R&D by OECD governments increased dramatically after the 1970 oil price increases, but has decreased steadily in real terms since the early 1980s; in some countries, the decrease has been as great as 75 per cent (IEA, 1997). Total 1995 expenditures were about US$10 billion. The total, however, hides some interesting shifts in the composition of energy R&D. The greatest declines in funding have been in the coal and nuclear sectors, while R&D for energy efficiency and renewable energy technologies has increased; these now form almost 20 per cent of total energy R&D or some US$2 billion. More recent OECD data (OECD, 1999a) suggest that spending on basic research has remained relatively stable, in part because research is seen by the Organisation (and its member governments) as the engine for long-term innovation and economic growth.

Overall there has been a shift in emphasis from longer-term technology options to meeting shorter-term needs (IEA, 1997), partly the result of increasing industrial competition and liberalisation in the utility industry. IEA (1999) notes that "many of the most promising technologies to achieve large reductions in carbon and emissions intensity require considerable applied R&D before they are commercially feasible. There are still others that are only at the conceptual stage but can be developed with further research."

The IEA has noted that statistics on energy technology R&D funding and deployment are weak, particularly for activities in the private sector. Most governments do not collect detailed information on private sector energy R&D (IEA, 1997), and the rates of deployment of new energy technology, for example, are the subject of "very rough and unsystematic estimates, if indeed they are measured at all" (IEA, 1997). The organisation notes that "...better information on the nature and level of private-sector efforts could help Member countries target their energy R&D and technology investments and programmes more advantageously; but such information is often considered sensitive or proprietary".

2.3.2   Core Technologies and Strategic Partnerships

International collaboration in energy R&D involving national governments, the private sector, and university research communities has been cited as important in creating energy technology options for the future and facilitating technology transfer (IEA, 1997, UNEP, 1998; see also section 4.3, 4.12 and 5.6 on the role of technology partnerships). A survey of 50 energy research institutions conducted for the Climate Technology Initiative found that respondents typically conducted less than 25 per cent of their climate change relevant research under international collaborative arrangements. Such efforts were focused on the planning and research stages rather than on development and deployment of technologies, possibly because governments were the funding source. The study identified a lack of collaboration between OECD countries and countries in Asia and Africa as a gap worth bridging. Furthermore, a feasibility study conducted by the Republic of Korea as part of the work programme of the Commission on Sustainable Development (Chung, 1998 and UNCSD, 1998) notes that there is little coordination of public R&D policies and ODA policies in most donor governments, and that partnership arrangements could yield significant results.

Looking beyond the energy sector, UNCTAD (1998) notes that the number of strategic R&D partnerships in core technologies, such as information technology and biotechnology, has also been rising steadily since 1990. Developing-country companies assumed a bigger role in strategic partnerships (three per cent in 1989 to 13 per cent in 1995), suggesting that these companies may have attained sufficient technological sophistication and capacity to make them worth having as partners (UNCTAD, 1998). Additionally, the OECD notes that R&D by foreign affiliates represented more than 12 per cent of total industrial R&D spending of the 15 OECD countries that account for 95 per cent of industrial R&D undertaken by Member countries (OECD, 1998b). Most of this, however, occurs in a few industrial sectors, among them the computer, pharmaceuticals, electronics, chemicals, and car industries. Parent companies often enter into contracts with their affiliates to carry out specific research, and vice versa. Public financing of the R&D of foreign affiliates is minimal in part because "restrictions remain on foreign participation in government-funded R&D and technology programmes" (OECD, 1999b).

2.4   Conclusions

This chapter has examined the characteristics and trends of different international financial flows that support technology transfer. None of these is a direct measure of technology transfer, nor is it clear how environmentally sound these flows are. The various financial flows are the best available indicators, however, and allow a comparison of technology transfer trends over time. In general, though, levels of both technology transfer and technology diffusion are difficult to measure.

Official Development Assistance experienced a general downward trend during the mid-1990s that was reversed in 1998, while foreign direct investment increased significantly during the decade. By 1997, private flows supplied more than three fourths of the total net resource flows from DAC member countries and multilateral agencies to developing countries, compared with one third in 1990. In large part this reflects the great increase in private sector investment in developing countries that occurred during the 1990s. Sources and amounts of development finance, some portion of which goes for technology transfer, vary widely from region to region. Countries in sub-Saharan Africa received in 1997 an average of some US$27 per capita of foreign aid and US$3 per capita of foreign direct investment. By contrast, countries in Latin America and the Caribbean received US$13 per capita of aid and US$62 per capita of foreign direct investment. Recent initiatives to spur development progress in Africa aim to respond to these disparities.

Total private flows to developing countries peaked in the first half of 1997 and then fell significantly in the wake of the financial crisis that started in Asia. Most of the decline was due to reduced bank lending by the private sector, although this remained robust to Latin America. Foreign direct investment in developing countries is estimated to have increased slightly. Most private transfer of improved technologies, however, still occurs between developed countries and that which occurs to developing countries is selective.

Grants provided by non-governmental and philanthropic organisations have remained steady during the 1990s. Despite their relatively modest amounts, like ODA many of these are directed at least developed countries and at capacity building efforts. NGOs are particularly important in identifying and promoting the transfer and diffusion of ESTs that are socially sound.

The role of governments in providing R&D funding for energy-related technologies deserves careful consideration; linking government-supported R&D programmes with bilateral aid efforts may hasten technology transfer to developing countries. Retrieving useful statistics on R&D funding, and deployment, however, is difficult, especially for private sector R&D activities.

The broad changes in transfer pathways have diminished the direct role of developed country governments in technology transfer and cooperation. The role of government is now viewed by many as helping put in place policies and measures that hasten the transfer of technologies by the private sector.

Technology transfer is difficult to measure, and little is actually known about how much climate-relevant technology is successfully transferred each year. Ambitious investigation and reporting over a number of years are needed to build a consensus on what is actually occurring, and to give governments better information on which to base policies and programmes.

References

Chung, Ray Kwon, 1998: *Report of the International Expert Meeting on the Role of Publicly-funded Research and Publicly-owned Technologies in the Transfer and Diffusion of Environmentally Sound Technologies (ESTs)*. Ministry of Foreign Affairs and Trade, Republic of Korea.

Felcke, R., 1997: *DAC Statistics and the Purpose of Aid: A Potential Tracking Tool for Aid in Support of Climate Friendly Technologies*. Presentation at the Expert Meeting on Terms of Transfer of Technology and Know-How; UNFC-CC, Bonn.

Gentry, B., 1999: Organisation for Economic Co-operation and Development (OECD). Working Paper No. CCNM/EMEF/EPOC/CIME(98)2 prepared for the Emerging Market Economy Forum, 28-29 January 1999, Paris.

Global Environment Facility (GEF), 1997: *Operational Programs*. Washington, DC.

Global Environment Facility GEF), 1996: *Operational Strategy*. Washington, DC.

Global Environment Facility (GEF), 1998a (June): *Operational Report on GEF Programs*. Washington, DC.

Global Environment Facility (GEF), 1998b (October): *Mainstreaming the Global Environment in World Bank Group Operations*. Washington, DC.

Global Environment Facility (GEF), 1999: *Project Implementation Review 1998*. Washington, DC.

Goldenman, G. 1999: Working Paper No. CCNM/EMEF/EPOC/CIME (98)3 prepared for the Emerging Market Economy Forum. Organisation for Economic Co-operation and Development (OECD), 28-29 January 1999, Paris.

International Chamber of Commerce, 1998: Background Note: The Financial Crisis in Asia *Analysis and Summary Tables*. Washington, DC.

World Bank, 1998c: *Balance of Payments* and Foreign Direct Investment. ICC, Paris.

International Energy Agency, 1997: *IEA Energy Technology R&D Statistics*. Paris.

International Energy Agency, 1999: The Role of Technology in Reducing Energy Related Greenhouse Gas Emissions. Meeting of the Governing Board at Ministerial Level, Committee on Energy Research and Technology, Paris.

Maurer, C., and R. Bhandari, 2000: *The Climate of Export Credit Agencies*. Climate Notes, World Resources Institute, Washington, DC.

Multilateral Fund Secretariat, 1998 (November): *Inventory of Approved Projects*. Montreal.

OPEC Fund for International Development, 1999: Development: Common Reference Paper. OECD Working Papers No. 75, OECD, Paris.

Organisation for Economic Co-operation and Development (OECD), 1997a: *Technology Cooperation in Support of Cleaner Technology in Developing Countries*. Synthesis Report, Paris.

Organisation for Economic Co-operation and Development (OECD), 1997b: *Recent Trends in Foreign Direct Investment*. Paris.

Organisation for Economic Co-operation and Development (OECD), 1997c: *Government Venture Capital for Technology-Based Firms*. Paris.

Organisation for Economic Co-operation and Development (OECD), 1998a: Aid and Private Flows Fell in 1997. News Release (18 June), OECD, Paris.

Organisation for Economic Co-operation and Development (OECD), 1998b: Internationalisation of Industrial R&D Patterns and Trends. OECD, Paris

Organisation for Economic Co-operation and Development (OECD), 1999a: *Science, Technology and Industry: Scoreboard of Indicators*. OECD, Paris.

Organisation for Economic Co-operation and Development (OECD), 1999b: *Fostering scientific and technology progress: OECD Policy Brief*. OECD, Paris.

Organisation for Economic Co-operation and Development (OECD), 1999c: Financial Flows to Developing Countries in 1998. Rise in Aid; Sharp Fall in Private Flows. News Release (10 June), OECD, Paris.

United Nations Commission on Sustainable Development (UNCSD). 1998. The Role of Publicly Funded Research and Publicly Owned Technologies in the Transfer and Diffusion of Environmentally Sound Technologies. Background Paper No. 22, DESA/DSA/1998/22, New York, NY.

United Nations Conference on Trade and Development (UNCTAD), 1996: *Wolrd Investment Report 1996: Investment, Trade and International Policy Arrangements*. UNCTAD, Geneva.

United Nations Conference on Trade and Development (UNCTAD), 1998: *World Investment Report 1997: Transnational Corporations, Market Structure, and Competition Policy*. UNCTAD, Geneva.

United Nations Environment Programme (UNEP), 1998: *Priority Issues which STAP Should Address in GEF Phase II*. UNEP, Nairobi.

United Nations Environment Programme (UNEP), 1997 (August): *Draft Study on the Terms of Technology Transfer Under the Multilateral Fund*. UNEP, Paris (unpublished study).

United Nations Framework Convention on Climate Change (UNFCCC), 1999: *Development and transfer of technologies*. Progress report; note by the Secretariat, FCCC/SBSTA/1999/2, UNFCCC, Bonn.

United Nations Industrial Development Organisation (UNIDO), 1997: *Industrial Development – Global Report of 1997*. Oxford University Press, Oxford.

United States Department of Commerce, 1998: *1990-1995: Survey of Current Business, Bureau of Economic Analysis*. Washington, DC.

World Bank, 1998a: *World Development Indices: Licensing & Royalty Payments in the WDI*. Washington, DC.

World Bank, 1998b: *Global Development Finance 1997: Statistics Yearbook. Washington, DC.*

World Bank, 1999: *Fuel for Thought: Environmental Strategy for the Energy Sector*. Washington, DC.

World Business Council for Sustainable Development, 1998 (June): Technology Cooperation for Sustainable Development. Position Brief, Geneva.

World Resources Institute. 1998. *Go with the flows? What the Changing Landscape of Development Finance Means for the Public Interest Community*. Frances Seymour. Washington, D.C., USA

Zarsky, L., 1999: Working Paper No. CCNM/EMEF/EPOC/CIME (98)5 prepared for the Emerging Market Economy Forum. Organisation for Economic Co-operation and Development (OECD), 28-29 January 1999, Paris.

3

International Agreements and Legal Structures

Coordinating Lead Authors:
MICHAEL GRUBB (UNITED KINGDOM) and
KILAPARTI RAMAKRISHNA (INDIA)

Lead Authors:
Raekwon Chung (South Korea), Jan Corfee-Morlot (USA), Michael Gollin (USA), Patricia Iturregui (Peru), Jorge Leiva (Chile), Atiq Rahman (Bangladesh), Terence Thorn (USA), Gerardo Trueba Gonzalez (Cuba)

Contributing Authors:
Marlena Castellanos (Cuba), Manuel Ruiz (Peru), Rajendra Shende (India), Jullio Torres (Cuba)

Review Editors:
Woodrow W. Clark, Jr. (USA), Richard Odingo (Kenya)

CONTENTS

EXECUTIVE SUMMARY

The topic of technology transfer has been an important part of international discussions on development policy for decades. In the context of environment and development, technology transfer was a key element in the United Nations Conference on Environment and Development (UNCED) agreements, including Chapter 34 of Agenda 21 and several Articles in the UNFCCC. These agreements note that effective technology transfer will be essential to meet the global challenges and to enable collaboration between developing and developed countries for the transfer of technologies. This will depend both upon local initiatives and regulations - many of which (such as efficiency improvements) could be beneficial anyway - and various international agreements that bear upon the technology transfer process.

Governments effectively own some important technologies, and these will require government-driven pathways. For private-sector pathways, harnessing the bulk of international investment, intellectual property rights (IPR) regimes are an important consideration. Overall the literature is diverse concerning the relationship between IPRs and technology transfer. Stronger IPRs, particularly resulting from the Trade Related Aspects of Intellectual Property (TRIP) agreement, may foster innovation and vertical technology transfer, but could impede horizontal dissemination of certain technologies through private-sector and community-driven pathways. Compliance with decisions reached in Agenda 21, UNGASS 1997 and other relevant fora is important in this regard, and the TRIP agreement allows for compulsory licensing as a last resort if other avenues are exhausted and with due compensation. A number of international mechanisms are available for bilateral and multilateral financial assistance including, for example, international financial assistance for licensing relevant technologies. This could be one way for the international regime to support technology transfer where market-driven, grants, equity investment, and joint venture solutions are not feasible. IPR regimes could be harnessed more widely to support innovation and dissemination of environmentally sound technologies (ESTs). Industry standards, including management standards developed through the International Standards Organization (ISO) and sectoral standards for some industries, could also play an important role in fostering global dissemination of ESTs.

Specific provisions on technology transfer form an important part of "positive measures" in several multilateral environmental agreements, which help to foster effectiveness and stability of these regimes. Apart from financial mechanisms, there is little empirical evidence concerning how technology transfer provisions have operated in practice. The most extensive experience is with the Montreal Protocol, where the Multilateral Fund (1998), working together with a multitude of supporting institutions and networks, has facilitated extensive technology transfer.

The UNFCCC itself spawned a number of technology-oriented initiatives. Attempts to set up Technology Assessment Panels failed primarily due to disagreements over representation. The Climate Technology Initiative is expanding as an important supporting endeavour though its impact is as yet difficult to evaluate. Ways of discouraging transfer of inferior technology need to be found. The text in the Kyoto Protocol recognises the need for cooperation and enhancement of supportive conditions, including for the private sector in developing countries, as well as the responsibilities of Annex II Parties. The project-level mechanisms of Joint Implementation (JI) and the Clean Development Mechanism (CDM) could be important avenues for furthering international technology transfer. This, however, does not absolve Annex II Parties from their commitments. The CoP4 decision on technology transfer represents an important new opportunity for furthering technology transfer under the Convention. Ultimately, however, it is what happens within countries, and companies, that will do the most to determine the pace and nature of technology transfer.

3.1 Introduction[1]

As noted in Chapter 1, technology transfer has many meanings, and many different terms are used for the processes involved in the development and international diffusion of technologies. Legally within the FCCC it relates to "transfer and access to environmentally sound technologies and know-how to implement" the Convention. Despite widely different perceptions and interests, all Parties agree that transfer and cooperation to advance the availability and use of environmentally sound technologies (ESTs) is critical to effective responses to climate change.

This chapter explores the international political context and actual or potential influence of international agreements and legal structures on technology transfer. The subsequent chapters explore how domestic circumstances and international programmes, including financial initiatives and institutions, can assist the process of technology transfer.

3.2 Technology Transfer and International Political Context

The legal, economic and political issues surrounding technology transfer in general have been around for a long time. Beginning with the establishment of the United Nations, this topic has invariably found its place in every international agreement that has anything to do with social, economic and environmental issues. With the release of the report of the World Commission on Environment and Development in 1987 the subject of sustainable development took centre stage in all of the international environmental and development initiatives. One major outcome of the report was the United Nations Conference on Environment and Development (UNCED) held in Rio de Janeiro in 1992. Among other things, the UNCED conference, also called the Rio "Earth Summit", adopted the "Agenda 21", an "action plan" for sustainable development.

Chapter 34 of Agenda 21 addressed the issue of technology transfer, and the United Nations Framework Convention on Climate Change, signed by many countries at Rio, contained more specific references in the context of climate change. These hard-fought instruments were based on the recognition that all countries needed to take action to achieve sustainable development and to address climate change, but that developed countries carried the responsibility to lead responses, and to assist developing countries towards sustainable development.

Consequently, Article 4.1 of the Convention, defining general commitments for all Parties, is complemented by Articles 4.3-4.5, which establish the financial responsibilities of countries in Annex II - essentially the OECD countries - to assist developing

countries to meet their obligations under the Convention, to help particularly vulnerable countries to adapt to climate change, and to aid these countries with "the transfer of, or access to, environmentally sound technologies and know-how." Article 4.7 of FCCC underlines this:

> "The extent to which developing country Parties will effectively implement their commitments under the Convention will depend on the effective implementation by developed country Parties of their commitments under the Convention related to financial resources and transfer of technology, and will take fully into account that economic and social development and poverty eradication are the first and overriding priorities of the developing country Parties".

This reflected a general perception in developing countries that implementing the Convention commitments could mean reducing one's economic growth. In practice, certain climate-change-related actions can be beneficial to developing countries. For example, measures to improve energy efficiency could support their economic growth, and widen the opportunities for transferring more advanced energy technologies that could bring multiple benefits, while also limiting their greenhouse gas emissions (some developing countries are already taking significant steps in this area, as indicated in Box 3.1, and later in this Report).

The UNFCCC recognises that economic and social development and poverty eradication are the first and overriding priorities of the developing country Parties of the Convention. All Parties have a right to, and should promote sustainable development. In implementing measures to address climate change and its adverse effects, the provision of the necessary financial resources and transfer of technology to developing countries is essential. Articles 4.3, 4.5, and 4.7 of the UNFCCC embody these commitments.

The UNFCCC also recognises that the climate change issue can be addressed through emission mitigation measures and adaptation measures. However, adaptation has played only a marginal part in the reports produced by IPCC so far. This is reflected on the attention given to adaptation technology development and its transfer. One view is that adaptation technologies help local people adapt because of the perceived local impact. Emphasis has therefore been on mitigation technologies which are seen as contributing to the solution to the global problem of climate change. All this is reflected in the little attention given to activating the CoP1 decisions on adaptation (see also section 4.10 in chapter 4 for a further discussion).

The diversity of issues means that attention at all levels of society is relevant. National Environmental Education Programmes emphasise public participation in the solutions to environmental problems. Just as in the National Innovation Systems and those of Environmental Impact Evaluation (which also offer possibilities to involve citizen participation), such programmes can identify and support actions related to technology transfer/climatic

---

[1] Relevant cases from the Case Studies Section, Chapter 16, bring out the experiences from international agreements such as the Montreal Protocol (cases 4, 17 and 23), and the AIJ process (case 12).

Box 3.1:	LEGISLATION FOR ENERGY EFFICIENCY IN DEVELOPING COUNTRIES

Many developing countries have adopted legislation for energy efficiency. Korea introduced an Energy Use Rationalisation Law (EURL)* in the 1980s and has been strengthening its implementation. This EURL was introduced to overcome energy dependency on foreign oil after the oil crisis, but it also meets objectives of the climate change convention and has contributed to reducing the rate of emissions growth. This was possible because it was motivated by the economic benefit of improving energy efficiency.

In Brazil a series of innovative energy efficiency programs are aimed toward increasing electricity efficiency. The Brazilian government is working with the national electric company, Electrobras, in an innovative programme called PROCEL, the National Program for Conservation of Electricity. PROCEL works with private enterprises to identify and implement options for reducing waste and saving power through efficient end-use technology. To date, PROCEL has implemented energy-saving programmes involving efficient motors with adjustable speed drives, improved water pumps and wastewater systems, efficient refrigerators, and compact fluorescent lightbulbs. These programmes allow Electrobras to defer investments that would produce greenhouse gas emissions in the future.

A major national programme of demand-side management (DSM) has been instituted in the Thai power sector. Coordinated by a government agency but with strong support from local utilities and NGOs, this programme has reduced the rate of growth in electricity demand significantly. Consequently, utilities have deferred power plant construction. One of the most successful areas of cooperation has been in the lighting sector. The Electricity Generating Agency of Thailand (EGAT) has worked with local manufacturers and distribution channels to expand the use of "thin-tube" high efficiency fluorescent lamps as an alternative to "fat-tube" conventional lamps. The new tubes reduce power demand, save fuel, decrease emissions, and, because they can be manufactured in-country, place a smaller burden on the scarce national reserves of hard currency. In the last four years, the Thai DSM programme has saved 1,063 gigawatt-hours (GWh), allowed the deferment of 242 Mwe of new power plants, and offset almost 800,000 tons of CO_2 in the lighting sector alone. The full national programme (including lighting, refrigeration, air conditioning, electric motors and commercial building applications) has led to the deferment of about 300 MWe of new power plants and offset almost 1.2 million tons of CO_2.

* For further details see Chapters 10 and 11 and the Case Studies in this Report.

change, referring also to recommendations of Agenda 21. Article 6 of the Convention specifically addresses the role education, training and public awareness-raising have in addressing climate change issues.

Internationally, the ISO 14000 series of environmental management standards can foster a greater ecological conscience in industrial sectors, and with this enhance the transfer and application of environmentally sound technologies (UN, 1997). In addition, sectoral initiatives and agreements can also be relevant to technology transfer and global dissemination, particularly for sectors (like vehicles and electric power generation) in which production of leading technology is concentrated in relatively few global companies.

3.3 Technology Transfer in International Environment and Development Agreements: An Overview

3.3.1 Technology Transfer in the Rio Declaration and Agenda 21

Global efforts to foster sustainable development came together at the Rio "Earth Summit" UNCED conference in 1992. At Rio, agreement on the Rio Declaration and Agenda 21 sought to define principles and an "action plan" for global sustainable development. It was aimed at fostering sustainable development through domestic and international actions. The following are of particular relevance:

Principle 9 of the Rio Declaration specifically addresses technology transfer: "States should cooperate to strengthen endogenous capacity-building for sustainable development ... through exchanges of scientific and technological knowledge, and by enhancing the development, adaptation, diffusion and transfer of technologies." The Declaration also stated that "States should cooperate to promote a supportive and open international economic system", and that "States should effectively cooperate to discourage or prevent the relocation and transfer to other States of any activities and substances that cause severe environmental degradation or are found to be harmful to human health". Several other of the Rio Declaration principles addressed requirements for states to develop domestic policies supportive of sustainable development, including participation, environmental legislation, liability and compensation for environmental damage, internalisation of environmental costs, environmental impact assessments, and international cooperation.

Agenda 21 supported these principles with more detailed proposals for action. Chapter 34 of Agenda 21 was devoted to technology transfer (Box 3.2). Unlike the two Treaties signed at Rio (UNFCCC, and the Convention on Biological Diversity), Agenda 21 does not carry legal force. The extent to which these recommendations have been implemented varies, and debate continues within the

Box 3.2:	EXTRACTS FROM

Environmentally sound technologies, including expertise, and related issues should be made available to developing countries. Developing countries, should have access to relevant information on technological choices, and the international information exchange systems and clearing houses should be developed. Access to and transfer of environmentally sound technology should be promoted "on favourable terms, including on concessional and preferential terms, as mutually agreed, taking into account the need to protect intellectual property rights, as well as the special needs of developing countries for the implementation of Agenda 21.

Commission on Sustainable Development. The rest of this section addresses issues of technology transfer associated with specific legally-binding multilateral environmental agreements.

3.3.2 Technology Transfer as a "Positive Measure" in Multilateral Environmental Agreements

The literature on global participation in Multilateral Environmental Agreements (MEAs) has increasingly emphasised the need to balance measures of enforcement with those of facilitation. Weak capacity can itself impede effective participation by developing countries in MEAs (including difficulties with the adoption and effective implementation of some kinds of commitments) There are a wide variety of instruments available for both enforcement and facilitation. The term "positive measures" is increasingly used for non-trade measures that facilitate participation in MEAs (UNCTAD, 1997b).

Osakwe (1998) presents a categorisation of such measures (Table 3.1) in which technology transfer is first on the list of such positive measures. He summarises the results of several years of work in the WTO Trade and Environment Committee, at UNCTAD, and elsewhere, illustrating the role of technology transfer while cautioning against simplistic treatment of "positive measures":

Positive measures tend to be considered as second order supplementary measures, offered as compromises only to weaker countries and firms lacking in capacity. Although developing countries may, on balance, need these non-restrictive measures more, this reasoning is incomplete. Positive measures are also required by developed countries and firms.

UNCTAD has proposed an agenda for positive measures and offers suggestions for implementing positive measures. UNCTAD also highlights three difficulties that hinder the implementation of positive measures. First, compared to trade measures, there are no implications which arise from the abandonment of commitments to positive measures [for the country abandoning them]. Second, the viewpoint expressed at several Conferences of Parties is restated: that exclusive rights granted under Intellectual Property Rights regimes could increase the cost of acquiring technologies authorised by MEAs. Third, apart from provisions for financial mechanisms in some MEAs, there are no explicit references to mechanisms for technology transfer, for instance, through the purchase of equipment, licensing, foreign investment, application of scientific research results in the public domain.

Balanced views should always be presented. For example, rebating energy taxes on weaker industries have produced unintended effects of increasing pollution levels . Also, long grace periods for adjustment by developing countries and weak producers have downsides. They could lead to high dependence, high phase-out costs for the developing countries, and the likelihood of the migration of environmentally harmful industries in the application of positive measures the "guilty" - polluters - should never be rewarded, weak capacities notwithstanding.

Table 3.1	The interaction of trade and positive measures: an indicative typology (Osakwe 1998)	
	TRADE MEASURES	**POSITIVE MEASURES**
Coercive	Bans / prohibitions Quotas Taxes / charges Mandatory labelling schemes	
Cooperative	Import / export permits Prior informed consent procedures Waivers Tradable emission permits	Transfer of environmentally friendly technologies Joint implementation of projects Project financing Funding incremental costs "Green loans" Credit guarantees Elimination of environmentally harmful subsidies "Green non-actionable subsidies" Grace periods within which to satisfy MEA commitments Market access and "green market access" Technical assistance for capacity-building

During the 1990s financial mechanisms in MEAs have become more widespread. The Global Environment Facility, obviously, has become a central actor for climate change and other agreements. Other specific experience of technology transfer and other positive measures in MEAs is limited to date, but growing.

Two major examples of MEAs, in addition to the UNFCCC, with strong components relating to technology transfer are the *Montreal Protocol* and the *Convention on Biological Diversity*.

3.3.3 The Montreal Protocol

The Montreal Protocol (MP) on Substances that Deplete the Ozone Layer is one of the more successful amongst international treaties related to global environmental problems (see also sec-

ENDA 21, CHAPTER 34 (AS SUMMARISED IN GRUBB ET AL, 1993)

ecific measures should include poli-
s and programmes to encourage
blic and private technology transfer
d regulatory measures, include sub-
lies and tax policies, and appropriate
chanisms for improved access and
nsfer of relevant technologies.
tional capacities, particularly of
veloping countries, should be built to
velop and manage environmentally
und technologies, including human
source development, and strengthen
earch and development capaci-
s."

e development of indigenous tech-
ogy and technology assessment

should be promoted, "a collaborative
network of research centres" should
be established and "programmes of
cooperation and assistance" should be
strengthened. The importance of technology transfer through business commerce is recognised, and while the availability to developing countries is of concern, "fair incentives to innovators" should also be provided; here the role of patent and property rights should be examined. Long-term partnerships between holders and users of environmentally sound technologies, and between companies in developed and developing countries as well as joint ventures should be promoted.

tion 5.5.7 in Chapter 5). It provides a good example of an international agreement that has been continuously amended in response to new scientific and technical assessments.

Signed by only 24 (mostly developed) countries in 1987, it has grown to almost global participation, thanks to a judicious combination of trade measures, with positive measures of financial and technology transfer provisions (Vossenaar and Jha, 1998). The overall commitments acquired by the international community to phase out ozone depleting substances (ODS) sent a "strong signal" to governments and enterprises to take actions to lower production and consumption of these chemicals and to search for adequate substitutes.

Furthermore, the initial reduction schedules were agreed to before adequate substitutes were in the marketplace. Therefore, the MP could be seen as speeding up the technological innovation and as a competitive process among leading enterprises. In this regard, the MP has created a broad range of business opportunities all around the world, and helped many developing countries create political and economic conditions that are more favourable to technology transfer.

However, the large scale of illegal trade in controlled chemicals to the US and the EU is an issue that threatens the Montreal Protocol achievements to protect the ozone layer (Environmental Investigation Agency, 1998). A number of authors have called for stronger intelligence and enforcement measures to prevent and eliminate smuggling of these substances.

The Multilateral Fund

The treaty established a Multilateral Fund (1998) to assist developing countries to comply with its commitments. Notwithstanding the fact that there are few formal technology transfer agreements under MP operations (as outlined in Chapter 2), the Fund has been a key factor that has facilitated technology transfer to developing countries.

The Fund operates under the authority of the Parties that appointed an Executive Committee (ExCom) that has to "develop and monitor the implementation of specific operational policies, guidelines and administrative arrangements, including the disbursement of resources, for the purpose of achieving the objectives of the Multilateral Fund (1998)". ExCom has 14 members divided equally between developed and developing countries, with annual renewal. The way that the ExCom works has allowed a wide participation of developing countries on an equitable basis, encouraging adoption of balanced decisions on many controversial and difficult issues. The practical operation and disbursement of the Fund to date has been summarised in Chapter 2 (Box 2.1).

The activities approved by the ExCom have helped developing countries in areas which can be called "hard technology transfer", carried out mainly by the World Bank, UNDP and UNIDO, and "soft technology transfer", directed mainly by UNEP. UNEP's activities provide various types of support including information exchange, networking, institutional strengthening, capacity-

building and training (Shende & Gorman, 1997). The World Bank, UNDP and UNIDO have focussed mainly on investment projects and country programme assistance. The implementing agencies have assisted developing countries in identifying projects, assessing national plans and policies, providing adequate technical support, and assistance for project development and implementation. At the same time the implementing agencies played a crucial role in assisting the members of the ExCom of the Multilateral Fund (1998) in solving difficult issues, through policy and technical papers.

In addition, the Multilateral Fund (1998) has funded and mandated UNEP to facilitate networks of National Ozone Units (NOUs), which enhance governments' abilities to manage ODS issues, and facilitate cooperative work among the implementing agencies, countries, technology suppliers and potential project proponents. Currently, there are eight regional networks consisting of 90 NOUs which meet regularly in their own region. Regional networking provides government ODS officers with a means of sharing their knowledge with their peers in developing and developed countries. Such networking has helped the countries to set up an enabling environment and policy settings for favourable technology transfer. Furthermore, the establishment of NOUs responsible for implementing the Montreal Protocol activities at country level have proved to be of critical importance to develop and promote the necessary changes to phase out ODS in developing countries.

A study requested by the Parties to evaluate the effectiveness of the Multilateral Fund (1998) indicates that the use of workshops as a primary mechanism to identify projects, even before a national plan is set-up and the decision to give priority to projects with high cost effectiveness, has favoured larger enterprises in developing countries (COWI, 1995). The facts that large conversion projects are becoming scarce and small, and medium size enterprises (SMEs) absorb most of the labour force in developing countries have made SMEs one of the main issues of the MP discussion, due to the fear that SMEs could become primary consumers of ODS in the medium term.

The problem of cost effectiveness

The main objective of the MP was to phase out as much ODS as possible in a minimum time frame. The cost-effectiveness index reveals both the incremental costs (operating and capital costs) covered by the Fund versus equivalent tons of ODS phased out, and does not constitute a measurement of the full costs to developing country societies of avoiding the use of ODS. Therefore, the costs of eliminating ODS are borne jointly by the Fund, the enterprises and consumers (COWI, 1995). It should be recognised that developing countries are not merely receiving the funds to eliminate ODS. They are active partners by contributing their own financial and technical resources to meet the overall responsibilities.

Meanwhile, the role and application of cost-effectiveness thresholds continue to evolve as the Executive Committee takes decisions on new concepts of investment projects to better assist SMEs.

3.3.4 Convention on Biological Diversity

The Convention on Biological Diversity (CBD) was adopted at the Rio Summit in 1992. It addresses one of the most pressing issues of the modern era, sustainable management of the living resources of the planet. It also, for the first time, adopts a comprehensive rather than a sectoral approach to the conservation of the planet's biodiversity and sustainable use of its biological resources. In the four sessions of the CoPs, the Parties developed a mechanism for the normative development of the Convention and its implementation. The Convention provides for "access to and transfer of technology" in Article 16. When this is read with Article 19 dealing with "handling of biotechnology and distribution of its benefits", it reflects the years of North-South debate in other fora over the issue of technology transfer and the issue of intellectual property rights (Glowka *et al.*, 1994).

The closely interlinked provisions along with Articles 16 and 19 are Articles 12 (research and training), 17 (exchange of information), and 18 (technical and scientific cooperation). Together they seek to "provide and/or facilitate access for and transfer to other Contracting Parties" of: technologies relevant to the conservation of biological diversity, technologies relevant to the sustainable use of its components or technologies that make use of genetic resources.

CBD includes provisions for biodiversity-rich developing countries to regulate access and, eventually, to the use of these resources. This approach is now being carried out in access to genetic resources and benefit-sharing agreements around the world (Reid *et al.*, 1993). A key component of the CBD supporting such mechanisms was the recognition of national sovereign rights over genetic resources.

The CBD specifically addresses access to and transfer of technology relevant to the conservation and sustainable use of biological diversity, including biotechnology. Such access and transfer to developing countries should be facilitated under fair and favourable terms, including on concessional and preferential terms where mutually agreed and in accordance with the financial mechanism (GEF). The situation calls for a careful balance between technology needs, financial requirements, appropriate regulatory frameworks and cooperation among countries. In this context – and for any other international agreement with technology transfer provisions – national strategies to determine technological needs play an important role.

The CBD also established that transfer should be under terms which recognise and are consistent with intellectual property rights over these technologies. Since the CBD is concerned directly with access to and use of genetic resources, IPRs have played a more central role than seems likely in the context of climate change. The debate has also been highly contentious, with some developing countries arguing that TRIP provisions may need to be amended or set aside in the context of the CBD, and a number of western governments and industry groups strongly contesting this position (*e.g.* UNICE, 1997). Technology transfer in the case of technologies related to genetic resources, particularly the biotechnology which is largely developed by the private sector, presents certain differences with respect to climate change related technology transfer.

3.4 Technology Transfer under the UNFCCC Agreements

As noted in the introduction to this chapter, the UNFCCC represents a global accord in which developing countries accept their need to be involved in global efforts to address climate change, and undertake the general provisions of Article 4.1, in return for assistance with finance and technology transfer that are set out in later sections of this Article. Article 4.7, referred to earlier, bares this point out in some detail (see Box 3.3).

Box 3.3: **KEY PROVISIONS RELATING TO TECHNOLOGY TRANSFER — UNFCCC**

Article 4.1.h (all Parties taking into account their common but differentiated responsibilities ...shall): Promote and cooperate in the full, open and prompt exchange of relevant scientific, technological, technical, socio-economic and legal information related to the climate system and climate change, and to the economic and social consequences of various response strategies.

Article 4.3 The developed country Parties .. in Annex II .. shall also provide such financial resources, including for the transfer of technology, needed by the developing country Parties to meet their agreed full incremental costs of implementing measures that are covered by paragraphs 1 of this Article and that are agreed ...

Article 4.4 (Annex II Parties) shall also assist the developing country Parties that are particularly vulnerable to the adverse effects of climate change in meeting the cost of adaptation to those adverse effects.

Article 4.5 The developed country Parties and other developed Parties included in Annex II shall take all practicable steps to promote, facilitate and finance, as appropriate, the transfer of, or access to, environmentally sound technologies and know-how to other Parties, particularly developing country Parties, to enable them to implement the provisions of the Convention. In this process, the developed country Parties shall support the development and enhancement of endogenous capacities and technologies of developing country Parties. Other Parties and organisations in a position to do so may also assist in facilitating the transfer of such technologies.

Article 4.7 The extent to which developing country Parties will effectively implement their commitments under the Convention will depend on the effective implementation by developed country Parties of their commitments under the Convention related to financial resources and transfer of technology ...

One important aspect of this was the financial mechanism. The GEF was entrusted with its operation, initially on an interim basis. Operating experience with the GEF and its role in financing the transfer of ESTs is discussed in Chapter 5 of this Report.

Until recently, the focus of the GEF has been primarily on mitigation of climate change, but following CoP3 and CoP4, adaptation has received increasing attention. However, major barriers to the international funding of adaptation activities remain to be overcome. The GEF Operational Strategy prescribes that activities need to have global benefits in order to be eligible for funding. Mitigation activities, aimed at reducing atmospheric greenhouse-gas concentrations, clearly have global benefits. For adaptation activities, on the other hand, it is difficult to imagine how global benefits can be produced. Adaptation takes place at the scale of an impacted system, which is regional at best, but mostly local.

At CoP4 in Buenos Aires, the decisions on the financial mechanism confirmed the Global Environmental Facility as an "entity entrusted with operation of the financial mechanism of the Convention" and broadened its scope. The mandate was broadened, with reference to the need for "flexibility to respond to changing circumstances". Its funding scope was formally expanded to include more wide-ranging support for building up the capacity of developing countries to address climate change issues, including support for adaptation technologies and full funding of their Second National Communications. The decision also established guidelines for the review of the financial mechanism every four years.

3.4.1 Follow-up to the Convention commitments on Technology Transfer

The first Conference of the Parties in Decision 13/CP.1 (UNFCCC, 1995) requested the Convention secretariat to prepare an itemised progress report on concrete measures taken by the Annex II Parties to fulfill their commitments related to the transfer of environmentally sound technologies and of the know-how necessary to mitigate and facilitate adequate adaptation to climate change. It also urged Parties listed in Annex II to include this information in their national communications in order to enable the Convention secretariat to compile, analyse, and submit the documents to each session of the Conference of the Parties.

Box 3.4 outlines the subsequent development of these issues in successive meetings of the SBSTA. It was evident from early discussions in SBSTA and the Subsidiary Body for Implementation (SBI) that while information from Annex II Parties is vital, the Parties recognised the complementarity of the roles of public and private sectors with technology transfer. From here on the emphasis was not just on Annex II Parties, but on urging all Parties to improve the enabling environment for private sector participation, and to support and promote the development of endogenous capacities and appropriate technology relevant to the objectives of the Convention (UNFCCC, 1996d). The work programme of

the secretariat, in addition to its work on the technology inventory database, included collecting information on: technology information centres; technology information needs; adaptation technology; terms of transfer of technology and know-how; and also private sector technology transfer. CoP2 took note of the progress made in implementing Decision 13/CP.1 by reviewing the Convention secretariat's documents on the subject, and expressed its concern at the slow progress in obtaining relevant information on this matter (Decision 7/CP.1 (UNFCCC, 1996e). The Decision also requested Annex II Parties to submit information so as to help Parties make informed decisions.

Box 3.4:	Development of reporting and analysis of

The secretariat, after reviewing information from 21 Annex II Parties in the aftermath of CoP1, organised information around multilateral, bilateral, and private sector cooperation (UNFCCC, 1996c). It found that four parties did not report on activities related to transfer of technology (most of these communications were submitted prior to the CoP1 decision). The secretariat found that the Annex II Parties submitted their communications as per the guidelines, which did not require the kind of detail sought by Decision 13/CP.1. Additionally, the secretariat's document points out that the part dealing with technology transfer could easily be subject to different interpretations by the Parties. The in-depth review process revealed that much more information is available than has been provided in the communications, leading the secretariat to the conclusion that a comprehensive picture of technology transfer activities was not available at that stage.

The secretariat prepared a report on the basis of 31 responses received to its request for information and supplied the same to the Subsidiary Body for Scientific and Technological Advice (SBSTA) (UNFCCC, 1996a). It was difficult for the secretariat to draw specific conclusions based on these responses. There was, however, a feeling that if more

time were to be available to the Parties and intergovernmental organisations, the many other valuable reports and information sources could be reviewed to help with the development of a long-term work programme. In response to a decision by SBSTA at its second session, Annex II Parties submitted information on technology inventory databases (UNFCCC, 1996 a and b)*.

The subsidiary bodies at their fourth and fifth sessions noted the progress made by the secretariat, and requested it to intensify and accelerate its efforts, and at the same time noted that the information provided by Annex II Parties differed considerably in format, comprehensiveness and level of detail. They urged the completion of an itemised progress report by its seventh session on access to, and the transfer of, environmentally sound technology, based on the national communications (UNFCCC, 1997a). By the time they met for the sixth session, the scope of work by the secretariat was enlarged to include specifically work with the World Bank, OECD, and the possibility of establishing intergovernmental technical advisory panels (ITAPs) at its seventh session. Moreover, there was support shown for the secretariat's plans to prepare reports on the role of the private sector, and on barriers and enabling

At Buenos Aires, Decision 4/CP.4 (UNFCCC, 1999a) recommended a series of practical steps to promote, facilitate and finance the transfer of, and access to, environmentally sound technologies and know-how. It outlined the importance of supporting the development and enhancement of endogenous capacities and technologies of developing country Parties, and identified areas where assistance in facilitating the transfer of environmentally sound technologies and know-how is required. It is addressed to all of the Parties in light of various provisions of the Convention (Articles 4.1, 4.3, 4.5, 4.7, 4.8, 4.9 and Articles 9.2, 11.1, 11.5). It asked Annex II Parties to provide a list of environmentally sound technologies and know-how related to adaptation to and mitigation of climate change that are publicly owned, and to report in their national communications steps taken to implement Article 4.5 of the Convention. It asked non-Annex I Parties to submit their prioritised technology needs. It invites all Parties and interested actors to identify projects and programmes that can serve as models for improving the diffusion and implementation of clean technologies under the Convention.

This survey illustrates the fact that technology transfer has been on the agenda of every session of SBSTA. Despite all of the discussions within and various papers produced by the secretariat, there has been limited progress on determining what constitutes technology transfer, the roles of the governments and private sector, and how compliance with Article 4.5 should be measured. As a note by the secretariat (UNFCCC, 1998c) pointed out, reaching agreement on these issues is of fundamental importance. It is vital that there be simple and agreed upon criteria for measuring whether Annex II Parties have met their commitments; the guidelines for national communications should clearly identify formats for reporting financial information (as an indication of the amount of technology transferred). In other words, as the secretariat pointed out, developing and reaching an agreement on what constitutes "success" might be an important issue for the Parties to consider.

Outside of formal discussions within UNFCCC, other initiatives on technology transfer are notable. Stemming from initiatives by Japan and Germany, the Climate Technology Initiative (CTI) was developed through the International Energy Agency (IEA). The CTI is still in evolution, but its activities include a variety of voluntary actions by IEA member states: setting up national advice and technological development plans; offering prizes for technological development; enhancing markets for emerging technologies; and promoting collaboration between states on technology research and development. In addition, the CTI has also established regional seminars on climate change related technology, and has worked in collaboration with the SBSTA.

In 1997, the IEA announced that it would increase its work with the CTI and seek new collaboration with existing bodies like the ISO. A new Global Remedy for the Environment and Energy Use (GREEN) initiative was also established, aiming to enhance the use of climate change mitigating technology in ODA and private investment.

However, the impacts of these official attempts to generate technology transfer are as yet unclear, and in terms of total investment are relatively limited compared with mainstream investment flows. Other complementary approaches have thus also been proposed. Du Pont *et al.* (1997) highlight that industrialised countries, at the same time as calling for stronger action in developing countries, frequently export to them second-hand equipment that is old, inefficient and polluting. Citing examples from vehicles, and from the pulp and paper industry, they note that facilities (cars or industrial plants) are exported that are essentially obsolete and no longer meet standards in industrialised countries.

TECHNOLOGY TRANSFER UNDER THE UNFCCC

activities of governments related to the transfer of technology.

The secretariat in preparation for the seventh session of the subsidiary bodies commissioned a number of papers and presented the key findings in its progress report on the topic (UNFCCC, 1997c). While the summaries are helpful in identifying the relative roles of different types of technologies and centres and networks, one should look elsewhere to find the results of the review of national communications of Annex II Parties (UNFCCC, 1997b). Reviewing these documents the SBSTA/SBI joint working group took note of the need for better information with respect to finance and transfer of technology, and agreed to consider at its ninth session what, if any, additions and/or amendments to the revised guidelines for the preparation of national communications by Annex II Parties would be required.

Decision 9/CP.3 (UNFCCC, 1998a) requested the secretariat to continue its work, consult with the Global Environment Facility and other relevant international organisations on international technology information centres and exploring the possibilities of funding the same. It urges Parties to create an enabling environment for the private sector to play its due role, as well as improve the reporting by Annex II Parties in their national communications.

In preparation for the eighth session of the subsidiary bodies, the secretariat produced another progress report (UNFCCC, 1998b). While this report touched on all of the various aspects of prior decisions, i.e. technology-information needs survey, adaptation technology, centres and networks, it could not say anything more on actions within Annex II, since any comment on this issue will need to come from a review of national communications and a review of available communications was already carried out earlier on.

---

*: Software packages include: IIASA's CO$_2$ *Technology Data Bank*, the *World Bank's Greenhouse Gas Assessment Methodology*, Germany's *Instruments for Reduction Strategies of Greenhouse Gases*, and the Netherlands *Information System on Conservation and Application of Resources Using a Sector Approach*. Reports include: IEA's *Comparing Energy Technologies*, the World Bank's *The Case for Solar Investments*, and USEPA's *A Guide for Methane Mitigation Projects, Gas-to-Energy at Landfills and Open Dumps*.

They suggest that such exports should be discouraged, for example, by using a tax credit in developed countries to accelerate equipment renewal (perhaps granted only if the obsolete technology is not exported) or by "negative emission credits" for export of second-hand technologies that increase greenhouse gas emissions compared with state-of-the-art technology (Du Pont et al., 1997). Procedures requiring "prior informed consent" from recipient countries, analogous to those in the international agreements on hazardous wastes and chemicals, might offer a "softer" alternative.

Overall, the need to better engage the private sector and to harness private investment, and much learning since the UNFCCC (including experience with AIJ projects summarised elsewhere in this Report), has implied rethinking the ways in which the UNFCCC has approached technology transfer. This has become clearer in the Kyoto Protocol and subsequent decisions.

3.4.2   *Technology Transfer in the Kyoto Protocol and Buenos Aires Decisions*

The wording on technology transfer in the Kyoto Protocol, compared to the UNFCCC, does lay greater emphasis upon the role of private investment and of actions by and in the developing countries themselves. The opening sentence in Article 10(c) of the Kyoto Protocol states that Parties should "Cooperate in .." contrasts with the Convention's onus upon developed country Parties only to "take all practicable steps". It also expands on similar language in the Convention by adding the general catch-all on "practices and processes pertinent to climate change". Perhaps most significantly, it formally recognises the role of the private sector and the need for an "enabling environment." All this reflects a significant evolution of thinking, but is very limited in terms of specific commitments (see 3.5).

In some ways, concerning technology transfer itself, the CoP4 follow-up in Buenos Aires was perhaps more significant than the Protocol itself. There was a much more focussed agreement that addressed the roles of all Parties. It called for Annex II Parties (largely OECD countries) to provide lists of environmentally sound technologies that were publicly owned, and for developing countries to submit prioritised technology needs, especially related to key technologies for addressing climate change. All Parties were urged to create an enabling environment to stimulate private sector investment, and to identify projects and programmes on cooperative approaches to technology transfer. Most importantly, the agreement called for a consultative process to be established to consider a list of 19 specific issues and associated questions, set out in an Annex (referred to in Chapter 1). One interpretation is that the Kyoto and subsequent Buenos Aires agreements on technology transfer 'embodied a considerable change of attitude in which the previous impasse has eased in favour of an approach which recognises that there are real problems that require real creative solutions to achieve global dissemination of environmentally sustainable technologies' (Grubb et al., 1999).

3.4.3   *Provisions for Enforcement and Compliance*

International mechanisms to monitor and review progress, to identify and respond to non-compliance, are relevant as a means to advance implementation of the technology transfer and cooperation provisions of the Protocol. The Convention already has in place an elaborate system for the collection and review of national information. The system aims to assess progress and facilitate national implementation of the international obligations under the Convention (OECD, 1998a). The Kyoto Protocol proposes to continue the monitoring, reporting and review functions, and calls for strengthening of the review and compliance assessment functions (OECD, 1999). Article 8 of the Protocol, in particular, requests the Climate Change Secretariat to identify questions related to implementation as part of the normal review of the performance of individual Parties (see Box 3.5).

At present UNFCCC reviews of national performance include an assessment of the Annex II Party implementation of technology transfer and cooperation provisions of the Convention. However, because these provisions are ambigu-

---

BOX 3.5:   KYOTO PROTOCOL - KEY PROVISIONS RELATING TO TECHNOLOGY TRANSFER AND REVIEW OF IMPLEMENTATION

Article 8.3: Review process shall provide a thorough and comprehensive technical assessment of all aspects of the implementation by Parties of the Protocol. The expert review teams shall prepare a report [to the CoP/MoP] assessing the implementation of the commitments of the Party and identifying any potential problems in, or factors influencing, the fulfillment of commitments. Such reports to be circulated to all Parties... ...secretariat shall list those questions of implementation indicated in such reports for further consideration by [CoP/MoP].

Article 8.5: [CoP/MoP] shall, with assistance of SBI and, as appropriate, the SBSTA, consider:

a) information reported under Art 7 and reports of the expert reviews thereon conducted under this Article; and

b) those questions of implementation listed by the secretariat[...] as well as any questions raised by Parties.

Article 8.6: Pursuant to its consideration of the information referred to in paragraph 5 above, [CoP/MoP] shall take decisions on any matter required for the implementation of the Protocol.

Article 10.c (All Parties, taking into account their common but differentiated responsibilities..., shall): Cooperate in the promotion of effective modalities for the development, application and diffusion of, and take all practicable steps to promote, facilitate and finance, as appropriate, the transfer of, or access to, environmentally sound technologies, know-how, practices and processes pertinent to climate change, in particular to developing countries, including the formulation of policies and programmes for the effective transfer of environmentally sound technologies that are publicly owned or in the public domain and the creation of an enabling environment for the private sector, to promote and enhance access to, and transfer of, environmentally sound technologies;

ous, it is difficult to assess performance in a compliance sense. An important step forward would be to identify a checklist of relevant activities or implementation actions that every Annex II Party should take. This would provide a point of departure for reviewing the performance of the Convention and the more recent Kyoto Protocol obligations. Because of the lack of clarity about these obligations, the UNFCCC review of national information tends to be a descriptive exercise rather than one that points out key areas of progress and shortfalls. Clarifying the expectations of the international community with respect to these obligations, for Annex II Parties as well as for all other Parties, would bring the technology transfer obligations of Parties fully into the international compliance system, and encourage open and transparent review of progress over time.

3.5 Technology Transfer and Intellectual Property Rights

The previous chapter has highlighted that international financial flows, including foreign investment, often bring some degree of technology transfer, have grown greatly during the 1990s and changed greatly in their structure. This has been due to many factors. Amongst these are the liberalisation of certain domestic sectors with the granting of access to foreign investors; development of domestic property laws, including on intellectual property; and developments in various international conferences such as the United Nations General Assembly Special Session of 1997 (UNGASS, 1997), and decisions adopted at the 6th session of the Commission on Sustainable Development (CSD) (UNCSD, 1998). Most notable among these agreements has been the completion of the GATT Uruguay Round and the associated establishment of the World Trade Organisation. In addition, there are a number of other associated agreements that have helped to encourage or facilitate international investment[2].

3.5.1 *Role of Government in R&D and the Transfer of Publicly-owned or Supported ESTs*

Governments play an important role in providing funding for public R&D programmes as part of their industrial policies or science and technology development strategy. These programmes are implemented either by government institutions or in joint partnership with the private sector. Over the past decade, about 40% of annual national R&D spending within a number of OECD countries was publicly funded (UNCTAD, 1998). To promote the development of ESTs that lack short-term commercial viability, government funding and public R&D programmes are vital. As pointed out in Chapter 2, while energy-related R&D by OECD governments has decreased (mainly in the coal and nuclear sectors) steadily in real terms since the early 1980s (Margolis and Kammen, 1999), R&D for energy efficiency and renewable technologies has been increased; they

now form almost 20%, some US$2 billion, of total energy R&D expenditures made in 1995.

In addition, governments sponsor a range of R&D that can underpin private sector investments in developing ESTs. OECD countries could influence flows of such technology directly through their influence on the private sector or on public institutes, which receive funding from government for their R&D to be more active in transferring technologies to developing countries (Chung, 1998). The issue of publicly owned technology transfer and the role that national regulatory frameworks play in creating demand and markets for ESTs (See Table 2, UNCTAD, 1998) was addressed both in the Rio Summit of 1992 and the UNGASS of 1997.

Chapter 34 para.34.18(a) of Agenda 21, adopted in 1992, states that "Governments and international organisations should promote the formulation of policies and programmes for the effective transfer of environmentally sound technologies that are publicly owned or in the public domain".

The programme for the further implementation of Agenda 21 adopted at the UN General Assembly at its 19th Special Session (UNGASS) in 1997 stated that "A proportion of technology is held or owned by Governments and public institutions or results from publicly funded research and development activities". To fully explore this issue, a feasibility study on "the role of publicly funded research and publicly owned technologies in the transfer and diffusion of ESTs" was undertaken in 1997 jointly by the UNCTAD , UNEP and UN/DESA (UNCTAD, 1998). The results of this study were reviewed at an expert meeting in February 1998 (Government of Korea, 1998) and were reported to the 5th session of CSD (UNCSD, 1997). CSD adopted a decision encouraging governments to undertake pilot projects to explore opportunities for sector-specific applications of the recommendations on transfer and commercialisation of publicly funded ESTs.

Major findings of the feasibility study done by the UNCTAD are that governments are funding public R&D programmes as part of their industrial policy aimed at improving their industrial competitiveness. Strong emphasis is placed on the commercialisation of those technologies developed from public R&D.

In many cases, co-financing with the private sector also plays an important role. Many governments either transfer or license the patents of the publicly funded technologies to the private sector as part of their industrial policy, and then the transferred patents follow the rules of the privately owned technologies and behave just like the other ordinary private IPRs. Thus, instead of focussing on the technologies held by the government, the focus should rather be on the public R&D programmes for ESTs in exploring the feasibility of transferring publicly owned technologies.

The country case studies of the feasibility study identified no specific policy measures were introduced in the countries reviewed to implement the agreement contained in Agenda 21 for the

---

[2] National legal institutions are discussed in section 4.8 of Chapter 4.

transfer of publicly owned technologies. No specific policy measures were implemented for the transfer of publicly owned technologies among 10 countries studied, except by the US which restricts by law the transfer of publicly owned technologies solely to companies manufacturing substantially in the US. The other countries do not have specific legal restriction or policies which restrict the transfer of publicly owned technologies to developing countries.

Based on the above findings, it could be summarised that governments have not yet reviewed this issue for implementation. As governments are funding public R&D programmes as part of their industrial policy, the challenge for the governments is whether they can integrate global environmental cooperation and commitment into their industrial policy.

As governments are the main driver of public R&D programmes for ESTs, governments can play a critical role in the promotion of the transfer and dissemination of the ESTs. However, the modalities for the transfer of these publicly funded ESTs are yet to be explored and discussed among the experts representing those governments which have the potential of transferring such technologies in order to build up consensus at a multilateral level.

3.5.2   *Private Investment and Intellectual Property Rights*

As illustrated in the previous chapter, the private sector plays an increasingly important role in international investment and technology development. This growing role has been supported by various domestic and international developments, including liberalisation of markets, development of stronger domestic legal and financial systems, and tariff reductions under the Uruguay Round of the GATT. In the context of technology transfer, a particularly important - and complex - set of issues are those relating to intellectual property rights (IPRs). IPRs may play an important role in ensuring economic returns to investors (including R&D resources they have devoted to developing and improving technologies), and to an extent enabling the transfer of and availability of protected technologies (see Box 3.6).

International development of IPRs and the TRIP agreement
IPRs were originally regulated exclusively at the national level and subject to national legislation. Regimes for IPR tended to vary widely, especially between developed and developing countries, due to differing interests, cultures and administrative capacities. Industrialised countries tend to see IPRs as a primary means for promoting technology development by offering inventors protection to reap profits from their labours. Developing countries tended to be more concerned to access existing technologies at affordable costs, and to make them more widely available. Not surprisingly, developing countries tended to have far weaker IPR laws than industrialised countries. The World Intellectual Property Organisation (WIPO) seeks to foster and harmonise IPR protection and disseminate information. Its mandate includes, subject to the competence of other organisations, promoting creative intellectual activity and facilitating the transfer of tech-

> **Box 3.6:**   **INTELLECTUAL PROPERTY RIGHTS AND THE ECONOMICS OF TECHNOLOGY PATENTS: A PRIMER (SOURCE: DERIVED FROM ACHANTA AND GHOSH (1993); BESEN AND J.BRASKIN (1991)**
>
> In very general terms, intellectual property rights protect human innovation and intellectual effort. IPRs include patents (utility patents, design patents, plant patents), plant breeder rights, trade secrets, trademarks and copyrights. Patents are specific rights granted by governments to inventors which enable the right holder - the inventor - to exclude third parties from utilising or exploiting or commercialising the protected invention in the countries where these are registered. To be protected under patent law, inventions need to be new, have an inventive step and industrial application.
>
> From an economic perspective, IPRs enable entrepreneurs to cover research and development expenses and ensure some profit from the use of the protected idea or invention returns to the innovator. From a sociological or anthropological perspective, IPRs serve to reflect those ideas most valued by a culture, although certainly this should not imply that IPR protected innovations are necessarily those most valuable to society.
>
> The standard economic justification for granting property rights over intellectual property is that this furnishes incentives for creative work. IPR regimes are premised on the belief that prospective financial returns in fact drive private creators of intellectual property. They do not necessarily guarantee that innovation takes place at least social cost, which will depend among other things upon the extent to which creators may borrow ideas or concepts from earlier work.
>
> From a societal perspective, one economic objective of an IPR regime could be to maintain a proper balance between creating and disseminating intellectual property. If innovations are not widely used, the net society benefits may be less than in a case where fewer resources are employed in creativity but the intellectual property is more widely disseminated. This focuses attention on the appropriate scope of protection: the optimal duration of IPR protection, and the optimal trade-off between the duration and breadth of IPR protection.
>
> Another way of looking at this is that a lack of IPR protection will enable everyone to access a technology, but this comes at a dynamic cost of deterring further innovation, and thus may stifle (positive) development.
>
> These interests may differ between developed and developing countries. Hence, it is unavoidable that there are disputes over the appropriate degree of protection when it comes to international transfers. These issues were debated very widely in the context of the GATT debates that led ultimately to the TRIP agreement.

nology related to industrial property.

Continuing differences in IPR treatment became a source of considerable international dispute in the 1980s, and the US especially threatened some developing countries with retaliatory trade action if they did not improve IPRs protection (Doane,

1994). At the same time, developing countries began to believe that stronger IPRs regulation could help them to attract more foreign technology and to develop stronger markets (Kwon, 1995). These developments have led to more common international standards for IPR protection during the 1990s.

In particular, the 1994 Agreement on Trade Related Aspects of Intellectual Property (TRIP), negotiated in the context of the Uruguay Round, is leading to increased homogeneity of laws around the world in accordance with minimum standards. According to Worthy (1994) the agreement "is a major breakthrough in the international protection of intellectual property rights, because of its substance and because of the wide measure of international acceptance it achieved". TRIP was adopted in the context of a trade negotiation and therefore was accepted as part of a "package". The main provisions are:

1. The establishment of minimum standards for the protection and enforcement of a wide variety of intellectual property rights, including the extension of copyright principles to computer code and certain kinds of databases;
2. A principle of "national treatment", preventing IPR discrimination in favour of domestic industries;
3. A principle of "most favoured nation", preventing IPR discrimination between investors from different signatories to the TRIP agreement.

The TRIP regime on patents is particularly relevant to technology transfer; it sets a minimum of 20 years for patent protection, it sets minimums for patentable subject matter, and it sets minimum standards as to the conditions that must be met for patents to be issued. It also establishes agreement on procedures that must be followed before countries can grant compulsory licences.

The TRIP agreement allowed developing countries four years (from January 1996) in which to make necessary changes, with an additional five years for least developed countries or others facing serious implementation difficulties.

Effects of stronger IPR protection
The benefits that may derive from legal systems having strong intellectual property rights include the following: an increase in innovation due to the incentive and reward that IPRs provide; fair treatment of innovators who can own the creative "sweat of the brow" and exert influence over how their technology is used; public disclosures of patented technologies, sharing of secrets under confidentiality agreements; ease of purchase, sale, or license; and enhanced investment due to the assurance that investors can recapture their investment in a technology subject to such protection.

Great IPR protection should give greater confidence for R&D investment in new technologies and processes. It should also lead to greater investment, including with new technology by western companies in developing countries, although empirical research suggests this may not be as significant as hoped for

(Kwon, 1995). Trebilcock and Howse (1995), in a survey of foreign investors, note that intellectual property protection was rated the least important of five factors affecting investment decisions in Thailand. Certainly, although in theory strong IPR systems should promote foreign investment, in the case of certain technologies it is not enough to have these systems in place. There is a need for the planning of technological development and better identification of technological needs. Furthermore, all countries do not necessarily need cutting edge technology to satisfy specific needs, particularly with respect to clean technologies.

In other respects, stronger IPR systems can impede technology development and transfer. The World Bank's 1998 World Development Report cautions that "there is now a risk of excessively strict IPRs adversely affecting follow-on innovations and actually slowing down the pace (of technological development)". The report goes on to identify patents which cover "not just products but broad areas of technology" as a particular concern. Concerning international transfer, a country that seeks to obtain a beneficial new technology for its inhabitants may find that the owner of the technology is unwilling to provide it on terms that the country (or host companies involved) can afford, whether or not there is IPR protection.

Thus, there is no absolute "right" degree of IPR protection, and notwithstanding the TRIP agreement, IPR regimes differ according to national circumstances and the agreements that governments have entered into, the technologies involved, and their national objectives. In addition, the appropriate response might be to negotiate specific guarantees with investors, rather than increasing intellectual property protection across the board (Trebilcock and Howse, 1995). In certain sectors and markets, including some energy and environmental sectors, the main advantages to investors may accrue simply from continued technological innovation and managerial expertise, in which case IPR issues may not be very important to them, and rigorous application of IPR may simply impede valuable technology transfer with little compensating benefit (Trumpy, 1997). In other cases, the reverse may be true.

International law recognises the right of a country to take legislative measures to provide for the granting of compulsory licenses to prevent the abuses that might result from the exercise of the exclusive rights conferred by the patent. While the Paris Convention for the Protection of Industrial Property of 1883, and WTO through TRIP referred above deal with the subject broadly, the North American Free Trade Agreement (NAFTA) of 1993 and OECD's proposal for a Multinational Agreement on Investments (MAI) severely restrict the use of compulsory licensing of patents. This issue has been addressed in several international fora, including UNCTAD, the Rio Summit, UNGASS and discussions within CSD. A survey conducted in 1999 indicated that a number of countries both among the Annex I and non – Annex I countries have legislation listing the circumstances under which provisions for compulsory licensing could be invoked (Health Care and IP, 1999).

IPRs and developing countries

As noted in the context of TRIP negotiations, developing countries have particular concerns about IPRs. The great majority of patents are owned and continue to be generated from the industrialised world; developing countries and their companies tend to have fewer resources to purchase licences and fear that stronger IPRs impede their access to such technologies. Indigenous companies or communities may find that traditional approaches are not familiar to investors, and will have difficulty competing with larger companies that have extensive experience obtaining and dealing with IPRs. The culture of competition may also make it difficult to obtain relevant data in the short term. In other words, it is probable that stronger IPR protection may to some degree enhance vertical technology transfer through foreign investment, but may in some circumstances impede horizontal dissemination of protected technologies through developing country societies. Forsyth (1999) emphasises this distinction, but argues that the climate change debate on technology transfer has tended to undervalue the potential contribution of vertical technology investment in its own right. Besides, international financial assistance could be made available where market-driven licensing is not feasible.

Ultimately, the impediments in any single case must be weighed against the overall societal advantages of IPR in promoting investment and innovation. Technology transfer is necessary to reach certain development goals, but developing countries argue that the balance of IPR weighs to the advantage of developed countries. In practice the incentive effects of IPR, weighed against the impediments they may raise to technology transfer, will differ according to the technology, sector, and country. While trade theory provides little basis for mandating uniform standards of intellectual property protection across all countries, intellectual property rights is an issue that is here to stay on the international trade agenda (Trebilcock and Howse, 1995).

Finally, in some circumstances, inadequate access or abuse of IPRs may be addressed through compulsory licensing procedures. Under Articles 30 and 31 of TRIP, member countries may provide for compulsory licensing of patented inventions, i.e. use of the invention without permission. Generally, compulsory licensing programmes require the user first to seek a license, and if no license is given, then a limited non-exclusive right to practice the invention domestically may be awarded by the government, with an obligation to pay reasonable compensation to the patent owner.

IPRs and the promotion of ESTs

The importance of IPRs needs to be set in context. Many of the technologies for addressing climate change may not be protected anyway. This may apply both to "soft" technologies, such as better energy management or agricultural practices, and "hard" technologies such as building insulation. Where there is no patent in force in the country seeking to acquire technology, the main barriers to technology transfer will be (1) inadequate technical expertise and know-how in the country, (2) the absence of professionals in the country able to negotiate a suitable transfer

agreement, and (3) the willingness of the technology owner to transfer the technology. Training and education are necessary to overcome the first two barriers. The third barrier may be overcome with financial support and encouragement by the technology owner's country, the technology recipient's country, or through bilateral or multilateral arrangements (*e.g.* the GEF).

In other cases, relevant technologies may indeed be protected. What steps can governments take to use IPRs to improve transfer and development of ESTs? IPR systems can be harnessed specifically towards environmental objectives in some circumstances (Gollin, 1991). Patents on environmentally friendly technologies have been increasingly common since before the main environmental statutes were implemented in the 1970s. In the US, patent regulations were amended in 1982 (37 C.F.R. q.102-c) to provide for faster processing of environmental patents there. The practical impact of such provisions is uncertain.

In certain circumstances, existing patents may affect ability to comply with domestic regulations. The US Clean Air Act of 1970 (42 USC 7608) permits compulsory licensing in such circumstances: if the Attorney General identifies that a patented technology is needed by others to comply and there are not reasonable alternatives, then the US Courts are authorised to order licensing, "on such reasonable terms and conditions as the courts, after hearing, may determine."

For some developing countries seeking access to patented technologies in connection with international environmental treaties, one option might be for license fees to be paid for by an international funding source such as the GEF and/or through bilateral or multilateral arrangements.

Another way a country may address the concern for an unlicensed patent is to charge increasing annual maintenance fees. If the fee becomes high enough by 5 to 10 years after patent issuance, the owner will let an uncommercialised patent lapse (Sherwood, 1990). It will then become part of public domain.

Trademarks for environmental products can be crucial to their success and help facilitate widespread acceptance of a product. So-called "green labelling" programmes employ trademark or related principles (*e.g.* A not-for-profit organisation allows a vendor to use an environmental seal of approval if certain requirements are satisfied).

This section has highlighted the complexity and specificity of the issues, so that generalisations may not be helpful. Governments can support the exchange of public domain information. And governments can provide incentives for private parties to implement technology transfer programmes or joint ventures. Under extraordinary cases of IPR abuse, compulsory licensing can be considered.

Quite apart from legal and IPR issues, cooperation plays a critical role in promoting technology transfer. Likewise, national planning and specifically identifying national needs becomes a pre-condition to any IPR-related discussion as it is only when the type of technology or expertise needed has been identified and the specific circumstance assessed that the IPR question becomes relevant.

Finally, while IPR and other developments have helped to facilitate combined foreign investment, they have not generally oriented such investment particularly towards sustainable development or environmentally sound technologies. That is the function of other international agreements.

IPRs and Restrictive Business Practices

In reviewing the actual practice of IPR, the issue of Restrictive Business Practice (RBP) of the private sector deserves adequate attention and analysis. Various types of RBPs ranging from refusal of licensing to attaching restrictive or even prohibitive conditions for royalty and equipment sales to maximise the monopolistic rent were reported. These RBPs are more often practiced in the initial stage of the innovation where there are not yet many other competitors. After the initial stage, other competitors may enter with similar technologies leading to the abolishment of RBP. Companies often quickly change their behaviour, and try to sell and license their own technology in order to maximise its market share. Businesses management theory identifies strategies for "competitive advantage" whereby companies will manipulate the market in any manner that can suit their global strategies (Porter, 1980; 1990; Porter *et al.*, 1995).

In order to maintain monopoly or competitive advantage, companies go after their potential competitors in other countries and register their patents to block similar technology development, and dump their products to drive out new competitors with similar technologies from the market. Several studies have been done that verify this strategy of using intellectual property rights as a market advantage and as a strategy to control markets, as well as dominate innovation within industrial sectors (Narin *et al.*, 1998, Mowery and Oxley, 1995; and Teece *et al.*, 1994)

According to a case study (UNCTAD, 1997a) on the experience of Korean companies importing foreign ESTs, the types of restrictive conditions are: non-exclusive basis, restriction on export, prohibition of consigning to a third party, sharing of improved technology, restriction on the licensee for dealing in competitive products or technologies. Among 523 technologies introduced in 1994, 122 (23.3%) were accompanied with restrictive conditions.

According to Korean firms and R&D institutions, there were cases where the private firms and even public institutions of industrialised countries refused to license such ESTs like HFC-134a, fuel cell and IGCC (Integrated Gasification Combined Cycle). Some private firms sell their equipment under the condition that the buyer cannot disassemble the equipment.

These RBPs have not been widely reported and sufficiently analysed. However, there are some cases of RBPs related to HFC-134a, fuel cell, and IGCC technologies that were identified in Korea. The case of HFC-134a technology draws particular attention. In the initial stage of HFC-134a technology introduction, companies with HFC technologies refused to license or sell the technology. When Korea launched its programme to develop HFC-134a technology, a (foreign) company already having this technology registered 40 process patents in Korea in 1993 in an attempt to block the development of similar technology. As Korea neared completion of its own HFC-134a technology development, this company changed its policy and approached a Korean company to sell its HFC-134a technology.

The experience is not confined to CFC technology. Among 168 Japanese technologies introduced to Korea in 1994, 15 (8.9%) were not allowed to be consigned to a third party, and, respectively, 13 (7.7%) were granted on a non-exclusive basis and on condition that improved technologies should be shared between two parties during the contract period. Seven (4.2%) were prohibited to be used for export products and three (1.8%) were granted on condition that the licensee cannot deal in competitive products or technologies. Among the 209 US technologies introduced to Korea in the same year, 10 (4.8%) were not allowed to be consigned to a third party, and 28 (13.4%) were granted on a non-exclusive basis and on condition that improved technologies should be shared (12 or 5.7% and 16 or 7.7% respectively). As suggested in the above statistics, Japan seems to impose more unfavourable conditions on technological transfer than the U.S. (UNCTAD, 1997a).

Even though at the time of writing this Report there were not many published reports or articles on such business practices (RBP), evidence is mounting. The case of Korea is only one among many. Some scholars have noted the problems at the company level (Lundvall, 1993), while others have documented how companies have prevented the introduction of new technologies into the marketplace (Shaynneran, 1996) in order to advance and retain their current own technological advantages (Clark and Paolucci, 1997a & b; and The Economist, 1999a and b). For a comprehensive economic historical overview of the problem, see Reinert (1995) on the abuses of the technological change and international competitiveness. More recent studies, including those reported in the business management literature begin to question such competitive strategies (Saxenian, 1994; Reinert, 1995 and Clark and Paolucci, 1997 a & b).

RBPs for other technologies could be simply regarded as the exercise of the intrinsic monopolistic power recognised by the patent system. But if these RBPs are exercised for ESTs in general and on the issue of EST transfer to developing countries, which would both have local as well as global environmental implications, then the implications of RBPs in hindering technology transfer deserve research and analysis.

3.6 Technology Transfer and Kyoto Mechanisms

The area in which the Kyoto Protocol itself may have greatest implications for technology transfer is in its establishment of the project based mechanisms, Joint Implementation (JI) (Article 6) and the Clean Development Mechanism (CDM) (Article 12).

These allow investments in projects that reduce or avoid emissions to generate emission credits (ERUs or CERs respectively), which may be used to contribute towards compliance by an Annex I Party. Features that distinguish these mechanisms from emissions trading are summarised in Table 3.2. This creates an incentive towards climate-friendly investments. In addition, the CDM specifies that a part of the proceeds shall be used to assist vulnerable developing countries adapt to the effects of climate change. In these respects, such mechanisms may offer significant contributions towards technology transfer.

The CDM, if well structured, could be a vehicle for transferring ESTs. Though no specific provision makes reference to technology transfer, a number of features make the CDM unique. First, the clean development mechanism invites Annex I Parties to work with developing countries to further sustainable development and the overall objectives of the Climate Convention. This is possible largely only by the transfer of ESTs. Assisting Annex I Parties to achieve their emission reduction obligations through a transfer of "credits" is another objective of the CDM.

Table 3.2	Main Characteristics of JI/CDM versus Emission Trading (Nondek, 1998)	
	EMISSIONS TRADING	**JI/CDM**
Compliance	Based on emissions inventories	Based on project baselines
Reference	National emissions limit	Project baseline
Transaction costs	Low	Possibly high
National implementationcost	Possibly high	Low
Implementation of reductions	Policies and measures	Direct technology investments
Emission reduction potential	Large	Limited
Time horizon	2008-2012	From 2000 (for CDM)
Commodity	Allowances	Credits

Secondly, the CDM is project based. Certified emission reductions may be generated through the investment in concrete projects and based on measurable, certifiable emission performance.

Thirdly, the provisions for governance of the CDM are more specific than for the other mechanisms. The CDM is to be supervised by an executive board, which is likely to be a sub-group of Parties to the Protocol, perhaps with inputs from other constituencies. This will provide oversight and guidance to the implementation of the CDM. The Protocol also calls for independent auditing and verification of project activities. These provisions reveal an effort to ensure transparency and credibility in the final results, and the need for agreement on standardised procedures of performance on which to base certification. They make the design of the mechanism clearly multilateral in nature, involving decisions and consensus among multiple Parties.

Some initial ideas on the CDM are starting to emerge from a variety of new literature and exchanges that have taken place since Kyoto (Aslam 1998 a & b; OECD, 1998a; 1999, Goldemberg, 1998, TERI 1998):

- **The CDM is a means to build trust and strengthen capacity**. It should strengthen working relations and understanding among partners, the private sector, non-

governmental organisations, and governments at various levels to enhance technology cooperation. Project-based crediting should lead to tangible investments and development of local capacity to maintain the performance of these investments. These projects should incrementally assist developing countries to achieve multiple sustainable development objectives (economic development, improvement of local environmental quality, minimisation of risk to human health from local pollutants, and reduction of greenhouse gases). Careful project screening and selection, including host community decision-makers, will assist in multiple benefits for all participants.

- **The need to design simple, unambiguous rules that ensure environmental performance in the context of sustainable development while also favouring investment**. The multilateral oversight and governance provisions of the mechanism, and the project basis of transactions, will raise the transaction costs of investment in CDM projects as compared to mitigation reduction through other more conventional means (*e.g.* local options or even within other Annex I countries). This increases the complexity of the transaction and removes a portion of the economic benefit that might otherwise be attained. Critical questions under this heading have to do with how to determine the emissions additionality of proposed projects as well as the baseline against which to assess performance and establish transferable "credits".

Much about the design and governance of the CDM remains to be resolved. One important distinction is between "bilateral" or "portfolio" approaches (Table 3.3). The bilateral approach is closest to joint implementation programmes where the host country negotiates directly with the investor about the terms of the contract. The portfolio approach would allow host (developing) countries to advance bundles of possible projects that fit with their own sustainable development objectives. Some authors (Siniscalco, 1998; Grubb *et al.*, 1999) argue that both models will be needed, depending on the type of project involved.

A practical understanding of one way the CDM may relate to technology transfer may be found in looking at progress made to date by some countries in establishing technology priorities for investments likely to emerge under the Convention and the Protocol. The Indian government, for example, has identified three different types of technologies as priorities for CDM investment (TERI, 1998): grid-connected photovoltaics; advanced fuel cells; and biomass for power generation. With respect to baselines, the Indian government has also identified three types: increases in energy efficiency, renovation and modernisation; introduction of new technologies; and projects that are currently subject to government incentives and subsidies (hence additional).

Presumably once a host country has its own vision of what sustainable development entails given its own national circumstances, it could develop broad guidance on types of CDM pro-

Table 3.3	Bilateral versus Portfolio Approach to the CDM (Source: Yamin, 1998)
BILATERAL APPROACH	**PORTFOLIO APPROACH**
Project by project	"Bundling" of projects in portfolios
Investor-led	Host country-led
Private sector emphasis	National sovereignty emphasis
Contribution to emission reductions emphasis	Contribution to sustainable development emphasis
Proceeds for adaptation unnecessary, seen as additional costs to achieve Article 3 compliance	Proceeds for adaptation seen as necessary to benefit all DCs to increase global participation in Protocol
May concentrate on countries already benefiting from Foreign Direct Investment	Could allow equity considerations to tailor portfolios to benefit all DCs' mitigation efforts
Primary purpose of CDM is clearing-house function	Primary purpose of CDM is to obtain best price for CERs, shield hosts from undue pressure; clearing-house function is a necessary feature

jects that would be acceptable to it. Such guidance could vary from country to country and possibly by region within a country, depending on a variety of different parameters (*e.g.* resource endowment, natural environment, geography, industrialisation, urbanisation and demographic trends).

This guidance might resemble a national "technology needs assessment," one of the ideas that has begun to emerge from discussions under the Convention on technology transfer (UNFCCC, 1998a and van Berkel *et al.*, 1998). Once a country has established a vision of its needs for sustainable development and technology, practical guidance on project types could be developed to fit these objectives. A portfolio of projects could be offered to international investors, giving more control to host countries to ensure the value of CDM projects over time (Yamin, 1998).

An international gathering hosted by the OECD offered some interesting perspectives and words of caution with respect to the CDM (OECD, 1999): Participants highlighted the potential role of the clean development mechanism for fostering climate-friendly technology cooperation and urged greater clarity on these issues. They also cautioned against the mechanism being used by investors to compete for the cheapest abatement projects, which could lead to uncoordinated and ultimately counter-productive efforts (OECD, 1998b).

3.7   Conclusions

This chapter has surveyed the role of international agreements and legal structures in facilitating technology transfer. Facilitating technology transfer is an intrinsic part of the evolving climate regime under the UNFCCC. So far steps towards this have proceeded slowly.

Government-driven technology and transfer pathways remain important, particularly in some regions. On the whole, however, the bulk of foreign investment is now through the private sector. Some relevant technologies (by no means all) are protected through intellectual property rights, particularly patents. The TRIP agreement embodies a considerable strengthening of IPRs in many developing countries. This may help to attract greater foreign investment (though IPRs do not appear to be a dominant consideration in most foreign investment decisions), and stimulate innovation. On the other hand, it may also impede the horizontal dis-

semination of technologies in some cases. More attention could be paid to the possibilities surrounding licensing (sometimes compulsory, where the TRIP procedures are followed and adequate justification is given). Where market-driven solutions are not feasible to address short-term access issues, international financial assistance should be considered. There may also be important roles for sector-based initiatives, standards and agreements.

The combined set of positive measures established under the Montreal Protocol, such as the participatory approach, where the relevant stakeholders are involved in the decision-making process for setting up national strategies, the globally agreed and binding targets to phase out ODS, industry-government partnerships and NGO involvement, plus direct investment projects, have helped in facilitating technology transfer of ozone friendly technologies, and also contributed towards endogenous capacity-building necessary for improving the regulatory structure in developing countries. A major obstacle for several of these countries is the inadequacy of the regulatory structure to support the phaseout process. The weak capacity of governments to institute and/or enforce laws, regulations, and policies is a common problem for them.

The Kyoto Protocol and the CoP4 decisions, including its establishment of the project-level crediting mechanisms, reflects significant changes and opens up important opportunities. Development of procedures relating to compliance could both highlight the performance of Annex II Parties with respect to technology transfer, and contribute to capacity-building in developing countries in ways that would also facilitate such transfer. In this context it is important to spell out the elements that constitute compliance with Article 4.5. These include agreeing on procedures to determine whether there is technology transfer, the expected roles of governments and the private sector, and finally on how to measure compliance with Article 4.5 of UNFCCC.

Ultimately, in all these respects, it is on-the-ground activities that do much to determine effective technology transfer, and the international process needs most to support the capacities and activities of states in creating stable and supportive environments for such activities. The recent Technical Paper prepared by the UNFCCC Secretariat (1998d) points out the need for:

- Stable macroeconomic and environmental policies to provide framework conditions conducive to investment by the private sector;
- Transparent and fair laws and public administration;
- Open trade and investment policies;
- Adequate infrastructure and human resources;
- Opportunities to use national legislation to provide incentives for EST;
- Educating and training people in the country to select, maintain and develop EST.

All these factors are relevant in creating the local conditions not only for compliance with commitments on reporting, finance, etc., but also for fostering adequate implementation of provisions on technology transfer, and for creating conducive national conditions in host countries. These conditions form the subject of the next two chapters.

References

Achanta, A.N.,and P.Ghosh, 1993: Technology transfer in the context of global environment issues. In *The Road from Rio: Environment and development policy issues in Asia*. P.Ghosh, A.Jaitly, (eds.), Tata Energy Research Institute, New Delhi.

Aslam, M. A., 1998a: *Endogenous Capacity Building for AIJ and CDM:Developing Country Needs*. Paper prepared for the UNFCCC Workshop on CapacityBuilding, September 1998, Abidjan, Cote D'Ivoire.

Aslam, M. A., 1998b: *International trading aspects of the CDM*. Paper for the UNCTAD International Working Group on the CDM, ENVORK, Islamabad.

Besen, S.M.,and J.Braskin, 1991: An introduction to the Law and Economics of Intellectual Property. *The Journal of Economic Perspectives*, **5**(1).

Clark, Jr. W W., and E. Paolucci, 1997a: Environmental Regulation and Product Development: issues for a new model of innovation. *Journal of International Product Development Management*.

Clark, W.W. Jr., and E.Paolucci, 1997b: An International Model for Technology Commercialization. *Journal of Technology Transfer*.

Chung, R. K., 1998: The role of government in the transfer of environmentally sound technology. In *Positive measures for technology transfer under the climate change Convention*. T.Forsyth (ed.), Royal Institute of International Affairs, London.

COWI Consult, and Gross Gilroy Inc, 1995: *Study on the Financial Mechanism of the Montreal Protocol*. UNEP, Nairobi.

Doane, M.L., 1994: TRIPs and International Intellectual Property Protection in an age of advancing technology. *American University Journal of International Law and Policy*, pp.465-497.

Du Pont, P., L.Velasco, and K.Opheim, 1997: *The problem of second-hand industrial equipment: reclaiming a missed opportunity* . Conference on "Towards a Sustainable Asia", Yokahama, Japan, 1997, International Institute for Energy Conservation, Washington / Bangkok.

Environmental Investigation Agency, 1998: *A Crime Against Nature. The Worldwide Illegal Trade In Ozone Depleting Substances*. A Second Report by the Environmental Investigation Agency (EIA).

Forsyth, T., 1999: *International Investment and climate change: energy technologies for develoipng countries*. Royal Institute of International Affairs/Earthscan, London.

Glowka, Lyle, F. B. Guilmin, and H. Synge, 1994: *A Guide to the Convention on Biological Diversity*. IUCN, Gland, Switzerland.

Goldemberg, J. (ed.), 1998: *Issues and options: the Clean Development Mechanism*, UNDP, New York.

Gollin, M., 1991: Using Intellectual Property to Improve Environmental Protection. *Harvard Journal of Law and Technology*, **4**, 193.

Government of Korea, 1998: *The role of publicly-funded research and publicly-owned technologies in the transfer and diffusion of ESTs*. International Expert Meeting, 4-6 February, 1998. Kyong Ju, Korea.

Grubb, M., M. Koch, A. Munson, F. Sullivan, and K.Thompson, 1993: *The Earth Summit Agreements: a Guide and Assessment*. Royal Institute of International Affairs/Earthscan, London.

Grubb, M., C. Vrolijk, D. Brack, 1999: *The Kyoto Protocol: a guide and assessment*. Royal Institute of International Affairs / Earthscan, London.

Health Care and IP, 1999: *Compulsory Licensing*. [Available at: http://WWW.cptech.org/ip/health/cl/].

Kwon, H.A., 1995: Patent protection and technology transfer in the developing world: the Thailand Experience. *George Washington Journal of International Law and Economics*, **28**, 568-605.

Lundvall, B-A., 1993: Explaining interfirm cooperation and innovation: Limits of the transaction-cost approach In *The Embedded Firm: on the socioeconomics of industrial networks*. Gernot Grabher, (ed.), Routledge, London.

Margolis, R. M., and D. M. Kammen, 1999: Underinvestment: The Energy Technology and R&D Policy Challenge. *Science*, **285** (30 July).

Mowery, D.C., and J.E. Oxley, 1995: Inward technology transfer and competitiveness: the role of national innovation systems. *Cambridge Journal of Economics*, **19** (1),.

Multilateral Fund, 1998: *Policies, Procedures, Guidelines and Criteria (as of July 1998)*. Multilateral Fund for the Implementation of the Montreal Protocol.

Narin, F., K. S. Hamilton, and D. Olivastro, 1998: The Increasing Linkage between U.S. Technology and Public Science. *Research Policy*, **March 17**, 1-17.

Nelson, R.R., 1994: Why do firms differ, and does it matter? In *Fundamental Issues in Strategy*. R.P. Rumelt, D.E. Schendel, D.J. Teece, (eds.), Harvard Business School Press, Boston, MA.

Nondek, L., 1998: Paper presented at the IEA Moscow workshop on Emission Trading and Joint Implementation, October 1998.

OECD, 1998a: *Ensuring Compliance with a Global Climate Agreement*. OECD Information Paper ENV/EPOC(98) 5/REVI, Paris. [Available at http://www.oecd.org/env/cc].

OECD, 1998b: *Moving Forward and Setting Priorities After Kyoto*. Highlights of the OECD and IEA Forum on Climate Change, 12-13 March 1998, Paris COM/ENV/EPOC/CDC/DAC/IEA/M(98)1).[Available at http://www.oecd.org/env/cc].

OECD, 1999: *Monitoring, Reporting and Review under the Kyoto Protocol*. OECD Information Paper ENV/EPOC(99) 20/FINAL, Paris. [Available at http://www.oecd.org/env/cc].

Osakwe, C., 1998: Finding new packages of acceptable combinations of trade and positive measure to improve the effectiveness of MEAs: a general framework. In *Trade and the Environment: Bridging the Gap*. A.Fijalkowski, J. Cameron, (eds.), Cameron May, London/T.M.C. Asser Institute, The Hague.

Porter, M E., 1980: *Competitive Strategy: Techniques for Analyzing Industries and Competitors*. Free Press, New York.

Porter, M E., 1990: *Competitive Advantage*. Free Press, New York.

Porter, M E. and C. van der Linde, 1995:Green and Competitive. *Harvard Business Review*, **September-October**.

Reid, W. et al., 1993: An Intellectual Property Rights Framework for Biodiversity Prospecting and The Convention on Biological Diversity. In *Biodiversity Prospecting: Using Genetic Resources for Sustainable Development*. World Resources Institute, Washington, DC.

Reinert, E. S., 1995: Competitiveness and its predecessors — a 500-year cross-national perspective. *Structural Change and Economic Dynamics*, *Vol. 6*. Elsevier Science, Amsterdam, pp. 23-42.

Saxenian, A., 1994: *Regional Advantage: Culture and Competition in Silicon Valley and Route 128*. Harvard University Press, Cambridge, MA.

Shaynneran, M., 1996: *The Car that Could: the inside story of the GM Impact*. Simon Schuster, New York.

Shende, R., and S. Gorman, 1997: *Lessons in Technology Transfer under the Montreal Protocol*. OECD Forum on Climate Change (Ozone Action Programme), United Nations Environment Programme, Industry and Environment Office, Paris.

Sherwood, 1990: *Intellectual Property and Economic Development*. Westview Press, USA..

Siniscalco, D., A.Goria, and J.Jasssen, 1998: Oustanding issues. In *Issues and options: the Clean Development Mechanism*. J. Goldemberg, (ed.), UNDP, New York.

Teece, D.J., J.E.L. Bercovitz, and J. M. De Figueiredo, 1994: *Firm Capabilities and Managerial Decision-making: A theory of Innovation Biases*. Haas School of Business, University of California, Berkeley, CA.

TERI, 1998: *Clean Development Mechanism: Issues and Modalities*. Tata Energy Research Institute, Multiplexus, New Delhi.

The Economist, 1999a: *Stepping on the gas: after many false starts, hydrogen power is at last in sight of commercial viability*, July 24, pp. 19-20.

The Economist, 1999b: Fuel Cells meet big business: A device that has been a technological curiosity for a century and half has suddenly become the centre of attention.. *The Economist*, July 24, 59-60.

Trebilcock, M.J., R. and Howse R., 1995: *The Regulation of Interna tional Trade*. Routledge, London.

Trumpy, T.B., 1997: Should needed energy and environmental technology be available "free" and when is it? *Haigh Yearbook of International Law*, pp.27-33.

UN, 1997: *Ranking procedures for environment, especially the ISO's series 14000: its effects for trade and investments in developing countries*. Board for Trade and Development. TD/b/COM.1/EM.42.

UNCTAD, 1997a: *Korea and the Montreal Protocol*. Sponsored by the UNCTAD in March 1997 and study carried out by the Korea Trade-Investment Promotion Agency.

UNCTAD, 1997b: *Positive measures to promote sustainable development in particular meeting the objectives of Multilateral Environmental Agreements*. Report by UNCTAD Secretariat, TD/B/Com.1/EM.3/2.

UNCTAD, 1998: *The role of publicly-funded research and publicly-owned technologies in the transfer and diffusion of ESTs*. UNCTAD, Geneva, January.

UNFCCC, 1995: Report of the Conference of the Parties on its First Session. Action taken by the Conference of the Parties at its first session. Addendum: Part Two, FCCC/CP/1995/7/Add.1, 6 June 1995. Berlin, 28 March - 7 April 1995.

UNFCCC, 1996a: *Technology inventory and assessment,* FCCC/SBSTA/1996/4, 2 February 1996.

UNFCCC, 1996b, *Technology inventory and assessment*. Initial report on an inventory and assessment of technologies to mitigate and adapt to climate change. Addendum: Additions to the technology inventory database. Note by the secretariat, FCCC/SBSTA/1996/4/Add.2, 31 May 1996.

UNFCCC, 1996c: *Transfer of technology*. Note by the secretariat., FCCC/SBI/1996/5, 2 February 1996.

UNFCCC, 1996d: *Development and transfer of technologies*. Review of the implementation of the Convention and of decisions of the first session of the Conference of the Parties. (Articles 4.1(c) and 4.5). Follow-up report on technological issues. Note by the secretariat, FCCC/CP/1996/11, 25 June 1996.

UNFCCC, 1996e: Report of the Conference of the Parties on its second session. Addendum, Part Two: Action taken by the Conference of the Parties at its second session Geneva, 8 to 19 July 1996, FCCC/CP/1996/15/Add.1, 29 October 1996.

UNFCCC, 1997a: Report of the Subsidiary Body for Scientific and Technological Advice on the work of its fifth session, Bonn, 25-28 February 1997, FCCC/SBSTA/1997/4, 7 April 1997.

UNFCCC, 1997b: Second communications from Parties included in Annex I to the Convention. Activities of Parties included in Annex II related to transfer of technology, FCCC/SBSTA/1997/13, 24 September 1997.

UNFCCC, 1997c: *Development and transfer of technologies*. Progress report, FCCC/SBSTA/1997/10, 30 September 1997.

UNFCCC, 1998a: Report of the Conference of the Parties on its third session, Kyoto, 1 to 11 December 1997. Addendum. Part two: Action taken by the Conference of the Parties at its third session, FCCC/CP/1997/7/Add.1, 18 March 1998.

UNFCCC, 1998b: *Development and transfer of technologies*. Progress report, FCCC/SBSTA/1998/5, 16 April 1998.

UNFCCC, 1998c: *Development and transfer of technologies (Decision 13/CP.1)*. Progress report on transfer of technology: draft work programme. Note by the Secretariat, FCCC/CP/1998/6, 22 September 1998.

UNFCCC, 1998d: *Barriers and Opportunities Related to the Transfer of Technology*. Second Technical Paper on terms of transfer of technology and know-how, FCCC/TP/1998/1, 13 October 1998.

UNFCCC, 1999: Action taken by the Conference of the Parties at its fourth session. Addendum, Part two: Report of the Conference of the Parties on its fourth session, Buenos Aires, 2 to 14 November 1998, FCCC/CP/1998/16/Add.1, 20 January 1999.

UNCSD, 1997: Report on the Fifth Session (7-25 April 1997), UNDoc. E/1997/29, E/CN.17/1997/25.

UNCSD, 1998: Report on the Sixth Session (20 April – 1 May 1998), UNDoc. E/1998/29, E/CN.17/1998/20.

UNGASS, 1997: Programme for the Further Implementation of Agenda 21 – Adopted by the General Assembly at its nineteenth special session (23-28 June 1997). UNGA Resolution A/RES/S-19/2, 19 September 1997.

UNICE, 1997: *TRIPS and the Environment*. Union of Industrial and Employers' Confederations of Europe, Brussels, September 1997.

Van Berkel, R. and E. Arkesteijn, 1998): *Transfer of Environmentally Sound Technologies and Practices under the Climate Convention: Survey of Experiences, Needs and Opportunities among Non-Annex II Countries*. IVAM Environmental Research, Unviersity of Amsterdam. Amsterdam.

Vossenaar, R. and V. Jha, 1998: Implementation of MEAs at the national level and the use of trade and non-trade-related measures: results of developing county case studies. In Fijalkowski & Cameron, 1998.

Worthy, J. 1994: Intellectual property protection after GATT. *EIPR*, **5**, 195-198.

Yamin, F., 1998: Unanswered Questions. In *Issues and options: the Clean Development Mechanism*, J. Goldemberg (ed.), UNDP, New York.

4

Enabling Environments for Technology Transfer

Coordinating Lead Authors:
MERYLYN MCKENZIE HEDGER (UNITED KINGDOM),
ERIC MARTINOT (USA), TONGROJ ONCHAN (THAILAND)

Lead Authors:
Dilip Ahuja (India), Weerawat Chantanakome (Thailand),
Michael Grubb (United Kingdom), Joyeeta Gupta (The Netherlands),
Thomas C. Heller (USA), Li Junfeng (China), Mark Mansley (United Kingdom),
Charles Mehl (USA), Bhaskhar Natarajan (India), Theodore Panayotou (USA),
John Turkson (Ghana), David Wallace (United Kingdom)

Contributing Author:
Richard J.T. Klein (the Netherlands/Germany)

Review Editor:
Karen R. Polenske (USA)

CONTENTS

EXECUTIVE SUMMARY

Governments and international agencies have a variety of policy tools for overcoming key barriers and creating enabling environments for technology transfer. Barriers and policy tools are discussed broadly in this chapter according to ten dimensions of enabling environments:

1. *National systems of innovation.* Technology transfers are influenced greatly by what have been called *national systems of innovation*--the institutional and organisational structures which support technological development and innovation. Governments can build or strengthen scientific and technical educational institutions and modify the form or operation of *technology networks*—the interrelated organisations generating, diffusing, and utilising technologies.

2. *Social infrastructure and participatory approaches.* Social movements, community organisations and non-governmental organisations (NGOs) contribute to the "social infrastructure" that plays an important and enabling role in many forms of technology transfer. Governments can devise participatory mechanisms and adopt processes to harness the networks, skills and knowledge of civil society, including community groups and NGOs, to better meet user needs, avoid delays and achieve greater success with technology transfer.

3. *Human and institutional capacities.* There are many failures of technology transfer that result from an absence of human and institutional capacity. Although much of the focus on capacity building has been on enhancing scientific and technical skills, other skills for selecting, managing, adapting, and financing technologies are equally important. Capacity building is a slow and complex process to which long-term commitments must be made. For adaptation there is a need: to strengthen scientific and policy institutions to enable the undertaking of assessments and, to access datasets, tools and techniques to produce outputs for nationally determined priorities.

4. *Macroeconomic policy frameworks.* Macroeconomic policies include direct and indirect financial support, energy tariff policies, trade and foreign investment policies, and financial sector regulation and strengthening.

5. *Sustainable markets.* Sustainable market approaches are important for renewable energy and energy efficiency technology transfer because these approaches promote replicable, ongoing technology transfers. Governments can conduct market transformation programmes that focus simultaneously on both technology supply (production technologies and product designs) and demand (subsidies, consumer education and marketing).

6. *National legal institutions.* National legal institutions are needed to secure intellectual property rights; reduce contract, property, and regulatory risks; and promote good governance and eliminate corruption. To these ends, governments can strengthen national legal institutions for intellectual property protection; and strengthen administrative and law processes to assure transparency, participation in regulatory policy-making, and independent review.

7. *Codes, standards and certification.* The importance and the need for technical standards, codes and certification have been well recognised by the technical community all over the world. If standards and codes are absent, transaction costs can increase as each buyer must ascertain the quality and functionality of potential technologies individually, raising transaction costs.

8. *Equity considerations.* Equity in technology transfer can be enhanced by devising analytical tools and providing training for social impact assessment, requiring social impact assessments before technology is selected, and creating compensatory mechanisms for 'losers'. Governments may also wish to develop criteria for ensuring that technology transfer projects do not disempower or negatively influence weaker social groups in a society.

9. *Rights to productive resources.* Rights to productive resources can be affected by technology transfer, including land (agriculture, forestry), natural resources (forests, water, coastal areas), factories, and other productive resources. Successful introduction of new technologies or modification of resource use often depends on a recognition of the existing forms of resource rights, or on taking steps to create an optimal resource rights regime.

10. *Research and technology development.* Developing countries' research and technology development efforts are often adaptive, following externally developed technology, thus suggesting the need for additional indigenous innovative capacity. Governments can develop science and educational infrastructure by building public research laboratories, providing targeted research grants, strengthening technical education, and directly investing in research and development.

4.1 Introduction

As discussed in Chapter 2, international financial flows have strongly influenced technology transfers between countries. Since UNCED in 1992, three major trends in finance for sustainable development have occurred. First, there has been increasing interest and activity in developing innovative domestic and international financial mechanisms. Second, both official development assistance and domestic resource mobilisation have fallen short of the commitments made. Third, private flows of financial resources from developed to developing countries have expanded enormously (Lin See Yan, 1997).

To improve the quality and efficacy of the transfer of environmentally sound technologies for the purposes of the UNFCCC, these trends need to be set in a broader context. That context - for promoting successful, sustainable technology transfer - implies multi-faceted enabling environments in both developed and developing countries. Enabling environments for technology transfer include national institutions for technology innovation, the involvement of social organisations, human and institutional capacities for selecting and managing technologies, macroeconomic policy frameworks, the underpinnings of sustainable markets for environmentally sound technologies, national legal institutions that reduce risk and protect intellectual property rights, codes and standards, research and technology development, and the means for addressing equity issues and respecting existing property rights. Enabling environments are inclusive, with all stakeholders potentially affected – governments, research institutions, national NGOs, technology developers and businesses. And actions to promote enabling environments are required in both developed and developing countries.

Any discussion of enabling environments inevitably conjures up the long-standing debate about "good government." Good government has been increasingly recognised as essential for promoting the environment and development agenda. Yet sharp disagreements have existed about what good government means; a wide range of actions have taken place in response to "simultaneously too much state and too little state" (Merquior, 1993). In both developed and developing countries in the 1980s and early 1990s, many governments committed themselves to market-oriented approaches for generating economic growth as a response to "too much state." Economic and political pressure forced governments to limit or eliminate economic interventions. This response to decades of "too much state" reduced the size, expenditures, and responsibilities of public sectors (Grindle, 1997). Reform priorities turned to the stabilisation of macroeconomic conditions, liberalisation of domestic and international trade, privatisation, and the reduction of state bureaucracies.

Over time it became clear to many that government also had to be strengthened to make it efficient, effective and responsive.

The most significant obstacle to the pursuit of social objectives is what may be described as the crisis of governance.

The crisis of governance includes an excessive degree of centralisation; overburdening and rigidity of the government machinery; the absence of local participation which can provide the requisite attention to detail; deterioration in the professionalism, competence and integrity of public functionaries and the weakening of judicial and quasi-judicial institutions (Banuri *et al.*, 1997, p. 9).

Governments not only had to have the capacity to manage macroeconomic policy but also had to be able to actively regulate market behaviour. So the trend towards market-oriented economic policy inevitably led to new (or rediscovered) roles for the state (Naim, 1995). In addition, reformers began to recognise that well-defined and functioning institutions of governance were important for the stability and legitimacy of new modes of public and community participation.

By the mid 1990s, good government had been added to the development agenda precisely because of greater awareness that neither markets nor democracies could function well--or perhaps function at all--unless governments were able to design and implement appropriate public policies, administer resources equitably, transparently and efficiently, and respond to the social welfare and economic claims of citizens. Although a general consensus developed on the imperative for good government, how to get good government has not been clearly understood (Grindle, 1997, p. 5).

Past experience with technology transfer in a variety of sectors can be used to suggest policy tools for providing enabling environments for the transfer of technologies for mitigation and adaptation to climate change that is supportive and sustainable. Evidence exists both of barriers and ways in which barriers can be avoided and overcome. Experience also shows that technology transfer offers many opportunities for sustainable economic and social development. Perceptions of the role of technology transfer in development have been transformed in the past twenty years, as the focus has switched from externally supplied hardware to people, processes, and the importance of local knowledge. Experience also has shown that technology transfer can be made more sustainable by ensuring that long-term learning and adaptation takes place, that conditions for market sustainability exist, that stakeholders continue to be involved, and that replicable financing mechanisms continue to function.

The sustainable use of environmentally sound technologies (ESTs) for climate change has to fulfill not only social but also economic and development objectives through a complex process of technological change. Chapters 4 and 5 explore what enabling conditions government can establish to facilitate this change in support of the Climate Change Convention, and how other stakeholders in the private sector, lending institutions, multilateral agencies, and non-governmental organisations can perform within this framework. This chapter identifies the key elements and broad relationships and Chapter 5 focuses on the crucial financial partnerships needed.

4.2   Policy Tools for Technology Transfer

Policy tools for technology transfer can be used to overcome a variety of barriers to technology transfer. The United Nations (UNCTAD, 1990) saw the lack of indigenous technological capabilities, infrastructure and institutions as key barriers to developing countries conducting technology transfer:

> [Developing countries] lack the strong knowledge base, integrated physical infrastructure and diversified economy required to weather shocks and recombine existing resources in new ways to adjust to what has become a continuous process of change. They also lack the institutional mechanisms and capabilities to perceive opportunities and constraints, and to translate these into effective policies for change. The financing and skills needed to innovate, to adapt and to diversify are also exceedingly rare in these countries. (p.3)

Many have recognised that barriers to technology transfer depend very much on the specific context in question:

> The limiting factors for technological transformation are not primarily technological but are instead part of the social, economic, political, and cultural milieus in which technologies are developed, diffused, and used. Market incentives, the structure of regulations, the content and quality of research and education, and social values and preferences all determine technological trajectories. (Heaton *et al.*, 1991, p. 21)

Governments and international agencies have a variety of policy tools for overcoming key barriers to technology transfer (see Table 4.1). Many of the barriers and tools are discussed broadly in this chapter and in Chapter 5, while barriers and policy tools more specific to individual sectors are discussed in Section II of this report. In addition, direct government financial support for technology transfer (such as grants, subsidies, loans, provision of equipment or services, and loan guarantees), as well as indirect financial support for private-sector innovation, are discussed further in Chapter 5.

Agenda 21, adopted in 1992, provided some of the earliest recommendations for public policies to promote technology transfer for environmental benefits (UN, 1993). These recommendations reflect not only the need for hardware, but also for building associated local capacities and for providing market intermediation. Strategies outlined in Agenda 21 include: (a) information networks and clearinghouses that disseminate information and provide advice and training; (b) government policies creating favourable conditions for both public-sector and private-sector transfers; (c) institutional support and training for assessing, developing, and managing new technologies; (d) collaborative networks of technology research and demonstration centres; (e) international programmes for cooperation and assistance in R&D and capacity building; (f) technology-assessment capabilities among international

organisations; and (g) long-term collaborative arrangements between private businesses for foreign direct investment and joint ventures.

In view of the considerable opportunities for action, some have argued for nothing less than a technological "transformation" towards more environmentally sustainable technologies (Heaton *et al.*, 1991). To promote such a transformation, successful technology transfer policies must make available financial resources, must reduce or eliminate barriers to technology transfer, must promote capacity building within developing countries, and must promote new forms of technology intermediation or "technology brokering." Capacity building should target technology acquisition, skills development, and local policies and institutions to support the technology transfer process. Market intermediation should include matching technologies with applications, brokering partnerships, and facilitating negotiations and financing packages.

A new role for Governments has been identified where they play an integrating role in managing knowledge on an economy-wide basis by making technology and innovation policy an integral part of overall economic policy (OECD, 1999). It should be recognised that fundamental social structures and personal values (*i.e.*, social factors) evolve along with a society's physical infrastructure, private and public institutions, and the technologies embodied within them. New (climate friendlier) technological trajectories for an economy can provide great opportunities. Governments can support the establishment of dynamic flexible learning mechanisms within the private sector and with the rest of civil society. Private companies can create greater environmental awareness. Greater participation across society in the charting of technological trajectories, and in the use of the products of goods and services they generate go hand in hand with raising awareness through education and the media.

4.3   National Systems of Innovation and Technology Infrastructure

Technology transfers are influenced greatly by what have been called "national systems of innovation," which are the networks of institutions that initiate, modify, import and diffuse new technologies (Freeman, 1987; OECD 1999). The term "national systems of innovation" started as an analytical tool but has also become a mobilising concept to drive policy. The term is given different emphases in the literature. One definition is: "the set of institutions to create, store and transfer the knowledge, skills and artifacts which define technological opportunities" (Metcalfe, 1995). National systems of innovation reflect a complex mixture of institutions (*e.g.*, financial; legal; scientific and technological; educational), public policies (regarding, *e.g.*, taxation; export/import promotion; science, technology and innovation), and business and social relationships.

National systems of innovation depend upon the development of so-called technology infrastructure, "a set of collectively supplied, specific, industry-relevant capabilities" such as technology cen-

Table 4.1   Policy Tools for Creating an Enabling Environment for Technology Transfer

POLICY TOOL	BARRIERS ADDRESSED	RELEVANCE
NATIONAL SYSTEMS OF INNOVATION AND TECHNOLOGY INFRASTRUCTURE (4.3)		
• Build firms' capabilities for innovation • Develop scientific and technical educational institutions • Facilitate technological innovation by modifying the form or operation of technology networks, including finance, marketing, organisation, training, and relationships between customers and suppliers	• Lack of technology development and adaptation centres • Lack of educational and skills development institutions • Lack of science, engineering and technical knowledge available to private industry • Lack of research and test facilities • Lack of information relevant for strategic planning and market development • Lack of forums for joint industry-government planning and collaboration	Primarily private-sector-driven pathways Primarily buildings, energy, and industrial sectors All stages
SOCIAL INFRASTRUCTURE AND RECOGNITION THROUGH PARTICIPATORY APPROACHES (4.4)		
• Increase the capacity of social organisations and NGOs to facilitate appropriate technology selection • Create new private-sector-focused social organisations and NGOs with the technical skills to support replication of technology transfers • Devise mechanisms and adopt processes to harness the networks, skills and knowledge of NGO movements	• Technology selection inappropriate to development priorities • Historical legacy of technology transfer in development • Problems of scaling cultural and language gaps and fostering long-term relationships	All pathways Particularly adaptation technologies, but applies to all sectors Particularly assessment, evaluation and replication stages, although NGOs are more and more participating in implementation stages
HUMAN AND INSTITUTIONAL CAPACITIES (4.5)		
• Build capacities of firms, non-governmental organisations, regulatory agencies, financial institutions, and consumers	• Inability to assess, select, import, develop and adapt appropriate technologies • Lack of information • Lack of management experience • Problems of scaling cultural and language gaps and fostering long-term relationships • Limited impact of technology because no long term capacity built to maintain innovation • Lack of joint venture capabilities for learning and integrating	All pathways All sectors Particularly assessment and implementation stages
MACROECONOMIC POLICY FRAMEWORKS (4.6)		
• Provide direct financial support like grants, subsidies, provision of equipment or services, loans and loan guarantees. • Provide indirect financial support, like investment tax credits • Raise energy tariffs to cover full long-run economic costs • Alter trade and foreign investment policies like trade agreements, tariffs, currency regulations, and joint venture regulations • Alter financial sector regulation (See also chapter 5 for further discussion of policy tools for financing technology transfer)	• Lack of access to capital • Lack of available long-term capital • Subsidised or average-cost (rather than marginal-cost) prices for energy • High import duties • High or uncertain inflation or interest rates • Uncertain stability of tax and tariff policies • Investment risk • Excessive banking regulation or inadequate banking supervision • Incentives for banks that are distorted against risk taking • Banks that are poorly capitalised • Risk of expropriation	Particularly private-sector-driven pathway, but relevant to all pathways Trade and foreign investment policies particularly relevant to private-sector-driven pathways Particularly assessment and repetition stages All sectors; energy tariffs relevant to buildings, industry, and energy sectors
SUSTAINABLE MARKETS FOR ENVIRONMENTALLY SOUND TECHNOLOGIES (4.7)		
• Conduct market transformation programmes that focus on both technology supply and demand simultaneous. • Develop capacity for technology adaptation by small- and medium-scale enterprises (SMEs) • Conduct consumer education and outreach campaigns • Targeted purchasing and demonstrations by public sector	• High transaction costs • Inadequate strength of smaller firms • Uncertainty of markets for technologies prevents manufacturers from producing them • Lack of consumer awareness and acceptance of technologies • Lack of confidence in the economic, commercial, or technical viability of a technology	Private-sector-driven pathways Buildings, industry, and energy sectors All stages
NATIONAL LEGAL INSTITUTIONS (4.8)		
• Strengthen national frameworks for intellectual property protection • Strengthen administrative and law processes to assure transparency, participation in regulatory policy-making, and independent review • Strengthen legal institutions to reduce risks	• Lack of intellectual property protection • Contract risk, property risk, and regulatory risk • Corruption	All pathways All sectors Particularly agreement stage
CODES, STANDARDS, AND CERTIFICATION (4.9)		
• Develop codes and standards and the institutional framework to enforce them. • Develop certification procedures, and institutions, including test and measurement facilities	• High user discount rates do not necessarily result in most efficient technologies • Lack of information about technology or producer quality and characteristics • Lack of government agency capability to regulate or promote technologies • Lack of technical standards and institutions for supporting the standards	All pathways Buildings, transport, industry, and energy sectors Assessment stage
EQUITY CONSIDERATIONS (4.10)		
• Devise analytical tools and provide training for social impact assessment. • Require social impact assessments before technology is selected • Create compensatory mechanisms for "losers"	• Social impacts not adequately considered • Some stakeholders may be made worse off by technology transfer	All pathways All sectors Assessment stage
RIGHTS TO PRODUCTIVE RESOURCES (4.11)		
• Investigate impacts of technology on property rights, test through participatory approaches, devise compensatory mechanisms for losers.	• Inadequately protected resource rights	All pathways Most sectors where land use is involved
RESEARCH AND TECHNOLOGY DEVELOPMENT (4.12)		
• Develop science and educational infrastructure by building public research laboratories, providing targeted research grants, and strengthening technical education system • Directly invest in research and development	• Insufficient investment in R&D • Inadequate science and educational infrastructure	Government-driven and community-driven pathways Assessment and replication stages Buildings, industry, energy, waste management and treatment sectors

tres and educational and skills development institutions (Justman and Teubal, 1995, p. 260). "Technology infrastructure consists of science, engineering and technical knowledge available to private industry. Such knowledge can be embodied in human, institutional, and facility forms. More specifically, technology infrastructure includes generic technologies, infratechnologies, technical information, and research and test facilities, as well as less technically-explicit areas including information relevant for strategic planning and market development, forums for joint industry-government planning and collaboration, and assignment of intellectual property rights. An important characteristic of technology infrastructure is that it depreciates slowly, but it requires considerable effort and long lead times to put in place and maintain." (Tassey, 1991, p.347)

Technology infrastructure may lack direct economic value to any one firm, and thus individual firms may lack adequate incentives to build technology infrastructure on their own. Thus policies to develop public technology infrastructure can build the capabilities that exist elsewhere but need to be imported, adapted, and absorbed in the local economy. Such policies include stimulating demand for technological capabilities through awareness-raising programmes and user-need determination; and building independent sources of such capabilities through learning-by-doing, training consultants and spinning off independent consulting services (Justman and Teubal, 1995, p.277). Institutional change and organisational development are critical to developing technology infrastructure, for example in promoting specific forms of networks through which capabilities can be acquired and disseminated.

National systems of innovation can influence a multitude of technologies simultaneously. But each technology may have it's own "network(s) of agents interacting in each specific technology area under a particular institutional *infrastructure* for the purpose of generating, diffusing, and utilising technology" (Carlsson and Stankiewicz, 1991, p.111). These networks may span across national boundaries, but are still subject to the influence of national culture, institutions, and policies (Dodgson and Bessant, 1996, p.25). "Empirical work...gives strong support to the view that it is the overall system and the quality of interconnections within it which affects successful innovation" (Dodgson and Bessant, 1996, p.54). And "the pattern of technological innovation depends on much more than the behaviour of individual firms." (Archibugi and Michie, 1997, p.291). In addition, there has to be social cohesion for the process to work well, as Lundvall (1999) states:

> today industrial and technology policies must be devised more broadly than has been the case- the societal framework is imperative for the effects of the policy. Learning is necessarily an interactive and socially-embedded process. Without a minimum of social cohesion the capability to master new technologies and new and more flexible forms of social organisation will be weak (Lundvall, 1999, p. 19-20).

Some explain this new approach in orthodox economic terms as an increased focus on "social capital", *i.e.* a recognition that the

ways in which actors organise themselves is important in explaining economic growth and development (World Bank, 1997b).

Policy interventions can address particular weaknesses in networks and support social cohesion. Policy objectives are not just to build firms' capabilities, but to facilitate the whole process of technological innovation by modifying the form or operation of networks (see Table 4.2). This can include considerations of finance, marketing, organisation, training, relationships with customers and suppliers, competitive positioning as well as the relationship between products and processes. Thus, much of the material in Chapters 4 and 5 relates indirectly back to the concepts of national systems of innovation discussed here.

Table 4.2	Typology of good policy practices for National Systems of Innovation (Source: OECD, 1999, p.70.)	
THEME	**POLICY AIM**	**MEANS**
Securing appropriate framework conditions	To develop human resources in S&T. To close market gaps in the financing of innovation	Reforms to post-secondary education. Increased government and industry support to professional education. Establishment of a legal framework for venture capital.
Building an innovation culture	To reduce asymmetry in information. To diffuse best practices in innovation management. To promote the creation of innovative firms.	Internet-based business information network. Funding greater use of benchmarking and diagnostic tools. Public investment in venture capital.
Enhancing technology diffusion	To increase firms' absorptive capacity. To improve linkages between SMEs and public research	Co-financing of consultants to upgrade firms' organisational ability. Co-financing of technology uptake via public/private partnerships
Promoting networking and clustering	To simulate the formation of innovative clusters of firms. To ensure a better match between the S&T infrastructure and industry needs.	Brokering and procurement policies. Competition among regions for funding of cluster initiatives. Co-funding of centres of excellence to facilitate university industry interactions. Building networks between public research actors and firms.
Leveraging research and development	To sustain technological opportunities in the long run. To increase economic return from public research	Increased government spending on basic R&D. Increased public support to R&D. Public-private partnerships. Technology foresight for policy setting. Regulatory reform (university-industry interface).
Responding to globalisation	To increase linkages between domestic and foreign-owned firms. To increase country's attractiveness as a location for knowledge-based activities.	Building networks of competitive domestic firms. Building innovative clusters (see above). Systemic upgrading of the S&T infrastructure
Improving policy making	To enhance policy coordination. To improve policy evaluation	Raising the coordination function to the highest policy level. Making evaluation obligatory. Developing new methodologies

Besides national-level systems of innovation, international (global) and subnational (regional) levels of innovation merit consideration (Freeman, 1999). One might think the trend towards globalisation might make national and regional systems of innovation less significant, but there is a complex interplay between these levels. New information technologies act as a powerful vehicle for the diffusion of information across distant communities. However, the process of generating and diffusing new technologies has been moulded and strengthened by flows of information and capital. Technology has facilitated globalisation and vice versa; technological change is both a factor in globalisation and one of its most important outcomes (Archibugi *et al.*, 1999). Globalisation facilitates transmission of best-practice techniques and is also a vehicle for the international flow of goods and services. Because location-specific advantages are still important to multinational corporations, as they seek market niches with competitive advantages, these corporations have a major influence on national systems of innovation. Globalisation is causing the integration of disparate national systems of innovation that are geographically dispersed and locally specialised (Cantwell, 1999). The role of the local business environment on the competitive advantages of firms was already recognised by Porter (1990).

The significance of the sub-national level (region) as a unit for economic analysis has long been recognised by economic geographers and has also been adopted in mainstream economics (Krugman, 1991; De Zooten-Dartenset, 1999). It is at the regional level that informal links between key personnel are formed and maintained and where economic decisions are most strongly influenced (Howells, 1999). Regional systems of innovation have been identified not to supplant the national systems of innovation approach, but rather to recognise that the national unit may be too large to allow a complete understanding of the dynamics of a technological system (Howells, 1999 after Metcalfe 1995). A concept similar to regional systems of innovation called "learning regions" has been used to elaborate how continuous social and technological learning can be established at the level of sub-national regions (Asheim, 1998; see Box 4.1). In a learning region, organisational (social) and institutional innovations to promote technological cooperation and learning can appear at a regional level rather than just at a company level. Such innovations can also serve as tools for regional development. In November 1999, the UK Government announced regional planning would identify innovative 'high-tech' cluster areas to boost business and encourage growth (Planning, 1999).

Technology transfers, both within a country and betweeen countries, are influenced greatly by national systems of innovation-- the institutional and organisational structures which support technological development and innovation. Governments can build or strengthen scientific and technical educational institutions and modify the form or operation of technology networks: the interrelated organisations generating, diffusing, and utilising technologies. The presence of regional and global systems of innovation interacting with national-level systems also has important implications for policy makers.

4.4 Social Infrastructure: the Need for Participatory Approaches

As mentioned in the chapter's introduction, the re-thinking of established approaches to the role of the state in relation to economic and governance agendas ("too much state" vs. "too little state") took a specific form at international and national levels in relation to development and technology transfer. The historical legacy of development efforts provoked a serious reassessment of the appropriate roles of the community, government and private sector in development and technology transfer.

Following the end of World War II and the period of decolonisation, orthodox views of development argued that "latecomers" in the development process should benefit from modern technologies. However, by the 1970s and 1980s, the effects of an overly simplistic approach to development through industrialisation, and the sometimes-harmful effect of economic and political motives by donor nations became noticeable. Increased unemployment, rural-urban migration, foreign debts, ineffective projects and growing technological dependency brought into question the appropriateness of the technology being transferred to developing countries. But by the 1990s, the changed geopolitical and economic environment finally allowed serious official attention to be given to approaches to development that emphasised the need for balance between state, market and civil organisations. This has had two consequences. First, participatory approaches have been introduced to mediate a better match

---

Box 4.1:   Learning Regions (Asheim, 1998).

The concept of learning regions should be looked upon as a policy framework or model for formulations of partnership-based development strategies. All definitions of the concept emphasise the role played by cooperation and collective learning in regional clusters and networks understood as regional development coalitions. By the concept "development coalition" is meant a bottom-up, horizontally based cooperation between different actors in a local or regional setting, such as workers and managers within firms or in a network of firms, but also generally the mobilisation of resources in a broader societal context as such, to initiate a learning-based process of innovation, change and improvement.

This definition of a learning region underlines the important role of innovation, understood as contextualised social processes of interactive learning, which highlights the significance of building social capital in order to foster cooperation, as well as transcending the artificial divide between high tech and low tech industries.

The concept of learning regions has become very popular with several international organisations, such as the OECD, as a strategy for institutional learning in order to promote regional development, as it at the same time promises growth and employment as well as social cohesion.

DG VI of the European Commission has launched learning regions as a policy instrument; and in Sweden three municipalities have been established as learning regions, understood as development coalitions, as a part of a bottom-up restructuring policy.

between host country requirements and donor assistance, and these have a broad resonance with the development of forms of good governance (OECD, 1997). There has been a massive shift in thinking (but only to a lesser extent in priorities), from things and infrastructure ("technology transfer") to people and capabilities (knowledge and learning; Chambers, 1997). Secondly, the *stated* development policy of governments and multilateral institutions has come to be dominated by what has been called a 'new Policy Agenda', where increasing amounts of aid are channelled to and through NGOs, rather than governments, in recipient countries (Edwards and Hulme, 1995).

4.4.1   The growth of NGOs[1]

Three stages of development in southern community environmental movements have been identified (Doyle and McEachern, 1998). In the sixties' "development" decade the pervasive optimism meant that there was little movement opposition within the countries of the South. In the seventies, some key environmental movements emerged such as the Green Belt Movement in Kenya, the Environment Liaison Centre International, Environment and Development Action in the Third World, and Sahabat Alam Malaysia. These movements fought for "people's" development but did not oppose northern development ideology. More recently, a gap has opened between those groups that work with government and international agencies, and more radical environmental protest movements. These latter movements criticise northern science and technology, the industrial practices of MNCs, national and northern governments and international aid agencies (Doyle and McEachern, 1998).

There is now a vast range and number of NGOs. On one count there were 530 NGOs in 45 countries in the African Environmental Network; 6,000 NGOs in Latin America and the Caribbean; about 12,000 development NGOs in India; 10,000 NGOs in Bangladesh and 18,000 in the Philippines (Princen and Finger, 1994).

In the environmental sphere, NGOs have varying tools and techniques for influencing the processes of development. Where they operate within a corporatist approach, they negotiate and undertake advocacy work for change. Where there is accountability, political leverage can be brought to bear. For example, commercial groups whose policies impact adversely on the environment can be vulnerable to consumer boycotts organised by NGOs (Potter and Taylor, 1996). In some cases, particularly in Latin America, groups are working on problems such as urban air pollution and have conducted successful consumer focused campaigns (Oviedo and Bossano, 1996). There may also be direct confrontation and opposition which can lead to projects being abandoned (Hirsch and Lohman, 1989). And there can also be negotiation to reach a consensus after confrontation. However, despite their increased roles and visibility, non-specialised NGOs are commonly thought to suffer from four sets of weaknesses:

---

[1]: See also section 2.2.7 in Chapter 2 on the financial flows stemming from NGOs.

limited technical capacity, limited scale, limited strategic capacity, and limited managerial capacity, so that capacity strengthening may be required (Chen, 1996).

Around the climate change issue, there is an increasing momentum for the development of organisations of "business" NGOs, that is groupings of businesses that support action in support of climate change and see commercial advantages in being there first. Two umbrella business groups have already started to develop networks. The Business Council for Sustainable Energy is working with partners in Latin America and Asia, while the Business Council for Sustainable Development is operating in Latin America (BCSD-LA) in Box 4.2).

BOX 4.2	THE BUSINESS COUNCIL FOR SUSTAINABLE DEVELOPMENT - LATIN AMERICA (BCSD-LA) *(SOURCE: SEPTEMBER 1999- MONTHLY NEWSLETTER PUBLISHED BY THE BCSD-LA ECO- EFFICIENCY PROGRAMME)*

The BCSD-LA has been operating for one year and has already compiled a book: "Global Climate Change - Foundations for Business and Strategy and Practice in Latin America". It includes a Latin American position towards the climate change problem, with the support of 410 companies. The document has support from 15 country organisations within the BCSD-LA network.

"This book also presents a collection of case studies documenting strategies implemented by Latin American industry for minimising the effects of climate change. This is proof that the commitment taken on by the Latin American private sector in 1998 is already providing real, tangible results. With this the Latin American private sector is once again setting a new and important precedent in the world, by showing its leadership and response strategy to the tremendous challenge presented by climate change, without waiting for the FCCC to require it."

4.4.2   Developing partnerships through participation

Participatory approaches have been shown to improve the quality, effectiveness, and sustainability of development projects, and strengthen ownership and commitment of government and stakeholders (World Bank, 1994a). For other (non-climate related) sustainable development policy objectives, there has been considerable experience with relevant climate mitigation technologies, ranging from large-scale electricity generating projects to small scale renewable technologies, to enhance energy security and rural economic development, and projects designed to promote energy efficiency. Technologies for adaptation to climate change impacts are required across all sectors, with the main ones being housing and settlement, health, agriculture, coastal zones, and forestry. From these projects, there is sufficient evidence to suggest that enhanced participation could improve the technology transfer system on climate-relevant technologies and indeed expand understanding of the way the system works.

Effective cooperation on climate-relevant technologies has been recognised to be dependent on being based on the actual need of the recipient country and adaptation to local circumstances through involvement of local stakeholders (OECD, 1998a; see also Chapter 5 on partnerships for technology transfer). It is therefore possible to make linkages between the increased use of participatory approaches and to improving technology transfer at all stages in the three pathways (government-driven, private-sector-driven, and community-driven) through participatory approaches.

4.4.3 How participation has developed

Participation received new attention as an explicit goal in development assistance in the late 1980s, due to increased emphasis on the sustainability of project benefits, institutional development and policy reform. Participation was initially promoted and applied mainly by NGOs and in small scale community development projects, then increasingly by multilateral organisations such as the FAO, ILO and UNRISD, as well as some bilateral agencies (OECD, 1997). Agenda 21 contained an innovative section on participation and responsibility which talked of the need for a social partnership to build environmental and economic security. Most of the innovations and accomplishments relating to participatory research and development have emerged from 'the third sector' (private organisations which are neither profit making nor affiliated to political parties) (Thompson, 1998). While these organisations are themselves attempting to spread and scale up their successes, they themselves recognise that experience is very limited: as Chambers points out, "*And, as usual with concepts which gain currency, rhetoric has run far, far ahead of understanding let alone practice.*" (Chambers, 1997). Mainstreaming of participation into operations - *i.e.* dialogue, planning, implementation, monitoring and evaluation of development activities- is still a major task ahead for most donors. In-country, public sector organisations are also taking an increasing interest in participatory approaches, for a variety of reasons connected with political and economic challenges in the public sector (Thompson, 1998). Consumer-oriented companies will engage in a quasi-consultation process through various business tools such as market research and focus groups. And, the involvement of business groups, trade associations and chambers of commerce can be a valuable way of disseminating a technology through a business community.

4.4.4 What is participation and why is it needed?

Participation is a process of complex social change (OECD, 1997). Definitions of participation have thus been developing along with the practice of it. Different dimensions and levels, degrees or types of participation can be analytically distinguished (Rudquist, 1987, in OECD, 1997). The terms are used differently depending as to where in the project cycle participation occurs (planning, implementation, monitoring and evaluation, takeover), as to the quality, intensity or extent of participation (as pas-

sive beneficiaries, informants, cost sharers, or as colleagues or counterparts with a voice in management, decision-making and control) and to societal levels (local, regional, national) (OECD, 1997). Current 'inclusive' approaches have several stages if they relate to all stages of a project cycle. Basically, participatory development stands upon a partnership that is built upon the basis of a dialogue among the various stakeholders. These stakeholders collaboratively:

1. Conduct the analysis and diagnosis at the outset;
2. Decide what is needed and set objectives;
3. Decide directions, priorities and institutional responsibilities to create a strategy, and
4. Oversee development of specifications, budgets and technologies to move from the present into the future, and formulate project tactics (World Bank, 1996; OECD, 1997).

Participation of the main stakeholders in the assessment stages can help establish a process that will produce a technology selection better matched to local needs:

- The current processes of technology selection often work against involvement and consultation of local communities. Social anthropological enquiry has long stressed the diversity and multiplicity of knowledge particularly related to the exploration of 'emics', *i.e.* indigenous concepts and categories, as opposed to 'etics', outsiders perspectives on how things are. This focus has become central to technological development, as it is important to comprehend how people themselves understand whatever technical issues are targeted for assistance (Fairbanks, 1992).

- Climate change-related problems may be perceived and defined differently by different social actors (Gupta, 1997). Solutions to these problems, including technology transfer efforts, should therefore be in keeping with community perceptions of local problems and should draw on local knowledge. There can be no universally applicable solutions. This approach is consistent with the viewpoint that there is a need for 'public interest science' and makes it "a factor located within environmental conflicts'. (Shiva and Bandyopadhyay, 1986, p. 87; cf. Grove-White *et al.*, 1992). Where problems involve high stakes, and are based on scientific uncertainty, as local climate change impacts undoubtedly are, and to which local responses are necessary, research into stakeholder perspectives is advocated in order to determine the most appropriate approach (Functovicz *et al.*, 1996).

- There are many technical options to reduce emissions of greenhouse gases (GHGs) at a reasonable economic cost. No single technology, nor small set of technical options, offers a "solution" to solving the climate change "problem" (Lashof and Tirpak, 1990). Because options available are so many and straddle so many different sectors of the economy, generalisations about how technologies are selected are difficult to make. The selection criteria for the technology depend upon, inter alia, the end product or service, the natural resources available at a site, the sources of financing and, of course, the cost of the technology for a particular application at a particular location. One generalisation appears to be safe—technology choices are value-laden rather than neutral (Madu and Jacob, 1989; Reddy, 1976).

- The source of financing has a bearing on the selection of a technology. Governments in developing countries shift budgetary allocations based on what is available from bilateral and multilateral sources. As discussed below, donors are biased in favour of companies based in their own countries. This could tend to favour certain technologies, sometimes even ESTs such as wind turbines. Also, multilateral assistance agencies prefer technologies that have a proven commercial track record. Their procurement policies often preclude support for the acquisition of the most advanced technologies.

4.4.5   *What can participation achieve?*

In relation to the technology transfer discussion, participation can be seen as essentially the opposite to top-down, external-expert determined approaches, whether at international or national levels. The process is broader than consultation and is usually dependent on capacity-building (see section 4.5) to be effective.

In many projects participation has taken place during implementation, but more seldom in project formulation, management, control over resources and distribution of benefits. Conversely, large MDBs have only recently made a commitment to supporting participatory approaches and their experience is mostly in the planning stages (World Bank, 1996). Participatory rural appraisal took off in India due to the involvement of senior government staff (Blackburn and Holland, 1998). It has been found to be a crucial tool to engage the 'main protagonists' rather than regarding them as 'targeted beneficiaries' and thus increased success (Malhotra *et al.*, 1998). Frequently, small-scale, community-specific projects aiming at social objectives are a common form of donor assistance to the promotion of participation.

Systematic evaluations of measurement of the costs and benefits of participation are scarce, but generally indicate that the costs, in terms of time and money spent, tend to be relatively higher for participatory projects in the course of their early phases (OECD, 1997). The initial investments in participation, however, tend to pay off in terms of increased efficiency and sustainability, and in saving time in subsequent phases.

Furthermore, it is widely recognised that people and civil society are key players to maintain long-term efforts on anti-corruption programmes (Heiman and Zucker Boswell, 1998;

Klitgaard, 1998). Public awareness campaigns can focus on the harm done by corruption, the misuse of public money, denied access to public services, and the public duty to complain when public officials act corruptly. Such campaigns can empower civil organisations to monitor, detect and reverse the activities of the public officials in their midst, by drawing on the expertise of accountants, lawyers, academics, non-governmental organisations, the private sector, religious leaders and ordinary citizens (Kindra, 1998).

There are a number of tools and methods which are used within participatory approaches depending on what stage participation is being used and whether poor or powerful stakeholders are being engaged. Table 5.2 summarises the main methods used.

Participation can thus achieve:

- Better choices and identification of possibilities and opportunities in local systems
- Better commitment to projects which improves implementation and sustainability
- Opportunities to negotiate conflicts
- Empowerment which raises awareness about the need for stakeholders to achieve solutions themselves.
- Access to additional resources for the project raised from the target project beneficiaries through payments, and time.

Some recent practice with participatory methods has shown a range of benefits in climate relevant technology (see Table 4.3).

Problems Arising with Participation
There are a number of problems that arise with participation:

- The commitment of considerable time and resources may not be available through completion of the project. Needs can also change. Over time stakeholders become capable of demanding and paying for goods and services from government and private sector agencies, and at this stage it is necessary to move from welfare-oriented approaches and to focus on building sustainable, market-based based financial systems and to strengthen local institutions. Capacity building for participation is essential and cannot be achieved over short time scales (Harkes, 1998).

- Considerable efforts are needed by all stakeholders to participate. The use of NGOs as intermediaries may have its own set of problems. The NGO label includes widely different kinds of organisations, some of which are working in a participatory manner while others do not. Additionally, NGO representativeness and accountability often remains unclear.

- There are constraints at a policy level, which impinge on the effectiveness of participation in relation to peoples' rights and access to information, and there may be resistance within bureaucracies to work in innovative ways.

Table 4.3 Methods and Tools for Participatory Development (Source: World Bank, 1996)	COMMENTS
COLLABORATIVE DECISION-MAKING	
Appreciation- Influence-Control (AIC). AIC is a work-shop based technique that encourages stakeholders to consider the social, political, and cultural f actors along with technical and economic aspects that influence a given project or policy.	• Stakeholders establish working relationships • Promotes ownership
Objectives-oriented project planning (ZOPP). The main purpose of ZOPP is to undertake participatory, objectives-oriented- planning that spans the life of the project or policy work, including implementation and monitoring, while building stakeholder team commitment and capacity with a series of workshops.	• Stakeholders establish rules of the game • Support may be needed for non-experienced participants
Team-Up. Team-Up builds on ZOPP but emphasises team building, it uses a computer software package (PC/Team-UP) . It enables teams to undertake participatory objectives-oriented planning and action while fostering a "learning-by-doing " atmosphere.	
COLLECTIVE DECISIONMAKING: COMMUNITY-BASED METHODS	
Participatory Rural Appraisal (PRA). PRA is a label given to a growing family of participatory approaches and methods that emphasis local knowledge and enable local people to do their own appraisal, analysis, and planning. PRA uses group animation and exercises to facilitate information sharing, analysis and action among stakeholders.	• Demystifies research and planning processes by drawing on everyday experience • Participants feel empowered by their participation.
SARAR. SARAR promotes five attributes: self-esteem, associative strengths, resourcefulness, action planning and responsibility. Its purpose is to (a) provide a multisectoral, multi-level approach to team building through training (b) encourage participants to learn from local experience rather than from external experts and (c) empower people at the community and agency levels to initiate action.	• Local communities have to be provided with decision-making authority in the project and/ or involvement in project management.
METHODS FOR STAKEHOLDER CONSULTATION	
Beneficiary Assessment (BA). BA is a systematic investigation of the perceptions of the poor and hard to reach beneficiaries, thereby highlighting constraints to beneficiary participation and (b) obtain feedback on development interventions.	• Field based needing time for regular consultation and interactions.
METHODS FOR SOCIAL ANALYSIS	
Social Assessment (SA). SA is the systematic assessment of the social processes and factors that affect development impacts and results. Objectives of SA are to (a) identify key stakeholders and establish the appropriate framework for their participation; (b) ensure that project objectives and incentives for change are appropriate and acceptable to beneficiaries, (c) assess social impacts and risks; and (d) minimise or mitigate adverse impacts.	• Provides a process for building information from local communities into plans and plans into action. • Need focused data analysis and needs experienced local consultants.
Gender Analysis (GA). GA focuses on understanding and documenting the differences in gender roles, activities, needs and opportunities in a given context.	

• There may be opposition from some stakeholders who stand to gain if alternative courses of action are pursued. To overcome this the active engagement of players keen for the process and project to succeed may be necessary by the project leaders. The participation of multiple stakeholders and social groups require negotiation processes that are difficult to influence and require a strong institutional focus (Neefjes, 1998). This can mean that participatory, process driven approaches re-create complex institutional structures. "Complexity itself can be the enemy of sustainability" (OECD, 1997).

• There may not always be a happy ending. Some stakeholders may withdraw if their views do not get general agreement. Parts of the project may need to be dropped in order for success in other areas (World Bank, 1996).

• The practical experience is rather limited. At a technical and procedural level there is a need to further develop evaluation frameworks, procedures and indicators that better accommodate participation processes. There is a key role for multilateral agencies. Further development of operational guidelines, evaluation frameworks, methods, procedures and indicators that better accommodate participatory processes are needed.

4.4.6 Participation and Climate-Relevant Technology transfer: Current discussions

Participatory development is now widely recognised as a way of achieving technology transfer at all levels of development endeavour. This has grown from a perceived need to move from donor driven technology transfer to national needs driven approaches. However, such a shift in approach is hampered by a lack of international consensus on what actions any particular country should take; how the international community should share the task of providing the resources for such actions; and, the disconnection in some developing countries between domestic politics and their stance in international fora. There is also a second bottleneck: the gap that exists between international agreement and domestic consensus in developing countries that is meant to follow international agreement (Gupta, 1997). Various of the work tasks initiated by the Subsidiary Body for Scientific and Technological Advice (SBSTA) reflect the perceived need of developing country Parties to benefit from capacity building so they are better able to identify their needs for technology transfer (for example, FCCC/SBSTA/1998/INF2, INF.5). In some ways the Technology Cooperation Agreements of the IEA are an attempt to build collaborative working relationships through a participatory forum of all significant actors with a role in technology cooperation activities relevant to the application of ESTs in a particular country (OECD, 1998b).

It has been suggested that if the FCCC wants to generate broader support, it may need to assist the Governments of developing countries to widen the base of domestic consensus among national stakeholders, investigating and responding to their needs (Van Berkel *et al.*, 1997). To help address the challenges, countries are encouraged to engage organised stakeholders representing different societal interests for the priority sectors into a capacity building process. Three "pillars" are proposed: creating an enabling environment for stakeholders' participation; implementing and evaluating mitigation and adaptation actions; and assessing mitigation and adaptation needs and opportunities. These all link to a roundtable process which engages stakeholders who represent different societal interests to produce and implement, in an organised way, a prioritised national action plan with wide buy-in (Van Berkel and Arkesteijn, 1998). Also, assistance is required for the organisation and facilitation of the participatory process that will result in sound proposals for climate relevant technology transfer and collaboration.

Table 4.4	Interest Groups
INTEREST GROUP	**DEFINITION**
Seller interest	Interests of the person selling the technology
Buyer interest	Interests of the purchaser of the technology
Third party supporter interest	Interests of those parties supporting the sale-purchase of the technology
Third party victim interest	Interests of those parties negatively affected by the sale-purchase of the technology
Payer interest	Interests of those parties that have a role in financing the transaction

4.5   Human and Institutional Capacities[2]

Increasing the use of participatory techniques also involves the strengthening of human and institutional capacities. Indeed the building, developing and strengthening of institutions' and people's skills is at the heart of making most dimensions of the enabling environment work better – to achieve technology transfer.

Several terms are in use: capacity building, capacity development and capacity strengthening, depending on whether the need is for creation, reform, or support of activities and structures. The term 'capacity strengthening' has been advocated because more often than not, organisations and institutions already exist and that capacity can often be increased more effectively by reinforcing existing structures rather than by building new ones (*et al.*, 1997. 'Capacity building' has also been criticised as implying an engineering approach to the creation of new capacity (OECD, 1995). However, as the term 'capacity-building' is now being used within the UNFCCC, it will be adopted here. Whilst there are linkages between capacity building for promoting the transfer of ESTs for mitigation and the adoption of technologies for adaptation, the emphases are different so they will be considered

separately here. Principally, the difference is that while there has been some experience with the transfer of energy efficient technologies and renewable energy technologies, principally for development objectives (so reasons for 'failure' are known), there is comparatively little experience with the assessment and transfer of adaptation technologies. In part this is due to the absence of systematic impact and vulnerability assessment which few countries, developed or developing, have undertaken.

4.5.1   *Capacity-building for mitigation - an overview*

There are many failures of technology transfer that result from an absence of human and institutional capacity. For example, inappropriate choice of technology can result from missing capabilities for searching, selecting and negotiating. Joint ventures may not produce long-term learning or absorption of technology or development of competence if capabilities for learning and integrating are missing. Missing implementation capabilities may result in unsuccessful employment of purchased technologies and unforeseen problems. Capacity is needed to assess, select, import, develop and adapt appropriate technologies. When existing technologies approach the end of their lifetime, replacement technology may require active market research for identifying and evaluating new technologies, which is generally limited because of the information costs and lack of indigenous capacity (Hoffman and Garvin, 1990).

Much of the focus on capacity building has been on enhancing scientific and technical skills, capabilities, and institutions in developing countries as a pre-condition for assessing, adapting, managing, and developing technologies (UNCTAD, 1995). But the need for enhanced skills and capabilities can also occur in the areas of technology selection, financing, marketing, maintenance, service, information dissemination, utility regulation, policy development, technology transfer, market intermediation, tax policies, macroeconomic policies, and property rights. Many studies acknowledge that capacity building needs vary greatly from country to country, and stress that case studies and other types of analyses should assess the needs of particular countries (see Mugabe, 1996; UNCSD, 1996; Van Berkel *et al.*, 1996).

4.5.2   *Capacity-building for assessment, agreement and implementation stages*

Many ways of developing capabilities for the assessment, agreement, and implementation stages of technology transfer are suggested by development experience: (1) formal training of employees, (2) technological gatekeeping, by keeping informed of technical literature, and forming links with other enterprises, professional and trade organisations, and research institutions; (3) learning-by-doing operational experience such as through twinning arrangements with other organisations. The experience with implementing the Montreal Protocol provides a useful example for capacity building within enterprises (see Chapters 3 and 5);

---

[2] Cases from the Case Studies Section, Chapter 16, that bring out the importance of institutional arrangements as an enabling environment include: DSM in Thailand (case 10), TIFAC (case 27).

a multilateral fund established by the treaty supports training, research and network building.

More fundamentally, there may also be a need to develop capacity to create a need for ESTs. This can be undertaken by developing a roundtable process, a national dialogue on climate change issues of organised stakeholders representing priority sectors and various private as well as public interests to consider mitigation and adaptation options (Van Berkel *et al.*, 1997).

Capacity for negotiation by recipient countries at the agreement stage can overcome some of the biases inherent in development assistance. Donors are often biased in favour of companies and technologies based in their own countries. International capital provided under preferential terms and conditions is often most important for financing technology transfer (Van Berkel and Arkestein, 1998). Multilateral assistance agencies prefer technologies that have a proven commercial track record. Their procurement policies, which encourage competitive bidding, often preclude support for the acquisition of the most advanced technologies that might be environmentally sounder, but carry additional technical risks. A survey of technology and technology needs carried out on behalf on SBSTA among non-Annex II Parties found that specific national and mitigation needs still have to be assessed in detail (Van Berkel and Arkestein, 1998).

There is also a need for increased publicity and awareness among consumers on the subject of energy efficiency. This becomes very relevant when considering domestic and commercial consumers, for whom there has been a felt need for large-scale awareness programmes as well as for demonstration projects. Consumer education could also be considered a form of capacity building that makes consumers more able to make intelligent purchases and investments that save them money. One example is the World Bank's Lithuania Energy Efficiency/Housing Pilot Project, in which the World Bank provides credit to condominium associations to improve the energy efficiency of their residential buildings (World Bank, 1996). The project provides extensive capacity building to allow homeowners to understand the technologies, costs and benefits, how to finance with credit, and the contracting mechanisms for implementing investments with private-sector firms.

4.5.3 Capacity-building for evaluation and adjustment, and replication stages

Too much policy attention has historically focused on market failures that prevent firms from providing adequate investment in research and development. But policy also needs to focus on the enhancement of competitive performance of firms and the promotion of structural economic change that allows firms to innovate and adopt new technologies. The OECD (1994) said: "technology policy has traditionally focused on the innovation end of the process....this approach has slowly been complemented by a parallel concern for an economic environment conducive to the diffusion of innovations... Policy needs to move towards recognising that, rather than two distinct activities, innovation and diffusion are two facets of the same process. Developing firms' ability to absorb and use new technology effectively also improves their ability to develop innovations themselves."

It is also important to consider the positive interrelationship of technology and innovation to create capacity for *autonomous* development. There is a widespread tendency to think of technology transfer in minimalist supply side terms—that the initial choice and acquisition of technology is the only critical factor (Brooks, 1995). Conceived this way, it leads to technology transfer's characterisation as a costly process that also may contribute to a perpetuation of technological underdevelopment, stagnation and dependency (Chantramondklasri, 1990). Newer, broader conceptions of technology transfer see it as a process of incremental and cumulative learning by which the results of the initial choice are internalised. Like R&D, it is an essential component of what is described as social learning (Brooks, 1995).

It is increasingly accepted that innovation is the single most important source of long-term economic growth in all countries (STAP, 1996) and that small incremental innovations are even more important to economic success in developing countries than in developed countries (Brooks, 1995). The objective of technology transfer should be to foster technological innovation in recipient firms so that not only do they master new processes, but also have the technical capability to generate improved processes and products (Stobaugh and Wells, 1984). This will depend not only on vendors but also necessarily and largely on the active technological behaviour of recipient firms (Chantramondklasri, 1990).

Despite the central importance of capability for technological innovation, the capacity to innovate and then replicate is poorly developed in developing countries (STAP, 1996). There are several factors accounting for this. First, leaders there are rarely interested in nurturing the development of the capacity to innovate. Second, even when R&D capacity exists, close involvement of industry is rare. Thirdly, bilateral and multilateral assistance targets only those technologies that have a proven track record in already industrialised countries. Moreover, they too provide little assistance aimed at nurturing the capacity to innovate (STAP, 1996).

The engineering and management skills required in acquiring the capacity to optimise and innovate are not trivial. Various kinds of high quality training are needed to embody in personnel of the receiving firm the skills, knowledge and expertise applicable to particular products and processes. Such training, both generic and specific, should be an important part of any technology transfer package and deliberately planned as a learning vehicle for the work force of the recipient firm (Imai, 1994). The transfer should not only be of specific know-how, but also of related systemic knowledge of the relevant technologies so that recipients can add value. This is an important consideration for developing countries, because it implies that the work force must experience continual cumulative learning, both from experience and formal training, in order to remain competitive in a world market where intense continual incremental improvement ("kaizen") is increasingly essential to sustained competitiveness (Brooks, 1995). In those regions of the developing world where existing capabilities are weak in specific technology areas, basic level of technological capability should be built via the establishment of regional institutes that provide training in the fundamentals of technology assessment and management. This approach can produce enabling capacity in a sustainable way and is already being explored within the FCCC process.

4.5.4 Capacity building: the needs in both developed and developing countries

While there is general agreement that capacity building is important, the record of capacity building in practice so far is mixed. There are many examples of attempts at capacity building that have failed (UNDP/HIID, 1996; UNDP, 1997; ODA, 1994), and fewer successes to chronicle (for some successful examples see for example Ch16 Case Study 5 "PV Solar Home Systems in Kenya" and Case study 16 "the Baltic Biomass Boilers". Part of the problem is that capacity building is neither easy nor quick:

> There is a great deal of uncertainty about precisely what capacities are needed and how they are developed.... Unfortunately, experience suggests that the necessary competencies can, at best, only be improved slowly, and that many of the requirements are cumulative, and involve tacit and uncodified knowledge that is difficult to purchase on the international market (Barnett, 1995, pp.15-16).

More often the donor agency's understanding or concepts of capacity development/building have frequently been understood to be that institutional and human capacity changes can be induced, that these capacities are weak and the human and institutional capacities can be developed or strengthened along the lines of management and organisational models of the donor countries. The experiences and results of over three decades of capacity and institutional building in sub-Saharan Africa, for instance, suggest that these assumptions or concepts do not necessarily hold (Berg, 1993). Increasingly, there is a consensus amongst both analysts and participants that: local involvement, in the form of traditional institutions, local organisations and individuals is critical to increasing national capacity. Further, less controlled styles of governing are required; central institutions need to be reorganised with attention given to creating space for them; and, local NGOs need institutional development (OECD, 1995). There is a need for extensive exchanges of experiences by the collaborating agencies. New relationships between donors and local actors are required (OECD, 1995).

Because much of donor understanding about institutional issues comes from *the lessons of failure* over the last three decades rather than success, more investment needs to be made in upgrading donor understanding of capacity development issues (OECD, 1995). Capacity strengthening is thus needed in developed countries. Due to low success rates in transfer of technology, implementation of programmes and sustainability, some donor and MDB programmes have undertaken capacity-strengthening in their own organisation to develop a more participative and less traditional approach (Wight, 1997). "Given the entrenched institutional bureaucracy and resistance to change in donor organisation, however, this (process facilitation) would require a major orientation and concerted change effort" (Wight, 1997). Local staff recruitment for projects in particular has been identified as a major issue (Gray, 1997)

4.5.5 Capacity building for adaptation

Increasing attention has been paid to the need to develop measures to cope with the climate change already contained within systems. How far such measures will need to go depends on: the magnitude and extent of climate change; the rates of accumulation of greenhouse gases; and, the level of stabilisation of the atmosphere to be achieved. Experiences with recent climate variability have given urgency to the adaptation agenda as it moves to its second stage in the FCCC process, particularly for the least developed countries and the small island states (SIS) amongst them because of their vulnerability. The SIS called for an acceleration of efforts at the UNGASS in October 1999. Few developing countries (or developed countries) have yet fully assessed their vulnerability to climate change and developed adaptation strategies.

Comparatively little consideration has been given in a systematic way to what capacity building is required for Annex II countries for adaptation. There are well developed methodologies as to how to assess the impacts from and potential adaptations to climate change (see, for example, Parry and Carter, 1998) but these may not be appropriate as a basis for assessing priorities for action by national governments of developing countries without modification and interpretation. The methods were originally developed for assessments of discrete climate impacts, not how to cope with future climate change at the local to regional level. And, the perspective of a national policymaker with limited resources, seeking to set priorities for research and then action, is inadequately covered. Existing approaches for evaluating adaptation often are methods and models designed for specific sectors or complex integrated assessment models (see, for example, UNEP and IVM, 1998). Skills, time and resources may not be available to apply such techniques and they may not produce required outputs. There may be too many players in 'traditional' studies: sponsors, researchers and the target community to produce an integrated package without coordination. Capacity-building, including substantive changes to social structures or planning paradigms, is often ignored in favour of evaluating discrete options, such as the cost-effectiveness of a higher sea wall. Weaknesses of these approaches have already been recognised in some national initiatives such as the UK Climate Impacts Programme, where emphasis is placed on facilitating stakeholder engagement to ensure policymakers get useful outputs.

The Programmes sponsored in host countries, such as the US Country studies programme, the UNEP/GEF Country case studies on Climate Change Impacts and Adaptations Assessment and the Netherlands Climate Change Assistance Programme have started work with many countries on impacts and adaptation assessment and in the case of the US Country studies programme, already provided significant capacity building for professionals and institutions (Dixon, 1998). In addition, UNDP's Support Programme for National Communications is starting work on vulnerability and adaptation assessments. In recognition of the gaps that exist, the SBSTA has asked the UNFCCC Secretariat to continue to develop its website relating to decision tools, models and methodologies to evaluate impacts and adaptation strategies (FCCC/SBSTA/1999/L.12). Within the adaptation rather than mitigation agenda, it would seem there is particular need for more support to ensure the scientific datasets, tools and skills are readily available in-country, for the undertaking of impact assessments, particularly climatological data and regional climate models. Impact studies are the basic building blocks for the identification of vulnerability and to enable the formulation of adaptation options. There may also be a need to establish internationally agreed approaches if funds are to be disbursed in the future on the basis of vulnerability indices.

4.5.6 Capacity Building: Current issues

The centrality and significance of the 'capacity-building' agenda has become recognised by the G77 and China group within the UNFCCC, which sponsored a decision for consideration at CoP5 (FCCC/SB/1999/CRP.9), which related to all the CoP4 decisions and was agreed with some modifications. If action follows at CoP6, and there is now a parallel text covering the countries with economies in transition (CEITs), it will provide an integrated framework on capacity building covering all dimensions of the Convention, not just technology transfer. The focus of required action is within the developing countries suggesting: the strengthening of national (and regional) focal points to cover training and human resources development; research activities; expertise on specialised aspects such as information technology; and a networking system between these components. The concept effectively is to develop National Systems of Innovation identified in section 4.3.

The need to place emphasis on strengthening capacity within developing countries to their agenda has been recognised by the OECD (1995).

> ... Capacity issues in all states including those in the Southern countries are embedded in political, cultural, and social dynamics of enormous complexity, a good number of which are likely to be beyond the understanding of the donor community. Raising the environmental performance of organisations and institutions in any society is a daunting task even for its own citizens. Assuming this can be done easily by outside interveners may be the first mistake in any capacity development programme (OECD, 1995, p.10).

Capacity-building is required at all stages in the process of technology transfer. It is a slow and complex process to which long-term commitments must be made for resources and to which the host country must also be committed if results are to bear fruit. Fundamental change requires an autonomous capacity to innovate, acquire and adapt technologies. For the mitigation agenda, there are specific needs for additional resources at the assessment and replication stages: (a) to ensure there is a broad national commitment to ESTs and that those most appropriate for national circumstances are selected; and (b) to ensure that indigenous capacity for ongoing innovation is developed to encourage North-north, South-North and South-South flows. For adaptation there is a need a) to strengthen scientific expertise and institutions capable of undertaking the relevant assessments and to promote linkages with this infrastructure with other parts of the public and private sectors; and b) to identify relevant tools and techniques to produce outputs for nationally determined priorities.

4.6 Macroeconomic Policy Framework

Barriers to technology transfer resulting from macroeconomic conditions include lack of access to capital, a poorly developed banking sector, lack of available long-term capital, high or uncertain inflation or interest rates, subsidised or average-cost (rather than marginal-cost) prices for energy, high import duties, uncertain stability of tax and tariff policies, investment risk (real and perceived), and risk of expropriation. Policy tools related to macroeconomic conditions to encourage foreign direct investment (FDI) and access to technology can include:

- Macroeconomic stability, such as low inflation and stable and realistic currency
- Deregulation of the investment regime, and free movement of private capital
- Foreign exchange convertibility and liberalisation of exchange restrictions
- Removal of (often prohibitive) restrictions on repatriation of profits and of capital
- Reduction of risk of expropriation
- Reduction of the role of the public sector in directly productive sectors, through privatisation of state enterprises and overall reduction of the share of state enterprises in total investment, by opening up the public utility sector and other public monopolies to private sector participation and foreign investment
- Provisions for the settlement of disputes ranging from direct negotiation among the disputing parties to third party arbitration (World Bank, 1997a)
- Removal of mandated local ownership requirements
- Promotion of the development of domestic institutional investors to assuage public fears about excessive foreign presence and to reduce the vulnerability of domestic capital markets to foreign investor herding. The presence of domestic institutional investors also reassures foreign investors about the host country's respect for corporate governance and property rights (World Bank 1997a)

- Reform of opaque regulations that leave much administrative discretion and scope for corruption which discourages investment flows (Shang-Jin Wei 1997).
- Streamlining and reduction of tax rates to competitive levels. Low tax rates have been found associated with high levels of direct foreign investment (Hines and Rice, 1994).
- Mobilisation of domestic resources through the gradual reduction of environmentally-damaging subsidies (Panayotou, 1997).

For the buildings, industrial, and energy sectors, the low price of conventional energy is a barrier that deters investment in alternatives. Different forms of levy to 'internalise' the environmental costs of fossil fuel use are now being tried to improve the competitiveness of the cleaner energy sources (Dale 1995), but this may increase perceived policy vulnerability. If the new technology is saving on scarce inputs or substitutes for dirty fuels which are subsidised, the technology may not be locally attractive even if it is globally preferable. Pricing of both the new technology and its complements and substitutes is just as important . The channels may be open but effective demand may be lacking because of inefficient pricing or other domestic distortions (Dale 1995).

The lack of financing for technology transfer can reflect the lack of capacity on the part of the financial sector in developing countries (Manas 1990). Further efforts are needed to convince banks and other lending institutions of the profitability of financing environmentally sound technologies and projects. Innovative financing mechanisms that allow reduced risks to lenders but allow profitable investments in environmentally sound technologies are one of the main impediments for greater energy efficiency in most of the developing countries (Pachauri and Bhandari, 1994). Most lending activities in developing countries are geared to corporate finance, not project finance.

The banking system plays a dominant role in the allocation of capital, and its health largely determines whether a country will be able to exploit the benefits of financial integration, including access to and transfer of technology. Banking systems in many countries exhibit characteristics that can adversely affect long-term investment, imports of capital goods and technology transfer, such as:

- Excessive banking regulation or inadequate banking supervision
- Incentives for banks that are distorted against risk taking
- Banks that are poorly capitalised
- Inadequate regulation and oversight
- Incentive distortions (*e.g.* excessive insurance)
- Surges in capital flows that expose the banking system to risk and vulnerability

These problems can be addressed through macroeconomic policies, increased reserve requirements and/or adoption of risk-weighted capital adequacy requirements (World Bank 1997a). It is necessary to build shock absorbers into the financial system and develop mechanisms to respond to instability by:

- Maintaining international reserves at levels commensurate with the variation of capital account; the lower the confidence of foreign investors, the larger the reserves need to be; they could also be
- Buttressed through contingent lines of credit (*e.g.* Argentina)
- Maintaining fiscal flexibility
- Building cushions in the banking system by using periods of credit booms to increase bank capitalisation.

Most modern economies comprise a mix between government and private sector investments which varies according to the nature of the economic system and circumstances, and it has varied over time. Traditionally governments, as well as their role in social infrastructure (education, health) have dominated investment in physical infrastructure and large-scale technology development (Rama 1997). In mixed economies, government finance has dominated where the benefits are not readily in the form of financial returns, or where such returns are likely to be very long term (as with technology development) (World Bank 1997a). In addition, governments have traditionally been heavily involved in sectors that are perceived as central to national economic security and development - such as the provision of energy. Private sector finance has been more readily involved in businesses and projects that provide a ready financial return over shorter timescales than would be required for government investment (Jacobson and Tarr, 1995).

The relationship between government and private finance has, however, changed considerably in recent years in many countries. Most relevant to climate change, this has been particularly clear in the energy sector. Faced with the rising costs and inadequate performance of state-funded energy developments, many governments around the world have privatised energy industries, are in the process of doing so, or are seeking to involve private finance even where energy systems remain primarily government owned (Dailami and Klein, 1998). For example, governments may set out terms of access to independent power generators. This has profound implications for financing the development and international transfer of clean energy technologies, and makes it all the more important to understand the operation of private finance and its relationship to the public sector in the funding of technology transfer (Blomstrom *et al.*, 1995).

4.7   Sustainable Markets for Environmentally Sound Technologies

Experience with development assistance for renewable energy projects in developing countries over the past three decades illustrates the importance of sustainable market approaches to technology transfer. In the 1970s and 1980s, development assistance agencies attempted to transfer many small-scale renewable-energy technologies like biogas, cooking stoves, wind turbines, and solar heaters. Many projects were considered failures because of poor technical performance, lack of attention to user needs and local conditions, and lack of replication of the original projects. Projects emphasised one-time technology demonstrations that failed to understand or provide incentive structures, failed to demonstrate institutional and commercial viability, failed to account for continuing maintenance requirements, failed to create a maintenance and service infrastructure, and in general failed to generate sustainable markets for the technologies demonstrated (Kozloff and Shobowale, 1994; Barnett, 1990; Hurst, 1990; Foley, 1993; Goldemberg and Johansson, 1995; GTZ, 1995).

These failures identify the need for an approach that promotes replicable, ongoing technology transfers rather than one-time transfers. Such an approach has been labelled a market transformation approach. In such an approach, public policy can consider what are the set of institutions which underlie markets and what are appropriate public interventions to shape those institutions. Many view the development process in a market-oriented context, in which technology transfer is intertwined with development assistance aimed at promoting functioning domestic commercial markets, including domestic production capability, access to financing, stakeholder partnerships, information channels, institutional capacities, and the removal of other market barriers. The need to support markets, market institutions, and entrepreneurs as the primary vehicles of technology transfer can be seen in alternative views of the development process that have come from schools of institutional and evolutionary economics (Hodgson *et al.*, 1994; Saviotti and Metcalfe, 1991).

The three central characteristics of a market are: *(i)* the number, nature, and capabilities of participants, *(ii)* the characteristics of the products and services, and *(iii)* the rules governing transactions (Feldman, 1994). Properly functioning markets generally require the availability of information, acceptable levels of risk, appropriate skills, a system of property definitions, quality and contractual norms or standards, oversight and intermediation bodies, decision-making autonomy for buyers and sellers, and stable political and legal regimes.

The scope of a "market" must be carefully defined. For example, we could speak of a market for grid-connected wind turbines, a market for wind farms, a market for independently generated electricity, and a market for electricity services like motive power and lighting. Each of these markets may face different sets of buyers, sellers and institutional constraints. A market approach draws attention past the producer to the consumer -- what decisions consumers make and why. But a market approach also highlights that consumers do not act alone, but as part of larger social groups. Thus, a market approach can bring into focus, for example, the way communities operate and how markets interweave with community structures and interrelationships. Governments define the property rights, contract enforcement mechanisms, and many of the rules for transactions that are necessary for markets to work well. Policies that build or facilitate markets can have a strong influence on the characteristics of those markets -- for example, the relative sales share of domestic vs. foreign products, the segments of consumers participating in the market, the ability of domestic producers to participate in the market, the

technologies available, and how regulations govern market behaviour. Markets that do not take account of externalities or technological path dependence can result in undesirable outcomes from a development and environmental point of view. For example, many promising technologies (*e.g.*, biomass gasification, efficient electric-wheel vehicles, and the next generation of basic materials manufacturing) may get "left out", because the market demand is in developing countries but the technology developers and financiers are still primarily in developed countries. Thus one important question from a market perspective is: will existing technology markets and incentives result in a transfer of the technologies that are most relevant to developing countries? Policies that promote technology transfer from a market perspective must clearly address the factors that drive technology choice in the marketplace, on both supplier and recipient sides.

The initial choice of the technology is not the only or the most critical factor in its diffusion, which is a dynamic process. Technology is often introduced in niche markets, later expanding into other markets, if supply is reliable, as its costs decline with increasing learning-by-doing and with economies of scale in manufacturing. In Kenya, the charcoal stove design originally adapted from a Thai design was introduced first for the urban market and then expanded to the rural market as well (Kammen, 1998a; also see Chapter 16, Case Study 1). Photovoltaic systems were first introduced for the rural affluent market and then smaller systems were introduced for the less affluent in the rural market (Kammen, 1998a; and see Chapter 16, Case Study 5). In Inner Mongolia, we see a reverse phenomenon. Windmills were initially adapted from a Swedish design to utilise the steady but low-speed wind resource prevalent in Inner Mongolia (see Chapter 16, Case Study 3). However, as incomes grew, adaptation was to larger systems (from 100 W to 300 W) and from intermittent wind generators to more reliable hybrid wind-PV systems that also provide electricity during low wind-speed but high insulation summer months.

There are many lessons to be learned from this experience for promoting technology transfer. Some of these lessons stem from the failures mentioned above. Others reflect the importance of technology adaptation, the need for enterprise and technological capability, the selection of compatible technology, and the need for a supportive and appropriate policy environment (Norberg-Bohm and Hart, 1995; Mugabe, 1996). Monitoring and verification protocols can also be useful (see Box 4.3).

The case of the failure of biomass gasifiers for the fueling of irrigation pumps in the Philippines is an example of the need to address the entire market (Bernardo and Kilayko, 1990). Some years after installation, only one per cent of the gasifiers remained in use. The gasifiers themselves were not to blame so much as the market institutions that would have supported them: inadequate user training and poorly funded service and spare parts infrastructure led to poor maintenance and equipment failure. The agency overseeing the programme lacked resources and installed gasifiers without sufficient testing. Users were not sufficiently convinced of the benefits of the gasifiers to invest further resources. Supplies of needed inputs (*i.e.*, charcoal) were difficult and expensive to obtain.

BOX 4.3 MONITORING AND VERIFICATION PROTOCOLS

Energy efficiency investments in the buildings, industrial, energy sectors have been constrained due to inconsistencies and uncertainties in their performance (i.e., actual energy savings achieved), and because financing for efficiency investments has been limited and inflexible. The existence of monitoring and verification protocols can help to reduce these inconsistencies and uncertainties. As a recent example, several dozen national organisations in 16 countries have developed industry best practices, including voluntary standard on implementation, measurement and verification of energy efficiency and called the International Performance Measurement and Verification Protocol (IPMVP) (U.S. Department of Energy, 1997). The IPMVP is being translated into 11 languages and is being widely adopted in countries ranging from Russia and Ukraine to South Korea, Brazil, Mexico and China. Multilateral Development Banks such as the World Bank are using the IPMVP as the technical basis for large scale energy efficiency financing. Use of the IPMVP results in higher and more persistent levels of energy efficiency savings, and in a standardised approach to contract development, implementation and monitoring. This uniform approach cuts transactions costs, allows project pooling and facilitates project financing. As a result of the rapidly increasing application of the IPMVP, there is increased efficiency project financing, with improved project performance, and increased availability of lower cost financing for energy efficiency projects.

Conversely, the case of improved cookstoves in East Africa shows a long and important history of how sustainable commercial markets evolved from what were initially one-time aid-based projects. The most popular improved stove, the Kenyan Ceramic Jiko, has become the mainstay of the Kenyan urban market, and is produced, sold, and serviced by a large and diffuse network of formal- and largely informal-sector artisans. The efficient use of the KCJ addresses economic, environmental, and public health challenges. The KCJ and the commercial network that supplies the technology has also become a model for direct emulation, or for adaptation to promote improved stoves in numerous nations. Comparisons with the successes and problems with these spin-off programmes highlight a number of critical issues in the feedback between design and use, and the limitations of economic analyses of some emerging technologies and informal markets.

Since the late 1970s, the Inner Mongolia Autonomous Region (IMAR) of China has achieved widespread dissemination of small, stand-alone wind electric systems among its rural herding population. This success has been attributed partly to actions taken by the IMAR government to create a local market for wind systems among individual household customers (including a modest subsidy provided directly to manufacturers), and the development of an extensive local manufacturing, sales, maintenance, and training infrastructure.

There is a growing literature on market transformation for energy efficiency (Geller and Nadel, 1994; Golove and Eto, 1996; Levine *et al.*, 1994). The National Association of Regulatory Utility Commissioners defines market transformation as "changing the types of products or services that are offered in the market, the basis on which purchase and behavioural decisions are made, the type or number of participants in the market, or in some other way altering this set of interactions in a self-sustaining way" (Hastie, 1995). The literature on market transformation has its roots in utility demand-side-management (DSM) approaches. Instead of targeting "participants" with rebates and other DSM programmes, a market transformation approach considers that utilities should act to transform the broader market in a sustainable manner, reduce market-barriers, and expand the role of energy-efficient products and services. The literature offers

many specific strategies for market transformation and regulatory approaches to encourage utilities to pursue these strategies (Geller and Nadel, 1994).

Successful market transformation programmes for energy efficiency have included the Poland Efficient Lighting Project (see Chapter 16, Case Study 2), where a combination of subsidies, consumer education, and marketing greatly expanded the market for compact fluorescent lights. The Thailand Demand-Side Management Project (see Chapter 16, Case Study 23) also contained market transformation elements for fluorescent lights that resulted in 100% of the market shifting to more efficient lighting designs.

A market transformation approach promotes replicable, ongoing technology transfers rather than one-time transfers. In such an approach, public policy can consider what are the set of institutions which underlie markets and what are appropriate public interventions to shape those institutions. It is useful to consider a broad brush of measures within a market transformation approach such as: technology adaptation, the need for enterprise technological capability, subsidies, consumer education, and marketing.

4.8   Legal Institutions and Frameworks

In addition to the macroeconomic policy frameworks discussed above, a range of national legal institutions are needed to secure intellectual property rights; reduce contract, property, and regulatory risks; and promote good governance and eliminate corruption. To these ends, governments can strengthen national legal institutions for intellectual property protection; and strengthen administrative and law processes to assure transparency, participation in regulatory policy-making, and independent review (see also section 3.5 on technology transfer and Intellectual Property Rights).

4.8.1   *Intellectual Property Rights*

Weak legal institutions in host countries can be a serious barrier to technology transfer agreements. The relevance of the law of intellectual property rights (IPRs) is of obvious importance to the volume of technology transfer. If actors with IPRs or their licensees cannot protect income flows associated with ESTs in whose development they have invested substantial resources, they will either avoid transferring leading technologies into jurisdictions with weak IP laws or enforcement systems, or will export only second line, more exposed technologies which put at risk less of their capital stock. Although IPRs, discussed further in Chapter 3, merit great attention, many other types of laws and legal institutions, often less emphasised, should receive equivalent concern.

Especially when technology is transferred between private firms through ongoing relationships such as joint ventures, wholly owned subsidiaries with local supply and distribution networks in the host country, or royalty based technology licenses, the risks associated with the investments will be affected by the state of law and legal institutions in the receiving state. This will also be the case for foreign portfolio investors, banks and export-import agencies who finance investments that bring ESTs into an economy. At times, the capacities of these national legal institutions may be supplemented by international treaties like TRIPS or the WTO or partially substituted by transnational arbitration (Dezalay and Garth, 1996). But, even if the role of national institutions is no more than the enforcement of foreign legal judgements, the ability of local law to mitigate risks without extensive cost or delay will have an impact on investment patterns and profiles of technologies selected (Clarke, 1996).

4.8.2 Managing contractual, property and regulatory risks

Three broad types of legal risk are likely to influence decisions to invest in advanced environmental technologies by foreign and domestic actors. Contract risk refers to the likelihood and costs of enforcing legal obligations with suppliers, partners, distributors, managers, labour forces, construction organisations or licensors (Lubman, 1998)]. Property risks refer both to more familiar risks associated with expropriation or other interference with asset ownership and to less visible, but essential questions of corporate governance including public shareholder rights that determine how decision making within the firm is divided or competition law which determines whether companies will be able to operate in open markets (Black *et al.*, 1999; Black, 1999). Clear property rights permit the evaluation of local companies, ownership of essential project assets including land, and outstanding or potential liabilities that determine the ability of firms to raise capital from banks and securities markets. Also essential are bankruptcy and commercial laws concerning security interests that allow creditors to know with relative certainty their rights versus other classes of creditors. Regulatory risk arises from the behaviour of public administrations which influence economic returns through licensing, tariff setting, taxation, and foreign exchange and trade controls (Corne, 1997). Well developed administrative law assures private actors relatively prompt and articulated regulatory decisions, an absence of excessive corruption, and coordination between multiple agencies with regulatory responsibilities. In addition, administrative processes can enhance investment by ensuring inclusive participation in regulatory policy making, transparent rules, and accessible independent review of regulations. Regulatory reform can also reduce impediments to the development of service sector organisations in law, accounting, business consulting, market evaluation, or investment rating that are important complements to foreign direct investments.

Different nations have used their laws to select the mechanisms through which private technology transfers have occurred. During its high growth years, Japan limited foreign direct investment by regulation and constrained equipment imports through foreign exchange controls. Technology licensing was encouraged by strong domestic IPR and contract enforcement. More recently, the People's Republic of China has favoured foreign direct investment of advanced technology industry with special economic incentives, while its technology licensing market is weakened by poor enforcement of the law (Oksenberg et. al., 1996). While there are no agreed propositions that the legal privileging of any particular mechanisms of private technology transfer is superior, there is coincidental evidence that the failure of a national system to manage legal risk can have unexpected consequences for technology development. Where capital assets are not secure, it is possible to discern concentrations of labour intensive production facilities that minimise risks of gradual expropriation (Huang, 1998). When the law does not afford certainty through the courts, there are incentives to engage in corruption to substitute personal and regulatory influence for contracts (Rose-Ackerman, 1996). When open and competitive markets are not assured, it is more likely firms will seek joint venture partners who retain monopoly power, and whose value they will then strive to protect. When IPRs are weak, obsolescing technologies will be more prevalent. In sum, to the extent that domestic legal institutions are deficient in managing contractual, property and regulatory risks, there will be incentives to distort technology choices and supporting financial flows in ways that discourage rapid international diffusion of ESTs.

4.8.3 Governance and corruption

Literature on good governance emphasises transparency, political stability, public audits, participation, accountability and fairness, the rule of law, and the absence of corruption (*Indian Journal of Public Administration*, 1998; Johnston, 1998; Khan, 1998; Theobald, 1999). Many of these aspects of good governance are discussed elsewhere in the chapter, particularly in the sections on human and institutional capacities and social infrastructure and participation. One of the most controversial elements of good governance, the absence of corruption, is discussed here. Although principles of good governance have been adopted in most democratic governments and corruption tends to be either limited to isolated incidents or to a few institutions, many developing countries have not adopted such principles or put them into practice.

Until the 1980s the literature on corruption tended to take a relativist tone and was functionalist in approach. Since then there has been a tendency to take a more critical approach to corruption which includes embezzlement and bribery. Corruption has been defined as "behaviour which deviates from the formal duties of a public role because of private regarding (personal, close family, private clique) pecuniary or status gains; or violates rules against the exercise of certain types of private regarding influence" (Nye, 1967, p. 444), or more recently as "the abuse of public roles or resources for private benefit" (Johnston, 1998, p. 89). Corruption can be incidental, institutional or systemic. Incidental corruption can be dealt with and controlled through existing legal mechanisms in most countries. However, institutional and systemic corruption is far more difficult to deal with. Khan (1998) argues that corruption occurs when capitalist accumulation in its early phases creates new classes of privileged property holders whose justifiable claim in relation to other potential contenders may be limited and this leads to political side payments. However, not all these payments deter economic growth. Hence, Khan (1998) argues that the economic problem is not corruption per se but the political structures which generate growth-retarding corruption. He argues that only to the extent that systemic corruption hinders development (and in this context – technology cooperation), should it be addressed as a priority.

Such corruption can be seen not only in the developing countries but also in the developed countries. Since 1977, when the Foreign Corrupt Practices Act was adopted in the USA, the US is trying to end tax deductibility in relation to bribes to foreign nationals in order to persuade them to buy certain technologies. European and Japanese companies, however, continued to bribe. When the US started to threaten sanctions on countries that condone bribery, the OECD Council on Bribery in International Business Transactions stated in 1994 that each member country should examine its own laws, and in 1998 the members signed an agreement to make bribery a punishable offence. However, this may still lead to bribery taking a more innovative form. On the basis of empirical research, Lambsdorff (1998) argues that Swedish exports to corrupt import countries are lower than those of Belgium, France, Italy, and the Netherlands.

Corruption in the developing countries tends to take the form of systemic and institutional corruption. Johnston (1998) argues that when there is systemic corruption, a long-term social solution is needed, and this calls for empowerment of the people via education and enhanced participation. This fits in very well with the analysis that increased participation of the local stakeholders in a technology transfer deal will increase the likelihood of successful technology transfer (see section 4.4). Galtung (1998) argues that the international dimension of such corruption can be partially tackled by social movements and environmental NGOs. For example, Transparency International, an international NGO with about 80 national chapters has introduced an Integrity Pact.

4.9   Codes, Standards and Certification

The importance and the need for technical standards and codes have been well recognised by the technical community all over the world. If standards and codes are absent, transaction costs can

increase as each buyer must ascertain the quality and functionality of potential technologies individually. Technology risks can increase because of the uncertain quality of technologies. Equipment purchasers or building architects may not be the ones paying the energy bills or using the equipment, so the interests of end-users (*i.e.*, for cost-effective levels of efficiency) are absent from purchasing or construction decisions. Non existent or inconsistent standards reduce confidence in the market for environmental technologies and deter investment. More, clear and harmonised standards, regulations and guidelines, enforced at regional and local levels, help alleviate this barrier.

4.9.1   Standards

Because equipment and supplies may come from a number of countries, most countries have some type of institution that looks after matters relating to standards. These official government agencies are invariably assisted by a number of industry working groups, NGOs, and research institutions in the business of standard setting. Although in some cases standards are voluntary (with the exception of safety related standards), many countries have realised the importance of making these standards mandatory with some form of penalties for noncompliance. Intermediary organisations such as the International Standards Organisation may have a role. It is an NGO federation of 130 national standards bodies through which capacity building is undertaken. International standardisation can be facilitated through this process. So far there are 12,000 international standards published which have been devised through Technical Committees using a consensus approach. The Environmental Management Standards 14000 series could be further applied to climate change issues. An important issue—and often the weakest link in policy -- is the enforcement of the standards and the availability of testing and technical institutions to provide the necessary technical assistance for standards setting and enforcement. Codes and standards are important because equipment purchasers or building architects may not be the ones paying the energy bills or using the equipment, so codes and standards provide a means for representing the interests (*i.e.*, to cost-effective levels of efficiency) of end-users who are absent from purchasing or construction decisions. Standards also provide another way to overcome information barriers, as they allow consumers to be less knowledgeable about the equipment they are purchasing. Standards can also create a market (*i.e.*, for high-efficiency equipment) where none existed (*i.e.*, perhaps buyers existed but suppliers did not). Eco-labels have been in place in a number of countries for energy-efficient and other environmentally sound products; the case of eco-labels in India is described in Box 4.4.

A number of countries have adopted laws requiring all manufacturers to follow energy efficiency standards. The European Union, for instance, has adopted a series of energy efficiency standards (see EU Directives: Directive 96/60/EC; Directive 97/17/EC; Directive 98/11/EC). Efficiency standards for commercial buildings are in place in countries such as Sweden, Germany, France and the UK, and also in some of the developing countries such as Pakistan and India.

Eco-labels have been used in India since 1991. They are given to products that have met specific environmental criteria related to the production process used, wastes arising from production, and resources consumed in production and use. A National Steering Committee and a Technical Advisory Committee have been formed to provide policy and technical guidance. So far 20 products have received labels, including toilet soaps, detergents, paper, and edible oils and coffee. Despite efforts by the Ministry of Environment and Forests, which is responsible for eco-labelling, only two companies have completed the requirements for getting an eco-label. Yet these companies have not used the eco-label in their consumer communications because they do not think that consumers are aware of the meaning and advantages of the labels and because the labels do not provide any direct financial incentive such as tax exemptions.

In contrast to eco-labels, energy-efficiency labels have yet to have an effect in India. Efficiency labels are more informative but also require a higher level of education to understand. On the other hand, there is a way to calculate the economic benefits of a product with an efficiency label, while this is not so easy for eco-labelled products. There is a need for research on the environmental benefits of eco-labelled products that might lead to a clearer understanding of financial benefits. There is also a need for greater consumer education and awareness about eco-labels and energy-efficiency labels.

In the United States and Europe, mandatory energy efficiency standards for appliances and buildings have produced significant energy savings (see Case Studies). In most of these countries, building standards are adopted at the state level or even at the town/city level and this is where implementation and enforcement occurs. In a number of countries, building codes do exist but they are only to act as a guideline for builders. In practice, substantial deviations do take place depending on the customer's requirement. In a significant move, reflecting globalisation of marketing, the EU (October 1999) agreed to adopt the US Energy Star programme for office equipment (see Box 4.5).

Renewable energy technology standards and certification have played important roles in building markets for dispersed, small-scale technologies where technologies are diverse, vendors are many, and consumers face high risks in evaluating and selecting technologies and suppliers. For example, in India, government performance standards for rural electrification have had a dramatic impact not only on technical system operation but also on the development of service/maintenance infrastructure (see Chapter 16, Case 14). In Inner Mongolia, technical standards for wind turbines have facilitated a market there (see Chapter 16, Case 3).

4.9.2   Certification

"Green power" marketing is another way to certify the qualities of an energy product (see Box 4.6). In the wake of liberalisation of markets, and more cross-boundary trading of electricity, cer-

tification issues increasingly need to be resolved at supra-national levels. Utility companies have now got an international working group which will start a pilot project, and there is a certification scheme included in the new draft renewables directive in the EU.

Even where developing countries have embarked on paths to create regulatory frameworks which will encourage the use of environmentally sound energy technologies, corruption can reduce their impact unless there are robust systems for control in place. Analysis of Inspection/Maintenance programmes has revealed that corruption or "bypassing" constitutes the major problem for such programmes in developing countries, and the impact of such programmes depends to a large extent on a successful scheme to prevent corruptive behaviour. There are ways to reduce this problem, on-road inspections are the only way to prevent all possible fraud with frequent controls and frequent fines (Grutter, 1996).

4.9.3   *Minimum environmental performance standards*

Transfers to developing countries of older technologies whose environmental performance is lower than that of average technologies used in developed countries have been called "technology dumping". Technology dumping is a controversial issue. It often occurs because of lower technology costs relative to newer technologies (including base costs, taxes, duties, and royalty payments). The literature distinguishes two types of dumping: (a) an exporting country firm sets export prices lower than prevailing prices in the exporting country (due to favourable or artificial

A voluntary labelling programme for office equipment has been identified as the most cost-effective action to achieve the potential energy savings; and a study proved that there is added value in an international coordination of labelling programmes to achieve the potential, as well as to ensure fair comparability between products of the same type. To facilitate international trade and to reduce the regulatory burden on the operators whilst maximising the energy savings, the European Commission decided that the best way of reducing the energy consumption of office equipment was to introduce the Energy Star Programme in the European Community. This would: 1) build upon the advantage of an already well-recognised logo in the Community market, 2) maximise manufacturers' participation and their active involvement in promoting energy savings; and 3) use a well-recognised logo to educate users about energy savings. Furthermore, this would result in a clear advantage for Community economic operators. The Energy Star logo was the "de facto" required standard for office equipment sold on the US market. In addition, the Energy Star requirements were becoming standard worldwide (in the Community as well), and this without any European input.

BOX 4.6	GREEN POWER CERTIFICATION

Green power schemes are now operating in many developed countries including Australia, USA, the Netherlands, UK and Germany. The schemes were encouraged by government policy to encourage renewable energy but growth prospects are considerable as a result of liberalisation of electricity markets whereby the opportunity exists for power companies to exploit consumers' environmental values. Operating frameworks require cooperation, though not necessarily formal regulation, between parties. Voluntary arrangements operating so far have avoided possible conflicts with international trade regulation. If premium prices are being paid, it is essential to establish a verification system, which involves an independent and credible certification procedure. Consumers are only prepared to pay a premium if they have confidence in the product.

Serious marketing issues have arisen with most schemes. How 'green' is green'? There has been resistance to the use of large scale hydro plants in Canada and Sweden. Energy recovery from incineration has been criticised for being classified as a renewable technology. Furthermore, schemes may not necessarily result in the creation of new capacity, and may merely result in the proportion of 'brown' power rising for the non-premium paying consumers. Plausible schemes are also thought to include requirements for energy efficiency advice. When schemes are certified, should the overall environmental performance of the company be assessed at the same time? Until there is general agreement, consumers can be confused by competing marketing information and these issues could delay internationalisation of these initiatives.

conditions prevalent through taxes and duties given by the exporting country); and (b) outdated technologies are purchased by users in developing countries because they are unable to afford new technologies (tax and duty structures may contribute to unaffordability of newer technologies).

Although industries trading older technologies may themselves be satisfied by the trade, environmental NGOs and governments have raised a variety of policy issues. Should old technologies with harmful environmental effects be sold to developing countries even if technologies meet the environmental standards of the recipient country? On the other hand, if newer technologies are not affordable to these countries and no financial assistance is available to them, are policies which deny them access to these technologies unfair?

Some ways forward have been suggested. Chapter 4 of the Habitat Agenda argues that "The international community should promote and facilitate the transfer of technology and expertise through…seeking to ensure that the process of technology transfer avoids the dumping of environmentally unsound technologies on the recipients." (UNHABITAT, 1996). There are existing examples of mandatory and/or voluntary minimum performance standards in developing countries. However, enforcing environmental standards requires some type of certification process, which can become complicated (*i.e.*, characterisation of degree of energy efficiency). It would be much easier to base standards on some known characteristic such as age or size (see Box 4.7). There is clearly a role for government and international assistance in setting and enforcing standards.

4.10   Equity Considerations

Any international regime that aims to address international problems will tend to influence the distribution of costs and benefits associated with the management of the regime, both internationally and nationally. To the extent that the transfer of technology is seen as an important operational tool for addressing the global climate change problem, it will also have a serious impact on distribution issues. The technology transfer issue is born out of the principle of equity and common but differentiated responsibilities of countries. The United Nations Framework Convention on Climate Change (UNFCCC) stated in Article 4.5 that the developed country Parties "shall take all practicable steps to promote, facilitate and finance, as appropriate, the transfer of, or access to, environmentally sound technologies and know-how to other Parties, particularly developing country Parties". However, the international equity aspect of climate change impacts and adaptation have received relatively little attention so far (Metz, 1999). So there is even less material to draw on to review the 'national' equity aspects of technology transfer. But it is evident that the execution of the international responsibility of developed country parties will have differential impacts in recipient countries. Within the domestic contexts of countries, there are also vast differences in emission and development levels between different groups of people. Technology transfer within countries is likely to affect some groups positively at the cost of other groups so clear distribution issues will become evident. If they are ignored there could be negative consequences on achieving technology transfer (see Box 4.8).

BOX 4.7: TECHNOLOGY IMPORT STANDARDS IN INDIA

India has recently announced a new import policy for second-hand technologies. Under this policy, imports of technologies that are more than 10 years old are prohibited. Technologies less than five years old can be imported at standard rates of duty. Technologies that are between 5-10 years old are decided on a case by case basis. The only criterium is the age of the technologies. Environmental performance is not considered, partly because any technologies will still have to meet existing environmental standards for manufacturing processes and equipment. This ignores the consideration that imports of older manufactured products could consume energy inefficiently and result in greater indirect environmental impacts (e.g., refrigerators, CFLs with inferior phosphors, chokes, and motors).

BOX 4.8	DEVELOPMENT CONFLICTS: FOR A NEW AGENDA
	(REDDY, 1998)

Environment-development conflicts can be minimised if not avoided. What is required is the implementation of a new agenda based on: a quantitative statement of what the project intends to achieve; a comprehensive listing of all the options (including megaproject and/ or mixes of smaller options) for achieving the objective; rigorous comparative real-costing of options, inclusive of externalities; determination of least-cost mixes; clear description of benefits according to region, location, income-group and gender; environmental impact assessment; universally accessible information and transparency; and democratic participation and decision-making along with the necessary institutional changes."

4.10.1   Equity and the Technology Transfer issue

The selection of technology for the purposes of transfer may have a serious impact on distribution issues:

Allocation of Emission entitlements
The climate change issue can be explained as the issue of how the rights to use and pollute resources will be shared among countries and people. The allocation of emission entitlements/targets is thus a crucial issue and will have an impact on technology choices. The initial distribution of rights both nationally and internationally may be a major bottleneck to the technology transfer process. Theoretically, a 'fair' allocation of emission allowances to all countries would allow the 'underusers' of such allowances to sell their allowances to 'overusers' of such allowances. Such an instrument would thus provide an incentive to 'overusers' to reduce their emission levels and help to redistribute resources to 'underusers' who generally would be the poorer countries. By allowing the market to function, the cheapest way of reducing the necessary global reductions in emissions can be achieved (UNCTAD, 1995; Agarwal and Narain, 1992). However, the critical issue is how should the initial allocations of emission allowances be undertaken? The recent literature indicates a proliferation of articles that discuss the definition of equity. These include allocation based rules which include, in addition to per capita principles, the principles of equal rights to pollute, the polluter pays principle, the basic needs approach, outcome-based rules such as equal abatement costs, equal percentage net GDP loss, compensation to net losers, etc. (Banuri *et al.*, 1996; Metz, 1999). Both legal precedent and the scientific literature do not point towards an optimum approach. Whatever agreements are made on modalities at the international level and on the decisions about approaches, these are likely to affect the outcomes of technology transfer within countries. The way the allocations of emission allowances are distributed would help to determine the potential for financing technology cooperation between countries (in terms of demand for CDM and JI, for example).

Adaptation versus mitigation
The climate change issue can be addressed through emission limitation measures and adaptation measures. Since available funds are limited, the issue of prioritising technology becomes central. Adaptation technologies tend to help local people adapt and may be seen as a local impact, and, until now the emphasis has been on limitation technologies which are seen as contributing to the solution to the global problem of climate change. Bodansky (1993) explains "adaptation measures generate primarily local benefits, so that developed countries have little incentive to fund adaptation measures". Until recently (see section 4.5), little attention had been given to activating the CoP1 decisions on adaptation and has adaptation played only a marginal part in the reports produced by IPCC. This reflects the limited attention given to adaptation by scientists as well as policymakers worldwide. In his elaborate review of the Working Group II volume of the IPCC Second Assessment Report, Kates (1997) suggested the reason for this lies in the existence of two distinct schools of thought about climate change, both of which have chosen not to encourage adaptation research and planning.

The "preventionist" school argues that the ongoing increase of atmospheric greenhouse-gas concentrations could be catastrophic, and that drastic action is required to reduce emissions. Preventionists fear that increased emphasis on adaptation will weaken society's willingness to reduce emissions and thus delay or diminish mitigation efforts. The "adaptationist" school, on the other hand, sees no need to focus on either adaptation nor mitigation. They argue that both natural and human systems have a long history of adapting naturally to changing circumstances and that active adaptation will bring with it high social costs (Kates, 1997).

The increasing awareness of climate change not as a theoretical phenomenon but as a genuine threat has led to the emergence of a third school of thought, which Klein and MacIver (in press) have labelled the "realist" school. The realist school regards climate change as a fact, but acknowledges that impacts are still uncertain. Furthermore, realists appreciate that the planning and implementation of effective adaptation options takes time. Therefore, they understand that a process must be set in motion to consider adaptation as a crucial and realistic response option along with mitigation (*e.g.*, Parry *et al.*, 1998; Pielke, 1998).

The least developing countries, and particularly the small island states, are putting increased emphasis on this side of the issue. Vulnerability indices are being developed, for example, by SPREP and UNEP. Adaptation technologies which involve institutional infrastructures (for example in agriculture, health and human settlement planning) could be integrated with other parts of efforts to alleviate poverty and promote development, but it is not yet clear if there is much joined-up thinking in evidence at the donor level.

Rural versus urban
Within the emission limitation context, when decisions are taken to support technology transfers to reduce emissions, the applica-

tion of the cost-effectiveness principle tends to imply that priority is given to technologies that can reduce emissions quickly and efficiently and where there is a strong vested interest. Inevitably, this implies that technologies that serve urban interests may get prioritised in the process. As the majority of technology transfer is from industrialised to developing countries, it might benefit a minority of elite rich: industrialists and investors in the advanced industrialised countries and a small group of wealthy political and economic elite in the recipient countries (Leys, 1975). The technology transferred is usually capital-intensive, and the products are directed toward urban middle and upper class consumption. With emphasis on export-oriented industrial production in many Third World countries, the development of domestic markets that would imply spreading greater wealth (and purchasing power) among the poorer classes is a low priority. Technology transfer in the agricultural sector also tends to benefit wealthier farmers with larger land holdings, often to the detriment of poorer, small landholders and landless tenants. The Green Revolution, touted as one of the great successes of technological development to improve agricultural production and food resources, was based on newly developed grain seeds that required high amounts of chemical fertiliser that only wealthier farmers could afford (Chambers, 1983; Pearse, 1980; Stewart, 1997). Even those whose concerns might be broader, who work in the public sector or for NGOs, have a strong urban bias. Planners, even those working on rural development and natural resource issues, rarely live in rural areas. Their offices and homes are in cities. Meetings and paperwork dominate their jobs. Their needs, and their perspectives, are tainted by their daily urban experiences (Chambers, 1983). It is difficult for many urban residents to communicate with rural people. The former look down upon the latter: it is easier to plan development activities FOR villagers than plan WITH them. The challenge is to find ways and means to ensure that technology transfers are also used to address relevant rural issues effectively.

Exclusion of poor social classes and ethnic minorities
Another bias that sometimes creeps in, is the bias against ethnic minorities. The relationship between dominant ethnic groups and non-dominant ethnic minorities in a country lends further complexity to the issue of technology transfer. Ethnic minorities are often viewed as "outsiders" to the national society, so little effort is made by members of the dominant ethnic group to communicate effectively with the minorities. This leads to frequent miscommunication on transfer of technologies with the ethnic minorities (Black, 1991). The unique identities of ethnic minorities are frequently tied to their use of "traditional" forms of technology. New technologies introduced by or through the dominant culture may be seen by ethnic minorities as "alien" to their own culture and identity (Bebbington, 1997).

Many analysts have noted the relationship between social class and the location of polluting activities, including those that emit GHGs. Many polluting activities are sited where poor, politically powerless, and socially deprived groups of people reside (Cable and Cable, 1995). The large thermoelectric plant in Catano, in Northeast Puerto Rico, is an example of environmental degradation of energy projects. The plant discharges toxic waste into the sea and sur-

rounding land, damages the health of nearby residents with its gas emissions, while producing electricity mainly for commercial and industrial uses in distant cities (Meyn, 1997). Plans to build a similar plant in another corner of the island have been met with extensive protests by local residents. A central authority made the plans, with no consultation with the people where the plant was to have been built. In 1993, after four years of protests and struggle, the local residents and other island-wide environmental groups forced the government to cancel the proposed thermoelectric plant. Several community organisations have since formed an environmental coalition to formulate alternative energy plans, by using renewable resources, through storage of excess generation during slack periods, and by increasing efficiency of existing plants (Meyn, 1997). The bias of technology transfer toward vested interests is reflected in transportation policies that favour roads and expanded use of private vehicles over public transportation. This implies that technologies transferred may often end up being sited in the poorer neighbourhoods, or alternatively that poor neighbourhoods grow around such technology sites and may have serious environmental impacts on the local people.

In serving certain vested interests, other interests may get neglected. Thus, large dams have led to huge displacements of the local people and of animals. Large-scale energy schemes and the extraction of natural resources (mining, logging) are rarely planned with the consensus of the local population. Rather than accruing some of the benefits to local residents, these schemes more often lead to considerable destruction of local environments, dispossession of people from their lands, and displacement of people. Dams and hydroelectric power stations lead to displacement – the Sardar Sarovar Dam on the Narmada will affect at least 100,000 people by submerging their homes with another 140,000 people to be adversely affected by the canal system[3] (Sen, 1995). Displacement and dispossession is possible when people do not offer resistance, and when they have not been consulted about the problem and included in the process to find a solution.

Gender bias
The selection of technologies often has a strong gender impact. As a result of the social position of a woman in a society, her needs may be shaped in different ways. Technology transfer may either lead to empowering women (Jain, 1985) by meeting needs of women in their productive, reproductive and community management roles, or it may lead to disempowering women by displacing their role in the society (Mies and Shiva, 1993). Thus, technology can be an opportunity for women when it provides them with greater income, reduced health risks and greater security and autonomy, or it can be a threat to the livelihoods of women, when the transfer of technology leads to processes in which men take over the traditional work of women (Everts, 1998). This may happen if women are not fully involved in the technological innovation process.

Formal interventions in the rural energy sector in India have traditionally been implemented in a manner that normally treats the actual users, *i.e.* local communities, as either 'data sources' or as 'beneficiaries', but rarely as full partners in the design, adaptation, and implementation processes. It is being recognised that one of the reasons for the inconsistent performance of these programmes is that the local communities and grassroots organisations are not adequately involved in the planning, implementation, and follow-up aspects of these programmes. The other serious shortcoming is the total absence of a role for women in these processes notwithstanding the fact that the women are usually the providers, users and managers of energy in rural areas (Malhotra *et al.*, 1998, p. 12-13). Specific evidence that interventions may actually have perverse effects is provided by the fuelwood crisis (Agarwal, 1991). Women in South Asia cook on fuel-wood which they cut from nearby forests and trees. Although governments have promoted social forestry schemes, if the ownership of such social forestry schemes is such that the poor women have no access to the wood, such schemes may not benefit the poor households nor the environment. Once fuelwood becomes part of the monetised economy, women are often excluded as men typically make decisions of a financial nature (Malhotra *et al.*, 1998). One can recommend an adaptation of Everts' (1998) approach to integrating gender (see Box 4.9).

Finally, one way forward is to specifically recognise that there are different stakeholders in a society and build assessment mechanisms to test for 'best social practices'. Although one acknowledges that it is the market that determines which technology is adopted, on the other hand, markets in developing countries are frequently very imperfect and tend to exclude social concerns. In order to better evaluate these concerns, the technology can be assessed for the community/private sector or government for which it is targeted. In such an approach the social actors are identified after the identification of the technology. Or, government/private sector/community needs are identified and appropriate technologies to meet those needs are sought out. It is important to recognise that there is a range of interest groups involved. Experiences from different social contexts need to be evaluated in order to generate generalisations of possible best social practices which can then be applied in relation to each individual case of technology transfer.

4.10.2 Problems associated with equity

Dealing with equity related issues is not always easy. Clearly if the key goal is to reduce GHG emissions, then selecting cost-effective technologies to reduce the most emissions with the least possible costs is a key issue. However, as illustrated above, such cost-effective and politically viable technology transfer may negatively affect some communities within a society. The challenge for the international community is to try and make these trade-offs in such a way that it also reduces existing imbalances in the society. This becomes even more

---

[3] Other sources report 85,000 families -originally in 1979 the government anticipated that 6,000 would be displaced, The Independent, 14th August 1999.

| BOX 4.9 | INTEGRATING GENDER INTO ASSESSMENT FOR TECHNOLOGY TRANSFER |

Most actors in the climate change and technology transfer debate are not very familiar with gender-related issues. The first step in such a process is to create gender awareness in the actors involved. This implies creating an understanding of how women as a social group benefit or lose from technology transfer and developmental initiatives, as opposed to assuming that men and women within specific social groups share the benefits and losses equally. The second step calls for undertaking the Efficient Gender Analysis (EGA) method. This method involves the formulation of 13 questions which aim to make explicit the implicit gender-specific objectives of a particular technology transfer project, to analyse the impacts on women and to formulate recommendations that would facilitate the effective inclusion of women in the process. The third step calls for the collection of context relevant specific gender-related information which can be included in the recommendations on how to minimise the adverse impacts on women, and to empower them in the process of transferring technologies to reduce the growth of greenhouse gas emissions.

complicated when one realises that empowering poor people, women and the generally neglected classes, does not immediately lead to a reduction of GHG emissions, but instead to an increase of the emissions as people move out of the poverty trap. This is why a key question is: whose development should technology transfer foster? Clearly, since there is a shortage of resources, the trade-off between environment and social issues is not an easy one.

However, the implementation of technology transfer has been hampered because of four reasons (Junne, cited in Gupta, 1997). First, the low GHG emitting "leap-frog technologies" are produced in few countries (USA, Japan and Germany), and technology transfer schemes facilitating transfer of the best technologies would lead to a North-North transfer of financial resources which few other developed countries would find acceptable. Second, in identifying the low GHG emitting technologies, new conflicts may arise. For example, if France were to promote the expansion of nuclear technology, and other countries with advanced coal technologies may support the promotion of clean coal technology. Third, cost-effectiveness arguments may lead to the transfer of technologies to the most polluting countries which may also be the rapidly industrialising countries with comparatively high growth rates, leading to a South-South inequity by helping those who need the help least at the cost of helping those who need it most. Fourth, ultimately, "leap-frog technologies" aim to assist developing countries develop quickly, a goal that industrialised countries may also not entirely support.

The domestic equity aspects of technology transfer, though central to the debates on appropriate technologies in the 1970s, have not really influenced the climate change technology transfer discussions since then. However, there is a good case to be made for learning from past mistakes and ensuring that technology transfer under the climate change regime enhances the sustainable development of the importing countries.

Finally, one way forward is to specifically recognise that there are different stakeholders in a society and build assessment mechanisms to test for what could be described as 'best social practices'. Such an approach would recognise that whilst the market determines technology adoption, markets in developing countries are very imperfect and tend to exclude social concerns. In order to better evaluate these concerns, the technology can be assessed for its impacts on intended users. Conversely, technology could be selected as matched to previously identified needs. Experiences from different social contexts could in time lead to general guidelines on 'best social practices' being formulated.

4.11   Rights to Productive Resources

In addition to the matter of intellectual property rights discussed in Chapter 3 - who owns (and benefits from the use of) the technologies to be transferred – there is a need to take into account the resource rights for the productive resources that would be used or affected by the transfer of any technology. These resources include land (certainly for agriculture and forestry), natural resources (forests, water, coastal areas), factories, and other productive resources.

The literature on technology transfer and development in agriculture, forestry, and coastal resources is rich with examples of the importance of resource rights. Certain programmes have failed because governments ignored resource rights, or ignored conflicts between legal resource rights and traditional or local resource systems. Other examples have shown how resource rights can affect the incentives to landowners to adopt new technologies.

If governments or others trying to introduce the technologies take into account the property relations of the land or resources affected by the technologies, they can adapt the technologies or take other measures that can ameliorate the potential disruptive social impacts of the new technologies. Better yet, technologies can be selected that are more appropriate to the particular property relations of the affected resources.

The distribution and control of property is a fundamental component of human relations: "property rights are but formal expressions of authority between persons and groups of persons." (Denman, 1978, p. 101). Property relations reflect the organisation of economic production and the allocation of wealth; property has traditionally been the foundation of political control and a key to social status (Agarwal, 1994).

The relation between technology transfer and property rights is multi-faceted. The adoption of a technology – by whom, and how extensive the adoption – is often contingent of the form of property relations. Forms of property rights can contribute to practices that degrade the environment (Thiesenhusen, 1991), and block the adoption of ameliorative practices or technologies.

Once a technology is adopted, the changes in production stimulated by the technology can often lead to shifts in property relations, and hence to shifts in economic and political power, from the household and local levels to the national level. If technologies are introduced that are more appropriate, more easily adopted, or otherwise appropriated by larger rather than smaller land owners, by owners rather than tenants, by men rather than women, or by one ethnic group rather than another, those introducing the technologies risk alienating and even further impoverishing already disadvantaged groups in the society (Agarwal, 1994; Carney and Watts, 1991).

4.11.1   *Experience from agricultural technologies*[4]

Past research on technology transfer in agriculture (particularly the Green Revolution) showed that tenure status of farmers had a great influence on the extent and timing of their acceptance of technologies. Size of landholding was another major influence in acceptance or rejection of technologies. These will have impact on the technology transfer issues raised in the chapter on agriculture. Technology can also affect the evolution or the status of property rights.
Adopting a new technology is a form of risk. Research in Thailand by Feder *et al.* (1988) found that farmers were more likely to make capital and technical improvements on their holdings if their land ownership is secure. This held true for farmers with small holdings as well as those with larger farms. Those without secure tenure were uncertain of their long-term status; even if given usufruct rights by the government, they were unable to obtain or did not want to invest the funds needed to make significant production changes.

The technologies known as the Green Revolution tended initially to benefit wealthier farmers with larger landholdings, and benefited landowners rather than tenants. Though production of rice, wheat, and other basic foods did increase remarkably, another effect of the Green Revolution was an increase in inequity. The newly developed grain seeds required high amounts of chemical fertiliser and pesticides that at first only the wealthier farmers could afford (Chambers, 1983). After a time-lag, small holders eventually adopted the technologies, with equal and often greater productivity than the larger farmers (Lipton and Longhurst, 1989).

Tenants – already among the poorest of farmers – suffered from the Green Revolution. The increased income from the more capital-intensive commercial production encouraged many landowners to cultivate their farms directly. Those who had been their tenants lost access to land and became further impoverished (Scott, 1985).

---

[4] See Chapter 11 for an extensive treatment of technology transfer in the agricultural sector.

The introduction of a new technology implies a reallocation of productive resources – whether of capital, labour, other inputs, or of the land itself – too often ignored when those introducing the technology focus mainly on the technological innovation. The introduction of irrigation schemes in the Gambia altered labour allocation and property rights among men and women, and appeared to increase tensions among ethnic groups with competing claims to the resources (Carney and Watts, 1991).

4.11.2 *Experience from forestry and use of other natural resources*[5]

As the literature on common property and other tenure relations involving natural resources has shown forests and other natural resources are controlled and managed under myriad forms of property relations. The legal *(de jure)* rights to these resources may reside with the state, but the actual *(de facto)* rights of ownership and control lie with local communities or individuals. These frequently traditional forms of property rights pre-date the state claims over the resources (National Research Council, 1986; Peluso, 1992a). These are of critical importance to the technology transfers recommended in the sections on Forestry and on Coastal Resources.

Public ownership of forests and other natural resources is difficult to enforce. When states attempt to revoke traditional community rights to forests by invoking state control, the result has often been a lack of control or management. The forests are open to overuse and encroachment. By clearing land, farmers indicate their ownership (Arnold and Campbell, 1986; Mueller *et al.*, 1994; Thiesenhusen, 1991).

The acceptance or rejection of a technology or of a mitigation strategy (tree planting, modification of coastal land use) will depend on who owns, who controls, and who manages the resources, both legally and in practice. States often claim rights over resources that are in fact controlled by local communities or individuals. Conflicts occur when the governments attempt to assert their legal claims over the traditional or local rights. (Lynch and Alcorn, 1994; Thomson *et al.*, 1986)

Since the late 1970s, governments and NGOs in Asia, Africa, and Latin America have come to realise the importance of alternative land management systems to give people greater rights to control and use their local resources. Management and usufruct rights to the forests and forest land are granted to communities (through community forests or joint community-state management) or to individuals (through leaseholds or other tenure arrangements), giving them the necessary security that encourages them to sustain forest resources (Peluso, 1992b; Poffenberger, 1990; Repetto, 1988).

The experience in agriculture, forestry, and use of other natural resources has shown that the successful introduction of new technologies or modification of resource use often depends on a recognition of the existing forms of property rights, or on taking steps to create an optimal property rights regime. With an understanding of the existing – legal and actual – forms of property rights, technologies or modified resource uses can be adapted to fit the property relations. If property issues are taken into account, those introducing new technologies or proposing modifications in land or resource use can be more assured of the support of the target populations or groups.

4.12 Research & Technology Development (RTD)

Technology capacity at both the inception/assessment and replication stages of the technology transfer process have to be underpinned by RTD. Central to this process are national systems of innovations and international cooperation between research institutions (public) and private sector entities in research and technology development. Innovation and technology development are the result of a complex set of relationships among actors in the system, which includes private-public enterprises, universities and government research institutions (OECD, 1997). The national systems of innovations are well developed in OECD countries compared to developing countries. Furthermore, there is more international cooperative research among OECD countries and the EU countries (see OECD, 1997; Guy, 1997).

4.12.1 *Publicly-funded research and publicly owned technology*[6]

Historically, governments have played a key role in supporting research and development through national laboratories, universities, and through international collaborative ventures. In the report of the international expert meeting on the role of publicly-funded research and publicly-owned technologies, it is stated that "…many governments emphasise the contribution that public support to R&D can make to economic competitiveness and the importance of commercialising publicly-funded R&D. The country case studies presented at this meeting indicate that public funding remains a major source for R&D activities in both industrialised and developing countries. Public funding of R&D, according to the UNCTAD, UNEP and UNDESA paper, usually takes two forms: general support to national R&D institutions and laboratories, and direct funding of specific projects according to set government priorities (UNCTAD *et al.*, 1998). Investment by OECD governments in energy expanded dramatically after the oil shocks, and was maintained at high levels during the 1970s, but overall government energy RTD has declined steadily since the early 1980s (Margolis and Kammen, 1999; Clark, 1999). However, some governments, especially in Europe, have invested steadily rising amounts towards renewable energy and other environmentally sound energy sources. The EU's fourth framework programme on research and development for 1994 to 1998 provided significant financial resources for non-nuclear activities. The framework programme also includes cli-

---

[5] See Chapter 12 for an extensive treatment of technology transfer in the forestry sector.

[6] See also section 3.5.1 in Chapter 3 on publicly-owned or supported ESTs.

mate-related research programmes such as EPOCH, and the science and technology for environmental protection (STEP) programme. Many governments are very much aware of the international dimension of global environmental degradation and the role of ESTs in addressing the problems. A recent OECD study concluded that industrial output from non-OECD countries will triple by the year 2010 as compared to 1990, and this would have direct implication for cleaner production technologies; therefore, OECD member countries must give attention to technology and information transfer (OECD, 1995).

There is a close relationship between the governments and the private sectors because RTD results 'spill over'. The literature on the returns to RTD spending consistently show that the social rate of return is higher than the private return (for example, Evenson *et al.*, 1979; Goto and Suzuki, 1989; Suzuki, 1993; Coe and Helpman, 1995; Bernstein, 1996; Griliches, 1992). This is because part of the return to RTD is a "public good" that cannot entirely be appropriated by the organisation making the investment. It has been estimated that optimal R&D investment is at least four times larger than actual RTD investment (Jones and Williams, 1998). Similarly, the "new growth theory" has drawn the attention of economists and development specialists to the spillovers from investments in education and capital formation that raise social productivity above the gains that can be realised by the private-sector agents making the investments (Romer, 1986a, 1986b, 1988, 1994; Lucas, 1988). These spill-overs constitute a strong justification for public policy to raise the level of private investment. These considerations apply to investments that transfer new and improved carbon-reduction technologies.

4.12.2   Policy and institutional environment

Some governments have developed clear policy guidelines not only on R&D financing, but also on the innovative activities required by recipients of such funding and specific areas requiring ESTs. In most countries, particularly in OECD countries, government policies play a role in the generation, development and transfer of technologies. They stimulate innovations in the areas of ESTs through regulatory policies that set environmental standards in areas such as emissions, and they also facilitate the process of encouraging public-private partnership in technology transfer and diffusion (see also section 5.5 in Chapter 5 on Public-Private Partnerships). Some countries have also developed incentive schemes such as tax incentives, preferential loans, financial grants and other incentive measures designed to support the process of transfer and commercialisation of ESTs (UNCTAD, 1998; Clark and Paolucci, 1997; Clark, 1999; Lanjouw and Mody, 1996). In some developing countries, the implementation of market liberalisation policies has created an environment for technology transfer through strategic technological alliances, mergers and acquisitions of foreign firms as the case of South Korea shows (Lee and Chung, 1997).

It is perhaps during the last 4 to 5 decades, that the private sector has taken up research and development activities, but these have been partly supported through government funding. The pri-

vate sector contribution has also come about, often due to policies of governments to encourage R&D, by giving certain incentives such as tax breaks, etc. It is the government that has taken on the role of underwriting the initial expenditure on research and development, on which returns are perhaps uncertain, and bear a significant amount of risk. Also, where it relates to technologies that have to be disseminated across consumer categories, it is again the government that has taken on this important role. This has been undertaken through exhibitions, mobile vans, and local dissemination centres, on which the government spends a substantial amount of funds. In ministries such as environment, renewables, rural development, etc., at least 10% of the total outlay, or more in some cases, is allocated for dissemination. Support is also provided by way of setting up new institutions and strengthening existing ones through capacity building for personnel, equipment and for dissemination (section 5.6 in Chapter 5 provides more information on technology intermediaries).

One of the major findings of this OECD expert meeting on publicly-funded R&D is that governments of several countries support programmes for the environment industry , however, coordinated efforts for the development and market access of environmental technologies and services is lacking.

In the light of the 1995 OECD study cited above, many OECD countries are evaluating the bilateral development cooperation polices with developing countries to facilitate the transfer of ESTs via their Official Development Assistance (ODA) programmes. Many of the new initiatives seek to increase North-South technology partnerships through encouragement of joint ventures, licensing, joint technology development and the creation of information clearinghouses and "matchmaking" services needed to make both suppliers and potential technology users aware of the opportunities in the area of ESTs (OECD, 1996).

In addition to the bilateral initiatives are multilateral initiatives. Typical examples of such multilateral initiatives are the Climate Technology Initiative (CTI) which was founded in 1995 as a voluntary initiative by 23 OECD/IEA member countries and the European Commission to support the technology-related climate objectives of the Framework Convention on Climate Change (OECD, 1996; OECD, 1997; EU Commission, 1997; CTI Press Release, 1997; Guy, 1997). In 1984 an international research cooperation network in the area of science and technology, linking 21 Spanish and Portuguese speaking countries from Europe and Latin America was created. This programme is supported by several international organisations and it involves different models of cooperation between universities, R&D centres and enterprises.

4.12.3   RTD and the innovation process

The private sector in developed countries invests substantial sums in RTD, and this figure is growing in developing countries as well. Thus, an understanding of firms' decisions to invest in RTD and the process through which decisions about new technologies are made are crucial given the importance of new technologies to improving productivity of labour and capital in organisations (Bolwijn and Kumpe, 1990).

The majority of traditional RTD innovation policies have been directed at closing the "resource" gap of firms -- that is, their available skills, investment capital, and science and technology base. But research has suggested that innovation fails at the level of capability - the ability to focus specific sets of resources in a particular way (*e.g.*, financial management, marketing, understanding user needs, etc.) - rather than because of inadequate resources.

The available literature available on the subject indicates that the degree of competitiveness of the industry the firm is in, and the extent to which a particular technology would enhance their competitive position or advantage determines the choice of technology the firm decides to produce (Baum, 1995). Georghiou (1998) argues along similar lines by suggesting that this process of technology choice by producers emphasises sources of competitive differentiation explicitly, and tries to organise efficient communication circles including all stakeholders who provide complementary knowledge. Early users of emerging technologies are identified. Reliable signals for value engineering and for technology evaluation and selection are transmitted back to the R&D community. Technology producers obviously require good R&D departments with the requisite expert staff. This is also determined by the relationship between public sector research institutions and innovating firms. Baum further suggests that choice of technology to produce be based on the enterprise model of user needs.

Such a model involves taking complete inventory of technology and the evaluation of technologies according to what best fits the current system of the firm. This is closely related to the changes in the way firms manage and organise R&D (Tschirky, 1994; Gerybadze, 1994). The current approach to R&D management plays a crucial role in the choice of technology. Gerybadze (1994) suggests three areas where R&D management plays such a crucial role, namely: *(i)* formulation and implementation of R&D strategy are considered as integral and vital elements of corporate strategy; *(ii)* balancing basic research and generic technologies with applied developments and with the needs of different business units; and *(iii)* implementation of a corporate-wide process of information exchange, strategy formulation and implementation.

The process of choice of technology by the user or recipient, on the other hand, depends to a large extent on the category of the recipient. Technology receiver/recipient may be a private company, a state-owned enterprise or local personnel in developing countries and will also include foreign affiliate (wholly or majority owned or joint venture) as well as supplier, customer and competitor enterprises. For foreign affiliates, technology choice is made through a parent organisation structure through its R&D or technology acquisition process, all with the explicit aim of improving its competitive position in foreign markets (Young and Lan, 1997). The experiences of Asian New Industrialising coun-

tries (NIC) showed the importance of accessing technology from abroad and utilising indigenous capacities to exploit the benefits of foreign technologies (Tolentino, 1993). State-owned enterprises normally go through a government-centralised bureaucracy to acquire technology. The choice is often made through a competitive bidding process. The local private sector, on the other hand, relies more on information about different technologies, and also with due recognition of the firm's organisational capacity to absorb and utilise the technology in question; an ability to determine the content of the technology in question is necessary to complement the existing capability of the recipient organisation (Lado and Vozikis, 1996).

From developing countries' standpoint, the willingness to obtain foreign technology also depends on the motives of the local firm or partner (in the case of partnership or joint venture), receiver's capabilities and ability to select from available technologies, market imported technologies and introduce a degree of novelty in products or processes (Young and Lan, 1997). The assimilation and adoption of outside technologies heavily depend on the development of in-house technological capabilities (Young and Lan, 1997; Lado and Vozikis, 1996). The role of government is critical to ensure sound management of the macro-economy, and also implementing policies on technology, trade, foreign direct investment (FDI), education and training and infrastructure development, and, finally, providing an enabling environment for technology acquisition to thrive (Dunning, 1992). The literature in technology transfer finds significant evidence of complementarity between imports of foreign technology and domestic R&D. Most R&D by developing countries' firms are adaptive in nature, *i.e.*, it adapts the foreign technology to the local conditions. Governments have been investing for three decades in RTD for environmentally sound energy technologies in the energy sector, often working with the private sector. There may be a case for seeing whether results from these processes have been used and disseminated sufficiently. Developing countries' RTD efforts are often adaptive, following externally developed technology, suggesting the need for additional resources to develop indigenous innovative capacity. RTD and the process of innovation are closely linked, but innovation has been found to fail at the level of capability - the ability to focus specific sets of resources in a particular way (*e.g.*, financial management, marketing, understanding user needs, etc.) - rather than because of inadequate resources or hardware. The process of dissemination depends closely on influencing key opinion formers at government departmental level, industry level, firm level or community level.

4.13 Conclusions

The discussion in this chapter on an enabling environment for technology transfer has produced several conclusions:

1. Governments can review whether their combination of laws concerning foreign investment, taxes, potential market growth and corruption would induce a firm to risk investment. Bureaucratic screening procedures, performance requirements and asset insecurity also act as barriers to for-

eign investment and could be addressed by policy changes that aim to attract foreign capital and technology.

2. Governments can devise participatory mechanisms and adopt processes which can harness the networks, skills and knowledge of community groups and non-governmental organisations to better meet user needs, avoid delays and achieve greater success with technology transfer.

3. Capacity-building is required at all stages in the process of technology transfer. It is a slow and complex process to which long-term commitments must be made for resources and to which the host country must also be committed if results are to bear fruit. Fundamental change requires an autonomous capacity to innovate, acquire and adapt technologies. Specific needs are for additional resources at the assessment and replication stages: (a) to ensure there is a broad national commitment to environmentally sound technologies and that those most appropriate for national circumstances are selected; and (b) to ensure that indigenous capacity for ongoing innovation is developed to encourage North-North, South-North and South-South flows. For adaptation there is a need to strengthen scientific expertise and institutions capable of undertaking the relevant assessments; to promote linkages with this infrastructure to other parts of the public and private sectors; and to identify relevant tools and techniques to produce outputs for nationally determined priorities.

4. Fundamental social structures and personal values (*i.e.*, social factors) evolve along with a society's physical infrastructure, private and public institutions, and the technologies embodied within them. New (climate friendlier) technological trajectories for an economy can provide great opportunities. Governments can support the establishment of dynamic flexible learning mechanisms within the private sector and with the rest of civil society, not only at national level -- national systems of innovation, but also at a level of subnational region –'the learning region.'

5. A market transformation approach promotes replicable, ongoing technology transfers rather than one-time transfers. In such an approach, public policy can consider what are the set of institutions which underlie markets and what are appropriate public interventions to shape those institutions. It is useful to consider a broad brush of measures within a market transformation approach such as: technology adaptation, the need for enterprise-technological capability, subsidies, consumer education, and marketing.

6. To reduce contract risk, property risk, and regulatory risks, particularly in private-sector-driven pathways, governments can strengthen national legal institutions for intellectual property protection; strengthen administrative and law processes to assure transparency, participation in regulatory policy-making, and independent review; and strengthen legal institutions to reduce risks.

7. Governments can work with the private sector and establish codes, standards and labels. This provides a framework which can work to the benefit of industry and consumers. This route can help build markets for dispersed, small-scale technologies where technologies are diverse, vendors are many, and consumers face high risks in evaluating and selecting technologies and suppliers. Codes and standards also provide a means for representing the interests (*i.e.*, to cost-effective levels of efficiency) of end-users who are absent from purchasing or construction decisions. Standards also allow consumers to be less knowledgeable about the equipment they are purchasing.

8. The principle of cost-effective reduction should not be uncritically applied in all cases, but rather each technology transfer case, whether government to government, private sector to private sector or community to community should be examined on a case by case basis to see what the social impacts will be and whether these are unavoidable or can be reduced. Governments may wish to develop criteria for ensuring that technology transfer projects do not disempower or negatively influence weaker social groups in a society. Such guidelines could draw from guidelines on integrating gender issues in technology development .

9. The experience in agriculture, forestry, and use of other natural resources has shown that the successful introduction of new technologies or modification of resource use often depends on a recognition of the existing forms of property rights, or on taking steps to create an optimal property rights regime. With an understanding of the existing – legal and actual – forms of property rights, technologies or modified resource uses can be adapted to fit the property relations. If property issues are taken into account, those introducing new technologies or proposing modifications in land or resource use can be more assured of the support of the target populations or groups.

10. Governments have been investing for three decades in RTD for environmentally sound energy technologies in the energy sector, often working with the private sector. There may be a case for seeing whether results from these processes have been used and disseminated sufficiently. Developing countries' RTD efforts are often adaptive, following externally developed technology, suggesting the need for additional resources to develop indigenous innovative capacity. RTD and the process of innovation are closely linked, but innovation has been found to fail at the level of capability— the ability to focus specific sets of resources in a particular way (*e.g.*, financial management, marketing, understanding user needs, etc.)—rather than because of inadequate resources or hardware. The process of dissemination depends closely on influencing key opinion formers, at governmentdepartmental level, industry level, firm level or community level.

References

Agarwal, B., 1991: Under the Cooking pot: The political economy of the domestic fuel crisis in rural south Asia. In *Women and the Environment: a Reader: Crises and Development in the Third World.*

Agarwal, A., and S. Narain, 1992: *Towards a Green World. Should Global Environmental Management be Built on Legal Conventions or Human Rights.* Centre for Science and Environment, New Delhi.

Agarwal, B., 1994. Gender and Command over Property: a Critical Gap in Economic Analysis and Policy in South Asia. *World Development*, **22**(10), 1455-1478.

Archibugi, D., and J. Michie (eds.), 1997: *Technology, Globalisation and Economic Performance.* Cambridge University Press, Cambridge, UK.

Archibugi, D., J. Howells, and J. Michie (eds.), 1999: *Innovation Policy in a Global Economy.* Cambridge University Press, Cambridge, UK.

Arnold, J.E.M., and J. G. Campbell, 1986: Collective Management of Hill Forests in Nepal: The Community Forestry Development Project. In *Common Property Resource Management.* National Research Council National Academy Press, Washington, DC.

Asheim, B.T., 1998 Learning Regions as Development Coalitions: Partnerships as Governance in European Welfare States. Paper to 2nd European Urban and Regional Studies Conference on Culture, Place and Space in Contemporary Europe, University of Durham, UK, 17-20 September 1998.

Banuri, T., K.Goran-Maler, M. Grubb, H.K. Jacobson, and F. Yamin, 1996: Equity and Social Considerations. In *Climate Change 1995: Economic and Social Dimensions of Climate Change.* Contribution of Working Group III to the Second Assessment Report of the Intergovernmental Panel on Climate Change, J. Bruce, H. Lee, E. Haites (ed.), Cambridge University Press, Cambridge, UK, pp. 79-124.

Banuri, T. *et al.*, 1997: Just Adjustment and Just Development. In *Just Development - Beyond Adjustment with a Human face.* T. J. Banuri, S. R. Khan, M. Mahmood (eds.), Oxford University Press, Oxford, UK, pp. 3-16.

Barnett, A., 1990: The diffusion of energy technology in the rural areas of developing countries: A synthesis of recent experience. *World Development*, **18**(4), 539-553.

Barnett A., 1995: *Do Environmental Imperatives Present Novel Problems and Opportunities for the International Transfer of Technology?* United Nations, New York, NY.

Baum, D., 1995: New technology burnout? Here is how to cope. *Datamation*, **15**(21), 45.

Bebbington, A., 1997: Debating 'indigenous' agricultural development: Indian organizations in the Central Andes of Ecuador. In *Green Guerrillas: Environmental Conflicts and Initiatives in Latin America and the Caribbean.* H. Collinson (ed.), Black Rose Books, Montreal.

Berg, E.J., 1993. *Rethinking Technical Co-operation*, UNDP, New York.

Bernardo, F.P., and G.U. Kilayko, 1990: Promoting Rural Energy Technology: The Case of Gassifiers in the Phillipines. *World Development*, **18**(4), 565-574.

Bernstein, J.I., 1996. International R&D spillovers between industries in Canada and the United States, social rates of return and productivity growth. *Canadian Journal of Economics*, Special Issue, **29** (April), S463-S467.

Black, B., R. Kraakman, and A. Tarassova, 1999: *Russian Privatization and Corporate Governance: What Went Wrong?* Working Paper #178, John M. Olin Program in Law and Economics, Stanford Law School. Available from the Social Science Research Network electronic library at http://papers.ssrn.com/paper.taf?abstract id=181348 .

Black, B., 1999: *The Legal and Institutional Preconditions for Strong Securities Markets: The Nontriviality of Securities Law.* Working Paper, September 1999. Available from the Social Science Research Network electronic library at http://papers.ssrn.com/paper?tafabstract id=182169.

Black, J.K., 1991: *Development in Theory and Practice: Bridging the Gap.* Westview Press, Boulder, CO.

Blackburn, J., and J. Holland, 1998: *Who Changes? Institutionalising participation in development.* Intermediate Technology Publications, London.

Blomstrom, M., A. Kokko, and M. Zejon., 1995: Host Country Competition and Technology Transfer by Multinationals. *Review of World Economics*, 521-533.

Bodansky, D. 1993: The United Nations Framework Convention on Climate Change: A Commentary. *Yale Journal of International Law*, **18**, 451-588.

Bolwijn, and Kumpe, 1990: Manufacturing in the 1990s - productivity and innovation. *Long Range Planning*, **21**, 44-57.

Brooks, H., 1995: What We Know and Do Not Know about Technology Transfer: Linking Knowledge to Action. In *Marshalling Technology for Development*, Proceedings of a symposium organised by the National Research Council and the World Bank, National Academy Press, Washington, DC, pp.83-96.

Cable, S. and C. Cable, 1995: *Environmental Problems, Grassroots Solutions: The Politics of Grassroots Environmental Conflict.* St. Martin's Press, New York, NY.

Cantwell, J., 1999: Innovation as the principal source of growth in the global economy. In *Innovation Policy in a Global Economy*. D. Archibugi, J. Howells, J. Michie, (eds.), Cambridge University Press, Cambridge, pp. 225-241.

Carlsson, B., and R. Stankiewicz, 1991. On the Nature, Function and Composition of Multinational Enterprise. In *The International Transfer of Technology: A Book of Readings*. R.D. Robinson, (ed.), Ballinger, Cambridge, MA..

Carney, J., and M.Watts, 1991. "Disciplining Women? Rice, Mechanization, and the Evolution of Mandinka Gender Relations in Senegambia. In *Signs*, **16** (4), 651-681.

Chambers, R., 1983: *Rural Development: Putting the Last First.* John Wiley and Sons, New York, NY.

Chambers, R., 1997: *Whose reality counts? Putting the last first.* Intermediate Technology Publications, London, pp. 297.

Chantramondklasri, N., 1990: The Development of Technological and Managerial Capability in the Developing Countries. In *Technology Transfer in the Developing Countries*. M. Chatterji, (ed.), St Martin's Press, New York, NY.

Chen, M.A., 1996: *Managing International Technology Transfer*, International Thomson Business Press, London

Chen, M.A., 1997: Building Research Capacity in the Non-Governmental Organization Sector. In *Getting Good Government - Capacity Building in the Public Sectors of Developing Countries*. M. S. Grindle, (ed.), Havard Institute for International Development, Cambridge, MA, pp. 229-253.

Clark, W.W., and E. Paolucci,1997: An industrial model for technology commercialisation: fuel cells into design manufacturing. Paper presented at the International Conference on Product Design and Manufacturing, Stockholm, May 1997.

Clark, W.W., 1999: Publicly-funded Research and Publicly-owned technologies in the Transfer and Diffusion of Environmentally Sound Technologies (ESTs): The case of the US. Paper prepared for the Division for Sustainable Development, Department of Economic and Social Affairs, July 1999.

Clark, W.W., 1999: Publicly-funded Research and Publicly-owned technologies in the Transfer and Diffusion of Environmentally Sound Technologies (ESTs): The case of the US. Paper prepared for the Division for Sustainable Development, Department of Economic and Social Affairs, July 1999.

Clarke, D.C., 1996: Power and Politics in the Chinese Court System: The Enforcement of Civil Judgements. *Columbia Journal of Asian Law*, **10**, 1-92.

Coe, D.T., and E. Helpman., 1995: International R&D spillovers. *European Economic Review*, **39**, 859-887.

Dailami, M. ,and M.Klein, 1998: Government Support to Private Infrastructure Project in Emerging Markets. World Bank Working Paper #1868.

Dale, G., 1995: Reforming the Energy Sector in Transition Economies: Selected Experience and Lessons. World Bank Discussion Paper #296, Washington DC.

De Zoeten-Dartenset, C.1999: *Towards a Methodological Framework for Trade Analysis.* Rotterdam School of Management, Rotterdam.

Denman, D.R., 1978: *The Place of Property.* Geographical Publications Limited, London.

Dezalay, Y., and B.G. Garth, 1996: *Dealing in Virtue: International Commercial Arbitration and Transnational Legal Order.* University of Chicago Press, Chicago.

Dixon R., 1998: Country Studies Overview: The US Country Studies Program. Paper presented at: National Assessment Results of Climate Change: Impacts and Responses, San Jose, Costa Rica March 25-28, 1998.

Dodgson, M., and J. Bessant, 1996: *Effective Innovation Policy: A New Approach.* International Thomson Business Press, London.

Doyle T., and D. McEachern, 1998: *Environment and Politics.* Routledge, London, pp. 206.

Dunning, J.H., 1992. The global economy, domestic governance, strategies and transnational corporations: interactions and policy implications. *Transnational Corporations,* **1**(3), 7-44.

Edwards, M., and D. Hulme, 1995: *NGO Performance and Accountability: Introduction and Overview.* Earthscan, London, p. 259.

Evenson, R.E., P.E. Waggoner, and V. W. Ruttan, 1979: Economic Benefits from Research: An Example from Agriculture. *Science,* **205** (14 September), 1101-1107.

Everts, S., 1998a: Technology and gender needs; an overview. In *Gender and Technology: Empowering women, engendering development.* S. Everts, (ed.), Zed Books, London, pp. 87-105.

Everts, S., 1998b: *Gender and Technology: Empowering women, engendering development.* Zed Books, London.

Fairbanks, J., 1992: *Indigenous technical knowledge and natural resource management a critical overview.* NRI, Chatham, UK.

Feder, G., T. Onchan, Y. Chalamwong, and C.Hongladarom, 1988: *Land Policies and Farm Productivity in Thailand.* A World Bank Research Publication. Johns Hopkins University Press, Baltimore.

Feenstra, J.F., I.Burton, J.B. Smith, and R.S.J. Tol, 1998: *Handbook on Methods for Climate Change Impact Assessment and Adaptation Strategies - Version 2.* United Nations Environment Programme and Institute for Environmental Studies.

Feldman, S., 1994: *Market Transformation: Hot Topic or Hot Air?* Proceedings of the 1994 ACEEE Summer Study on Energy Efficiency in Buildings, American Council for an Energy-Efficient Economy. Washington, DC.

Foley, G., 1993: Renewable energy in Third World development assistance— learning from experience. In *Renewable Energy: Prospects for Implementation.* T.Jackson, (ed.), Stockholm Environment Institute, Stockholm, pp.193-202.

Freeman C., 1999: Foreword. In *Innovation Policy in a Global Economy.* D, Archibugi, J. Howells, J. Michie. (eds.), Cambridge University Press, Cambridge, pp. xiii - xiv.

Freeman, C., 1987: *Technology Policy and Economic Performance: Lessons from Japan.* New York, NY.

Geller, H., and S. Nadel, 1994: Market Transformation Strategies to Promote End-Use Efficiency. *Annual Review of Energy and the Environment,* **19**, 301-346.

Georghiou, L., 1998: Issues in the evaluation of Innovation and Technology Policy. *Evaluation,* **4**(1), 37-51.

Gerybadze, A., 1994: Technology Forecasting as a process of organisational intelligence. *R&D Management,* 24(2), 131-141.

Goldemberg, J., T.B. Johansson (eds.), 1995. *Energy as an Instrument for Socio-Economic Development.* United Nations Development Program, New York, NY.

Golove, W.H, and J.H. Eto, 1996: *Market Barriers to Energy Efficiency: A Critical Reappraisal of the Rationale for Public Policies to Promote Energy Efficiency,* LBL-38059. Lawrence Berkeley National Laboratory Berkeley, CA.

Goto, A., and K. Suzuki, 1989: R&D Capital, Rate of Return on R&D Investment and Spillover of R&D in Japanese Manufacturing Industries. *The Review of Economics and Statistics,* **71**(4 -, (November), 555-564.

Greenpeace, 1997a: Benefits of methyl bromide phase out in Article 5 countries, http://www.greenpeace.org/~ozone/greenfreeze/env_imp/11benefit.html.

Greenpeace, 1997b: Greenpeace exposes import of dirty technology to Israel, http://www.greepeace.org/majordomo/index-pres-releases/1997/msg00021.html.

Griliches, Z., 1992: The Search for R&D Spillovers. *Scandinavian Journal of Economics,* **94** (Supplement), S29-S47.

Grindle, M.S., 1997: The Good Government Imperative:Human Resources, Organizations, and Institutions. In *Getting Good Government - Capacity Building in the Public Sectors of Developing Countries.* M.S. Grindel, (ed.), Havard Institute for International Development, Havard University, Boston, MA, pp.3-30.

Grove-White, R., S.Kapuitza, and V. Shiva, 1992: Public Awareness, Science and the Environment.. In *An Agenda of Science for Environment and Development into the 21st Century.* J.C.I. Dooge, G.T. Goodman, J.W.M. la Riviere, J. Marton-Lefevre, T.O'Riordan, F. Praderie, Cambridge University Press, Cambridge.

Grutter, J., 1996: *How to prevent corruption in pollution control programs, Vol. 1.* In Proceedings of the World Congress on Air Pollution in Developing Countries, Costa Rica, October 21-26 1996.

GTZ, 1995: *Basic Electrification for Rural Households: Experience with the Dissemination of Small-Scale Photovoltaic Systems.* Eschborn, Germany.

Gupta, J.,1997: *The Climate Change Convention and Developing Countries: from Conflict to Consensus?* Kluwer Academic Publishers, Dordrecht.

Guy, K., 1997: *Enhancing Foreign Access to Technology Programmes: Trends and Issues.*

Harkes, I.H.T., 1998: Project Success: different perspectives. Paper to Conference on Participatory Rural Resource Management, Mansfield College, University of Oxford, April, 1998.

Hastie, S.M., 1995: Market Transformation in a Changing Utility Environment. Prepared for the National Association of Regulatory Utility Commissioners (NARUC), Synergic Resources Corp., Bala Cynwyd, PA.

Heaton, G., R. Repetto, and R. Sobin, 1991: *Transforming Technology: An Agenda for Environmentally Sustainable Growth in the 21st Century.* World Resources Institute, Washington, DC.

Heiman F., and N. Zucker Boswell, 1998: The OECD Convention: Milestone on the Road to Reform. In *New Perspectives on Combating Corruption.* Publication prepared for Transparency International's Annual General Meeting held in Kuala Lumpur, Malaysia, 11-16 September 1998 and the IMF -World Bank Annual Meetings held in Washington, DC, 6-8 October 1998. Transparency International and Economic Development Institute, pp. 65-74.

Hines, J.R., and E.M. Rice, 1994: Fiscal Paradise: Foreign Tax Havens and American Business. *Quarterly Journal of Economics,* **109,** 149-182.

Hirsch, P., and L.Lohman, 1989: Contemporary Politics of the Environment in Thailand. *Asian Survey,* **XXIX**, (4 - April), 439-451.

Hoffman, K., and N.Garvin, 1990: *Managing International Technology Transfer; A Strategic Approach for Developing Countries.* IDRC-MR259e, International Developing Research Center, Ottawa.

Hodgson, G.M., W.J. Samuels, and M.R. Tool (eds.), 1994: *The Elgar Companion to Institutional and Evolutionary Economics, 2 vols.* Elgar Publishing Co., Hants, UK.

Howells, J., 1999: Regional Systems of Innovation. In *Innovation Policy in a Global Economy,* D. Archibugi, J. Howells, J, Michie, (eds.), Cambridge University Press, Cambridge, pp. 67-93.

Huang, Y., 1998: *FDI in China.* Chinese University Press, Hong Kong.

Hurst, C. 1990: Establishing new markets for mature energy equipment in developing countries: Experience with windmills, hydro-powered mills and solar water heaters.*World Development,* **18**(4), 5-615.

Imai, M. 1997: *Gemba Kaizen; a common sense, low cost approach to management.* McGraw Hill, New York, NY.

Indian Journal of Public Administration, 1998: Towards Good Governance. Special Issue, **44**(3).

Jacobson, C.D., and J.A. Tarr, 1995: Ownership and Financing of Infrastructure. Policy Research Working Paper #1466, World Bank.

Jain, S.C., 1985: *Women and Technology.* Rawat Publications, Jaipur, India.

Johnston, M., 1998: Fighting systemic corruption: social foundations for Institutional Reform. *The European Journal of Development Research,* **10**(1), 85-104.

Jones, Charles I., and J.C. Williams, 1998: Measuring the social return to R&D. *Quarterly Journal of Economics,* **113**(4), 1119-1135.

Justman, M., and M. Teubal, 1995 Technology Infrastructure Policy (TIP): Creating Capabilities and Building Markets. *Research Policy,* **24**, 259-281.

Kammen, D.M., 1998a: The Commercialization of Photovoltaic Systems in Kenya: Technology Transfer a few Watts at a time. Case Study 5.

Kammen, D.M., 1998b: Research, Development and Commercialization of the Kenya Ceramic Jiko and other Improved Biomass Stoves in Africa. Case Study 7.

Kates, R.W., 1997: Climate change 1995-impacts, adaptations, and mitigation. *Environment,* **39**(9), 29-33.

Khan, M.H., 1998: Patron-client networks and the economic effects of corruption in Asia. *The European Journal of Development Research,* **10**(1), 15-39.

Kindra, G., 1998: Social marketing strategies to fight corruption. In *New Perspectives on Combating Corruption.* Publication prepared for Transparency International's Annual General Meeting held in Kuala Lumpur, Malaysia, 11-16 September 1998 and the IMF –World Bank Annual Meetings, held in Washington, DC, 6-8 October 1998. Transparency International and Economic Development Institute, pp. 75-86.

Klein, R. J. T., and D.C. MacIver, Adaptation to climate variability and change: methodological issues. *Mitigation and Adaptation Strategies for Global Change,* in press.

Klitgaard, R. 1998: International cooperation against corruption. In *New Perspectives on Combating Corruption.* Publication prepared for Transparency International's Annual General Meeting held in Kuala Lumpur, Malaysia, 11-16 September 1998 and the IMF- World Bank Annual Meetings, held in Washington, DC, 6-8 October 1998. Transparency International and Economic Development Institute.

Kozloff, K., and O. Shobowale, 1994: *Rethinking Development Assistance for Renewable Electricity.* World Resources Institute, Washington, DC.

Krugman, P. R., 1991: *Geography and Trade.* The MIT Press, Boston, MA.

Lado, A.A., and G.S. Vozikis, 1996: Transfer of Technology to promote entrepreneurship in developing countries: An integration and proposed framework.*Entrepreneurship Theory and Practice,* **21** (2 - Winter), 55-72..

Lambsdorff, J.G., 1998: An Empirical Investigation of Bribery in International Trade. *The European Journal of Development Research,* **10**(1), 40 - 59.

Lanjouw, J.O., and A. Mody, 1996: Innovation and the international diffusion of environmentally responsive technologies. *Research Policy,* **25** (4 – June).

Lashof, D., and D. Tirpak, 1990: *Policy Options for Stabilizing Global Climate.* Report to Congress, (Main Report), USEPA, Washington, DC.

Lee, M.J., and S.C. Chung, 1997: *Globalisation of Industrial Research and Development: The Korean Experience.*

Levine, M.D., E. Hirst , J.G. Koomey, J.E. McMahon, and A.H. Sanstad, 1994: *Energy Efficiency, Market Failures, and Government Policy.* Lawrence Berkeley Natl. Laboratory, Berkeley, CA.

Leys, C., 1975: *Underdevelopment in Kenya: The Political Economy of Neo-Colonialism.* The University of California Press, Berekely.

Lin See-Yan, 1997: Chairman's Summary. In *Finance for Sustainable Development - The Road Ahead.* Proceedings of the Fourth Group Meeting on Financial Issues of Agenda 21, Satiago, Chile. United Nations, New York NY.

Lipton M., and R. Longhurst, 1989: New Seeds and Poor People. *Johns Hopkins Studies in Development.* Johns Hopkins University Press, Baltimore, MA.

Lubman, S.B., 1998: *The Legal and Policy Environment for Foreign Direct Investment in China: Past Accomplishments, Future Uncertainties.* Symposium Private Investments Abroad 1997, Matthew Bender & Co., New York, NY.

Lucas, R.E. Jr., 1988: On the Mechanics of Economic Development. *Journal of Monetary Economics,* **22**(1 - July), 3-42.

Lundvall, B.-A. 1999: Technology policy in the learning economy. In *Innovation Policy in a Global Economy.* D. Archibugi, J. Howells, J. Michie, (eds.), Cambridge University Press,Cambridge, pp.19-34.

Lynch, O., and J. Alcorn, 1994: Tenurial Rights and Community-based Conservation. In *Natural Connections.* D. Western, R. M. Wright, (eds.), Island Press, Washington, DC, pp. 378-380.

Madu, C. N., and R. Jacob, 1989: Strategic Planning in Technology Transfer: A Dialectical Approach. *Technological Forecasting and Social Change,* **35**, 327-338.

Malhotra, P., S.Dutta, and V. Ramana, 1998: *Participatory Rural Energy Planning - A Handbook.* Tata Energy Research Institute, New Delhi.

Manas, C. (ed.), 1990: *Technology Transfer in Developing Countries.* St. Martin Press, New York, NY.

Margolis, R.M., and D.M. Kammen, 1999: Under-investment: The energy technology and R&D policy challenge. *Science,* **285** (July)..

Merquior, J. G., 1993: A Panoramic View of the Rebirth of Liberalism. *World Development,* **21** (8).

Metcalfe, J.S., 1995: Technology systems and technology policy in an evolutionary framework. *Cambridge Journal of Economics,* **19,** 25-46. Reprinted in Archibugi and Michie, (eds.),1997b.

Metz, B., 1999: International equity in climate change policy: an overview of summary and submitted papers. EFIEA Policy Workshop, Integrating Climate Change Policies Costs and Opportunities, March 4-6, 1999.

Meyn, M., 1997: Puerto Rico's energy fix. In *Green Guerrillas: Environmental Conflicts and Initiatives in Latin America and the Caribbean: A Reader.* H. Collinson, (ed.), Black Rose Books, Montreal, pp. 168-77.

Mies, M., and V. Shiva, 1993: *Ecofeminism.* Zed Books, London.

Morse, B., and T.R. Berger, 1992: *Sardar Sarovar,* Report of the Independent Commission, Resource Futures International, Ottawa.

Mueller, B., A. Lee, G.D. Libecap, and R. Schneider, 1994: Land, Property Rights and Privatisation in Brazil. *The Quarterly Review of Economics and Finance,* Special Issue, **34** (Summer), 261-280.

Mugabe, J., 1996: Technology and Sustainable Development in Africa – Building policy and institutional capacities for needs assessment. Background paper prepared for the UN Division for Policy Co-ordination and Sustainable Development, New York, NY.

National Research Council, 1986: *Common Property Resource Management.* National Academy Press, Washington, DC.

Nelson, R.R. (ed). 1993: *National Innovation Systems: A Comparative Analysis.* Oxford University Press, New York, NY.

Neefjes, K., 1998: Learning on Participatory Environmental Impact Assessment of Community Centred Development; the Oxfam experience. Paper to Conference on Participatory Rural Resource Management, Mansfield College, University of Oxford, April, 1998.

Norberg-Bohm, V., and D. Hart., 1995: Technological Cooperation: Lessons from Development Experience. In *Shaping National Responses to Climate Change: A Post-Rio Guide.* Henry Lee, (ed.), Island Press, Washington, DC, pp.261-288.

Nye, J.S., 1967: Corruption: a cost - benefit analysis. *American Political Science Review,* **LXI** (2).

ODA, 1994: A synthesis study of Institutional Capacity Development Projects for Overseas Development Administration.

Oksenberg, M., P. Potter, and W. Abnett, 1996: *Advancing Intellectual Property Rights: Information Technologies and the Course of Economic Development in China.* Annual Papers 7, Number 4, National Bureau of Asian Research, Seattle, WA.

Organisation for Economic Co-operation and Development (OECD), 1994: *Effective Technology Transfer, Co-operation, and Capacity Building for Sustainable Development: Common Reference Paper.* OECD Working Paper No. 75. Paris.

Organisation for Economic Co-operation and Development (OECD), 1995: *Developing Environmental Capacity - A Framework for Donor Involvement.* OECD Publications, Paris

Organisation for Economic Co-operation and Development (OECD), 1996: Climate Technology Initiative – Inventory of Activities, Paris.

Organisation for Economic Co-operation and Development (OECD), 1997: *Evaluation of Programs Promoting Participatory development and Good Governance. Synthesis report.* DAC Expert group on aid evaluation.

Organisation for Economic Co-operation and Development (OECD), 1998a: Development Cooperation and the Response to Kyoto. OECD/IEA Forum on Climate Change, 12-13 March 1998, Paris.

Organisation for Economic Co-operation and Development (OECD), 1998b: *Development Co-operation and the Response to Kyoto,* COM/ENV/EPOC/DCD/DAC//IEA (98)3, OECD, Paris, 10 pp.

Organisation for Economic Co-operation and Development (OECD), 1999: Managing National Innovation Systems, Paris.

Oviedo , and F. Bossano, 1996: *Development of Fuel Policies and Technical Norms for new Vehicles in Ecuador.* Proceedings (vol. 1) of the World Congress on Air Pollution in Developing Countries, Costa Rica, October 1996, pp.361-364.

Pachauri, R.K., P. Bhandori (eds.), 1994: *Climate Change in Asia and Brazil: the role of Technology Transfer,* Energy Research Institute.

Panayotou, T., 1997. Taking stock of trends in sustainable development since Rio. In *Finance for Sustainable Development - The Road Ahead.* J. Holst, P. Koudal, J. Vincent, United Nations, New York, NY, pp. 35-74.

Parry, M., N. Arnell, M. Hulme, R. Nicholls, and M. Livermore, 1998: Adapting to the inevitable. *Nature, 395,* 741.

Parry, M., and T. Carter, 1998: *Climate Impact and Adaptation Assessment.* Earthscan Publications Ltd, London.

Pearse, A., 1980: *Seeds of Plenty, Seeds of Want: Social and Economic Implications of the Green Revolution.* Oxford University Press, New York, NY.

Peet, R., and M. Watts, 1996: Liberation Ecology. In *Liberation Ecologies: environment, development, social movements.* R. Peet, M. Watts (eds.), Routledge, London and New York, pp.1-45.

Peluso, N.L., 1992a: The Ironwood Problem: (Mis)Management and Development of an Extractive Rainforest Product. *Conservation Biology,* 6(2), 1-10.

Peluso, N.L., 1992b: The Political Ecology of Extraction and Extractive Reserves in East Kalimantan, Indonesia. *Development and Change,* 23(4), 49-74.

Pielke, R.A. Jr., 1998: Rethinking the role of adaptation in climate policy. *Global Environmental Change,* 8(2), 159-170.

Planning, 1999: *The Journal of the Royal Town Planning Institute,* 12 November.

Poffenberger, M. (ed.), 1990: *Keepers of the Forest: Land Management Alternatives in Southeast Asia.* Kumarian Press, West, Hartford, CN.

Porter, M.E., 1990: *The competitive advantage of nations.* MacMillan, London.

Potter, D., and Taylor, A, 1996: Introduction. In *NGOs and Environmental Policies : Asia and Africa.* D. Potter, (ed.), Frank Cass, London.

Princen, T., and Finger, M., 1994: *Environmental NGOs in World Politics: Linking the Local to the Global.* Routledge, London and New York.

Rama, M., 1997: Efficient Public Sector Downsizing. Policy Research Working Paper #1840, Policy Research Department, World Bank, Washington, DC.

Reddy, A.K.N., 1976: The Trojan Horse. *Ceres 50,* 9(2), 40-43.

Reddy, A.K.N., 1998: Development Conflicts for a New Agenda. In *The Hindu Survey of the Environment,* pp. 9-13.

Repetto, R.C., 1988: *The forest for the trees? : government policies and the misuse of forest resources.* World Resources Institute, Washington, DC.

Romer, P.M., 1986a: Increasing Returns and Long-Run Growth. *Journal of Political Economy,* 94(5), 1002-1037.

Romer, P.M., 1986b: Cake Eating, Chattering, and Jumps: Existence Results for Variational Problems. *Econometrica,* 54 (July), 897-908.

Romer, P. M., 1988: Capital Accumulation and Long-Run Growth. In *Modern Business Cycle Theory.* R. Barro, (ed.), Harvard University Press, Cambridge, MA.

Romer, P.M., 1994: The Origins of Endogenous Growth. *Journal of Economic Perspectives,* 8(1 - Winter), 3-22.

Rose-Ackerman, S., 1996: Democracy and 'Grand' Corruption. *International Social Science Journal,* 48, 365-380.

Saviotti, P., and S. Metcalfe (eds.), 1991: *Evolutionary Theories of Economic and Technological Change.* Harwood, London.

Scott, J., 1985: *Weapons of the Weak: Everyday Forms of Peasant Resistance.* Yale University Press, New Haven, CN.

Sen, G., 1995: National Development and local environmental action – the case of the River Narmada. In *The North the South and the Environment.* V. Bhaskar, A. Glyn, (eds.), Earthscan/ United Nations University Press, Tokyo, pp. 184-200.

Shah, A., and J. Slemrod, 1995: *Do Taxes matter for Foreign Direct Investment?* Oxford University Press for the World Bank. Collection Volume Article.

Shang, Jin Wei., 1997: How Taxing is Corruption on International Investors? NBER Working Paper #W6030.

Shiva, V., and J. Bandyopadhyay, 1986: Environmental Conflicts and Public Interest Science. *Economic and Political Weekly,* 21(2), 84-90.

STAP (Science and Technical Advisory Panel of the Global Environment Facility), 1996: *International Industrial Collaboration for accelerated adoption of environmentally sound energy technologies in developing countries.* Report of a STAP Workshop, Amsterdam, 7-8 June 1996.

Stewart, S., 1997: The Prices of a Perfect Flower: environmental destruction and health hazards in the Colombian flower industry. In *Green Guerrillas: Environmental Conflicts and Initiatives in Latin America and the Caribbean, a Reader.* H. Collinson (ed.), Black Rose Books, Montreal, pp.132-39.

Suzuki, K., 1993: R&D spillovers and technology transfer among and within vertical keiretsu groups: Evidence from the Japanese electrical machinery industry. *International Journal of Industrial Organisation,* 11, 573-591.

Tassey, G., 1991: The Functions of Technology Infrastructure in a Competitive Economy. *Research Policy,* 20, 345-361.

Theobald, R.1999: So what really is the problem about corruption? *Third World Quarterly,* 20(3), 491-502.

Thiesenhusen, W.C., 1991: Implications of the Rural Land Tenure System for the Environmental Debate: Three Scenarios. *The Journal of Developing Areas,* 26(1), 1-24.

Thompson, J., 1998: Participatory Approaches in Government Bureaucracies: facilitating institutional change. In *Who Changes? Institutionalising participation in development.* J. Blackburn, J. Holland, (eds.), Intermediate Technology Publications, London.

Thomson, J.T., D.H. Feeny, and R.J. Oakerson, 1986: Institutional Dynamics: The Evolution and Dissolution of Common Property Resource Management. In *Common Property Resource Management.* National Research Council, National Academy Press, Washington, DC.

Tolentino, P.A.E., 1993: *Technological Innovation and Third World Multinationals.* Routledge, London and New York.

Tschirky, H.P., 1994: The role of technology forecasting and assessment in technology management. *R&D Management,* 24 (2), p.121.

UNCTAD, 1995: Controlling carbom dioxide emissions: the tradeable permit system. UNCTAD, Geneva, UNCTAD/GID/11.

UNDP/HIID, 1996: Building Sustainable Capacity: Challenges for the Public Sector

UNHABITAT, 1996: The Habitat Agenda, http://www.unhabitat.org/agenda/ch-4e-4html.

United Nations Conference on Trade and Development (UNCTAD), 1990: *Transfer and Development of Technology in the Least Developed Countries: An Assessment of Major Policy Issues.* Report UNCTAD/ITP/TEC/12, United Nations, New York, NY.

United Nations Conference on Trade and Development (UNCTAD), 1995: *Technological Capacity-Building and Technology Partnership: Field Findings, Country Experiences and Program,* New York and Geneva.

United Nations Conference on Trade and Development (UNCTAD), 1997: *Promoting the Transfer and Use of Environmentally Sound Technologies: A Review of Policies,* New York, NY.

United Nations Conference on Trade and Development (UNCTAD). 1998: The role of publicly-funded research and publicly-owned technologies in the transfer and diffusion of environmentally sound technologies. Prepared by the UNCTAD secretariat in co-operation with UNEP and DESA as background to the International Expert Meeting, January 1998.

United Nations Economic and Social Council, Commission on Sustainable Development (UNCSD – First session), 1993: *Progress Achieved in Facilitating and Promoting the Transfer of Environmentally Sound Technology, Co-operation and Capacity-Building: Report of the Secretary-General,* E/CN.17/1993/10, New York, NY.

United Nations, 1993: *Agenda 21: Program of Action for Sustainable Development.* United Nations, New York, NY.

Van Berkel, R., C. Westra, F. Verspeek, P. Lasschuit, and L. Pietersen, 1996: *National Needs Assessment for Environmentally Sound Technologies for Developing Countries: Methodologies and Comparative Evaluation.* University of Amsterdam, Amsterdam.

Van Berkel, R., H. Blonk, C. Westra, and L. Pietersen, 1997: *A Primer on Climate Relevant Technology Transfer.* IVAM Environmental Research/Report 1997-3, Amsterdam, 48pp.

Van Berkel R., and E. Arkesteijn, 1998: *Participatory Approach for Enhancing Climate-relevant Technology Transfer: Opportunities for International Collaboration.* IVAM Environmental Research, Amsterdam, 12pp.

World Bank, 1994a: *Evaluating Capacity Development.* Report of the task force.

World Bank, 1994b: *World Development Report - Infrastructure for Development*, Oxford University Press, New York, NY.

World Bank, 1996: *World Bank Participation Sourcebook.* Environment Department Papers 019, February 1996.

World Bank, 1997a: Social capital: The Missing Link? In *Monitoring Environmental Progress-expanding the measure of wealth.* Indicators and Environmental Valuation Unit, Environment Department, The World Bank, January 1997.

World Bank, 1997b: *World Development Report - The State in a Changing World,* Oxford University Press, New York, NY.

Young, S., and P. Lan, 1997: Technology Transfer to China through Foreign Direct Investment. *Regional Studies*

5

Financing and Partnerships for Technology Transfer

Coordinating Lead Authors:
MARK MANSLEY (UNITED KINGDOM), ERIC MARTINOT (USA)

Lead Authors:
Dilip Ahuja (India), Weerawat Chantanakome (Thailand), Stephen DeCanio (USA), Michael Grubb (United Kingdom), Joyeeta Gupta (The Netherlands), Li Junfeng (China), Merylyn McKenzie Hedger (United Kingdom), Bhaskhar Natarajan (India), John Turkson (Ghana), David Wallace (United Kingdom)

Contributing Authors:
Ron Benioff (USA), Ibrahim Abdel Gelil (Egypt)

Review Editor:
Karen R. Polenske (USA)

CONTENTS

EXECUTIVE SUMMARY

Financing is an important dimension of environmentally sound technology transfer. This chapter looks at the practical issues involved in conducting technology transfers, and explores a wide range of mechanisms and approaches. Initiatives need to be carefully tailored to the relevant circumstances and objectives. Many environmentally sound technologies (ESTs) are essentially new – often requiring change and innovation in the relevant institutions to support their transfer, such as new partnerships, new financing mechanisms, new information distribution, and new models for participation.

Public finance has a crucial role in supporting the transfer of ESTs, especially in the absence of pricing that incorporates environmental costs. Public finance has different roles than private finance, which can vary by investment type and sector. For example, it is more important for long-term and infrastructure investments. Public finance remains central in the coastal zone adaptation and transport sectors, and still plays a large role in the energy sector despite growing private finance in some countries. There has been increasing interest in opening public infrastructure development to the private sector, for example, by privatising state-owned companies, opening markets to competition, and opening projects to private finance.

Official Development Assistance (ODA) is still significant for the economies of the poorest developing countries. There is increasing recognition that ODA can best be focused on mobilising and multiplying additional financial resources. ODA can also assist the improvement of policy frameworks and take on long-term commitments to capacity building. Donor coordination by the host country is key to avoiding distortions such as those induced by tied aid, which can be detrimental to the technology transfer process, preventing the establishment of the institutions to support technology choice, financing, operation and management, etc. More generally, trade support (*e.g.*, export credits) rarely takes account of environmental factors and may in many respects be biased against environmentally sound technologies.

Because of the ongoing shift in many countries from the public to the private sector as a principal source of finance, maximising the support for technology transfer may require a new degree of financial innovation and an increased emphasis on new or different forms of finance such as microcredit, leasing and venture capital. While the private sector has started a number of environmental initiatives, there is scope for governments to enhance these mechanisms through partnerships.

There is a need for adequate resources to enable adequate project preparation in the agreement stage of all pathways, yet this is precisely the area where funding can be most difficult to obtain, particularly for private-sector-driven and community-driven pathways. Project preparation needs to consider issues of financing and participation. There is wide scope for governments, both industrialised and developing, as well as multilateral organisations, to provide support directly for the project preparation process in private-sector-driven and community-driven pathways.

Public-private partnerships are becoming increasingly important, because the relationship between government and private finance has changed considerably in recent years in many countries. These partnerships can involve a mixture of governments at national and local levels, private firms (companies), private financial institutions, and non-governmental organisations. Examples include voluntary agreements, technology partnerships, information dissemination to the financial sector and support for the development of innovative financial instruments. There have been a number of examples in these areas, many of them funded by the multilateral development banks and the Global Environment Facility (GEF).

Technology intermediaries are an important form of financing and/or partnership that can overcome barriers associated with information, management, technology, and financing. Information clearinghouses are simple forms of technology intermediaries, but policies can create an environment that encourages more sophisticated forms of technology intermediaries, such as technology-specific national-level institutions, energy-service companies, and electric power utilities.

There is scope for governments to more formally organise, develop and report on the practical initiatives they undertake in support of environmentally sound technology transfer. A formal programme could monitor activities, disseminate best practices, and develop new ideas and initiatives. These initiatives could encompass a variety of action-oriented interventions that support technology transfer, typically based on addressing specific problems, and incorporating both private- and public-sector involvement.

5.1   Introduction

Finance is a critical aspect of technology transfer. This chapter reviews various funding sources and financial mechanisms for conducting EST transfers, and the types of partnerships and stakeholder relationships that can support technology transfers. The chapter looks at the practical issues involved and explores a range of mechanisms and approaches. As discussed in Chapter 4, greater emphasis is being placed on more participatory models of technology transfer and creation of "social infrastructure." This chapter continues this discussion with a review of different forms of public-private partnerships.

Introduction of a new technology into a country usually requires investment, as does the diffusion of existing technologies within a country. Technology adaptation may also require substantial investments in design and/or production. Financing is also often required (and particularly difficult to obtain) in the early (developmental) phases of a technology transfer project or business. Without financing, very little technology investment or transfer takes place. The provision of financing depends upon those who have financial resources—whether governments or the private sector—being convinced that projects and the businesses that run them will justify the financial support or investment. And investment in ESTs and businesses will depend upon governments and private investors being convinced that these will justify—by whichever criteria they apply—the expenditure. This is the financial reality that underpins all technology investment and transfer processes. However, financing perspectives may differ enormously not only according to the project, technology and business, but according to the investor. Thus governments may offer a range of financing possibilities that differ radically from the private sector—and each contains enormous diversity itself.

Table 5.1 summarises the policy tools available related to financing and partnerships, and the barriers these are designed to overcome. The chapter begins by considering the role and scope of participatory techniques to help promote stakeholder dialogues and partnership. The chapter then considers the investment decisions made by private firms that bear on climate-change mitigation and technology transfer for private-sector-driven pathways. Public-sector finance and investment, which is of key significance for many forms of transfer, is considered in terms of direct government finance, official development assistance, and multilateral development bank lending. The section on private-sector finance and investment, which is becoming increasingly important in both the national and international diffusion of technology, discusses a broad range of financial mechanisms and modalities for finance within the private sector. Because the relationship between the public and private sectors has already changed markedly in recent years in many countries, public-private partnerships are discussed. Public-private partnerships can combine the positive attributes of both sectors and provide increasingly impor-

tant opportunities to promote technology transfer. Finally, technology intermediaries are discussed as important mechanisms to overcome barriers associated with information, management, technology, and financing.

5.2   Public-Sector Finance and Investment

Public sector finance inevitably has a substantial role in investing in environmentally sound technologies and otherwise supporting the transfer of ESTs (see also section 2.2 in Chapter 2 on the public sector contribution in international financial flows)[1]. At a fundamental level, much of this involvement arises because the public sector has direct responsibility for managing public and common goods, and investing in their protection and conservation. The role of public sector finance becomes particularly important in supporting the development and dissemination of ESTs in the absence of efficient pricing mechanisms or other policies to incorporate environmental costs, when the private sector finance will be unable to operate efficiently.

The public sector typically directly invests in a range of infrastructure, although this is changing. There has been increasing interest in opening public infrastructure development to the private sector, for example, by privatising state owned companies, opening markets to competition, and opening projects to private finance.

The public sector can also provide various incentives (tax benefits, grants, subsidies, etc.) to private firms to encourage investment in ESTs – these can cover R&D grants, project subsidies, support for information dissemination and support for trade activities (note that several of these are covered primarily in Chapter 4).

The public sector is a major purchaser of goods and services, and can use its purchasing power to buy ESTs.

Public finance has different roles from private finance, being more important with respect to long-term and infrastructure investment, and assumes different roles in different sectors. For example, it remains central in the coastal-zone adaptation and transport sectors, and still plays a large role in energy alongside rapidly growing private finance.

---

[1] A large number of case studies in the Case Studies Section, Chapter 16, address the use of subsidies to promote market development. These include: Wind in Inner Mongolia (case 3), Butane gas stove by TOTAL (case 7), renewables in Ladakh (case 14). Innovative private sector initiatives include: Mobil (case 13), Green Lights (case 2), PV in Kenya (case 5), micro-hydro in Peru (case 25), GEF in India (case 22). Public-private partnerships are illustrated by cases 4, 17, 22 and 23.

Table 5.1. Policy Tools for Financing and Participation (Source: Mansley et al., 1997a, b)		
POLICY TOOL	**BARRIERS ADDRESSED**	**RELEVANCE**
PUBLIC-SECTOR FINANCE AND INVESTMENT (5.2)		
• Provide direct finance • Provide official development assistance • Provide multilateral development bank finance	• Lack of confidence in "unproven" technology • Lack of access to capital • User acceptability • Costs of developing new public infrastructure • Lack of policy harmonisation • Uncertain future energy prices	Government-driven and community-driven pathways All sectors All stages
PRIVATE-SECTOR FINANCE AND INVESTMENT (5.3)		
• Support, through a variety of policy tools, private-sector financing mechanisms such as microcredit, leasing, venture capital, project finance, "green" finance, and a range of other private-sector financing initiatives. • Reduce perceived risks through consistent policies and transparent regulatory frameworks (see Ch. 4)	• Lack of access to capital • High transaction costs • High front-end capital costs • High user discount rates • Inadequate financial strength of smaller firms • Lack of information • Lack of confidence in "unproven" technology	Private-sector-driven pathways All sectors Agreement and implementation stages
PRIVATE-FIRM INVESTMENTS (5.4)		
• Create incentives for firms to make environmentally sound investments, such as energy taxes, investment tax credits, and emissions fees (see Ch. 4) • Engage firms in public-private partnerships, as discussed in 5.6, particularly to overcome managerial misincentives	• Managerial misincentives • Competing purchase decision criteria • Sunk investments in existing equipment and infrastructure • Energy-supply-side financing bias • Lowest-cost equipment favoured	Private-sector-driven pathways Buildings, transport, industry, and energy sectors All stages
PUBLIC-PRIVATE PARTNERSHIPS (5.5)		
• Enter into voluntary agreements with the private-sector • Develop technical partnership programmes • Conduct informational initiatives • Provide tax incentives, guarantees, and trade finance • Promote new financial initiatives	Barriers are similar to those for public-sector and private-sector financing and investment, plus the following: • Uncertain markets for technologies • Import duties • Utility acceptance of renewable energy technologies • Permit risk • Environmental externalities • Shortage of trained personnel	All pathways All sectors All stages
TECHNOLOGY INTERMEDIARIES (5.6)		
• Create information networks, advisory centres, specialist libraries, databases, liaison services • Create and support technology intermediaries like energy service companies and national-level institutions	• High transactions costs • Lack of available information about technology costs and benefits among potential purchasers and partners • Missing connections between potential partners and credible information about partners • Disaggregated opportunities that do not provide sufficient benefits for individual firms to capture on their own • Lack of capacity to contract and conduct technology transfers	Private-sector driven pathways Buildings, industry, and energy sectors Primarily assessment and agreement stages

5.2.1. *Government Finance in Climate-Change-Related Projects*

Governments raise finance from tax revenues and through borrowing from domestic and international financial markets or from multilateral organisations, and use the funds for government spending, including on projects that are perceived, or assumed, to be justified in terms of the public interest. Traditionally, governments have been the principal suppliers of finance for infrastructure projects, which are seen as being in the public interest. This encompasses many sectors of relevance to climate change such as energy, transport, agriculture, water and waste, and coastal defences.

Such finance can be provided as part of the capital expenditure programmes of state or local governments, through the investment activities of state owned industries, or through the lending of government-owned financial institutions, such as national development banks. While there has been a trend in recent years to increase the involvement of the private sector in such activities, public sector finance remains a very important source of finance in many areas, both in the developed and developing world.

While the allocation of government finance is subject to a number of influences, such as political pressure and central spending limits, the principal method used for many public sector projects by governments and government-controlled companies is through the internal rates of return. Because such businesses are backed by government and/or by a monopoly customer base (as with many electricity systems), the risk is perceived to be very low, and low rates of return are required. Financial rates of return in the range 3-8%/yr, set by governments according to macroeconomic and other factors, have been typical.

To expand the scope of this approach, sometime governments have sought to expand the definition of benefits beyond financial returns, to include other factors such as environmental benefits based on estimates of quantified 'external costs.' This results in sophisticated and extensive cost-benefit evaluation of approaches against a range of criteria (Anderson, 1979). Such an approach has provided the dominant criteria for public sector financing decisions in many countries over the past few decades, both nationally and - to some extent - in the area of foreign aid. It is also possible to incorporate non-financial factors into the decision process by means of multi-criteria analysis, which takes different non-monetary considerations into account and makes them comparable using a system of non-monetary weights. The external costs may also be "internalised" by measures to make Coasian bargaining possible[2] or by targeted policy measures in line with the Polluter Pays Principle. It should be noted that such cost-benefit analysis is largely the preserve of the public sector – the commercial private sector cannot include non-financial considerations into its analysis, unless measures are taken to monetise them.

Such approaches to quantitative evaluation are also applied in developing countries, and often indicate that government funding of programmes with positive climate change impacts are worthwhile in their own right. This has been seen particularly in the area of energy efficiency, and developing countries are increasingly turning to energy efficiency investment as a means to provide energy services rapidly with limited capital resources. They are doing this by enabling more work to be done and more services to be provided with less energy input, reduced capital expenditure, and minimal environmental impact. Economic planners in some developing countries seek to employ demand side management (DSM) as a cornerstone of sustainable economic expansion. The government of Thailand, for example, has committed US$60 million per year to an Energy Conservation Fund. In addition, the Electricity Generating Authority of Thailand has adopted a five-year, US$189 million DSM programme focused on commercial and industrial energy savings. In Mexico, the national electric utility has begun a move toward DSM with a programme to procure and sell two million compact fluorescent lamps (CFLs) for residential applications in two cities. The Mexican government is also committed to promoting energy efficiency in its federal buildings, and in municipal services such as street lighting and water pumping. Similar initiatives are emerging in the Philippines, Indonesia, Poland, the Caribbean, and China, among others.

With continuing pressures to reduce taxation and government expenditure, governments are increasingly seeking to justify expenditure on public infrastructure and to consider alternatives. Thus, in many cases there has been increasing interest in opening public infrastructure development to the private sector, for example, by privatising state-owned companies, opening markets to competition, and opening projects to private finance. This increasing role of the private sector in areas such as electricity supply has tended to increase the required rate of return. While this might at first glance appear to increase the cost of the services to be provided by the project, in many cases this is expected to be more than offset by the gains in efficiency. However, this can create problems for environmentally sound technologies if, as is often the case, they involve increased capital costs (in return for reduced operating costs). Such problems can be exacerbated by the fact that the private sector will not be able to take account of external costs/benefits in the same way as public entities. These factors are not insurmountable, and there are structural options to help direct private finance both at the macro level (*e.g.*, envi-

ronmental charges) and micro level (*e.g.*, the public-private partnerships in section 5.6) to help overcome them. However, governments should be aware of the potential climate change drawbacks in shifting from public to private sector finance.

One clear example of the consequences of this shift is the impact it has on DSM programmes above. As the energy market is deregulated and privatised it becomes increasingly difficult to support formal DSM programmes. As such attention has shifted to alternative mechanisms for encouraging energy efficiency, both through macroeconomic measures and through specific activities such as energy service companies or ESCOs (see section 5.7.3)

Since the beginning of the 1990s, several countries in Central and Eastern Europe (CEE) and the Newly Independent States (NIS) have explored the creation of public environmental funds with the specific purpose of investing in environmental infrastructure, technology and conservation. These funds are financed by earmarked revenues from charges and fines for pollution and use of natural resources or environmentally harmful products. The significant advantages of these funds are that the resources are dedicated for environmental purposes and not such to competition with other demand, and that they are off-budgetary. Although in NIS (countries) these funds remain insignificant and somehow a flawed source of financing environmental investments, they have been able to mobilise significant resources and play an essential role in maintaining high levels of environmental investments in the economy of some CEE countries (in particular Poland and the Czech Republic). Such funds could potentially be developed in other economies. (See OECD 1995a, OECD 1995b, OECD 1999a, Peszko 1995; Peszko and Zylicz 1998; Mullins *et al.*, 1997).

5.2.2   *Official Development Assistance*

It is increasingly recognised that ODA should not be seen as a leading source for investment in environmentally sound and cleaner technologies, but rather be used to address the fundamental determinants of development, which include a sound policy environment, strong investment in human capital, well-functioning institutions and governance systems and environmental sustainability (Killick, 1997; OECD, 1998a). This realisation has arisen partly from the policy view among donors that aid should not go to sectors where the private sector can take the lead role, and also from the mixed experience with development aid programmes. The dominance of certain interests in donor governments led most industrialised countries to promote economic and geo-political goals – *e.g.*, contracts for their domestic firms, support for friendly political regimes – that often ran contrary to the fundamental development objectives. This has resulted in development aid and, in particular, to tied bilateral aid having a very mixed record. Problems range from the controversies over major projects such as big hydro-electricity projects, to the disappointments and failures of some programmes to support the transfer of smaller-scale renewable energy. ODA is still

---

² If actors will be able to solve the problem among themselves through appropriate distribution of property rights, symmetric information and low negotiation transaction costs (see Coase, 1960). [If adopted insert ref.: Coase R.H., 1960, The problem of social cost, Journal of Law and Economics, pp.355-378.

significant for the poorest developing countries, where it accounts for up to 20% of gross domestic product (GDP), with external private flows accounting for 3-4% on average (OECD, 1998b; see also section 2.2.2 in Chapter 2 on ODA flows).

The extent to which aid is tied to being donor-country supplied reveals the persistence of the tendency for development to be subordinated to other goals. Accurate estimates of the extent to which aid is tied are difficult to come by because of the multitude of ways it can be hidden in mixed credit financing. The Organization for Economic Cooperation and Development (OECD) estimates that in 1996 tied aid accounted for around US$22 billion of a total of US$52 billion of official development assistance. Another estimate is that only half of the top 20 donor countries tie less than half their aid (Jain, 1996).

Tied aid started in part because of industry pressure in donor countries. In the 1970s and 1980s, the effects of an overly simplistic approach to development through industrialisation, and the harmful effect of self-serving economic and political motives in the donor nations became noticeable. Increased unemployment, white elephant projects, rural-urban migration, foreign debts and growing technological dependency brought into question the appropriateness of the technology being transferred to developing countries (Chambers, 1997):

Early beliefs of the 1950s and 1960s in linear and convergent development through stages of growth, in central planning, in unlimited growth, in industrialisation as the key to development, in the feasibility of a continuous improvement in levels of living for all these have now been exposed as misconceived and, with the easy wisdom of hindsight, naive. Hundreds of millions of people are now worse off than twenty years ago.

In the 1990s, the apparent absence of sustained economic and social improvement in many recipient countries, the end of the Cold War as a motive for providing aid to friendly nations, and budgetary constraints in donor countries have led to increased questioning of the effectiveness of overseas development assistance and to a sharp decline in aid flows (Graham and O'Hanlon, 1997). Funding might have declined even further in some countries were it not for industry in donor countries arguing for subsidies for exports, to protect jobs and to match subsidies provided by other donor nations to their firms (Morrissey, 1992), although in other countries there is more general public support for aid, partly based on its effectiveness. Thus, while aid has always been tied to some extent, the proportion of tied aid may have lately increased in importance[3].

Tied aid is less likely to promote economic growth in recipient countries than untied aid. Empirical studies suggest that tied aid increases costs for the acquiring country anywhere from 10 to 50% (Morrissey, 1992). In principle tied aid is better than no aid, and could be a positive sum transaction; in practice it often ends up that neither side benefits. The importation of more expensive, capital intensive, and inappropriate technologies creates a dependency for maintenance and spare parts. In general, the technology being imported may be a low national priority for the recipient country.

The consequences of tied aid go beyond the distortion of technology choice. It inhibits the development of domestic capacity in selecting technology – technology choice becomes a matter of finding the biggest subsidy rather than the most appropriate technology. It can crowd out good technologies and viable business models. It also acts to prevent private financial institutions from becoming involved in supporting technology transfer and developing appropriate expertise, notably when tied aid finance is provided on greatly subsidised terms in order to [win/secure/procure exports. For example, there are few cases where aid finance has been useful in helping to mobilise private capital into technology transfer, or to support financial innovation and new forms of financing for technology transfer - most such work is being done by the multilateral development banks. The challenges of tied aid have been recognised by the donor community, and Development Assistance Committee (DAC) donors have made important efforts to limit tied aid on the grounds that it limits the effectiveness of aid. Specifically, the "OECD 1992 tied aid discipline" prohibits subsidised finance (*e.g.*, to support manufacturing or power investment) to developing countries except the least developed countries or LDCs (OECD, 1998c). Nonetheless, there remains substantial scope for the abuse of tied aid. Transparency has been advocated as a way of reducing donor's use of tied aid, recipients' use of aid for short-term political and economic gains, and temptations to divert aid to private pockets (Lin See-Yan, 1997).

Increasingly, there is recognition that aid can be more effective and useful to development if it is focused less on core financing of specific projects and more on areas such as capacity-building, in providing incentives for direct investment (public or private) or in supporting the public private partnerships discussed later in this chapter (OECD, 1998a):

[3.] In order to increase the effectiveness of aid, both increased recipient participation (including NGO) and reduced donor control are required. Today, aid is increasingly seen as a resource to help ensure the sustainable and efficient use of domestic resources in recipient countries. One approach to increase the effectiveness of aid has been the call to link, or tie, aid to performance. To the extent this is perceived as another conditionality, and leads to a reduction in country ownership, it could be counter-productive. Untying of aid has advantages for both recipients and donors. In 1992, OECD Member countries agreed that tied aid should be extended only to projects that are not commercially viable and that are unable to attract commercial financing. To the extent that this commitment is respected, it could prove beneficial to several climate-friendly renewable energy technologies, which face difficulty in attracting financing even when they are least-cost options for certain applications.

Support for the dissemination of technological know-how must concentrate on developing the necessary human, scientific, technological, organisational, institutional and resource capabilities to underpin the long-term application of new technologies.

Specifically, this can include supporting development of the right policy mix, direct support for investment in appropriate technologies, or support for project preparation and development. That is to say that such support should be provided in the abstract – in many cases it will be more effective if linked to specific projects or programmes. Also, it is recognised that adapting assistance to local needs requires establishing working relationships among the various external and domestic actors involved and that coordination under the leadership of the host country is key (OECD, 1998a)

5.2.3   Trade Finance and Export Credit Agencies

The largest source of public sector support for cross-border finance is trade finance in its various forms, where a government agency provides a guarantee on loans to support exports (see also section 2.2.7 in Chapter 2 on international ECA flows). Export credit is a massive area – in 1996 export credit agencies (ECAs) supported exports totalling US$432.2 billion (Berne Union Yearbook, 1998). While much of this was short-term cover, approximately US$100 billion was for medium and long-term transactions (over one year). Such guarantees normally cover political and sovereign risks only and not commercial risks, and will usually require that the business is either a state entity or is backed by a local bank guarantee. Trade finance is also particularly relevant in that it normally operates in conjunction with the private financial sector, and, for example, has increasingly played a critical role in supporting project finance. Many deals would not be possible without the support of ECAs.

A key aspect of trade finance however is that it does not focus particularly on clean technology, and indeed the ECAs appear to be heavily involved in supporting activities which contribute to climate change. One study found that the two U.S. export credit/investment insurance agencies (ECAs), the Overseas Private Investment Corporation (OPIC) and the Export-Import Bank of the United States (Ex-Im), underwrote US$23.2 billion in financing for oil, gas and coal projects around the world between 1992 and 1998. These projects will, over their lifetimes, release 29.3 billion tonnes of CO_2 (Institute for Policy Studies *et al.*, 1999). The experience of the US ECAs is unlikely to be atypical and around 60% of ECA activity may be related to carbon or energy intensive exports or investments.

Given this, it is not surprising to find that most ECAs have no environmental or climate change policies. This is because the mandate of ECAs is not developmental or environmental, but to support the trading activities of the host nations.

Furthermore, they are traditionally secretive in their operations and policy, failing to disclose their activities openly and act accountably. There are some exceptions: the US Ex-Im and OPIC mentioned above have some modest environmental requirements. Even these have been under pressure, as they have led to trade tensions, notably over the Three Gorges Dam Project (in China), which Ex-Im refused to back, whereas the German Hermesbuergschaften and other ECAs were prepared to provide support. This has lead to the Ex-Im Bank coming under pressure to relax its standards. In the absence of harmonisation among ECAs, any ECA seeking to develop environmental standards will be penalised. (Cornerhouse, 1999)

The fact that no standardised environmental requirements exist among ECAs is in many cases at odds with the commitments many countries have made in multilateral agreements such as the Climate Change Convention. It reflects the fact that ECAs generally report to the trade ministry, rather than the environment or development ministries. It would clearly appear desirable to develop some harmonised environmental standards, probably based on World Bank standards, with a particular emphasis on avoiding technology dumping and (supporting) undesirable projects, and possibly considering giving special support for transfer of ESTs. Rather than "a race to the bottom" there should be procedures for upwards harmonisation. Furthermore, in keeping with the discussions on participation in Chapter 4, it would appear important that such mechanisms are developed in an open and accountable manner, with participation from all interested parties.

Increasingly there is international recognition of the need for environmental standards within ECAs, and it has been placed on the international policy agenda. The final communiqué of the G8 summit in Koln stated "We will work within the OECD towards common environmental guidelines for export finance agencies. We aim to complete this work by the 2001 G8 Summit". The OECD Export Credit Group has been attempting to share information and coordinate between ECAs for some time now, but relatively little progress has been made, with reluctance from some ECAs to take action, although with the emergence of increased political pressure it may be that the process will gain new impetus.

With environmental policies, Export Credit Agencies will be better placed to focus on issues particularly relevant to the transfer of ESTs. This will including moving away from the focus on exports (the most limited form of technology transfer in that there is no capacity building or value added in the host country) as well as the emphasis on large deals. It may also involve looking at the potential for specific activities in climate change related areas (for example, ECAs may have a role to play in the operation of the Clean Development Mechanism).

To date there have been few examples of activities which have sought to focus particularly on the problems of environmental technology transfer. One interesting exception, which illustrates the potential, is the creation of the private-sector Global Environmental Fund (not to be confused with the Global Environment Facility) in the United States, which invests in environmental projects and businesses worldwide. Its formation was greatly facilitated by the provision of an investment guarantee from the OPIC.

5.2.4. *Financing by Multilateral Development Banks*

The Multilateral Development Banks (MDB) have seen technology transfer as part of their mission to encourage development. More recently they started to focus on the challenges of the environment and the specific problems involved in transferring environmental technology. In response many have started to develop a range of initiatives and activities.

In particular, development banks have become aware of the role they can play in helping to mobilise private capital to help meet the needs of sustainable development and the environment, and of the potential to use financial innovation to encourage environmental projects and initiatives. While much of the earlier work they did was sporadic, the private sector arms of the MDBs are now seeking to identify ways they can work with international private capital to help address the environmental and developmental needs and are discussed in section 5.5.

- The World Bank itself is limited by its charter to only working with governments and quasi government organisations, although it is increasingly developing mechanisms to deal with private and quasi-private sector entities. Its affiliate, the International Finance Corporation (IFC), is most directly involved in private sector investment. However, the World Bank has developed a number of initiatives with the potential to support environmental technology transfer (Asad, 1997) . These include financing a number of environmental lending programmes at domestic financial institutions, which will then lend to industry. An important new World Bank initiative is the proposed US$60-$150 million Prototype Carbon Fund. This vehicle will provide additional finance for CO_2 mitigating projects in return for carbon offsets, *i.e.* the right to transfer the credit for the CO_2 saved to the investor. It is expected to have a substantial private sector financing and project execution. The World Bank has also prepared a major environmental strategy for the energy sector called Fuel for Thought.

- The Global Environmental Facility is a financial mechanism that was established prior to the 1992 Earth Summit (see Box 5.2) to provide grant and concessional finance to recipient countries for projects and activities that aim to protect the global environment.

BOX 5.1   ECAs: THE TOOLS OF THEIR TRADE (SOURCE: MAURER AND BHANDARI, 2000)

ECAs are bilateral organisations such as investment promotion agencies or Export Import banks that offer a variety of financing options for foreign export and investment. Most advanced industrialised countries in the OECD have ECAs that are dedicated to promoting their economic and business interests overseas. Most of their incentives are directed toward companies trying to enter or compete in emerging market economies of developing countries and economies in transition (newly independent states). ECAs use a variety of financial instruments to give their private sector clients a leg up on foreign competitors. The following are the most common instruments used by ECAs:

Export credits or loans: loans to buyers or suppliers of export goods usually of a short term nature (maturities of less than a year or two) including letters of credit and banker's acceptances. These are provided on favourable terms that are not as easily available from private commercial banks.

Import credits or loans: essentially the same as export credits, but they are provided to overseas purchasers of domestic goods and services.

Project financing: direct or indirect loans for overseas projects with the significant participation of a domestic company, including joint-ventures. These are provided on favourable terms, such as extended maturities, that are not as readily provided by private commercial banks for politically risky markets.

Guarantees: agreement by a sovereign entity (usually a government) to cover or insure a domestic investor against any losses suffered on an investments or export that results from civil unrest, expropriation of property, or nationalisation (political risk), the inability to convert local currency into hard currency (currency transfer risk), from a breach of contract by the host country government (partial risk guarantee), and back a loan provided by a commercial bank against a borrowers default (loan guarantee).

Insurance: this is very similar to guarantees, the difference being that the coverage against political, currency transfer or loan defaults are purchased as an insurance premium.

Equity: these are direct investments into a project or an equity fund that in turn invests directly in development, infrastructure or other projects in the recipient country. Essentially, this equity buys down credit risk and permits private funds to raise additional financing more easily.

The instruments most commonly used by ECAs are export and import credits (also known as trade financing), project financing and various forms of guarantees and insurance. The use of equity funds is a fairly recent phenomenon, but it is growing more common as various types of financing instruments are increasingly packaged together (equity, bonds, loans, guarantees) to assemble sufficient capital to get a project off the ground.

There are three official implementing agents, the UNDP, UNEP and the World Bank. The World Bank also acts as trustee of the GEF Trust Fund. The focus of the GEF is on providing incremental funding for projects that would not be viable on the basis of domestic considerations alone. The GEF has supported a growing, but proportionately small number of private sector investment operations. As part of its private sector portfolio, the GEF has made US$110 million in commitments to the IFC for climate change mitigation initiatives. A requirement for funding by the GEF is the demonstration of incremental global environmental benefits. This can be a time consuming and expensive process that increases transaction costs, although these problems are more marked with the Biodiversity Convention than support provided under the FCCC. The Buenos Aires Conference (CoP4) also officially expanded the scope of the GEF, for example, to include adaptation.

- The European Bank for Reconstruction and Development (EBRD), as the newest MDB, is the only one to have sustainable development incorporated in its charter. It is also much more active in working with the private sector than the other regional MDBs. Thus it is not surprising that it has shown a fair amount of leadership in helping to encourage technology transfer. It has built up a large and successful portfolio of loans in areas such as private energy service companies and municipal environmental infrastructure in developing countries. In addition, it has also worked with intermediary banks to educate them on environmental issues and the potential of clean technology.

- Other MDBs, such as the regional development banks (ADB, IDB, AfDB) all play an important role in regional investment, and most have given some attention to issues of sustainable development, in varied ways. The European Investment Bank, while differing in that it is an institution of the European Union (and thus has a limited membership), is also a huge investor with an increasing focus on lending outside the EU, particularly to the Asia-Pacific-Caribbean countries under the Lome Convention, and also to Central and Eastern Europe and the countries of the Mediterranean basin. Its investment programmes, however, have not generally been coordinated with the EU's goals on sustainable development and climate change.

- Various international agencies also harness considerable expenditure. Most notably, the UN Development Programme makes substantial grants in the area of institutional capacity-building. It has also directly supported the development of various financial mechanisms to encourage environmental technology transfer, such as supporting the feasibility work for an environmental venture capital fund.

5.3 Private-Sector Finance and Investment

Private sector finance is increasingly important in both the national and international diffusion of technology, and the relationship between public and private is particularly important in the context of technology transfer. It will have a major role in private-sector-driven pathways and often a role in community-driven pathways. This section thus discusses the criteria used by private sector finance and the forms it can take, and then identifies some of the financing mechanisms most relevant to environmentally sound technology transfer, and then looks at the some of the important initiatives taking place in private finance.

While such initiatives can be successful, they clearly will not work if the macroeconomic and environmental framework is not adequately supporting ESTs so that they are financially viable. Even if this is the case, they may not be sufficient to cover other concerns of the financial markets such as the significance of climate change to their business or the risks of getting involved in this area. Governments can be a source of risks themselves in the way they develop policy and consistent, consensual policy development can help reduce risks.

It is also important to distinguish between investment and financial products from private sector financial institutions, which is the real focus of this section, and investment by private businesses as part of their business development, which will be discussed in section 5.4 on private-firm perspectives. Most foreign direct investment is by its nature in this latter category.

5.3.1 Private-Sector Finance: Criteria and Forms

Private financial institutions invest in businesses - or specific projects - which can generate a financial return. However, they have no particular interest in any individual business and can typically choose from a very wide range of investments available to them. In selecting investments most financiers will focus on the two criteria of risk and return - higher perceived risk results in higher expected return, with the level being primarily set by the market. Different financiers will have different preferences for risk and return. An impact of the emphasis on risk, and compensation for that risk through increased return, is that the private sector will find it most difficult to finance high risk, longer term projects. Many environmentally sound technologies are essentially of this nature with low operating costs and high up front expenditure.

In trying to alter the behaviour of the financial markets, governments can choose regulation or persuasion. Regulation is unpopular with financial institutions, particularly internationally. Persuasion is difficult to initiate, but can ultimately be more successful, offering major benefits for relatively small outlays and gives rise to opportunities for private-public partnerships (discussed in section 5.6). In seeking to persuade the

way the financial markets allocate capital, governments can focus on four key aspects: the perception of risk; the calculation of expected return; the structuring of the financial package; and the transaction costs associated with that investment. However, to be relevant these aspects have to be looked at in the context of an individual financing problem or type of finance.

Understanding the role of private finance in technology transfer necessitates some understanding of the detail of different forms of finance in technology transfer. There is a very wide range of types of finance which are potentially relevant in financing technology transfer. Their relevance depends on the specific opportunity under consideration. Key factors are the scale of the investment and whether the investment is a venture (a business intending to grow and develop) or a project (a stand-alone specific entity – *e.g.*, a power plant). Scale can be roughly divided into large (roughly at least US\$20 million), medium (over \$500,000), small (\$10,000 to \$500,000), or micro (say less than \$10,000, but mostly from \$10 to \$100), although the size is to some extent subjective and, for example, will depend on the level of economic development.

BOX 5.2:   TECHNOLOGY TRANSFER AND MARKET DEVELOPMENT PROMOTED BY THE GLOBAL ENVIRONMENT FACILITY

Since its inception in 1991, the Global Environment Facility (GEF) has promoted technology transfer of energy efficiency and renewable energy technologies through a series of projects in developing countries. Following a three-year pilot phase, the GEF in 1996 adopted an operational strategy and three long-term operational programmes for promoting energy efficiency and renewable energy technologies by reducing barriers, implementation costs, and long-term technology costs. A significant aim of these programmes is to catalyse sustainable markets and enable the private sector to transfer technologies.

From 1991 to mid-1999 the GEF approved grants totalling US\$706 million for 72 energy efficiency and renewable energy projects in 45 countries. The total cost of these projects exceeds US\$5 billion, because the GEF has leveraged financing through loans and other resources from governments, other donor agencies, the private sector, and the three GEF project-implementing agencies (UN Development Programme, UN Environment Programme and World Bank Group). An additional US\$180 million in grants for enabling activities and short-term response measures have been approved for climate change.

GEF projects are testing and demonstrating a variety of financing and institutional models for promoting technology diffusion. For example, fourteen projects diffuse photovoltaic (PV) technologies in rural areas through a variety of mechanisms: financial intermediaries (India and Sri Lanka), local photovoltaic dealers/entrepreneurs (Peru, China, Zimbabwe and Indonesia), and rural energy-service concessions (Argentina).

Several other projects assist public and private project developers to install grid-based wind, biomass and geothermal technologies (China, India, Philippines, Sri Lanka, Indonesia, Mauritania, Mauritius). For energy-efficiency technologies, projects promote technology diffusion through energy-service companies (China), utility-based demand-side management (Thailand, Mexico and Jamaica), private-sector sales of efficient lighting products (Poland), technical assistance and capacity building (China), and regulatory frameworks for municipal heating markets in formerly planned economies (Bulgaria, Romania, Russia). In addition, projects provide direct assistance to manufacturers for developing and marketing more efficient refrigerators and industrial boilers through foreign technology transfer (China).

The achieved energy savings and renewable-energy capacity installed through GEF-supported projects are small but not insignificant relative to world markets. For example, wind-power capacity directly installed or planned for approved projects is 350 MW, relative to an installed base of 1,200 MW in developing countries in 1997. The GEF has approved close to 500 MW of geothermal projects, which compares with over 1,100 MW installed worldwide from 1991 to 1996. There are an estimated 250 to 500 thousand solar home systems now installed in developing countries and approved GEF projects would add up to one million additional systems to this total in the next several years. Replication or "indirect" effects are also key aspects of GEF project designs; through demonstrations, new institutional models, policy changes, stakeholder dialogues, and other project activities, GEF projects have provided an important stimulus for technology transfer beyond these direct project impacts.

Capacity-building is a central feature of most GEF projects and is resulting in indirect impacts on host countries' abilities to master, absorb and diffuse technologies. Projects build the human resources and institutional capacities that are widely recognised as important conditions for technology adoption and diffusion. For example, the China Energy Conservation project is building capacities of private-sector energy service companies, as well as those of public agencies to disseminate information, experience and best practices. In West Africa, a GEF project is helping develop regulatory frameworks, standards, tariff structures, and technical capacity for more efficient buildings.

Several GEF projects are designed to directly mobilise private-sector finance. For example, in the IFC/GEF Poland Efficient Lighting project (Case Study 2, Chapter 16), a US\$6 retail price reduction for energy-efficient lamps was possible with only a US\$2 grant because of manufacturer contributions, and 1.6 million lamps were installed. Through the International Finance Corporation (IFC), four GEF projects—the Renewable Energy/Energy Efficiency Fund, the Photovoltaics Market Transformation Initiative, the Solar Development Corporation, and the Hungary Energy Efficiency Co-financing programme are designed to leverage US\$490 million in private-sector financing for technology transfer with US\$105 million in GEF grants.
(Sources: GEF 1996, 1997, 1998; Martinot and McDoom 1999)

Table 5.2 Applicability of types of finance

TYPE OF FINANCE / DEBT	APPROPRIATE SCALE				APPROPRIATE TYPE		COST
	LARGE	MEDIUM	SMALL	MICRO	VENTURE	PROJECT	
Personal Loans	-	-	++	++	+		medium
Bank Overdraft	-	+	++	?	?		medium
Secured Loans	++	++	+	-	+	++	low
Leasing	+	++	++	+	+	++	low
Export Finance	++	+	-	-	+	++	low
Securitised Debt	++	+	-	-	+	+	low
EQUITY							
Personal	-	+	++	++	+		N/a
Private Investors		+	++	+	+	?	High?
Venture Capital	+	++	+		++		V High
Strategic Investors	+	++	+		+	++	High
Institutional	++	+			+	++	High

KEY: ++ MOST RELEVANT DEVELOPING COUNTRIES + QUITE RELEVANT ? OCCASIONAL RELEVANCE - NOT RELEVANT DITTO

It is also important to note that in many cases there may be two stages of financing required: *e.g.*, the financing to establish the business of manufacturing/distributing the technology, and the financing for the end-users of the technology to enable them to purchase the technology. For example, financing the establishment of a PV module factory in the developing world has different challenges from providing finance to households to enable them to purchase the solar home systems produced by the plant. Table 5.2 summarises the applicability of different forms of finance to different scales and types of business.

5.3.2 Initiatives within the Private Financial Sector

While on one level the private financial sector has no special reason to consider environmental issues, many are beginning to realise, like much of the rest of business, that environment is a strategic issue for them, and a particular focus has been climate change. It is with adaptation to the impacts of climate change that progress has been most rapid as financial institutions recognise that climate change could directly affect their business. Insurance companies are increasingly aware that climate change could increase their losses on property and general insurance (*e.g.* from increased sea level rises and storm damage). Banks could also see the undermining of the security behind much of their lending. As a result, some insurance companies have become increasingly active, and have been working with others to develop and transfer technology in this area. Measures taken have included: adjusting premiums to reflect risks (where they are permitted to do so), thus sending a clearer signal about the dangers of climate change to owners and developers; working with local authorities on preventative measures such as enforcing building codes and zoning; and developing disaster recovery measures such as improved telephone support.

In the other areas of climate change the mainstream financial industry has had a less direct impact, and to date most financial institutions have made only a modest commitment to supporting the development and use of mitigation technologies, and particularly to overcoming some of the barriers identified in section 5.1 as preventing greater investment. However, progress has been made in some areas – for example, bankers no longer regard wind energy technologies as being a particularly high risk.

Certain financial institutions have been prepared to innovate and show leadership in finance related to the environment. For example, some banks have also been active in working with smaller businesses to improve their environmental impact, often with a focus on energy efficiency, through providing advice and information. In doing this they hope to improve the credit standing of their clients, as well as secure general environmental benefits. Some banks have also instituted lending programmes with more favourable terms than in ordinary lending for businesses seeking to reduce their environmental impact.

Other initiatives within the private financial sector include:

Green financial institutions. While most mainstream financial institutions have paid only modest attention to the environment, a number of smaller organisations or groups within organisation have made it a major feature of their activities. These "green financiers" are usually driven by, firstly, the growing number of investors with concerns about the environment and a desire to see their money invested to take account of these concerns and, secondly, a high level of personal commitment by the professionals involved. These green financial organisation are much more prepared to work to overcome some of the problems identified earlier, either independently or in conjunction with the public sector. Many of these "green" financiers are involved in some of the activities above. They include:

- a number of Ecological or Social Banks (typically very small, although growing) that focus on providing fairly low cost lending for environmental and other worthy projects, and have strong links to the micro-credit movement (see Box 5.3);
- environmental equity funds which invest in listed "green" companies (many billions of dollars are now invested in such funds, but they are limited in the extent to which they can provide money to new ventures);
- a few environmental venture funds and specialist corporate financiers which provide support to new environmental businesses.

While these green financiers are still small as a proportion of the overall financial markets, they are providing a very useful pathfinder role in developing new concepts and ideas. There

**BOX 5.3:   ENVIRONMENTAL PROTECTION BANK
                   IN POLAND**

An example of a successful green financial institution is the Polish
Environmental Protection Bank. Established at the beginning of the
1990s, it has received substantial equity investments from the Polish
National Fund for Environmental Protection. Share capital has also
been raised several times from strategic investors and from the private sector. The bank was listed on the Warsaw stock exchange in
1997 and became the world's first publicly traded bank specialising
in environmental protection financing. In 1997, the bank granted
over 27,000 individual credits and loans worth PLZ 1,431 million (363
million Euros). The bank lends primarily to businesses (54 %), municipalities and other public-sector entities (26 %) and individuals.
Specific environmental investments make up two thirds of the
bank's portfolio. The bank held a one-per cent share of the banking
market in Poland (in terms of total assets of all commercial banks),
and it has built up a good reputation for quality of services as well as
for financial performance.

appears to be substantial merit in the public sector finding ways
to support them and work with them to encourage their work.

Collective initiatives and organisations. In recent years a number
of initiatives and organisations have been created to bring together industry participants to look at environmental issues as a collective basis. The most notable have been the UNEP initiatives, where
banks and insurance companies have signed a statement on the environment, and subsequently the signatories have formed organisations to develop further activities. Other organisations have developed, mostly at a national level, to further the cause of environmental
investment, such as the Social Investment Forum (USA), UK
Social Investment Forum, VfU (Verein fur Umweltmanagement in
Banken und Versicherung) in Germany/Switzerland and the Social
Venture Network (USA). These organisations provide forums for
networking, information gathering and sharing experiences. They
also have been involved in lobbying for change and encouraging
investment and green finance.

5.3.3   *Potential Financial Solutions*

Slow diffusion together with the consideration of cost and availability of finance suggest that there is potential for innovation and
focus to help support and accelerate the transfer of environmentally sound technology. In particular, certain types of finance
appear to offer particular potential for helping to finance the
transfer of technology, although they may require adaptation to
the specific issue; it is worth considering these in more detail.
(Mansely *et al.*, 1997a and b).The public-private partnerships discussed in section 5.6 can play a role in developing and implementing these solutions.

Project finance. Project finance is the packaging of investment
into specific, stand-alone projects. Notably there is only limited

recourse to other parties (*e.g.*, the promoters and financiers) if the
project runs into difficulties, so the project has to stand on its own
merits. Of particular relevance to climate change are energy projects, which are frequently financed this way. Project finance uses
a range of finance instruments and typically consists of a mixture
of debt (normally secured loans) and equity (strategic investors
and institutions). Project finance aims to reduce risks and thus
financing costs through a series of robust contracts, notably to
charge for the services provided (*e.g.*, power). Negotiating these
contracts can be difficult and time consuming. Often there can be
some flexibility over ownership of the facility - such as build-
operate-transfer (BOT) structures. One key issue for climate
change technologies is the need to achieve a certain scale. To justify the transaction costs involved, project finance normally
requires sums of US$20 million and above. This can restrict its
applicability in many areas such as renewable energy and energy efficiency where only few projects in certain sectors reach this
size. Project finance is particularly relevant to government-driven pathways, and to the transport, energy, solid waste and
coastal zone adaptation sectors.

Leasing. Leasing is a highly flexible form of finance used
throughout business to finance everything from photocopiers to
aircraft. In 1994 over US$350 billion of new equipment, machin-
ery and vehicles were financed through leasing, and some US$44
billion in developing economies. It is often packaged as a form
of sales financing - *i.e.*, it helps customers of a company buy that
company's equipment. Despite higher spreads than conventional lending, leasing offers several advantages such as simplified
security arrangements, convenience and speed, flexibility, low
transaction costs and frequently tax advantages. The principle
constraint on the development of leasing has been access to local
currency medium-term lending. MDBs, notably the IFC, have
been active in promoting leasing businesses and have found it to
be a successful form of investment (Carter, 1996). Leasing offers
potential to be a major source of finance for the transfer of EST,
particularly to the business community (private-sector-driven
pathways). Leasing has been used to buy various types of environmental technology from monitoring equipment to wind turbines, although it has not been possible to identify any leasing
company focusing specifically on environmental technology.
There is scope to encourage leasing companies to support the
transfer of EST through selective tax incentives, information
sharing and bringing together environmental entrepreneurs and
leasing companies. Leasing is particularly relevant to the private-
sector-driven pathways, although it can be used by governments,
and especially the industrial sector, although it can be used in several other sectors.

An example of an established and successful leasing company that
focuses specifically on environmental technology is Towarzystwo
Inwestycyjno-Leasingowe Ekoleasing S.A. (joint stock company Investment and Leasing Society Ekoleasing) in Poland. It
was established in 1993. Currently the share capital is almost
US$1 million. Over 40% of the total value of PLZ 33 million
(about US$9 million) of contracts concluded in 1998 was leasing
of specifically environmental technologies.

Private equity from strategic investors. Strategic investors, often in the form of multinational corporations, have the potential to be major investors in technology transfer. As large organisations they have ready access to finance. As well as using capital internally they can also act as external investors, investing in projects or businesses. They look for a financial return but also usually expect other business benefits, such as a role as supplier to a project or investing in a joint venture as a way to gain access to new markets. They frequently bring additional skills and expertise as well as finance. They are major suppliers of equity to many energy projects already, and with many major companies becoming increasingly interested in climate change (such as Enron, BP and Shell) are increasingly investing in renewable and clean energy internationally. This type of investor is also the most interested in the flexibility mechanisms of the Kyoto Protocol, as indicated by their participation in the various precursor instruments such as joint implementation and the Carbon Investment Fund, probably because they can see benefits beyond the strictly financial ones.

Portfolio investment. For listed companies in developing countries, issuing new stock is an option for raising capital that can be attractive. Doing so enables risk capital to be raised, without involving loss of control, for example to overseas partners, and enables investment in several needed areas, which may include energy efficiency or environmental technologies, without changes to management or business structure. However, the vast majority of portfolio investors will place little direct importance on the investment in environmental technologies, and will instead consider more general aspects of the firm and management's track record when deciding to purchase the new shares. Furthermore, the ability to raise such capital cannot be guaranteed, and depends on market conditions and on the company's performance at the time.

Venture capital. Venture capital is particularly relevant to the development and transfer of new technologies. Venture capitalists are prepared to back risky investments in return for high returns and will invest in small companies, such as those who have developed new technology and/or have difficulties raising capital from most other investors. Venture capitalists have a relatively long-term focus, aiming to hold companies for several years before selling them, and have a more active approach than most other types of investors, in terms of participating in management of the company. This means they can play an active role in supporting technology transfer if it forms part of the business development plans of their investee companies. Venture capital is largest in the USA but has grown recently in the rest of the world, including in developing countries where multilateral institutions have provided substantial support. Venture capitalists have tended to focus on high-return sectors such as computer software and biotechnology, and to date only a relatively small amount of finance has gone into environmental business, and only a few funds focus on environmental ventures. Indeed, the environmental sector has had a very mixed track record in delivering returns to investors. However, there is growing interest in venture capital funds with an environmental focus and a number are expected to be launched in coming years. Venture capital is predominantly relevant to private-sector-driven pathways, and especially important in the industrial sector, with some relevance in the transport and energy sectors. It does require a relatively sophisticated financial infrastructure.

Micro-credit. Micro-credit is the provision of small amounts of finance to individuals. While the basic concept is the same as traditional banking, the attitude to risk is radically different, because micro-credit institutions are prepared to lend to those ignored by conventional financial institutions – those on low incomes or with no assets. A particular emphasis is on enabling access. It is often provided by non-conventional financial intermediaries such as cooperatives, farmers' associations and distributors. Micro-credit has been successful in many areas now and is receiving increasing attention from Multilateral Financial Institutions as a way of encouraging development (Ledgerwood, 1999). Many believe there is substantial scope for adapting and focusing micro-credit to finance the uptake of ESTs at the household level. However, others have argued that micro-credit is generally not suitable for environmental technologies. because the credit is usually short-term (less than 1 year), comes with high interest rates, is limited to small amounts (US$100, whereas a solar home system might cost US$600) and is not granted for capital investments (van Berkel and Bouma, 1999).

5.4 Private-Firm Investment Decisions and Foreign Direct Investment

While section 5.3 has looked at external private sector investment, in many areas the most significant investment decisions are those made within firms. Within a single country, such investment decisions have an important role to play in the diffusion of technology. Across borders such investment forms Foreign Direct Investment (FDI), increasingly seen as one of the largest and most important financial flows (see section 2.2.2 in Chapter 2).

Particularly in the buildings, industrial, transport, and energy sectors, investment decisions made by private firms can significantly affect greenhouse gas (GHG) emissions in firms' processes and products, and in firms' conduct of environmentally sound technology transfer along private-sector-driven pathways. These decisions are often at variance with simple economic models that assume universal optimisation, because of barriers that can exist within firms to technology transfer. Private-sector investment comes in many varieties (see section 5.3), but whatever form the financing mechanism takes, obstacles originating in the organisational structures and decision-making procedures of firms may limit adoption of the most environmentally sound technologies. Internal barriers within a firm are those that slow down the adoption of technologies that would be in the firm's own interest given prevailing market prices, external macroeconomic conditions, and regulatory requirements. Both local and multinational firms are subject to internal barriers, but the ways in which technology transfer is impeded by the barriers may differ across classes of firms, depending on the nature of the barriers.

5.4.1   Multinational Corporations and Foreign Direct Investment

Concern has been expressed that multinational corporations' (MNCs) direct foreign direct investment gravitates toward countries with lower environmental standards or lax enforcement (pollution havens), as MNCs seek to avoid the high cost of environmental compliance in their original bases of operation. There is debate about differences between local and foreign firms with limited empirical evidence available (Jun and Brewer, 1997). However, according to some sources, there is growing evidence that foreign-owned or joint ventures tend to be cleaner ("halos") than local firms for the following reasons: they use the usually higher standards of the developed countries embedded in the overseas subsidiary; they export to environmentally sensitive markets, and the parent firms do not want their image to be tarnished by irresponsible overseas operations (as has happened) (Panayotou, 1997).

Multinational corporations (MNCs) have significantly expanded their environmental management capacity. In general the environmental effects on the host economy depend also on the policies of the host government as well as their practises. Domestic regulatory policies can increase the contribution of private capital flows to sustainable development (Jun and Brewer, 1997). Environmental sustainability, including mitigating and adapting to climate change, can then be seen not as a barrier to growth, but as a boundary condition that could stimulate the emergence of a sustainable industrial economy, a process in which technology transfer is likely to play a major role. Host governments can require enforcement of environmental regulations, transparency in reporting and pre-screening of projects before commencement. Home governments could potentially screen projects for their environmental effects before granting them political risk guarantees (see the discussion on ECAs, section 5.2.3). MNCs themselves have in some cases assumed responsibilities to minimise the detrimental consequences of FDI projects and guidelines have been developed by both private-sector and public sector organisations, such as the Business Charter for Sustainable Development developed by the ICC (Jun and Brewer, 1997). Presently under discussion is a revision of the OECD Guidelines for Multinational Enterprises. Capacity building is required in developing countries to strengthen regulatory reform and monitoring.

Western multinational corporations are sometimes at the leading edge of lean production techniques and new ways of working with the community. Transfer of these approaches through foreign direct investment by MNCs can be a critical pathway for developing countries to acquire these essential building blocks for sustainable industrialisation. Today's dominant industrial paradigm – lean production – tends towards minimising raw mater-

ial needs (by reducing waste and unwanted stock, Womack *et al.*, 1990). It is associated with participation and individual responsibility of workers throughout the chain of production, and is, therefore, potentially a building block for an integrated system of industrial production within an environmentally sustainable economy (Wallace, 1996). At the same time, possible elements of a future sustainable industrial economy can be seen in developed economies. Some essential elements have already been suggested, including: new relationships with workers and the community (Hawken, 1993; Silverstein, 1993); decentralised production of a wide variety of goods (leading to greatly reduced pollution, lower costs with a transfer of wealth from large corporations to local communities; Hawken, 1993); and the integration of production, consumption and waste streams into a single "ecological" system (Socolow, 1994). In the real world, limited application of industrial ecology principles is being demonstrated in the Kalundborg, Denmark, eco-industrial site (Hawken, 1993).

However, barriers and obstacles do exist with modern multinational corporations to the transfer of technology and the reduction of GHG emissions. It is in the nature of such large organisations that they must be divided into semi-autonomous divisions, and that central coordination can only imperfectly control the actions of the far-flung branches. Even if the central headquarters of a MNC were to decide on a particular course of technology transfer or diffusion, it would require effort and management attention to make that decision happen. Control from the centre can be exercised only imperfectly, so it would be unusual if all branches performed up to the same standard in every dimension. Thus, there should be no reason to presume that all segments of a globally dispersed MNC have optimised their technological choices. Rather, there will always be opportunities for profitable transfers of knowledge and technique.[4] A great deal of foreign direct investment takes the form of expansion of operations by MNCs, so internal or trans-divisional decisions regarding technology choices, in addition to decisions about the location and direction of capital flows, are important in determining the pace of transfer of ESTs

Decisions within firms may not be made according to rational decision-making models, and a more open process can help the private sector decision-maker. Traditional economic theory would tend to predict that the choice of a technology would be chosen on the bases of cost minimisation and perfect information. Even where such adherence to theory would be expected to be most prevalent, such as in the choice of manufacturing methods, empirical studies have shown this to be inadequate to explain technology choice decisions actually made by managers. Observers have noted that firms in the same country use quite different technologies to manufacture ostensibly identical products. Flouting theory, labour-intensive and capital-intensive plants survive side by side (Stobaugh and Wells, 1984). Factors that do influence choice include the ratio of manufacturing costs to total costs, the price elasticity of demand, competition faced by the firm, flexibility in changing output, quality of the product, whether it was being produced for the domestic market or export, etc. Moreover, lack of information is an important determinant of choice (Stobaugh and Wells, 1984), and is often accompanied by a lack of initiative in searching for information other than that provided by existing suppliers of technology (Chantramonklasri, 1990).

It is also well known that the present configuration of technology choices depends to a significant degree on past choices (Arthur, 1994). This sort of "path dependence" of technological progress may result from economies of scale that cause technologies with larger market shares to exhibit lower costs than newer, potentially superior, technologies. Alternatively, path dependence can arise in the normal course of a firm's evolution, because it can be easier in the short run to make marginal changes in existing methods than to switch to an entirely new way of doing things. In the case of energy technologies, the past choices were made without taking account of the climate externality. Thus, existing technologies enjoy advantages over newer, lower-emission technologies, in terms of having learned by doing, having learned by using, having already realised scale economies, having established information channels and user confidence, and having established inter-related technologies. These factors can perpetuate continued "lock-in" of the fossil fuel technologies that have grown around the historic availability of cheap fossil fuels and historic neglect of the greenhouse gas externality. Policy intervention may be required to replace locked-in technological choices that are no longer globally optimal.

Probably the most general reason large firms fail to take advantage of opportunities to adopt profitable new technologies that would reduce GHG emissions is that doing so is not a strategic priority. Investment in energy technologies usually is not seen as central to the firm's growth and survival, so this type of decision receives a lower level of attention from top management than other concerns. The mechanisms and controls that serve to maintain accountability and control principal/agent problems across different layers in the organisation's hierarchy in relation to its mainstream activity can themselves become barriers to action in relation to energy efficiency. These considerations suggest that an effective way for a corporation to achieve emissions reductions would be an organisational change that would make a clearly identifiable person or group within the firm responsible for the monitoring and abatement of greenhouse gas emissions. Such an internal organisational unit could be charged with identifying and implementing profitable energy-saving investments (with the managers of the group rewarded accordingly), and could be self-financing to the degree that such opportunities exist. Other regulatory or policy regimes, such as carbon taxes, cap-and-trade systems, or proportional abatement obligations (PAOs) could play an important role in focusing the attention of management and making GHG emissions reductions a measurable objective for the firm.

---

4. Even if a MNC could be optimised at a given moment in time, the rapid rate of change of the market, regulatory, and environmental factors with which it must contend would guarantee the emergence of profit opportunities.

5.4.2 Incentives and Barriers to Energy-Saving Investments

At one level, it may see sensible for a private firm to invest in GHG reductions only if there is an obvious financial benefit. But for many firms, greenhouse gas reductions can result from corporate decisions that are taken in response to direct economic incentives—like reduced costs, increased profits, and increased market share. For example, the World Business Council for Sustainable Development is promoting the view that corporate efforts for eco-efficiency are a sign of management competence, and hence will increase shareholder value (*e.g.*, ease raising of equity) as well as credit-worthiness (*e.g.*, ease raising of debt financing) (Schmidheiny and Zorraquin, 1996). A "green" corporate image can become a corporate asset, and some preliminary research in the United States has shown that better environmental management systems and environmental performance tend to reduce firms' risk profile (controlling for other factors), with the expectation of a positive impact on the stock price of the greener firms because of their reduced risk (Feldman *et al.*, 1997). The value of a company's equity ultimately is determined by the present value of the entire expected future stream of earnings (suitably discounted), so long-run concerns including environmental performance should have an effect on the market evaluation of a firm's net worth.

Internal barriers are often overlooked, partly because economic models typically make the simplifying assumption that all market agents are fully maximising their objectives. Yet several strands of recent research have shown that private sector firms do not take full advantage of all the cost-effective investments in energy efficiency and other cleaner energy technologies that are available. This evidence comes from "bottom-up" studies (Interlaboratory Working Group, 1997; Energy Innovations 1997; National Laboratory Directors 1997; IPCC, 1996, Table 9.8 and the studies cited therein), statistical tests of the maximisation hypothesis (DeCanio and Watkins, 1998a; DeCanio, 1998), and theoretical and empirical studies (Koomey, 1990; Ayers, 1993; Lovins and Lovins, 1991; Jaffe and Stavins, 1993; Koomey *et al.*, 1996; DeCanio and Watkins 1998b; Porter and van der Linde, 1995a, 1995b).

A number of specific intra-firm barriers to the adoption and diffusion of profitable energy saving and other cleaner energy technologies have been identified. In addition to the long-recognised tension between the goals of shareholders and management, managers at different levels within a firm may have conflicting incentives. It is often the case that data that could be used for energy auditing and control either is not available or is scattered through the organisation in such a way as to make cost-saving investments in energy efficiency more difficult. Capital budgeting procedures that are put in place to control principal/agent problems within the organisation may have the unintended side effect of screening out profitable energy-saving investments. Managers can be inappropriately risk averse because of the way their performance is evaluated, and their incentives to pursue energy efficiency blunted by frequent turnover or switching of positions within the company. Managers rarely have incentives to make the long-run decisions that will benefit their successors at the expense of their own performance in the short run (DeCanio, 1993, 1994).

Furthermore, many firms in NIS countries (mainly Russia and Ukraine) operate under perverse microeconomic conditions that encourage them to under-report or hide revenues, expenses, and profits. Barter transactions, which make up a substantial percentage of economic activity in these countries, along with corruption in many forms, complicate matters further. Under these conditions, judging the financial condition of enterprises become problematic, and adds risk and uncertainty to energy efficiency and other types of otherwise profitable investments. In general, issues of corporate governance in non-monetary, distorted economies of some large NIS countries can represent significant barriers to environmentally sound technology transfer (OECD, 1997; EBRD, 1998; Commander and Mumssen, 1998).

Firms are not unitary entities having a mind and will of their own. Instead, they are made up of a multitude of individuals, each of whom has their own individual interests and objectives. The decisions of firms are thus the result of collective action, and it has long been understood that collective action may not yield optimal outcomes, even if all the individuals taking part in the decision-making process are perfectly ratio-

nal (Olson, 1965). This problem can manifest itself in the operation of for-profit firms just as it does in voting models of group choice. In both cases, the (possibly divergent) interests of individuals have to be 'aggregated' into organisational decisions.[5] The task of management is to bring about as much correspondence as possible between the interests of the individuals making up the organisation and its formal goals, and this task is neither straightforward nor simple. The modern theory of the firm is based on an exploration of the multitude of ways in which agency problems, asymmetric information, and incentive incompatibility can create a gulf between the formal objectives of the firm (maximisation of profits or of stock value) and the behaviour of its employees. It should come as no surprise that perfect maximisation is rarely achieved, and in particular that it is not realised in the realm of energy efficiency.

---

[5] Of course, the assumption of individual rationality has itself been questioned (Zey, 1992; Etzioni, 1987). For studies dealing specifically with energy technology choices, see Stern and Gardner (1981), Dennis *et al.* (1990), Stern (1992), Geller (1992), and Crabb (1992). Arrow (1951) gives the classical rigorous treatment of the problem of reconciling individual and social choice. The general problem of governance and of the efficiency of collective action pertains not only to voting rules for public decisions and the operation of capital-controlled firms, but also to cooperatives and worker-managed firms.

BOX 5.4A: SMALL- AND MEDIUM-SIZE ENTERPRISES IN CHINA

Small and medium size enterprises (SMEs) in China consist of community enterprises (mainly owned by townships and villages), multiple cooperative enterprises, joint ventures, and individual and private enterprises. SMEs produce a significant share of China's GDP in a number of industrial sectors. In 1995, there were about 22 million SMEs in China employing 129 million people. SMEs in China face a number of constraints to engaging in technology transfer, such as for producing more energy-efficient products or investing in more-energy-efficient processes, including:

Information. SMEs lack contact with technology manufacturers and customers so information about technology availability and customer demands is lacking. The evolution of industrial SMEs from non-sector-specific commune-based enterprises made SME rely on low-grade technologies and gave them little access to formal information and training channels. SMEs learn largely by visiting and copying other firms in the same sector. This constraint on information acquisition is

especially true of what might be called organisational technologies such as project analysis, financial methods, or studies of market developments and factor price forecasts. SMEs have limited interchange with government ministries that might be in a position to advise them on technology choices.

Rural customer demand. Rural customers show little appreciation for product quality (such as energy-efficiency). Competition is based solely on price and regulatory initiatives to promote product quality do not exist. In cases where some product quality standards do exist (i.e., minimum heat efficiency of bricks), they are usually not enforced. Even when customers do appreciate quality, they are often not able to pay for higher up-front capital expenditures because of severe capital constraints. And there are usually few marketing activities or product labelling initiatives to better inform customers, and encourage them to distinguish between higher quality products and lower quality products.

Financing. SMEs do not possess the financial means to invest in more advanced technologies. On the other end, technology manufacturers are not in the position and other intermediaries do not exist to provide financial mechanisms encouraging technology supply push. Financial institutions are reluctant to lend for such investments to SMEs.

Market competition: SMEs face little competitive pressure in their rural markets. All local producers operate under the control of the SMEs and local markets are highly segregated. SMEs are integrated into a spatial network of enterprises supplying largely to local markets and not in a product oriented network. For this reason, inter-local distribution networks are weak or not existing, and opportunities to exploit existing economics of scale in production are limited. Product pricing is somewhat arbitrary and an SME is not driven out of the market when its profitability is too low. So far, SMEs have no experience with market/competition based regulation.

BOX 5.4B: **SMALL- AND MEDIUM-SIZE ENTERPRISES IN THAILAND**

SMEs are the real backbone of the Thai economy. A major barrier in transferring environmentally sound technologies to SMEs is insufficient financial resources. However, the difficulty in improving technology transfer capacity is not merely a financial problem. To prop up Thailand's industrial strength in the long run, Thailand has to modernise SME management and international marketing, and enhance industrial science and technology capability and labour skills. This means that:

(1) Government policies for restructuring Thai industries, including SMEs, need to be spelled out more clearly. Even though SME assistance programmes have been launched recently, resource allocation and priorities are not well managed and cannot efficiently meet the needs of SMEs.

(2) SMEs' pool of skilled personnel and ability to attract skilled R&D staff are still comparatively weak and must be strengthened. Academia-industry linkages that target SMEs would be one potential remedy.

(3) The application of voluntary environmental management in SMEs is urgently needed. Good examples are the implementation of ISO 14000 standards for medium-scale industries and the promotion of cleaner production to enhance competitiveness and environmental sustainability. It is necessary that the adoption of the environmentally sound technology transfer be on a "willingness to accept" basis.

(Source: Chantanakome, 1999)

5.4.3 Small- and Medium-Size Enterprises

Many of the incentives to embody best practice (in both capital equipment and in the products produced) that exist in multinational corporations also exist in small- and medium-size enterprises (SMEs). Indeed, SMEs can be highly innovative and competitive. But organisational difficulties and the lack of scale economies can diminish the ability of SMEs to make the best economic and environmental decisions (Lin See-Yan, 1997). Even if an SME in a developing country understands (or receives a policy signal) that it would be a good choice to invest in environmentally sound technology, it may still fail to do so simply because of a lack of information, skilled personnel or financial resources (see Box 5.4a for the example of SMEs in China and Box 5.4b for Thailand).

SMEs may also face additional barriers to technology adoption because of language differences or a lack of scientific and technical training on the part of their personnel. Even though much of the current stock of technical knowledge is available in the open scientific and engineering literature, this literature is not easily accessible in all parts of the world.

This last possibility illustrates that SMEs may be handicapped by the lack of infrastructure. The emission-reducing benefits of cogeneration in manufacturing cannot be obtained if there is no hook-up to the electrical grid, or if the scope and stability of the grid is limited so that there is no steady demand for cogenerated power. Absence of adequate infrastructure may place an additional constraint on technology choice that leads to less energy-efficient methods being chosen, as when, for example, production techniques have to be designed to work around the likelihood of power outages or brownouts.

5.5 Public-Private Partnerships

Public-private partnerships are increasingly seen as an effective way in which the public sector can achieve public policy objectives by working with the private sector. For the public sector such partnerships have the potential of harnessing the efficiency of the private sector, as well as overcoming budget restrictions and leveraging limited public funds. For the private sector, they aim to help overcome some of the internal and external barriers which prevent appropriate technology transfer taking place, and to create interesting business opportunities. Central to the concept is the recognition of corporate self-interest, and the opportunity to harness this self-interest to achieve goals such as greenhouse gas reduction.

Public private partnerships can take many forms and involve different entities. From the public sector, they can involve central government departments, agencies, multinational organisations or local government. On the private side, they can involve technology suppliers, technology user or private financiers.

5.5.1 Build Operate Transfer Projects

The Build-Operate-Transfer (BOT) structure for projects has gained considerable popularity as a form of private - public partnership that enables private participation in the development of public infrastructure. The essence of the BOT structure is that the private sector takes responsibility for the detailed design, construction, commissioning and operation of a particular project. In return it receives a payment for providing the services once operational, either from the public sector or from users, in the form of a long-term contract. After an agreed period (typically of between 10 and 30 years) the project is transferred back to the public sector. Finance for BOT projects is usually provided through project finance, discussed earlier. Equity investors usually include strategic investors such as the private construction or equipment companies, and public sector partners may also take an equity holding.

BOT and related arrangements(*e.g.* BOOT, Build Own Operate Transfer and BOO, Build Own Operate) have been very successful in opening up public infrastructure to the private sector finance, and they have a number of advantages. They bring private sector disciplines to the project development and design process, and the public sector is far less exposed to risks of cost overruns or below expected performance. By enabling focus on a particular facility they can be more readily financed and do not impose a burden on public funds, and have thus probably helped encourage financial flows. The costs of financing are relatively low, with the bulk of the capital cost being financed by bank lending at modest margins. (UNIDO, 1996; World Bank, 1994; World Bank Group, 1994)

However, there is increasing recognition of some of the limitations of BOTs. The costs are increased by high project arrangement fees, and with private sector finance being more costly than public finance, the overall costs of the output can be higher than, *e.g.*, a well managed project by a public sector utility. The long term contract typically provided to the project by the public sector is inflexible and can become a burden to the public spending if for example demand falls or prices change, particularly if certain risks have been passed through to the public sector (*e.g.*, fuel costs). Finally, BOT projects can lead to a project based focus (*e.g.*, what is the cheapest power project at present), to the detriment of broader considerations (*e.g.*, what other power options, such as energy efficiency, are best). Thus, in certain areas attention is moving away from BOTs to other mechanisms (such as merchant power plants which sell power in a deregulated market).

From the perspective of ESTs, BOT projects usually encourage the use of modern equipment and there is an incentive to be efficient, with climate benefits. However, it is worth noting that the BOT mechanism is heavily used in sectors with major climate impacts (power, transport, oil & gas). Thus, the success of the mechanism itself could be seen to be contributing to increasing greenhouse gas production. More significantly, there is some indication that the BOT mechanism favours climate unfriendly technologies (*e.g.*, coal power generation) over alternatives (such as wind power), because of the need for proven technology, the preference for established project paradigms, the large scale of BOTs and the failure of governments to internalise environmental costs.

5.5.2   *Voluntary Agreements*

Traditional legislation creates a legal sanction for desired activities by the private sector and imposes penalties for non-compliance. Alternatively, regulations may specify contractual obligations between parties, including targets, time schedules, monitoring and evaluation efforts, etc. More recently, self-regulation, or so-called voluntary agreements have gained prominence. Under voluntary agreements, industry and government get together and come to some form of understanding and commitments on certain targets and achievements, and agree to undertake their own monitoring and reporting.

One example of a voluntary programme is the Top Management Commitment Programme (TMCP) in the UK. This programme,

implemented by the Energy Efficiency Office (EEO), is targeted at the top management of companies in the UK. The programme attempts to elicit a formal commitment from the Chief Executive Officer (CEOs) of a company, requiring them to state their commitment to energy efficiency and display the same all over the company. The commitment was signed by the CEO and the Minister in charge, and it requires the company to formalise their commitment to energy efficiency. The company would undertake to report on energy consumption and efficiency, set up working groups and/or councils, suggestion schemes, etc. This programme has already been joined by over 1,000 companies, including IBM, Shell, Ford, BA, etc. Follow-up surveys carried out by the EEO indicated that the success on energy conservation efforts was significantly higher in the companies that signed on than in the companies that were not part of the TMCP.

The U.S. EPA's Green Lights Programme is another example of a voluntary agreement (see Case Study 2). Companies participating in the programme agree to invest in energy-efficient lighting retrofits in exchange for technical expertise and public relations benefits from the programme. China has started its own Green Lights programme. Numerous successful corporate voluntary programmes for ozone-depleting-substance phaseouts have also occurred (see Case Study 17).

In developing countries, one example of an especially successful voluntary programme was part of the Thailand Promotion of Electricity Efficiency Project by the Thai national electric utility (EGAT), partially financed by the Global Environment Facility (Martinot and Borg, 1999). EGAT wanted to rely on voluntary agreements and market mechanisms, and elicited a voluntary agreement with all five Thai manufacturers and the sole importer of T-12 fluorescent tubes. Under the voluntary agreement, the manufacturers and importer of T-12 lamps agreed that they would switch to producing and importing more-efficient T-8 lamps instead of the less-efficient T-12 lamps. In return, EGAT engaged in an extensive public education and information campaign to educate consumers about the switch and make the switch acceptable to the market. By 1995, all lamp manufacturers and importers had complied with the agreement, and virtually all T-12 lamps were eliminated from the Thai market. Success was aided by a zero net cost to manufacturers (reduced T-8 production costs paid for the production conversion), T-8 retail prices similar to those for the T-12 lamps, and luminaire compatibility. Success was also attributed to cultural factors ; the utility stated that the public considered such voluntary agreements more desirable and fairer than price incentives like rebates or subsidies.

There is much scope for voluntary agreements and other types of voluntary pollution prevention programmes, particularly for reduction of GHG emissions (Aloisi de Larderel, 1997). Berry (1995) suggests that industry and government should together take steps to refine knowledge of technologies and to educate users and manufacturers. Bringing the cost of non-standard technologies such as photovoltaics (PV) systems down implies reaping economies of scale in manufacturing the PV systems. To bring the transaction barrier down, Berry further suggests more knowledgeable buyers, sellers, and suppliers, risk taking on the part of investors in large manufacturing plants, efforts at cooperation by major users of PV systems, and opportunities for risk sharing.

However, as identified in a recent report (OECD, 1999b), there can be problems with voluntary agreements, and the overall experience has been more mixed. In particular, an essential prerequisite for voluntary agreements is an underlying ability and willingness by policymakers to develop and enforce environmental regulations, so that the threat of alternative measures is credible. In countries where such conditions do not exist, the use of voluntary agreements could be at best ineffective and potentially very damaging for environmental objectives.

New examples of unilateral corporate commitments are also emerging, involving senior executives of the company voluntarily making a clear commitment to addressing environmental issues. An example of leadership has been the recent clear announcement of BP Amoco to reduce CO_2 emissions by 20%, combined with the establishment of an internal trading system to achieve that end.

5.5.3 Technology Partnership Programmes

Technology transfer aimed at fostering mitigation and adaptation responses to climate change will be most effective where it engages all key stakeholders in designing and implementing technology transfer actions. These key stakeholders include in-country and international private businesses and investors, government agencies, and bilateral and multilateral donor organisa-tions. While private businesses play the key role in implementing most technology transfer activities, national governments and international donor agencies can help remove market barriers and set conditions to ensure effective private sector participation in technology transfer. Technology transfer activities will be most effective where businesses, governments, and donor organisations collaborate in designing and implementing these activities to make the most productive use of their respective resources and authorities.

Since climate change is not explicitly considered in most development plans, climate change considerations are not fully integrated into the development plans that shape markets for new technologies. In many cases, consideration of climate change issues will only require marginal adjustment to development plans, but this process of adjustment and review is critical in ensuring that development programmes contribute to climate change goals. Therefore, it is important for climate change technology transfer activities to respond to developing country determination of what type of technology transfer will best contribute to their development needs while also addressing climate change. Once these technology transfer priorities are well understood, developing countries can work with the private sector and the international donor community to facilitate technology transfer activities to respond to these priorities.

Technology partnerships can also be exclusively at a firm (company) level. The United Nations (1996) has suggested that "firms in many developing countries—the least developed ones in particular—often do not have the funds, trained human resources or infrastructure to pursue a technology-based innovation process

---

BOX 5.5   TECHNOLOGY COOPERATION AGREEMENT PILOT PROJECT (TCAPP)

In 1997, the U.S. Government launched the Technology Cooperation Agreement Pilot Project (TCAPP) to provide a model for a collaborative approach to foster technology cooperation for climate change mitigation technologies. Under TCAPP, the Governments of Brazil, China, Kazakhstan, Mexico, and the Philippines are currently working with the private sector and bilateral and international donor organisations to attract private investment in clean energy technologies in their countries. Many other donor initiatives have also adopted similar collaborative approaches between country officials, businesses, and donors in fostering private investment. However, TCAPP is one of the few initiatives that has engaged climate change officials in this collaborative process to lead to actions that address both development needs and climate change goals.

TCAPP has two basic phases of activities. In the first phase, the participating countries have developed technology cooperation frameworks that define their climate change technology cooperation priorities and the actions necessary to attract private investment in these priorities. These actions include efforts aimed at capturing immediate investment opportunities (e.g., issuance of investment solicitations, investment financing, business matchmaking and capacity building, etc.) and longer-term efforts to remove market barriers. In the second phase, TCAPP assists the country teams in securing the private sector, in-country, and donor participation and support necessary to successfully implement these actions. This second phase of activities includes two major types of activities:

1   Attracting direct private investment in immediate market opportunities. This includes helping the countries develop and issue investment solicitations for large-scale opportunities and business matchmaking and financing activities. TCAPP has established an international business network to help guide the design and implementation of these activities.

2   Securing support for actions to address market barriers. These actions range from business capacity building to policy reform. This includes development of domestic implementation plans for these actions and securing necessary donor support to assist with implementation of these plans. TCAPP assists the countries in preparing implementation plans and donor proposals and in matching country needs with donor programmes.

(Sources: NREL, 1998, UNFCCC, 1999)

on their own. In such cases a level of cooperation is needed that is qualitatively different from that associated with traditional technology transfer. Technology partnership (TP) is one opportunity for participation by developing countries' firms in the emerging forms of technological alliances and cooperation."

The essential characteristics of technology partnerships between enterprises from industrial and developing countries are typically the following: (a) they are long-term arrangements; (b) they are mutually beneficial; (c) they contain an explicit commitment to cooperation; (d) they have as one of their central goals the learning process of both partners; (e) they occur within a technology system and within specific economic relations; (f) they enhance the level and depth of both partners' technological capabilities (UN, 1996).

One recent example of a technology partnership programme is the Technology Cooperation Agreement Pilot Project by the U.S. government (see Box 5.5). Another example is the Technology Partnership Initiative, run in the United Kingdom by the Joint Environmental Markets Unit of the DTI and DETR. Joint demonstration projects of many kinds represent another form of technology partnership; one example is the project in Brazil by a consortium of twelve companies, including private and public Brazilian enterprises and multinational firms, to develop a biomass gasifier/gas turbine power plant designed to use wood chips as fuel. The consortium was created by the joint entrepreurship of foundations, industry associations, and government entities (Norberg-Bohm and Hart, 1995). And the multilateral "Climate Technology Initiative" has established a programme called "Technology Cooperation Implementation Plans" (TCIP) to carry out a number of partnerships (UNFCCC, 1999).

5.5.4   *Informational Initiatives in Private Finance*

The public sector can aim to encourage private finance to be more active in the development, dissemination and transfer of environmentally sound technology through a variety of activities which aim to remove some of the obstacles identified above, without spending large sums of public expenditure. A wide range of initiatives is possible here such as:

- Providing particular support to environmental businesses on accessing finance to help them in their dialogues with financiers. The USA, U.K. and Canada have all had initiatives in this area.

- Raising awareness within the financial sector on climate change, through seminars and working groups. Some countries have actively tried to get the financial sector involved in discussions on the appropriate response to climate change.

- Supporting the activity of some of the industry organisation identified above, and involving them in government consultations.

- Looking at industry training and regulation to ensure that there is a level of environmental management. Most

active in this area have been the multilateral development banks such as the EBRD, who have required banks to develop environmental procedures as part of their financial markets' development programmes. While such activities will tend to emphasise avoiding the worse rather than actively encouraging the use of the best technology. it can still encourage a change in attitude among financial institutions.

- The lower risks are perceived to be, the cheaper finance will be. The public sector can help reduce risk perceptions by supporting information dissemination, *e.g.* of successful case studies.

5.5.5   *Fiscal Measures: Tax Incentives and Guarantees*

Governments can take a number of fiscal actions to encourage the uptake of environmental technology. While these can be highly effective, it should be noted that in many cases these may be second best alternatives to more consistent measures to internalise environmental costs or remove subsidies, which may not be politically acceptable (essentially offsetting one subsidy with another). In other cases they may be useful to initiate a market, but should not become permanent features.

Investor tax incentives. Governments can seek to encourage investment by supplying tax incentives for investors in certain types of companies or investments. One example is in the U.K. where private investors buying venture capital funds and new shares in unlisted businesses (not necessarily environmental) can partially offset the investment against income thus reducing tax liabilities. This has helped to encourage investors to put capital out into new technologies, although it has proved difficult to ensure that investors back the kind of risky investment the schemes are aimed at (rather than creatively packed less risky alternatives). An alternative approach has been taken in the Netherlands where the Government has given a tax-free status to returns on investments in approved environmental funds. These have succeeded in attracting substantial amounts of capital and in reducing the cost of finance for appropriate environmental projects. Perhaps most importantly they have helped encourage financial institutions to find and help develop new "green projects". However, they do not provide risk capital, and there have been problems of definition. While such projects normally support domestic businesses and ventures, the Dutch government recently extended its investor tax incentives to the selected projects in developing countries, with the potential to support technology transfer. In theory, the sort of tax support given to domestic venture capital could also be extended to selective venture capital opportunities in developing countries which might be an effective way of encouraging appropriate technology transfer.

Capital expenditure tax incentives. An effective way to encourage the uptake of new technology is by providing accelerated capital depreciation on certain equipment – in extreme cases all in the first year, or alternatively on a faster schedule than normally used. This has been used in a number of markets to help accelerate the uptake of renewable energy (California, India) and is currently being used in the Netherlands to encourage businesses to install certain types of environmental technology. One advantage of accelerated capital allowances is that they can usually be combined with leasing to provide an accessible and flexible sort of financing.

Loan guarantee schemes. In order to help support new business development, a number of governments have introduced loan guarantee schemes to support domestic small business development. They consist of the central government guaranteeing loans made by domestic banks to the small business sector to encourage the development of that sector. In most cases only a partial guarantee is provided so the participating private sector banks have an incentive to lend prudently. Other organisations have introduced schemes more specifically targeted at lending for environmental projects (*e.g.*, the European Investment Bank). Such loan guarantees do depend on the existence of a strong, independent financial system.

5.5.6 Partnering and Sponsorship for New Financial Initiatives

Another area in which the public sector can encourage the financial sector to become involved in the transfer of environmentally sound technology is through partnering and sponsoring new financial initiatives. This can reduce the costs and risks for private financial institutions in developing new products and instruments, and can help them give such initiatives a higher priority.

Support for such initiatives could come from a number of areas within the public sector: domestic industry or environment departments, bilateral assistance agencies, bilateral development banks, multilateral development agencies, and multilateral development banks. While some other examples do exist, it is the multilateral development banks that have been by far the most prominent in this sort of activity.

In particular, the World Bank's IFC, through its environmental projects unit, is now aiming to find ways to develop and support innovative mechanisms which help address environmental challenges and also encourage private financial sector participation, thereby using limited concessional finance efficiently. While in many cases it is proving more challenging and time consuming than originally expected to pull together and develop such initiatives, the IFC has started to accumulate a portfolio of activities (Asad, 1997). These include:

Finance for SME environment business. IFC's Environmental Projects Unit delivers a GEF-funded Small and Medium Enterprises (SME) programme which is designed to channel concessional funds through intermediaries to SMEs for renewable energy, eco-tourism, energy efficiency, sustainable forestry and agriculture. The SME activity needs to address the objectives of GEF programmes involving climate change and conservation of biodiversity. Intermediaries have included private companies, NGOs, financial institutions and a venture capital fund. These intermediaries can benefit from low interest rate loans and incentives, along with limited amounts of technical assistance to assume the business risk and invest in SME enterprises. The use of intermediaries by the GEF/IFC SME Programme helps overcome the obstacles of scale and of transaction costs identified above when dealing with the SME sector. The EBRD has developed similar programmes to encourage finance through intermediaries.

Emerging sector and market funds. The IFC is helping to create Sector and Market Investment Funds to assist professional and institutional investors to look at biodiversity, renewable energy and energy efficiency. Of greatest relevance is a major fund, co-financed with the Global Environment Facility (see Box 5.2), which will invest in renewable energy and energy efficiency projects, namely the proposed Renewable Energy and Energy Efficiency Fund for Emerging Markets (REEF). In addition, the IFC is also encouraging the development of environmental funds for a particular region or country, such as the proposed MENA Environmental Fund – while such funds invest in a variety of environmental projects, energy and climate-change-related investments are a significant proportion of the total. With these funds, the IFC contributes some of the capital but the aim is to attract funds from outside, particularly from mainstream financial investors. Here the IFC, as a supporter of the fund, can play a valuable role in reassuring investors about new markets. In addition, the IFC can help reduce the costs and risks of developing such funds.

Transforming inefficient or non-existent environmental markets. IFC's Market Transforming Initiatives recognise that while new environmental markets may offer potential, there are significant barriers to their development which other market players cannot address on their own. Examples relevant to climate change technologies include the Photovoltaic Market Transformation Initiative (PVMTI), and the Poland Efficient Lighting Project (PELP, see Case Study 2). The Initiatives aim to minimise the risks of developing them by providing concessional funds for innovative solutions to market development, with the objective of taking the markets to the point where fully commercial operations are viable, or to accelerate the penetration of commercial technology. Although concessional, the funds are intended to operate in many ways like private sector funds, and projects will be judged on the basis of current and future ability to leverage additional private sector finance, trigger market growth potential and promote longer term sustainability and replicability. The World Bank is also actively developing new market transformation approaches

These initiatives are to be welcomed. However, it should be noted that the close cooperation between the public and private sectors that these initiatives are based on can often entail tension in areas such as cost-sharing, timescales and objectives. For example, REEF has taken several years longer to develop than originally anticipated. There is a need to learn from experience, for increased education of the private sectors and possibility some more flexibility and pragmatism from the sponsors of such initiatives to encourage private sector participation.

5.5.7   *The Case of the Montreal Protocol*

The possibilities for partnerships and financing for climate change mitigation can be better understood through an examination of the historical experience in phasing out ozone-depleting substances (ODSs) under the Montreal Protocol on Substances that Deplete the Ozone Layer (see also section 3.3.3 in Chapter 3). Evaluations of the economics of the phaseout process that have been made since the Protocol was signed in 1987 have concluded that the speed of the phaseout has been faster, and the cost lower, than had been anticipated when the Protocol was negotiated (Hammitt, 1997; Cook, 1996; Economic Options Committee, 1991, 1994, 1998). This happy surprise is attributable largely to the unusual and unexpected channels for technology transfer that emerged once the Protocol was in place.

The signing of the Montreal Protocol meant that significant cutbacks in ozone-depleting substances had become a strategic business necessity. Industrial leaders' recognition of this fact may in part have been simply an acknowledgement of the legal reality of the Protocol, but there is ample reason to believe that their support of the Protocol was based on an understanding of the science as well. Given the existence of this sort of consensus, undertaking the kind of organisational changes needed to eliminate ODSs followed. Participation in the ozone protection effort became a basis for career advancement and a source of personal pride for individuals within companies. Firms found that by redesigning processes to reduce their need for ODSs, they could realise previously unforeseen productivity gains, as when it was discovered that printed circuit boards could be manufactured without having to clean off soldering residues with a CFC solvent (Iman and Lichtenberg, 1993; Wexler, 1996a).

A variety of cooperative arrangements evolved under the umbrella of the Montreal Protocol that facilitated technology transfer. Soon after the Protocol was signed, a number of large corporations with major electronics interests formed ICOLP, the Industry Cooperative for Ozone Layer Protection, to share information, discoveries, and procedures for eliminating ODSs in their manufacturing processes. A similar consortium was founded in Japan, the JICOP (Japan Industrial Conference for Ozone Layer Protection). Company-to-company deals, multifaceted agreements involving both companies and governments, and both formal and informal information exchanges characterised the process. Examples include the trilateral agreement between Thailand, Japan MITI, and the U.S. EPA (see Case Study 23) and the cooperation between the Government of Mexico, Camara Nacional de la Industria de la Transformacion, the Canadian telecommunication company Nortel (Northern Telecom), the

International Cooperative for Ozone Layer Protection, and the U.S. EPA. (Economic Options Committee 1994; also see Case Study 17)).

Governments played a supportive role in the formation of information-exchange networks. In the United States, for example, a vital clearinghouse role was filled by the U.S. Environmental Protection Agency's Stratospheric Ozone Protection Division. The U.S. EPA, along with the industry-based Alliance for Responsible Atmospheric Policy (formerly the Alliance for Responsible CFC Policy), Environment Canada, and the United Nations Environment Programme, co-sponsored an annual meeting in which industry practitioners presented papers detailing the progress they had made in eliminating ODSs in their own operations. The culture of this well-attended meeting was akin to a scientific symposium or a scholarly conference, with an emphasis on free exchange of ideas. Part of the conference space was devoted to a trade fair, in which the most recent advances in technology to replace ODSs were on display. By 1998 the conference had evolved into the "Earth Technologies Forum" covering climate protection technologies as well as ODS elimination.

The Multilateral Fund, set up under the Montreal Protocol to assist developing countries in defraying the "incremental costs" of compliance with the Montreal Protocol, also played an important role. The Fund has financed projects ranging from the development of individual country ODS-elimination programmes to the building of large-scale industrial facilities that use alternate technologies. In some respects, the Fund may be seen as a precursor and proving ground for the functioning and organisation of the Global Environmental Facility (see Box 5.2 on the GEF). It has been the objective of this Multilateral Fund to provide development assistance that is "additional" to other aid funds. Through establishment of the Multilateral Fund, the Parties to the Montreal Protocol have addressed the equity concerns that for a time retarded full participation in the ozone protection process by all the key countries (neither China nor India signed the Montreal Protocol until the Multilateral Fund was established at the London meeting of the Parties in 1990).

Government policies were able to exert a positive influence on technology development in other ways. A major stumbling block to the elimination of ODSs in electronics manufacture was removed when the U.S. Department of Defense changed its requirement that CFC-113 be used to clean soldered electronic assemblies to a performance standard (Wexler, 1996b). Governments have helped to develop standards for recycling CFCs, and have supported the establishment of Halon "banks" to get full use from the stock of ODSs that have already been produced (Economic Options Committee, 1994). The United States imposed both an excise tax on 'new' ODSs and a 'floor tax' on inventories beginning in 1990. The effect of such taxes is to make new chemicals and technologies more attractive, and to encourage reclamation and recycling (Economic Options Committee, 1994). The Ninth Meeting of the Parties in Montreal, 1997, decided to require Parties to the Protocol to establish licensing systems to control the import and export of new, used, recycled and reclaimed substances, in order to reduce and eventually eliminate illegal trade in controlled ODSs (Economic Options Committee, 1998).

5.6 Technology Intermediaries

The World Bank and many other agencies have recognised that technology intermediation is needed to reduce barriers to technology transfer associated with information, management, technology, and financing (World Bank, 1993; Martinot *et al.*, 1997; Heaton *et al.*, 1994). Research on technology innovation also highlights the role of intermediaries in the innovation process (Dodgson and Bessant, 1996, p.54; see also section 4.3 on National Systems of Innovation and Technology Infrastructure). Examples of technology intermediaries include specialised government agencies, energy-service companies, non-governmental organisations, university liaison departments, regional technology centres, research and technology organisations, electric power utilities, and cross-national networks. Non-governmental organisations in particular are playing a greater role in technology intermediation; for example, there are many cases where technology intermediation by NGOs played a key role in the success of particular technology transfer efforts for renewable energy (Kozloff and Shobowale, 1994). The functions of technology intermediaries can include:

- articulation of specific technology needs and selection of appropriate options
- education, information dissemination, and communication
- identification of skill and human resource needs
- selection, training, and development of personnel
- investment feasibility, appraisal and business plan development
- development of business and innovation strategies
- locating key sources of new knowledge
- building linkages with the external sources of information
- creating and/or operating new dealer and service networks
- project management and organisational developmentreferrals
- training and consulting
- energy audits
- matching potential supplier and recipient firms
- feasibility, evaluation, and packaging of projects for public or private financing
- translating, compiling, vetting, and endorsing information

In general, there are seven key questions for policies that promote technology intermediaries:

1. What are the needs of users?
2. Who are the suppliers of technology?
3. What are the needs of technology suppliers?
4. What is an appropriate role for intermediaries?
5. What kinds of agency can help bridge the gap between suppliers and users as an intermediary?
6. What are the mechanisms whereby such intervention can take place?
7. What can public policy do to enable or assist the process of intermediation?

5.6.1 The Value of Technology Intermediaries

The need for intermediation to overcome transaction barriers is often discussed in the context of technology development, both internationally and in purely domestic contexts. The World Resources Institute (Heaton *et al.*, 1994) has proposed sector-specific intermediation as an important policy goal for greater international technology transfer, development and cooperation:

In intermediation, third parties create linkages, transmit knowledge, and expedite other transactions for the principals. The greater the barriers that separate parties who could create relationships of mutual benefit, the greater the need for intermediation. In technology development, the value of intermediation is well recognised. (p.20)

Evidence to-date with institutions that perform some intermediary functions shows that sector-specific intermediaries have advantages over broad, general-purpose intermediaries, because the technologies and applications involved are simply too diverse. A consequence is that intermediaries for energy efficiency and renewable energy should to some extent be specialised. In this view, many or most or the actors are already in existence and working, but communication and new, and more specific, problem-solving capacities are required.

Others have called technology intermediaries bridging institutions. Dodgson and Bessant (1996) highlight the importance of the intermediaries that operate between users and suppliers of technology and which help to create the links within networks and systems. They say that "bridging institutions...encourage interaction within the system, assisting with undertaking search, evaluation and dissemination tasks. They ensure that technological know-how is broadly dispersed within the system and can provide a compensating mechanism for weaknesses or 'holes' in the system" (p. 26). Innovation agents are another name for intermediaries(Dodgson and Bessant, 1996, p.186).

The high value of technology intermediation is illustrated by many of the case studies presented throughout this report. In the Baltic States, the Swedish government aid agency NUTEK promoted conversion of heating boilers to biomass by bringing boiler operators and manufacturers of conversion equipment together, by providing financing for the conversions, and by providing assistance to boiler operators in financial and technical analysis, competitive procurement, and contracting (see Case Study 18). In Mexico, the national electric utility played an intermediary role by marketing and selling efficient lighting through its offices, and by reducing the retail price through bulk procurement (see 5.2.1). In East Africa, the promotion and dissemination of improved efficiency cookstoves was facilitated by small-scale informal-sector entrepreneurs providing sales and service (see Case Study 1).

In countries with economies in transition (CEITs), technology intermediaries are important ways to overcome the lack of business, financing and marketing skills among firms whose managers never learned these skills in the centrally planned economy (because these skills were not needed). In particular, energy service companies, financial intermediaries and information centres have been playing important technology intermediation roles for climate-friendly technologies in many CEITs (discussed in more detail in the sections below). In CEITs, because of generally well developed technical skills among enterprises, intermediaries can focus on business, information, and financing services (Evans and Legro, 1997; Martinot *et al.*, 1997; Martinot, 1998; Marousek *et al.*, 1998).

In many rural photovoltaic programmes, a local or foreign intermediary provides critical marketing activities, education, financing or leasing mechanisms, sales and service infrastructure that helps to create a market. In the Dominican Republic, an innovative leasing programme by SOLUZ has successfully transferred PV technology. In Kenya, a network of dealers, along with education and training programmes have resulted in 80,000 solar home systems in use (see Case Study 5). In Bangladesh, the Grameen bank has successfully provided micro-credit for solar photovoltaic home systems (see Box 5.3). The World Bank and GEF have recently incorporated innovative intermediary mechanisms into solar PV home system projects in China, Indonesia, and Argentina, reflecting the World Bank's increased focus on rural energy for development (World Bank, 1997).

Technology intermediaries also can play an important role in strengthening the enabling environments for technology transfer discussed in Chapter 4. In particular, they can help to establish codes and standards locally or nationally, they can help facilitate programmes that create sustainable markets for environmentally sound technologies, and they can influence regulatory conditions and macroeconomic policies.

5.6.2   *Information Clearinghouses and Technology Transfer Agencies*

In order for technology transfer transactions to take place, parties must know about each other and understand the costs and benefits of different technology transfer pathways. Often projects, particularly to introduce the new energy technologies, are conceived without proper understanding of the needs and priorities of the targeted users (Mapako, 1997). Consumers or purchasers must be aware that technologies exist, must know their performance characteristics, reliability, capital costs, operating costs, and economic benefits, and must know how to maintain and service technologies or know of firms who can. While in most of the developed countries there are a multitude of information sources, the same is not the situation in the developing countries. Interviews with more than a hundred negotiators and policy-

makers in developing countries can be summed up in the words of one interviewee: "we do not know what is available and what we really need" (Gupta, 1997, p.89).

Information clearinghouses and technology transfer agencies are specific forms of technology intermediaries that have been proposed by UN and other public agencies. These agencies point out that numerous public and private environmental information systems already exist. Improving these existing systems and linking them through clearinghouses can be a first step towards establishing an international network of technological information. A number of international information networks and databases that specifically address climate-mitigation technologies already exist (UN, 1997).

Although governments commonly set up information centres, in some countries, national or sector specific industry associations have also set up information centres. The information centres are of two types: information of a highly technical nature, required by larger energy consumers, and general information as would be required by households and small commercial establishments. Traditionally, schools and colleges, science centres, and museums have also been common vehicles for providing general information.

But beyond the simple supply of information, more sophisticated technology transfer agencies can actively promote knowledge transfer through a number of activities:

*   conducting workshops, seminars, and conferences
*   assisting technology producers in marketing their technology and understanding markets
*   providing training and assistance in preparing business plans
*   matching potential joint venture partners
*   securing intellectual property rights and assisting in creating licensing agreements
*   educating financiers about specific technologies and channelling investment proposals

Notwithstanding the UN initiatives in this direction, many countries in the developing world and CEITs have initiated systems to provide information on different technologies. In a recent climate technology and technology-information-needs survey among developing countries under the auspices of the Subsidiary Body of Scientific and Technology Advice, 60% of the respondents pinpointed to national technology information centres as an important vehicle for dissemination of climate relevant technologies and practices. In over 75% of these respondents' home countries at least two such technology information centres exist (van Berkel & Arkesteijn, 1998). Successful energy-efficiency centres in several economies in transition (China, Russia, Bulgaria, Romania, and the Czech Republic) are good examples of technology intermediaries that have been established with international assistance (Chandler *et al.*, 1996). As another example, Box 5.6 describes national-level technology intermediaries in India for energy efficiency and renewable energy technologies.

BOX 5.6   NATIONAL TECHNOLOGY INTERMEDIARIES IN INDIA

National-level government agencies acting as intermediaries can also be important in creating incentives and facilitating a market for cleaner technologies. The Energy Management Centre (EMC), an autonomous agency, under the Ministry of Power, Government of India, is an example of a technology intermediary for energy efficiency. EMC has been carrying out a number of initiatives to promote energy conservation and efficiency in India. To begin with, EMC set up and trained 25 agencies (public, private, NGOs), to provide specialised energy auditing and management to consumers in India. Each of these agencies are carrying out an average of 10-12 energy audits annually, and the feedback from the industry is that there is an urgent need for many more such professional agencies to be able to serve the consumers in the country. EMC also carried out a number of studies in the area of technologies for energy efficiency, issues relating to standards and labelling, as well as implementing a nation-wide energy conservation awareness project. EMC annually organises, through industry associations, about 20-25 training programmes and workshops for wider dissemination of information on ener-

gy conservation in the country. To date, it is reported that over 5,000 professionals have been provided training in different aspects of energy efficiency. Regular feedback carried out indicated that the participants have actually implemented energy efficiency projects in their organisations. EMC was the executing agency for international cooperation projects with Germany, the European Union, and the Department of Energy (USA), among others.

The initiatives of the Indian Government implemented through the EMC have resulted in a significant rise in the exposure and awareness on energy conservation technologies. It is reported that there are proposals to introduce standards for appliance and energy consuming devices and these would be mandatory. Penalties for non-compliance would be enforced once the law is passed by the Indian parliament. Under a collaborative programme with the EU, EMC has set up an Information Service on Energy Efficiency (ISEE), jointly with a national industry association. The database established is expected to contain information on technologies, guide books, manuals, best

practice programmes, a list of manufacturers, etc. and is expected to fill the gap in information for energy consumers.

The Technology Information, Forecasting and Assessment Council (TIFAC) in India was established as an autonomous organisation of the Indian Department of Science and Technology, and has been particularly successful in making the public-private sector linkages, providing information on patent issues, and supporting start-up ventures. Each of these activities provides important examples for other similar, knowledge-based technology transfer policy offices.

The Ministry of Non-conventional Energy Sources (MNES) in the nodal ministry responsible for providing the overall thrust and direction for increased adoption and installation of renewable energy devices in the country. MNES implements the programmes through the state governments and through state energy nodal agencies. MNES has separate programmes for biogas, solar thermal, solar PV, biomass gasifier, and for new technologies.

Most developing countries have a large proportion of small and medium-scale industrial concerns whose outputs constitute a significant portion of the GDP of these countries. The technology information needs of these enterprises may not be the same as that of big MNCs operating in these countries. Nevertheless the technology information centres springing up in developing countries do or can not serve MNCs. While the private sector has often been disdainful of such information programmes, many representatives of developing countries report that they find vendor

information biased and confusing and need help to evaluate competing vendor claims. Such technology information centres can provide such help. The small and medium scale industries are generally short in management and technological capabilities, and can make effective use of information clearinghouses on ESTs. Having said that one should also realise that the small-scale industries or the so-called informal sector are far removed from such initiatives, thus it is desirable that agencies use intermediaries to reach them with this information.

BOX 5.7   REGULATING ELECTRIC POWER UTILITIES TO BE TECHNOLOGY INTERMEDIARIES

The key role played by electric utilities as technology intermediaries in promoting energy efficiency has been well established in scientific literature. Historically, utilities in the US began offering energy efficiency or demand side management (DSM) programmes to consumers after regulators made this a policy goal. The utilities were compensated by the utility commissions for any loss of revenue that may have occurred in this process, after accounting for other savings due to reduced fuel costs, etc. Investment on DSM is reported to exceed US $2 billion, and this is expected to account for about 14% of the new investment in the power sector in the US in 1994. Utilities in Germany, Denmark, Canada and other countries followed. Now DSM programmes are being taken up in Thailand, the Philippines, Mexico, Jamaica, Brazil and other developing countries as well (see Case Studies 10 and 23 on DSM in Chapter 16). These programmes are in their early phase, and they account for a small portion of the activities on energy efficiency in these countries. Utilities are essentially playing the role of information provider to start with; then they assist consumers to achieve energy efficiency at the consumers' premises. Several utilities have set up independent companies, 100% owned by the utility, which are outside the control of the regulator, since energy efficiency business was essentially unregulated.

Regulating electric power utilities to perform so-called Demand-Side Management (DSM) is a form of technology intermediation that became well established in the United States in the 1980s (see Box 5.7). Utilities in Germany, Denmark, Canada and other developing countries have followed. Now DSM programmes are being taken up in Thailand, the Philippines, Mexico, Jamaica, Brazil and other developing countries as well (see also Case Studies 10 and 23 in Chapter 16).

5.6.3 Energy Service Companies

Energy service companies (ESCOs) are a specific form of technology intermediary that has gained widespread acceptance in developed and developing countries and countries in transition. Due to the previously discussed barriers to conventional financing, innovative financing schemes are needed. A financing scheme is a particular institutional arrangement that determines who pays what to whom and who bears the risks of the transaction(s). An energy service company "ESCO" addresses the financial capability, and other institutional market issues. An ESCO is a company that offers energy services to customers with performance guarantees. Typical performance contracting arrangements provide customers with feasible means of improving their competitiveness by reducing energy consumption costs. Additionally, companies' cash flows are enhanced, which add value to their financial worth.

Historically, ESCOs evolved in three broad categories as follows:

- Technology based (technology suppliers).
- Financial and legally based, with sub-contracting for the technical aspects of projects
- Technically based such as engineering consultancy firms

Two common ESCO approaches in the United States are guaranteed savings and shared savings approaches. In the guaranteed savings structure, the end-user finances the project's initial investment costs from a third financier and, in turn, the ESCO guarantees that the energy savings will at least cover the debt services. Then the ESCO receives a share of the net savings after debt services and the operations and maintenance costs. However, if the savings fall short of the customer's financial obligations as stated in the performance contract the ESCO assumes the shortfall. In this respect, the ESCO assumes all the risks associated with the project's performance and the third party financier assumes the end-user's credit risk. In the second approach, the shared savings structure, the ESCO finances the project's initial investment costs, usually by borrowing from a third party financier. In turn, the ESCO is compensated by a higher share of the project savings. Given the current market situation in most of the developing countries, ESCOs are most likely to evolve in one of the following forms:

- Local engineering consultancy firms expanding their portfolio of services to include energy efficiency as one of their activities.

- Local engineering consultancy firms entering into joint ventures with foreign technical partners.
- Local equipment suppliers expanding their services to include energy efficiency services.
- Financial and legal firms creating specific companies for this purpose.

ESCOs have been successful in many developed countries, in particular the United States, United Kingdom, France, Germany, Australia and Brazil. However, the risks and the absence of clear success in developing countries are still issues. While there have been one or two successes in the former communist countries with specific injection of bilateral grant funds, the ESCO concept is still emerging. Nevertheless, energy service companies are operating or being formed in several developing countries and countries in transition, including Brazil, Mexico, China, Thailand, India, Russia, Hungary, and the Czech Republic (see Case Studies, Ch. 16). With assistance from USAID, two or three ESCOs are now operating in India, with efforts underway to increase the number of operations as well as to sensitise the consumers to take advantage of the services provided by the ESCOs. A pilot project by the World Bank and the Global Environment Facility is pioneering parastatal ESCOs in China by developing standard contractual models, providing financing and technical assistance to a group of pilot government-owned ESCOs (which may be privatised over time), and disseminating information about energy efficiency measures to industry (World Bank, 1998).

Energy service business associations have been recently formed in Egypt and Brazil. They represent groups of private companies offering energy efficiency products and services. Their members share the common goal of providing solutions that reduce energy costs, improve productivity, and enhance operating conditions of energy users. The associations will address current market barriers facing the energy service business community, and will provide a forum for energy efficiency development. Members of the associations include companies providing turn-key services as energy service companies (ESCOs), equipment vendors and service suppliers, support vendors such as legal firms and consultants, and other interested organisations.

5.7 Conclusions

ODA is still significant for the poorest developing countries. There is increasing recognition that ODA can best be focused on mobilising and multiplying additional financial resources, assist the improvement of policy frameworks and be based on long-term commitments to capacity-development. However, the advantages of public-sector finance may be offset by assistance in the form of tied aid, which can be detrimental to the longer-term prospects for indigenous technology development by preventing the establishment of the institutions to support technology choice, financing, and operation and management. Tied aid is more useful at targeting areas such as capacity building and project preparation.

There are several channels for international public finance. Bilateral development aid is the largest, with very mixed records relating to technology transfer and sustainable development that, furthermore, varies widely between countries. The MDBs are another major route, and in many cases these now have more developed environmental criteria and sources. Agencies such as UNDP may also strongly influence technology transfer through their programmes. On the other hand, trade support, such as export credits, rarely takes account of environmental factors and may in many respects be biased against environmentally sound technologies.

The choice between different international financing routes is determined by many factors. One major issue regarding the effective transfer of ESTs through public finance is the fact that foreign aid expenditure tends to be institutionally divorced from other powerful agencies that also have a huge influence on technology choice and investment patterns, such as trade ministries. It is not uncommon to find governments pursuing seemingly contradictory international financial policies, and one of the strongest recommendations in this area is for greater institutional coherence within donor governments.

An important trend is the shift away from the public sector to the private sector as the principal source of finance. Transfers of ESTs are influenced by this trend, partly because the private sector requires relatively high rates of return and does not monetise externalities. Although the private sector has begun to recognise the importance of climate change, governments can enhance private involvement through various types of initiatives and partnerships. Private support for climate-friendly technology transfer may also require financial innovations and emphasis on different forms of finance such as micro-credit, leasing and venture capital.

There is a wide variety of types of traditional private sector debt and equity finance available depending on the scale and type of the project. The most flexible way to finance debt is secured loan and leasing. The transfer of ESTs to developing countries will also involve increased use of innovation to structure existing financial products to new markets and to develop new ones as appropriate. Just as support for scientific and technical innovation is seen as an appropriate use of public funds, so is support for financial innovation. A number of worthwhile initiatives have been undertaken to date (such as micro-credit, project finance, green finance and also the use of strategic investors), and there is scope to replicate and extend these as well as develop new concepts.

Although the private-sector pathway is one of the key channels for the transfer of EST, it should not be assumed that the search for economic gains on the part of individuals and firms will guarantee adoption of best-practice techniques. A number of obstacles that are internal to firms can retard the diffusion of pollution-reducing innovations even when such innovations would be profitable. These obstacles are not instances of "market failure" in the traditional sense, because they originate within firms rather than arising from the strategic interactions between firms or from the existence of externalities or public goods. These barriers can also retard the transfer of environmentally beneficial technologies between firms. Policy measures, public-private partnerships, and internal organisational improvements can help overcome these barriers and promote the interests of all stakeholders.

Public-private partnerships are increasingly seen as an effective way in which the public sector can achieve public policy objectives by working with the private sector. For the public sector they have the potential of harnessing the efficiency of the private sector, as well as overcoming budget restrictions and leveraging limited public funds. For the private sector, they aim to help overcome some of the internal and external barriers which prevent appropriate technology transfer from taking place, and to create interesting business opportunities. There have been a number of examples in this area, many of them funded by the multilateral development banks, and the support of the GEF has been useful in many cases – its flexibility and adaptability has been a major strength.

In order to overcome information barriers, technology information centres have been widely advocated, but existing experience and literature does not provide enough certainty about exactly what is required for key stakeholders at critical stages. The value of other forms of technology intermediaries is better established, such as national-level technology transfer agencies, electric utilities, and energy service companies. ESCOs in particular have gained widespread acceptance to stimulate innovative financing schemes; they address the financial, capability, and other institutional market issues. An ESCO is a firm that offers energy services to customers with performance guarantees. Typical performance contracting arrangements provide customers with feasible means of improving their competitiveness by reducing energy consumption costs. Governments and other public-sector entities can develop technology intermediaries through direct support and other interventions.

References

Aloisi de Larderel, J. 1997: *Reducing greenhouse gas emissions: the role of voluntary programmes*. New York: United Nations Environment Programme, Industry and Environment Programme Activity Centre, and United States Environmental Protection Agency, Atmospheric Pollution Prevention Division, New York/Washington.

Arrow, K.J., 1951: *Social Choice and Individual Values, 2nd edition*. John Wiley & Sons, Inc.

Arthur, W.B., 1994: *Increasing Returns and Path Dependence in the Economy*. Ann Arbor: University of Michigan Press.

Asad, M., 1997:Innovative Financial Instruments for Global Environmental Management. In *Environment Matters*. The World Bank, Washington, Winter / Spring, pp. 12-13.

Ayers, R.U., 1993: On Economic Disequilibrium and Free Lunch. Centre for the Management of Environmental Resources Working Paper 93/45/EPS. INSEAD, Fontainebleau, France.

Berne Union, 1998: *Berne Union Yearbook*, Berne, Switzerland.

Berry, D., 1995. You've got to pay to play: Photovoltaics and transaction costs. *Electricity Journal* **8**(2), 42-49.

Carter, L. W. , 1996: *Leasing in Emerging markets: the IFC's experience in promoting leasing in emerging economies*. Washington, DC: International Finance Corporation.

Chambers, R., 1997: *Whose reality counts? Putting the last first*. Intermediate Technology Publications, London, pp. 297.

Chandler, W. U., J.W. Parker, I. Bashmakov, J. Marousek, S.Pasierb, and Z.Dadi. 1996: *Energy Efficiency Centres: Experiences in the Transition Economies*. Pacific Northwest National Laboratory, Washingotn DC..

Chantanakome, W. 1999: *Thailand strategic industries: positioning of economy and technology*. Thailand National Research Council.

Chantramonklasri, N., 1990: The Development of Technological and Managerial Capability in the Developing Countries. In *Technology Transfer in the Developing Countries*. M. Chatterji, (ed.), St. Martin's Press, New York.

Commander, S., and.Ch. Mumssen, 1998: Understanding Barter in Russia. EBRD Working paper No. 37. European Bank for Reconstruction and Development, .

Cook, E, (ed.), 1996: *Ozone Protection in the United States: Elements of Success*. Washington DC: World Resources Institute, Washington, DC.

Cornerhouse, 1999: *Snouts in the Trough, Export Credit Agencies, Corporate Welfare and Policy Incoherence*. Sturminister Newton, Dorset

Crabb, Peter B. , 1992: Effective Control of Energy-Depleting Behaviour. *American Psychologist*, **47**(6 - June), 815-816.

De Groot, W. T. (1992) *Environmental Science Theory: Concepts and Methods in a One World, Problem- Oriented Paradigm*. Elsevier Science Publications, Amsterdam.

DeCanio, S.J. 1993: Barriers within firm to energy-efficient investments, *Energy Policy*, **21**(9), 906-914.

DeCanio, S. J. 1994: Agency and Control Problems in US Corporations: The Case of Energy-efficient Investment Projects. *Journal of the Economics of Business*, **1**(1): 105-123.

DeCanio, S.J. 1998: The efficiency paradox: bureaucratic and organisational barriers to profitable energy-saving investments. *Energy Policy* **26**(5), 441-454.

DeCanio, S. J., and W.E. Watkins, 1998a: Investment in Energy Efficiency: Do the Characteristics of Firms Matter? *The Review of Economics and Statistics* **80**(1), 95-107.

DeCanio, S.J., and W.E. Watkins, 1998b: Information processing and organisational structure, *Journal of Economic Behaviour and Organisation*. **36**(3), 275-294.

Dennis, M.L., E. J. Soderstrom, W.S. Koncinski Jr., and B. Cavanaugh.' 1990. Effective Dissemination of Energy-Related Information. *American Psychologist*, **45** (10 – October), 1109-1117.

Dodgson, M., and J.Bessant., 1996: *Effective Innovation Policy: A New Approach*. International Thomson Business Press, London.

EBRD, 1998. *Transition Report*. London.

Economic Options Committee, 1991: *1991 Economic Assessment Report*. United Nations Environment Programme, Nairobi.

Economic Options Committee, 1994: *1994 Report of the UNEP Economic Options Committee*. United Nations Environment Programme, Nairobi.

Economic Options Committee, 1998: *1998 Assessment Report of the UNEP TEAP Economic Options Committee*. United Nations Environment Programme, Nairobi.

Energy Innovations, 1997: *Energy Innovations: A Prosperous Path to a Clean Environment*. Alliance to Save Energy, Council for an Energy-Efficient Economy, Natural Resources Defense Council, Tellus Institute, and Union of Concerned Scientists, Washington, DC.

Etzioni, A., 1987: How Rational Are We? *Sociological Forum*, **2** (1), 1-20.

Evans, M., and S. Legro, 1997: *Business Planning: A Key to Energy Efficiency in Russia*. In Proceedings from the ECEEE 1997 Summer Study on Sustainable Energy Opportunities for a Greater Europe, European Council for an Energy-Efficient Economy, Prague/Copenhagen.

Feldman, S.J., P.A. Soyka, and P. Ameer, 1997: Does Improving a Firm's Environment Management System and Environmental Performance Result in a Higher Stock Price? *Journal of Investing* ,Winter, 87-97.

Geller, E.S., 1992: It Takes More Than Information to Save Energy. *American Psychologist*, **47**(6 - June), 814-815.

Global Environment Facility, 1996: Operational Strategy. Washington, DC.

Global Environment Facility, 1997: Operational Programs. Washington. DC.

Global Environment Facility, 1998: Operational Report on GEF Programs (June). Washington, DC.

Graham C., and M. O'Hanlon, 1997: Making Foreign Aid Work. *Foreign Affairs*,. July /August 1997, pp. 96-104.

Gupta, J., 1997: *The Climate Change Convention and Developing Countries: from Conflict to Consensus?*

Kluwer Academic Publishers, Dordrecht, the Netherlands.

Hammitt, J.K., 1997: *Are the Costs of Proposed Environmental Regulations Overestimated? Evidence from the CFC Phaseout*. Centre for Risk Analysis and Department of Health Policy and Management, Harvard School of Public Health, Cambridge, MA (May).

Hawken, P., 1993: *The Ecology of Commerce: How Business can Save the Planet*. Weidenfeld and Nicolson, London.

Heaton, G.R. Jr., R. D.Banks, and D. W. Ditz, 1994: *Missing Links: Technology and Environmental Improvement in the Industrializing World*, World Resources Institute, Washington, DC.

Iman, R.L., and L. R. Lichtenberg, 1993: Clean Connections—Putting Innovation to Work for the Environment. *Machine Design*, **65**(8 - April 23), 62-64.

Institute for Policy Studies, Friends of the Earth, and the International Trade Information Service, 1999: *OPIC, Ex-Im, and Climate Change: Business as Usual: An Analysis of OPIC and Ex-Im Support for Fossil Fueled Development Abroad, 1992-98*.

Intergovernmental Panel on Climate Change, 1996: *Climate Change 1995: Economic and Social Dimensions of Climate Change*. Contribution of Working Group III to the Second Assessment Report of the Intergovernmental Panel on Climate Change. Cambridge University Press, New York.

Interlaboratory Working Group, 1997: *Scenarios of U.S. Carbon Reductions: Potential Impacts of Energy Technologies by 2010 and Beyond*, LBNL-40533 and ORNL-444. Lawrence Berkeley National Laboratory/ Oak Ridge National Laboratory, Berkeley, CA/Oak Ridge, TN.(Available at http://www.ornl.gov/ORNL/Energy_Eff/CON444).

Jaffe, A. B., and R. N. Stavins, 1993: *The Energy Paradox and the Diffusion of Conservation Technology*, Faculty Research Working Paper R93-23. J. F. Kennedy School of Government, , Harvard University, Cambridge, MA.

Jain, R. K., 1996: European Developments and Aid Policies in the 1990s: Implications for India. *India Quarterly*, 31-52.

Koomey, J.G. 1990: Energy Efficiency Choices in New Office Buildings: An Investigation of Market Failures and Corrective Policies. *Energy and Resources* (PhD Thesis), University of California at Berkeley, CA.

Koomey, J.G., A.H. Sanstad, and L. J. Shown, 1996: Energy-Efficient Lighting: Market Data, Market Imperfections, and Policy Success. *Contemporary Economic Policy*, **14**, 98-111.

Kotabe, M., A. Sahay, and P.S. Aulakh, 1996: Emerging role of technology licencing in the development of global product strategy: Conceptual framework and research propositions. *Journal of Marketing*, **60**(1), 73-88.

Kozloff, K., O. Shobowale, 1994: Rethinking Development Assistance for Renewable Electricity. World Resources Institute, Washington, DC.

Ledgerwood, J., 1999: *Microfinance Handbook: An Institutional and Financial Perspective.* World Bank, Washington, DC.

Lovins, A.B., and L.H. Lovins, 1991: Least-Cost Climatic Stabilization. *Annual Review of Energy and the Environment,* **16,** 433-531.

Mansley, M., 1996; The Transfer of Environmentally Sound Technologies: a Financing Perspective. Prepared for UNCSD. Delphi International, London.

Mansley, M., et al., 1997: The Role of Financial Institutions in Achieving Sustainable Development. Prepared for the European Commission. Delphi International and Ecologic GmbH, London.

Mansley, M., et al., 1997; Going International – A Guide to Finance. Prepared under contract to ETSU for the DTI. Delphi International / Impax Capital, London.

Mapako, M.C., 1997: Promotion of Renewable Energy Technologies; Some Thoughts. *Renewable Energy for Development,* **1** (1- March).

Martinot, E., 1998: Energy Efficiency and Renewable Energy in Russia: Transaction Barriers, Market Intermediation, and Capacity Building. *Energy Policy,* **26**(11), 905-915.

Martinot, E. and N. Borg, 1999: Energy Efficient Lighting Programs: Experience and Lessons from Eight Countries. *Energy Policy,* **26**(14), 1071-1081.

Martinot, E., and O. McDoom, 1999: Promoting Energy Efficiency and Renewable Energy: GEF Climate Change Projects and Impacts. Pre-Publication Draft, October. Washington, DC: Global Environmental Facility, Washington, DC.

Martinot, E., J.Sinton, and B. Haddad, 1997: International Technology Transfer for Climate Change Mitigation and the Cases of Russia and China. *Annual Review of Energy and the Environment,* **22,** 357-401.

Marousek, J., M.Dasek, S. Legro, B. Schwarzkopf, and M. Havlickova, 1998: Climate Change Mitigation: Case Studies from the Czech Republic: Pacific Northwest National Laboratory. Washington, DC., 32 pp. (Available at http://www.pnl.gov/aisu/czechcase.pdf).

Maurer, C., and R. Bhandari, 2000: *The Climate of Export Credit Agencies.* Climate Notes, World Resources Institute, Washington, DC.

Morrissey, O., 1992: *British aid and international trade: aid policy making, 1979-89.* Open University Press, Buckingham, UK.

Mullins, F., S. Boyle, C. Bates, and I. Househam, 1997: Policies and Measures for Financing Energy Efficiency in Countries with Economies in Transition: Policies and Measures for Common Action. Annex I Expert Group of the UNFCCC, Working Paper 10. OECD, Paris.

National Laboratory Directors, 1997. Technology Opportunities to Reduce U.S. Greenhouse Gas Emissions.. U.S. Department of Energy, Washington DC. (Available at http://www.ornl.gov/climate_change).

NREL, 1998: *Technology Co-operation Agreement Pilot Project: Development-Friendly Greenhouse Gas Reduction.* Status Report, NREL, Golden, CO.

Norberg-Bohm, V., and D. Hart, 1995: *Technological Cooperation: Lessons from Development Experience. In Shaping National Responses to Climate Change: A Post-Rio Guide,* H. Lee, (ed.), Island Press, Washington, DC, pp. 261-288.

OECD, 1995a: Environmental Funds in Economies in Transition, OECD, Paris.

OECD, 1995b: The St. Petersburg Guidelines' on Environmental Funds in the Transition to a Market Economy, CCET/ENV/EAP(94)13. OECD, Paris,

OECD, 1997: Economic Survey. Russian Federation, Paris.

OECD, 1998b: Application of Climate-Friendly Technologies in Developing Countries: the Role of International Collaboration. OECD/IEA Forum on Climate Change, March 1998.

OECD, 1998c: The Export Credit Arrangements, Achievements and Challenges 1978-1998

OECD,1999a: forthcoming, Sourcebook on Environmental Funds in Countries in Transition (working title). Prepared by the EAP Task Force Secretariat in co-operation with EU's Phare Programme. OECD, Paris.

OECD, 1999b: Voluntary approaches for environmental policy in OECD countries: an assessment, ENV/EPOC/GEEI(98)30/REV1.

Olson, M. , 1965: *The Logic of Collective Action: Public Goods and the Theory of Groups.* Harvard University Press, Cambridge, MA.

Panayotou, T., 1997: Taking stock of trends in sustainable development since Rio. In *Finance for Sustainable Development - The Road Ahead.* J. Holst, P. Koudal, J. Vincent (eds.), United Nations, New York, pp. 35-74.

Peszko G., 1995: Environmental Funds and Other Mechanisms of Financing Environmental Investments in Central and Eastern Europe and the Former Soviet Union. In *Effective Financing of Environmentally Sustainable Development,* ESD Proceeding Series No. 10, The World Bank, Washington, D.C.

Peszko G., and T. Zylicz. 1998: Environmental Financing in European Economies in Transition. *Environmental and Resource Economics,* **11**(3-4), 521-538.

Porter, M.E., and C. van der Linde, 1995a;: Green and Competitive: Breaking the Stalemate. H*arvard Business Review* (**September-October**), 120-134.

Porter, M.E., and C. van der Linde. 1995b: Toward a New Conception of the Environment-Competitiveness Relationship. *Journal of Economic Perspectives,* **9**(4 - Fall), 97-118.

Schmidheiny, S., and F. Zorraquin, 1996: *Financing Change: the financial community, eco-efficiency and sustainable development.* MIT Press, Cambridge, MA.

Schmidheiny, S., and F.J. Zorraquin, 1996: Financing Change. MIT Press, Cambridge, MA.

Silverstein, M., 1993: *The Environmental Economic Revolution: How Business will Thrive and the Earth Survive in Years to Come.* St. Martin's Press, New York.

Socolow, R.C., 1994: Six Perspectives from Industrial Ecology. In *Industrial Ecology and Global Change.* R. Socolow, C. Andrews, F. Berkhout, V. Thomas, (eds.), Cambridge University Press, Cambridge, MA.

Stern, P. C., 1992: What Psychology Knows About Energy Conservation. *American Psychologist,* **47**(10 - October), 1224-1232.

Stern, P.C., and G. T. Gardner, 1981: Psychological Research and Energy Policy. *American Psychologist,* **36**(4 -April), 329-342.

Stobaugh, R., and L. T. Wells (eds.), 1984: *Technology Crossing Borders: The Choice, Transfer, and Management of International Technology Flows.* Harvard Business School Press, Boston.

UNFCCC, 1999: Development and Transfer of Technologies (projects and programs incorporating cooperative approaches to the transfer of technologies), FCCC/SBSTA/1999/Misc.5. UNFCCC, Bonn.(Available at www.unfccc.de/resources/docs/1999/sbsta/misc05.pdf).

UNIDO, 1996: Guidelines for Infrastructure Development through Build-Operate-Transfer (BOT) Projects. Vienna.

United Nations, 1996: Conference on Trade and Development (UNCTAD). Emerging Forms of Technological Cooperation: The Case for Technology Partnership. UN, New York.

Van Berkel R., and E. Arkesteijn, 1998: *Participatory Approach for Enhancing Climate-relevant Technology Transfer: Opportunities for International Collaboration.* IVAM Environmental Research, Amsterdam, 12pp.

Van Berkel, R., and J. Bouma, 1999: *Promoting Cleaner production Investments in Developing Countries: a Status Report on Key Issues and Possible Strategies.* UNEP Industry and Environment Department, Paris.

Wallace, D., 1996: *Sustainable Industrialisatio*n. Royal Institute of International Affairs/Earthscan, London.

Wexler, P., 1996a: Saying Yes to Ozone? Clean. In *Ozone Protection in the United States: Elements of Success.* E. Cook, (ed.), World Resources Institute, Washington, DC.

Wexler, P., 1996b: New Marching Orders. In *Ozone Protection in the United States: Elements of Success.* E. Cook, (ed.), World Resources Institute, Washington, DC.

Womack, J.P., D.T. Jones, and D. Ross, 1990: The Machine that Changed the World. Rawson Associates, New York.

World Bank., 1993: *Energy Efficiency and Conservation in the Developing World: The World Bank's Role.* Washington, DC.

World Bank Group, 1994: Submission and Evaluation of Proposals for Private Power Generation Projects in Developming Countries (Discussion paper Nr. 250), P. Cordukes, Washington, DC.

World Bank, 1994: *World Development Report – Infrastructure for Development.* Oxford University Press, Oxford.

World Bank, 1997: *Rural Energy and Development: Improving Energy Supplies for Two Billion People.* Washington, DC.

World Bank, 1998: *China Energy Conservation Project: Project Appraisal Report.* World Bank Report No. 17030-CHA. Washington, DC.

Zey, M. (ed.), 1992: *Decision Making: Alternative to Rational Choice Models.* Sage Publications, Newbury Park.

Section II

Technology Transfer: A Sectoral Analysis

Section Coordinators:
JAYANT SATHAYE (USA), YOUBA SOKONA (SENEGAL),
OGUNLADE DAVIDSON (SIERRA LEONE), WILLIAM CHANDLER (USA)

6

Introduction to Section II

Coordinating Lead Author:
JAYANT SATHAYE (USA)

Lead Authors:
William Chandler (USA), John Christensen (Denmark),
Ogunlade Davidson (Sierra Leone), Youba Sokona (Senegal)

CONTENTS

6.1   Introduction

Article 4 of the United Nations Framework Convention on Climate Change (UNFCCC) calls for the transfer of technologies, including those for adaptation, from developed to developing countries (Climate Change Secretariat, 1992). Under various sub-articles, it lays out ways by which such transfers could be supported by the developed countries. Furthermore, at the Third session of the Conference of the Parties to the UNFCCC in Kyoto, Japan, in December 1997, three new mechanisms of cooperative implementation were established (Climate Change Secretariat, 1998). The new mechanisms include transactions among Annex I Parties, international emissions trading (IET), which provides for cooperation among the Annex B Parties, and the clean development mechanism (CDM), which extends the scope of cooperation to non-Annex I Parties.

Domestic actions, and those taken in cooperation with other countries, will require an increased market penetration of environmentally sound technologies, many of which are particularly important in their application to each sector. What is the potential for the penetration of mitigation and adaptation technologies? What barriers exist to the increased market penetration of such technologies? Can these be overcome through the implementation of a mix of judicious policies, programmes and other measures? What can we learn from past experience in promoting these, or similar technologies? Is it better to intervene at the R&D stage or during the end-use of fuels and technology? The chapters in this section (Section II) address these questions using examples specific to each sector. Technology transfer activities may be evaluated at three levels – macro or national, sector-specific and project-specific. Many of the options explored in the Section II chapters are at the latter two levels. We present criteria that authors have used for the evaluation of what might constitute effective technology transfer activities.

Greenhouse gas emissions from some sectors described in Section II are larger than those from other sectors, and the importance of each greenhouse gas varies across sectors and countries as well. Methane, for instance, is a much bigger contributor to emissions from agricultural activity than, for instance, from the industry sector. Table 6.1 shows the carbon emissions from energy use in 1995. Emissions from electricity generation are allocated to the respective consuming sector. Carbon emissions from the industrial sector clearly constitute the largest share, while those derived from agricultural energy use comprise the smallest share. In terms of growth rates of carbon emissions, however, the fastest growing sectors are transport and buildings. With rapid urbanisation promoting an increased use of fossil fuels for mobility and habitation, these two sectors are likely to continue to grow faster than others in the future. Carbon emissions from fossil fuels used to generate electricity amounted to 1,762 MtC of the total for all sectors in Table 6.1. Chapters 7-9 examine technology transfer opportunities in the energy demand sectors, and Chapter 10 focuses on energy supply options.

SECTOR	CARBON EMISSIONS AND(%SHARE) 1995	AVARAGE ANNUAL GROWTH RATE (%)	
		(1971-90)	(1990-95)
Industry	2370 (43%)	1.7	0.4
Buildings			
• Residential	1172 (21%)	1.8	1.0
• Commercial	584 (10%)	2.2	1.0
Transport	1227 (22%)	2.6	2.4
Agriculture	223 (4%)	3.8	0.8
All Sectors	5577 (100%)	2.0	1.0
Electricity Generation*	1762 (32%)	2.3	1.7

Table 6.1   Carbon emissions from fossil fuel combustion in Mt C (Price *et al.*, 1998)

NOTE: EMISSIONS FROM ENERGY USE ONLY; DOES NOT INCLUDE FEEDSTOCKS OR CARBON DIOXIDE FROM CALCINATION IN CEMENT PRODUCTION. BIOMASS = NO EMISSIONS.

* INCLUDES EMISSIONS ONLY FROM FUELS USED FOR ELECTRICITY GENERATION. OTHER ENERGY PRODUCTION AND TRANSFORMATION ACTIVITIES DISCUSSED IN CHAPTER 10 ARE NOT INCLUDED.

Carbon emissions from the forestry sector were estimated in the IPCC Second Assessment Report at 0.9 +- 0.5 MtC for the 1980s (IPCC, 1996a). Tropical forests, as a whole, are estimated to be net emitters, but temperate and boreal forests are net sequesters of carbon. Estimates of emissions from the agricultural sector are not available. Carbon equivalent emissions from waste disposal amounted to between 335-535 MtC. Chapters 11 and 12 focus on the technology transfer opportunities in the agricultural and forestry sectors, and Chapter 13 focuses on the waste disposal sector.

Changes in atmospheric concentrations of greenhouse gases and aerosols are projected to lead to regional and global changes in temperature, precipitation, and other climate variables, such as soil moisture, an increase in global mean sea level, and prospects for more severe extreme high-temperature events, floods, and droughts in some places (IPCC, 1998). Climate models based on alternative IPCC emissions scenarios project that the mean annual global surface temperature will increase by 1-3.5 degrees Celsius, and that the global mean sea level will rise by 15-95 cm (IPCC, 1996a).

Climate change represents an additional stress on systems already affected by increased resource demands. In coastal areas, where a large part of the global population lives, climate change can cause inundation of wetlands and lowlands, erosion and degradation of shorelines and coral reefs, increased flooding and salinisation of estuaries and freshwater aquifers. Health care systems may be further stressed as diseases spread beyond their current domains, and vectors migrate to other parts of the world and to different altitudes. Model projections show that at the upper end of the range of projected temperature increase (3-5 degrees Celsius), the world's population exposed to malaria will increase from 45% to 60% by the latter half of the next century. Heat-stress mortality and air pollution will create additional problems for health systems, particularly those in urban areas. Technology transfer options for adapting to these consequences are discussed in Chapters 14 and 15.

Technology transfer includes both within and between countries by actors who are engaged in promoting the use of a particular technology along one or more pathways. The market penetration of a technology may proceed from research, development, and demonstration (RD&D), adoption, adaptation, replication and development. At a project-specific level, the elements of the pathway are different, and may proceed from project formulation, feasibility studies, loan appraisals, implementation, monitoring, and evaluation and verification of carbon benefits. The pathways may include many actors, starting with laboratories for RD&D, manufacturers, financiers and project developers, and eventually the customer whose welfare is presumably enhanced through their use. This presumption needs to be carefully established through an assessment of the technology needs of the consumer. A poor needs-assessment can result in barriers to technology transfer that could have been avoided had the assessment fully captured the social and other attributes of the technology. The actors may make specific types of arrangements – joint ventures, public companies, licensing, etc. that are mutually beneficial. These arrangements will define the particular pathway chosen for technology transfer.

The transfer of a particular technology may proceed along one or more pathways, as it evolves from R&D towards commercial application. The importance of actors may change over time, as activities that were carried out earlier by governments are turned over to private industry or to communities. On the other hand, in times of crisis, the government role may become more prominent as national or international interests become the primary drivers for taking action.

The spread of a technology may occur through transfer within a country and then transfer to other countries, both may occur simultaneously, or transfer between countries may precede that within a country. Generally, the spread of a technology is more likely to proceed along the first option rather than the other two, since the transfer of technologies to markets within a country is likely to be less expensive given the proximity to the market, and lower barriers to the penetration of that technology in the indigenous markets. Transfer of technology from one country to another will generally face trade and other barriers, both in the initiating and recipient country, which may dissuade manufacturers and suppliers from implementing such transfer.

Many market barriers prevent the adoption of cost-effective mitigation options in developing countries. In the energy sector these barriers include the high initial cost of equipment, a lack of information on new technologies, the presence of subsidies for electricity and fuels, and high tariffs on imported energy technologies. In the forestry sector, barriers include pressures on land availability for mitigation; absence of institutions to promote participation of local communities, farmers and industry; risk of drought, fire, and pests; inadequate research and development capacity in countries; and poorly developed reforestation and sustainable forestry practices. Both sectors also suffer from an absence of appropriate methods and institutions to monitor and verify carbon flows (IPCC, 1996b).

What conditions and policies are necessary to overcome these barriers and successfully implement GHG mitigation options? The combination of barriers and actors in each country creates a unique set of conditions, requiring "custom" implementation strategies for mitigation options. Each chapter in this Section discusses the barriers that are particularly important to a sector, such as fuel and electricity price subsidies, weak institutional and legal frameworks, lack of trained personnel, etc. Each chapter also provides examples and case studies to highlight the barriers, and policies, programmes and measures that were used, or could be developed, to overcome them.

6.2 Criteria for Determining Effective Technology Transfer

Evaluation of technology transfer requires criteria that are specific to the manufacture and use of technologies, and to the process of transferring these technologies. The IPCC Technical Paper I included a discussion of the criteria that could be used to evaluate technologies and measures (IPCC, 1996). These criteria are also useful to determine the effectiveness of technologies that are transferred. The criteria for evaluating the effectiveness of a technology are grouped into three categories as shown below, and in more detail in Annex 1-2, Chapter 1:

1. GHG and other environmental criteria
2. Economic and social criteria
3. Administrative, institutional and political criteria

Some of these criteria, such as the GHG reduction potential, are amenable to quantitative evaluation, while others such as political considerations are at best evaluated qualitatively. An example of the application of these criteria would be the market penetration of wind power technology within a country and/or between countries. It would be possible for instance to quantitatively evaluate the transfer of this technology with respect to criteria 1 and 2, and qualitatively with respect to the third criterion. In addition, the case example would identify and describe the policies, programmes and measures that were used to overcome barriers to the transfer of the technology, and how these affected one or more of the aforementioned criteria. For example, the programme may have resulted in a more equitable distribution of jobs, increased government subsidy, or to easier replication of wind turbine projects.

In addition to the above criteria one needs to consider criteria to evaluate the effectiveness of the process of technology transfer. These process-related criteria include:

4. Rate and geographic extent of technology transfer
5. Long-term institutional capacity building
6. Monitoring and evaluation considerations
7. Leakage that reduces the impact of a programme or measure

6.3 Chapter Outline and Content

The sectoral chapters in this section follow a common outline to the extent that this was feasible given the material that was reviewed for each chapter. The common outline is shown in Box 6.1. The substantive material in each chapter will include a discussion of the relevant mitigation and adaptation technologies, but since the mitigation technologies, policies and measures are already discussed in the IPCC, 1996b report, and the adaptation technologies were addressed in the IPCC, 1996a report, the focus of this report will be on the barriers to the transfer of technologies within a country and between countries. To the extent technologies are not addressed in these IPCC reports, they are addressed in respective chapters. In addition, each chapter will discuss the policies, programmes, and measures that might be used to overcome these barriers.

The discussion in each chapter is to be organised primarily by barriers and policy tools, because these are felt to be common to each sector and are not expected to vary by each technology within a sector. Since the audience for each chapter will be the many actors engaged in influencing the discussions on climate change, each chapter concludes its discussion by organising the material by the roles that these actors could play in fostering the spread of technology. The material is then condensed in an executive summary focused towards government experts providing information or themselves engaged in the negotiations on climate change.

Some of the substantive material spans across sectoral boundaries. Bioenergy, for instance, gets treated in both the energy supply and forestry chapters from different perspectives. In the energy supply chapter, the focus is on the technologies used to convert biomass into energy, while that in forestry is more towards the growing of biomass for use in generating energy. Likewise, the spread of cook-stove technologies that use biomass is covered in the buildings chapter. Technology transfer for rural energy use is treated separately in the buildings chapter.

Another topic is the manufacture of mass-produced goods, such as refrigerators and motor vehicles, which could belong in the industry chapter or in the buildings and transportation chapters respectively. The literature on improving energy efficiency of refrigerators or motor vehicles, however, is largely in the buildings and transportation sectors, and thus it is appropriate to treat the material in the two respective chapters.

BOX 6-1: **TEMPLATE FOR SECTION II: CHAPTERS 7–15**

1. *Executive summary* – Focuses on the role of governments and UNFCC mechanisms to foster technology transfer. Highlights roles of other essential actors.

2. *Introduction*– Discusses current development patterns as a context.

3. *Climate mitigation and adaptation technologies* - This section lists and describes technologies that are climate friendly and climate safe with a near-term time horizon in the subjective chapter. To the extent possible, it provides the costs and potential for reducing GHG emissions when compared with baseline technologies (See IPCC Technical Paper 1 (TP1) for list, description and GHG impacts (IPCC, 1996b); see also UNFCCC paper on adaptation for initial description.

4. *Magnitude of current and future technology transfer using various systems (pathway, process, stakeholder) and their limitations.* Estimates the extent to which each transfer is comprised of mitigation and adaptation technologies.

5. *Technology transfer within a country (includes a discussion of barriers, evaluates with respect to criteria in 2 above, and draws on case studies)* The primary focus of this section is on barriers and the policies, programmes and measures to overcome them (see Section 1 for a list of generic barriers and policy tools). The discussion is built around drivers/incentives, pathways, and stakeholders/actors.

6. Technology transfer between countries (includes a discussion of barriers, evaluates with respect to criteria in 2 above, and draws on case studies) - The primary focus of this section is on barriers and the policies, programmes and measures to overcome them. The discussion is built around drivers/incentives, pathways, and stakeholders/actors.

7. Lessons learned from 5 and 6 above – This section discusses changes in future technology transfer systems regarding interventions suggested in 5 and 6. It also analyses options available to different actors to change the current trajectory of technology transfer discussed in 4 above. It is organised by actors such as national governments, international governmental organisations, non-governmental organisations, industry, community organisations, etc.

References

Climate Change Secretariat, 1992: *United Nations Framework Convention on Climate Change*. United Nations Environment Programme, Information Unit for Conventions/UNEP/IUC publication, Geneva.

Climate Change Secretariat, 1998: *The Kyoto Protocol to the Convention on Climate Change*. Published with the support of the UN Environment Programme, Information Unit for Conventions/UNEP/IUC/98/2.

IPCC 1996a: *Climate Change 1995: Scientific-Technical Analyses of Impacts, Adaptations, and Mitigation of Climate Change: The IPCC Second Assessment Report, Vol. 2*. Watson, R.T., M.C. Zinyowera, and R.H. Moss (eds.), Cambridge University Press, Cambridge, UK.

IPCC 1996b: *Technologies, Policies and Measures for Mitigating Climate Change. IPCC Working Group II Technical Paper I*: Watson, R.T., M.C. Zinyowera, and R.H. Moss (eds.), Cambridge University Press, Cambridge, UK. 85 pp.

IPCC 1998: *The Regional Impacts of Climate Change: An Assessment of Vulnerability*. Watson, R.T., M.C. Zinyowera, and R.H. Moss (eds.), Cambridge University Press, Cambridge, UK, 516 pp.

Price, L., L.Michaelis, E. Worrell, and M. Khrushch, 1998: Sectoral Trends and Driving Forces of Global Energy Use and Greenhouse Gas Emissions. *Mitigation and Adaptation Strategies for Global Change,* **3**(2).

7

Residential, Commercial, and Institutional Buildings Sector

Coordinating Lead Author:
JOHN MILLHONE (USA)

Lead Authors:
Odon de Buen R. (Mexico), Gautam Dutt (Argentina), Tom Otiti (Uganda), Yuri Tabunschikov (Russia), Tu Fengxiang (China), Mark Zimmermann (Switzerland)

Contributing Authors:
Marilyn Brown (USA), Jim Crawford (USA), Howard Geller (USA), Joe Huang (USA), Martin Liddament (United Kingdom), Eric Martinot (USA), John Novak (USA)

Review Editor:
Ewaryst Hille (Poland)

CONTENTS

EXECUTIVE SUMMARY

The residential, commercial, and institutional buildings sector accounted for about one-third of the global energy used in 1990 and roughly one-third of the associated CO_2 emissions. The sector's share of energy consumption is higher in developed countries than in developing and transition countries. Energy is used to heat and cool buildings, and to provide lighting, as well as services ranging from cooking to computers. The emissions from the building sector includes those from the direct use of fossil fuels in buildings, and emissions from the fuels used to furnish electricity and heat to buildings. About two-thirds of these emissions were from residences; the other one-third from commercial and institutional buildings (IPCC, 1996a). The achievable reductions below baseline projections are estimated at 10-15% in 2010, 15-20% in 2020, and 20-50% in 2050, relative to the IS92 scenarios.

To achieve these reductions requires technology transfer programmes that work rapidly and effectively to diffuse the best environmentally sound technologies (ESTs). The buildings sector is more atomistic and decentralised than the industrial, energy, and transportation sectors, making it more difficult to transfer technology and transform markets. The most successful government-driven pathways include (mandatory) energy and environmental standards for new buildings and equipment; information, education and labelling programmes; and government-supported research, development, and demonstration (RD&D) programmes. Governments also have a key role in creating a market environment for successful private sector-driven technology transfer through decisions on financing, taxes, regulations, and customs and duties. Governments, particularly local governments, can encourage successful community programmes by proactively identifying community-level needs, and by encouraging and responding to community initiatives.

In the near term, the most successful technology transfer programmes will not be driven by their environmentally sound benefits alone, but because they also meet other human needs and desires. Examples include new energy-efficient buildings that are more comfortable and provide more services, yet have lower energy costs and lower greenhouse gas (GHG) emissions. The most successful technology programmes focus on new products and techniques that have multiple benefits.

7.1   Introduction[1]

Those responsible for the transfer of ESTs to residential, commercial and institutional buildings face two challenges. First, they must find ways to advance the best technologies from a great range of new technologies available to the buildings sector. And second, they must advance them rapidly to meet international climate change goals. Buildings are long lasting and community development patterns have even longer lives. The incremental costs of the best technologies is slight at time of construction, compared with the cost of replacing energy-wasteful buildings and equipment. The technologies themselves are varied and powerful. The IPCC Technical Report I found an existing technical potential to meet the sector's global energy needs through 2050 with no increase in energy use from the 1990 level (IPCC, 1996b).

Yet, the transfer of these technologies poses special problems. Buildings vary greatly in their size, shape, function, equipment, climate, and ownership--all of which affects the mix of technologies needed to improve their performance. In some countries, the energy used in housing is "free" or subsidised to the occupants. When they do not have to pay the full cost of the energy they use, they have less incentive to use it wisely. Where a large portion of the occupants would have great difficulty paying the full costs immediately, there is strong political pressure to continue the subsidies. Governments and energy suppliers find it simpler and more predictable to invest in increasing energy supplies than in reducing the energy demands of millions of building owners and operators.

The nature of the buildings sector is changing. Urbanisation is having a great impact on development choices, particularly in developing countries, causing a rapid expansion of the housing and commercial building sectors. Due to an unmet demand for adequate housing in many countries, this trend is expected to continue, especially in some developing countries. Driven by these changes, a growing share of the GHG emissions in many countries is coming from the buildings sector. The related government decisions on environmentally sound land use planning and energy, water, and wastewater infrastructure will have long-term effects on population density and ecology systems.

The challenge is to identify and implement technologies that meet these changes and also lower GHG emissions. Fortunately, the same investments can achieve multiple goals. Investments in energy efficient buildings also lower future energy costs; produce more comfortable and healthy indoor environments; create more productive work places; achieve other environmental improvements; and acquire more durable, long-term investments.

Successful technology transfer strategies link climate change goals with measures that produce these companion benefits. See Figure 7.1.

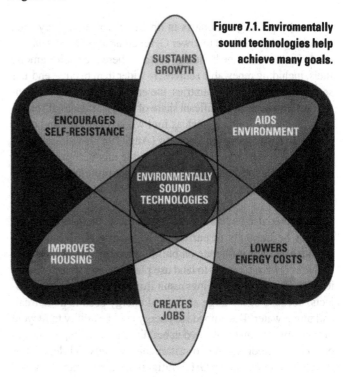

Figure 7.1. Enviromentally sound technologies help achieve many goals.

For each country, the desired mix of new technologies will be different, depending upon its unique climate, building stock, energy sources, stage of development, and social, economic, and political priorities. These values will be reflected in its assessment of its technology transfer priorities, which is the first stage in the technology transfer process. Following this assessment, the next stages will be obtaining agreement on the technology transfer programme and then its implementation, evaluation and adjustment, and replication. (See also Figure 1.2, Chapter 1.)

This chapter will provide a brief description of ESTs to illustrate their nature and potential. The current technology transfer processes are described, including their limitations, focusing on the barriers to change, the different pathways for overcoming these barriers, and the roles of the different stakeholders. Governments play a prominent role in the buildings sector through their programmes and through decisions that affect private-sector stakeholders and community groups. The bulk of the chapter describes experiences with different technology transfer programmes, both national and international, analysing what has worked, what has not, and the lessons to be learned.

The primary focus of this chapter is buildings. The boundary between the buildings sector and the energy supply sector is the building envelope, where electricity, thermal energy and other energy resources are delivered to buildings. Renewable systems that deliver energy directly to buildings, such as photovoltaic

---

[1] Relevant cases from the Cases Studies Section, Chapter 16, for Chapter 7 are: Cookstoves (case 1), Green Lights (case 2), Inner Mongolia Wind (case 3), PV in Kenya (case 5), Butane in Senegal (case 7), Ladakh renewables (case 14), CFC-free refrigerators in Thailand (case 23).

arrays, are considered part of the buildings sector. When an industrial process dominates a building's energy consumption, it becomes part of the industrial sector.

While the primary emphasis is on increased efficiency, fuel switching also can lead to lower GHG emissions. This is particularly important in the buildings sector, where the choice among fuels, including renewable sources, is wider than in other end-use sectors. In developing countries, the energy used in the residential sector includes a significant share of "non-commercial" or traditional energy sources, such as fuel wood, charcoal, and other biomass, particularly for cooking (Ang, 1986). The changing patterns in energy sources and the possible mix of future sources will influence a country's selection of its preferred buildings sector technologies.

This chapter also covers some of the subjects in the chapters on Human Settlements in earlier IPCC reports (IPCC, 1996a). Because of their influence on building energy use, the chapter includes brief references to land use planning and water use topics. Significant energy savings result from water conservation projects, because of the large amount of energy used to heat, treat, and pump water. The embedded energy and the ability to recycle the construction materials used in buildings also are important and provide an important link to sustainable systems. While most of the attention is given to GHG mitigation, this chapter includes some adaptation measures, *i.e.* water management, sewage systems, and building codes and standards.

7.2   Climate Mitigation and Adaptation Technologies

An extensive body of literature describes building and municipal technologies that improve energy efficiency, apply renewable energy resources, reduce GHG emissions, and adapt to potential climate change. For much of our history, energy was viewed as so abundant and inexpensive that insufficient scientific and technical attention was given to how to use it wisely. During the last 25 years, this has changed. Scientists have focused on energy and found fertile ground in virtually every area of energy use. The diversity of the building sector has made it a favourite field for technology innovation. Since, however, this Report focuses on how to transfer the technologies to new users, only a summary description is provided of this rich field.

7.2.1   Residential Buildings

In the residential sector, GHG mitigation technologies can be divided into three groupings: building envelope strategies, building equipment strategies and renewable energy strategies. Building envelope strategies address the size, shape, orientation, and thermal integrity of the residential unit. Examples of mitigation technologies include increased wall and roof insulation, advanced window technologies, roof coatings, and reduced or controlled infiltration. Building equipment strategies improve the space heating and cooling, lighting, cooking, refrigerators, water

heating, clothes washing and drying, air conditioning and other household appliances used in homes. Examples include such advanced technologies as condensing furnaces, compact fluorescent lamps and advanced refrigerator compressors. The renewable energy strategies include passive solar building designs and active solar water and space heating systems, ground-source heat pumps, daylighting strategies, and photovoltaic systems. In the residential subsector, the choice of technologies will vary greatly due to climate, between single-family residences and multi-family apartment buildings, and between urban and rural (traditional) communities.

7.2.2   Commercial and Institutional Buildings

This sub-section includes different building types, such as offices, retail stores, schools, hospitals, hotels, warehouses, theatres, and places of worship. However, within each building type, such as office buildings, the buildings are often similar in both developed and developing countries, inviting similar energy-saving strategies. Electricity is the dominant energy source, providing 70% of the resource energy demand in the industrialised countries (EIA, 1994). However, energy sources vary greatly among countries, e.g. coal is the dominant heating source for commercial and institutional buildings in China, while other sources are dominant in other countries.

As with residences, the mitigation technologies for commercial buildings can be divided into three categories. Building envelope strategies vary, depending upon the size and type of building and the climate. Wall and roof insulation is important in many building types. Modern commercial office buildings have higher internal heat loads from equipment and people, decreasing the importance of insulation and raising the importance of window and glazing systems. Building equipment strategies emphasise heating and cooling, efficient lighting, energy management control systems, and office equipment efficiency. Renewable technology strategies include photovoltaics, active and passive systems and daylighting. Too often overlooked, renewable strategies are most effective when integrated into the building orientation, shape, and design, and can be important in constraining the growth of energy consumption in urban settings. In the near future, the growing use of Internet-based information systems may change the shape of the workplace with dispersed and at-home work stations. The restructuring of the electric power industry is placing more attention on time-of-day pricing and encouraging the incorporation of load-shedding by agreement and energy storage systems within commercial buildings.

7.2.3   Adaptation Technologies.

Many of the technologies that mitigate GHG emissions also help adapt to the potential effects of climate change. For example, the ability of local governments to provide effective land use planning is essential to address many environmental problems. With such authority, local governments can cluster higher density residential and commercial land use to improve the system efficiency of combined heat and power systems. A city's streets and building lots can be laid out to optimise the potential use of solar energy. By limiting developments on flood plains or potential mud slide zones, a city can adapt to both current and anticipated future flooding. The minimisation of paved surfaces and the use of trees can reduce flooding, moderate the urban heat-island effect and reduce the energy required for air conditioning. Water using equipment, such as clothes washers, can be developed and marketed that are both energy efficient and use less water. Building codes and standards reduce energy consumption and also reduce the damage to buildings from destructive weather anomalies. A systems, or whole-building approach, can achieve both mitigation and adaptation objectives through the optimal integration of land use, building design, equipment and material choices and recycling strategies.

A fuller description of these technologies can be found in the IPCC's Second Assessment Report, Working Group II (IPCC, 1996a) and the IPCC's Technical Paper I, *Technologies, Policies and Measures for Mitigating Climate Change* (IPCC, 1996b). Other sources are included in the references (Interlaboratory Working Group, 1997; CADDET, 1997; Worrell, 1996). Adaptation strategies vary between developed and developing countries. While the published literature deals primarily with new technologies, indigenous technologies using thermal mass, convective air movement and night radiation use no energy, and could be more widely used in both developing and developed countries.

7.3 Current and Future Technology Transfer Systems

The barriers to the rapid transfer of these ESTs include the lack of information about new technologies, their higher initial cost, the presence of subsidies for electricity and fuels, the absence of delivery and maintenance services, a diversity of building and equipment codes, different performance testing methods, public procurement practices, restrictions in building materials and limited recycling (See Chapter 6). This report analyses the role of different stakeholders in the technology transfer process, the primary pathways they use, and the stages in this process. The key stakeholders include the developers, owners, and suppliers of technologies, buyers of technologies such as private firms, state enterprises, and individual consumers, financiers and donors, governments, international institutions, non-governmental organisations (NGOs) and community groups. These roles of these stakeholders are intertwined in each stage of the major strategies that are used to accelerate the transfer of ESTs in the building sector. Developed countries have a very important role to play in technology transfer, since most advanced technologies in the buildings sector are developed within those countries.

Governments play an important leadership role in the transfer of "climate friendly" technologies, which reduce GHG emissions from the buildings sector. The major pathways include information and education programmes, the use of cost-based energy prices, energy and environmental labels, building energy codes, appliance and equipment efficiency standards, leading by example in government buildings and purchases, and government support for RD&D. Governments also play the leading role in the transfer of "climate safe" technologies, which reduce vulnerability to climate changes, through land use planning and infrastructure developments.

The primary role of the private-sector stakeholders is to meet the consumer demand for the shelter and services provided by the buildings sector. While consumer surveys show support for environmental goals, this support may not be expressed in their purchasing decisions (Federal Environment Agency, Germany, 1998). Education programmes that draw this connection are gaining popularity and are starting to influence private-sector decisions. Governmental policies affect the marketplace through subsidy and taxation programmes, the regulation of energy tariffs, import and export controls and laws covering intellectual property rights.

The role of community groups is of great importance for the buildings sector, but is less well characterised in the technology transfer literature than the other pathways. Decisions about land-use, building materials and intensity, energy and water services are made within communities. These decisions are driven by immediate priorities, yet they have long-term environmental impacts. The rapid urbanisation in many developing countries underscores the importance of finding ways to use sustainable development pathways in cities.

It is important to recognise that traditional technologies have an important role in providing building energy services. Natural ventilation provides comfortable building environments in both hot humid and hot dry climates, including India and the Middle East. Traditional methods of heating are used in Korea (floor heating) and Japan (under-table heating). These traditional approaches may be enhanced through modern scientific re-investigation, measurement technologies, and computer simulations. The resulting guidelines for building design based on local conditions and using local craftsmen could minimise the cost and environmental impact of providing energy services. The combination of traditional and new technologies in buildings offers promising results, which only recently are beginning to draw some attention.

The flexibility mechanisms in the Kyoto Protocol could give these stakeholders powerful new tools to advance the dissemination of ESTs. The Clean Development Mechanism is a potential tool for the transfer of ESTs to the growing building sectors of non-Annex I countries. Joint Implementation projects are particularly attractive for reducing the GHG emissions of the buildings sectors of countries with economies in transition (CEITs). By monetising GHG emissions, emissions trading would add value and flexibility to environmentally sound investments.

7.4 Technology Transfer Strategies

The strategies for accelerating the transfer of ESTs are illustrated in Figure 7.2. The top curve shows the distribution curve of the efficiency of products acquired by consumers prior to the start of a technology transfer programme. The curve could be applied to any product that uses energy or produces GHG emissions, such as furnaces, cookstoves, or housing. The slope on the right side of the curve shows the slow penetration of more efficient technologies.

The bottom curve in Figure 7.2 shows the combined effect of a successful technology transfer programme. Building codes and appliance standards eliminate the sale of energy-wasteful products on the left side of the curve. RD&D introduces advanced technologies on the right side of the curve. Market transformation strategies encourage consumers to select more ESTs. Working together, they can produce major changes.

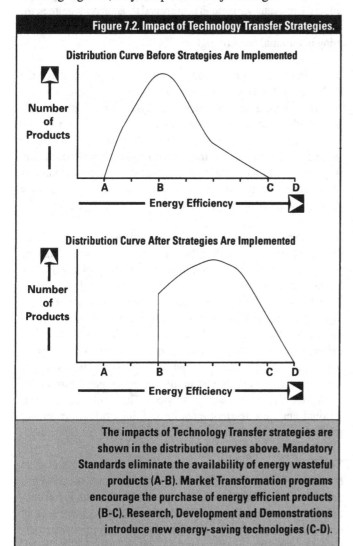

Figure 7.2. Impact of Technology Transfer Strategies.

Distribution Curve Before Strategies Are Implemented

Number of Products

Energy Efficiency

Distribution Curve After Strategies Are Implemented

Number of Products

Energy Efficiency

The impacts of Technology Transfer strategies are shown in the distribution curves above. Mandatory Standards eliminate the availability of energy wasteful products (A-B). Market Transformation programs encourage the purchase of energy efficient products (B-C). Research, Development and Demonstrations introduce new energy-saving technologies (C-D).

International interests are playing an increasingly important role both in government-driven and private sector-driven technology transfer. The recognition of a common global interest in reducing GHG emissions has become a powerful additional motive for international programmes. The growing role of multinational corporations has led to an increasing global market for new goods and services. For each of the national technology transfer pathways, described above, there are parallel international pathways.

7.5   Role of Governments in Technology Transfer[2]

7.5.1   *Information and Education Programmes*

Within Countries
The most pervasive barrier to increased energy efficiency and environmentally sound practices is simply the lack of information about the impact of our decisions and how they can be improved. Many families do not realise what they can do to end (or reduce) energy waste in their own homes. Building owners and operators do not know how operation and maintenance and retrofit decisions can reduce their energy costs. Entrepreneurs do not recognise the potential market for new energy-saving products and services. Because of the large number and diversity of the decision-makers in the buildings sector, this barrier requires special education and information programmes and the establishment of a permanent consulting infrastructure. Governments can play an influential role through Best Practices guides, school curricula and public education campaigns.

There are many audiences for these programmes. Homeowner guides provide practical, money-saving tips. Adult training programmes teach building engineers how to manage and operate the new energy systems of commercial buildings. Business and finance classes show how to develop bankable projects. Innovative school curricula combine lessons on energy sources and environmental issues with how to read meters and calculate utility bills.

At a deeper level (of involvement), these programmes encounter behavioural barriers, such as the limited ability of individuals to deal with life cycle cost minimisation due to the complexity of this concept, lack of data, and the low priority given to energy use, which remains a relatively small fraction of the total costs of owning and operating buildings in many countries. Another behavioural barrier is a short time horizon--consumers often demand two or three year paybacks (i.e., high implicit discount rates) even though there is a societal interest in accepting longer paybacks on efficient and renewable energy measures. To a greater or lesser degree in all countries, poverty is a barrier. Impoverished consumers often are forced to buy the cheapest product available, even if this means higher future energy, environmental and social costs in the long term.

When confronting these barriers, does the provision of free technical information really influence human behaviour? A recent study by the Resources for the Future seeks to answer this question as it applies to a private firm by modelling the many factors that go into a company's decisions on whether to invest in energy-efficient lighting. The study found that information programmes make a significant contribution to the transfer of efficient lighting in commercial buildings, although these programmes are less important than the basic price signals (Morgenstern, 1996).

---

[2] See also chapter 4 on the role of governments in creating enabling environments for technology transfer.

Among Countries

International programmes offer a relatively low cost and highly important mechanism for exchanging information on ESTs. For example, the Centre for the Analysis and Dissemination of Demonstrated Energy Technologies (CADDET) of the International Energy Agency (IEA) provides a large and growing computer database of more than 1,600 energy efficiency and renewable energy demonstration projects (CADDET, 1998). CADDET also provides valuable analyses of technology applications. Designed to complement national programmes, CADDET is now seeking to provide information to persons in developing countries.

Although the transfer of information from developed to developing countries is important, a higher priority also needs to be given to the transfer of information about the energy-using patterns and opportunities in developing countries to developed countries, and the exchange of information among developing countries and countries with economies in transition.

7.5.2   *Energy Pricing*

Within Countries

For the market to stimulate investments in energy efficiency, energy prices must reflect the full cost of providing energy to end users. While this is true for all end-use sectors, it is particularly relevant in the buildings sector, because of the political interest in keeping energy prices low to homeowners in countries with administered price systems. Where energy prices are subsidised, homeowners and commercial building owners and managers receive muted signals on the benefits of investing in efficiency, undercutting the potential market for energy-efficient products. In countries with administered price systems, the industrial sector often subsidises the housing sector. These cross-sector subsidizes can make industries less competitive. If the housing subsidies were paid by governments, taxes would need to be increased or funds diverted from other public services, such as education or health care.

A World Bank study of the effect of energy price increases in six countries--Columbia, Ghana, Indonesia, Malaysia, Turkey and Zimbabwe--found that eliminating subsidies does not cause disproportionate hardship for the poor, lower economic growth, create inflation or reduce industrial competitiveness, but does improve public revenues (Hope, 1995). The consumption of commercial fuel increases greatly with income, according to the study, so energy subsidies largely benefit non-poor, urban households. A study of subsidised household energy prices in transition economies also concluded that they benefited the rich more than the poor (Freund, 1995). One approach is a staged removal of energy subsidies, which creates a market for investments in energy-saving products and services.

The landlord-tenant relationship can create problems unique to the buildings sector when the landlord pays the energy costs, but has little control over the energy-using practices of the tenant. For tenants, if energy services are free, there is no incentive to use energy wisely. Technical problems may also make it difficult to make homeowners and tenants responsible for the energy they use. In many high-rise multifamily buildings, heat is delivered to apartments through vertical pipes, making it difficult and costly to try to measure the energy used by each individual apartment unit. In new buildings, this problem can be addressed by building codes that require that heating systems be designed to serve individual, metered apartment units.

Even in market economies, energy prices rarely include the full societal costs of related environmental externalities. These costs are reflected in adverse health impacts and environmental degradation. If the market mechanism is to exercise its full potential for achieving environmental goals, the price of energy needs to incorporate these environmental externalities.

Among Countries

The lessons learned from in-country technology transfer programmes also need to be recognised in international programmes. This is particularly important for countries moving toward market economies. During this transition, it is often useful to ask: Who is paying the energy costs?. Someone is. It may be municipalities through subsidies, industries through cross-sector subsidises, or energy supply industries through unrecovered costs. Whoever is paying these costs will have an interest in reforms. It is also useful to track the flow of energy from source to end use, to identify the changes in ownership, to metre the quantity of energy in each transaction, to measure the performance of each owner, and to move toward a system of rewards and penalties that improves the overall efficiency of the energy delivery system.

Multilateral and bilateral assistance programmes can encourage the movement to full-cost energy pricing by ensuring that any projects that are supported incorporate progress toward market reforms. The Russian Enterprise Housing Divestiture Project of the World Bank exemplifies this approach. The Bank is providing loans of US$300 million for basic energy efficiency measures in a total of 3,500 residential apartment buildings in six Russian cities. The participating cities are undertaking policy reforms designed to reduce housing maintenance and utility costs. The World Bank is also initiating a similar project in Lithuania.

7.5.3   *Energy and Environmental Labels*

Within Countries

To create more informed consumers, a number of product labelling programs have been initiated. At least 11 countries and the European Union have initiated mandatory or voluntary programmes that have products labelled with descriptions of their energy performance (Casey-McCabe, 1995). The United States requires labels on furnaces, water heaters, refrigerators, central and room air conditioners, clothes washers, dishwashers and lamp ballasts (USDOE, 1996). The European Union has initiated a programme under the SAVE (Specific Actions for Vigorous Energy Efficiency) Programme that requires labels for refrigerators and freezers, washing machines and clothes dryers, and is being phased in for other appliances.

Worldwide, more than 30 products are covered by one or more labelling programme, including the major energy-using appliances, such as refrigerators, furnaces, clothes washers and dryers, and ovens. Labelling programmes may be mandatory or voluntary and comparison or endorsement in type. Comparison labels describe the performance of a product with others in the same class. Endorsement labels identify a product that meets a high efficiency standard. Most programmes use comparison labels, are mandatory, and are operated by government agencies. Only the United States, Canada, and the European Union also have programmes with endorsement labels, which may be operated by government agencies or NGOs.

Stakeholder cooperation is illustrated by the window labelling programme of the National Fenestration Rating Council (NFRC), a coalition of window manufacturers, governments, utilities, and consumer groups. The labels, which identify the energy performance of windows, doors and skylights, help consumers select high performance products in an area of rapid technology change. Manufacturers pay a fee to have their products tested in NFRC-accredited laboratories. Since 1993, the NFRC has certified and labelled 12,000 products made by more than 160 manufacturers. NFRC is working with the International Organization for Standardization (ISO) on an internationally recognised testing and labelling programme.

The record of labelling programmes is mixed. The initial U.S. Energy Guide labels were largely ineffective, because they were difficult to understand, yet went unchanged for 10 years. The initiation of harmonised energy labels in Europe was slow—17 years between the initiation and implementation of common levels for refrigerator-freezers. On the other hand, two years after the initiation of the Energy Star computer programme, 50 per cent of the computers and 80 per cent of the printers were meeting its standard. The average power requirements for personal computers fell from 75-80 watts to 35-45 watts (Duffy, 1996).

The policy objectives of the programmes make a significant difference in its results, according to a comparison between the U.S. and Thai household appliance labelling programmes. The objective of the 20-year-old U.S. programme is to provide customers with information to assist them with their purchasing decisions. By contrast, the objective of the three-year-old Thai programme is to persuade customers to buy more efficient appliances that save money and protect the environment, an objective that is backed by a massive, nation-wide advertising campaign. Energy efficiency was reported among the top three purchase priorities by 28 per cent of the Thai customers, compared with only 11 per cent of the U.S. customers (du Pont, 1998).

Among Countries
Since many appliances and other energy using equipment are produced and sold worldwide, it would be beneficial to have a uniform international labelling system, rather than separate nation-

al systems. The initial move in this direction might be through regional programmes, such as those being implemented by the European Union. Multinational efforts are already underway through the ISO to harmonise the test procedures that underlie national labelling and standard programs (CADDET, 1997).

An international approach faces formidable obstacles, including the standardisation of testing protocols, the treatment of different product designs, and non-tariff barriers. Even so, some international approaches are moving forward. The Green Lights and Energy Star computer programmes, initiated by the U.S. Environmental Protection Agency, have been transferred successfully to other countries (Case Study 2, Chapter 16).

7.5.4 Mandatory Standards

Within Countries
For many years, governments have been responsible for setting standards for new buildings to protect the health and safety of their occupants. After the 1973 oil embargo created a growing public awareness of the cost and security risks of wasting energy and rising energy imports, many countries expanded these regulatory programmes to ensure that new and renovated buildings were designed to avoid squandering energy. Without standards, architects and builders are under pressure to minimise investments in efficiency to hold down the initial cost of the buildings, even when the additional investments would be repaid rapidly through lower energy costs. These standards now have the additional advantage of reducing GHG emissions from the burning of fossil fuels.

Building energy codes have become widespread. A survey of 57 countries found 31 of them with codes for both residential and non-residential buildings, nine countries with codes for non-residential buildings only, four countries with codes for residential buildings only, and 13 countries without any building codes (Janda and Busch, 1994). Many countries modelled their building codes on those in other countries. The most often cited codes were those of the American Society of Heating, Refrigeration and Air-Conditioning Engineers (ASHRAE); 11 countries used the ASHRAE standards.

The rate of compliance with energy codes varies widely among countries (Duffy, 1996). While most codes are adopted nationally, local agencies often are responsible for their enforcement. Buildings differ in size, function, and location. An effective enforcement system requires trained local inspectors, who periodically visit the construction sites of complex buildings. Studies in the United States report rates of non-compliance of 50 per cent or more. By contrast, in Singapore compliance with the Energy Code is reported to be very high (Alliance to Save Energy, 1997). After building standards laws are approved, it is necessary to have strong implementation, training, and enforcement programmes to realise their potential benefits.

Some governments have also moved lately to mandate appliance and equipment energy efficiency standards. A recent international survey found that at least nine countries have energy efficiency standards for household appliances (Duffy, 1996). The United States and Canada cover the most appliances, followed by Switzerland and China. The programmes cover 18 different household appliances, the most common being refrigerators, air conditioners, clothes washers and clothes dryers. The standards are mandatory in all nine countries, except Japan and Switzerland. Japan's standards are voluntary, but largely met. Switzerland's standards are target values; however, the Swiss legislation indicates that if the target values are not met the government intends to mandate standards.

The survey found that at least six countries have standards for commercial and industrial equipment. The United States and Canadian programmes cover the most types of commercial equipment, including fluorescent lamps and ballasts, incandescent lamps, electric motors, and commercial air conditioning/heat pumps, furnaces/boilers, water heaters and water chillers. Japan and Korea cover lamps. China covers electric motors and furnaces/boilers. Malaysia covers electric motors. In the United States and Canada, the standards cover an estimated 70 per cent of the energy used in commercial buildings.

Among Countries

The same factors that led to national energy standards are creating pressure for international standards, i.e. the public interest in increased efficiency and decreased emissions of pollutants, the market failure when low initial cost dominates the selection of products with long lifetimes, and the benefits of harmonised standards throughout the growing international market. The standards programmes in developed countries cannot be simply extended to other countries. The standards need to be modified to reflect the energy uses and preferences of a country. New infrastructure investments may be needed, including new laboratory testing and certification facilities. Domestic industries may need support to upgrade their products to meet the international standards. However, the potential benefits are large. A harmonised international approach to standards would widen the market for energy-saving products, lower the cost of such products, increase their market penetration, and encourage manufacturers to produce only the more efficient units (CADDET, 1997).

7.5.5 Leading by Example

Within Countries

The government is the largest single consumer of energy in most countries. Governments can reduce their own energy costs through the operation of their buildings, through the design features of new buildings, and through the efficiency of the energy-using products they buy. Through environmentally sound decisions, governments can also provide an example to those who own, rent, and operate privately-owned residential, commercial and institutional buildings. Government policies are also important in myriad additional ways. For example, governments can use their purchasing power to create a market for energy-efficient products. Government leaders can stimulate the demand for

---

Box 7.1	APPLIANCE SUCCESS STORY

Appliance standards were first initiated in the United States in California in 1974, a year after the 1973 oil embargo. During the 1970s, California extended standards to cover 15 products. Other States also initiated their own standard programmes. Meanwhile, little was happening at the national level. A 1978 law created a U.S. appliance standards programme, but opposition delayed its implementation. The impasse was broken in 1986 with the passage of the National Appliance Energy Conservation Act, which established minimum U.S. efficiency requirements for 12 types of residential appliances. The Act was approved due to a remarkable--and instructive--collaboration between energy and environmental NGOs and appliance manufacturers. NGOs had long championed standards. Manufacturers came to recognise that uniform national standards were preferable to the growing number of different State standards.

This political process contains lessons for other national and international standard initiatives. The command and control strategy was used, but its impact was moderated to recognise industry's interests. In setting standards, the U.S. Department of Energy is required by law to consider their economic consequences, including any adverse impacts on manufacturers. The standards cannot be changed more frequently than every five years and manufacturers are given at least three years to meet any new standard levels. While this schedule makes the standards more acceptable to industry, it also delays the resulting energy savings and reduced GHG emissions.

The U.S. standards programme continued to be politically controversial, but was expanded by the 1992 Energy Policy Act. The results are significant. The average refrigerator sold in 1994 used about 653 kWh per year, down from about 1725 kWh per year in 1972. In 2001, the next iterations of the standards will lower this average to 475 kWh per year. The standards already adopted are expected to save 1.3 EJ[6] of primary energy in 2000, rising to 3.4 EJ in 2015 (Geller, 1997). An analysis shows that for every US$1 increase in the price of products due to the standards, consumers save an average of $3.20 during the life of the product (Goldstein, 1996).

---

[6] 1 EJ (Exa Joule) = 10^{18} joules

"Green" products through public recognition for voluntary industry efforts to market ESTs.

Few governments are taking advantage of the opportunity to show leadership in their management of their own energy consumption, according to a recent survey of 25 countries in Europe, North America, Latin America and Asia (Borg *et al.*, 1997). The survey questionnaire was distributed to knowledgeable government people in 25 countries. The most active programmes were reported in Canada, the Netherlands, Switzerland and the United States. The common government programmes included setting energy-saving targets, tracking progress toward the targets, recognising successes, requiring the purchasing of efficient equipment, providing information and training, as well as financing schemes, and conducting energy audits and demonstration projects. Additional countries have adopted Climate Change programmes in their building sectors, including Australia, the United Kingdom, Japan, New Zealand, France and Germany (OECD, 1999)

Among Countries

Leading by example has an appealing potential for transferring one country's success to another in the climate change field, where countries are striving for a common, international goal. The buildings sector offers highly visible opportunities to demonstrate this leadership, for example, by integrating climate friendly policies into government operations, housing and education programmes. Annex I countries have a special responsibility to lead by example in order to stimulate adoption by developing countries and CEITs.

A potentially powerful form of international cooperation is present among the countries of a region that share common resources, climates, languages, traditions and aspirations. For example, renewable energy resources such as solar radiation, hydropower, wind, biomass, and geothermal resources are distributed regionally without regard to political boundaries. The demand for heating and cooling and the available construction materials are common to multi-country regions. Within these regions, the technology opportunities are similar. The successful deployment of a new technology can spread rapidly. The enhanced regional demand for climate friendly technologies can attract development and investments that otherwise would be slow to respond to efforts of a single country.

7.5.6   Research, Development and Demonstration

Within Countries

Government and private sector research programmes play a critical role by developing and demonstrating advanced technologies that meet human needs more effectively, at lower costs and with fewer adverse environmental impacts. The technologies that are making a difference now are the results of past research. While this Report focuses on existing and near-term technologies, it is important to recognise the role that today's RD&D programmes will have in developing the ESTs for the technology transfer programmes of the future.

For example, the U.S. Department of Energy played a key RD&D role in the introduction--among other new products--of low-emissivity windows, electronic ballasts, and high efficiency supermarket refrigeration systems. These three technologies, alone, have provided U.S. manufacturers with US$3.5 billion in cumulative sales and are delivering 250 Tbtu (264×10^{15} J) /yr of primary energy savings worth US$1.5 billion a year (Alliance to Save Energy, 1997).

Although not often recognised, a high degree of scientific skill is also required to develop improved products in developing countries, such as the Jiko cookstove in Kenya (See Case Study 1, Chapter 16). In sub-Saharan Africa, where household cooking accounts for more than 60 per cent of total energy use in some countries, this is a high priority. Inefficient combustion of traditional fuels has also resulted in high concentrations of pollutants and acute respiratory infections. For 20 years, international aid organisations have tried to develop improved cookstoves, but have encountered a complex tangle of combustion, convection, conduction, cost, and acceptance problems. The Jiko cookstove is a collaboration between scientists, local craftspeople, and potential users. Today, hundreds of local craftspeople manufacture some 20,000 stoves a month and more than 1 million are in use throughout Kenya. Each stove cost roughly US$2 , uses 1,300 pounds less fuel per month, and saves urban households as much as US$65 a year (one-fifth of the average annual income) (Kammen and Dove, 1997). Another example is in Senegal where the butane cookstove was re-engineered to meet local conditions (See Case Study 7, Chapter 16).

The demonstration component of RD&D can be important in countries with little experience in the application of technologies used elsewhere. This is the situation in Russia and other transition countries with limited experience in modern space heating technologies. Space and hot water heating dominate the energy use of the building sector in Russia, accounting for two-thirds to three-fourths of total residential energy consumption. Most of the space heating is provided to multifamily buildings supplied by district heating systems. The buildings suffer from high energy losses due to heating intensities--the energy required for indoor comfort adjusted to different climate conditions--that are one to two times higher than in Western countries (Martinot, 1997). In typical apartment units, if households paid the actual cost of the space heat and hot water they receive, this would represent 40% of their monthly wages. National and municipal governments face the challenge of addressing this problem, which requires a combination of technical, financial, institutional and social measures. The challenges include forming homeowners associations, developing consumption-based metering, creating utility regulations that encourage energy efficiency investments, providing long-term financing, and increasing the number and capabilities of local design and construction firms.

Among Countries

Research activities among countries fall into two categories. The first area is multilateral and bilateral RD&D programmes that can give countries access to research advances at a lower cost than through separate national programmes and enlarges the pool of researchers, which can lead to more creative approaches and more significant results. An example is the collaborative energy RD&D programme of the International Energy Agency.

A second area is adaptive RD&D, which examines how the advances in one country might be adapted to the needs of another. For example, the RD&D in a developed country might lead to the commercial introduction of a highly efficient 20-cubic-foot refrigerator. In other countries, there may be little interest in such a refrigerator. However, the technologies embedded in the refrigerator—the advanced insulation, seals, compressor, and controls—may be adapted to different refrigerator models for a wide variety of different international markets.

7.6 Role of the Private Sector in Technology Transfer[3]

In the buildings sector, the roles of governments and the roles of private sector stakeholders often overlap; they merge in patterns that vary from country to country. The private sector is defined here as including two quite different types of organisations: 1) businesses and industries that are motivated primarily to improve their profit-making positions and long-term economic viability, and 2) NGOs, including charitable and church organisations, that are motivated by environmental, humanitarian, and religious missions, or organisations which are advocates of certain interest groups.

7.6.1 Business and Industry

Within Countries
While government decisions create an enabling environment for environmentally sound investments, the bulk of investment funds are coming increasingly from the private sector (See Chapter 2). The developing countries are expected to be the largest markets for energy efficiency products in the 21st Century, driven by a growth in population, economic activity, and energy demand that far outstrips that of industrialised countries. For example, the developing countries' current rate of growth in annual energy demand is more than double that of the OECD countries, 3.7 per cent versus 1.7 per cent (IIEC, 1996). To respond to this trend in ways that minimise economic, environmental, and social costs will create an attractive opportunity for new, environmentally sound products and services. While these changes are often characterised as opportunities for international exports, they also invite the growth of domestic production and joint ventures. To grasp these opportunities, participating governments could remove any artificial trade, regulatory, taxation, or commercial barriers that hinder the diffusion of advanced technologies (World Energy Council, 1998).

Among Countries
The largest market for ESTs in the 21st Century will be in the developing countries and CEITs. This will offer an opportunity for the domestic industries of these countries. It also will also be an export opportunity for international industries. A 1995 study by Hagler Bailly Consulting, Inc. estimates the annual international market for energy efficiency will rise from about US$40 billion currently to US$125 billion in 2015 (Hagler Bailly, 1995). A recent assessment of the export market identified building technologies among the major items, including building envi-

ronment controls; heating, ventilation and air conditioning equipment; lighting; household appliances; and building materials (IIEC, 1996). International strategies can capitalise on this demand by identifying the international corporations that are targeting this market and by encouraging them in the rapid deployment of the best ESTs.

7.6.2 Financing Programmes

Within Countries
The buildings sector faces inherent problems when it seeks to attract environmentally sound investments. The size of the investment in any one building application is relatively small. As a result, the acquisition cost for a building project is a relatively high share of the total project cost when compared with other investment opportunities. In addition, the number of parties involved in a project is large and diverse, including the architects and engineers, who design the project; the technicians, who install the measures; and the owners and occupants, who operate and maintain the new systems. As a result, there is an increased risk that the project will not achieve its expected benefits. To address these complications, financing programmes need to be developed that lower administrative costs and reduce risks.

The risks are perceived as being particularly high in countries with limited experience in energy efficient investments. In such settings, there is little experience in preparing, reviewing, approving, financing, implementing, evaluating, and replicating climate friendly projects. Because of these barriers, a market transformation strategy needs to consider approaches that will attract environmentally sound investments. Direct approaches have included tax credits for energy efficiency or renewable energy investments and partial subsidies of the project costs. Special funds have been created to cover all or part of the cost of such investments. Energy-saving performance contracts have been used, where the investment costs are paid back from the energy savings. A new business venture has emerged, the energy service companies (ESCOs), which deliver energy performance contracts and a broad range of contract energy services (see also section 5.6.3 on ESCOs). Lending institutions can encourage efficiency in new buildings by allowing the additional administrative costs to be included in the normal financing agreement.

Among Countries
The lack of investment by developed countries in developing countries and in CEITs is often cited as the greatest obstacle to the deployment of mitigation and adaptation technologies. In addition, developed countries, when they undertake investments in developing countries, do not always bring in the latest technologies in which they have invested in their own home countries. This financial support by developed countries is crucial for increasing the transfer of advanced technologies to developing countries. Host countries can make such investments more attractive by taking appropriate, supportive public policy decisions. These include the elimination of trade barriers, the avoidance of

---

[3.] See also Chapter 5 on the role of private sector finance, investment and public-private partnerships in the technology transfer process.]

punitive taxation, the lifting of import and export restrictions, and the adoption of fair and expedient procedures for resolving disputes.

After reviewing data from 52 countries, a World Bank Policy Paper has recommended major reforms in financing programmes in the building sector (World Bank, 1994). The reforms would shift government policy from producing small-scale public housing toward managing the housing sector as a whole. On the demand side, it suggests ways to develop property rights, increase mortgage finance, and target housing subsidies. On the supply side, it shows how to regulate land and housing development and organise the building industry for maximum productivity. The framework for initiating the reforms brings together public agencies, NGOs, community groups, and the private sector. Different strategies are recommended for low-income countries, highly indebted middle-income countries, other middle-income countries, and CEITs.

The creation of special funds to finance energy-saving and climate-friendly investments has also been done successfully in a number of countries. An international example is the use of German Coal Aid to create the Hungarian Environment and Energy Service Co. (EESCO). The fund has made more than 200 revolving loans totaling more than 7 billion HUF (28 million USD[4] and been producing savings of more than 110 thousand tonnes of oil per year. A significant portion of these savings is being returned to replenish the fund.

7.6.3   Utility Programmes

Within Countries
Under some conditions, utilities—the electricity, natural gas, and thermal energy supply organisations—can play significant roles in market transformation. Public utilities are given special powers, such as limited monopolies in their service territories, and the right to use eminent domain to obtain easements for their distribution networks. In turn, they are subject to varying forms of regulation, including the approval of the tariff rates that they charge to different classes of customers. The role of utilities places them in a unique position to deliver energy efficiency and renewable energy programmes to their customers. They are a source of technical expertise in the supply and use of energy; are in regular contact with their customers; and are in a strategic position to aggregate customer demand to introduce new technologies.

Despite these advantages, utilities historically have had a disincentive to encourage energy efficiency. In the past—and continuing in many jurisdictions—the profits earned by utilities have been based on the volume of their energy sales. In this regulatory climate, if a utility encourages efficiency successfully, its sales and earnings decline. To correct this disincentive, a number of national and state governments are adopting utility reforms.

Regulatory programmes are being changed to require utilities to carry out energy efficiency programmes and to allow them to earn a profit on these services—the traditional demand-side management (DSM) model. Where utility restructuring is taking place, utilities are under competitive pressure to reduce costs, which has reduced their investments in DSM programmes. To preserve these programmes, some restructuring legislation is experimenting with mandatory "line" charges–surcharges on each kWh of electricity carried on a transmission line–to create special funds for DSM services, including energy efficiency and renewable energy investments and subsidies for low-income customers.

Among Countries
The features that made utility DSM programmes successful--the technical expertise of utilities and their customer contacts--have lead to the transfer of utility programmes among countries. In Brazil, a comprehensive national electricity conservation programme, PROCEL, conducts R&D, energy audits, equipment testing and rating, and educational campaigns (Geller, 1997). In Thailand, a DSM programme has been initiated through cooperation between the utility and manufacturers. In China, DSM is integrated into a Sustainable Future programme. The rapidly increasing cross-boundary investments in utilities is increasing the potential of this form of technology transfer.

7.6.4   Non-Governmental Organisations

Within Countries
During the last quarter century, the growing attention given to energy efficiency and renewable energy and environmental issues has seen the creation and growth of NGOs, which are playing an increasingly influential role in all forms of technology transfer. NGOs provide an organisational focus for public concerns about energy and environmental issues; influence public policies at the local, state and national levels; represent their members' priorities in interaction with governments, businesses and industry; and sometimes are able to initiate programmes more rapidly and at lower costs than the traditional government programme.

Among Countries
The influence of NGOs within developed countries has been transferred by international NGOs and government programmes to developing and transitioning countries. With this support, energy conservation centres have been created in many countries, including Russia, China, Poland, the Czech Republic, Hungary, Bulgaria, Ukraine, India, Indonesia, Pakistan, South Korea and Thailand. The centres perform numerous functions, including public education, energy audits, professional training, development of model legislation, demonstration projects, and innovative financing schemes (IIEC, 1996).

7.7   Role of Community Groups in Technology Transfer[5]

---

[4] This (rounded-off) figure is based on the exchange rate from November 1999.

[5] See also section on the role of participatory approaches and community groups in the technology transfer process.

Within Countries

Technology transfer strategies are recognising alternatives to the old paradigm that envisioned all change as being mandated by top-down directives. The pervasive influence of community programmes has been recognised; these programmes often involve a vibrant mix of stakeholders, representing international, national, regional, city and community organisations (See Figure 7.3). With increasing frequency, programme responsibilities are being delegated by national governments to regional and local governments and community organisations. In Russia, for example, the 89 regional governments are moving toward a federal structure with a more flexible fiscal system that is more responsive to the country's changing needs (Wallich, 1994). In Columbia, municipalities are developing the capacity needed to take control of responsibilities formerly belonging to the national government (World Bank, 1995). The city of Curitba has earned the nickname, "Ecology Capital of Brazil", through its innovative public transportation system, garbage recycling programme, and large number of trees, parks and green spaces (Dobbs, 1995; Fiszbein, 1999).

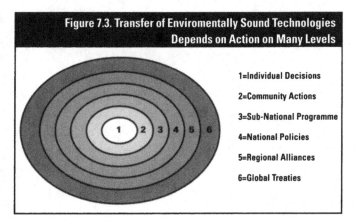

Figure 7.3. Transfer of Enviromentally Sound Technologies Depends on Action on Many Levels

1=Individual Decisions

2=Community Actions

3=Sub-National Programme

4=National Policies

5=Regional Alliances

6=Global Treaties

Within this context, there is an increased recognition of the role of communities where citizens' organisations, such as villages, neighbourhoods, and grassroots organisations, are the initiators in defining the need for new technologies through a high degree of collective decision-making (See Chapter 1). The development of the Jiko cookstove, mentioned above, is an example of a community-driven pathway (Case Study 1, Chapter 16). Many other examples are reported in a growing Best Practices Database maintained by the Together Foundation, an NGO, and the United Nations Centre for Human Settlements (UNCHS)-Habitat (Together Foundation, 1999).

Some of these examples:

- In Benavides, a small Argentine village north of Buenos Aires, a non-profit civic association created a 173-home neighbourhood at 40 per cent of the cost of private real estate developers;
- In El-Tadamon Village in upper Egypt, the inhabitants of two villages destroyed by floods—one Moslem and one Christian—were relocated into a new village on higher ground. The new inhabitants worked together to build 102 houses, using a new "healthy house" design, and a textile and handcraft;

- In Rajasthan, India, the Academy for a Better World, by the Brahman Kumaris, constructed a new village for 800 people on barren land. The complex is using renewable energy and recycling its water to reforest the adjoining land.

An analysis by the Inter-American Development Bank of the role that community organisations have played in alleviating poverty throughout Latin American suggests that governments could recognise the potential of such organisations (Navarro, 1994). The objective of most community-driven initiatives has been to alleviate poverty, which often includes improved housing. However, community leaders have also shown a sensitivity to environmental issues (Tietenberg and Wheeler, 1998). The community-driven pathway appears open to playing a larger role in the diffusion of ESTs.

Among Countries

Government-to-government programmes can empower community-driven technology transfer in two ways. The first is by simply informing communities about what other communities have done and how they did it. NGOs, such as the Together Foundation, can provide this information exchange, as it does through its Best Practices Database. Through their support for the Together Foundation, the UNCHS-Habitat and the European Union make this information exchange possible (Together Foundation, 1999). Other international NGOs also are recognising the potential of the community-driven pathway, such as the International Institute for Energy Conservation, Alliance to Save Energy, Natural Resources Defense Council, Environmental Defense Fund, and Habitat for Humanity.

The second way to empower these community-based initiatives is to recognise their potential role in the design of international programmes. The multinational development banks and multilateral assistance agencies are starting to do this in many of their analyses (Bamberger and Aziz, 1993; World Bank, 1994, 1995; Fiszbein, 1999; Navarro, 1994). The challenge now is to work with developing and transitioning countries to incorporate these insights into their lending and technology assistance programmes.

7.8   Lessons Learned

- The lessons learned from the experiences gained in the transfer of ESTs in the buildings sector include:
- Buildings vary greatly in their function, size, shape, climate, ownership, lifetimes, equipment, construction material, culture, quality and cost. The mixture of characteristics also varies among countries and regions. Technology transfer strategies need to respect these differences.
- National governments have a central responsibility for promoting successful programmes directly through government-driven programmes, and indirectly through the creation of national environments that attract private-sector-driven programmes and encourage community-driven programmes.

- National governments could begin by identifying the technologies that are most important in achieving its social, environmental, economic, and energy goals for their buildings sector.
- The most effective way to advance these technologies is through an integrated programme that includes information and education programmes, full-cost energy pricing, energy and environmental labels, building and equipment standards, leading by example, and support for RD&D.
- The largest source of funding for ESTs will be from the private sector. To attract these funds, a country needs to remove any artificial trade, regulatory, taxation, or commercial barriers that discourage investments.
- Community organisations have an essential role to play as part of a national strategy. Direct citizen participation in identifying priorities, barriers, and pathways is especially important in the design and implementation of housing reform programmes.
- International linkage and regional alliances are necessary to identify the technology needs of the buildings sector, to stimulate the development of these technologies, and to facilitate their transfer.

References

Alliance to Save Energy, American Council for an Energy-Efficient Economy, Natural Resources Defense Council, Tellus Institute, Union of Concerned Scientists, 1997: *Energy Innovations: A Prosperous Path to a Clean Environment.* Washington, DC. 171 pp.

Ang, B.F., 1986: A method for estimating non-commercial energy consumption in the household sector of developing countries. *The International Journal,* **11**(3), 315-325.

Bamberger, M., and A. Aziz, 1993: *The Design and Management of Sustainable Projects to Alleviate Poverty in South Asia.* World Bank EDI Seminar Report, Washington, DC.

Borg, N., E. Mills, N. Martin, and J. Harris, 1997: *Energy Management in the Government Sector–An International Review.* Proceedings of the 1997 European Council for an Energy Efficient Economy, Summer Study, Spindelruv Mlyn, Czech Republic.

CADDET, 1997: *Saving Energy With Appliance Labelling,* Maxi Brochure 09. Sittard, The Netherlands, 16 pp.

CADDET, 1998: The Centre for the Analysis and Dissemination of Demonstrated Energy Technologies is an information activity of the International Energy Agency. It maintains three databases, which are available on the Internet: CADDET-Energy Efficiency— http://www.caddet-ee.org; CADDET-Renewable Energy—http://www.caddet-re.org; and GREENTIE—http://www.greentie.org.

Casey-McCabe, N., and J. Harris, 1995: *Energy Labelling: A Comparison of Existing Programs.* European Council for an Energy-Efficient Economy, Mandelieu, France.

Dobbs, F. (producer), 1995: *Curitba: City of the Future* (video). World Bank, Washington, DC.

Du Pont, P., 1998: *Energy Policy and Consumer Reality: The Role of Energy in the Purchase of Household Appliances in the U.S. and Thailand.* International Institute for Energy Efficiency, Washington, DC.

Duffy, J., 1996: *Energy Labelling, Standards, and Building Codes: A Global Survey and Assessment for Selected Developing Countries.* International Institute for Energy Conservation, Washington, DC.

Energy Information Administration, 1994: *Annual Energy Outlook 1994.* U.S. Department of Energy, Washington, DC, 185 pp.

Federal Environmental Agency (Federal Republic of Germany), 1998: *Sustainable Development in Germany–Progress and Prospects.* Erich Schmidt, Berlin, 344 pp.

Fiszbein, A., and P.Lowden, 1999: *Working Together for Change: Government, Business, and Civic Partnerships for Poverty Reduction in Latin America and the Caribbean.* Economic Development Institute, Washington, DC.

Freund, C.L., and C.I. Wallich, 1995: *Rising Household Energy Prices in Poland: Who Gains? Who Loses?* World Bank, Washington, DC.

Geller, H., 1997: National appliance efficiency standards in the USA: cost-effective federal regulations. *Energy and Buildings,* **26**,101-109.

Goldstein, D.B., 1996: Appliance Efficiency Standards: A Success Story of Economically Beneficial Environmental Protection. Testimony to the House Commerce Committee's Sub-Committee on Energy and Power. Washington, DC.

Hagler Bailly Consulting, Inc., 1995: *The Future Market for Energy Efficiency Technologies and Services.* Arlington, VA.

Hope, E., and B. Singh, 1995: *Energy Price Increases in Developing Countries: Case Studies of Columbia, Ghana, Indonesia, Malaysia, Turkey, and Zimbabwe.* World Bank, Washington, DC.

IIEC, 1996: *The Export Market for the Energy Efficiency and Renewable Energy Industry.* International Institute for Energy Conservation, Washington, DC.

Interlaboratory Working Group, 1997: *Scenarios of U.S. Carbon Reductions; Potential: Impacts of Energy Technologies by 2010 and Beyond.* LBNL-40533 and ORNL-444, Lawrence Berkeley National Laboratory, Berkeley, CA and Oak Ridge National Laboratory, Oak Ridge, TN.

IPCC 1996a: *Climate Change 1995: The IPCC Second Assessment Report, Volume 2: Scientific-Technical Analyses of Impacts, Adaptations, and Mitigation of Climate Change.* Watson, R.T., M.C. Zinyowera, and R.H. Moss (eds.), Cambridge University Press, Cambridge, UK.

IPCC 1996b: *Technologies, Policies and Measures for Mitigating Climate Change.* Watson, R.T., M.C. Zinyowera, and R.H. Moss (eds.), IPCC Working Group II Technical Paper I., 85 pp.

Janda, K., and J.Bush, 1994: Worldwide Status of Energy Standards for Buildings. *Energy,* **19** (1), 27-44.

Kammen, D., and M.R. Dove, 1997: The Virtues of Mundane Science. *Environment,* **July/August**,10-15; 38-40.

Martinot, E., 1997: *Investments to Improve the Energy Efficiency of Existing Residential Buildings in Countries of the Former Soviet Union.* World Bank Studies of Economies in Transformation 24, Washington, DC.

Morgenstern, R.D., 1996: *Does the Provision of Free Technical Information Really Influence Firm Behavior?* Resources for the Future, Washington, DC.

Navarro, J.C., 1994: *Community Organizations in Latin America.* Inter-American Development Bank, Washington, DC.

OECD, 1999: Summary Report on Sustainable Building Policy. OECD/ENV/EPOC/PPC/RD(99), Paris.

Tietenberg, T., and D. Wheeler, 1998: *Empowering the Community: Information Strategies for Pollution Control.* Presented at the Frontiers of Environmental Economics Conference, Airlie House, Richmond, VA.

Together Foundation, 1999: Best Practices Database: Tools You Can Use Today. Together Foundation and UNCHS-Habitat. New York, NY. Available via website: http://www.bestpractices.org.

USDOE, 1996 (July): *Policies and Measures for Reducing Energy Related Greenhouse Gas Emissions: Lessons From Recent Literature.* U.S. Department of Energy, Washington, DC.

Wallich, C.I., 1994: *Russia and the Challenge of Fiscal Federalism.* World Bank Regional and Sectoral Studies, Washington, DC.

World Bank, 1994: Housing: Enabling Markets to Work. World Bank Policy Paper, Washington, DC.

World Bank, 1995: Local Government Capacity in Columbia: Beyond Technical Assistance. World Bank Country Study, Washington, DC.

World Energy Council, 1998: *Energy Efficiency Policies.* The report of a Working Group of the Council, London.

Worrell, E., M. Levine, L. Price, N. Martin, R. van den Broeck, and C. Blok, 1996: *Potential and Policy Implications of Energy and Material Efficiency Improvements.* A Report to the United Nations Division of Sustainable Development. Ministry of Economic Affairs, The Netherlands.

8

Transportation

Coordinating Lead Author:
OGUNLADE DAVIDSON (SIERRA LEONE)

Lead Authors:
Oyuko Mbeche, (Kenya), Laurie Michealis, (United Kingdom), Lee Schipper, (USA),
Suzana Kahn Ribeiro, (Brazil), Romeo Pacudan, (Philippines), Mike Walsh, (USA),
Yang Honghian (China)

Review Editor:
Lars Sjöstedt (Sweden)

CONTENTS

EXECUTIVE SUMMARY

Transport-related greenhouse gas (GHG) emissions are the second-fastest growing worldwide, but the transport sector is the least flexible to change due to its almost complete dependence on petroleum-based fuels and current entrenched travel lifestyles. Transportation is growing worldwide, and so are carbon emissions. Over 65% of the growth is by Annex I countries, but the share by non-Annex I countries will increase faster in the future as they satisfy their development needs (IPCC, 1995). The overall expected growth in transport-related GHG emissions will be huge and reducing it to meet the demands of the climate convention will require major changes because of the limits of technology changes and entrenched lifestyles. Significant change in current use patterns and lifestyles people in the transport sector of developed countries are needed, and efforts should be made to avoid repetition in developing countries. These changes are very challenging for the transport sector.

Efforts mainly driven by other concerns than climate change have led to technological options (improved technology design and maintenance, alternative fuels, vehicle use change, and modal shifts) and non-technical options (transport reduction, and improved management systems) that can reduce GHG emissions significantly. Similarly, there are non-transport options such as urban planning, and transport substitution such as telematics and improved telecommunications. Some options are low cost such as vehicle and aircraft maintenance, stringent enforceable regulatory systems along with inspection and testing, improved driving, fleet control, improved signalling, and better road signs. Others that involve changes in infrastructure such as modal change, dedicated lanes for different systems, and convenient walkways can be expensive and have long lead times. Greater use of light and heavy rail transport for both passenger and freight transport can result in reduction of GHGs, but initial cost can be high. Reducing transport intensities such as more use of public transport modes, re-organisation of local markets, regionalisation of production, and use of new logistics systems for freight travel can lead to substantial GHG reductions, but may require change in lifestyles. Disparities in technological conditions, socio-economic and historical factors necessitate the transfer of these options through market and non-market oriented paths within and between countries.

Significant barriers exist in the transfer of these options such as lack of a suitable business environment and technological capacities in technology recipient countries; in addition, there is a lack of a stimulating environment for transfer in technology supplier countries. However, policies such as promoting cooperative technology agreements between companies of different countries can result in joint R&D and other activities leading to transfer of some of these transport options and overcome some implementing barriers. Similarly, joint information networks can lead to transfer of improved technical and management skills. Creating an enabling environment in countries worldwide will stimulate technology outflows, and increase technology inflows to and from these countries with greater participation of the private sector. In addition, specialised training programmes through technical assistance can enhance local technical capacities in technology recipient countries. Adoption of appropriate standards and regulations can stimulate and facilitate technology transfer within and between countries. Partnership between government and the private sector and among countries can also help promote technology transfer within and between countries.

8.1 Introduction[1]

In 1995, the transport sector accounted for 26.5% of total final energy consumption and 22% of carbon emissions from energy use (OECD, 1997; IPCC, 1996). Transport-related carbon emissions, including international aviation bunkers but excluding marine bunkers, was the fastest growing between 1990 to 1995, with an annual rate of 2.4%. This growth occurred globally but not evenly. In 1995, Annex II countries accounted for 66% of the total, though emissions from developing countries are increasing their share. Similarly, per capita transport emissions are uneven, in 1995 it was 3.33 t CO_2/capita in Annex II as compared to 0.29 t CO_2/capita in non-Annex I countries (OECD, 1997).

Transport activities, and thus carbon emissions, because of the almost total dependence (about 97%) on petroleum-based fuels, have been growing dramatically since 1950. This is mainly due to decline in relative travel cost while at the same time incomes were rising. The dominant source of these carbon emissions is road travel. Vehicle ownership rose from 50 million in 1950 to 770 million in 1997, of which 78% were cars (Philpott, 1997). Vehicles on the road are estimated to rise by 50 million annually leading to a billion of them within 12-15 years time without direct intervention (UNEP, 1998). This growth varies among the different regions of the world. The industrialised regions with 21% of global population have 60% and 80% of commercial and passenger vehicles, while developing countries with 79% of the world's population have 40 and 20% respectively (Hilling, 1996). This disparity has a clear impact on their transport-related carbon emissions as shown in Table.8.1. Since 1970, phenomenal growth in private vehicle ownership is occurring worldwide as income rises; this is shown for cities of selected countries in Figure 8.1. This brings about a modal shift in passenger transport as can be seen for EU countries in Figure 8.2. Similarly, freight travel by road is growing faster than other modes with a significant shift from rail to road. These changes are shown in Figure 8.3 for European Union (EUROSTAT, 1997). Air travel has been growing faster than road travel. Between 1960 and 1990, air travel (passenger-km) grew by 9.5% annually and airfreight by a faster rate of 11.7% annually (ICAO, 1992). Though in recent times the overall annual growth has been about 5-6% and there has been reduction in energy intensity, this growth still surpasses economic growth. Controlling emissions from aviation is proving difficult.

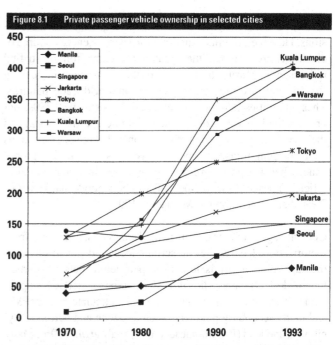

Figure 8.1   Private passenger vehicle ownership in selected cities

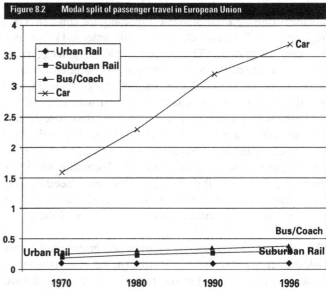

Figure 8.2   Modal split of passenger travel in European Union

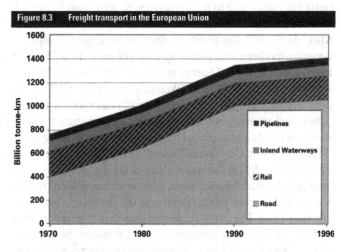

Figure 8.3   Freight transport in the European Union

WORLD REGION	GENERAL AND SELECTED TRANSPORT DATA FOR 1990 (% OF WORLD TOTAL)				DIRECT TRANSPORT CARBON EMISSIONS (MTC)		
	AREA	POPULATION	COMM.VEHS.	PASS.VEHS.	1971	1991	2010
North America	16	5.2	40.1	35.3	1201	1588	2133
Europe	4	9.3	18.2	37.4	475	867	1180
Former USSR	16	5.2	0.2	4.8	215	429	736
OECD Pacific	7	2.5	1.9	2.0	150	300	460
Latin America	15	8.4	8.7	6.7	77	337	552
Africa	22	12.4	3.5	2.0	52	110	215
Asia (ex Japan)	20	57.0	27.4	11.8	136	522	1747

Table 8.1   Vehicle use and carbon emissions for world regions
Source: United Nations Statistical Yearbook; IEA (1996)

• OECD PACIFIC REFERS TO JAPAN, AUSTRALIA AND NEW ZEALAND

[1] Relevant cases from the Case Studies Section, Chapter 16, are Brazilian Ethanol programme (case 8), Transport in Uganda (case 11).

Emissions from rail transport have been declining mainly due to shifts in fuel use and modal changes in the overall activities in the transport sector. Fuel changes, mostly from coal to electricity and diesel, have had a significant impact on reducing carbon emissions. However, if emissions from the primary energy source of electricity are included, there may not be gains from electrified rail systems. Marine transport activities have been increasing steadily because they are dependent on world trade patterns. These activities declined in the late 1970s till the 1980s, but grew in the late 1980s due to the economic upturn in Asia, and then declined again as result of recession in that region. Emissions from shipping transport have followed this pattern, though not exactly because goods are now being moved in larger and more efficient marine vessels, especially dry bulk goods that form 60-70% of such movements (Michaelis, 1996).

Generally, growth in transport-related emissions, especially from motorisation, has resulted in serious local and global environmental problems. Emissions from petroleum fuels are carbon monoxide (CO), carbon dioxide (CO_2), hydrocarbons (HC), sulphur oxides (SO_x), nitrogen oxides (NO_x) and particulate matter (PM), and many secondary pollutants such as ozone and peroxyacetyl nitrates, and lead (Pb) from leaded fuels (Faiz *et al.*, 1996). Many of these are hazardous to human health as irritants, allergens or carcinogens causing local problems especially for children. On a regional scale, SO_x and NO_x can cause air, water and land contamination; and globally, CO_2 contributes to global warming (Walsh, 1997). In addition, growing road congestion in large cities worldwide is causing local pollution leading to high economic and social costs. The example of Santiago, Chile in which the children tend to have 25% higher incidence of coughing than those in other Chilean cities illustrates this (WRI *et al.*, 1996). Stopping and starting of engines in congested traffic consumes more than three times more fuel than free-flowing traffic (Jourmad *et al*, 1990). Emissions from aircraft engines can be severe, especially CO_2, which has the same effect as that at ground level, while the effects of other gases may be affected by height (Michaelis, 1996).

Transport related carbon emissions would grow significantly unless there is a major change in lifestyle and travel behaviour. It has been predicted that global transport energy use and, thus, GHG emissions (due to their linkage with petroleum fuels) will grow by 2.2-2.6% yearly from 1993 to 2010 (OECD, 1996). Also, since 1971 the transport share of total petroleum consumption has been rising steadily, and will grow from 55.6% in 1993 to 60% in 2010 if energy use continue to grow by 2.6% as it did between 1971 to 1993 (Wohlgemuth, 1997). GHG emissions from this growth will be substantial, and expected gains in vehicle fuel economy is estimated as 1 to 2% per annum based on historical trends and expected technical changes (Grubler *et al.*, 1993; IEA, 1993, Greene and Duleep, 1993; DeCicco and Ross, 1993; Walsh, 1993). The major share of the growth is expected from private road transportation. Also, road transport of freight is expected to continue to dominate freight travel, as goods become lighter. The overall expected growth in transport related GHG emissions will be huge, and reducing it to meet the demands of the climate convention will require major changes because of the limits of technology changes. OECD estimated that projected technology improvements would only meet half of the effort required to achieve the demands of the UNFCCC (OECD, 1995,

OECD 1997). Coping with expected transport growth is a challenge for the transport sector.

However, a number of policies and measures exist which if used while recognising local specificities could reduce transport related GHG emissions (IPCC, 1996). Similarly, good management practices are available which can moderate the growth in aviation fuel consumption and so reduce GHG emissions. These policies and measures could be transferred between and across regions if certain implementing barriers can be overcome. This chapter will concentrate on looking at the potential to transfer some of these options within and between countries. The chapter will largely be based on road transport due to its dominance in current and future GHG emissions in the transport sector.

8.2   Transport Mitigation Technologies

Significant achievements have been made in developing transport systems that reduce GHG emissions though the development of other concerns such as performance gains, safety, and energy intensity improvements has been paramount in their development (IPCC, 1996). A large number of potential GHG emission-reduction options are technically feasible but only some are economically feasible. Some are cost-effective but vary among users, resource availability, know-how, institutional capacity and local market conditions (IPCC, 1996). Energy savings, performance improvements and reduction in private costs also affect their cost effectiveness. These options with potential to mitigate GHG emissions can be generally put into four categories: vehicle technology improvements, including for aircraft and marine vessels; fuel technology improvements, including alternative fuels; non-motorised systems; and infrastructure and system changes. Introducing these options may require justification of other objectives other than GHG mitigation such as competitiveness, security concerns, and improvement of quality of life or local environment improvement.

8.2.1   *Vehicle Technology Improvements*

Vehicle technology improvements normally involve proper maintenance, improving the engine or vehicle body, or reducing inertia with the main aim of reducing the energy intensity (energy use per useful product) and so reducing carbon emissions. Regular servicing, including regular tire and oil checks, and engine tuning can lead to fuel savings of 2-10% (Davidson, 1992; Pischinger and Hausberger, 1993). Use of three-way catalytic converters along with electronic fuel injecting systems can result in reduction of ozone precursors (unburned HC, CO, NO_x) emitted from gasoline cars and heavy-duty vehicles, but the effect on global warming is uncertain because the impact of fuel consumption is also uncertain (IPCC, 1996). Improved combustion by use of gas turbines and low-heat-rejection engines can potentially result in higher efficiency and, thus, in lower emissions, but there will be a need for high temperature materials along with compatible high temperature lubricating systems. Also, direct-ignition stratified-charge engines can be more efficient because of their ignition-enhancing qualities. Details of these

Table 8.2   Technical and potential combustion control technologies (Source: IEA/OECD, 1998)

TECHNOLOGY	EXAMPLES	STATUS	TECHNICAL FEASIBILITY	CONVERSION EFFICIENCY	ENVIRONMENTAL IMPACT	MARKET POTENTIAL TIME FRAME
ICE CONTROL						
1. Improved Exhaust Treatment	• Catalyst traps, exhaust gas recirculation (EGR) • Intake and exhaust systems • Advanced emissions abatement in heavy-duty vehicles	• Deployed in autos • Limited diesel application	• Continuing after treatment improvement • Allows continued use of ICE	• Slight decrease for significant O_x reduction • Increased back pressure reduces efficiency in diesels	• Up to 97% control for HC and CO • Up to 85% control for O_x • Up to 85% control for particulate	• 0-5 years
2. Improved Combustion	• Ceramic components • Ignition systems • Flow dynamics variable valves • Turbine engine	• Incremental improvements	• Good variety of technology • Available technology must integrate with current ICE	• 5-10% engine efficiency gains	• O_x particulate and CO_2 reduction	• 0-10 years
3. Fast Warm-up	• Thin wall engines • Start/stop with flywheel storage	• Incremental improvements	• Transient time decreased by 50%	• Average efficiency gains of <5%	• <10% average reduction • <30% reduction in first 60-120 seconds	• 0-10 years

potential reductions are given in Table 8.2. The potential exists for increasing vehicle mileage and, therefore, energy intensity by reducing the aerodynamic drag and rolling resistance leading to improved efficiency and, thereby, reducing the emissions (ETSU, 1994; DeCicco and Ross, 1993). Similarly, through size reduction, material substitution or component redesign, the inertia can be reduced and so lower the fuel consumption (DeCicco and Ross, 1993). Improving the transmission system to electronically allow for optimal speed and load conditions can result in energy savings and reduced emissions (Tanja *et al.*, 1992; NRC, 1992). More details of these potential reductions are summarised in Table 8.3.

Trends show that if priorities shift among manufacturers and users, improvements of 10-25% in energy intensity may be achievable on cars by 2020 at a higher cost, but the potential for commercial vehicles will be smaller. However, fuel savings and environmental gains may be offset by the increase in number of vehicles and driving (Wootton and Poulton, 1993).

The trend in buses for higher level of comfort and safety, and more powerful engines has tended to increase fuel consumption per seat compared with old buses, but this can be reduced by using advanced composite materials and turbo-compound diesel engines. Electric buses are in use as minibuses in urban areas, but they have higher GHG emissions than diesel buses when the primary emissions than diesel buses when the primary energy used is from fossil sources. Hybrid buses (diesel/electric) are now being tested because they can save up to 30% in energy if the motor/generator efficiency is about 85%. Alternative fuels (CNG, alcohol fuels and vegetable oils) are used in buses and when rapeseed methyl ester is used as substitute for diesel, life-cycle GHG emissions can be reduced by 25-50% (IEA/OECD, 1994). Use of turbo-charging and charge cooling in engines of trucks improves the fuel economy and so reduces GHG emissions, but retarding fuel injection worsens the fuel economy. Potential exists for improvement in fuel economy based on developments of new engine materials (IEA/OECD, 1993). Fuel economy can also be

Table 8.3   Technical and potential vehicle improvements options (Source: IEA/OECD, 1998)

TECHNOLOGY	EXAMPLES	STATUS	TECHNICAL FEASIBILITY	CONVERSION EFFICIENCY	ENVIRONMENTAL IMPACT	MARKET POTENTIAL TIME FRAME
VEHICLE IMPROVEMENTS						
1. Drag and rolling resistance reduction	• Drag coefficient reduction • Reduced rolling resistance • Reduced bearings friction	• Commercial potential for improvement in low-friction bearings and lubrications • Low-friction tyres to be tested	• Continuation of improvements dependent on material properties & cost of manufacture • Study on basic physics	• Speed sensitive benefits • Gains of 1-5% possible	• Reduction of all emissions in proportion to efficiency gains	• 0-10 years
2. Structural weight	• Light structures • Bonded/composite structures • Light powertrains	• Commercial/ demonstrated • Bonded structures in limited use • Composite materials in most vehicles	• Continuation of improvements • Limited by material properties and relative cost of manufacture	• 0.2 to 0.4% gain for every 1% weight reduction	• Reduction of all emissions in proportion to efficiency gains • Greater effect on acceleration emissions (urban traffic) as vehicle inertia is diminished	• 0-10 years
3. Transmission	• Electronic shift • Multistep lock-up • Continuously variable transmission (CVT) electric drives • Drivelines and suspensions	• Commercial/ demonstrated technology • CVT available • High power CVT in prototype • Lock-up and electronic control	• CVT/IVT in widespread use in next decade • Hybrid powertrains feasible with CVT/IVT	• 10-15% gain over manual with CVT or IVT • Electronic drives could further increase this conversion efficiency	• Reduction of all emissions in proportion to efficiency gains • Engine operation optimised, decreasing emissions even more than efficiency improvement	• 0-10 years
4. Accessories	• On-board electronic controls • Constant speed drives • Efficient components	• Demand responsive systems gaining preference • Constant speed systems in demonstrations	• Highly feasible for constant speed • High efficiency accessory systems	• <5% efficiency gain	• Emissions reduction facilitated by on-board electronic controls and sensors	• 0-10 years

Box 8.1   COMPRESSED NATURAL GAS AS A
                 TRANSPORT FUEL

Compressed natural gas (CNG) can be an attractive alternative transport fuel to gasoline because of its environmental benefits including reduction of GHGs. It is more useful for countries with natural gas resources and a relatively good gas distribution system. The use of CNG as a transport fuel started in the 1930s but failed to increase its share because, as with most alternatives, petroleum was a preferred fuel due to cost advantages. However, the current threat of climate change has increased the focus on alternative transport fuels including CNG. Countries with programmes on the use of CNG as a transport fuel include the USA, Canada, UK, Thailand, New Zealand, Argentina and Pakistan. CNG is used in both private vehicles and transport fleets. It is estimated that about 250 million vehicles are using this fuel worldwide, and its use is on the increase, representing 2% of total global transport fuel use. The advantages to using CNG, beyond environmental ones, include reduced engine maintenance cost, and improved engine and fuel efficiency. Disadvantages include power loss, limited range of storage (100-150 km), and high cost of conversion. The environmental benefits relating to climate change are given in the Table below:

EMISSIONS	CNG	REG. GASOLINE	SUPER GASOLINE	DIESEL
CO	1	10.4	9.0	1.2
Unburned HC	1	2.0	1.4	1.2
NOx	1	1.2	1.4	1.1
Particulates	neg.	present	present	very high
SO2	neg.	neg.	neg.	very high
Lead	nil	declining	declining	nil

A case is given below to illustrate transfer of CNG technology between Pakistan and New Zealand.

Pakistan has proven reserves of natural gas in excess of approximately 850 billion m3 (90% methane, sp. gr. of 0.56, and octane of 130). The country embarked on using CNG as a transport fuel in 1980 with officials of the Hydrocarbon Development Institute of Pakistan (HDIP) visiting Italy and New Zealand for two years to gain experience with CNG technology. A pilot phase was first introduced in 1982 in Karachi, and all the compressors and conversion equipment was purchased in Italy and New Zealand. In 1992, the government, through the Ministry of Petroleum and Natural Resources, promulgated the CNG Rules of 1992 that have commercialised CNG as a transport fuel in Pakistan. Six years later, 25 CNG stations became operational and another 25 were at various stages of completion. Along with the Rules of 1992, the Gazette of Pakistan Extra of July 28, 1992 provided guidelines for the safe practices of CNG relating to storage, filing and distribution. Under a UNDP/ESCAP programme in 1991, HDIP in collaboration with Liquid Fuels Management Group (LFMG) of New Zealand undertook a detailed six-month field test of completely retrofitted buses, partially converted buses and diesel buses. The results revealed that CNG is environmentally better, because it was lead free, had no particulate matter, and very low smoke density, but the levels of CO and NOx emitted were higher than that of diesel. It was found that CNG is more economically and technically suitable for conversion from spark ignition engines, but will require major modifications for high compression diesel engines. Originally, all conversions were done in New Zealand, but now Pakistan only receives kits from there and the conversions are done by local technicians. Though the cost differential between CNG and gasoline is small, it is estimated that about 25,000 gasoline vehicles have been converted to CNG.   (Source: Sarwar et al., 1999)

improved in the design of trains. About 5-10% savings is possible in diesel locomotives and up to 30% if a regenerative braking system is used in urban metro systems; 15% savings could be realised in suburban train systems and 5-10% for inter-city systems.

Energy intensity in aircraft can be improved with engine modifications and new engine designs. Future improved supersonic engines that are expected after 2010 may lead to an increase in energy efficiency and lower emissions, but this improvement could lead to increase traffic movements (Balashov and Smith, 1992).

Energy intensity for boats can be improved by modifying marine engines by making improvements in the hull and propeller designs that could yield to higher energy gains. The use of vertical-axis turbines as sails can assist the engine and result in energy savings (CEC, 1992).

8.2.2   *Fuel Technology Improvements*

Gasoline and diesel can be improved by chemical reformulation that can lead to decrease in ozone-forming pollutants and carbon monoxide emissions per km travelled, but will be greater for non-catalyst controlled vehicles (IEA/OECD, 1998). Performance problems, cold-start ability, smooth operation and avoidance of vapour lock are disadvantages of using reformulated fuels. Alternative fuels to petroleum include compressed natural gas (CNG); liquefied petroleum gas (LPG); methanol from natural gas, coal or biomass; ethanol from biomass; electricity and hydrogen. The use of these options in reducing GHGs will depend on ease of use, performance and cost, however, CNG, LPG and ethanol are now used in niche markets (high mileage and urban travel) in both developed and developing countries (Sanwar *et al.*, 1999). On a full-cycle basis, use of LPG can result in 20-25% reduction in GHG emissions as compared to petrol, while emission benefits from CNG are smaller - about 15%. Although CNG emits less CO than petrol, gains from CNG depend on the amount of associated methane emissions from gas recovery, transmission, distribution, and use. Life cycle GHG emissions from alcohol fuels, such as methanol and ethanol, depend on the source and conversion technology (DeLuchi, 1993; IEA, 1993). GHG emissions from methanol made from coal will be double that of petrol, whereas methanol from natural gas will be the same, and from wood will be lower. Ethanol from maize, wheat and sugar beet will result in GHG emissions of 20-110% of that of petrol depending on fertiliser inputs and fuel used for conversion. Ethanol from sugar yields 80% GHG emission benefits in comparison to petrol, and almost 100% if baggase is used instead of coal in conversion (Goldemberg and Macedo, 1994). The use of ethanol and CNG as transport fuel is shown in Boxes 8.1 and 8.2.

Electric vehicles have the potential of having significant life-cycle GHG reductions depending on the primary energy source; the vehicle technology and method of use (DeLuchi, 1993; ETSU, 1994; Martin and Michaelis, 1992). Its widespread use will depend on battery charge/discharge efficiencies at high current, motor and controller efficiencies at high load, and improvements in vehicle design.

Also, electric powered systems can be costly and inflexible. Hydrogen is a clean transport fuel but requires high energy input and has serious storage and cost problems. Development of alternative aviation fuel, such as liquefied natural gas (LNG) and liquid hydrogen, is going on and can result in up to 20% lower carbon emissions than kerosene for LNG, but is not expected to be commercial in the next 10 years. Details of alternative fuels are in Table 8.4.

Recent increasing interest in the development of hybrid vehicles (combination of gasoline engine and battery-motor system) is yielding positive results that can have a positive impact on GHG reduction. Making optimum use of the different power supply system to suit the demand, and with an engine switch-off system during short breaks, up to 30% fuel efficiency gains can be achieved, as has been shown for models by Mitsubishi and Toyota (Tsuchiya, 1997). However, there are some disadvantages such as higher weight and slightly higher cost, but current declining cost could assist its commercial success. Use of hybrid vehicles with fuel cells and batteries is actively being considered by many manufacturers, but the cost is still very high for commercial application and may be in market early next century (OECD, 1997).

Box 8.2	Ethanol as a transport fuel in developing countries (Source: Goldemberg, and Macedo, 1994)

Ethanol can be important in helping to reduce GHG emissions. The energy derived from biomass, and in this case, from a renewable, "clean" source, i.e., from sugar cane, has the unquestionable advantage of permitting the almost complete re-absorption of CO_2 emitted through the combustion of ethanol. This closed cycle allows, in principle, to increase the global energy supply, essential for sustained economic growth, without creating hazards for the environment. The relevance of fuel alcohol in connection with the global efforts for reducing CO_2 emissions is singled out as one of the major contributors to the reduction of the greenhouse effect. It is important to note that, with technological advances and research, the price of alcohol can be made competitive with gasoline in the long run, but with the added advantage of providing a clean and renewable source of energy.

Ethanol in Brazil

The National Alcohol Programme (PROALCOOL), launched in November 1975, in Brazil, appeared to be the answer to the dangers of oil shortages. The programme's objectives were to guarantee the steady supply of fuel in the country; substitute a motor vehicle fuel from a renewable energy source for imported gasoline; use the sugar cane production to its full potential, especially in view of the drop in the world sugar prices; diminish regional inequalities and promote greater rural employment; and to encourage technological development in connection with the production of sugar cane and alcohol.

The programme benefited from a combination of favourable circumstances: the availability of adequate technology for the production of alcohol; the ability of the sugar sector to adjust quickly to the production of alcohol; the expansion of the distilleries; and the low international price for sugar, due to the general crisis of the sector, in part caused by overproduction.

Until 1979, the first phase of the Programme, alcohol production concentrated on anhydrous alcohol (99.33% ethanol) for blending with gasoline. The proportions of the mixtures varied. During this period, the programme benefited greatly from the expanded installed capacity of the distilleries annexed to the sugar mills. The second phase began with the second oil crisis (1979), placing considerably more ambitious goals before PROALCOOL. A change occurred, resulting in the predominance of hydrated alcohol, used in pure form as car fuel. The car factories in Brazil began to design vehicles using fuel alcohol exclusively. The industry appeared to welcome the new fuel and invested in research and development of alcohol-run cars; given the high oil prices, fuel alcohol would permit increasing production. The first cars run solely on fuel alcohol were produced in 1979. By December 1984, the number of cars run on pure hydrated alcohol reached 1,800,000, *i.e.*, 17% of the country's car fleet.

Fuel alcohol is a "clean" fuel because the local effects of gas emissions are less damaging to the environment, generally speaking. On the average, alcohol-run vehicles emit less carbon monoxide, hydrocarbons and sulphur. Another advantage of alcohol over gasoline is that alcohol replaces tetraethyl lead, which is hazardous to health and the environment, and is used as an additive to increase gasoline octane level. In fact, in the United States, alcohol is added to gasoline to diminish the high index of environmental pollution.

Ethanol in Zimbabwe

Zimbabwe, in Southern Africa, also operates an ethanol plant that was locally planned and is producing 40 million litres annually. As a lot of other developing countries, Zimbabwe had energy security problems in the 1970s. Petroleum products accounted for 14% of energy consumption and besides that the country was an exporter of sugar. At that time the international price of sugar was very low, so the conversion of sugar to ethanol was both economic and strategic.

In 1975, The Triangle, a private enterprise, decided to use surplus molasses from up to 40,000 tonnes of sugar for ethanol production and started production in 1979. The German Company Gebr. Hermman supplied the plant design and was willing to provide a "turn-key" project. The Zimbabwean firm only purchased the plants (at a reduced price) while the German firm supervised its activities. Adaptation was necessary and involved discarding many automatic controls in favour of manual operation to suit the capabilities of the local workforce. Local material was utilized up to 60%, substituting stainless steel used in the distillation columns with approval of and supervision by the German firm. The final cost was US$ 6.4 million for a plant capable of 40 million litres a year.

The ethanol produced is sold to National Oil Corporation of Zimbabwe and they resell it to various oil companies. Ethanol as a fuel with 13% blend was the only petrol available for a long period in Zimbabwe, with very few modifications. The plant has been operating for over 18 years with a lot of environmental benefits, skill transfer and technological adaptation as well.

8.2.3   Non-Motorised Systems

Non-motorised transport systems (bicycles, rickshaws, push-carts, etc.) have been used to meet transport needs of urban poor and rural dwellers in developing countries for a long time (Dimitriou, 1990; WRI *et al*, 1996). Recently, because of its environmental benefits, cycling is seen as an option to meeting the growing demand for urban travel provided its associated infrastructure is available (WRI *et al*, 1996). Some European cities have instituted bicycle-friendly measures such as dedicated lanes, improved signalling, etc. that have resulted in increased use and, thereby, in an improvement of the environment. Delft and Groningen in the Netherlands and Copenhagen in Denmark are examples. In the latter, cycle trips increased by 50% in five years after instituting a cycle infrastructure (PTRC, 1991). Cycling can be useful for short urban trips for both passenger and freight trips in cities of developing countries (Dimitriou, 1990). Its use varies widely among developing countries depending on the terrain, cost and safety. It represents 30-50% of urban trips in China, 15-20% in India, 3.5% in Africa and 16% in Latin America, though in the latter it is used more for recreational purposes. However, safety is a major problem; in New Delhi and Bangkok, 14% and 6% of fatal accidents were suffered by cyclists (Gate/GTZ, 1998). Despite this, for many developing countries with economic constraints cycling can be an option if they can satisfy associated infrastructure needs. Studies show that investments in infrastructure needs (dedicated lanes, parking facilities, inter-modal facilities, special signalling) could be recovered between 1 to 4 years with good planning (Simon, 1996). Walking forms a substantial share of movements in many countries, and can be a viable alternative for short distances if the associated infrastructure such as footbridges, attractive and convenient walkways and proper signalling are instituted.

8.2.4   Infrastructure and System Changes

Travel trips and choice, and freight volume and modal choice can be affected by several factors such as urban density, transport infrastructure and the design of transport systems. Recent major innovations in infrastructure design such as linking urban transport to land-use patterns, zoning, increase access to jobs and shops, comprehensive and integrated planning strategies have lead to reduction of urban pollution with possible climate change benefits as they reduce the reliance on car (Newman and Kenworthy, 1989). Designing clusters under the mixed land model, where homes, jobs and stores are together reduces the trips significantly (ECMT/OECD, 1995). However, designing infrastructure to suit transport demands is difficult because of the multiple needs of the different transport modes. Also, an integrated strategy requires coordination between many stakeholders which is not always easy (Gakenheimer, 1993).

Traffic and fleet management aimed at reducing road congestion and increasing traffic flows for different modes is gaining prominence because energy savings up to 10% in urban areas can be achieved, resulting in reduction of GHG emissions. Advanced Traffic Control Systems (ATCS), which include well-designed and coordinated control systems, have led to improvements in many cities of developed countries such as Cambridge in UK, and Oslo and Bergen in Norway (OECD, 1995). Singapore is one of the few countries that have instituted several demand management measures aimed at reducing demand for cars and restraining car use. These include fitting smart cards on vehicles as a means of paying tolls and use of road sensors (Pezoa, 1995), vehicle quota system based on traffic and electronic road pricing to reduce traffic jams in peak hours (Ang, 1996a) and the area licensing scheme that is continuously being reviewed (Ang, 1996b; see Box 8.3). Modal shifts from road to rail can yield energy savings of 0-50% resulting in GHG reductions and even higher in non-fossil fuel electricity powered electric trains, but these savings vary among regions and current trends of moving from rail to road travel can negate such benefits.

Increasing occupancy of travel trips by promoting mass public transport systems can result in substantial energy savings and GHG reductions because the emissions per passenger are lower especially with well organised routing and stops. Energy efficiencies in passenger-km for such systems can be up to 4 times better than private cars (Hilling, 1996). Potential reduction in emissions from using such systems in comparison to personal transport can be up to 99% in HC, 97% in CO, 85% in NO_x, 46% in SO_x, and 27% in PM (Hilling, 1996). Also, public transport systems especially well coordinated bus systems can prove very effective as shown in the case of Curitiba (see Box 8.4). Urban rail systems can prove beneficial in

Box 8.3	**Urban transport demand management: a case (Source: Ang, 1993; Ang, 1995; Ang, 1996 a)**

Many cities in fast developing countries are adversely affected by serious traffic congestion leading to increased travel time and fuel consumption, environmental degradation and productivity losses. Singapore is one of the few countries in the world that has adopted measures to moderate demand for cars and to restrain their use.

Singapore, a small and densely populated state, has being going through rapid economic development in the past 30 years resulting in significant increases in road transportation. The country embarked on land transport policy with technical, regulatory and policy measures to control transport growth. A recent white paper on transportation clearly identified two demand management measures that are aimed at reducing demand and use of cars. These are

a vehicle quota system and a road pricing system.

The vehicle quota system (VQS) was introduced on May 1, 1990 after several fiscal measures such as import duties, registration fees, and road taxes had been tried with very little success. As an example, purchase price of new cars increased by almost 200% in real terms between 1974 and 1990, but the car population doubled in the same period. Rising incomes was believed to have accounted for this growth and there was no indication that it will slow down by further increase in taxes. Hence, the VQS was designed to link vehicle population to road capacity. Vehicle registration is fixed by the government based on current traffic conditions and the market determines the vehicle price. It was applied to all vehicles except scheduled buses and school buses. New

Table 8.4   Technical and potential transport fuel technologies

TECHNOLOGY	EXAMPLES	STATUS	TECHNICAL FEASIBILITY	CONVERSION EFFICIENCY	ENVIRONMENTAL IMPACT	MARKET POTENTIAL TIME FRAME
FUEL SUBSTITUTION						
1. Improved Gasoline and Diesel Fuel	• Reformulated gasoline • Reduced Reid vapour pressure (RVP) Σ	• Within current refining techniques	• Chemistry available • Refinery balance and the need to produce more light ends Σ	• <2% gain for vehicles • Increase in refinery energy efficiency	• Significant VOC reduction • Limitations on fuel additives	• 0-10 years
2. Natural Gas and LPG	• On-board storage • System integration	• Demonstration fleets • Field trials • Fuels commercially available Σ	• Range extension needed • System cost abatement	• Close to gasoline with engine adaptation	• VOC, CO2 and particulate reduction	• 0-15 years
3. Alcohol Fuels (in an ICE)	• Neat methanol • Neat ethanol M85	• Demonstration fleets • Field trials in large vehicles • Commercial availability of blends	• Supply limitation and cost needs • Change in OEM design • Low-cost emissions control option • Multiple feedstocks	• 15% improvement	• VOC and CO2 reductions	• 0-20 years
4. Hydrogen (in an ICE)	• Neat H2 in ICE storage systems Σ	• R&D and prototypes	• Renewable source of supply • Distillation & production hurdles	• Dependent on feedstock and storage system	• Substantial reductions in all pollutants	• 30 years
5. Energy Regeneration	• Hydraulic and kinetic storage • Engine management and electric storage	• R&D and prototypes • Large vehicles with electric trains • Demonstration bus fleets	• Energy storage systems • Transmission of power requires availability of CVT to blend power	• Dependent on mission profile but 15-20% improvement possible	• Reduction of all emissions in proportion to efficiency gain	• 10-20 years
6. Electric and Hybrid Vehicles	• Electric batteries • Fuel cells • Solar photovoltaic cells • Σ Hybrid systems	• Demonstration fleets • Field trials in niche markets Σ	• Range and cost limitations may limit market • Adopting hybrid drives may increase use options	• Dependent on base fuel with 20-40% gain possible	• Reduction to zero of all vehicle emissions • Environmental benefit to be gauged against overall fuel cycle	• 10 years

SOURCE : IEA/OECD, 1996.

STUDY OF SINGAPORE
AND B; SINGAPORE LAND TRANSPORT AUTHORITY, 1996)

vehicles are put on a tender that is held in specific periods and cars can be used for a limited time without further taxes. Although the system is constantly under review to reduce its weaknesses, some changes have been achieved as shown in Table below.

The Area Licensing Scheme (ALS) was introduced in 1975 to reduce traffic congestion in designated restricted zones in the city centre during peak hours. The scheme included components such as carpooling and park-and-ride for weekdays. Special public buses for commuters and car parks were introduced for commuters, Area licenses were made available in convenient public outlets, and compliance was enforced. As a result of this scheme, the number of vehicles entering the restricted areas fell by a third and cars by 61%. In addition, commuters changed trip times and modal shift was observed with car owners moving to buses. Noise and air pollution was also reduced. Between 1975 and 1989, employment in the restricted zones grew by a third in size. With the introduction of the rail-based Rapid Transit, all vehicles had to pay ALS fees except public buses, motorcycles and essential vehicles; car pool exemption was banned; and evening ALS was introduced. Also in 1994, a whole ALS system was introduced that led to a reduction of traffic by about 9% and increase in traffic speed.

The Electronic Road Pricing (ERP) system was introduced to replace the ALS, which was a manual pricing system that had limited coverage after a lengthy process of design and evaluation, as this was the first such system in the world. In 1993, the government offered three multinational consortia to develop a prototype, which they did for about a year. After modifications and improvements on the designs, the ERP started in late 1997. The system uses a smart card that is slotted into a unit attached to the windscreen. A detector automatically deducts charges when a gantry is crossed. Vehicles without cards or with cards of insufficient value are photo-recorded. These cards can also be used for phone calls and supermarket purchases.

On the whole, no doubt thanks to the introduction of these systems, Singapore car ownership is 11 per 100 people, which is very low compared with other countries with the same economic output. In 1995, the average road transport fuel consumption was 0.34 tonne of oil equivalent per capita resulting in 0.29 tonne of carbon emission per capita which is also low in comparison. Also, over 80% of the cars are less than 1600 cc and are well maintained, with an efficiency of 9 litres of gasoline per 100 km.

Table	Annual Average Growth Rates of Road Vehicles during the introduction of VSQ	
	1980-90	1990-94
Cars	5.9%	4.4%
Other road vehicles	2.2%	1.4%
All road vehicles	3.9%	3.1%
Real GDP	7.0%	8.1%

Curitiba, a Brazilian city of 1.6 million people, underwent a major restructuring of its bus system in order to reduce congestion and save journey time. This restructuring is a good example of the integration of transport with land-use planning. The main features of restructuring are the reserved bus lanes, the priority of buses over other models at intersections, a hierarchy of frequency, interconnecting routes, an easy ticketing system to avoid time wastage, and shelters at bus stops. Besides those features, various innovations, such as level boarding tubes, assure the speed of bus loading and unloading, therefore allowing the buses to operate like a surface subway. Feeder buses provide services to the five main corridors from more outlying areas. As a result, the use of buses in Curitiba has increased significantly.

Meanwhile, three factors made Curitiba a unique case: 1) a continued local policy, supported for many years, and favourable to the use of a public transport; 2) an urban plan compatible to the solution of public transport; and finally 3) an implementation of an integrated net transport.

The implementation of the integrated net was facilitated mainly by the interest of the Volvo Manufacturing Company, which was already established in that region, and the geographic characteristics of the city that allow the elaboration of an urban plan suitable to the net.

The rationalisation of the integrated transport system, the land-use legislation, and the fact of meeting human needs were seen as the driving forces for this system. However, this system is yet to be repeated elsewhere.

reduction of GHG emissions but experience shows that they are expensive especially for cities with less than 5 million people. The number of tracks also seriously affects its viability (PTRC, 1991)

With a well-developed communication system, use of telephones and telecommuting can assist to reduce travel trips leading to reduction in GHG emissions (Davidson, 1994). As satellite-based communication systems develops, more comprehensive routing of different modes will be possible leading to reduction in GHG emissions. More details are available in Table 8.5. However, these options can be expensive.

8.3   Technology Transfer Pathways

In general, the transfer of transport technology goes through the five different stages outlined in Chapter 1 (assessment, agreement, implementation, evaluation and adjustment, and replication). Also, the pathways for technology transfer involves the three main pathways (government driven, private sector driven, and community driven) mentioned in the first chapter; however, government and private sector driven pathways are more dominant than community driven pathways. In addition, the transfer of transport technologies can either be market-oriented, including recent trends such as globalisation and liberalisation, international production networks, and mergers and acquisitions (M&A); occur via regional networks; or be non-market-oriented. In general, the path of transfer differs according to the type of technology. Market-oriented private sector pathways are used on transactions involving vehicle and fuel technologies, while bilateral and multilateral treaties are common for transport infrastructure that are government driven pathways. The relatively more recent introduction of privatisation in infrastructure development and

Table 8.5   Technical and potential transport productivity options

TECHNOLOGY	EXAMPLES	STATUS	TECHNICAL FEASIBILITY	CONVERSION EFFICIENCY	ENVIRONMENTAL IMPACT	MARKET POTENTIAL TIME FRAME
TRANSPORT PRODUCTIVITY						
1. Traffic Control	• Electric rolls • Road information systems • High occupancy vehicles • Telecommunications	• Commercial/ demonstrated • Interactive signals and controls deployed	• Interactive signals and controls feasible • Inter-vehicle controls • High occupancy vehicle	• Significant local gains with the elimination of stop & go traffic (>10%) • Increased capacity of highways	• Major pollutant reduction (30-50%) at local sites • Reduction of emissions in proportion to efficiency gains	• 0-20 years
2. Travel Substitution	• Software development • On-board navigation systems • Electronic displays and interactive warning	• Commercial/ demonstrated • Telecommuting in trials • High occupancy vehicle programmes • Electronics use expanding rapidly	• Dependent on modal shift of transport market and flexibility of work arrangements	• Move to more efficient energy conversion modes • Higher overall transport productivity (10-30% possible)	• Reduction of emissions and change of emission sources • Reduction potential is approximately the same as fuel reductions	• 0-20 years
3. Route Planning		• Commercial/ demonstrated • Satellite-based communication positioning systems in marine/air use • Demonstration on highway • On-board maps and interactive reporting	• Anti-congestion measures feasible • Advanced communication technology • Staggered work hours	• Gains depend upon interaction between vehicle and overall road transport system	• Reduction of emissions and change of emission sources • Reduction potential is approximately the same as energy reductions	• 0-10 years

SOURCE : IEA/OECD, 1998.

maintenance is attracting market transactions such as joint ventures, management contracting and concession arrangements in such transactions. More recently, technology licensing has become a preferred strategy for market oriented pathways, because of its capacity to be a component of a technology-based global product strategy, especially for more matured technologies.

In the transport sector, significant technology is being transferred through M&A and Foreign Direct Investment (FDI) flows. More recently, growth in inter-firm technology agreements has had an impact in the transport sector. FDI has being growing substantially for the past seven years, surpassing growth in ODA, and the transport sector has been part of that growth, especially among major transport MNCs in both industrialised and developing countries. Cross–border M&A in the automobile industry such as General Motors (GM) acquiring Isuzu and Suzuki of Japan, Vauxhall from the U.K and Opel from Germany is an example of transport technology transaction across countries. Two-way inter-firm technology partnerships, which involve joint research and development agreements and the creation of joint R&D ventures with specific research programmes, are increasing, and the automobile industry is the third largest with such type agreements (UNCTAD, 1998). This creates a major pathway for transferring technology across firms, because as R&D become more knowledge intensive firms tend to concentrate their efforts to remain competitive. Hence, concentration of R&D efforts may increase in the future.

8.4 Technology Transfer within Countries

In the transport sector, technology transfer within a country is comparatively smaller than transfer between countries except for very large countries such as the USA, Brazil and China. The transfer of vehicle technologies is limited to subcontracting between major firms and components manufacturers, especially among automobile manufacturers. The development of vertical disintegration and new forms of supplier-client relationships in the automobile industry, which were triggered by the entry of Japan, has led to the growth of local companies that can produce specialised components with the desired quality and precision required. As a result, by 1996 only Ford and Volvo among the major car companies manufactured more than 60% of their components in-house. Others depend on outside units such as Volkswagen that used only 43% in-house, Mercedes-Benz 38%, GM 37%, BMW, Renault and Peugeot 33%, Fiat 30% and Honda, Nissan and Toyota 25% (UNCTAD, 1998). This change has led to flows of technology within certain countries such as Japan, which depends almost entirely on local Japanese companies for such transactions. Companies in other countries depend on other developed countries; for example, Ford of the USA depends on suppliers from Germany, the U.K., Japan, and Italy, in addition to those of the USA (UNCTAD, 1998). Another area in the transport sector that involves sub-contracting is infrastructure construction and maintenance. Local companies are used for

the supply of services to the main contractors and designers in very high quality jobs, especially jobs that involve international financing. This provides opportunities for the transfer of skills within countries.

The transfer of low cost measures such as management practices, fleet control, modal shifts to less carbon intensive modes, improved maintenance, better road signs and signalling is possible within countries but social, cultural and institutional barriers may need to be analysed (Pacudan, 1999). The transfer of fuel technology, such as use of LPG in taxis and CNG in buses, is possible provided the associated infrastructure and distribution networks such as filling stations, repair shops, qualified technical personnel for vehicle conversion and installation competence are available (Sanwar et al., 1999).

8.4.1 *Barriers to Technology Transfer within Countries*

There are barriers specific to the transport sector to the transfer of transport technology within a country. The first is lack of suitable local companies with the required technical skills and competence to supply technology components and services suited to the detailed and precise requirements of major firms. Also in countries where these capabilities exist, the number of such companies is not large enough to encourage competition which can enhance quality of output (UNCTAD, 1998). Another barrier is lack of technical, business and general information within a country, which can lead to poor links between the different local stakeholders. An organised local information network with necessary local data and inter-linkages to external networks is needed for effective technology flows. Access to capital by local companies to build or strengthen technical and other capacities can be a financial barrier, especially if local financial institutions do not perceive such need as a priority. An overriding barrier is the lack of political will among the different government authorities to transfer technology within a country.

8.4.2 *Policies and Programmes for Technology Transfer within Countries*

Important policies for transport technology transfers within a country are promotional policies for the development and strengthening of small and medium-scale sub-contracting companies. Also, major firms should be encouraged to enter into technology agreements with local sub-contracting companies. These policies could include incentives, loans and grants (UNCTAD, 1998). Providing grants or other incentives for institutional building and strengthening could assist small and medium-scale businesses to achieve the necessary capacities required by major firms. Promoting specialised training programmes to suit required skills would help to enhance local capacities. Creation of and support for a locally organised information network using professional societies will assist to improve the knowledge of local companies on the necessary local capabilities available.

Demonstration of technical options along with a reward system as incentive can boost technology flows between countries. Encouraging indigenous R&D and continuing education programmes, and supporting the necessary linkage with local educational institutions will create the opportunity for the development of new skills and innovation. Setting up of stringent standards with the complimentary compliance mechanisms can assist to transfer technology within a country (Pacudan, 1999). Furthermore, standards set up by a region in a country can influence activities in other regions and such standards get replicated. Standards, such as those for emissions, can stimulate technology change in the process of compliance.

8.5 Technology Transfer between Countries

Several programmes and policies exist that can promote the transfer of transport technologies between countries, because most of such activities are between countries rather than within countries. There are also barriers that may impede transfers between countries, because of the varying conditions among the different countries and regions of the world.

8.5.1 Barriers to Technology Transfer between Countries

The barriers to technology transfer of transport options between countries discussed above can be categorised into technological, financial, institutional, information and social. These should be seen along with the generic barriers already discussed in Chapter 5 of this report. In the transport sector, an overriding barrier that require emphasis is the lack of an enabling business environment for both technology supplier and technology recipient countries to promote technology transfer. Industrialised countries, which are mostly technology suppliers, can institute economic and fiscal measures and regulations with the necessary compliance regimes that can stimulate the private sector to transfer transport technologies. Technology recipients, which are mostly developing countries, need to create the enabling environment that is receptive to transport technologies (UNEP, 1998). Lack of a suitable enabling environment is particularly absent in low-income and capital constrained countries. In general, technology recipient countries need to build an effective business environment to attract involvement of the private sector, which is now increasing its role in transport technology flows, especially in transport infrastructure.

An important technical barrier to technology inflows to any country is lack of the necessary manufacturing capabilities, especially in technology recipient countries. Additionally, a lack of companies to undertake sub-contracting, as may be required by large transport companies, and the absence of suitable facilities for training and RD&D can create serious problems for technology development and transfer. An important financial barrier is access to capital, because most of the transport options are very expensive and involve long lead times such as building or modifying highways and bridges. These activities may involve significant capital outlay and many institutions with different interests.

Harmonising and optimising these interests can prove to be challenging (Pacudan, 1999). Also, implementing some non-motorised measures such as wider use of cycling can be expensive, because of the need for dedicated lanes and other support infrastructure, which would be a barrier for many countries. Lack of compliance and arbitration institutions can be a barrier for effective private sector participation. Lack of knowledge of the existence and development of environmentally friendly transport options, including their weaknesses and benefits, will be a major barrier in adopting them. This is common among technology recipients. Differences in social and cultural systems among countries can be a barrier, because some transport options are sensitive to these differences. Adopting cycling may require certain lifestyle changes as well as some other non-motorised systems. Similarly, adoption of recently smaller and more fuel-efficient cars that is being manufactured by many of the major manufacturers may not be acceptable to many countries because of their transport needs. Political will by respective governments for technology transfer is needed and so can be a major obstacle if absent.

8.5.2 Policies and Programmes for Technology Transfer between Countries

Creating an enabling business environment and building suitable indigenous technological capability in technology recipient countries, and creating the enabling environment for stimulating technology outflows are basic to transport technology transfer. Hence, these are briefly commented on in the following two paragraphs before continuing with a description of specific technology transfer policies and measures.

Building the necessary human resource for technology transfer and development is important to the transfer process, and this may involve programmes and policies for an adequate S&T human resource base, promotion of general S&T literacy, ensuring active participation of the indigenous S&T community and utilising local materials and other resources. A modern computerised system for information storage, retrieval and use with linkages to relevant national and global systems is very important. Equally important is a well-organised and coordinated institutional framework performing functions ranging from regulating, skill development, information development and dissemination, and to financial management. Suitable legislation for technology imports (choice, selection, evaluation and monitoring), foreign investments, and project assessment, approval and monitoring are needed, as well as institutions for law enforcement and arbitration (Davidson, 1998).

Creating a favourable business environment is important in addition to a stable macro-economy for technology flows. This includes having clear and well-defined operating guidelines for business operations, organised financial institutions, market information development and dissemination, and support training programmes and associated technologies. These capacities will not only enhance the absorption of technologies being transferred, but create the capacity for generating new technologies and

Table 8.6	Transfer of vehicle technology: a case study of South Korea				
TECHNOLOGY	ASSEMBLY	LOCALISATION	INTERNALISATION (1)	INTERNALISATION (2)	GENERATION
	1962-1967	1968-1974	1975-1981	1982-1990	1991-present
Passenger Cars (Hyundai and Daewoo)	• Semi-knocked manufacturing of Japanese Models	• Complete knocked down manufacturing of American and European models	• Mass production of local model	• Restyling and front wheel drive car model development	• Engine development
	• 21% local content	• 30% local content	• 85% local content	• 97% local content	• 100% local content
	• Suppliers of spare parts	• Technology licensing and joint ventures	• Technology licensing and joint ventures	• Technology licensing, joint ventures and in-house R&D	• In-house R&D and technology licensing
	• Acquisition of assembly and operational skills	• *Acquisition of inspection and production management skills	• *New model quality control and EEC tests	• Mass production of front wheel drive designed vehicles.	• Acquisition of engine design skills
				• FMVSS tests	

so offer new transfer opportunities. As an example, these capacities were crucial in the transfer of technologies to the Asian countries, which have enjoyed significant economic progress recently (World Bank, 1996).

Stimulating technology flows from technology supplier countries is needed for the transfer of transport technologies. This will require instituting certain measures such as informing local companies about potential markets, financial support for joint ventures with companies in recipient countries, grants for investments with transfer-of-transport options, preferential treatment to firms that transfer and sponsor expertise to live and work with recipients. In addition, supplier countries should ensure that options for transfer have been adequately tested domestically to build the confidence and respect of the recipient (UNCTAD, 1999).

An important specific policy for technology transfer is the will and commitment of governments involved in the transfer and development of transport technology. This was the most single important factor for the success of technology transfer of the automobile industry in South Korea as shown in Table 8.6 (Pacudan, 1998). This commitment can be from the highest possible authority and be demonstrated by long-term strategic plans of technology growth, while linking it to other development policies and programmes.

Another specific policy is the promotion of R&D in the transport sector. Globally, industrial R&D in the transport sector have undergone major changes in the last two decades. First, there has been a substantial increase in R&D funds by vehicle manufacturers in industrialised countries, especially in the US, Europe and Japan; second, a change in direction and scope of R&D funding; and third, a change in the management and organisation (Gerybadze, 1994). R&D management has changed from centralisation with no explicit strategies in the 1970s, to project planning, strategic management and decentralisation in the 1980s, and now to integration of R&D strategy to corporate strategy, balancing between basic and applied research programmes within networks. Governments can institute new policies and programmes to exploit these changes, because despite production still being concentrated in a few companies, opportunities exist for technology recipients with vehicle assembly plants to encourage joint R&D activities with the main firms. The case of increased imports of cars produced in Mexico by the U.S. automobile manufacturers illustrates this point. During the devaluation of the Mexican peso, U.S. companies easily imported high quality produced cars from Mexico back to the U.S. for distribution because they were cheaper. It shows the importance of building capabilities in recipient countries. Another aspect of R&D programmes is to enhance local R&D through promotion of contract R&D programmes between public and private sectors within and between countries. This can promote relevance and stimulate interest in local R&D, as has been the case of Malaysia where companies have been able to receive a 200% tax exemption for R&D expenditure (Idrus, 1988). Setting up a revolving fund for R&D promotion and commercialisation of results as being done in countries such as South Korea and Japan can help to promote R&D. Due to the rapidity in innovations, and the expensive and skill demanding nature of modern R&D globally, countries need to participate in international and regional R&D networks. Joint R&D will allow technology recipient countries access to very expensive equipment, laboratories, and high quality skills, which are normally absent in many technology recipient countries.

In general, technology cooperative agreements among firms can form a very good basis for the transfer of transport technologies. Joint ventures in equity and no-equity firms offer opportunities for technology transfer. In equity form, the recipient will be fully involved in decision-making in the capitalisation of technology, royalty fees and organisational arrangements, and depending on the capability of the recipient, significant technical skills such as procurement of equipment, R&D, quality control and marketing skills can be transferred through joint programmes. Joint R&D programmes could be an efficient means of transferring knowledge and stimulating activities especially among resource-constrained countries. Financial, marketing and management cooperation agreements can be used to access technology through non-equity joint ventures. The use of joint agreements in other forms can be used effectively to promote information access and exchange. However, in joint agreements, partners need to be prepared to share the risks and costs of programmes. These agreements could be used to stimulate the transfer of smaller and fuel-efficient cars between parent and subsidiary firms, hence encouraging transfer between countries.

Bilateral cooperative agreements between countries which could lead to twining or linking of cities can provide opportunities for cities with successful urban planning systems that had yielded GHG benefits to transfer such experiences to other cities. Exchange of personnel, targeted visits, joint implementation of programmes are some of activities which could result in the transfer of skills and knowledge (Figueroa *et al*, 1998). "Integrated transport planning" which involves a set of policies that matches the environmental and transport agendas using experiences worldwide can improve the local and global environment of many cities. This can include the harmonisation of a wide range of policy objectives such as wider use of public transport, new road infrastructures, pricing policies, and environmental quality monitoring (IIEC, 1996).

Technical assistance programmes can offer technology recipient countries opportunity to access specific skills such as technical advice on operation, maintenance, and quality control, but not proprietary technical information. Consultancy services can be used to transfer specific technical skills. Management contracts could be used to acquire selected skills such as organisational and technology-sourcing skills, but problems could arise from different perceptions of project objectives between the two parties, unnecessary extension of contracts, and level of involvement of indigenous employees.

Promotion of standards and regulations can significantly be used to control emissions, especially from road vehicles as was demonstrated in the USA with the 'Clean Air Act'. Nearly all countries have standards for vehicle emissions, but many technology recipient countries have weak enforcement mechanisms, hence compliance is generally weak. Applying standards depends on several factors such as the manufacturing base of the country, capacity to enforce the standards and facilities to comply with the standards. Using similar standards in countries with assembly plants as those of the parent companies will ensure that updated technology will be transferred between both countries. Countries without manufacturing capacities can institute strict standards, but they need to have the capacity to enforce them such as testing and inspection facilities. Strict standards can also be used to control second-hand imports.

Policies that can promote local technology development if the basic technical capacities are present include discouraging imports of completely built up units, providing incentives to local assemblers to increase local content in production (tax reduction, subsidies), and regulations to ensure local content in varying percentages. These policies, which were instituted in the development of automobile manufacture in South Korea, can provide significant lessons for technology recipient countries in the acquisition of many GHG reduction options in the transport sector (see Table 8.6). Some countries have embarked on national programmes for transport technology development, though most of them are for automobile promotion as was the case of South Korea and India. More recently, some African countries are doing the same. Nigeria and Kenya are examples of such moves. In the case of Nigeria the prototype has been demonstrated, while in the case of Kenya the prototype is yet to be adequately tested.

Policies and measures to promote non-motorised systems mainly involve the construction of the associated infrastructure such as dedicated lanes with the supporting signalling systems, and parking facilities for cycles. Creating local manufacturing facilities for bicycles to increase public access will be useful to improve their use by poorer members of the society. On the whole, these policies and measures should be integrated into the overall transport policy, as has been done in some European cities.

Multilateral agencies can be useful in facilitating the transfer of transport technologies through provision of specific information, support for local information networks, provision of technical advice and providing training programmes. Also, these organisations can be used to provide unbiased information regarding technology products and services. The Global Environment Facility (GEF) that is concerned with providing funds for abating GHG emissions will hopefully approve a transport programme soon that is aimed at funding non-motorised transport, modal shifts, electric and hybrid vehicles, and biomass transport fuel. GEF is expected to be restructured to include support for technology transfer activities (GEF, 1999).

8.6   Lessons Learned

The past experiences on the transfer of transport technologies provide some useful lessons, which could help future technology transactions between technology suppliers and technology recipients.

The need for governments' commitment in the transfer and development of technology as shown in the different cases studied in this chapter is vital for the transfer process. This is true for both technology suppliers and technology recipients. Technology suppliers need to demonstrate their commitment by providing conditions that stimulate technology outflows, while technology recipients have to demonstrate their will by providing the enabling environment for the private sector's participation. Building indigenous technological capacities increases the technology receptivity of a country. This activity is extremely important for technology recipients, as is the improvement of their overall business environment. These attributes will attract foreign and local investments along with technology inflows. Establishing a partnership between government and the private sector can contribute to useful technological relationships involving transfer. High quality manpower with a supporting industrial infrastructure is the main factor that drives technology growth in a country. This provides the capacity for "un-packaging" technology imports, effective R&D policy, and the drive to select vital technology imports for further development. The important role of government in the process is summarised below:

- Formulating the national transport development strategies, including encouraging or restricting certain transport options.
- Providing market development support for increased investments in the transfer of existing environmentally sound transport technologies and fuels, and support for their further development.
- Supporting R&D and demonstration projects related to the advantages of environmentally sound transport options.
- Setting and enforcing standards for improved efficiency and GHG emission controls
- Giving preferential support through fiscal and other measures for environmentally sounds transport modes.
- Promoting the expanded use of information networks to foster technology transfer

- Using regional market opportunities to stimulate closer inter-government interests and government-industry relationships and interactions
- Promoting local and external investments including co-financing programmes
- Providing policy and regulatory frameworks for fiscal incentives for alternative fuel use and distribution infrastructure, and mechanisms for internalising social costs of transport.
- Promoting international cooperation in making available necessary financial resources through regional and international financial institutions for programmes such as collaborative R&D and demonstration, and information exchange.

A significant number of transport options exist now and will be developed in the future that will reduce GHG emissions significantly, but they need to be transferred to where they are needed once certain implementing barriers can be overcome. Similarly, policy and measures are available that can transfer some of these options using the different stakeholders. Creating an enabling environment for effective participation of the private sector is very important for both technology outflows and inflows. In addition, improving the overall business environment will enhance investments and technology inflows. Exploiting the facilitating role of multilateral agencies will assist the adoption of the transport options discussed. The will and political commitment of all governments is crucial for the process of technology transfer and development.

The inflexibility in the transport sector due to its fossil fuel dependency and entrenched interests will need to change by departing from present lifestyles and travel patterns, if the current growth in transport-related GHG emissions is to be decreased.

References

APO, 1994: *Technology Development, Adaptation and Assimilation Strategies at Cooperative Level*. Survey report, Asian Productivity Organization, Japan.

Ang, B.W., 1993: An Energy and Environmentally-sound Urban Transport system: A case of Singapore. *International Journal of Vehicle Design*, **14** (5/6), 431-444.

Ang, B.W., 1995: Relieving Urban Traffic Congestion: The Singapore Area Lincensing Scheme. *The Journal of Urban Technology*, **2**(3),1-18.

Ang, B.W., 1996a: *Transport Demand Management in Singapore and its Energy and Air Pollution Impacts*. Proceedings on High-Level Panel on Transportation, Urban Pollution and Energy Efficiency, Fourth Session of the Commission on Sustainable Development, United Nations, New York, NY, 23 April.

Ang, B.W., 1996b: Urban Transportation Management and Energy Savings: A case of Singapore. *International Journal of Vehicle Design*, **17**, 1-12.

Balashov, B., and A. Smith, 1992: CAO Analyses Trends in Fuel Consumption by World's Airlines. *ICAO Journal*, August.

Davidson, O.R. 1992: *Transport Energy in Sub-Saharan Africa: Options for a Low-Emission Future*. Report No.267, Princeton, NJ.

Davidson, O.R, 1994: Opportunities for Energy Efficiency in the Transport Sector, Chap 8. In *Energy Options for Africa*, Zed Press.

Davidson,O.R, 1998: Transfer of Renewable Energy Technologies for Sustainable Development: Opportunities for Developing Countries. The European Network for Energy Economics Research (ENR) *ENER Bulletin*, **22**.

DeCicco, J., and M. Ross, 1993: *An Updated Assessment of the Near-Term Potential for Improving Automotive Fuel Economy*. American Council for an Energy-Efficient. Washington, DC/Berkeley, CA.

DeLuchi, M.A., 1993: *Emissions from Greenhouse Gases from the Use of Transportation Fuels and Electricity, Vol.2*, Appendixes A-S, ANL/ESD/TM-22. Center for Transportation Research, Argonne National Laboratory, IL.

Dimitriou, H., 1990: *Transport Planning for Third World Cities*.Routledge, London, UK.

ECMT/OECD, 1995: *Urban Travel and Sustainable Development*. Organisation for Economic Co-operation and Development and European conference of Ministers of Transport. Paris.

Ellis J. & Teanton, 1997: Recent trends in energy-related CO_2 emissions. *Energy Policy*, **26**(3), 159-166.

ETSU, 1994: Appraisal of UK Energy Research, development, Demonstration and Dissemination. *Energy Technology Support Unit, Vol.15*. Transport Sector. HMSO, London.

EUROSTAT, 1998: *European Union Direct Investment Data, 1998 edition*. Office for Official Publication of the European Communities.

Faiz, A., C.Weaver, and M. Walsh, 1996: *Air Pollution from Motor Vehicles: Standards and Technologies for Controlling Emissions*. World Bank, Washington, DC.

Gakenheimer, R., 1993: Land Use/TransportationPlanning: New Possibilities for Developing and Developed Countries. *Transportation Quarterly*, Eno Transportation Foundation, Landsdowne, VI.

Gate/GTZ, 1998: Gate technology and development: Mobility for the Majority. D 13548 F No.3/1998, July-September.

GEF, 1999: Promoting Environmentally Sustainable Transport. September.

Gerybadze, A.,1994: Technology forecasting as a process of organizational intelligence. *R&D Management*, April, V24 N2, 131 (10).

Goldemberg, J., and I.C. Macedo, 1994: Brazilian alcohol program; an overview. *Energy for Sustainable Development*, **1**(1), 17-22.

Gougeon, P., and J. Gupta (eds.), 1997: *Contemporary Issues in Technology Transfer: Theories and Practice from a Management Perspective*. ESKA Publishing, Paris.

Greene, D.L., and K.G. Duleep, 1993: Costs and benefits of automotive fuel economy improvement: a partial analysis. *Transportation Research A*, **27**A(3), 217-235.

Grieb, H.,and B. Simon, 1990).Pollutant emissions of existing and future engines for commercial aircraft. In *Air Traffic and the Environment-Background, Tendencies and Potential Global Atmospheric Effects*. U. Schumann, (ed) , Springer Verlag, Berlin.

Grubler, A., S.Messner, L. Schrattenholze, and A.Schafer, 1993: Emission Reduction at Global Level. *Energy*, **18**(5), 539-581.

Hilling, D., 1996: *Transport and Developing Countries*. Routledge, London and New York

ICAO, 1992: Outlook for Air Transport to the year 2001. Circular 237-AT/96, ICAO, Montreal.

IEA, 1993: *Electric Vehicles: Technology, Performance and Potential*. OECD, Paris.

IEA, 1996: *World Energy Outlook 1996*. IEA/OECD, Paris.

IEA/OECD, 1994: Energy and Environmental Technologies to respond to Global Climate Change Concerns Concerns, Paris.

IEA/OECD, 1994: Energy Technologies to Reduce CO_2 Emissions in Europe: Prospects, Competition, Synergy. Conference in Petten, The Netherlands, 11-12 April, 1994.

IEA/OECD, 1997: Energy Technologies for the 21st Century. Paris.

IIEC, 1996: *Integrated Transport Planning: Beginner's Handbook*. The International Institute for Energy Conservation Transport Program., January.

IPCC, 1996: Technologies, Policies and Measures for Mitigating Climate Change. IPCC Technical Paper 1. Intergovernmental Panel on Climate Change, Geneva.

Jourmad, R., L.Paturel, R Vidon, J-P Guitton, A-I Saber, and E.Combet, 1990: *Emissions Unitaires de Polluants des Vehicules Legers, INRETS* Report No. 116. Institut National de Recherche sur les Transports et leur Securite, Bron, France.

Kuhfeld, H., 1998: Global rise in fuel consumption by the transport sector undermines aim of reducing CO2 emissions.

Lee, H., and Kim, J.,1994: The Role of Technology Transfer in Abating CO_2 Emissions: The Case of the Republic of Korea. In *Climate Change in Asia and Brazil*. R.K.Pachauri, Bhandari, (eds.), Tata Energy Research Institute, India.

Magezi, S.A.K., 1998: Mitigating Transport Sector GHG Emissions: Options for Uganda. African Energy Policy Research Network (AFREPREN), Working Paper No. 163. Nairobi.

Martin, D.J., and L.A.Michaelis, 1992: *Research and Technology Strategy to Help Overcome the Environmental Problems in Relation to Transport. SAST Project No.3. Global Pollution Study*. Report EUR-14713-EN. Commission of the European Communities Directorate-General for Science, Research and Development. Brussels and Luxembourg.

Michaelis, M., 1996: The Transport Sector in OECD Europe. In *Energy Technologies to reduce CO_2 emissions in Europe*. IEA/OECD, Paris.

Michaelis, L., and O.R. Davidson, 1996: GHG mitigation in the Transport Sector. *Energy Policy*, **24** (10/11).

Nakicenovic, N., and D.Victor, 1993: Technology Transfer to Developing Countries. *Energy*, **18**, (5), 523-538.

Newman, P., 1992: Policies to Influence Urban Travel Demand. Paper at OECD Project on Urban and Sustainable Development. OECD, Paris.

Newman, P., and J. Kenworthy, 1989: *Cities and Automobile Dependence: An International Sourcebook*. Gower Publishing Co., UK.

NRC, 1992: *Automotive Fuel Economy: How far should we go?* National Academy Press, Washington, DC.

OECD, 1995: *Motor Vehicle Pollution: Reduction Strategies beyond 2010*. OECD, Paris.

OECD, 1997: *Report on Phase II of the OECD Project on Environmentally Sustainable Tr*ansport. OECD, Paris.

OECD/IEA, 1997: CO_2 Emissions from Fuel Combustion: 1972-1995. OECD/IEA, Paris.

Rabinovitch, J., and J. Leitman, 1996: Urban Planning in Curitiba. *Scientific American*, March 1996.

Pacudan, R., 1998: Transport, Energy and the Environment in Singapore: An Institutional Perspective. *Cities*, **14**,165-168.

Pacudan, R.B., 1999: Transportation, Greenhouse Gas Emissions and Institutions in Asia. *World Resources Review1999*.

Pezoa, P., 1995: Telepeaje, el Gran Cobrador. *La Nacion* (March 14) Santiago, Chile.

Philpott, J., 1997: Transport: Lets move toward a cleaner path, E-Notes. *Newsletter of the International Institute for Energy Efficiency (IIEC)*, December 1997.

Pischinger, R., and S. Hausberger, 1993: Measures to Reduce Greenhouse Gas Emissions in the Transport Sector . Institute for Internal Combustion Engines and Thermodynamics, Technical University, Graz, Austria.

PTRC, 1991: Urban Transport in Developing Countries: Lessons in innovation. PTRC Education & Research Services Ltd., UK.

Sarwar, S.N., M.A. Quoddus, and M.I.Khan, 1999: Environmental Impact of Compressed Natural Gas (CNG) Used as Transport Fuel. *Asia Pacific Monitor*, **16**(1), Jan-Feb.

Simon, D., 1996: *Transport and Development in Third World*. Routledge, London and New York.

Schipper, L. and Marie-Lilliu, 1998: Transportation and CO2 Emissions: Flexing the link A path for the World Bank. World Bank Document.

Tanja, P.T, W. Clerx, C.G.van Ham, T.J.de Ligt, A.A. Mulders, R.C. Rijkeboer, and P. van Sloten, 1992: *EC Policy Measures aiming at Reducing CO_2 Emissions in Transport*. TNO Policy Research., Delft.

Tsuchiya, H., 1997: *Key Technology Policies to Reduce CO_2 Emission in Japan*. WWF (Worldwide Fund for Nature), Japan.

UNCTAD, 1998: *The Financial Crisis in Asia and Foreign Direct Investment*. A joint report by the United Nations Conference on Trade and Development (UNCTAD) and the International Chamber of Commerce.

UNCTAD, 1999: *World Investment Report 1999: Foreign Direct Investment and the Challenge of Development*. United Nations Conference on Trade and Development (UNCTAD), Geneva.

UNEP, 1998: Sustainable Business: Economic development and environmentally sound technologies. Regency Corporation Limited/United Nations Environment Programme, Nairobi, Kenya.

United Nations, *United Nations Statistical Yearbook* (various issues).

Walsh, M.P., 1993: Global Transport Scenarios: Technical Annex. In *Towards a Fossil Free Energy Future: The next Energy Transitio*n. Greenpeace International, Amsterdam..

Wootton, H.J., and M.L. Poulton, 1993*: Reducing carbon dioxide emissions from passenger cars to 1990 levels*. TRL Project Report PA 3016, TRL Crowthorne, UK.

World Bank, 1992: *World Development Report 1992: Development and the Environment*. World Bank, Washington, DC.

WRI/UNEP/UNDP/World Bank, 1996: *World Resources 1996-97*. World Resources Institute, United Nations Environment Programme, United Nations Development Programme and The World Bank. Oxford University Press, New York, NY.

9

Industry

Coordinating Lead Authors:
ERNST WORRELL (THE NETHERLANDS), MARK LEVINE (USA)

Lead Authors:
Rene van Berkel (The Netherlands), Zhou Fengqi (China),
Christoph Menke (Germany), Roberto Schaeffer (Brazil), Robert O. Williams (United
Kingdom)

Contributing Authors:
Sanghoon Joo (South Korea), Xiulian Hu (China)

Review Editors:
Prosanto Pal (India), Doug McKay (United Kingdom)

CONTENTS

EXECUTIVE SUMMARY

The industrial sector is extremely diverse and involves a wide range of activities. Aggregate energy use and emissions depend on the structure of industry, and the energy and carbon intensity of each of the activities. The structure of industry may depend on the development of the economy, as well as factors like resource availability and historical factors. In 1995, industry accounted for 41% (133 EJ[1]) of global energy use and up to 43% of global CO_2 emissions. Besides CO_2 industry also emits various other GHGs. Although the efficiency of industrial processes has increased greatly during the past decades, energy efficiency improvements remain the major opportunity to reduce CO_2 emissions. Potentials for efficiency improvement and emission reduction are found in all processes and sectors. In the short term, energy efficiency improvement is the major GHG reduction measure. Fundamentally new process schemes, resource efficiency, substitution of materials, changes in design and manufacture of products resulting in less material use and increased recycling can lead to substantial reductions in GHG emissions. Future reductions in GHG emissions are technologically feasible for the industrial sector of OECD countries if technologies comparable to that of efficient industrial facilities are adopted during stock turnover. For Annex I countries with economies in transition (CEITs), GHG reducing options are intimately tied to the economic redevelopment choices and the form that industrial restructuring takes. In developing countries large potentials for adoption of energy efficient technologies exist as the role of industry is expanding in the economy.

In industry, GHG emission reduction is often the result of investments in modern equipment, stressing the attention to sound and environmentally benign investment policies. Industrialisation may affect the environment adversely, stressing the need for the transfer of clean technologies to developing countries. Technology transfer is a process involving assessment, agreement, implementation, evaluation and adjustment, and replication. Institutional barriers and policies influence the transaction process, as well as the efficiency of the transfer process. Developing countries suffer from all barriers that inhibit technology transfer in industrialised countries plus a multitude of other problems. Investments in industrial technology (i.e. hardware and software) are dominated by the private sector. Foreign direct investment is increasing, although concentrated on a small number of rapidly industrialising countries. These countries may impact regional industrial development patterns, as seen in Southeast Asia. Private investment in other developing regions is still limited, although increasing. Public funding (in industrialised and developing countries) for technology development

and transfer, although still important, is decreasing. Funding for science and technology development is important to support industrial development, especially in developing countries. Public funding in the industrial sector, although small in comparison to private funding, remains important.

It is essential that policies provide a clear framework for technology transfer. An effective process for technology transfer will require interactivity between various users, producers and developers of technology. The variety of stakeholders makes it necessary to have a clear policy framework as part of an industrial policy for technology transfer and cooperation, both for a technology donor and recipient or user. Such a framework may include environmental, energy, (international) trade, taxation and patent legislation, as well as a variety of well-aimed incentives. Policymakers are responsible for developing such a comprehensive framework. The interactive and dynamic character of technology transfer stresses the need for innovative and flexible approaches, through partnerships between various stakeholders, including public-private partnerships. There is a strong need to develop the public and private capacities to assess and select technologies, in particular for state-owned and small and medium-sized industries. Stakeholders (policymakers, private investors, financing institutions) in developing countries have even more difficult access to technology information, stressing the need for a clearinghouse of information on climate change abatement technology, well integrated in the policy framework. To be successful, long-term support for capacity building is essential, stressing the need for public support and cooperation of technology suppliers and users.

Adaptation of technology to local conditions is essential, but practices vary widely. Countries that spend on average more on adaptation seem to be more successful in technology transfer. As countries industrialise the technological capabilities increase rapidly, accelerating the speed of technology diffusion and development, and demonstrating that successful technology transfer includes transfer of technological capabilities.

9.1 Introduction[2]

The industrial sector is extremely diverse and involves a wide range of activities including the extraction of natural resources, conversion into raw materials, and manufacture of finished products. We define the industrial sector as industry including the minerals processing industries. The sub-sectors that account for roughly 45% of all industrial energy consumption are iron and steel, chemicals, petroleum refining, pulp and paper, and cement. These industries are generally concerned with the transformation of raw material inputs (e.g. iron ore, crude oil, wood) into usable materials and products for an economy. Due to the wide variety in activities, energy demand and GHG emissions vary widely. Hence, the aggregate energy use and emissions depend on the structure (or specific set of activities) of industry, and the energy and carbon intensity of each of the activities. The structure of industry may depend on the phase of the economy, as well as many other factors like resource and technology[2] availability as well as historical factors.

In 1995 industry accounted for 41% (131 EJ) (Price *et al.*, 1998) of global energy use and up to 47% of global CO_2 emissions (IPCC, 1996a). Besides CO_2, industry also emits various other GHGs, i.e. CFCs, HFCs, HCFCs, CH_4, N_2O, PFCs, CF_4, C_2F_6, and SF_6 (IPCC, 1996a). Between 1971 and 1990, industrial energy use grew at a rate of 2.1% per year, slightly less than the world energy demand growth of 2.5% per year. This growth rate of industrial energy use has slowed in recent years, falling to an annual average growth of 0.2% between 1990 and 1995, primarily because of declines in industrial output in the CEITs. Energy use in the industrial sector is dominated by the OECD countries, which account for 44% of world industrial energy use. Non-Annex I countries and CEITs used 37% and 20% of 1995 world industrial energy, respectively. Industrial production is growing at a fast rate in non-Annex I countries. The trends in industrial energy use and CO_2 emissions are depicted in Table 9.1.

Industrial production is an important engine to increase the economic activity, generate employment, and build up the infrastructure in developing countries. Investment in industry seems to have a stronger relation with economic growth than investments in other sectors (UNIDO, 1997). This can also be observed from the growing importance of industry in, and its contribution to the growth of a developing nation's economy (UNIDO, 1997). High industrial growth also promotes technological change (UNIDO, 1997). Capital investment in industry is important to achieve economic welfare in developing countries. Capital relates to physical (e.g. equipment), human (e.g. education) and technological capital (e.g. science, R&D). Industrialisation builds on the contribution of science and technology, as is evidenced by the Chinese economic development in the past decades (Song, 1997). However, industrial technology should fit the needs of the users in developing economies. Technologies developed for a specific industrial infrastructure (e.g. raw materials used (UNEP, 1997), relative shares of production costs) may not always be the right choice for another one, as is shown by examples of industrial technology applied in Tanzania (Yhdego, 1995) and India (Schumacher and Sathaye, 1998). Adaptation and development of technology to suit the needs is an essential step in the successful transfer of technology. Hence, technology transfer is a process, involving the trade and investment in technology, the selection (e.g. new, second-hand), adoption, adaptation, and dissemination of industrial technology, and, last but not least, capacity building, as science and technology are strongly related (Song, 1997) in the development of an industrial infrastructure.

Future growth of basic industries will, to a large extent, occur in developing countries. While developing countries are the most important markets for new and energy efficient processes, technology is still primarily developed in industrialised countries, despite the fact that the absolute demand for such technologies is stagnating or relatively low. Industrialised countries will be less favourable theatres for the innovation of technologies in the primary materials process-

Table 9.1.	Historical Energy Use in Industry (EJ). Primary energy consumption is calculated using a 33% conversion efficiency for electricity generation for all years and regions. Source: Price et al. (1998)							
REGION	**TOTAL INDUSTRIAL ENERGY USE**					**AVARAGE ANUAL GROWTH RATE (%/ANNUM)**		
	1960	1971	1980	1990	1995	1960/1990	1971/1990	1990/1995
OECD	28	49	55	54	57	2.3	0.6	0.9
EE-FSU		26	34	38	26		2.0	-7.3
Developing Countries		13	24	37	48		5.4	5.0
World		88	114	129	131		2.1	0.2

ing industries, if there are limited applications for such in industrialised countries. This development stresses the need both for technology adaptation to the prevailing conditions in developing countries, and intensified collaboration between suppliers and users of new industrial processes. Technology transfer needs to be studied within these perspectives. However, it seems that environmentally sound technologies do not transfer as rapidly as e.g. information technology, particularly with regard to developing countries. Also, the rapidly increasing role of transnational companies and foreign direct investment (UNCTAD, 1997) may change the patterns of technology transfer (see section 9.3). These issues warrant a specific study of transfer of environmentally sound industrial technology, with an emphasis on GHG abatement technologies.

---

[2] Cases relevant from the Case Studies Section, Chapter 16, for Chapter 9 fall in the category of technologies, new innovations and corporate leadership. Cases reflecting a technological initiative are Ecofrig in India (case 4), heat recovery in Chinese steel industry (case 15), ODS in Mexico (case 17), Biomass boilers in the Baltics (case 18), CFC-free refrigerators in Thailand (case 23). Mobil (case 13) reflects corporate leadership. The Bamboo reinforced cement case (case 9) represents the new innovations in technologies and an application from North-South transfer to South-South transfer.

In this chapter we describe the experiences with various forms of technology transfer. After a brief summary of the technologies for GHG mitigation, mainly based on previous IPCC reports, we discuss the trends in technology transfer from a 'macro' perspective (section 9.3). In section 9.3 we describe the trends from an economic perspective, and study magnitude and directions, as well as sources of investment and technology. In sections 9.4 and 9.5 we study the processes of technology transfer between and within countries, based on case-study material and other literature sources. Next, there is an evaluation of the analysed material and a description of the main lessons learned in section 9.6, and this is followed by a summary.

9.2   Climate Mitigation and Adaptation Technologies

Future reductions in CO_2 emissions are technologically feasible for the industrial sector of OECD countries if technologies comparable to the present generation of efficient industrial facilities are adopted during regular stock turnover (replacement) (IPCC, 1996a). For Annex I countries with economies in transition, GHG reducing options are intimately tied to the economic redevelopment choices and the form that industrial restructuring takes. In developing countries large potentials for adoption of energy and resource efficient technologies exist as the role of

industry is expanding in the economy. Although the efficiency of industrial processes has increased greatly during the past decades, energy efficiency improvements remain the major opportunity (IPCC SAR, 1996) for reducing CO_2 emissions. Efficient use of materials may also offer significant potential for reduction of GHG emissions (Gielen, 1998; Worrell *et al.*, 1997) (see Table 9.2). Much of the potential for improvement in technical energy efficiencies in industrial processes depends on how closely such processes have approached their thermodynamic limit. For industrial processes that require moderate temperatures and pressures, such as those in the pulp and paper industry, there exists long-term potential to maintain strong annual intensity reductions. For those processes that require very high temperatures or pressures, such as crude steel production, the opportunities for continued improvement are more limited using existing processes. Fundamentally new process schemes, resource efficiency, substitution of materials, changes in design and manufacture of products resulting in less material use and increased recycling can lead to substantial reduction in energy intensity. Furthermore, switching to less carbon-intensive industrial fuels, such as natural gas, can reduce GHG emissions in a cost-effective way (IPCC, 1996a; Worrell *et al.*, 1997). In addition to stock replacement, which is an excellent opportunity to save energy, there are many low cost actions that can be adopted as part of good management practices. Table 9.2 provides categories and examples of tech-

Table 9.2	Categories and selected examples of practices and technologies to mitigate GHG emissions in the industrial sector, based on IPCC (1996 a), WEC (1995), Worrell et al. (1997), and IPCC (1996 b).			
OPTION	**MEASURES**	**CLIMATE AND OTHER ENVIRONMENTAL EFFECTS**	**ECONOMIC AND SOCIAL EFFECTS**	**ADMIONISTRATIVE, INSTITUTIONAL AND POLITICAL CONSIDIRATIONS**
END USE	National Cleaner	National Cleaner		National Cleaner
Energy Efficiency Gains • more efficient end uses • reduction of energy losses	• Market Mechanisms • Voluntary Agreements • Energy Price Reform • Information programmes • International Corporation	• Savings on CO_2 emissions • Reduction of air pollution	• Highly cost-effective • Restructuring tax system to taxing resource use • Equity issues in providing energy services	• Major effort from industry • Change regulatory and tax systems • Coordination • International coordination and monitoring
Process Improvement • process integration • reduction non-CO_2 emission	• Voluntary Agreements • Regulatory Measures	• Savings on CO_2 and non-CO_2 GHG emissions • Reduction of air pollution	• Highly cost-effective	• Major effort from industry • See above
New Technologies and Processes • new production technologies, e.g. steel, chemicals, pulp	• RD&D • International Corporation	• Savings on CO_2 and non-CO_2 GHG emissions • Reduction of air pollution	• R&D investments • Cost-effective on the long-term • Transform industrial infrastructure and basis	• Funding • Industry, academic and government labs • Modest changes in administrative factors
CONVERSION				
Cogeneration • CHP using gas turbines, fuel cells	• Voluntary Agreements • Regulatory Measures • Market Mechanisms • RD&D	• Reduction in CO_2 emissions • Reduction in air pollution	• Highly cost-effective • Some industry restructuring (PPI)	• Major effort from industry • Changes in regulatory regimes • Siting for optimal use
Fuel Switching • natural gas • biomass • solar (drying, water heating)	• Regulatory Measures	• Reduction in CO_2 emissions • Reduction in air pollution	• Highly cost-effective • Internalizing external costs may hasten shift • Trade-off with other uses (e.g. biomass)	• Major effort from industry • Opposition of producers fuels being displaced
MATERIAL USE				
Efficient Material Use • efficient design • substitution • recycling • material quality cascading	• Voluntary Agreements • Market Mechanisms • Regulatory Measures • RD&D	• Reduction in CO_2 emissions • Reduction in air pollution • Reduction in solid waste and primary resource use	• Highly cost-effective • Decreased use of primary resources • Dislocations in existing industry • Job creation near product users	• Major effort from industry • Engage all actors in problem solving • Regulatory changes • Opposition to regulatory changes

nologies and practices to mitigate GHG emissions in the industrial sector (based on IPCC (1996a), WEC (1995), and Worrell *et al.* (1997)). This summary is by no means comprehensive, but rather an indication of the wide range of possibilities that exist within and among industrial sectors for reducing GHG emissions. For more specific technologies and information, the reader is referred to a wide body of literature, as has been described in the references mentioned above.

The sensitivity of industry to climate change is widely believed to be low, compared to that of natural ecosystems (IPCC, 1996a). Climate change, however, may have (local and regional) impacts on availability of resources to industry as a result of changes in average temperature, precipitation patterns and weather disaster frequencies, in particular, availability of water (as a resource, energy source or for cooling) and renewable inputs (industrial and food crops) may be affected. Industry thus also needs to adapt to climate change, depending on local conditions, e.g. by improving its water efficiency, by strengthening its flexibility to cope with fluctuations in input availability, by reducing the vulnerability of production for weather conditions, and through proper siting and adaptations of industrial facilities. This may include a wide variety of measures such as protecting industrial sewage cleaning installations from flooding by storm water, reducing dependence on water use for various purposes, and siting away from vulnerable coastal areas. Fluctuating water levels at sea or rivers may also affect the steady supply of resources to industrial facilities, as evidenced by the impact of extremely high water levels on river bulk transport on the Rhine river system. There are already examples in which water scarcity has driven innovation into water efficient industrial technologies, which have significant energy efficiency improvements (and hence GHG mitigation potential) as spin off. For example, water scarcity was identified as a potential threat to the textile industry in Surat (India) in the early 1990s. This incited a local engineering firm to invest in the development of dyeing machines customised for local fabric quality. Water and energy consumption are only approximately 1/3 of the water and energy consumption of comparable dyeing machines available on the international market, while the investment is much lower due to local industry. Several hundreds of dyeing machines are now being installed annually in the Surat region, and efforts are underway to market the technology in other regions and abroad (Van Berkel *et al.*, 1996).

9.3 Magnitude of Current Technology Transfer

As countries develop from an agrarian society to an industrial urban economy the economic structure of a developing nation goes through a transition process, as described by Kuznets (1971). The structure of the economy is strongly dependent on the stage of development, and hence the technology needs. A World Bank study confirmed the transition patterns for a large number of economies (Syrquin and Chenery, 1988). The transition process may not be smooth (especially in short periods), and may follow various paths. Syrquin and Chenery (1988) showed that the performance of the economy is associated with large size, a manufacturing orientation and with a higher degree of openness. The

smaller the economy, the more it relies on the open character of the economy. However, alternative paths may be successful too, as evidenced by the development of economies as in Korea, where industries matured under economic protection (Lee, 1997).

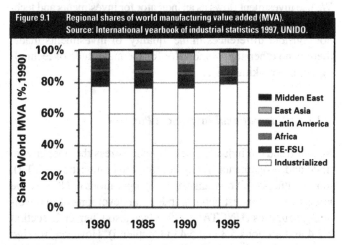

Figure 9.1   Regional shares of world manufacturing value added (MVA). Source: International yearbook of industrial statistics 1997, UNIDO.

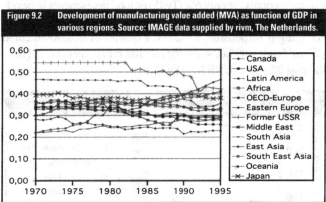

Figure 9.2   Development of manufacturing value added (MVA) as function of GDP in various regions. Source: IMAGE data supplied by rivm, The Netherlands.

The rate of technological change strongly affects the rate of investment and the productivity and *vice versa*. Investment in modern equipment, evidenced by the economic growth in newly industrialised economies in East Asia, is seen as a more important contribution to growth than other investments (UNIDO, 1997). The growing industrial production in Asia, especially China (5% of world manufacturing value added (MVA) in 1995), is shown in Figure 9.1. Figure 9.1 also shows that world MVA is still dominated by the industrialised countries, while MVA of the economies in transition has decreased. The share of MVA from other regions of the world has remained stable. The regional development of the industrial sector is depicted in Figure 9.2. It shows that the importance of the industrial sector in the regional economy is increasing in most developing regions.

Although the growth pattern of the industrial sector may differ between countries, generally the growth is associated with the use of capital intensive technology such as in the raw material based industries. China, India (Kaplinsky, 1997) and Korea are examples of this pattern, though in different stages of development. There is, however, considerable debate about the importance of the industrial sector in the economic development process. The growing importance of the services sector in some developing economies (Asia, Latin America) generates an increasingly larger part of total economic growth (World Bank, 1998). Different investment patterns influence industrial growth, structure and technology adoption.

We use investment flows as an indicator for investments and technology transfer. While, recognising that investment flows do not consider differences in the 'quality' of investments made, there is no other simple indicator for the magnitude of technology transfer taking place.

9.3.1   International Investment Patterns

Recent trends in industrial development stress the openness in trade and investments. Today, foreign direct investment (FDI), joint ventures (JV) by transnational corporations (TNC) are the largest foreign investments in industrial development in developing countries (UNCTAD, 1997) (see also Chapter 2, section 2.2.4 and sections 5.3 and 5.4 in Chapter 5). However, foreign capital accounts for only 6% (1995) of total investments in developing countries (UNIDO, 1997). In developing countries public spending is responsible for about a quarter of national income (World Bank, 1997a), while the role has relatively declined over the past 25 years (UNIDO, 1997).

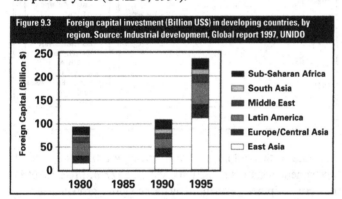

Figure 9.3   Foreign capital investment (Billion US$) in developing countries, by region. Source: Industrial development, Global report 1997, UNIDO

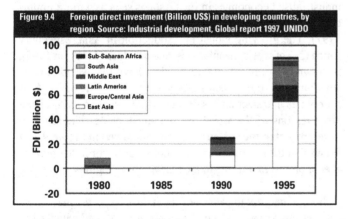

Figure 9.4   Foreign direct investment (Billion US$) in developing countries, by region. Source: Industrial development, Global report 1997, UNIDO

Transnational corporations' spending in international investments increased from less than US$100 billion (B$) in the early 1970s to over US$1.4 trillion in 1996 (UNCTAD, 1997). The majority of the funds is still spent in industrialised countries, but an increasing part is spent in developing countries. Foreign industrial investment in developing countries has increased sub-

stantially, especially since 1990, as shown in Figure 9.3. Figure 9.3 shows that foreign industrial capital spending is concentrated in two regions, East Asia and Latin America. These regions have experienced successful industrial growth in the last decade, although concentrated in a small number of countries. Foreign direct investment, a part of the international investments, has grown to 350 B$ in 1996 (UNCTAD, 1997), of which 34% was invested in developing countries (see Figure 9.4).

Previous periods of high growth in FDI were mainly directed to oil producing countries. The current growth of FDI seems to be more diverse, although there is a strong geographic concentration in current FDI. Of the 129 B$ FDI in developing countries, 42 B$ was spent in China, followed by 10 B$ in Brazil (UNCTAD, 1997). Favoured regions are Asia and Latin America, and there are signs of increasing FDI in Africa, although still limited. Important is the increasing FDI from developing countries, especially Asia, which increased to 52 B$ (UNCTAD, 1997). In Asia regional investments seem to be the main driver for industrialisation. FDI in the former Central and Eastern Europe are relatively constant at 12-14 B$, but also concentrated in a few countries (Poland and the Czech Republic). Also, FDI is concentrated in a relatively small number of TNCs. Only a few (from Korea and Venezuela) of the top 100 TNCs are based in developing countries, yet TNCs from developing countries are growing in importance. TNCs seem to be most important in the electronics, automotive, and chemical industries, as well as petroleum and mining. TNCs seem to be more productive than domestic companies (UNIDO, 1997), which may be partly due to more efficient production technologies and practices used. The role of TNCs in industrial development is generally seen as positive, although negative effects may arise from TNC involvement if the market power of the TNC is high.

It appears that future trends in FDI will be sustained, as international trade seems to gain in importance, and as countries are liberalising trade and investment. FDI aims at accessing and developing markets, whereas portfolio equity investment (PEI) is more directed to participating in local enterprises. Following the globalisation trend PEI is also growing, but tends to be more centred on developed markets and to be more fluid. PEI is estimated at 45 B$ (1995) (UNCTAD, 1997).

Small and medium sized enterprises (SMEs) have less access to international financing, and hence rely more on domestic capital and public spending. Even small investments in cleaner production and GHG abatement projects in SMEs are often not done, due to lack of capital, poorly developed banking systems, lack of appropriate financing mechanisms, lack of knowledge (both within the industrial and the financial sectors), technology risks, and management's unwillingness to borrow funds (Van Berkel and Bouma, 1999). These barriers reduce the availability of capital, stimulating investors to keep investment costs low, which may result in the purchasing of second-hand equipment, low quality products, or equipment without modern controls and instrumentation. This may lead to higher operating costs, and environmental impacts. Lack of access to capital and credit is seen as the strongest barrier to the development of SMEs (UNIDO, 1997). Various developing countries have experimented and applied financing schemes for SMEs, e.g. Ecuador, Indonesia, Korea, Malaysia, Pakistan and Tanzania, with varying rates of success (UNIDO, 1997). Trends in foreign investments are relatively easy to monitor. However, domestic capital spending in developing countries/CEITs, especially by SMEs, is more difficult to monitor. Research in some developing countries shows that especially SMEs contribute for a large part to industrial employment, and that in LDCs industrial employment is found in rural areas (Little, 1987; Putterman, 1997; UNIDO, 1997). However, this does not necessarily mean that SMEs are more efficient with regard to capital and resource use (Little, 1987). There is growing evidence that SMEs in some countries may be less efficient with respect to resource use (World Bank, 1997b). Sound market conditions are crucial to create a competitive market in which innovation by SMEs in process technology is stimulated.

The above trends in industrial investments are difficult to translate to technology choice and transfer. It is obvious, though, that increasing international investments influences the rate of technology transfer, although it gives no information on the way and on what technology is transferred. Generally, the majority of investments in many developing countries seem to be in low-technology industries, though the share of high-technology industries is increasing (UNIDO, 1997). Also, there is no hard information available on the role of the markets for environmentally sustainable technologies (including greenhouse gas abatement) (Luken and Freij, 1995).

9.3.2 Official Development Assistance and Other Flows

Annual official development assistance (bilateral and multilateral) has averaged about 60 B$ since 1990 (UNIDO, 1990), of which only a small part is invested in industrial development (see also section 2.2.2 in Chapter 2 and 5.2 in Chapter 5 on ODA and public sector finance). In the 1970s about 10% of foreign aid was invested in industrial infrastructure, but this decreased sharply during the 1980s and 1990s. Approximately 2% of bilateral and 6% of multilateral aid is spent on industrial development (UNIDO, 1997), or US$820 million (M$). Major recipients of development aid earmarked for industry are low to medium income countries, e.g. Bangladesh, China and Indonesia. In 1995 45% of the total budget was spent in these three countries (UNIDO, 1997). Official development assistance funds have been reduced in real terms in the past decade. Future trends in development aid are unclear.

Other financial flows include development loans and export credits, used primarily to finance the export of capital goods and equipment. In 1994 the lending of export credit agencies to developing countries and CEITs has increased to 420 B$ (UNIDO, 1997), but it is unclear what part is spent on industrial development and technology. Export credits and loans seem to be heavily concentrated in large low-income but creditworthy countries, and increasingly in countries that have (gained) access

to international financial markets. However, the majority of low income developing countries have no access to these funds. Finally, the lending by multilateral financing banks to industry has decreased from 8.5 B$ in 1990 to about 4 B$ in 1994 (UNIDO, 1997). The reduction seems to be due to the reduced role of project lending by these banks, as well as the increased access of developing countries/CEITs to international capital markets (UNIDO, 1997).

9.3.3 Role of Research and Development

Scientific and technical capability are crucial to the economic and industrial development of developing countries (Rama Rao, 1997; Song, 1997; Suttmeier, 1997; also section 2.3 in Chapter 2 and sections 4.3 and 4.12 in Chapter 4 discuss the role of R&D). Technology transfer is defined as the transfer and development of "hardware" and "software". The "software" may include scientific and engineering knowledge as well as managerial and operational skills. Direct investment in industrial R&D may be included in the investment figures discussed above. In industrialised countries the private sector is often the largest investor in R&D. However, in developing and transitional countries, the public sector is the largest contributor, e.g. in China (Song, 1997), the Czech Republic (Moldan, 1997) and India (Rama Rao, 1997; Tripathy, 1997). Although difficult to estimate, the R&D funds allocated to environmental technology are only a small part of the total industrial technology R&D budget. Energy R&D budgets of OECD countries have declined in past decades (Williams and Goldemberg, 1995). Less than 6% of the total energy R&D budget in IEA countries was spent on energy efficiency (incl. industrial technology), whereas most is allocated to nuclear technology R&D (IEA, 1994). Scientific knowledge and R&D are getting more and more internationally oriented, as evidenced by foreign direct investment in R&D. It is estimated that foreign corporation spending in the U.S. in 1994 amounted to 15 B$, or 15% of total industrial R&D spending (Florida, 1997). Generally, FDI in R&D is comparatively small, mostly directed to support local industry. However, FDI in R&D is growing rapidly, particularly in the U.S., and also the focus is changing to developing new products, obtaining information on local scientific developments and access to local human capital (Florida, 1997). International R&D collaboration can be an effective means of technology transfer (see, for example, Case Study 4, Chapter 16), and recent initiatives like the Climate Technology Initiative (CTI) can enhance this collaboration. Preliminary analysis seems to suggest that newly industrialised countries seem to increase the generation of scientific and technological knowledge within their countries, although the majority of knowledge is still generated in the industrialised world (Amsden and Mourshed, 1997). The type of scientific output and knowledge may vary by country. In India, in 1994 to 1995, total research expenditure is estimated to have been 0.8% of GDP (Rama Rao, 1997) and in China it was estimated at 0.5% in 1995 (Song, 1997), while total spending in science and technology development was estimated at 1.5% of GDP. The figures are slightly lower than the years before. However, no accurate information is available on the global role of and investments in scientific knowledge in developing countries.

9.4 Programmes and Policies for Technology Transfer within Countries

Technology transfer is a process involving assessment, agreement, implementation, evaluation and adjustment, and replication. Although technology transfer is often seen as a private interaction between two companies or trade partners, institutional barriers and policies influence the transaction process, as well as the efficiency of the transfer process. In this section we will first discuss the barriers to technology transfer, followed by a discussion of existing programmes and policies within countries. A wide body of literature discusses the barriers and policies that affect implementation and diffusion of technologies (see, *e.g.*, Worrell *et al.*, 1997). We will concentrate on the experiences of programmes with respect to environmental and energy efficient technologies in developing countries and CEITs. Developing countries and CEITs suffer from all barriers that inhibit technology transfer plus a multitude of other problems. Potential conflicts in policies and goals between sectors can act as a barrier. For example, energy costs in industrialised countries often do not reflect the total costs, but the problem is especially serious in some developing countries and CEITs, where energy is considerably underpriced, with the government providing the energy supply industries (especially electric power producers) subsidies. Recently, subsidies in many countries have been reduced, possibly due to deregulation of the energy sector. Deregulation of the power sector may help to remove energy subsidies. Rigid hierarchical structure of organisations and the paucity of organisations occupying the few niches in a given area, lead to strong and closed networks of decision makers who are often strongly wedded to the benefits they receive from the status quo (see Gadgil and Sastry (1994) for an example of efficient lighting systems).

9.4.1 Barriers to Technology Transfer

Under perfect market conditions all additional needs for energy services are provided by the lowest cost measures, whether energy supply increases or energy demand decreases. There is considerable evidence that substantial energy efficiency investments that are lower in cost than marginal energy supply are not made in real markets, suggesting that market barriers exist. We first discuss barriers to the transfer of climate change technologies that apply to all economies, followed by a discussion of additional barriers that are of particular importance to developing nations.

Decision-making processes in companies are a function of their rules of procedure, business climate, corporate culture, managers' personalities and perception of the firm's energy efficiency (DeCanio, 1993; OTA, 1993). Energy awareness as a means to reduce production costs seems not to be a high priority in many companies, despite a number of excellent examples in industry worldwide (e.g. Nelson, 1994). Cost-effective energy efficiency measures are often not undertaken as a result of *lack of information* on the part of the consumer, or a lack of confidence in the information, or high transaction costs for obtaining reliable information (Reddy, 1991; OTA, 1993; Levine *et al.*, 1995; Sioshansi, 1991). Information collection and processing consumes time and resources, which is especially difficult for small companies (Gruber and Brand, 1991; Velthuijsen, 1995). Especially in many developing countries and CEITs, public capacity for information dissemination is lacking, which suggests the importance of training in these countries, and is seen as a major barrier for technology transfer (TERI, 1997). The problem of the information gap concerns not only consumers of end-use equipment but all aspects of the market (Reddy, 1991). Many producers of end-use equipment have little knowledge of ways to make their products energy efficient, and even less access to the technology for producing the improved products. End-use providers are often unacquainted with efficient technology. In addition to a lack of information at least two other factors may be important: a focus on market and production expansion, which may be more effective than efficiency improvements to generate profit maximisation; and the lack of adequate management tools, techniques and procedures to account for economic benefits of efficiency improvements.

Limited capital availability will lead to high hurdle rates for energy efficiency investments, because capital is used for competing investment priorities. Capital rationing is often used within companies as an allocation means for investments, leading to even higher hurdle rates, especially for small projects with rates of return from 35 to 60%, much higher than the cost of capital (~15%) (Ross, 1986). In many developing countries cost of capital for domestic enterprises is generally in the range of up to 30-40%. When energy prices do not reflect the real costs of energy (without subsidies or externalities) then consumers will necessarily underinvest in energy efficiency. Especially for SMEs, capital availability may be a major hurdle in investing in energy efficiency improvement technologies due to limited access to banking and financing mechanisms, as was also shown in the evaluation of a Japanese energy audit programme for SMEs (Oshima, 1998). Energy prices, and hence the profitability of an investment, are also subject to large fluctuations. The uncertainty about the energy price, especially in the short term, seems to be an important barrier (Velthuijsen, 1995). The uncertainties often lead to higher perceived risks, and therefore to more stringent investment criteria and a higher hurdle rate. Lack of skilled personnel, especially for small and medium sized enterprises (SME), leads to difficulties installing new energy-efficient equipment compared to the simplicity of buying energy (Reddy, 1991; Velthuijsen, 1995).

In many companies (especially with the current development toward *lean* companies) there is often a shortage of trained technical personnel, as most personnel are busy maintaining production (OTA, 1993). In CEITs the disintegration of the industrial conglomerates may lead to loss of expertise and hence similar implementation problems. In most developing countries there is hardly any knowledge infrastructure available that is easily accessible for SMEs. In Brazil, the SEBRAE programme provides institutional and technical assistance for SMEs, financed through a federal industry tax. SMEs are often a large part of the economy in developing countries. Special programmes may alleviate this barrier (see below).

In addition to the problems identified above, other important barriers include (1) the "invisibility" of energy efficiency measures and the difficulty of demonstrating and quantifying their impacts; (2) lack of inclusion of external costs of energy production and use in the price of energy; and (3) slow diffusion of innovative technology into markets (Levine *et al.*, 1994; Fisher and Rothkopf, 1989; Sanstad and Howarth, 1994). Regulation can, sometimes indirectly, be a barrier to implementation of low GHG emitting practices. A specific example is industrial cogeneration, which may be hindered by the lack of clear policies for buy-back of excess power, regulation for standby power, and wheeling of power to other users. Cogeneration in the Indian sugar industry was hindered by the lack of these regulations (WWF, 1996), while the existence of clear policies can be a driver for diffusion and expansion of industrial cogeneration, as is evidenced by the development of industrial cogeneration in the Netherlands (Blok, 1993). In addition, alternative models may be found important in focusing public policy on the need to raise end-user awareness and the priority to increase energy efficiency. This is likely to be an effective route to ensuring industry takes a comprehensive view of energy efficiency.

9.4.2 Programmes and Policies for Technology Transfer

In this section we will follow the steps in the transfer process, using experiences reported in the literature, as well as case studies (see Table 9.3). The steps we follow are assessment, agreement, implementation, evaluation and adjustment, and replication. Various programmes try to lower the barriers simultaneously in some steps. A wide array of policies to increase the implementation rate of new technologies has been used and tested in the industrial sector in industrialised countries (Worrell *et al.*, 1997), with varying success rates. We will not discuss general programmes and policies (e.g. taxation, subsidies, integrated resource planning, regulation and guidelines, voluntary programmes and information programmes; see Chapter 4), but rather concentrate on specific examples in the industrial sector, with an emphasis on developing countries' experiences. With respect to technology diffusion policies there is no single instrument to reduce barriers; instead, an integrated policy accounting for the characteristics of technologies, stakeholders and countries addressed is needed. Technology diffusion is also influenced by many parameters, including capital costs, resources, productivity and resource efficiency.

Assessment

Selection of technology is a crucial step in any technology transfer. Information programmes are designed to assist energy con-

sumers in understanding and employing technologies and practices to use energy more efficiently. These programmes aim to increase consumers' awareness, acceptance, and use of particular technologies or utility energy conservation programmes. Examples of information programmes include educational brochures, hotlines, videos, audits, and design-assistance, energy use feedback and labelling programmes. Information needs are strongly determined by the situation of the actor. Therefore, successful programmes should be tailored to meet these needs. Surveys in Germany (Gruber and Brand, 1991) and the Netherlands (Velthuijsen, 1995) showed that trade literature, personal information from equipment manufacturers and exchange between colleagues are important information sources. In the United Kingdom, the "Best Practice" programme aims to improve information on energy efficient technologies, by demonstration projects, demonstrating technologies in various industrial environments and conditions, information dissemination and benchmarking (see section 5.6 in Chapter 5 on information dissemination through intermediaries). The programme has been effective in achieving cost-effective energy savings, and is now replicated in various countries (Collingwood and Goult, 1998). In developing countries and CEITs technology information is more difficult to obtain. The case studies in India (TERI,1997; Van Berkel,1998a), see Table 9.3, show various efforts to organise technology users and to collect and distribute data. These efforts seem to be successful, and have even lead to the establishment of visions on technology development (TERI, 1997). In China, visions on technology needs have also been developed.

BOX 9.1. INTERNATIONAL COOPERATION FOR ENERGY AUDITING

Information and methods to identify and assess opportunities for greenhouse gas emission abatement and energy efficiency are essential steps in the successful implementation of these practices and technologies. Energy audits for industries have been used as a tool to bridge this information gap. In India, energy audits for industry had a bad history, as historically these were often subsidised and provided at almost no cost. Often the quality of the audits was very low. Consequently, recommendations were seldom implemented by the recipient. The cooperation between Tata Energy Research Institute (TERI, New Delhi), India, and the German organisation for Technical Cooperation (GTZ) aims to strengthen the capabilities of the TERI Bangalore Centre, to provide energy audits for industry and to strengthen the capabilities to offer high quality advise to industry. The Indo-German project provided various forms of training, established an energy information centre, provided improved measuring instruments for energy audits, helped to re-organise the institution by building specialised teams for the various industrial sectors, and helped to establish South-South cooperation. The energy audit centre in Bangalore has established itself, now has nine years of experience in providing energy audits to industry in India, and has expanded from having eight to more than 25 energy experts. This has provided the critical mass for the success of the project. It is planned to replicate this process in other parts of India and other countries. Currently the Jordan-German Rational Use of Energy Project is an attempt to replicate the positive experiences from India, by twinning the Jordan RSS Institute with TERI (Menke, 1998).

Energy audit programmes are a more targeted type of information transaction than simple advertising. Industrial customers that received audits reduced their electricity use by an average of 2 to 8%, with the higher savings rates achieved when utilities followed up their initial recommendations with strong marketing, repeated follow-up visits, and financial incentives to implement the recommended measures (Nadel, 1990; Nadel, 1991; Oshima, 1998). Energy audit programmes exist in numerous developing countries, and an evaluation of programmes in 11 different countries found that on average 56% of the recommended measures were implemented by audit recipients (Nadel *et al.*, 1991). The Indo-German energy audit project (see Box 9.1) in Indian industries (Menke, 1998) confirms that 50-60% of the recommendations were implemented, resulting in energy savings of 5-15%. Moreover, energy auditing proved to be a viable self-sustaining business opportunity, as the Indian partner was well equipped and motivated.

Agreement and Implementation
Actual implementation of technologies and practices depends on the motivation of management and personnel, external driving forces, e.g. legislation and standard setting, economics (i.e. profitability), availability of financial and human resources, and other external driving forces (e.g. voluntary agreements). Environmental *legislation* can be a driving force in the adoption of new technologies, as evidenced by the case studies for India (TERI, 1997) and the process for uptake of environmental technologies in the U.S. (Clark, 1997). Energy prices often do not reflect the full costs of energy production. Higher energy prices can increase the implementation rate of efficient practices, as evidenced by the Russian case study (Avdiushin *et al.*, 1997). Market deregulation can lead to higher energy prices in developing countries and CEITs (Worrell *et al.*, 1997), although efficiency gains may lead to lower prices for some consumers. Small energy or carbon taxes have been implemented for small energy users (incl. industry) in Denmark and the Netherlands, but it is too early to evaluate the effect on GHG emissions. Energy intensive industries operating in export-oriented markets are often exempted from such taxation schemes. The Czech case study shows a scheme, somewhat similar to a "feebate", where funds from pollution fines are used to finance pollution prevention projects (Marousek *et al.*, 1998).

Direct subsidies and tax credits or other favourable tax treatments (to raise end-use energy efficiency) have been a traditional approach for promoting activities that are thought to be socially desirable. Incentive programmes need to be carefully justified to assure that social benefits exceed cost. Direct subsidies might also suffer from the "free rider" problem, where subsidies are used for investments that would be made anyway. Estimates of the share of "free riders" in Europe range from 50 to 80% (Farla and Blok, 1995), although evaluation is often difficult. An example of a financial incentive programme that has had a very large impact on energy efficiency is the energy conservation loan programme that China instituted in 1980.

This loan programme is the largest energy efficiency investment programme ever undertaken by any developing country, and currently commits 7% to 8% of total energy invest-

ment to efficiency, primarily in heavy industry. The programme not only funded projects that on average had a cost of conserved energy well below the cost of new supply, it also stimulated widespread adoption of efficient technologies beyond the relatively small pool of project fund recipients (Levine and Liu, 1990; Liu *et al.*, 1994). The programme contributed to the remarkable decline in the energy intensity of China's economy. Since 1980 energy consumption has grown at an average rate of 4.8% per year (compared to 7.5% in the 1970s) while GDP has grown twice as fast (9.5% per year), mainly due to falling industrial sector energy intensity. Of the apparent intensity drop in industry in the 1980s, about 10% can be attributed directly to the efficiency investment programme (Sinton and Levine, 1994), and a larger amount from unsubsidised efficiency investments, efficiency improvements incidental to other investments, and housekeeping measures. Economic reforms in many countries opened China's economy, which has favoured growth of light industries over heavy industries. The industrial structure has thus changed remarkably, in favour of less energy intensive sectors (World Bank, 1997b).

New approaches to industrial energy efficiency improvement in industrialised countries include voluntary agreements (VA). A VA generally is a contract between the government (or another regulating agency) and a private company, association of companies, or other institution (VA and other forms of public-private partnerships are also discussed in section 5.5 of Chapter 5). The content of the agreement may vary. The private partners may promise to attain a certain degree of energy efficiency improvement, emission reduction target, or at least try to do so. The government partner may promise to financially support this endeavour, or promise to refrain from other regulating activities. Various countries have adopted VAs directed at energy efficiency improvement (IEA, 1997). No thorough evaluations of VA schemes have been published yet. Experiences with early environmental VAs varied strongly - from successful actions to very limited impacts (Worrell *et al.*, 1997). In some cases the result of a voluntary agreement may come close to those of regulation. Voluntary agreements can have some apparent advantages above regulation, in that they may be easier and faster to implement, and may lead to more cost-effective solutions. Some NICs, e.g. Korea, also consider the use of VAs (Kim, 1998), while the Global Semiconductor Partnership is an example of an international voluntary agreement by TNCs to reduce PFC emissions, to avoid regulation (Andersen, 1998a).

Evaluation and Adjustment
Every industrial facility is unique in the process equipment used, lay-out, resources used, and organisation. Translation from a generic technology level into practical solutions within a country, sector or individual plant is needed. In UNIDO's National Cleaner Production Programme, it was found that investors only accepted the results of a technology demonstration if these are generated in a situation similar to theirs (Van Berkel, 1998b). Among other activities, the

"Best Practice" programme in the UK (and replicated in China (Dadi *et al.*, 1997), Brazil, Australia and New Zealand) demonstrates a technology in different industrial applications. Various countries have subsidy programmes under which new applications of technologies are eligible. Unless the capacity to adapt technology to the specific circumstances is developed, either in industry or technical assistance providers, investments in clean and energy efficient technology will not be successful.

Replication

Research and development can have various goals, depending on the barriers to be tackled to implement a technology. Blok *et al.* (1995) differentiate between technical development of a technology, improving the technology to reduce costs, and exploration and alleviation of barriers to the implementation of a technology. The challenge of climate change is to achieve substantial GHG emission reductions over time, which can only be reached by building (technological) capacity through sustained RD&D efforts. Large potential efficiency improvements do exist in the long term (Blok *et al.*, 1995). A recent US study (DOE, 1995) quotes many successes of energy RD&D. There is consensus among economists that R&D has a payback that is higher than

many other investments, and the success of R&D has been shown in fields like civilian aerospace, agriculture and electronics (Nelson, 1982). Still the private sector has a propensity to under invest in RD&D, because it cannot appropriate the full benefits of RD&D investments, due to "free riders" (Cohen and Noll, 1994). Companies will also under invest in RD&D that reduces costs not reflected in market prices (Williams and Goldemberg, 1995), such as air pollution damages and climate change. The example of the Waste Minimisation Cycles in India (Van Berkel, 1998a) demonstrates further development of technologies to improve performance, through a network of industries from the same industry sector to reduce some of the barriers. The Brazilian Alcohol programme is an example of indigenous technology development. Although seen as expensive due to lower oil prices since 1986 (Oliveira, 1991; Weiss, 1990), it is seen as a success in the field of technology development. Development has decreased the production costs of alcohol considerably (Goldemberg and Macedo, 1994; Macedo, 1998). Copersucar, a cooperative of sugar and alcohol producers, operates a (leading) joint research centre for agricultural and technology development (Macedo, 1998), as well as training. The centre also maintains a benchmarking programme to monitor and improve performance among members.

Table 9.3 Summary of case studies on technology diffusion programmes and policies within countries

CASE STUDY	COUNTRIES /ORGANISATION	TECHNOLOGY	TYPE				REFERENCE
			ASSESMENT	AGREEMENT& IMPLEMENTATION	EVALUATION& ADAPTATION	REPETITION	
Energy Management in Metal Manufacturing Plant	Russia	Monitoring & Control	•	•	•		Avdiushin et al.,1997
Waste Heat Recovery & District Heating	Czech Republic	Waste Heat Recovery at Rolling Mill & Distribution	•	•	•		Marousek et al.,
Energy Conservation Audit Programme for SMEs	Japan	Energy Auditing	•				Oshima, 1998
Waste Minimisation Circles	India	Improved Operation, Maintenance and Management Practices	•		•	•	Van Berkel, 1998a.
Technology Information, Forecasting and Assessment Council	India	Information Collection, Assessment and Promotion on Technologies	•				TERI, 1997

Table 9.4 Summary of case studies on technology diffusion programmes and policies between countries

CASE STUDY	COUNTRIES /ORGANISATION	TECHNOLOGY	TYPE				REFERENCE
			ASSESMENT	AGREEMENT& IMPLEMENTATION	EVALUATION& ADAPTATION	REPETITION	
National Cleaner Production Programme	UNIDO & various host countries	Training & Facilitation of Cleaner Production	•		•	•	Van Berkel, 1998b
Energy Efficiency for Large Industry as Business	Germany India	Energy Auditing & Training	•			•	Menke, 1998
COREX Smelt Reduction	Austria Korea	Advanced Ironmaking Process Technology	•	•			Joo, 1998a
Development of the FINEX Process	Austria Korea	Joint Development of new Ironmaking Process			•	•	Joo, 1998b
Pulverised Coal Injection for Blast Furnaces	USA Korea	Coal Grinding and Injection Equipment	•	•		•	Joo,1998c
Global Semiconductor Partnership	Global	Technology Development to Reduce PFC Emissions	•	•	•	•	Andersen,1998a
Vietnam Leadership Initiative	Vietnam TNCs	Technology Cooperation to Phase Out CFC Use	•	•	•		Andersen,1998b
Mexico Solvent Partnership	Mexico U.S.	Phasing out CFC use in Mexican Industry	•	•			Andersen,1998c
Dry Coke Quenching	China, Japan	Dry Coke Quenching		•			Hu et al.,1998

9.5   Programmes and Policies for Technology Transfer between Countries

As in the previous section, here we will follow the steps in the transfer process, using experiences reported in the literature, as well as case studies (see Table 9.4). We focus on the transfer of technology between countries. The steps we follow are: assessment, agreement, implementation, evaluation and adjustment, and replication. In this regard we give strong emphasis to the adaptation, assimilation and replication of technologies in developing countries and CEITs.

9.5.1   Barriers to Technology Transfer between Countries

Developing countries and CEITs suffer from the same factors that inhibit transfer of environmentally sound technologies as in industrialised countries (see section 9.4.1), plus a multitude of other problems. The problems also hinder transfer between countries.

High inflation rates in developing countries/CEITs and lack of sufficient infrastructure increase the risks for domestic and foreign investors and limit the availability of capital. Lack of capital may result in the purchasing of used industrial equipment (Sturm *et al.*, 1997), resulting in higher energy use and/or GHG emissions, as well as higher production costs. Trade in second-hand industrial equipment to developing countries and CEITs is quite common in most industrial sectors, e.g. cement, chemical, pulp & paper and steel industries. National trade and investment policies may limit the inflow of foreign capital. This might be a barrier to technology transfer (see also section 9.3). Recent liberalisation of investment regimes, in e.g. the mining industry, is seen as a way to transfer and acquire new technologies and reduce environmental damage (Warhurst and Bridge, 1997). This also applies to the role of TNCs and their role in technology transfer (see e.g. Case Study 13, Chapter 16). The technology cooperation to phase out the use of PFCs in the manufacture of semiconductors in the Global Semiconductor Partnership provides an example of cooperation between TNCs as a way to improve access of knowledge and technologies (Andersen, 1998a) within a more liberalised market, and a way to avoid command and control regulations.

Information about and assessment of technologies provided by foreign suppliers is more difficult for local investors in developing economies. Dependence on foreign suppliers may also induce risks in the case of technological support. For almost all industries the major suppliers can be found in the industrialised world, although some developing countries (e.g. China, India) or sectors (e.g. sugar cane processing) develop and supply indigenous and even advanced technologies (e.g. Korea) as well. Experience has shown that environmental considerations should be more carefully integrated into development and corporation policies. The policies in technology producing countries for transfer of environmentally sound technologies to developing countries seem to be inadequate (UN, 1998). In developing countries and

CEITs a lack of protection of intellectual property rights (IPR) may exist, which is seen as a barrier by technology suppliers (UN, 1998; see also section 3.5 on IPR). Also, technology licensing procedures may be time consuming, leading to high transaction costs. Besides the problems with technology selection and supply, inadequate environmental policies, or implementation thereof, in developing countries and CEITs may reduce the demand for such technologies.

Basically, similar problems affect the international transfer of technology, but even more severely. This illuminates the need for closer collaboration between industrialised and developing countries as well as CEITs, especially in the areas of technological innovation, strengthening of local capacity, and increased training and information. In the next section we will discuss international experiences with technology transfer, based on case studies and available literature.

9.5.2   Programmes and Policies for Technology Transfer between Countries

Energy efficiency and GHG emission abatement could be viewed as an integral component of national and international development policies. Energy efficiency is commonly much less expensive to incorporate in the design process in new projects than as an afterthought or a retrofit. In the environmental domain, we have learned that "end of pipe" technologies for pollutant cleanup are often significantly more expensive than project redesign for pollution prevention, leading to widespread use of pre-project environmental impact statements to address these issues in the planning phase. Energy efficiency should also be incorporated into the planning and design processes wherever there are direct or indirect impacts on energy use such as in the design of industrial facilities, reducing the costs for energy supply and reducing the risks of local air pollution. This has not always been the case, as shown by Callin *et al.* (1991) for the investment in a new paper mill in Tanzania. Local circumstances often limit even the small investments needed for cleaner production and GHG abatement, due to lack of capital, poorly developed banking systems, lack of appropriate financing mechanisms, lack of knowledge (both within the industrial and financial sectors), technology risks, and management's unwillingness to borrow funds (Van Berkel and Bouma, 1999). These barriers reduce the availability of capital, stimulating investors to keep investment costs low, which may result in selection and purchase of inappropriate technologies.

Most policies and programmes for the transfer of environmentally sound and greenhouse gas abatement technologies are national, and only a few are internationally oriented. Examples of the latter are the Greenhouse Gas Technology Information Exchange (GREENTIE) of the OECD/IEA, the PHARE programme of the European Union with Central and Eastern-Europe, and various bilateral programmes, e.g. US-AEP (U.S. and various Asian countries), Green Initiative (Japan), and the Technology Partnership Initiative of the UK. Most industrialised (donor) countries have policies in place, but strongly connected to (technology) interests of the donor country. Joint Implementation or Activities Implemented Jointly (JI/AIJ) may also be a useful energy efficiency promotion instrument. JI (see also Chapter 3) involves a bi- or multi-lateral agreement, in which (donor) countries with high greenhouse gas abatement costs in implementing mitigation measures in a (host) country with lower costs receive credit for (part of) the resulting reduction in emissions. Under CoP3 the Clean Development Mechanism (CDM) (see also Chapter 3) has been introduced as a means to accelerate emissions reduction and credit emission reductions from project activities in non-Annex I countries to Annex I countries. The criteria for JI/CDM are still in the process of development (Goldemberg, 1998). Most likely the projects should fit in the scope of sustainable development of the host country (without reducing national autonomy and with cooperation of the national government), have multiple (environmental) benefits, be selected using strict criteria and be limited to a part of the abatement obligations of a donor country (Jepma, 1995; Pearce, 1995; Jackson, 1995). Determination (and crediting) of the net emission reductions is a problem that stresses the need of well-developed baseline emissions (La Rovere, 1998), i.e. emissions that would occur in the absence of the project (Jackson, 1995). JI/CDM can prove to be a viable financing instrument to accelerate developments in CEITs and in developing countries, if implemented according to specific criteria (Goldemberg, 1998). Comprehensive evaluation of pilot projects is necessary to formulate and adapt these criteria, including the issue of crediting.

Assessment

Technology assessment and selection is very important. However, often the capacity is missing, or the selected technology is determined by a donor country or by available financing (e.g. bilateral export loans or tight aid). This may lead to suboptimal technology choices (Schumacher and Sathaye, 1998, Yhdego, 1995). An important arena for cooperation between the industrialised and developing countries therefore involves the development and strengthening of local technical and policy-making capacity, for example, for an assessment of (technical) needs. Large companies may be able to access information or resources or hire engineering companies more easily, like in the chemical industry (Hassan, 1997). SMEs and local companies have generally less easy access to external resources. Project-oriented agencies eager to show results commonly pay inadequate attention to the development of institutional capacity and technical and managerial skills needed to make and implement energy efficiency policy.

The Japanese Green Assistance Plan aims at supporting Japanese exports of energy efficient technologies to other Asian countries, including China and Thailand (Sasaki and Asuka-Zhang, 1997). It is not always clear how the technologies supported under this programme are selected. Hu *et al.* (1998) made a report on the transfer of dry coke quenching technology from Japan to China, as part of the Japanese Green Assistance programme and JI/AIJ. The payback period under current Chinese

conditions is 7 years (Hu *et al.*, 1998). The recipient, Capital Steel, had no choice in the technology selection, as the transfer was the product of cooperation between both governments. Projects in India (Menke, 1998; Van Berkel, 1998b), as well as Leadership Programmes under the Montreal Treaty in Thailand and Vietnam aimed at the development of the needed capacity (Andersen, 1998b; see also Box 2.1 in Chapter 2 and section 3.4 in Chapter 3 on the Montreal Protocol). The Indian projects proved to be successful, in the sense that they built active capacity assessing needs and opportunities for energy efficiency improvement and clean technologies for industries in various regions (see also Box 9.2). Formal recognition of the acquired skills in knowledge transfer seems to be important to improve the status of a programme (Van Berkel, 1998a). International partnerships of firms can be a successful tool to transfer technologies, as shown in the Vietnam Leadership Programme between various TNCs active in Vietnam and government agencies to phase out the use of CFCs in the Vietnamese electronics industry (Andersen, 1998b). The example of bilateral cooperation between U.S. electronics manufacturers and Mexican suppliers helped to overcome some of the barriers in information supply and access to technology and financing (Andersen, 1998c).

BOX 9.2. THE NATIONAL CLEANER PRODUCTION CENTRE PROGRAMME

The basis of successful technology transfer is the capacity to adapt, operate and integrate a new technology. The National Cleaner Production Centre (NCPC) Programme is a global project managed by UNIDO, together with UNEP. The Programme aims to facilitate the application of cleaner production in industry and the incorporation of the concept in policies of developing countries and economies in transition. In collaboration with a host institution, the programme establishes a unit (called NCPC) that provides continuous support to cleaner production initiatives in companies, business organisations, and local and national governments. An NCPC undertakes four sets of activities: in-plant demonstrations, training, information dissemination and policy advice. These activities can differ in intensity and form, depending on the situation in a country. The programme has established NCPCs in Brazil, China, Costa Rica, India, Czech and Slovak Republics, El Salvador, Guatemala, Honduras, Hungary, Mexico, Nicaragua, Tanzania, Tunisia, and Zimbabwe. New centres are being established in Slovenia, Croatia, Vietnam, and Morocco. Experiences with the NCPCs have showed that disseminating knowledge on cleaner production and showing the gains were not sufficient to spur the demand in industry. The programme will need to improve the identification of the needs of companies and responding to these needs. There is a need to formalise the process, e.g. by linking cleaner production concepts to certification systems like ISO 14000. Replication needs several prerequisites to be successful including: effective environmental policy, regulation and enforcement, environmentally sound behaviour (embedded in society); the use of operational, accounting and management systems for data collection in industries; and a relation between cost of inputs, waste and emissions and the proceeds of the output. Access to adequate financing is also necessary to enable industry to invest (Van Berkel, 1998b).

As industrial development increases, capabilities for technology assessment and selection improve, as evidenced by the case study of pulverised coal injection for blast furnaces in the steel industry in Korea (Joo, 1998c), as well as by investment projects in new cement plants in Mexico (Turley, 1995) and Chinese Taipei (Chang, 1994). It is stressed that development of technical capabilities is a continuous process, because it takes large resources to build up a knowledge infrastructure, and the key to success is so-called "tacit knowledge" (unwritten knowledge obtained by experience) (Dosi, 1988), which is easily lost. The greater the existing capability, the greater the opportunities are for gaining knowledge from industrial collaboration and technology transfer (Chantramonklasri, 1990). Finally, language can be a barrier in successful transfer of a technology, especially when working with local contractors or suppliers (Hassan, 1997).

Agreement and Implementation
As in adoption of technology and practices within countries, adoption across countries depends on the motivation of management and personnel, external driving forces, e.g. legislation and standard setting, economics (i.e. profitability), availability of financial and human resources, and other external driving forces (e.g. voluntary agreements). Financing in particular may be more difficult, hindered by high inflation rates, and needing hard currencies to acquire technologies. Budgets of multilateral financing institutes are relatively small, while bilateral financial assistance schemes may influence the technology selection (see above). The example of the Montreal Protocol Multilateral Fund shows that efficient and effective financing mechanisms can be deployed, although specific barriers may delay the financing schemes, as happened in Mexico (Andersen, 1998c). The case studies have shown that financing schemes for small companies, e.g. soft-loans, subsidies and tax credits, may help to improve the adoption rate (TERI, 1997). Large companies in NICs seem to have easier access to capital, as shown by the case studies for the steel industry in Korea (Joo, 1998a,b and c). Trade barriers, such as import taxes, can influence the economic assessment, and hence technology selection and implementation.

Evaluation and Adjustment
Adaptation of technologies to local conditions is crucial. There is a great need for technological innovation for energy efficiency in the developing countries and CEITs. The technical operating environment in these countries is often different from that of industrialised countries. For example, different raw material qualities, lower labour costs, poorer power quality, higher environmental dust loads, and higher temperatures and humidities require different energy efficiency solutions than successful solutions in industrialised country conditions. Technologies that have matured and been perfected for the scale of production, market, and conditions in the industrialised countries may not be the best choice for the smaller scale of production, raw materials used or different operating environments often encountered in a developing country. Transferred technologies seldom reach the designed operational efficiencies, and often deteriorate over their productive life (TERI, 1997) due to several reasons. Improper maintenance, inadequate availability of spare parts and

incomplete transfer of "software" are some of the problems. This stresses the need for effective adaptation strategies, including transfer of technical and managerial skills (see also Box 9.3). Technical training is a very important aspect of a technology transfer (Hassan, 1997), and should preferably be done in the local language.

Box 9.3   DEVELOPMENT OF EFFICIENT FLUIDISED BED BOILERS IN CHINA

Much of China's coal consumption is in inefficient polluting equipment. Coal burning is a major contributor to air pollution in many Chinese cities. The average boiler efficiency of small and medium capacity industrial boilers, which consume approximately 1/3 of China's annual coal production, is only 60 to 65% (LHV, Lower Heating Value). In China there are already about 2000 fluidised bed boilers burning low grade coal. However, almost all of them are bubbling fluidised bed combustion (BFBC) boilers that have performance disadvantages and development limitations. In OECD countries, a new generation of circulating fluidised bed combustion (CFBC) concept has been developed. CFBC addresses the problems of combustion efficiency and air pollutant emissions. It was decided to demonstrate imported CFBC technology to China's coal users. Ahlstrom Pyropower was selected as the technology supplier. The project aimed to demonstrate CFBC technology at an existing industrial site, and enhance the capacity of China to design, manufacture, install and operate CFBC systems in various sizes with the flexibility to burn numerous coal types. The planned project costs of US$8.5 million (M$) were exceeded by 2 M$. UN funds provided 2 M$ and the Chinese Government provided 8.5 M$. Government input in kind was estimated at RMB 292 million (35.3 M$[3]) to meet other costs in China. The cost overrun was due to additional auxiliary equipment that needed to be imported. Eight training groups consisting of 16 researchers and engineers were trained in OECD countries, while over 174 Chinese engineers participated in a training workshop held in China. The R&D facilities provide a necessary tool for CFBC technology development in China. At least seven domestic boilermakers are now involved in CFBC design and construction, with a total of over 200 units either in operation, construction or under contract (Williams, 1998).

In practice, adaptation practices vary widely in various countries. For example, Chinese enterprises have spent, on average, only 9 (US) cents on assimilation for every (US) dollar on foreign technology. In contrast to countries as Korea and Japan where the amounts spent on assimilation were greater than those spent on technology itself (Suttmeier, 1997). Countries in a later stage of industrialisation may be better equipped for adapting technologies to the local industrial environment, while countries or companies in an earlier stage may (have to) rely more on the foreign suppliers of technology. Equipment suppliers may license part of the construction or parts supply to local firms. This is illustrated by the construction of an advanced steel plant in Korea, which

---

[3] This figure is based on a currency exchange rate from November 1999.

was partly done by Samsung Heavy Industries (Worrell, 1998), as well as examples in the construction of cement plants in India (Somani and Kothari, 1997), Mexico (Turley, 1995) and Chinese Taipei (Chang, 1994). The examples in Korea, Mexico and Chinese Taipei show a heavy involvement in technology procurement, design and management. The Korean and Mexican firms belong to the largest producers in the world of respectively steel and cement.

Replication

Replication and further development of practices and technologies in developing countries and CEITs is needed. It is also a heavily debated issue involving intellectual property rights (see Chapter 3), and dependence on (foreign) technology suppliers. Many industrial technologies are privately owned, although (part of) the (pre-competitive) research may have been publicly funded. When transferring dry coke quenching technology to China the proprietary rights stayed with the Japanese technology providers for a period of 10 years, avoiding replication in China for a long period (Hu *et al.*, 1998). A clear (legal) framework is needed to improve adaptation and replication of technology (ESETT, 1991). Technology transfer projects need continued support from the technology supplier. This is beneficial to both the technology user and supplier. The user can benefit from experience from other licensees, and the licensor gets an opportunity to gain further market entrance. Experience has shown that reasonable plant performance will improve future business opportunities (Hassan, 1997). However, technology owners may be hesitant to share all parts of a technology, including "software", without sufficient legal protection in the country of the user (see Chapter 3).

Various concepts of replication and development are demonstrated by other case studies. Waste Minimisation Circles were started in a few regions in India, and are now replicated in other sectors and regions (Van Berkel, 1998a). UNIDO/UNEP replicated National Cleaner Production Centres in various developing countries and CEITs (Van Berkel, 1998b). Replication of programmes and experiences as a form of South-South cooperation is demonstrated by the transfer of the Indian auditing programme to Jordan (Menke, 1998). The examples of furnace technology development for SMEs in India through joint organisations (e.g. research institutes, NGOs) demonstrate the benefits of combining the experiences and strengths of various partners in innovative development and implementation schemes (TERI, 1997). Countries possessing a higher technical capability are faster to replicate and develop a technology. The first implementation of pulverised coal injection in a blast furnace in Korea made it possible to replicate the technology in another plant (Joo, 1998c) of the same company. The examples of the FINEX (fine-ore-based smelt reduction) process development, as well as the development of the HYL direct reduction process in Mexico (Zervas *et al.*, 1996), illustrate the capability of companies in NICs to develop a new process. The advanced FINEX project is an example of technology cooperation between the Austrian supplier and the Korean industry (Joo, 1998b). The steel sector is an industry with relatively frequent and open communication. In other sectors, e.g. the chemicals industry, process and technology knowl-

edge is proprietary, limiting replication and development for developing countries and CEITs. Licensors and contractors are interested in the successful transfer of proprietary technology to secure future sales (Hassan, 1997).

9.6 Lessons Learned

The industrial sector is extremely diverse and involves a wide range of activities including the extraction of natural resources, conversion into raw materials, and manufacture of finished products. Due to the wide variety in activities, energy demand and GHG emissions vary widely. Hence, the aggregate energy use and emissions depend on the structure (or specific set of activities) of industry, and the energy and carbon intensity of each of the activities. The structure of industry may depend on the phase of the economy, as well as many other factors like resource availability and historical factors. Industrial production and GHG emissions are still dominated by industrialised countries, but the role of developing countries in world industrial production, especially Southeast Asia, is increasing. Cost-effective potentials and opportunities for GHG emission abatement exist in all regions and industrial sectors. A wide variety of practices and technologies to reduce GHG emissions are available (see Table 9.2), often with high paybacks.

In industry, energy efficiency is often the result of investments in modern equipment, stressing the attention to sound and environmentally benign investment policies. Investments in technology (including hardware and software) in the industrial sector are dominated by the private sector. Recent trends in globalisation of industry seem to affect the international transfer of investments and technology. Foreign direct investment (FDI) is rapidly increasing, although concentrated on a small number of rapidly industrialising countries. These countries may have an impact on regional industrial development patterns, as seen in Asia. Private investment in other developing regions is still limited, although increasing. FDI is dominated by transnational companies, while SMEs in industrialised, developing countries and CEITs have less access to (international) financial markets and technologies. Although difficult to measure, domestic investments in developing countries are still larger than FDI. Official development assistance, although earmarked for low to medium income countries, is also concentrated on a few countries. Public funding (in industrialised, developing countries and CEITs) for technology development and transfer, although still important, is decreasing. Funding for science and technology development is important to support industrial development, especially in developing countries. Public funding in the industrial sector, although small in comparison to private funding, remains important but its future role may be changing. Regular evaluation of the goals of public funding is needed for industrial development with respect to the role of cleaner technologies and with respect to the role of private funds.

Barriers limit the uptake of more efficient technologies. These barriers may include the (un)willingness to invest in (new) technologies, the level of information and transaction costs, the lack of effective financing (e.g. lack of sufficient funds, high interest), the lack of skilled personnel and a variety of other barriers, e.g. the "invisibility" of energy and CO_2 emission savings and the lack of inclusion of external costs. Developing countries and CEITs suffer from all of these factors that inhibit market acceptance of technologies plus a multitude of other market problems. Consumers often have no knowledge of energy efficiency (technologies) or cannot afford increases in equipment costs, due to a limited ability to pay increased initial costs, limited foreign currency and high inflation rates. A well developed banking system and existence of appropriate financing mechanisms are essential for the uptake of efficient and cleaner technologies in industry.

Traditionally, technology transfer is seen as a private transaction between two enterprises. However, innovation and technology transfer is an interactive and iterative process, involving many different parties. An effective process for technology transfer will require interactivity between various users, producers and adaptors of technology. The variety of stakeholders makes it necessary to have a clear policy framework as part of an industrial policy for technology transfer and cooperation, both for a technology donor and recipient or user. Such a framework may include environmental, energy, (international) trade, taxation and patent legislation, as well as a variety of well-aimed incentives. The framework may help to give the right signals to all parties, as well as help to develop innovative concepts for technology assessment, financing, procurement, adaptation, replication and development. Policymakers are responsible for developing such a comprehensive framework. The interactive and dynamic character of technology transfer stresses the need for innovative and flexible approaches, through (long-term) partnerships between various stakeholders, including public-private partnerships.

The case studies and the literature demonstrate clearly that there is a strong need to develop the capacity to assess and select technologies. Stakeholders (policymakers, private investors, financing institutions) in developing countries and CEITs have even more difficult access to technology information, stressing the need for a clearinghouse for information on climate abatement technology. Various innovative policy concepts, including networking and joint research and information organisations, were found to be successful. To increase the likelihood of success, long term support for capacity building is essential, stressing the need for public support for capacity building and cooperation of technology suppliers and users.

Adaptation of technology to local conditions is essential, but practices vary widely. Countries that spend on average more on adaptation seem to be more successful in technology transfer. As countries industrialise the technological capabilities increase rapidly, accelerating the speed of technology diffusion and development. This demonstrates that successful technology transfer includes transfer of technological capabilities, which may be beneficial to both the supplier and user. Technology users, suppliers as well as financial institutions and governments could give attention to adaptation as an essential and integral part of technology procurement.

The introduction and diffusion of clean or low-GHG technologies in the industrial sector needs a sound environmental and economic policy, stressing the need for long term goals and commitment by policymakers. This also means that technology transfer needs to be incorporated in R&D strategies, as many (public) environmental sound technologies "remain on the shelves" and are not brought into the market as rapidly as may be expected. Several countries and equipment suppliers envisage that environmentally sound product development can enhance the future competitive position of domestic suppliers, making technology transfer (through strengthening local capacity and demonstration of technology) a way to open new export markets. Subsequently, policies to support the development of new technologies and markets could be used in these countries as part of economic and trade policies.

References

Amsden, A.H., and M. Mourshed, 1997: Scientific Publications, Patents and Technological Capabilities in Late-Industrializing Countries. *Technology Analysis & Strategic Management*, **9** (3), 343-359.

Andersen, 1998a: Global Semiconductor Partnership to Reduce PFC Emissions. Case-Study performed for the *IPCC Special Report on Methodological and Technological Issues in Technology Transfer*.

Andersen, 1998b: Vietnam Leadership Initiative. Case-Study performed for the *IPCC Special Report on Methodological and Technological Issues in Technology Transfer*.

Andersen, 1998c: Mexico Solvents Partnership. Case-Study performed for the *IPCC Special Report on Methodological and Technological Issues in Technology Transfer*.

Avdiushin, S., I. Grtisevich, and S. Legro, 1997: *Climate Change Mitigation: Case Studies from Russia Pacific Northwest*. National Laboratory, Washington, DC.

Blok, K., 1993: The Development of Industrial CHP in the Netherlands. *Energy Policy*, **21** (2), 58-175.

Blok, K., W.C. Turkenburg, W. Eichhammer, U. Farinelli,and T.B. Johansson (eds.), 1995: *Overview of Energy RD&D Options for a Sustainable Future*. European Commission, DG-XII, Brussels/Luxembourg.

Callin, J., B. Svennesson, and E. White, 1991: *Energy and Industrialization, The Choice of Technology in the Paper and Pulp Industry in Tanzania Dept. of Environmental and Energy Systems Studies*. Lund University, Lund, Sweden.

Chang, T.H., 1994: Successful Operation of the World's Only 5000 tpd Short Rotary Kiln. *World Cement*, **25**(9), 4-9.

Chantramonklasri, N., 1990: The Development of Technological and Managerial Capability in the Developing Countries. In *Technology Transfer in the Developing Countires*. M. Chatterji, (ed.), The Macmillan Press Ltd., Houndsmill, Hampshire/London, pp. 36-50.

Clark, W.W., 1997: *The Role of Publicly-Funded Research and Publicly-Owned Technologies in the Transfer and Diffusion of Environmentally Sound Technologies. The Case Study of the United States of America*. In Proceedings International Expert Meeting of CSD on the Role of Publicly Funded Research and Publicly Owned Technologies in the Transfer and Diffusion of Environmentally Sound Technologies, Ministry of Foreign Affairs, Republic of Korea, February, 4-6, 1998.

Cohen, L.R., and R.G.Noll, 1994: Privatizing Public Research. *Scientific American*,**September**, 72-77.

Collingwood, J., and D. Goult, 1998: The UK Energy Efficiency Best Practice Programme: Evaluation Methods & Impact 1989–1998. Paper presented at Industrial Energy Efficiency Policies: Understanding Success and Failure, Utrecht, The Netherlands, 11-12 June 1998.

Dadi, Z., Z. Fengqi, Y. Cong, S. Yingyi, and J. Logan, 1997: Climate Change Mitigation. Case Studies from China Pacific Northwest National Laboratory, Washington, DC.

DeCanio, S.J., 1993: Barriers within Firms to Energy-Efficient Investments, *Energy Policy*, **21,** 906-914.

De Oliveira, A., 1991: Reassessing the Brazilian Alcohol Programme. *Energy Policy*, **1** (19),47-55.

Department of Energy, 1995: *Energy R&D: Shaping our Nation's Future in a Competitive World*. Task Force on Strategic Energy Research and Development, Department of Energy, Washington, DC.

Dosi, G., 1988: The Nature of the Innovative Process. In *Technical Change and Economic Theory*. G.Dosi, C. Freeman, R. Nelson, G. Silverberg, L. Soete, (eds.), Pinter Publishers, London.

ESETT, 1991: *Symposium Report*. International Symposium on Environmentally Sound Energy Technologies and their Transfer to Developing Countries and European Economies in Transition, Milan, 21-25 October 1991.

Farla, J.C.M., and K. Blok, 1995: Energy Conservation Investment of Firms: Analysis of Investments in Energy Efficiency in the Netherlands in the 1980s. *ACEEE 1995 Summer Study on Energy Efficiency in Industry Proceedings*, Washington, DC.

Fisher, A. C. R., and M. Rothkopf, 1989: Market failure and Energy Policy. *Energy Policy*, **17**, 397-406.

Florida, R., 1997: The Globalization of R&D: Results of a Survey of Foreign-Affiliated R&D Laboratories in the USA. *Research Policy*, **26**, 85-93.

Gadgil, A., and A. Sastry, 1994: Stalled on the Road to Market: Lessons from a Project Promoting Lighting Efficiency in India. *Energy Policy*, **22**, 151-162.

Gielen, D., 1998: Western European Materials as Sources and Sinks of CO_2. *Journal of Industrial Ecology*, **2** (2).

Goldemberg, J. (ed.), 1998: *The Clean Development Mechanism: Issues and Options*. United Nations Development Programme, New York.

Goldemberg, J., and I.C. Macedo, 1994: Brazilian Alcohol Program: An Overview. *Energy for Sustainable Development*, **1** (1), 17-22.

Gruber, E., and M. Brand, 1991: Promoting Energy Conservation in Small- and Medium-Sized Companies. *Energy Policy*, **19**, 279-287.

Hassan, N, 1997: Successfully Transfer HPI Proprietary Technology. *Hydrocarbon Processing*, **76** (2), 91-99.

Hu, X., K. Jiang, and S. Zheng, 1998: *The Potential of Energy Efficiency Improvement in China's Iron and Steel Industry and a Case Study on Technology Transfer*. Proceedings Workshop on Technology Transfer and Innovation in the Energy Sector, STAP/GEF, 19-20 January 1998, Amsterdam.

IPCC 1996a: *Climate Change 1995. Impacts, Adaptations and Mitigation of Climate Change: Scientific-Technical Analysis*. R.T. Watson, M.C. Zinyowera, R.H. Moss, (eds.), Intergovernmental Panel on Climate Change - Working Group II, Cambridge University Press, Cambridge.

IPCC 1996b: Technologies, Policies and Measures for Mitigating Climate Change. IPCC Technical Paper, Intergovernmental Panel on Climate Change, Watson, R.T., M.C. Zinyowera, and R.H. Moss (eds.), Cambridge University Press, Cambridge and New York.

International Energy Agency, 1994: Energy Policies of IEA Countries. *1993 Review*, IEA/OECD, Paris.

International Energy Agency, 1997: *Voluntary Actions for Energy-Related CO_2 Abatement* . IEA/OECD, Paris.

Jackson, T., 1995: Joint Implementation and Cost-Effectiveness under the Framework Convention on Climate Change. *Energy Policy*, **23,** 117-138.

Jepma, C.J. (ed.), 1995: *The Feasibility of Joint Implementation*. Kluwer Academic Publishers, Dordrecht, The Netherlands.

Joo, S.H., 1998a: The COREX C-2000 Plant of POSCO. Case-Study performed for the *IPCC Special Report on Methodological and Technological Issues in Technology Transfer*.

Joo, S.H., 1998b: FINEX Development Project in POSCO. Case-Study performed for the *IPCC Special Report on Methodological and Technological Issues in Technology Transfer*.

Joo, S.H., 1998c: Pohang No.3 Blast Furnace Pulverized Coal Injection Plant of POSCO. Case-Study performed for the *IPCC Special Report on Methodological and Technological Issues in Technology Transfer*.

Kaplinsky, R., 1997: India's Industrial Development: An Interpretative Survey. *World Development*, **25** (5), 681-694.

Kim, J-I., 1998: *Industry's Effort to Mitigate Global Warming: Options for Voluntary Agreements in Korea*. Proceedings 1998 Seoul Conference on Energy use in Manufacturing: Energy savings and CO_2 Mitigation Policy Analysis, Korea Energy Economics Institute, Korea Resource Economics Association, Seoul, 19-20 May 1998.

Kuznets, S., 1971: *Economic Growth of Nations: Total Output and Production Structure*. Harvard University Press, Cambridge, MA.

La Rovere, E.L., 1998: The Challenge of Limiting Greenhouse Gas Emissions through Activities Implemented Jointly in Developing Countries: A Brazilian Perspective. Federal University of Rio de Janeiro, Rio de Janeiro (draft).

Lee, J., 1997: The Maturation and Growth of Infant Industries: The Case of Korea. *World Development*, **25** (8), 1271-1281.

Levine, M.D. , and X. Liu, 1990: *Energy Conservation Programs in the People's Republic of China*. Lawrence Berkeley Laboratory, Berkeley, CA.

Levine, M.D., E. Hirst., J.G. Koomey, J.E. McMahon, and A.H. Sanstad, 1994: *Energy Efficiency, Market Failures, and Government Policy*. Lawrence Berkeley Lab./Oak Ridge National Laboratory, Berkeley/Oak Ridge, USA.

Levine, M. D., J.G. Koomey, L.K. Price, H. Geller, and S. Nadel, 1995: Electricity and end-use efficiency: experience with technologies, markets, and policies throughout the world. *Energy*, **20**, 37-65.

Little, I.M.D., 1987: Small Manufacturing Enterprises in Developing Countries. *World Bank Economic Review,* **1** (2).

Liu, Z., J.E. Sinton, F. Yang, M.D.Levine, and M.Ting, 1994: *Industrial Sector Energy Conservation Programs in the People's Republic of China during the Seventh Five-Year Plan (1986-1990).* Lawrence Berkeley National Laboratory, Berkeley, CA.

Luken, R.A., and A-C. Freij, 1995: Cleaner Industrial Production in Developing Countries: Market Opportunities for Developed Countries. *Journal Cleaner Production,* **3** (1-2), 71-78.

Macedo, I.C., 1998: *The Role of Copersucar in Improving Technology for Ethanol Production from Sugar Cane in Sao Paulo.* Proceedings Workshop on Technology Transfer and Innovation in the Energy Sector, STAP/GEF, 19-20 January 1998, Amsterdam.

Marousek, J., M. Dasek, , S. Legro, B. Schwarzkopf, and M. Havlickova, 1998: *Climate Change Mitigation. Case Studies from the Czech Republic.* Pacific Northwest National Laboratory, Washington, DC.

Menke, C., 1998: Energy Auditing for Large Industry as Sustainable Business Case-Study performed for the *IPCC Special Report on Methodological and Technological Issues in Technology Transfer.*

Moldan, B., 1997: *The Role of Publicly Funded Research and Publicly Owned Technologies in the Transfer and Diffusion of Environmentally Sound Technologies, A Case Study on the Czech Republic.* In Proceedings International Expert Meeting of CSD on the Role of Publicly Funded Research and Publicly Owned Technologies in the Transfer and Diffusion of Environmentally Sound Technologies. Ministry of Foreign Affairs, Republic of Korea, February, 4-6, 1998.

Nadel, S., 1990: *Lessons Learned: A Review of Utility Experience with Conservation and Load Management Programs for Commercial and Industrial Customers.* Final Report prepared for the New York State Energy Research and Development Authority (NYSERDA), the New York State Energy Office, and the Niagara Mohawk Power Corporation, New York.

Nadel, S., 1991: Electric Utility Conservation Programs: A Review of the Lessons Taught by a Decade of Program Experience. In *State of the Art of Energy Efficiency: Future Directions.* E.Vine, D. Crawley, (eds.), American Council for an Energy-Efficient Economy, Washington, DC and Berkeley, CA.

Nadel, S., V. Kothari, and S. Gopinath, 1991: *Opportunities for Improving End-Use Electricity Efficiency in India.* American Council for an Energy-Efficient Economy, Washington, DC.

Nelson, R.R. (ed.), 1982: *Government and Technical Progress.* Pergamon Press, New York.

Nelson, K., 1994: Finding and Implementing Projects that Reduce Waste. In *Industrial Ecology and Global Change.* R.H. Socolow, C. Andrews, F. Berkhout, V. Thomas, (eds.), Cambridge University Press, Cambridge.

Office of Technology Assessment, 1993: *Industrial Energy Efficiency.* US Government Printing Office, Washington, DC.

Oshima, T., 1998: Energy Conservation Audit Project for Small and Medium Scale Industries in Japan. Case-Study performed for the *IPCC Special Report on Methodological and Technological Issues in Technology Transfer.*

Pearce, D.,1995: Joint Implementation, A General Overview. In *The Feasibility of Joint Implementation.* C.J. Jepma, (ed.), Kluwer Academic Publishers, Dordrecht, The Netherlands.

Price, L., L. Michaelis, E. Worrell, and M. Khrusch, 1998: Sectoral Trends and Driving Forces of Global Energy Use and Greenhouse Gas Emissions. *Mitigation and Adaptation Strategies for Global Change,* **3,** 263-319.

Putterman, L., 1997: On the Past and Future of China's Township and Village-Owned Enterprises. *World Development,* **25**(9), 1639-1655.

Rama Rao, P., 1997: India: Science and Technology from Ancient Time to Today. *Technology in Society,* **19** (3/4), 415-447.

Reddy, A.K.N., 1991: Barriers to Improvements in Energy Efficiency. *Energy Policy,* **19,** 953-961.

Ross, M.H., 1986: Capital Budgeting Practices of Twelve Large Manufacturers. *Financial Management,* **Winter,**15-22.

Sanstad, A. H., and R.B. Howarth, 1994: 'Normal' Markets, Market Imperfections and Energy Efficiency. *Energy Policy,* **22,** 811-818.

Sasaki, S., and S. Asuka-Zhang, 1997: *The Role of Publicly Funded Research and Publicly Owned Technologies in the Transfer and Diffusion of Environmentally Sound Technologies, The Japanese Study.* In Proceedings International Expert Meeting of CSD on the Role of Publicly-Funded Research and Publicly-Owned Technologies in the Transfer and Diffusion of Environmentally Sound Technologies, Ministry of Foreign Affairs, Republic of Korea, February, 4-6, 1998.

Schumacher, K. and J. Sathaye, 1998: *India's Pulp and Paper Industry: Evaluation of Productivity Growth through Econometric, Statistical and Engineering Analysis,* Report No. 41843, Lawrence Berkeley National Laboratory, Berkeley, CA.

Sinton, J.E., and Levine, M.D.,1994: Changing Energy Intensity in Chinese Industry. *Energy Policy,* **22,** 239-255.

Sioshansi, F.P.,1991: The Myths and Facts of Energy Efficiency. *Energy Policy,* **19,** 231-243.

Somani, R.A., and S.S. Kothari, 1997: "Die Neue Zementlinie bei Rajashree Cement in Malkhed/Indien". *ZKG International,* **50**(8), 430-436.

Song, J., 1997: Science and Technology in China: The Engine of Rapid Economic Development. *Technology in Society,* **19** (3/4), 281-294.

Sturm, R., K. Opheim, and P. Kelly-Detwiler, 1997: *The Problem of Second-Hand Industrial Equipment: Reclaiming a Missed Opportunity.* International Institute for Energy Conservation, Washington, DC.

Suttmeier, R.P., 1997: Emerging Innovation Networks and Changing Strategies for Industrial Technology in China: Some Observations, *Technology in Society,* **19** (3/4), 305-323.

Syrquin, M., and H.B. Chenery, 1988: Patterns of Development, 1950-1983. World Bank Discussion Papers No.41, The World Bank, Washington, DC.

Tata Energy Research Institute (TERI), 1997: *Capacity Building for Technology Transfer in the Context of Climate Change.* TERI, New Delhi.

Tripathy, U., 1997: *The Role of Publicly-Funded Research and Publicly-Owned Technologies in the Transfer and Diffusion of Environmentally Sound Technologies in India: A Feasibility Study.* In Proceedings International Expert Meeting of CSD on the Role of Publicly-Funded Research and Publicly-Owned Technologies in the Transfer and Diffusion of Environmentally Sound Technologies, Ministry of Foreign Affairs, Republic of Korea, February, 4-6, 1998.

Turley, W., 1995: Mexican Mammoth. *Rock Products,* **September,** 21-25.

United Nations, 1998: *The Role of Publicly Funded Research and Publicly Owned Technologies in the Transfer and Diffusion of Environmentally Sound Technologies,* UNCTAD, UNEP, DSD, New York.

United Nations Conference on Trade and Development, 1997: *World Investment Report 1997, Transnational Corporations, Market Structure and Competition Policy,* United Nations, New York and Geneva.

United Nations Industrial Development Organization, 1997: *Industrial Development - Global Report 1997,* Oxford University Press, Oxford.

Van Berkel, R., 1998a: Waste Minimisation Circles. Case Study performed for the *IPCC Special Report on Methodological and Technological Issues in Technology Transfer.*

Van Berkel, R., 1998b: National Cleaner Production Programme. Case Study performed for the *IPCC Special Report on Methodological and Technological Issues in Technology Transfer.*

Van Berkel, R., and J. Bouma, 1999: *Promoting Cleaner Production Investments in Developing Countries.* A Status report on Key Issues and Potential Strategies, United Nations Environment Programme, Paris.

Van Berkel. R., S. Chandak, P. Sethi and R. Luken, 1996: *From Waste to Profits: Towards Financial and Environmental Dividends from Waste Minimisation in Small Scale Industries in India.* Final Report of Project DESIRE (Demonstrations in Small Industries for Reducing Waste), UNIDO Environment and Energy Branch, Vienna.

Velthuijsen, J.W., 1995: *Determinants of Investment in Energy Conservation.* SEO, University of Amsterdam, Amsterdam.

Warhurst, A., and G. Bridge, 1997: Economic Liberalisation, Innovation, and Technology Trasnfer: Opportunities for Cleaner Production in the Minerals Industry. *Natural Resources Forum,* **21**(1), 1-12.

Weiss, C., 1990: Ethyl Alcohol as a Motor Fuel in Brazil, A Case Study in Industrial Policy. *Technology in Society,* **12**(3), 255-282.

Williams, R.H., and J. Goldemberg, 1995: *A Small Carbon Users' Fee for Accelerating Energy Innovation.* Center for Energy and Environmental Studies, Princeton University, Princeton, NJ.

Williams, R.O., 1998: Development of Circulating Fluidized Bed Combustion Technology in China. Case-Study performed for the *IPCC Special Report on Methodological and Technological Issues in Technology Transfer.*

World Bank, 1997a*: The State in a Changing World, World Development Report 1997*, Oxford University Press, Oxford.

World Bank, 1997b: Clear Water, Blue Skies, China's Environment in the Next Century. *China 2020 Series.* The World Bank, Washington, DC.

World Bank, 1998*: World Development Indicators 1998.* The World Bank, Washington, DC.

World Energy Council, 1995: *Energy Efficiency Improvement Utilising High Technology, An Assessment of Energy Use in Industry and Buildings.* World Energy Council, London.

World Wildlife Fund, 1996: *Sustainable Energy Technology in the South.* A Report to WWF by Institute of Environmental Studies , Amsterdam, and Tata Energy Research Institute, New Delhi.

Worrell, E., 1998: *Energy Efficient Technologies in the Cement and Steel Industry – Experiences in Developing Countries. Proceedings Workshop on Technology Transfer and Innovation in the Energy Sector*, STAP/GEF, 19-20 January 1998, Amsterdam

Worrell, E., M. Levine, L. Price, N. Martin, R. Van Den Broek, and K. Blok, 1997: *Potentials and Policy Implications of Energy and Material Efficiency Improvement.* United Nations Division for Sustainable Development, UN, New York.

Yhdego, M., 1995: Environmental Pollution Management for Tanzania: Towards Pollution Prevention. *Journal Cleaner Production,* **3** (3), 143-151.

Zervas, T., J.T. McMullan, and B.C. Williams, 1996: Gas-Based Direct Reduction Processes for Iron and Steel Production. *International Journal of Energy Research,* **20** (2), 157-185.

10

Energy Supply

Coordinating Lead Author:
JOSE ROBERTO MOREIRA (BRAZIL)

Lead Authors:
Jos Bruggink (The Netherlands), Hisashi Ishitani (Japan), P.R. Shukla (India),
Katia J. Simeonova (Bulgaria), John J. Wise (USA)

Contributing Authors:
Youba Sokona (Senegal), Helena Li Chum (USA), Eric Martinot (USA)

Review Editors:
R. S. Agarwal (India), Steven Bernow (USA)

CONTENTS

EXECUTIVE SUMMARY

General

Major objectives of the current energy supply sector are economic development and international competitiveness. Climate change objectives, particularly the reduction of carbon dioxide (CO_2) emissions, do not play a significant role. Nevertheless, significant opportunities exist to reduce CO_2 emissions (see Table 10.1) as well as other greenhouse gas (GHG) emissions.

Technology transfer in the oil, natural gas and electricity sectors is mainly driven by the private sector. On the other hand, coal, nuclear and renewable energy sources are often dependent on government to preserve or increase their presence in the market. Technology transfer, as presently understood, is a relatively new process since historically it was used as a euphemism for large-scale power projects financed by multilateral banks, or for limited knowledge transfer from the international oil and gas companies to national industries. The oil crises in the 1970s changed the contractual terms in the oil and gas sectors and enabled powerful national oil companies to negotiate technology transfer on terms that were more favourable to them. In the early 1990s, the process of market globalisation and the availability of private capital on a global scale triggered investments, and hence technology transfer opportunities in the electricity sector also. The private sector is now playing a bigger role in electric power generation in concert with a new set of regulations and standards.

In the energy supply sector, technology transfer comes with investment. One of the keys to the transfer of technology is to promote investment through an appropriate economic and institutional framework. For some clean energy supply alternatives technology improvement is necessary for large-scale commercialisation. In general, economic and institutional barriers rather than technology availability are more apt to be the cause of failure to transfer technology. In all energy sectors, the role of government in facilitating technology transfer is critical. Annex II countries could develop more effective policies to stimulate and finance private investments in clean energy sources in developing countries and countries with economies in transition (CEITs). Governments could initiate policies for liberalising the energy supply market, fostering and ensuring conditions to allow international financing, promoting infrastructure development, eliminating unnecessary regulatory and trade barriers, educating and training local workforces, protecting intellectual property rights, and for strengthening local research and development (R&D) and environmental management regimes. Table 10.2 lists for each major technology group the existing barriers preventing their transfer, as well as the possible policies to overcome such barriers.

It is expected that markets will respond to whatever regulatory policies are adopted to promote GHG emissions reduction. Technology transfer will be stimulated as investments are made in response to the price signals when uncertainties on policies to secure environmental goals are settled.

Fossil Fuel and Nuclear

Technology transfer in the fossil fuel sector is mature, and well-established mechanisms are in place. Technology is readily available from a wide variety of sources, such as the oil, gas, and coal industries, engineering contractors, equipment vendors, etc. Barriers to technology transfer, therefore, are primarily economic and institutional.

Technology transfer in the nuclear power sector for water-cooled and water moderated reactors is mature and has well-established mechanisms. To further promote the successful transfer of nuclear technology, major government involvement is needed. The large capital costs, public acceptance, availability of cheap domestic fossil fuel, and safety and waste disposal concerns provide significant barriers to the use of nuclear energy. In many cases, nuclear proliferation issues are also a major problem to be addressed by governments and other international institutions.

Specific measures and policies could include:
- Create and/or enlarge the market for clean energy technologies by, for example:
- Using high efficiency technology and procedures in electric power generation;
- Switching to lower carbon fuels (eg. substituting natural gas for coal);
- Using combined heating (cooling) and power generation where feasible;
- Using sequestration of CO_2 where feasible; and
- Using high efficiency in electric power generation.
- Enhance regional cooperation in development and transportation of natural gas and electricity across national boundaries.
- Increase efficiency by promoting dissemination of best practices.
- Educate the public on the benefits and risks of nuclear power.
- Develop infrastructure and trained personnel required to insure the highest possible level of nuclear safety.

Renewables

In the renewables sector, with the exception of large scale hydropower, technology transfer has been constrained by the

EXECUTIVE SUMMARY

lack of investment and high costs. Investment has been generally limited to niche (such as solar photovoltaics (PV) or wind power) or protected markets because of technical, institutional and economic barriers. Governments could promote the development of improved and more cost effective renewable technologies, provide incentives for investment and remove policies which hinder the application of renewable energy.

Measures to improve renewable technologies could include:
- *Biomass* - Promote RD&D on biomass crop selection, conversion technology, and integrated systems.
- *Solar* - Promote RD&D on PV integrated systems.
- Wind - Develop performance standards for wind turbines.
- *Hydro* - Promote development of hydropower consistent with environmentally and socially sustainable development.

Measures to promote investment and broaden the market for renewables could include:
- *General*
- Create political and regulatory frameworks to allow full cost pricing, and recognise the indirect benefits of renewables such as the creation of more local jobs, improvement of the environment, balance of trade, etc.
- Develop human and institutional capacities.
- Foster joint research and technology development.
- Promote assessment of the potential of renewables in appropriate countries.
- Involve local communities, mainly in small size energy supply projects.

- *Biomass* - Develop biomass resource assessment (including agricultural residues) for countries with biomass potential.

- *Solar* - Promote market development through government procurement and low cost financing and in-country technology transfer.

- *Wind* - Develop resource assessment to allow market to judge potential.

10.1 Introduction[1]

The purpose of this chapter is to provide information about methodological and technological issues in climate friendly technology transfer for the energy supply sector. The discussion is restricted to technologies for climate change mitigation, since technologies for adaptation to climate change (so-called climate safe) in the energy supply sector are not yet discussed in the literature and have little potential for the nearby future.

The total primary energy supply including non-commercial biomass reached 398 EJ in 1996, of which renewables (excluding hydro) represent 10.9% (43.4 EJ) and total final consumption reached 283 EJ (IEA, 1998). According to other authors (Craig and Overend, 1996) total primary energy supply should be higher (around 435 EJ), since non-commercial biomass supply is probably near to 80 EJ. Global annual energy consumption has grown at an average amount of 2% for almost two centuries (IPCC, 1996a; IPCC, 1996d). According to five notable global energy scenarios exercises (IPCC, 1992; WEC/IIASA, 1995; Kassler, 1994; Shell, 1995; IPCC, 1996b) forecasts are quite different, ranging by a factor of 2, which means an average growth of 1.96% per year for the most energy intensive scenario to 0.73% per year for the lowest one. The large difference is due to assumptions regarding improvements in energy efficiency, change in user habits, change in the profile of available energy sources and the capacity for development of different regions of the globe. Even so, a significant conclusion is that there are opportunities to improve the standard of living of the world's population with an increase of energy needs below the historical trend.

Energy related CO_2 emissions are projected to increase at a slower rate than energy consumption. Historically, CO_2 emission intensity of both economic and energy activities has been decreasing by 1.3%/year since the mid 1800s (1% decline in energy intensity per unit of economic value added and 0.3%/year due the replacement of fuels with high carbon content (e.g. coal) by those with lower carbon content (e.g. natural gas) or those with zero carbon content (e.g. nuclear power; sustainable biomass) (Nakicenovic *et al.*, 1993; Nakicenovic *et al.*, 1996). This is a trend with no inducement from environmental concerns. However, with environmental issues gaining momentum, CO_2 emission intensity should decline at a faster rate. CO_2 emission in the IPCC - IS 92 scenarios from the energy sector moves from the 2.3 GtC/year

value of 1990 to 2.3-4.1 GtC/year in 2020 and 1.6-6.4 GtC/year in 2050 (IPCC, 1996f).

The energy supply sector is also responsible for substantial methane (CH_4) emissions. Out of a total of 535 TgC CH_4 emissions per year 100 TgC or roughly 20% is related to fossil fuel use (Prather *et al.*,1994). Of the fossil fuel related part, two thirds originates in the oil and gas industry (pipe and compressor leakages, venting and flaring) and the remaining in the coal industry (mine leakages and incomplete combustion) (IEA, 1997b; IEA, 1996).

CO_2 emissions caused by the energy supply sector can be reduced with the use of some or all of the following options (IPCC, 1996a; IPCC, 1996b):

- More Efficient Conversion of Fossil Fuels;
- Switching to Low-Carbon Fossil Fuels;
- Decarbonisation of Flue Gases and Fuels, and CO_2 Storage and Sequestering;
- Switching to Nuclear Power; and
- Switching to Renewable Sources of Energy.

A recent IPCC publication (1996b) provides a complete discussion of the many available technologies for each option, while another one (IPCC, 1996f) makes an effort to quantify the amount of CO_2 abated for each option. Table 10.1 lists the technical CO_2 reduction potential for six different mitigation technologies, by the year 2020. According to Table 10.1 CO_2 abatements of more than 10% can be obtained (a more precise figure is difficult to quote since the mitigation potential of the individual options identified are not additive, because the realisation of some option is mutually exclusive or may involve double-counting). The technical potential of each mitigation strategy, in the long term, is much higher. Only replacement of coal by natural gas can reduce overall emission per unit of electricity generated in the range of 50%. The main question is how far CO_2 abatement in the energy supply sector is cost effective when compared with other possible actions able to achieve the same reduction. A global evaluation of the optimal mixture of options for 2.4 GtC emission reduction concludes half of it should be performed through increasing carbon sinks (forestry), and a fraction of the other 25% should be obtained by improvements in energy conservation (Jepma and Lee, 1995). Major constraints considered in the figures of Table 10.1 are the long average life of energy conversion plants, and the rate of technology transfer. Most of the growth in the energy sector will occur in the developing countries and these usually are the last to adopt new technologies, due a number of issues, but particularly the absence of binding commitments regarding GHG emissions.

10.2 Climate Mitigation Technology

About 75% of the world's energy supply comes from burning non-renewable fossil fuels (coal, oil, and natural gas) (IPCC, 1996f). The balance comes from nuclear or renewable energy. Since fossil fuels are expected to be a major source of energy for

---

[1] Cases cover decentralised and centralised applications. Decentralised applications include:, Inner Mongolia Wind (case 3), PV in Kenya (case 5), Butane gas in Senegal (case 7). Centralised applications include micro-hydro in Peru (case 24) and Clean coal power plant in China (case 6). Other cases include natural gas production in Indonesia (case 13), alternative energy development in India (case 22).

[2] 1 EJ = 1018 Joules.

Table 10.1	Technical CO$_2$ Reduction Potential Based on IS92a Scenario (and Range for IS92e to IS92c) (Source: IPCC, 1996F)		
MITIGATION	**Gt C**	**% of Annex I**	**% of World**
Replacing Coal with Natural Gas for Electricity Generation in Annex I Countries	0.25 (0.01 - 0.4)	4.0 (2.0 - 6.0)	2.5 (1.0 - 4.0)
Flue Gas Decarbonisation (with de-NO$_x$ and de-SO$_x$) for Coal in Electricity Generation in Annex I Countries	0.35 (0.1 - 0.6)	6.0 (3.0 - 8.0)	3.5 (1.5 - 5.0)
Flue Gas Decarbonsation (with de-NO$_x$) for Natural Gas Electricity Generation in Annex I Countries	0.015 (0.0 - 0.05)	0.5 (0.0 - 0.5)	0.15 (0.0 - 0.45)
CO$_2$ Removal from Coal Before Combustion for Electricity Generation in Annex I Countries	0.35 (0.1 - 0.6)	6.0 (3.0 - 8.0)	3.5 (1.5 - 5.0)
Replacing Natural Gas and Coal with Nuclear Power for Electricity Generation in Annex I Countries	0.4 (0.15 - 0.65)	7.0 (3.0 - 9.5)	4.0 (2.0 - 5.5)
Replacing Coal with Biomass (In Electricity Generation, Synfuel Production and Direct End Use) in Annex I Countries(a)	0.55 (0.25 - 0.85)	9.5 (5.5 - 12.0)	5.5 (3.0 - 7.0)

NOTE: (A) THE BIOMASS REQUIREMENTS WOULD AMOUNT TO 9-34 EJ/YR, WHICH IS LESS THAN THE RANGE OF 72-187 EJ FOR THE BIOMASS POTENTIAL BY 2020 TO 2025 (SAR II, B.3.3.2, 1996). THESE FIGURES ARE HIGHER THAN THOSE ASSESSED IN THE SAR CHAPTER ON AGRICULTURE (SAR II, 23, 1996), AND CAN BE ACHIEVED ONLY THROUGH ACTIONS WHICH GO BEYOND AGRICULTURAL MEASURES (E.G. LARGE EFFICIENCY IMPROVEMENT IN THE CONVERSION PRIMARY ENERGY/ELECTRICITY, POSSIBLE LARGE MARKET FOR BYPRODUCTS, INCLUSION OF EXTERNALITIES IN THE COST OF ENERGY).

the near future, reducing GHG emissions in the energy supply sector in the relatively short term must come about primarily by reducing the emissions of GHGs during the production and transformation of fossil fuels. Some technologies with both near and longer term potential to economically reduce GHG emissions from fossil fuel production are listed below, along with comments on technology transfer.

10.2.1   Oil

Crude petroleum is refined into many products, primarily gasoline, jet, diesel, fuel oil, and lubricants. In the refining of crude to produce finished fuels, 5-10% of the oil is consumed during processing in a typical refinery. Potential exists in some refineries to improve efficiency (Phylipsen *et al.*, 1998) and thereby reduce energy consumption in the process by up to 28% (Larsen, 1990). Among the measures and technologies to improve energy efficiency in refining are:

- Dissemination of "best practices", such as improved operating procedures and strategies to bring low-efficiency refineries up to the level of high-efficiency refineries.

- Use of newer more energy-efficient technologies, such as catalytic dewaxing, membrane separations, supercritical solvent extraction, simulated moving bed chromatographic separations, moving bed naphtha reforming, heat integration, co-generation, and advanced control and real-time optimisation systems.

- The greatest potential to reduce GHG emissions in the oil-producing sector is to mitigate the release of GHGs pro-

duced as a byproduct of oil production. These are mainly CO$_2$ and CH$_4$. Methods to mitigate CO$_2$ and CH$_4$ are discussed below.

10.2.2   Natural gas

The most important contribution of natural gas related to carbon emission reduction is the substitution of gas for coal. For example, this could be very significant in China where natural gas is desirable for replacing direct coal combustion for industry and heating in urban areas, where it causes serious air pollution problems.

Nevertheless, natural gas transportation is always a source of potential methane emission. There are large opportunities to reduce methane emissions from this source by promoting the more widespread use of available technologies and practices in the production, transmission, storage and distribution of natural gas (US EPA, 1993). The largest emission reductions can occur from using or reinjecting gas associated with oil production, rather than venting or flaring this gas. Flaring is preferable to venting because even an inefficient flare will convert the majority of the methane (usually more than 98 per cent) to a less harmful greenhouse gas (CO$_2$). Around 5 per cent of world natural gas production goes to flaring and venting (80% and 20% respectively). A similar amount, 6 to 7 per cent of world production, is thought to be reinjected. Encouraging the use of natural gas and developing the infrastructure to do so can help reduce emissions from oil production. Estimates are that a 50 per cent reduction in emissions from venting and flaring is readily achievable (US EPA, 1993). Thus, technologies are readily available to reinject gas or to efficiently flare when necessary. Large oil corporations could show leadership and actively stimulate the use and dissemination of these technologies (see section 5.4).

Other measures to reduce losses during natural gas treating and transport have been developed. For example, the Natural Gas Star Program at the US EPA has documented a variety of best practices such as (US EPA, 1997):
- Use of dry seals in centrifugal compressors;
- Detailed inspections and maintenance at compression stations;
- Reduction of glycol circulation in dehydrators;
- Vapour recovery on crude storage; and
- Replacement of high bleed pneumatic devices with low or no bleed devices

An important issue regarding natural gas transportation is pipeline leakage. This is a serious problem in the Former Soviet Union (FSU), particularly in Russia and the Ukraine, through which passes a significant share of European gas from Russia, but these leakages can be mitigated (Strategic Development of the Russian Gas Industry, 1998).

Another source of greenhouse gases is the CO$_2$ that is co-produced with oil or natural gas. The CO$_2$ is usually removed by amine scrubbing. Rather than release the CO$_2$ to the atmosphere, CO$_2$ can be reinjected. Excess CO_2, extracted from natural gas, is being injected into a shallow underground aquifer in the Sleipner Field in offshore Norway (Baklid and Korbol, 1996; US DOE, 1997). In this development, 1 TgC per year are being re-injected.

10.2.3 Coal

Methane is often vented to the atmosphere in the degassing of coal mines. Technologies to economically recover methane for fuel use, which could reduce emissions by 30-90% (IPCC, 1996f), have been developed and are commercially available. Pre-mining degasification or enhanced recovery using vertical or horizontal in-mine boreholes can economically recover methane produced via drainage (which represents 30% of all coal mine methane). Nevertheless, the major fraction (70%) is produced via the ventilation systems and being very dilute (< 1% in air) is presently vented (US EPA, 1993). Technology should focus on the utilisation of this dilute stream. Estimate for equivalent CO_2 emission reduction by recovering and utilising CH_4 from one mine reaches 1 TgC/year. Capturing all CH_4 from coal mine and burning is estimated to have the potential to reduce greenhouse gases by about 5% of the CO_2 emitted from coal burning (APEC, 1997).

10.2.4 Nuclear

Water-cooled and moderated nuclear fission is a power source that does not directly generate GHGs. Indirect GHG emissions do occur and are accounted for by life cycle analysis (IAEA, 1994). Nevertheless, these emissions are at least one order of magnitude lower when compared with fossil fueled plants of the same electric capacity (Uchiyama, 1996; Taylor, 1994; Dones *et al.*, 1994). Major reasons for the significant slowdown in construction of these plants are the huge capital investment required for the construction of the plant, the handling of nuclear fuel materials which can be used for military purposes, the large volume of radioactive materials which can cause significant social and environmental disasters if improperly handled, and issues related to nuclear waste storage (WEC, 1998). Considering the positive aspect of low GHG emission, the technology could be considered provided it can contribute to sustainable development that is advocated in Article 12 of the Kyoto Protocol.

10.2.5 Power Generation

Electric power generation is one of the biggest single sectors emitting CO_2. Major options to reduce GHG emissions are summarised below:

- Replacement of fossil fuelled plant by non-fossil fuel systems with low GHG emissions. On a life cycle basis renewables and nuclear systems could release up to 2 orders of magnitude less CO_2 emissions than those of fossil fuelled systems (CRIEPI,1995; Uchiyama,1996).

- Fuel switching to less carbon-intensive fuel. Fuel switching from coal to petroleum or natural gas or from petroleum to natural gas can contribute to reducing CO_2 emissions (OECD/IPCC, 1991; 1995). Switching from fossil fuels to nuclear power can significantly reduce CO_2 emissions. For instance, CO_2 emissions avoided by nuclear generation in Japan amounted to 66 TgC, an amount equal to 20% of the overall CO_2 country emission in 1995 (FEPC,1998).

- Improvement of conversion efficiency by using advanced fossil fuel based technologies, such as combined cycle or retrofitting inefficient fossil fuel plants. According to IEA statistics, current average power conversion efficiency is around 30%, whereas that of most efficient commercial plants with natural gas combined cycle systems already reach over 55% (IEA,1998a).

- Improvement of thermal efficiency by use of cogeneration to supply process or district heat. Depending on the circumstances, this can increase thermal efficiency substantially. District cooling systems can also improve overall thermal efficiency in megacities with dense population and stable demand.

- Improvement of efficiency of transmission line by increasing busbar voltage and/or using DC. This could improve transmission efficiency up to 10% in some situations. More localised power production will also bring less transmission losses and contribute to local and regional development.

- Improvement of efficiency by maintenance and modification of existing systems. For example, rehabilitating hydropower plants or recovery of capacity of reservoirs by dredging.

- CO_2 capture and sequestration from power plants has the potential to substantially reduce CO_2 emissions, but more R&D is needed to make it economically viable and assure that environmental impacts are negligible.

- Utilisation of fuel cell technology when commercially viable. Fuel cells are able to convert hydrogen to electricity at higher efficiency than through direct combustion. Hydrogen can be produced from fossil fuels, renewable fuels or by electrolysis of water. It is also an effective way to cogenerate heat and power in relatively small quantities. By improving efficiency, less CO_2 emission will be produced for the same amount of electricity generation.

10.2.6 Biomass and Small Scale Renewables

Renewable sources of energy can provide energy in the final form required by users while emitting significantly less GHGs than fossil fuels. Carbon emissions are present through the use of fossil fuels during the process of planting, harvesting and storage of the renewable resource and its transformation into a commercial form of secondary energy.

Technology improvements in biomass productivity, harvesting and collection, and conversion are expected to further reduce GHG emissions by:

- Increasing the yield of biomass forests and plantations;
- Gasifying biomass as an intermediate step for gas turbine based electricity generation or cogeneration or other high efficiency combined cycle concepts;
- Converting ligno-cellulosic materials to ethanol for use as a motor fuel. This is one of the few possibilities in the near term to displace fossil fuel in the transportation sector because of existing distribution infrastructure (Interlaboratory Working Group, 1997).

When regarding other small scale renewable technologies it is very important to consider the historic technological improvements. Wind propelled electric generators and photovoltaics have been able to reduce their cost for generating electricity, but costs are still relatively high as compared to small-scale hydroelectricity and geothermal generation. These are options with lower unit costs, but for which cost reduction potential is more difficult to identify. Technical feedback and increasing economy of scale will lead to further cost reductions. Opportunities for commercial-grid-connected wind and photovoltaic-based electricity production systems will allow these technologies to be part of the large-scale renewables framework, today primarily consisting of hydro and biomass-based electricity generation. Use of small-scale hydroelectricity and geothermal generation are other options.

10.2.7   Cost and Potential

The fossil fuel-producing industries are quite mature and their technologies are well developed. Biomass and small-scale renewables have potential in niche markets but generally cannot compete with fossil fuels on a direct cost basis. A serious obstacle is the lack of inclusion of external costs regarding social, environmental, and even indirect economic gains which can affect country trade balance involving hard currency for developing countries and CEITs (Moreira and Goldemberg, 1999).

Fossil fuel technology is readily available in most instances from commercial and governmental sources and can be, in most cases, readily licensed. For example, technology to catalytically remove wax from lubricating oil was developed by an oil company (Hydrocarbon Asia,1994). This technology replaced the costly and energy intensive solvent dewaxing process and reduced energy consumption and therefore CO_2 emissions by about 85%. This technology is now used both in developed and developing countries. In some countries, to facilitate technology transfers, this technology is offered through a partnership with local institutions. This is one of many technologies that together can result in significant improvement in energy efficiency and greenhouse gas reductions.

Deployment of new technology, improvement of operating efficiencies, and other best practices to reduce GHG emissions are currently driven primarily by economic opportunity. However, developing countries are not always aware of the opportunities, and education and awareness programs should be encouraged. In the energy supply sector, economic, educational and institutional barriers, rather than technology availability, are more apt to be the cause for the failure to transfer technology.

10.3   Current Transfer of Climate Change Mitigation Technology

10.3.1   Current energy supply technology transfer

Energy supply technology transfer in the past
Current pathways and actors of energy technology differ substantially from historical patterns. In the 1950s and 1960s large scale power projects and rural electrification schemes financed by multilateral banks and operated by national public monopolies dominated the electricity sector, while the international oil and gas companies ruled over the exploration, development and production of hydrocarbons. Technology transfer was in fact a euphemistic term for large-scale public investments based on foreign technology and soft loans with minimal knowledge transfer and domestic capacity building. The oil crises of the seventies caused a revolution by stimulating the domestic search for oil substitution and shifting interests from supply side issues to demand side issues. The international oil and gas companies now had to negotiate with demanding governments of exporting nations and their emerging national oil companies. Requirements for training of local talent and purchasing of domestic equipment were increasingly included in new contracts. At the same time new ventures in small-scale renewable technology and energy conservation began to influence views on technology transfer. The incentives for large-scale hydro and nuclear were weakened due to environmental and social concerns. Technical assistance, including training and R&D activities, became a standard component of project financing by multilateral agencies. The emphasis gradually shifted from isolated hardware package deals to more integrated, process-oriented approaches including proper incentive structures for relevant actors. Moreover, at the same time it became clear that the diversity among developing countries was growing fast in terms of economic development potential and technological absorption capacity, which implied a growing divergence in the needs for technology transfer in the developing world.

Energy supply technology transfer in transition
In the early 1990s a second revolution followed when processes of market globalisation accelerated and energy demand growth in the industrialised countries remained low. The availability of private capital on a global scale increased greatly and competition between global vendors of technology became intense. In addition, market restructuring through liberalisation and privatisation spread from the industrialised to the developing world. This restructur-

ing process will ultimately affect the role of government in guiding technology transfer in a major way. At the same time, the strong growth of energy demand in some parts of the world increased the range of potentially profitable technologies of interest to developing countries. At low levels of demand it does not pay to build complex refineries and long-distance pipelines. These changes are also affecting the traditional actors on the global energy scene, in particular multilateral banks and oil and gas multinationals, who are no longer concentrating on their traditional roles. Multilateral banks are keen on diversifying their portfolios of energy supply projects and have included environmental objectives besides traditional economic goals (see also section 5.2). Special funds such as the Global Environmental Facility (GEF; see Box 5.2 in Chapter 5) have been created. Oil and gas multinationals are venturing outside their traditional domain of hydrocarbon development and production, and are involved in power projects and renewable energy technologies. They now share the global stage with major equipment vendors, emerging domestic firms, private financing institutions, independent project developers and major energy clients.

From a policy perspective these changes have important consequences. In today's dynamic market environment government policy on the level of specific technologies and isolated actors is no longer very effective and could even be detrimental. It is becoming more and more difficult to pick specific winners and losers. While formerly governments played an active role as recipients in the technology transfer process, they now concentrate more and more on regulating the rules of the game and promoting enabling policies of a general nature. This involves more attention for the proper economic incentives, the technological consequences of trade regulations and the legal aspects of innovation policies. They are now keen on facilitating technology transfer in general through creation of an adequate institutional infrastructure with high quality engineering education, promotion of R&D activities, adequate industrial standards and flexible market intermediation services.

Climate change and energy supply technology transfer
Current energy supply technology transfer is primarily driven by objectives of economic development and international competitiveness. Climate change objectives and in particular the reduction of CO_2 emissions do not play a significant role. This does not imply that energy supply technology transfer has no effect on climate change, but that such effects are coincidental rather than intended. This situation could change in the next few years because of the binding commitments for industrialised countries and CEITs under the Kyoto Protocol, and also because of the opportunity to use, in addition to the measures implemented domestically, the provision for the flexible mechanisms under the protocol. At the Buenos Aires Fourth Conference of Parties (CoP₄) further steps were agreed upon towards implementation of these instruments. Article 6 of the Kyoto Protocol allows for Joint Implementation (JI) projects among Annex I Parties (developed countries), which implies that emission reduction units from specific projects can be transferred between countries. Article 17 allows opportunities for international emission trading

among Annex I Parties. These mechanisms could be of major interest to CEITs in Eastern Europe. Finally, Article 12 defines the new Clean Development Mechanism (CDM) as a multilateral mechanism to assist Non-Annex I Parties (developing countries) in achieving sustainable development while allowing Annex I Parties to comply with their reduction commitments[3]. The intentions of the CDM instrument are particularly relevant for initiatives in the energy supply sector of developing countries, and raise the question how energy supply technology transfer can be guided in a direction which will both enhance economic development and reduce potential CO_2 emissions. For this purpose it is useful to concentrate on the technologies which are inherently beneficial from both the economic and climate change perspective Section 10.2. has already discussed the categories of technology options as presented in IPCC (1996f) and listed in Table 10.2. The six categories distinguished include: efficiency improvement, switching to low-carbon fuels, removing and storing CO_2, nuclear power, biomass resources and small-scale renewables.

10.3.2 Stakeholders and pathways for energy supply technology transfer

Efficiency improvement
The actors involved in the four sectors listed under efficiency improvements in Table 10.2 are diverse and require different approaches for purposes of technology transfer policy. Yet problems of energy efficiency improvement in developing and transitional economies share some basic characteristics with efficiency improvements in other areas of resource use. Few technologies are operated at design specifications right from the start, and their performance generally tends to diminish as time goes by. This observation is particularly relevant for advanced technology with low operational tolerance levels. The reasons have to do with system-wide deficiencies such as inadequate maintenance discipline and lack of spare parts. They are not unique for certain technologies. A brief outline of the actors and pathways involved in the four areas of potential efficiency gain follows:

- The oil and gas sector is a mature, globally oriented industry involving not only the large multinationals, but also a complex network of specialised suppliers of equipment and services. In many developing and transitional economies local companies are tied into this network. The preferred pathway for technology transfer is clearly through private sector contracts in many forms, from licensing to foreign direct investment. The government role is primarily limited to enabling actions having to do with establishing a stable and competitive economic climate. Given the huge investments in oil and gas production, refineries and pipelines, geopolitical considerations play a major role and the risks of not recovering investments is potentially high. Major gains in efficiency

---

[3] For more information on Buenos Aires decisions, the Kyoto Protocol, including JI and CDM see sections 3.4 and 3.6 in Chapter 3.

improvement are possible given the right incentive structure for private firms.

- In contrast to the oil and gas industry, coal industries, with some notable exceptions, are more nationally oriented even in Europe, and technology transfer flows primarily through government driven pathways. The industry also has a long historical record in some key developing countries, particularly in India and China. Indeed, where the lack of vested interests may hamper efficiency improvement in the oil and gas sector, the opposite may be the case where coal is concerned. The position of domestic firms in emerging economies is strong, but investments in conversion efficiency improvements and technological innovation are insufficient to reach internationally accepted standards of performance. In those economies large efficiency gains throughout the chain from mining and bulk transportation to steam and power generation are potentially available. Sector restructuring is in many cases necessary to create competitive conditions and improve performance. Incentives for domestic mining companies and equipment manufacturers to actively seek cooperation with foreign firms are needed. Local coal R&D activities form a necessary component of technology transfer policies. Stimulation of joint ventures and foreign direct investment would involve better protection of proprietary information, support for licensing agreements, removal of import restrictions, and adequate enforcement of environmental standards.

- Like the coal industry, power generation has been an almost exclusively domestic affair in most countries for a long time. The past decade has seen a major change in the orientation of this sector because of a global wave of liberalisation and privatisation. At the same time electricity demand growth in the OECD has been slow (Price *et al.*, 1998). These factors have led to an increasing international orientation of the major players in the industry, ranging from equipment suppliers to former public utilities. On the other hand, electricity supply (both in terms of total load and reliability) in most developing countries cannot keep up with demand, and this is becoming a major constraint on economic development. The gradual decoupling of public policies and political interests from utility regulation and electricity markets will have a profound impact on the future performance of the sector. The initial drive in this process must come from the government. Restructured national utilities whether in private or public hands can then involve the private sector through independent power producers or other forms of partnerships in efforts to improve efficiencies in generation, transport and distribution.

- Cogeneration (combined heat and power application) has had a major impact on energy efficiency improvements in many OECD countries (IEA, 1997). Without a proper regulatory framework and special tariff construc-

tions, cogeneration will be difficult to implement in many cases. Although cogeneration projects are based on private sector initiatives and pathways, the public sector has to play a major enabling role.

Switching to low-carbon fuels

The major option for switching to low-carbon fuels is replacing coal or oil with natural gas. The natural gas sector is dominated by the oil and gas multinationals, global pipeline construction companies and major gas equipment manufacturers. It is a mature industry characterised by gradual technological innovation in the last decade primarily on the production and conversion ends of the resource flow (off-shore operations and gas turbine technology). Core technologies in this chain are typically R&D intensive and logistically complex, and thus not easily transferred to domestic firms. There are, however, substantial opportunities for peripheral supplies during construction, and once installed operation and maintenance can be transferred relatively easily. The projects involved are politically highly visible and of the front page news type. This makes a stable gas market regime essential for attracting capital at reasonable conditions. Besides technical skills of construction and operation such projects require entrepreneurial and bargaining skills at the recipient end to balance interests competently. Natural gas markets and applications are a relatively new phenomenon in most developing countries, and thus need considerable policy attention when it comes to building up required skills and supplier industries. Regional cooperation is an essential element of success in transnational gas ventures, because a large part of the supply costs are in infrastructural investments with high risks of recovery.

Decarbonisation of Fuels or Flue Gases, and CO_2 Storage

The actors involved in CO_2 removal and storage will first be oil and gas companies and perhaps later electric utilities. Because CO_2 becomes available in pure form in some petrochemical complexes (hydrogen, methanol and ammonia production), and because CO_2 stripping from natural gas fields is necessary to meet commercial fuel specifications, oil and gas companies are facing just the costs of storage. If, in addition, the injected CO_2 improves recovery rates in production from natural gas and coal beds, niche applications can even be attractive when carbon emission taxes are low. Removing CO_2 from natural gas (e.g. steam reforming) or from flue gas in power plants is an expensive add-on technology, while new integrated technologies like coal gasification or synfuels with undiluted CO_2 flows are not commercial. Technology transfer concerning this option could involve both North-North and North-South partnerships. At the same time large-scale CO_2 storage arrangements without commercial benefits will require government-driven technology transfer pathways.

Nuclear power

Nuclear planned capacity additions in the past few years were located primarily in Asia. Moreover, considerations of operational safety and waste management have led to considerable involvement of industrial countries in nuclear power plant upgrading in transitionary economies. From the point of view of technology transfer these developments cannot be ignored. Governmental organisations at the national and international level are the most important stakeholders in this respect, as are the major engineering and construction companies involved in nuclear energy. In general, nuclear power requires an elaborate national regulatory and technical infrastructure, and affects key international political issues. The major pathways for technology transfer in this area are thus strongly government-driven and embedded in international agreements. With respect to operation of existing nuclear power plants, measures to strengthen and improve technology transfer in the areas of plant safety, personnel training, and the nuclear fuel cycle are needed.

Biomass

Biomass technology for energy generation or fuel production is the most complex cluster of the six major options listed in Table 10.2. First of all, biomass technology is still evolving, which makes it difficult to decide what exactly should be transferred in terms of knowledge and techniques. Secondly, biomass technology requires an interconnecting series of difficult technological choices concerning biomass sources and production, biomass handling and transportation, and biomass conversion and end use. These choices are to a large degree area-specific and cannot realistically be addressed on a generic level. Finally, there are a multitude of actors who potentially could become crucial players in global markets. Nevertheless, at least for some developing countries in Latin America, Asia and Africa, biomass energy may become the most important opportunity on a community level for economic development in an environmentally conscious world. The Brazilian alcohol programme (see Case Study 8, Chapter 16) testifies to this observation despite its present economic difficulties (Moreira and Goldemberg, 1999). Biomass technology transfer under current conditions is mostly dependent on government driven pathways, such as active involvement in R&D activities, demonstration projects financed locally or internationally, and government sponsored programmes to determine the resource availability. An example is the joint USA-China effort to develop a biomass resource database (Zhu, 1998). Such efforts are necessary to prepare the ground for large-scale involvement at a later stage[4]. The development of biomass energy options can also promote incremental carbon sequestration.

Hydroelectricity

Hydroelectricity is the largest source of renewable energy now being used. Technology transfer is occurring as shown by the intensive programme of hydro plant construction in several developing countries. Unfortunately, local impacts due to the use of rivers for other purposes, and the social problems related with population displacement for water storage are making it difficult to justify large-scale hydroelectricity as environmentally sustainable, unless several complementary measures are added to the projects (Liebenthal *et al.*, 1996). Hydroelectricity generation,

like most renewable energy technologies, is capital intensive which can be an important financial barrier. The electrical sector is, however, now searching for low cost alternatives because of economic pressures due to de-regulation or to privatisation. Run-of-the-river, small-scale hydro and pumped-storage hydros are being considered as more sustainable alternatives to the use of large scale hydroelectricity, despite reducing significantly the available economic potential (Moreira and Poole, 1993).

Small-scale renewables

Small-scale sources for renewable electricity based on wind or PV have been popular items of technology transfer programmes since the early 1970s. Only in recent years have these led to impressive success stories such as the penetration of wind parks in India and Mongolia (see Case Study 3, Chapter 16) or the penetration of solar home systems in Kenya (see Case Study 5, Chapter 16). These technologies are to a large extent dependent on specific niche markets created through government intervention, combined with the entrepreneurial spirit of the involved communities. Yet they hold great potential for the immediate future. In general, the main actors in the world market are equipment manufacturers from industrialised countries, who try to penetrate worldwide through a variety of cooperative agreements with counterparts in developing countries and strong reliance on international aid funds. Their role is increasingly challenged by domestic manufacturers. Because these technologies are generally purchased by end users rather than power producers, arrangements with respect to marketing, financing, and after-sales services on the local community level are just as important as technical performance and manufacturing capability. Without competent intermediaries the chances of successful market penetration are low no matter the origin and performance of the product. The necessary involvement of a large number of people distributed over a large area makes technology transfer for renewable electricity difficult and requires continuous government intervention to increase awareness and institutional commitment, and to stimulate appropriate education and technical facilities.

10.4 Programmes and Policies for Technology Transfer within Countries

In some areas, like energy, the opportunity to transform knowledge advances in commercial products and services is enhanced due to the large amount of investments made every year. Expenditures in energy are approximately 6% of the GDP for OECD countries (Baron, 1996). Unfortunately, even technologies which provide costs reductions for the energy provider are not easily transferred due do barriers which are present in all stages of the transference process - assessment, agreement, implementation, evaluation and adjustment, and replication, (see Chapter 1, and Figure 1.3).

Examples of barriers and opportunities for the transfer of environmentally sound technologies (ESTs), and the participation of different actors in the process are presented in this section.

---

[4] This effort estimated 100GW equivalent potential from the biomass resources available as agriculture residues.

Table 10.2	Major Transfer Options, Stakeholders, Pathways, Barriers and Policies in the Energy Supply Sector					
MAJOR TRANSFER OPTIONS	KEY STAKEHOLDERS	KEY PATHWAYS	NATIONAL BARRIERS	INT'L BARRIERS	NATIONAL POLICIES	INT'L POLICIES
Efficiency improvement						
Oil refinery/gas transport	Oil & Gas multinationals National oil companies Engineering contractors	Private sector	Lack of competitive conditions	Lack of competitive conditions	Promote FDI Voluntary agreements Promote best practices	Promote FDI Trade Policies Prom. best practice.
Coal mining	Coal mining companies Equipment manufacturers Labour unions	Private sector	Lack of competitive conditions	Lack of competitive conditions	Promote FDI Voluntary agreement Promote best practices Capacity building Regulatory policies	Promote FDI Trade Policies Prom. best practice. Promote FDI Capacity building
Power Generation	Utilities Regional agencies Equipment manufacturers	Private sector Government driven	Domestic technology and Management skills		Regulatory policies Promote FDI	Promote FDI Capacity building
Cogeneration	Industries Utilities Regional agencies	Private sector Government driven	Economic feasibility Level of heat demand	Economic feasibility Scale		
Switching to Low C fuels						
Natural gas development	Oil & Gas multinationals National Gas companies Regional agencies	Private And Public sector	Political commitment Large scale markets	Political commitment Transportation costs	Regulatory policies	Regional cooperation.
Decarbonisation of Flue Gases and Fuels and CO2 Storage	Oil & Gas companies Electric utilities		Economic cost	Economic cost	Regulatory policies	Regulatory policies
Switching to Nuclear Power	National governments International agencies Equipment builders	Government to Government driven	Public acceptance Economic cost	Nuclear proliferation	Regulatory issues Energy supply security	Promote best practices Capacity building Financial policies
Biomass						
Biomass resources	Extension agencies Food, fodder, Fiber industries	Private and Community driven	Logistic infrastructure Land use competition		RD&D policies Infrastructure policies	
Biomass conversion	Utilities Equipment manufacturers	Private and Community driven	Immature technologies Economic costs Fuel standards	Economic costs	RD&D policies Financial policies	Financial policies
Small-scale renewable						
Wind	Domestic manufacturers International component manufacturers	Private, public and community driven	Economic cost Operation and management skills	Economic cost	Resource assessment policies	Financial policies
Solar	Utilities/Private International components manufacturers. Community	Utilities ,private And community driven	Utilities	Economic cost	Green electricity regulation	Financial policies
Small hydro	Utilities/Private Community	Utilities private And community driven	Utilities	Economic cost	Green electricity regulation	Financial policies

Discussion will be centred on issues and opportunities for promotion of technology transfer within the country. Most of the discussion shows that developing countries are less prepared to absorb energy technologies due the presence of barriers, some of which are intrinsic to the development process (shortage of information and finance), and others which can be more easily removed through political and administrative decisions.

10.4.1 Barriers to Technology Transfer

Many opportunities to reduce carbon emissions in the energy supply sector are well documented. A series of barriers precludes actions which can be undertaken to use already available technologies. A detailed list of barriers limiting energy efficiency diffusion and more intensive use of renewables is provided in Chapters 4 and 5. Instead of listing all of them we prefer to discuss a few, particularly the ones important for conventional energy sources.

More efficient conversion of fossil fuels can be obtained through identification of the necessary technology and its utilisation to improve oil refineries processing efficiency, coal mining processing, power generation and the widespread use of cogeneration, to name the most important ones.

Efficiency improvements in oil refineries are limited mainly by the lack of competitive financial conditions. Due to the low price of oil, capital investments in refinery upgrades often would not provide an economical return. The same is true for coal processing upgrades. Low price of fossil fuels are consequences of many traditional policies like direct and indirect subsidies (IPCC, 1996b), non-inclusion of external costs (Harou *et al.*, 1998) associated with their production and use (environmental and social costs), large scale of consumption and long time presence in the market which allowed the creation of an optimised production, transportation and commercialisation structure.

Opportunity to reduce emissions while reducing economic costs exists in the FSU, where the efficiency of long-distance, high-pressure natural gas transmission pumping is only 28-30 per cent in the compressor turbines. As far as technology transfer is concerned, Ukraine and Russia have developed 35 per cent efficient systems, but they cannot implement them due to the financial crises.

Efficiency improvements in power generation are already demonstrated and there is significant potential to increase present world average thermal efficiency from 30% to 60% (IPCC, 1996f; Willians and Zeh,1995). Diffusion of thermal plants with 60% energy efficiency is limited in industrialised countries to the small growth in electricity demand and by the long lifetime of existing conventional power plants. In developing countries, where opportunities for supply expansion are frequent, diffusion of the most efficient plants is occurring but limited by the availability of natural gas and capital constraints for the construction of gas pipelines. As result, efficient coal plants are being built but constrained by capital availability.

Cogeneration or sequential production of power and heat is a much more efficient process than the production of each one of these energies alone. Major obstacles are shortage of capital and lack of regulatory policies to allow commercialisation of the excess electricity produced through access to the existing grid systems.

Energy efficiency improvements in power generation and cogeneration also face barriers due to:

- Lack of incentive of the major utilities: Many electric utilities sell electricity through a regulatory review process that allows the utility to recover all operating expenses, including taxes and a fair return for its investments. This could give insufficient incentive to improve efficiency (US DOE, 1996).

- Deregulation creating uncertainties in the power generation business. Deregulation may not necessarily lead to the most environmentally friendly outcome unless the proper institutional policies are in place.

- Lack of human qualification in developing countries. Without investment in capacity building the existing electricity service is of lower quality compared with industrialised countries, information gathering is not a priority, and new technologies, which may be less costly or more environmental friendly, are seldom taken into account.

Switching to low carbon fuels is an important way of abating GHG emissions. In particular, replacement of coal and oil by natural gas as the primary energy source in power generation is an excellent solution. Conversion to natural gas is constrained by the long lifetime of the existing coal and oil power plants in operation in industrialised countries and by the costs associated with installation of pipelines and other infrastructure in the developing countries. The trend to power deregulation could inhibit fuel switching in the absence of complementary policies due to the existence of cheaper alternatives such as coal.

Lack of a consistent and comprehensive framework for the evaluation of energy costs from different energy sources is another serious barrier to technology transfer. For such an evaluation it is necessary to include the complete energy cycle analysis. In the case for biomass the amount of land needed, cost of collection and competition with farming to produce food must be considered. For nuclear energy costs for security, for handling and storing radioactive wastes and for disassembling the plant after its operational life are barriers to private sector involvement.

Uncertainties in the economic systems discourage long term investments, including sustainable energy. Most multilateral and international lending institutions are technologically risk averse. As a result, governments may be reluctant to invest in high-tech projects that entail high capital costs (ECOSOC, 1994). Unfortunately, most ESTs are characterised by large up-front investments. In effect, the pollution abatement advantage is paid in advance. This is also a serious obstacle for nuclear energy. Reduction in nuclear unit scale may be a way of widening its market, but must be initially tested commercially in industrialised countries to attract further economic interests.

Another concern is the lack of continuous energy supply from some renewable sources. Energy supply intermittence may require another energy source in a hybrid system or a storage mechanism to guarantee continuous supply. This adds cost or limits the maximum share of intermittent renewables in an integrated electric system (Ishitani *et al.*, 1996).

10.4.2 Programmes and Policies for Technology Transfer

Assessment
Assessment of the most efficient technology to abate GHG emissions while requiring the minimum amount of investment is an important aspect of the technology transfer mechanism.

Traditionally government participation in the innovation process has occurred as a main promoter of R&D and buyer of new technologies. It is important to note that government's role is also important in the area of information, regulation, and in the definition of market-based initiatives (see Chapter 4 on the role of enabling environments for technology transfer).

Information and education programmes can be conceived as efforts for increasing public awareness regarding the availability of climate-friendly investments and also directed to reeducate experts who are trained in conventional technologies and procedures. Programmes directed to both of these purposes can be very effective. In Brazil hydroelectricity has been almost the only source of electricity generation for many decades, and most of the population were not aware of other alternatives which could bring many advantages (low initial cost, short implantation time, few social issues, etc.), as well as many difficulties (GHG emissions, electricity cost dependence on oil prices, etc.). Information and education programmes were mostly carried out with knowledge acquired abroad through non-financial technology transfer services (intelligence gathering and dissemination, technical assistance, independent validation and testing, and brokering activities performed by economic development agencies, and non-governmental organisations), and diffused later on through the same mechanism.

The fact that some ESTs are in the public domain[5] provides opportunities for the countries which want to develop them locally (Korea, 1998). Hence, the priority task is to maximise the diffusion of information on these technologies. Increasing the use of computers and all their presently available associated software for communication (the Internet) can provide a very good opportunity for most countries. Better education is seen as an opportunity for developing countries to have the required trained human resources for developing and utilising these technologies, as well as to absorb and take advantage of the ones available from IC by any possible transfer mechanism (Davidson *et al.*, 1991). Regulation, such as setting minimum efficiency requirements for the energy supply sector, could be important for inducing use of more efficient technology.

Agreement and Implementation

Once assessment of the technology is completed it is important to examine how the technology transfer can evolve to its next step, which is the agreement between parties and project implementation.

Market availability is a very important condition for the success of a new technology (OTA, 1995a; Rothwell, 1974; Langrish *et al.*, 1972). Market pull is important since private entrepreneurs with short-term interests will be resistant to investments in new technologies unless a market exists and uncertainties are low. Usually this market is initiated by pilot projects, government investments or acquisitions, importation of products and services from abroad (FCCC, 1998), or consumers willing to pay a premium price. As they gain acceptance rising public interest, private investors from the country or abroad will understand that a

potential market is available, and through actions on maintenance services and on better prices it can be enlarged. The creation of a market has been a responsibility assumed quite often by the government through procurement of new technologies for public missions like energy production and distribution, and provision of incentives for its development, including grants, low interest loans, import duty exemption, income tax exemption, and competitively determined subsidies. Grants are a very common incentive used to stimulate adoption of a new technology in industrialised countries (e.g. The Clean Coal Program in US – see the Compendium on Clean Coal – http://www.lanl.gov/projects/cctc/; The Salix Consortium – Biomass for Rural Development (http://www.eren.doe.gov/biopower/newyork.html) and in developing countries (see Box 10-1, and Case Study 6, Chapter 16). An example of the use of income tax exemption is discussed for The Netherlands (see below). Competitively determined subsidies have been employed in industrialised countries – the Non-Fossil-Fuel Obligation (NFFO) in the UK, and the electricity feed law (EFL) in Europe (Grubb and Vigotti, 1997; Brower *et al.*, 1997; Mitchell, 1995).

BOX 10.1   WIND POWER TECHNOLOGY IN INDIA

India has a history of policy-driven technology promotion for emerging technologies. Prior to the economic reforms initiated early this decade, the new and renewable energy technologies were promoted in the initial stages through the target-oriented-supply-push approach. Government agencies identified the investors as well as suppliers and fixed the technology price. Financial support was limited to target capacity without any price signal since market had no role.

Policies in the post-economic reform have allowed a greater role for the market. Project promoters accessed the financial incentives in addition to their own investments. The government provides tax concession through accelerated depreciation of capital equipment. Under the market dynamics, the technology transfer occurred via the market, as companies competed to provide the technology. The success of this pathway is evident in the rising deployment of wind power in India in past five years (Ramana and Shukla, 1998).

Market-based incentives set up by the government are useful approaches that are helping renewables to be commercialised in small niches. Three examples of this approach - the Government of India's effort to promote wind power technology (see Box 10.1), fiscal support for renewables in the Netherlands and the carbon tax in Norway (see below) are presented.

- In the Netherlands the government actively supports renewable energy through various financial measures. Fiscal policies allow accelerated depreciation and investment cost allowances for companies investing in renewables. The generated income of private investors in so-called Green Funds are tax-exempt. In addition, the distributing utilities are allowed to retain the standard reg-

---

[5] E.g. small hydro plants, natural gas flaring, municipal solid waste (MSW) burning and/or biodigestion, biofuels (from sugarcane, starches, and vegetable oils), solar thermal heating, biomass residues combustion (including co-firing), biodigestors, coal and biomass gasification, combined heat and power production, and wind mills.

ulatory energy tax to support renewable energy producers and may also sell Green Electricity at a premium price to environmentally conscious customers. In a pilot market scheme for Green Certificates, a trading mechanism (spot and forward market) has been established for renewably generated electricity, which makes it feasible to efficiently reach quantified targets for renewable energy introduction.

- Natural gas industry investments in most of the industrialised countries are essentially done by the private sector. But the presence of the government is important to create an enabling environment for private-sector driven technology. CO_2 extracted from natural gas wells and injected into a shallow acquifer (see 10.2.2) has turned economically feasible in Norway due the existence of a carbon tax. The factors that make this development different from the already practical injection of CO_2 for improved oil recovery is the shallow depth and low injection pressure, the wet CO_2, and the need for the system to dispose of available CO_2 at any time. By publishing the results of this development, the technology is made available to anyone who wishes to sequester CO_2 in an aquifer under similar geological and operating conditions.

A mix of policy instruments appears effective in many countries in promoting renewables. In Germany, for example, the success in promoting renewable energy in the last few years is attributed to a large extent to the Electricity Feed Law, which requires utilities to buy electricity from renewable sources at premium rate, and to direct subsidies for renewables, which amounted to DM 55 (approx. US\$ 29[6]) million in 1995/1996. The most remarkable aspect is the success of wind energy in Germany, whose capacity increased from 61 MW in 1990 to about 1,545 MW at the end of 1996. The process has been supplemented by an aggressive information campaign, which helps to disseminate information on current application and the most recent research available on renewable energy (IEA, 1997).

Technology implementation is facilitated if it fits in with country capital-labour mix. Under this vision, the GHG abatement option of increasing the use of renewable energy is welcome, because of the large number of permanent jobs created. Considering this measure, biomass-based electricity combined with heat production is the most favourable option, followed by PV. Hydro and mini-hydro plants also create a large number of jobs; yet only during the construction period (Moreira and Poole, 1993). The Brazilian Alcohol Programme has survived during the first half of the 1990s with arguments based strongly on the large number of permanent jobs created, and the synergism between the low qualification requirement of the cane harvester and the large number of low-literacy workers. This particular match was an important factor for the rapid diffusion of ethanol

production in Brazil (Moreira and Goldemberg, 1999). A similar situation is occurring in the United States where the creation of numerous work opportunities for corn farmers are used as an important argument for the promotion of this liquid fuel (Evans, 1997).

Another way to guarantee the implementation of technologies is the use of the legal structure. The Brazilian Alcohol Programme was introduced in such way. A law was passed to strengthen the national sugar industry, thereby reducing hard currency expenditures on oil imports. Thus, the Programme was motivated by potential economic benefits and not to meet GHG abatement targets. Another opportunity for energy technology transfer is the recognition by decision makers that availability of power, even in small amounts, provides a significant improvement in the quality of life (see Case Study 3, Chapter 16). This is pushing the market for small PV units. International experience has shown a positive correlation between access to energy and electricity services and educational attainment and literacy among both the rural and urban poor. Families lacking adequate energy supplies will tend to limit children's time spent on schoolwork and reading; in extreme cases, families may withdraw children from the school systems to spend time on firewood and dung collection. Worldwide, female children are disproportionately affected.

In stand-alone applications, remote from the electrical grid (for lighting, water pumping and refrigeration), PV has been competitive for several years (IPCC, 1996b). In such applications, PV systems are often competitive with presently used kerosene, candles and dry-cell batteries (assuming that low cost money is available), but typically there is no infrastructures to provide people with access to this technology (see Case Study 5, Chapter 16).

Evaluation & Adjustment
Even considering that the energy industry is quite homogeneous throughout the world, several opportunities for evaluation and adjustment of technologies exist in particular countries or regions. The operation efficiency of the electricity generating equipment in developing countries - with some notable exceptions - is often substantially below that achieved in the industrialised countries, despite the fact that the basic technology is the same (Maya and Churie, 1996). Thermal efficiency and forced outage rate, which are important measures of maintenance efficiency, should improve (OTA, 1992). Over the past two decades, the cost effectiveness of generation system rehabilitation has become recognised in the United States and Europe, where a great deal of attention has been placed on what has become known as "life extension" or "life optimisation" (OTA, 1992). Efforts of this type require hard and soft technologies, since most of the technological barriers leading to such poor performance are credited to inadequate training.

One of the adjustment processes is on-the-job training. Successful innovations require that entrepreneurs assemble a team of well trained scientists, engineers, technicians, managers and marketers who develop new technologies and incorporate them into products; manufacture them in a way that is timely, cost effective,

---

[6] This figure was calculated based on the currency exchange rate in November 1999.

and responsive to the market; and sell them. Training workers with these diverse skills is the responsibility of different institutions, both public and private (OTA, 1995a). In many universities, graduates often receive little training in manufacturing processes, product design and teamwork. Nevertheless, workplace education supplements formal education, as workers learn through experience and formal training programmes. For emerging technologies in particular, many of the skills needed for commercial success are not available in the formal education systems, but are developed instead by companies engaged in proprietary R&D programmes (OTA, 1995). Labour force skills are expanded through industry conferences, technical committees, and trade publications and technical journals, which provide an opportunity for industry participants to exchange ideas and share knowledge.

In the case of biomass technologies the issue of adjustment is magnified. Biomass yields are sensitive to climate and soil characteristics, and any successful project in one site has to be properly adapted to provide similar results in new sites[7]. This is exemplified in the discussion of Case Study 8, Chapter 16.

Replication
Even in the developed world R&D represent only a small fraction, 10 to 15%, of the resources required to bring to market a new product that incorporates substantially new technology. The other 85-90% is so-called downstream investment: design, manufacturing, applications engineering, and human resource development (OTA, 1995). Developing a highly qualified human infrastructure in the downstream investment sector, will facilitate in dealing with one of the two faces of technology development -- "problems in search of solutions" -- which is what industry, society, and design engineers encounter in practice. Even with poor investments in the second face of technology development -- "solutions in search of problems" -- a country can reach a high level of technological development (Greene and Hallberg, 1995).

Assuming R&D conducted at universities and research centres is sufficient, it is very important to establish linkages between networks of research laboratories and the private sector to facilitate the transfer of technologies to industry, and for the users of technology to channel their feedback to the generators of knowledge. In the USA, since the 1980s, Congress and the executive branch began to supplement this approach within a series of programmatic efforts aimed at helping specific industries. Some of these efforts encourage government and private industry to (OTA,1995a; DOE, 1999):

1. Share the cost of strengthening the supplier base of some important industries

2. Share the cost of pre-competitive research projects, and,
3. Disseminate best-practices to manufacturing firms, many of which are unfamiliar with the most advanced manufacturing technologies and practices.

The US Government is committed to ensuring the full implementation of technology transfer policies and legislation in all federal agencies. For the DOE technology transfer is a priority mission at all levels of department management. The EPA and Department of Commerce also have made significant improvements in their technology transfer programmes (National Energy Strategy, 1991). This attention on technology transfer builds in a strong base of legislative and policy mandates. In recent years, the US Congress and the Administration have cooperated on legislation specifically directed towards increasing the transfer of federally developed technology to US entities (National Energy Strategy, 1991). States have also funded RD&D. California funds certain public interest RD&D projects to "advance science or technology not adequately provided by competitive or regulated markets". A surcharge on electric rates is the source of funds. Focus areas include renewables, efficiency, advanced power generation, power system reliability, and environmental research (Tanton, 1998; http:/www.energy.ca.gov).

Clusters of research-production-marketing activities, such as applied research parks or university-industry parks will facilitate the connection of research to production and the marketing of the results. Rather than dispersing assets, the parks offer a synergistic concentration of knowledge, workers, and facilities (Greene and Hallberg, 1995).

Although there is a consensus among economists that over the long-run innovation is the single most important source of long-term economic growth, and that returns on investment in research and development are several time as high as the returns on other forms of investment, private firms generally tend to underinvest in research and development (Cohen and Nell, 1991). Overall, government support for energy research and development in International Energy Agency member countries fell by 1/3 in absolute terms and by half a percentage of GDP in the decade ending in 1992 (see Table 5.1; Reddy *et al.*, 1997). Not only has overall energy research and development spending been declining, but also only a modest fraction of R&D spending has been committed to the sustainable energy technologies.

10.5   Technology Transfer between Countries

Technology transfer between countries adds another dimension of complexity to the issues traditionally addressed in the context of technology transfer within countries. This includes, but is not limited to, a difference in the price, taxation and tariff policy, difference in technology and manufacturing cultures and standards, and barriers to imports. Additionally, there are just a few examples of programmes for technology transfer that have an explicit aim to facilitate the transfer of climate friendly technologies, and include an analysis of barriers, approaches on how to overcome

---

[7] Efforts in this direction help diffusion of publicly owned technologies. A significant number of them are quite sensitive to transportation costs (e.g., small hydro, biomass) or are more demanded in rural areas than in large cities (e.g., small hydro, wind).

them, and the specified roles of different stakeholders. One such example is the programme for technical cooperation agreement launched by the US (see Box 5.5 in Chapter 5).

The status of private power in developing countries is growing. Developing Countries are at various stages in allowing private investments in power production. Brazil, Chile, Costa Rica, the Dominican Republic, India, Jamaica, Mauritius, Mexico, Pakistan, the Philippines, Tanzania, Thailand and Turkey all have legislation governing the private production of power either pending or in force (Kozloff and Shobodale, 1994). Private sector share in the energy supply is growing in the CEITs too, as result of the deregulation of the electricity market, with Hungary, Poland and the Czech Republic being the leaders in this process. This implies new and emerging opportunities for privately-driven technology transfer.

10.5.1 Barriers to Technology Transfer between Countries

Many barriers exist for technology transfer between countries:

- Access to capital is limited. The capital costs of ESTs are generally higher than those of conventional technologies. Also, owing to the risks perceived for new technologies, financing costs will tend to be higher. Moreover, the availability of FDI is limited and unevenly distributed around the world.

- Although many countries are revising their trade policies in order to liberalise markets, substantial tariff barriers remain in many cases for imports of foreign technologies including energy supply equipment. This limits exposure to energy efficiency improvement pressures from foreign competition on domestic suppliers and prevents early introduction of sustainable energy innovations from abroad. Where foreign exchange limitations and public revenue considerations make across-the-board tariff removal difficult, preferential treatment of ESTs could be a realistic option.

- National interest groups such as powerful extraction and construction companies can influence technology choices in favour of conventional technologies.

- Institutional and administrative difficulties exist to develop technology transfer contracts, which can be a necessity to qualify regional construction companies as potential partners of the entrepreneurship. There is a need for greater regional cooperation among developing countries, both in R&D work and in the international commercial contracting network.

- Developing countries have in general poor access to information. It is one thing to recognise that the information and technology desired are available, but is quite another issue to gain access to them. That will require that developing countries strengthen their linkages with the rest of the world by investing in the infrastructure needed to receive and transfer information. In this undertaking, partnerships are key: between research institution in developed and developing countries, between domestic and foreign firms, and between research institutions and the private sector. A modern information and communication infrastructure provides up-to-date technical information and publications, and allows instant communication among scientists around the world. Technical information services linked to worldwide information networks distribute knowledge quickly and cheaply to the producing sector. Thanks to satellites, the developing world can leapfrog immediately to advanced telecommunications capabilities, by passing the long road already travelled by the industrialised countries (Greene and Hallberg, 1995).

- A major requirement for successful agreement in technology transfer is the guarantee of intellectual property rights (IPR). Without an IPR law that is effectively enforced, there is little incentive for private companies to share their technology. The cost of IPR is usually quite small when compared to the capital investments and risks that are involved. In the energy sector, a well-developed mechanism exists for sharing IPR. It is the production-sharing contract. Under this agreement, private firms contract with local parties, usually state-owned companies or governments, to invest and share technology with them in return for a share of the products produced. This practice has proven very successful in the international oil and gas sectors, and could be a model for other energy supply areas (see section 3.5 in Chapter 3 for an elaborative treatment of IPRs).

- Needs of the developing countries are quite different to those of the developed countries. Developing countries are generally still focused on large capacities of cheap, reliable power with low technical risk, and have new technologies as a lower priority. Industrialised countries have the infrastructure to assimilate riskier new technologies, whereas developing countries do not. However, this approach prevents developing countries from technological leapfrogging.

- Economic incentives for donors are weak mainly when energy demand is scarce and scattered. This barrier can be minimised by the additional potential value gained through JI/CDM schemes.

10.5.2 Programmes and Policies for Technology Transfer between Countries

Assessment

In most developing countries, government used to be the major actor in the energy supply sector. With the emerging privatisation wave public utilities are losing their monopolistic advantage and opening opportunities for private entrepreneurs. The government is losing importance as an energy supplier, but still has an important role in the sector as controller, and as the responsible body for the definition of policies which can shape the energy market (e.g. regulation, market-based incentives). Government interest is the major single item able to define the country attitude regarding GHG emission control. For example, under the auspices of the US EPA Coal Bed Methane Program, technology transfer to other major coal mining states is being promoted. For example, Coal Bed Methane Centres or Clearinghouses have been set up to help overcome economic and institutional barriers to technology transfer in China and Russia. These centres promote information exchange, demonstration projects, and commercial development. Also, the Asian Pacific Economic Council (APEC) is starting a coal mine gas project to recover and utilise methane from deep coal mines in Southeast Asian countries.

The choice of generating technologies for capacity expansion is deeply influenced by the planning tools available. Energy plans rely strongly on the knowledge and manpower from utilities, which greatly prefer the old technologies they are familiar with, and often distrust new technologies. Quite often their actions optimise only the energy supply side and do not take demand into account. The integrated resource planning (IRP)[9] approach is a valuable tool to examine the potential and barriers for new technology transfer, and thus to stimulate the analysis of the interventions governments could prescribe to overcome these barriers. In many industrialised countries, IRP is introduced by law. In Denmark, for example, the electric utilities are required to present such plans and to update them on a regular basis. In developing countries IRP is being done when required, but it also can provide substantive information on the technology priorities for development of the energy system. In China, for example, an extensive study on the energy sector development using the IRP approach was conducted recently. The study explored options, effects and costs of GHG mitigation in the energy demand versus energy supply. Hydro, nuclear and wind options were considered (IEA, 1997). The role of IRP is changing under the conditions of a liberalised market. It provides a basis for a dialog between companies and the government with respect to the objectives of the future technology development and deployment, and the policy framework necessary to support these objectives. The presence of the government, which is likely have a better and broader view of the country's requirements, is imperative when energy plans are under development (Greene and Hallberg,1995):

Multilateral Development Banks (MDBs) are financing many activities necessary to ensure the transfer of technology. For example, they finance building institutional capacity, establishing research centres and funding demonstration projects. The

World Bank funding of ESTs is estimated to be approximately US$ 2 to 3 billion per year. Acting as a lead investor, however, these banks have influenced a significantly larger proportion of the total investment in ESTs. Environmental loans totalled US$ 1.63 billion and MDBs leveraged another US$ 1.64 billion in fiscal year 1996 (United Nations, 1997). ODA has addressed about 3% (US$ 1.3 billion) of total reported bilateral energy assistance to renewable projects (except large hydro) between 1979 and 1991 (Kozloff and Shobowale, 1994)[10].

Regional organisations have an important function at this stage of technology transfer, since they have a number of generic strategies that can promote and/or facilitate the implementation of some of the emission reduction options described in this Report. These strategies (Von Hippel, 1996) could include, but are certainly not limited to, the following:

Provide Information and General Training to Government Officials - Getting initiatives such as industrial energy efficiency, and utility boiler emissions control programmes, and fuel switching/renewable energy initiatives off the ground in most developing countries (particularly China and India) will be impossible without top officials embracing the concept. Consequently, the advantages and local/international opportunities provided by the measures and technologies must be presented to top officials in a manner that is both forceful and forthright[11].

Provide Specific Information and Training to Local Actors - Training of a very specific and practical nature must be provided to personnel at the local level. Examples include workshops and courses for factory energy plant managers, and power plant and heating system operators; and new job classifications such as energy-efficiency and pollution control equipment installers, energy auditors, and environmental officials.

Encourage the Implementation and Enforcement of Energy and Environmental Standards - Although some countries of Northeast Asia and Latin America have general policies supporting energy efficiency and environmental sustainability, not all have a well-defined, quantitative set of standards in place to codify these general policies. Where standards exist, furthermore, they may not be stringent enough to satisfy GHG emissions reduction targets. Once standards are set, it will be necessary to create the capability to enforce them by recruiting and training enforcement personnel and supplying them with the tools necessary to do their job, as well as the high-level administrative support needed for credible implementation of sanctions. Setting up these regulations and support structures is an area where international assistance may be valuable for some countries.

---

[9] IRP treats energy supply and demand on equal terms, and searches for the most economical solution to provide the final user the energy services he needs (cooking, lighting, transportation, etc.).

---

[10] See also chapter 5 on MDB finance and other sources of private and public sector finance and Chapter 2 on international financial resource flows.

[11] Tokyo Electric Power Company's educational programme is a good example of this strategy (TEPCO,1999).

Establish Programmes of Grants and Concessional Loans - Experience in China has shown that such a programme in itself can have a significant positive impact in overall sectorial energy efficiency (Liu *et al.*, 1994). The benefits of institutionalising support for pollution control and energy efficiency, however, would go beyond those obtained through the various individual projects themselves. Creating government agencies or corporations with their own budgets would signal a strong commitment to acid gas and GHG emissions reduction on the part of the government, and would create a constituency within official circles for promoting EST (see Case Study 3, Chapter 16). Moreover, by establishing a pool of funds for which government ministries, sectors, and/or individual enterprises could compete, it would stimulate at all levels awareness of the potential, methods, and technologies for reducing local, regional and global gas emissions (see Case Study 24, Chapter 16).

Promote Joint Ventures and Licensing Agreements - The growth in the need for pollution control and energy efficient equipment could be met by domestic production through joint ventures and licensing agreements between governmental or private organisations in developing countries and foreign companies with the necessary expertise to produce the needed equipment. For example, a wide variety of efficient industrial equipment and controls - including adjustable speed drives, higher-efficiency electric motors, and improved industrial boilers - have already been introduced to China through commercial channels and are being or will be manufactured there.

The effort of such organisations in promoting EST cannot be neglected. The United Nations Development Programme (UNDP) over the last two decades committed over US$ 400 million to energy sector activities, and funded more than 900 projects in energy planning, energy efficiency and conservation, conventional energy, electric power, nuclear energy, and new and renewable energy. Such UNDP funding helped energy-endowed developing countries develop their resources and train national staff in cost-effective and environmentally cleaner methods of producing and using energy resources. Despite its success, UNDP energy assistance was fragmented, reflecting the constantly shifting international consensus about what is needed (renewables in the 1970s; conservation in the 1980s and concern about the environment in the 1990s) (Gururaja,1995). Also, the amount of financial support (on average US$ 20 million/yr in the last two decades) is quite small to induce significant results in the energy sector[12]. Most of the contribution has been in providing information on feasible projects (ESMAP activities), provision of small-scale lending for renewable energy (FINESSE initiatives) and acting as an intermediary to obtain loans from multilateral lenders. One of the areas most benefited by UNDP actions has been the use of renewable sources of energy, promoted as early as 1981 in the UN Conference on New and Renewable Sources of Energy in Nairobi. Another very useful contribution of many regional organisations has been the promotion of non-commercial energy technology transfer through many international conferences, employment of personnel through technical assistance programmes, education and training of host country's labour force and information databases.

Agreement and Implementation

Availability of capital is the major obstacle for the implementation of a selected technology. The gap between the foreign exchange needs of developing countries in the power sector and total aid flows from abroad implies that total investments will have to be reduced, foreign aid increased, domestic finance increased, or private foreign investment increased even more[13]. All these options play a role in the oil sector and in the power sector expansion, but the last is the most likely to dominate (Barnett, 1992).

Oil and natural gas are being produced in many developing countries through one or several of the following arrangements: concession, production-sharing contracts, risk service contracts, service contracts, joint ventures and nationalisation. Excluding nationalisation, all other arrangements are made between an International Oil Company (IOC) and the Government, usually through the National Oil Company (NOC). This means that agreement and implementation procedures for technology transfer are well established. With the emerging privatisation wave, opportunities for FDI are growing (see Figure 9.4, Chapter 9).

Laws allowing increased presence of the private sector in the oil and gas markets, as well as laws allowing private electric power sales, are no guarantee that power generation markets will develop or that EST will be used. Nevertheless, improvements in the efficiency of energy supply should be achieved through improving efficiency in oil exploration and exploitation, in power generation, transmission and distribution, and increased investments in system-interconnections. Through competition it is expected that improved efficiency will occur. Global conversion losses in these areas are responsible for 85 EJ or 22 per cent of global commercial energy used in 1990 (IPCC, 1996d).

For energy production from conventional energy sources, usually based in large centralised plants, FDI can be the preferred option. The investment required may be unavailable in the host country, so FDI can very quickly create new job opportunities, raise technological and educational levels of the labour force, transfer technical and management skills, and help the local R&D institutions. Nevertheless, it is important to remember that through FDI various elements of the technology are supplied in "bundled" or "packaged" form (Able-Thomas, 1996). FDI enables the technology supplier to retain ownership and to exercise control over the new production plant which is inconvenient to the host country from a technology transfer viewpoint, but provides a flow of productive capital from one country to another (Able-Thomas, 1996). Thus, FDI is suitable for the energy sector where techno-

---

[12] The electricity sector of developing countries requires investments of the order of US$ 100-200 billion per year (Reddy *et al.*, 1997; IIASA/WEC, 1998).

[13] Aggregate net long-term resource flows to low and middle income countries has increased substantially from US$ 100 billion in 1990 to 284 billion in 1996 (World Bank, 1997).]

logical progress and spin-off opportunities are less common, as is the case of mature technologies like hydroelectricity, or for the manufacturing of solar cells and PV modules where technological improvements are occurring so fast that developing countries have little chance of recovering commercial investments from the technology before it becomes obsolete (Able-Thomas, 1996).

Government has a significant role as a promoter of ESTs, since a large part of the R&D money used is coming from public sources. Policies for the diffusion and accelerated transfer of publicly funded ESTs are available (Korea, 1998), and include financial incentives for their transfer to developing countries and promotion of the transfer of uncommercialised publicly funded R&D results to enhance capacity building in developing countries and CEITs. For other energy supply technologies with small commercial markets, which are still in the development stage, or are publicly owned, joint ventures and licensing may be a more appropriate way for technology transfer. Joint ventures are dependent on a reasonable national partner with similar technological capabilities and complementary skills and resources, and necessary where relations with host-country governments and institutions would be difficult for foreign companies (partners) alone (Chowdhury, 1992). At the other extreme comes licensing, which is of value if the transfer is limited to product and processes that the nation is capable of producing or using at internationally competitive costs, or products serving purely domestic markets, for which neither imports or exports are an option (Able-Thomas, 1996).

Even considering the strong and emerging presence of the private sector in energy supply there are other important actors involved in technology transfer.

MDBs influence private financial markets to support technology transfer. At the macro level, this is done by encouraging financial reforms in banking, privatisation, and stock market development. At the micro level, influence comes through developing, demonstrating and transferring the innovative financial instruments needed to accelerate access to technology (United Nations, 1997). In the short term the most effective contribution of MDBs regarding GHG abatement is through the money used to fund and leverage other loans in the energy sector (see Case Study 14, Chapter 16). MDBs have significant leadership in this sector, and in the period 1991 to 1995, 44% of all the external financing of infrastructure in developing countries has been addressed to this sector (United Nations, 1997). Funding has impacted GHG emission through the large share which financed hydroelectric plants and the share invested in promoting sectorial reforms in favour of competition in the developing countries' power sector (World Bank, 1994). These actions abated GHG through the options of switching to renewable sources of energy and more efficient conversion of fossil fuels to power and heat. Loans to thermoelectric plants and oil refineries also impacted GHG emission, since it is assumed that under the MDB management, projects are more technically appropriate to the country economy with implications on efficiency optimisation of plants (United Nations, 1997).

Since the establishment of the pilot phase of the Global Environmental Facility (GEF)[14] a total US$ 5.2 billion have been allocated to climate change activities. Of this sum, US$ 775 million was provided in grants from the GEF Trust Fund. An additional US$ 4.4 billion was contributed through co-financing (GEF, 1998). About one third of these projects are to enhance penetration of advanced renewable and cogeneration technologies in the developing countries' energy markets, such as PV, solar-thermal hybrid technology, micro-hydro and biomass. Another objective of these projects is to improve energy facilities operation, production management and marketing capabilities (GEF, 1997). A very interesting example is the US$ 6 million joint USAID/India and GEF project to prevent GHG emissions, which addresses the major market, financial and institutional barriers that hinder penetration of the modern energy supply technologies. The first, near-term component of the project aims at introducing cost-effective conversion measures and state-of-the-art coal technologies at the existing power plants. The second, long-term component promotes more efficient bagasse (sugarcane processing residue) cogeneration technologies based on high pressure multi-fuel boilers and high-efficiency turbines. This component will provide incremental cost support for six demonstration projects in addition to the technical assistance, training and cost-shared research grants for developing cane trash and other biomass fuels. (Climate Action Report, 1997).

In the same way as the World Bank has exhibited leadership in requiring developing countries' hydroelectric projects to take into account social and local environmental issues, MDBs have the capability to enforce international regulations on GHG emissions and promote the use of International Standards (ISO-series), in addition to their traditional role of addressing funding and verifying the feasibility of EST projects. One difficulty associated with these actions is the necessity to invest in projects based on new technologies that conflict with the traditional posture of most of these institutions, which are technologically risk adverse (ECOSOC, 1994).

Another multilateral source of support for transfer of climate friendly energy supply technologies to CEITs is the European Bank for Reconstruction and Development (EBRD). In fact, EBRD is the first international financial institution with a proactive environmental mandate, having sustainable development goals among its highest priorities. Its objective is to foster the transition of the CEITs to a market driven economy and to promote entrepreneurial initiative.

Bilateral organisations have special programmes with technology transfer as an objective. For example the Japanese government provides technical assistance in the electric power field through studies of project development, training of overseas technicians, sending industry experts overseas, etc. Development studies consists of initial master-plan development and preliminary best-plan scenario development, and feasibility studies requested from partner-countries. Total loans from the Japanese government

---

[14] See Box 5.2 in Chapter 5, for information about GEF.

equal nearly ¥ 1,000 (approx. US$ 9.5) billion in 1991. Of this total, ¥ 100 (approx. US$ 0.95) billion is in the electric power field. Of this aid, ¥ 40 (approx. US$ 0.38) billion is for hydro-electric power development assistance. In the period 1959 to 1991, the total amount of loans the Japanese government offered in the hydropower field was ¥ 700 (approx. US$ 6.6) billion (Fujino, 1994). Another example is the Swedish government programme for biomass boiler conversion in the Baltic states (see Case Study 18, Chapter 16).

As is well known, an important objective of many bilateral aid programmes is to promote donor country exports of goods and services (Kozloff and Shobowale, 1994). Although the distinction between development assistance and export promotion is frequently blurred, it is no accident that bilateral donors often direct assistance to technologies and products in which they have a comparative advantage in domestic world markets (e.g. The Energy Policy Act, 1998). Under this approach, more sophisticated alternative technologies are being promoted in the energy supply sector. Most of the resources have been addressed to increasing the use of renewables through geothermal energy, wind resources and photovoltaic cells. Also, technology transfer is mostly provided through provision of equipment. There is little support for developing human capabilities for environmentally sustainable energy technology, which is strongly needed by developing countries (der Werff *et al.*, 1994).

Considering the emerging presence of the private capital in the energy sector of the developing countries and the relatively small amount of incentives provided by the GEF, World Bank, Regional Development Organisations and Bilateral Organisations compared to the amount of investment needed in the energy sector, as extensively discussed in this section, it is necessary to recognise that incentives are almost non existent for the governments of these countries to stop using their cheap natural resource of coal and oil in favour of more expensive, imported ecologically-friendly fuels such as LNG or pipeline gas.

Evaluation & Adjustment
Adjustment is almost always needed due the difference in infrastructure, social, and economic system of the technology provider and recipient countries (see Case Studies 3 and 5, Chapter 16). Creation of labour opportunities is a high priority in most developing countries. Due to the large investments in the energy supply sector, the potential number of job opportunities are often underestimated. Adjustment in favour of manpower versus automation is important. Heavy automation implies more importation of goods and more hard currency expenditures, which is seriously constrained in developing countries and CEITs.

Evaluation of all potential gains in the technology recipient country must be performed with care, including short, medium, and long-term views, checking all aspects considered important for the country's development by extrapolating the energy boundary. As an example, in the oil sector, there is already a well-established system for technology transfer from International Oil Companies (IOCs) and the Government, usually through the National Oil Companies (NOCs)[15] (Barrows, 1993). Technology transfer occurs through provision of state of the art technology and capacity building, such as training. Most legislation or contracts today contain provisions obliging the IOCs to provide training of nationals of the host country (personnel from NOCs, the Government, and relevant government agencies (Barrows, 1993). Transfer of more detailed technology is required in some countries (such as Norway), and this necessitates a special agreement to assure patent protection and secrecy regarding technical processes (Barrows, 1993). On the other side, the combination of business and social objectives is a feature of the public-private partnership requiring disbursement of IOC money in the economic and social development of host countries (See Box 10.2 and Case Study 13, Chapter 16). Indirect technology transfer is induced through agreements on Preference for Use of Domestic Goods and Services. Generally the legislation obliges the oil companies to purchase domestically produced goods and services, if they are available at competitive terms; in particular the World Bank financed projects allow a cost differential of up to 15% in favour of local suppliers. Because of potential demand by the IOCs, the national suppliers of goods are stimulated to search and make arrangements for acquisition of new technologies which can make their products and services more competitive.

From the above example it is clear that technology transfer has been used by the recipient country as a way to get fuel, job opportunities, commercial and industrial activities, hard-currency savings or gains, human and institutional capacity building, and social and economic development. All of these gains must be properly evaluated by the country's society. Lessons must be derived from past experiences and results used for better agreements able to promote even more country development.

BOX 10.2   LIQUEFIED NATURAL GAS (CASE STUDY 13 IN CHAPTER 16)

The development of the remote Arun natural gas field in North Sumatra by Mobil Oil and Pertamina, the state oil company of Indonesia, is a good example of capacity building and technology transfer. The Arun field is located in a rural area lacking industrial infrastructure and a trained workforce.

Capacity building included building bridges, roads, water supply, schools, mosques, training, etc. Training in operation, health, safety, and environmental issues was provided for the local workforce. Now 98% of the nearly 3000 workers are Indonesians.

Liquefied natural gas is supplied to Japan and Korea for use in power generation, displacing the alternative of heavy oil or coal. The technology transfer was state of the art and has been constantly updated as new developments emerge. Lessons learned in Indonesia are now being applied to other LNG projects in developing countries.

---

[15] This system is not working in the FSU due to market distortion.

Replication

Opportunities for replication of the new technology are favourable in developing countries due the large and rapidly growing energy supply markets. Developing Countries now have the opportunity to promote innovative energy technologies that would be helpful in meeting their sustainable development objectives (Reddy *et al.*, 1997). Adoption of advanced technologies not only provides increases in the energy supply, but can also provide opportunities to fulfil one of the characteristics of technology transfer. This is putting it into operation in its new environment, and through incremental innovation (small improvements in existing products and process, or relatively small extensions of the scope of existing applications of the product design or technology (Greene and Halberg,1995)) producing major changes on behalf of the country's economy. Incremental innovations generate new business opportunities that do not require the same level of advanced knowledge necessary for radical innovations. Developing countries with a minimum level of basic education in their work force and minimum industrial experience may be able to capitalise on these opportunities quite successfully, often at lower cost than industrialised countries.

Trade, which is an important system for technology transfer[16], has a direct impact on the energy supply system of developing countries and CEITs. In general, trade liberalisation can have a positive effect on the environment, by helping to allocate resources more efficiently, provided effective environmental policies are implemented (OECD, 1994; OECD, 1995a). Since energy is an ingredient always present in products and services, international competition poses price constraints on it, stimulating more efficient use of primary fuels in the energy sector. However, search for lower energy prices can stimulate the use of cheap fuels (e.g. coal), which are intrinsically environmentally aggressive, or can be as such, due to the shortage of investments in end-of-pipe cleaning equipment. Thus the enforcement of effective environmental policies is a necessary condition to increase the positive and decrease the negative environmental effects of trade liberalisation. A quite serious problem that needs to be addressed is conflict between existing trade rules and environmental issues (see Box10.3).

Shortage of money is a serious obstacle for any project development. This chronic issue in developing countries is even aggravated by the difficulty that innovative projects are usually undertaken by small enterprises. Larger enterprises traditionally finance their projects through corporate financing, where the assets of the corporation are used as a guarantee. Small enterprises must rely mostly on project finance where the guarantee for a loan is provided by the cash-flow of the project.

---

[16] It has been estimated that 75% of international technology transfer arises from trade flows (OECD, 1995a). International trade in capital equipment is the most direct of these channels, since a large proportion of trade is in "producer" goods (i.e. goods which are used in the production of other goods and which are therefore important determinants of the production technologies) (Coe and Helpman, 1995).

Innovative third party financing can be obtained through multi-lateral organisations (i.e. GEF), but with limited availability of funds. Joint Implementation (JI) and the Clean Development Mechanism (CDM) can be a way of directing investment to CEITs and developing countries. Money from the private sector may be more available than from government organisations depending on the incentives.

The participation of non-governmental organisations (NGOs) has become important to the growth of simple distributed energy supply facilities which have low energy output and relatively low capital cost. For example, an important intervention programme to alleviate the rural energy crisis in India has been the dissemination of biogas technology among rural households through which 2.5 million individual biogas plants were constructed up until March 1997. The diffusion of such biomass technology has relied on the cooperation among government agencies, private firms, and NGOs. Technological exchange with China led to the adaptation from the floating dome biogas technology to the less experienced fixed dome biogas technology, mainly because capital subsidy remained with the organisations - mainly the NGOs - which were constructing the plants. Most technological improvements were incremental and resulted from local innovations (Ramana and Shukla, 1998).

Commercialisation of new technologies is an important part of the innovation process, and policies to facilitate these actions are urgently needed in developing countries and CEITs. Although developing countries' enterprises may be developing renewable energy technologies and have advanced scientific and technical capabilities and skilled workforces, the translation of these capa-

---

BOX 10.3   CONFLICTS BETWEEN TRADE RULES AND ENVIRONMENTAL ISSUES

Trade issues became more complex when the various stages of the product's life cycle are at issue (i.e. processes and production methods – PPMs). For example, recent packaging and recycling requirements which address environmental issues, but which are associated with the disposal stage of the product life cycle, have caused some trade concerns. This is because waste reduction policies tend to have a national focus, and can thereby impose relatively higher costs on importers, amounting to de facto protection for domestic products (OECD, 1997). To date, trade rules have been interpreted as precluding policy differentiation on the basis of non-product related PPM requirements (the ones that do not affect product characteristics, but generate an environmental impact at the production stage, such as emissions). Similarly, adjusting the price of imported products at the border, to account for the additional cost incurred by domestic industry in complying with non-product related PPM requirements has generally been considered to be incompatible with existing trade rules (OECD, 1997).

An interesting case is import taxes set to a certain level for products, without any consideration of the implicit environmental benefits.

bilities into commercial products is still a major problem. The associated market-oriented skills and institutions to take full advantage of these technological capabilities are still poor.

In Russia these market-related deficiencies are the persistent legacy of the former Soviet paradigm of central economic planning and development. Central planning avoided the need for many market-oriented skills and created a variety of disincentives and structural economic conditions that stifled innovation, creativity, efficiency, and quality (Cooper, 1991; Nove, 1986; Martinson and Valdemars, 1992). Key underdeveloped capabilities are business management, finance systems, marketing, creative product development and innovation, quality assurance and economic analysis (like cost-benefit and life cycle analyses); and legal, contracting, and accounting skills. Quality assurance, statistical quality control, management for quality, and other methods common in the West are uncommon in Russian industries (but not in the military sector), because incentives in the Soviet economic system emphasised quantity over quality. Another problem is the excessive presence of monopolies in Russia. In Poland, the Czech Republic, and Hungary markets are being developed while these countries have also been out of the market economy for 50 years. Thus, business planning and training will be helpful, but only if there is a market environment to use them (Martinot, 1999).

Other important aspects of commercialisation are continuous product improvement and cost reduction. As an example, product improvement was often intentionally avoided in the former Soviet economic system because enterprise incentives encouraged quantity over quality, and because design changes could mean changes in needed inputs that might not be available. For example, in the West, extensive experience with operating and maintaining wind turbines through commercial markets over the past 15 years has led to a refinement of designs and cost reductions. This experience has been critical to the current success of modern wind turbines.

A promising process for speeding up the commercialisation of new and renewable sources of energy is through the acquisition of small companies with good products in the energy supply sector by large ones with well established positions in the energy market. Small companies find it difficult to commercialise their products worldwide, while this can be an easier process for large companies. Several energy multinationals are taking up the commercialisation challenge of renewables by drastically expanding their investments and acquisitions in the area of solar, biomass and wind technologies. With the exception of the two Japanese companies, Kyocera and Sharp, all companies with a global PV market share above 10% are owned by large energy companies such as Siemens, BP, Amoco and Enron. Shell has recently established a fifth business group under the name of Shell International Renewables. Although still dwarfed by the other groups in every respect, it signifies a strong commitment to renewables. Half a billion (US) dollars of investment will be spent in the first five years of operation. Concentrating on PV, forestry and biomass power, Shell intends to obtain a strong position in rural energy development by coupling biomass plantations with local power production. Similar trends are evident in the wind power industry. Although independent producers are still strong, mergers and acquisitions are increasing.

10.6 Lessons Learned

General Comments: Technology transfer as presently understood in the energy sector is a relative new process, since historically it was used as a euphemistic term for large–scale power projects financed by multilateral banks or for a limited knowledge transferred from the international oil and gas companies to the national industries. The oil crises in the 70s changed this tendency in the oil and gas sector, when powerful national oil companies were able to negotiate technology transfer with better terms. In the early 90s the process of market globalisation and the availability of private capital on a global scale triggered technology transfer opportunities in the electric sector also.

Major objectives of the current energy supply technology transfer is economic development and international competitiveness. Climate change objectives and in particular the reduction of CO_2 emission do not play a significant role.

Even without strong interest allocated to climate change, energy efficiency improvements and new and renewable energy sources are gaining influence but at a lower than desirable pace. Most of the projects in non-conventional energy sources are dependent on grants or subsidies, since they are small in size and unable to compete with the more economic fossil fuel based energy suppliers. Also, the number of major investors are quite small due the large capital requirement and they are quite resistant to significant changes in the sector.

The role of government in facilitating the technology transfer of GHG reducing technologies will continue to be important. Especially important is liberalization of the energy market combined with appropriate environmental regulation. Useful measures include creating economic incentives, adopting policies to ensure international financing, promoting infrastructure development, eliminating unnecessary regulatory and trade barriers, educating and training local workforces, protecting intellectual property and strengthening R&D, amongst others.

Fossil Fuels: Technology transfer in the fossil fuel sector is mature, and well-established mechanisms are in place in most countries. Technology is readily available from a wide variety of sources, such as the oil and gas industry, engineering contractors, equipment vendors, etc. Technology transfer comes with investment. Although, there is no technology transfer "magic bullet". The key to transfer of technology in the fossil fuel sector is to promote investment. Incremental improvements in technology transfer are being made and should continue to be encouraged by using policy instruments, such as education and awareness programmes and voluntary agreements. Nevertheless, it is necessary to identify what incentive is there for governments of coal rich developing countries (e.g., China and India) to neglect using their cheap natural resource in favour of more expensive imported fuels such as LNG or pipeline gas.

Nuclear: Technology transfer in the nuclear power sector for water-cooled/water-moderated reactors has a well-established mechanism. To promote successful transfer of nuclear technology, major government involvement is needed. The large capital costs, public acceptance, availability of cheap domestic fossil fuel and the resolution of safety and waste disposal concerns provide significant barriers to the use of nuclear. In many cases, nuclear proliferation issues are a major problem to be addressed by governments and other international institutions.

Renewables: Technology transfer for new and renewable energy forms are not yet fully commercially established. Several pilot and demonstration projects, as well as a few commercial projects have been performed under the umbrella of the government, regional organisations, and NGOs. Examples of technology transfer within countries and even between countries exist, showing that the technology process can work provided a market is available for the products. But, in comparison with conventional sources of energy we have to recognise that technology transfer is immature. In general, renewables with the exception of hydro cannot economically compete with fossil fuels, except in some special markets (such as wind power) or niche markets (such as solar photovoltaic). Externalities (clean environment, job creation, social development, etc.) should be considered. Unfortunately, the low economic value attributed to some of the benefits provided by renewables are a serious obstacle for their supply expansion. Also, most of these projects have been implemented in developing countries and CEITs where experience with innovations are not well developed and it is more difficult to add marginal technology advances – thus there is a need to promote market development.

References

Able-Thomas, U., 1996: Models of Renewable Energy Technology Transfer to Developing Countries. In *Renewable Energy, Energy Efficiency and the Environment.* AA. Sayigh, (ed.), Pergamon Press, NY, pp. 1104-7.

APEC, 1997: *APEC Coal Mine Gas Project in the People's Republic of China.* Interim site survey report, May.

APEC, 1997a: *Environmentally Sound Infrastructure in APEC Electricity Sectors.* A Report to the APEC Energy Working Group, Apogee Research , August

Baklid, A., and R. Korbol, 1996: Sleipner Vest CO_2 disposal, CO_2 injection in a shallow underground aquifer. SPE 366000, Society of Petroleum Engineers, Richardson, TX.

Barnett, A., 1992: The Financing of Electric Power Projects in Developing Countries. *Energy Policy,* **20**(4), 326-34.

Baron, R., 1996: Economic/Fiscal Instruments: Taxation (i.e. Carbon/Energy). Annex I, Policies and Measures for Common Action Working Paper, Expert Group on the UNFCCC.

Barrows, G.H., 1993: *World Review: Analysis of Recent Trends in Structures Used for Oil / Gas Exploration / Production Contracts.* Proceedings of the International Energy Conference on Natural Resource Management: Crude Oil Sector, Moscow, 23-25 November, OECD/IEA, pp. 281-348.

Brower, M., S. D. Thomas, and C. M. Mitchell, 1997: Lessons from the British restructuring experience. *Electricity Journal* (April), 40-51.

Chowdhury, J., 1992: Performance of international joint ventures and wholly owned foreign subsidiaries: a comparative perspective. *Management International Review,* 32(2), 115-123.

Climate Action Report, 1997: 1997 Submission of the United States of America under the UNFCCC, Department of State Publication # 10496, Bureau of Oceans and International Environmental Scientific Affairs, Office of Global Change, Washington DC.

Coe, D.T., and E. Helpman, 1995: International R&D Spillovers. *European Economic Review,* **39**, 859 – 887.

Cohen, L.R., and R.G. Nell, 1991: *The Technology Pork Barrel,* Brooking Institution, Washington, DC.

Cooper J., 1991: Soviet technology and the potential of joint-ventures. In *International Joint-Ventures: Soviet and Western Perspectives.* A.B. Sheer, I.S. Korolev, I.P. Faminsky, T.M. Artemova, E.L. Yakovevla, (eds.), Quorum Books, New York, NY, pp. 37-56.

Craig, K., and R.P. Overend, 1996: *Biomass Energy Resource Enhancement: The Move to Modern Secondary Energy Forms. Vol. 1: Thematic Papers.* In Proceedings of the Symposium on Development and Utilization of Biomass Energy Resources in Developing Countries, 11-14 December 1995, Vienna, Austria, United Nations Industrial Development Organization (UNIDO), pp. 109-121; NICH Report No.TP-430-20511.

CRIEPI, 1995: Life Cycle Analysis of Power Generation Systems. CRIEPI Report No. Y9400.

Data, D.K., 1988: International Joint Ventures: a framework for analysis. *Journal of General Management,* **14**(2),78-91

Davidson, O., L. Kristoferson, and Y. Sinyak, 1991: *Technology Transfer to Developing Countries and European Economies in Transition.* Symposium Report, ESETT'91 International Symposium on Environmentally Sound Energy Technologies and their Transfer to Developing Countries and European Economies in Transition, Milan, October 21-25, 71 pp.

Der Werf, P., R. Heintz, S. Barathan, P. Bhatia, and G. Sethi, 1994: *Sustainable Energy Technology in the South.* A Report to WWF, prepared by the Institute of Environmental Studies and Tata Energy Research Institute, Word Wide Fund for Nature, Switzerland, pp. 27.

Dones, R., S. Hirshberg, I. Knoepfel, 1994: *Greenhouse gas emission inventory based on full energy chain analysis.* Proceedings of the IAEA Advisory Group meeting/Workshop, Beijing, October.

ECOSOC, 1994: *Transfer of Environmentally Sound Technology, Cooperation and Capacity Building.* Report of the Secretary General of the Commission on Sustainable Development, Intersessional ad hoc working group on technology transfer and cooperation, 23-25 February 1994.

Evans, M.K., 1997: The Economic Impact of the Demand for Ethanol. Midwestern Governors' Conference, Lombard, IL February - http://www.ethanolrfa.org/docs/evans.html.

FCCC, 1998: *Barriers and Opportunities Related to the Transfer of Technology – Terms of transfer of technology and know-how.* Second Technical Paper, FCCC/TP/1998/1, August.

FEPC, 1998: Measures Against Global Climate Change. Japanese Electric Power Industry.

Fujino, K., 1994: International Cooperation for Environmental Coexistence and Technology Transfer of Hydropower Development in Developing Countries, Energy & Environment, **5**(2), 159-171.

GEF, 1998: Report of the GEF to the Fourth Session of the Conference of the Parties to the United Nations Framework Convention on Climate Change (Draft, to be finalised in October).

Goldemberg, J., 1997: The Importance of the Cooperation Between Industrialized and Developing Countries for the Prevention of Global Warming. International Environment and Technology Forum, December 5-6, Kyoto, Japan.

Greene, M., and K. Hallberg, 1995: *The Globalization of Knowledge and Technology.* Proceedings of Symposium Marshaling Technology for Development, November 28-30, 1994, Irvine, CA, National Academic Press, Washington, DC, pp.17-28.

Grubb, M., and R. Vigotti, 1997: *Renewable Energy Strategies for Europe, Vol II: Electricity Systems and Primary Electricity Sources.* Energy and Environmental Programme, The Royal Institute of International Affairs, Earthscan, London.

Gururaja, J. 1995: *Lessons Learned: UNDP Assistance to the Energy Sector.* An Ex-Post Evaluation Study, Office of Evaluation and Strategic Planning, UNDP, New York, NY, August.

Harou, P. M., L. A. Bellu, and V. Cistulli, 1998: Environmental Economics and Environmental Policy: A Workbook, World Bank, Washington, DC.

IEA -International Atomic Energy Agency, 1994: Net Energy Analysis of Different Electricity Generation Systems. IAEA-TECDOC-753, IAEA, Vienna.

IEA- International Energy Agency, 1996: *Methane Emissions from Coal Mining.* IEA Greenhouse Gas R&D Programme, Stoke Orchard, Cheltenham, UK

IAE- International Energy Agency, 1997: *Renewable Energy Policy in IEA Countries. Volume I: Overview.* Paris

IEA-International Energy Agency, 1997a: *Electric Technologies – Bridge to the 21st Century and a Sustainable Future.* Report of an IEA Workshop, OECD/IEA.

IEA-International Energy Agency, 1997b: *Methane Emission from the Oil and Gas Industry.* IEA Greenhouse Gas R&D Programme, Stoke Orchard, Cheltenham, UK.

IEA-International Energy Agency, 1998: Key World Energy Statistics 1998, IEA, Paris (http://www.iea.org/stats/files/keyatats/stats_98.html).

IEA-International Energy Agency, 1998a: World Energy Outlook. International Institute for Applied Systems Analysis and World Energy Council. *Global Energy Perspectives.* N. Nakicenovic, A. Grubler, A McDonald, (eds.), Cambridge University Press, Australia.

Interlaboratory Working Group, 1997: *Scenarios of US Carbon Reduction: Potential Impacts on Energy Efficient and Low-Carbon Technologies by 2010 and Beyond.* Berkeley, CA: Lawrence Berkeley National Laboratory, Berkeley, CA (LBNL-40533) and Oak Ridge National Laboratory (ORNL-444).

IPCC, 1992: *The Supplementary Report to the IPCC Scientific Assessment Report of Working Group I.* J.T. Houghton, B.A. Calander, S.K. Varney, (eds.), Cambridge University Press, Cambridge, UK.

IPCC, 1996a: Summary for Policymakers. In *Impacts, Adaptation and Mitigation of Climate Change: Scientific - Technical Analyses.* Contribution of Working Group II to the Second Assessment Report of the Intergovernmental Panel on Climate Change, R.T. Watson, M.C. Zinyowera, R.H. Moss (eds.), Cambridge University Press, Cambridge and New York, NY.

IPCC, 1996b: Energy Supply mitigation options (Chapter 19). In *Impacts, Adaptation and Mitigation of Climate Change: Scientific - Technical Analyses.* Contribution of Working Group II to the Second Assessment Report of the Intergovernmental Panel on Climate Change, R.T. Watson, M.C. Zinyowera, R.H. Moss (eds.), Cambridge University Press, Cambridge and New York, NY.

IPCC, 1996c: Agricultural Options for Mitigation of Greenhouse Gas

Emissions (Chapter 23). In *Impacts, Adaptation and Mitigation of Climate Change: Scientific - Technical Analyses*. Contribution of Working Group II to the Second Assessment Report of the Intergovernmental Panel on Climate Change, R.T. Watson, M.C. Zinyowera, R.H. Moss (eds.), Cambridge University Press, Cambridge and New York, NY.

IPCC, 1996d: Energy Primer (Chapter B1). In *Impacts, Adaptation and Mitigation of Climate Change: Scientific - Technical Analyses*. Contribution of Working Group II to the Second Assessment Report of the Intergovernmental Panel on Climate Change, R.T. Watson, M.C. Zinyowera, R.H. Moss (eds.), Cambridge University Press, Cambridge and New York, NY.

IPCC, 1996e: Energy Primer (Chapter B1 - B3.3.2). In *Impacts, Adaptation and Mitigation of Climate Change: Scientific - Technical Analyses*. Contribution of Working Group II to the Second Assessment Report of the Intergovernmental Panel on Climate Change, R.T. Watson, M.C. Zinyowera, R.H. Moss (eds.), Cambridge University Press, Cambridge and New York, NY.

IPCC, 1996f: *Technologies, Policies and Measures for Mitigating Climate Change. IPCC Technical Paper I*, R.T. Watson, M.C. Zinyowera, R.H. Moss, R.H. Moss (eds.), Cambridge University Press, Cambridge and New York, NY

Ishitani, H., T. B. Johnsson, S. Al-Khouli, H. Audus, E. Bertel, E.Bravo, J.Edmonds, S. Frandsen, D. Hall, K. Heinloth, M. Jefferson, P. de Laquil, J.R. Moreira, N. Nakicenovic, Y. Ogawa, R. Pachauri, A Riedacker, H.H. Rogner, K. Saviharju, B. Sorensen, G. Steves, W. C. Turkenburg, R. H. Williams, and Zhou Fengqi, 1996: Energy Supply Mitigation Options. In *Climate Change 1995 – Impacts, Adaptations and Mitigation of Climate Change: Scientific-Technical Analyses*. R.T. Watson, M.C. Zinyowera, R.H. Moss, (eds.), Cambridge University Press.

Jepma, C.J., and C.W. Lee, 1995: Carbon dioxide emissions : A cost-effective approach. In *The Feasibility of Joint Implementation*. C.J. Jepma, (ed.), Kluwer, Groningen, pp. 57-68.

Kassler, P., 1994: Energy for Development. Shell Selected Paper, Shell International Petroleum Company, London, November, 11p

Korea, 1998: *Report from the International Expert Meeting on the Role of Publicly-Funded-Research and Publicly-Owned Technologies in the Transfer and Diffusion and Environmentally Sound Technologies (ESTs)*. Kyongju, Republic of Korea, February 4-6.

Kozloff, K., and O. Shobowale, 1994: *Rethinking Development Assistance for Renewable Electricity*. WRI, November, 57 pp.

Langrish, J. *et al*, 1972: *Wealth from Knowledge: A Study of Innovation in Industry*. Halsted/John Wiley, New York, NY.

Larsen, 1990: Heat Recovery and Energy Conservation in Petroleum Refining. PhD.Thesis, University of Michigan.

Liebenthal, A., *et al*, 1996: *The World's Bank Experience With Large Dams: A preliminary Review of Impacts*, Operation evaluation Department. The World Bank, Washington, DC.

Liu, Z.P., J.E.Sinton, F.Q. Yang, M.D. Levine, and M.K.Ting, 1994: *Industrial Sector Energy Conservation Programs in the People's Republic of China during the Seventh -Five-Year Plan (1986-1990)*. Lawrence Berkeley Laboratory, Berkeley, CA and Energy Research Institute, Peoples Republic of China (LBL – 36395).

Martinot, E. 1999: Renewable Energy in Russia: Markets, Development, and Technology Transfer. *Renewable & Sustainable Energy Reviews*, **3**, 49-75

Martinson, M.G., and K. Valdemars, 1992: Technology and Innovation Management in the Soviet Enterprise. *International Journal of Technology Management*, **7** (4/5), pp.359-369.

Maya E.S., and A.N. Churie, 1996: An Assessment of the Overall National Perspective on JI and Potential Areas of JI Involvement for Zimbabwe. In *Joint Implementation: Carbon Colonies or Business Opportunities?* S. Maya, J. Gupta, (eds.), Southern Centre Publications, Harare.

Mitchell, C., 1995: The Renewable NFFO: a Review. *Energy Policy*, **23**(12), 1077-1091.

Moreira, J.R., and A.D. Poole, 1993: Hydropower and its Constraints. In *Renewable Energy - Sources for Fuels and Electricity*, T.B.Johansson, H. Kelly, A.K.N. Reddy, R.H. Williams and L. Burnham, (eds.), Island Press, Washington DC., pp. 73 – 120.

Moreira, J.R., and J.Goldemberg, 1999: The alcohol program. *Energy Policy*, **27**, 229-245.

Nakicenovic, N.A., Grübler, A. Inaba, S. Messner, S. Nilsson, Y.Nishimura, H.H.Rogner, A. Schaffer, L. Scharattenholzer, M. Strubegger, J.Swisher, D. Victor, and D. Wilson, 1993: Long term strategies for mitigating Globa Warming. *Energy*, **18**(5), 401-609.

Nakicenovic, N., A Grubler, H. Ishitani, T. Johansson, J. R. Moreira, and H.-H. Rogner, 1996: Energy primer. In *Climate Change 1995: Impacts, Adaptations and Mitigations of Climate Change_ Scientific and Technical Analysis*. Contribution of Working Group II to the Second Assessment Report of the IPCC. Cambridge University Press, Cambridge, pp 75-92

National Energy Strategy, 1991: *National Energy Strategy - Powerful Ideas for America, 1991*. National Technical Information Service, US Department of Commerce, Springfield, VA, February.

Nove, A. 1986: The Soviet Economic System. Allen & Unwin, Boston, MA.

OECD, 1994: The Environmental Effects of Trade, OECD, Paris.

OECD, 1995a: *Report on Trade and Environment to the OECD Council at Ministerial Level*. OECD, Paris.

OECD, 1997: *Economic Globalisation and the Environment*. OECD, Paris, pp. 88.

OECD/IPCC, 1991: *Estimation of Greenhouse Gas Emissions and Sinks*. Final Report from OECD Experts Meeting, 18-21 February. Prepared for Intergovernmental Panel on Climate Changes, (Revised) OECD, Paris.

OECD/IPCC, 1995: *IPCC Guidelines for National Greenhouse Gas Inventories. Vol. 3*. Greenhouse Gas Inventory Reference Manual, OECD and IPCC, UK Meteorological Office Bracknell, UK.

OTA 1992: Fuelling Development - Energy Technologies for Developing Countries. Congress of the United States, Office of Technology Assessment, April, Washington, DC.

OTA, 1995: Renewing Our Energy Future, Congress of the United States, Office of Technology Assessment, September, Washington DC.

OTA, 1995a: Office of Technology Assessment, US Congress, Innovation and Commercialization of Emergency Technologies, OTA-BP-IT6-165, US Congress, Washington, DC., September.

Phylipsen, G.J.M., K. Block and E. Worrel, 1998: *Handbook on International Comparisons of Energy Efficiency in the Manufacturing Industry*, Department of Science, Technology and Society, Utrecht University, The Netherlands.

Prather, M., R. Derwent, D. Erhalt, P. Fraser, E. Sanhueza and X. Zhou, 1994: Other Trace Gases and Atmospheric Chemistry, in *Climate Change 1994 – Radiative Forcing of Climate Change*. IPCC. Cambridge University Press, Cambridge and New York, NY

Price, L.K., L. Michaelis, E. Worrel, and M.Khrushch, 1998: Sectoral trends and driving forces of global energy use and greenhouse gas emissions. *Mitigation and Adaptation Strategies for Global Change*, **3** , 261-319.

Ramana, P.V., and P.R. Shukla, 1998: Climate Change Policies and Long-Term Rural Energy Transitions in India. In *Climate Change Mitigations, Shaping the Indian Strategy*. P.R. Shukla (ed.), Allied Publishers, New Delhi.

Reddy, A.K.N., R.H. Williams, and T.B. Johansson, 1997: *Energy After Rio - Prospects and Challenges*. United Nations Development Program and Stockholm Environment Institute, UNEP, New York, NY.

Rothwell, R. *et al*, 1974: SAPPHO Updated Project SAPPHO Phase 11. *Research Policy*, November.

Shell International Petroleum Company, 1995: The Evolution of the World's Energy System 1860-2060. Shell Centre, London, December.

Strategic Development of the Russian Gas Industry, 1998: Gasoil Press, Moscow, 231pp.

Tanton, T., 1998: Energy and Environmental Workshop, Aalburg University, October.

Taylor, M., 1994: *Greenhouse gas emissions from the nuclear fuel cycle*. Proceedings of the IAEA Advisory Group meeting/Workshop, Beijing, October.

The Energy Policy Act, 1998: TITLE XII – Renewable Energy, Sec 1211. Innovative Renewable Energy Technology Transfer Program, US-DOE.

TEPCO-Tokyo Electric Power Company, 1999: TEPCO's Commitment to Energy and Environmental Issues, Chapter 10: International Cooperation. Environmental Action Report, Tokyo, pp 146-149 July (http:www.tepco.co.jp/plant-sit-env/environment/99report-e/menu-e.html).

Uchiyama, Y., 1996: *Life Cycle Analysis of Electricity and Supply Systems: Net energy analysis and greenhouse gas emission.* Proceedings of a Symposium on Electricity, Health and the Environment: Comparative Assessment in Support of Decision Making, IAEA-SM-38/33, Vienna.

United Nations, 1997: Trends of Financial Flows and Terms and Conditions Employed by Multilateral Institutions. First Technical Paper on Term of Transfer of Technology and know-how , FCCC / TP / 1997 - I , 25/07/97, United Nations, 41 pp.

US DOE, 1996: *Policies and Measures for Reducing Energy Related Greenhouse Gas Emission, Lessons from Recent Literature.* Office of Policy and International Affairs, USDOE, Washington, DC., July

US DOE, 1997: *Carbon Management: Assessment of Fundamental Research Needs, Appendix a.* Workshop for the Preliminary Identification of Basic Science Needs and Opportunities for the Safe and Economic Capture and Sequestration of CO_2, Office of Energy Research, Department of Energy, Washington DC, August.

US DOE, 1999: http://www.oit.doe.gov/techdeliv.shtml.

US EPA, 1993: *Options for Reducing Methane Emissions Internationally. Volume* I (430-R-93-006), July.

US EPA, 1997: Note # 430-B-97-009 Oct.; Note # 430-B-97-011 Oct.; Note # 430-B-97-014 May; Note # 430-B-97-032 Oct.

Von Hippel, D., 1996: Technology Alternatives to Reduce Acid Gas and Related Emissions from Energy - Sector Activities in Northern Asia - Energy, Security and Environment in Northeast Asia, ESENA, November (www.nautilus.org/esena/papers/techannex1).

WEC/IIASA, 1995: *Global Energy Perspectives to 2050 and Beyond.* World Energy Council / International Institute for Advanced Systems Analysis, London and Laxemburg.

WEC-World Energy Council, 1998: *The role of nuclear power in a sustainable future.* World Energy Council Conference, September 13-17.

Willians, M.C., and C.M. Zeh (technical coordinators), 1995: Proceedings of the Workshop on Very High Efficiency Fuel Cell / Gas Turbine Power Cycles. Morgantown Energy Technology Center, Office of Fossil Energy, US Department of Energy, DOE / METC - 96/1024, October 19.

World Bank, 1994: *World Development Report, 1994,* Oxford University Press, New York.

World Bank, 1997: *Global Development Finance, 1997.* World Bank, Washington, DC.

Zhu, D. (ed.), 1998: *Assessment of Biomass Resource Availability in China.* PRC Ministry of Agriculture and US. Department of Energy Project Expert Team, China Environmental Science Press.

11

Agricultural Sector

Coordinating Lead Author:
LIN ERDA (CHINA)

Lead Authors:
*Carlos Clemente Cerri (Brazil), George Frisvold (USA), Katsuyuki Minami (Japan),
Otto Doering (USA), Neil Sampson (USA), Paul Waggoner (USA)*

Contributing Authors:
*Don Plucknet (USA), Heinz Ulrich Neue (Germany), Karim Makarim (Indonesia),
Kenneth Hubbard (USA), Li Jiusheng (China), Li Yu'e (China), Vernon Ruttan (USA)*

Review Editor:
Walter Baethgen (Uruguay)

CONTENTS

EXECUTIVE SUMMARY

Agriculture is a worldwide critical strategic resource expected to double its production in 30 years to feed the world. Yet, agriculture is most directly affected by climate change through increased variability as well as temperature and moisture changes. Agriculture's adaptation to climate change will require new genetic stocks, improved irrigation efficiency, improved nutrient use efficiency, and improved risk management and production management techniques.

Agriculture can contribute modestly to mitigation through carbon sequestration in soils and in other ways. Emissions from manure can be turned into methane (CH_4) fuel. Methane from ruminants can be reduced through straw ammoniation and increased feed efficiency, and methane from rice paddies can be mitigated. Better nutrient management can reduce emissions of nitrous oxide (NO_x).

Adaptation to uncertainty such as climate change requires assembling a diverse portfolio of technologies; and keeping the flexibility to transfer and adopt needed technology. However, small farms and related businesses are risk-averse. Lack of information, financial and human capital, and transportation; as well as temporary (land) tenure and unreliable equipment and supplies discourage transfer of technology. These hindrances can rarely be surmounted unless the transferred technology profitably has high chances of quickly solving an evident problem.

The effectiveness of technology transfer in the agricultural sector in the context of climate change response strategies would depend to a great extent on the suitability of transferred technologies to the socio-economic and cultural context of the recipients, considering development, equity, and sustainability issues. This is particularly relevant when applied to North-South technology transfers in this sector.

Governments can facilitate transfer with incentives by regulating and by improving institutions. For example, farmers need stronger incentives to adapt to a changing environment. However, if subsidies and distortions of markets make farmers less susceptible to fluctuations in yield and price, they will also tone down signals of climate change, and thus slow technology transfer. For example, cereal prices have been almost exempt of the 5 to 6 times rise of prices for almost everything in the course of the 55 years post wartime period.

The success of a response to an actual climate demonstrates that necessary technology can be developed, transferred and adopted. A new rice variety was developed in Sierra Leone to exploit seasonal rain and require less pesticide. Once success of the variety became apparent, farmers themselves transferred it to others. This transfer demonstrates the success of a policy that responds to present need, concentrates researchers, devises cheap technology, and promptly benefits farmers.

Some technologies will not be so easily transferred. Irrigation, a preeminent adaptation to climate, costs millions and requires communities to take up unfamiliar crops and methods. Nations must deal with scarce water and environmental impacts, marshal capital to construct the dams and canals, and assist in the marketing of new crops. On the other hand, new crops (*e.g.*, oil seeds), may require new processing plants located near the production site, or else they require considerable transportation to existing processing plants. To facilitate the transfer, banks could extend credit to farmers and research and training could turn to irrigation design, new crops, water use efficiency and prevention of salinity. Only an integrated national effort that extends to the farm level succeeds.

Centuries of experience, much of it governmental, have demonstrated the value of new plants and the useful genes of established ones. Breeding, testing and demonstrating in the diverse locales and climates where farmers must cope with drought, pests, and different lengths of season have had a high payoff and are essential for adaptation to climate change.

Trade free of barriers facilitates technology transfer. Trade has exploited comparative advantages among climates and will continue to do so if climate changes. Free movement of innovations, knowledge and trained work forces will speed technology transfer. This was demonstrated in th past, during the Green Revolution which showed the power of free movement of money, technicians, genes and information. As governments and donors have economised, business has stepped forward in research and innovation, but merely focusing on profitable areas.

Education lies at the heart of technological transfer. The public role in education is pre-eminent and must be supported. Policies of practical demonstration have proven the most helpful form of education. Private business can announce technology in advertisements or demonstrate it at fairs. Governments have a role in monitoring claims and educating broadly. For example, the "training and visit" transfer requires both training technicians and getting them into the field to educate farmers.

Although the transfer of new varieties proceeds quickly and easily, transferring systems of management requires persistence.

For example, America established the Conservation Technology Information Center in 1982 to encourage conservation tillage. Great progress has been made, but after 17 years, adoption is still ongoing.

Uncertainties cloud the favourable outlook for the transfer of agricultural technology. Because people transfer technology most readily to solve a problem that is serious, certain and imminent, the uncertainty of climate change hinders transfer. When, for example, climatologists assessing the climate for the next few years cannot agree whether it will be wetter or drier, no farmer will pay to mitigate emissions or invest in irrigation or drainage.

The growing role of private business and intellectual property rights introduces unfamiliar rules. At the same time, slowing public research adds to the problem. While the annual financial growth of the up-to-now-successful international research establishment (CGIAR), which stimulated and initiated yearly crop yield increases and aided in outgrowing the additional demand by the rising world population, has fallen from 14% to near 0%, it has spread its activities over new areas. Growth of national establishments has also slowed.

Despite uncertainties, past successes, established by embattled researchers and trainers, and the incentive of feeding the world encourage hope. The aim is reliable and adequate agricultural growth with less emission and more storage of greenhouse gases (GHGs).

11.1 Introduction[1]

11.1.1 Special Character of Agriculture

The United Nations Framework Convention on Climate Change (UNFCCC) challenges the world community to take actions that will stabilise GHGs in the atmosphere within the framework of sustainable development. This places a special responsibility on agriculture, which must support an expected doubling of the world's food and fibre needs in the next 30 years, while adapting to any climate change effects experienced, and doing its proportional share to reduce net GHG emissions.

The ability of world agriculture to meet the needs of an expanding population has, for the latter part of the 20th century, been due to the development and adoption of new technologies, rather than to the expansion of cultivated land (IPCC, 1996a). In the face of limits to unused arable land or additional water for irrigation, it is realistic to anticipate that meeting future agricultural needs in the face of uncertain and, perhaps, changing climate conditions, will rest even more heavily upon technology development and transfer. The exposure of crops and animals outdoors to the vagaries of weather heightens this challenge. Increasing social demand for environmentally sound technologies (ESTs), and the need to address the opportunities for reducing GHG emissions and increasing the sequestration and long-term storage of carbon within agricultural soils, heightens the challenge.
Enhancing the role of agriculture in meeting the goals of the UNFCCC will require attention to speeding the development and transfer of technologies that will:

- Reduce the consumption, or improve the production efficiency, of inputs based on fossil fuels;
- Reduce the emission of GHGs from soil deterioration, animal metabolism or waste, nutrient cycling, or water management;
- Increase the sustainable production of bio-energy crops to replace fossil fuels; and,
- Improve the capacity of agricultural systems to sequester carbon in soil compounds and other forms of long-term terrestrial storage.

Policymakers should recognise, however, that the opportunities described above are relatively modest, in the face of the total obligations under the UNFCCC, and that they are significantly different in magnitude. It is estimated, for example, that the potential opportunities in terms of total annual climate forcing impact, as measured by million metric tons of carbon equivalent (TgC), are in the following orders of magnitude:

- Production of bio-energy crops 400 to 1,460 TgC (IPCC, 1996b);
- Increased sequestration into soil carbon compounds in US - 75 to 208 TgC (Lal *et al.*, 1998);
- Reduced emissions from soils, animals, etc. - 576 to 1,386 TgC (IPCC, 1996b); and,

- Reduced fossil fuel consumption in agriculture - 10 to 50 TgC (IPCC, 1996b).

These compare with the total estimated annual GHG emissions of 6,000 TgC on a global basis, and the current obligations for GHG emission reductions of 100-200 TgC under the Kyoto Protocol. These comparisons indicate that agriculture can contribute to the goals of the UNFCCC, but only in a partial manner. The situation facing individual Parties may be substantially different than the global estimates indicate, and there may be little or no opportunity for improvement available under some circumstances.

11.1.2 Objectives of this Chapter

As agriculture around the world has been challenged to meet the growing need for food accompanied by the competition for arable land that has been generated by expanding human populations, it has relied upon a large and complex cadre of scientific researchers, communicators, and technical advisors from both the public and private sectors. The experience in agricultural technology transfer is therefore both wide and deep, providing a useful basis upon which to build an expanded capacity to meet climate change-related challenges.

Drawing from the extensive research and experience of agricultural advisors, this chapter seeks to provide insight into the potential for improved or accelerated technology transfer to help agriculture meet the new challenges identified in the UNFCCC. Brief case studies illustrate different policies and programmes that have affected technology transfer in a variety of situations. The cases illustrate the role of government, industry and farmers themselves. In the end, this chapter shows both the uncertainties and ways how countries can adapt to climate change to achieve reliable and adequate agricultural growth while emitting less and storing more GHGs.

11.2 Climate Adaptation and Mitigation Technologies

Sustainable agricultural development is an ongoing priority for all countries. While the outlook for meeting the growth rates of production required to meet projected food needs by 2010 to 2030 appears reasonable with wider and more efficient use of existing technology and technology transfer for rice and maize or animal production, it seems less hopeful for crops that have a high proportion of their area in difficult ecoregions, that is, wheat, barley, sorghum, and millet (Oram and Hojjati, 1995). The challenges to adaptation will be significantly affected by the manner in which climate change effects are experienced. A slow change in climate over decades, without a significant change in variation and weather extremes, will facilitate adaptation by farmers themselves, and may demand little new technology transfer in itself. If the future effects include an increase in short-term weather extremes, the ability of farmers to adapt will be severely

---

[1] Relevant case from the Case Studies Section, Chapter 16, for Chapter 11: mangrove rice variety (case 29).

challenged, and improved technology transfer, such as more reliable long-term weather forecasting, will become critical to their ability to adapt.

It is conceivable that a changed climate will increase agricultural production in some regions. However, more effective technology transfer to developing countries and improved prices of grains could encourage farmers to adopt known adaptation and mitigation technology more widely than the present very low levels or, in some countries, to expand the area cropped.

The Green Revolution technologies and associated policies have played a major role in bringing about a rapid increase in crop yields. These technologies involve the use of high-yielding varieties, fertilisers, irrigation, and plant protection materials. In spite of yield improvements at a global level, crop yields vary considerably among regions and countries. For example, wheat yield varied between 1.5 tons per hectare in Africa and 4.8 tons per hectare in Europe. Similarly, rice yields also varied between 2.0 tons per hectare in Africa and 7.4 tons per hectare in Oceania. These differences indicate that the available technologies are not equally adapted to all economic, cultural, or environmental conditions.

The large variations in crop yields among different countries and gaps between farmers' yields and experimental-station yields suggest that a considerable untapped potential exists to increase crop yields in developing countries and countries with economies in transition (CEITs). However, the realisation of this untapped potential depends on the existence of a stable and conducive policy environment. Technology transfer should be a high priority.

11.2.1 Adaptation Technologies

Adaptations - such as changes in crops and crop varieties, improved water management and irrigation systems, and changes in planting schedules and tillage practices - will be important in limiting negative effects and taking advantage of beneficial changes in climate. The extent of adaptation depends on the affordability of such measures, particularly in developing countries: access to know-how and technology, the rate of climate change and biophysical constraints such as water availability, soil characteristics and crop genetics. (IPCC, 1996b)

Many adaptation opportunities suitable for climate change have already been applied by farmers. Table11.1 provides a list of currently available adaptation opportunities that can be applied at the farm or farmer community level. Most available options take advantage of the general flexibility of agricultural systems related with the short management cycles involved. It is likely that autonomous adjustment by farmers will continue to be important as climate changes, provided that farmers have access to the right information and tools. However, some agricultural systems are less flexible, for example because they are constrained by soil quality or water availability, or because they face economic, technological, institutional or cultural barriers. In such

cases, autonomous adjustments may not be implemented in time because of lack of awareness (of both problems and solutions), and anticipatory planned adaptation would be required to provide the right conditions (*i.e.*, information and tools) to farmers for autonomous adjustment (Klein and Tol, 1997).

Anticipatory strategies for adaptation to climate change and climate variability aim to increase flexibility so as to allow the type of adjustments shown in Table 11.1. For example, increasing the variety of crops may require the introduction of new knowledge and machinery to a farming community. However, as climate changes, the technologies listed in Table 11.1 may not be sufficient, and the need may arise for the development of new technologies to allow farmers to cope better with anticipated climate-change impacts, and to reduce the costs of adaptation (Klein and Tol, 1997).

Table 11.1	Examples of adaptation opportunities to climate-change impacts on agricultural systems (Smit, 1993).
RESPONSE STRATEGY	**ADAPTATION OPTIONS**
Use different crops or varieties to match changing water supply and temperature conditions	• Conduct research to develop new crop varieties • Improve distribution networks
Change land topography to reduce runoff, improve water uptake and reduce wind erosion	• Subdivide large fields • Grass waterways • Land leveling • Waterway-leveled pans • Bench terracing • Tied ridges • Deep plowing • Roughen land surface • Use windbreaks
Introduce systems to improve water use and availability and control soil erosion	• Low-cost pumps and water supplies • Dormant season irrigation • Line canals or install pipes • Use brackish water where possible • Concentrate irrigation water during peak-growth period • Level fields, recycle tailwater, irrigate alternate furrows • Drip-irrigation systems • Diversions
Change farming practices to conserve soil moisture and nutrients, reduce runoff and control soil erosion	• Conventional bare fallow • Stubble/straw mulching • Minimum tillage • Crop rotation • Contour cropping to slope • Avoid monocropping - • Chisel up soil clods • Use lower planting densities
Change timing of farm operations to better fit new climatic conditions	• Advance sowing dates to offset moisture stress during warm period

11.2.2 Mitigation Technologies

Agriculture is the main source of methane (CH_4) and nitrous oxide (N_2O) emissions. CH_4 emissions from domestic ruminants, animal waste and rice fields were estimated to be 65-100 (IPCC, 1992; Hogan, 1993), 20-30 (Safley *et al.*, 1992) and 25.4-54 (IPCC, 1996b) Tg/yr. N_2O is produced primarily by microbial processes in the soil (Bouwman, 1990; Duxbury and Mosier, 1993). It was estimated that more than 75% of the anthropogenic N2O sources are derived from agriculture, the total amount was 4.2Tg N2O-N/yr (Mosier *et al.*, 1998b). A significant fraction of the CH4 and N2O emitted from agricultural systems could be avoided if some combination of agricultural management practices listed in Table 11.2 were adopted worldwide (IPCC, 1996b).

Table 11.2	List of practices to reduce CH_4 and N_2O emissions from agricultural systems (IPCC, 1996)	
MITIGATION PRACTICE	**ESTIMATED DECREASE DUE TO PRACTICE (TG CH4 OR N2O-N/YR)**	
Ruminant livestock	*29.0 (12-44)*	
• Improve diet quality and nutrient balance	25.0 (10-35)	
• Increase feed digestibility	2.0 (1-3)	
• Production-enhancing agents	2.0 (1-3)	
• Improve animal genetics	-	
• Improve reproduction efficiency	-	
Livestock manure	*5.1 (2.6-8.7)*	
• Covered lagoons	3.4 (2-6.8)	
• Small digesters	1.7 (0.6-1.9)	
• Large digesters	-	
Flooded rice	*20.0 (8-35)*	
• Irrigation management	5.0 (3.3-9.9)	
• Nutrient management	10.0 (2.5-15)	
• New cultivars and other cultural practices	5.0 (2.5-10)	
Biomass burning	*6.0 (3-9)*	
• Incorporate crop residues into soil	-	
• Increase the productivity of lands	-	
• Lengthen the rotation time	-	
Agricultural soil	*0.68*	
• Match N supply with crop demand	0.24	
• Tighten N flow cycles	0.14	
• Use advanced fertiliser techniques	0.15	
• Optimise tillage, irrigation and drainage	0.15	

Beyond the use of biomass fuels to displace fossil fuels, the management of forests, agricultural lands and rangelands can play an important role in reducing current emissions of CO_2, CH_4 and N_2O, and in enhancing carbon sinks. A number of measures could conserve and sequester substantial amounts of methane (approximately 24-92 Tg/yr, 15-65% of current levels) over the next century (Mosier *et al.*, 1998a). A total potential reduction of global N_2O emissions from agricultural soils is thus 0.7 (0.36 to 1.1) Tg N_2O-N/yr or 9-26% of current emissions from agriculture (Mosier *et al.*, 1998b).

11.2.3 Transferable Technologies for both Adaptation and Mitigation

Mitigation options are available that could result in a significant decrease in GHG emissions or increase carbon sequestration into agricultural soils. If implemented, most of them are more likely to increase rather than decrease crop and animal productivity. Considerable progress has been made in evaluating the potential effects of climate change on global agriculture, but significant uncertainties remain, so agricultural policies are specifically appropriate for adapting to climate change. A range of adaptation options can be employed to increase the flexibility and adaptability of vulnerable systems, and reverse trends that increase vulnerability. Many of these attempts to abate climate change will be of immediate benefit, and can therefore be considered "no-regret" technologies.

In order to achieve these objectives, technology transfer must occur more rapidly, and with a more intense focus on those technologies that further sustainable development. Table 11.3 summarises some key technology examples, catalogued by objectives:

There are many barriers that may be encountered (Table 11.4). Some of these options require more labour and some need more capital investment, which may represent the main constraints slowing adoption of the technologies.

In most cases the transfer of technologies needs to take into account, *e.g.*, the link between technology adoption and diffusion, and enhancing income in technology receiving countries. This will be critical in achieving adoption and wide-scale diffusion of alternative technologies in these countries. The lack of financial incentives will be a major obstacle to the adoption of some of these practices. Otherwise, the system depends on subsidies of some sort (*e.g.*, cost shares) to speed adoption and diffusion, and when the subsidy runs out, the practice is dropped.

Table 11.3	Examples of transferable technologies catalogued by objectives and specific technology/objectives							
OBJECTIVES	**TECHNOLOGY**	**MITIGATION**	**ADAPTATION**	**POTENTIAL IMPACT**	**RELATIVE COST**	**TIME PATH**	**FOOD SAFE**	**REGIONAL APPLICATION**
CO2 Sequestering in Soils	Conservation tillage	Yes	Yes	M	H	Decades	H	South America
	Improve irrigation	Yes	Yes	H	H	Years	H	Developing Countries
	Yield improvement	Yes	Yes	M	M	Decades	H	All
Higher Yields	Genetics	Yes	Yes	H	M	Decades	H	All
	Improve inputs	Yes	Yes	M	M	Decades	M	All
	Pest Control	Yes	Yes	M	M	Decades	M	All
Reducing Emissions	Improve animal agriculture	Yes	Yes	H	M	Years	M	All
	Lower GHGs	Yes	Yes	H	Not sure	Years	H	All
	Improve feed efficiency	Yes	Yes	M	M	Years	H	Developing countries
	Concentrating on best lands	Yes	Yes	M	M	Decades	H	All
	Improve nitrogen efficiency	Yes	Yes	M	M	Years	H	All

M = MEDIUM   H = HIGH

Table 11.4	Barriers for adoption of the selected mitigation technologies	
NO.	OPTIONS	CONSTRAINTS
1	Irrigation efficiency	Requires large investments and national technology and assessment commitment Requires technology transfer to the farm level Requires cooperative community action
2	Direct seeding of rice	Requires intensive weed control
3	Substitution of traditional varieties by improved varieties	Less preferred grain quality New pest problems in certain areas Changed management
4	Conservation tillage	Risk of reduction of yield Different machinery needs, crop varieties, soil moisture and temperature conditions Requires intensive weed control
5	Ammoniation of straw for animal feed	Ammonium sulphate is more expensive than urea
6	Large scale biogas digester	More investments, more complex to operate and maintain

11.3 Magnitude of Current and Future Technology Transfer

11.3.1 Global System for Agricultural Technology Transfer Already in Place

The agricultural sector is unique in that a global network of research, development and technology transfer has already been in place for a number of decades. Before World War II, little cooperation existed between countries in agricultural research or technology transfer. Most countries developed agricultural technologies in relative isolation. That is no longer the case. Today major elements of what can be described as a global agricultural research system are in place, where any country can link its research efforts to the international system to help solve important problems.

The global system is made up of three major players: NARSs (National Agricultural Research Systems) of developing countries, IARCs (International Agricultural Research Centers), and advanced laboratories and institutions in developed countries. These players interact in a variety of ways, including bilateral agreements, multilateral agreements, contracts, and research networks. The global system, being informal, depends largely on the meshing of perceived needs of numerous research organisations. With its growth and development, it has become the world's largest and most collaborative scientific enterprise. Almost every country is involved in some way and has invested some of its own funds, mostly at home, to participate. The community-driven technology transfers often proceed in the system.

Since the 1960s the Consultative Group on International Agricultural Research (CGIAR) has established 16 IARCs. The initial centres focused their research on the major food crops grown in developing countries - rice, wheat, maize, potatoes and cassava. These were joined in the 1970s by centres focussing on livestock production, animal disease and genetic resources, on arid and semiarid areas, food policy and the capacity of the national research system. During 1990 to 1992, five new centres were added to the CGIAR system, focussing on areas such as irrigation management, agro-forestry and aquatic research. This expansion was not accompanied by an expansion of the resources available to the system. In contrast with expanded missions and increasing worldwide demand for agricultural technology, support for the CGIAR system has actually declined in real terms.

11.3.2 Yield Growth Critical to Climate Change Mitigation

Maintaining yield growth is critical to climate change mitigation in the agricultural sector since the conversion of forestland to crop and livestock production is the single largest source of CO_2 and overall GHG emissions from agriculture. If growing demands for agricultural goods are not to be met by expanding acreage, yields must rise, and post-harvest technology must improve to reduce loss and spoilage.

Continued yield growth is a necessary, but not sufficient, condition for higher agricultural output without land conversion. While higher yields imply that more food can be produced without land expansion, higher yields do not necessarily reduce pressures for land expansion (Mundlak, 1997; Larson, 1994). Some types of technologies may increase demand for land or displace labour, encouraging migration to frontier areas (De Janvry, 1988; Fearnside, 1987). Much research is needed to identify the conditions under which different types of yield-increasing technologies reduce or increase demands for land expansion.

The international system of agricultural research and technology transfer has been highly successful in raising global agricultural productivity, albeit unevenly across regions. Between the beginning of the 1960s to the end of the 1980s, global production of major cereals doubled (Table 11.5). About 92 per cent of this doubling can be attributed to yield increases with only 8 per cent coming from expansion of agricultural land. Yet, in some parts of the developing world, particularly Latin America and sub-Saharan Africa, agricultural land expansion remains an important source of agricultural growth (Table 11.5). In the 1980s, about 8 per cent of the world's tropical forests were cut down. Three-quarters of this occurred in Latin America and sub-Saharan Africa.

Table 11.5	Contribution of increases in area and in yields to growth of cereals production in developing regions and high income countries between 1961 to 1963 and 1988 to 1990		
Country Group	Percent Growth in Cereal Production Attributable to Increased Area	Percent Growth in Cereal Production Attributable to Increased Yield	30-year Growth Rate of Cereal Production (per cent)
World	8	92	100
High Income Countries	2	98	67
Developing Countries	8	92	118
East Asia	6	94	189
South Asia	14	86	114
Latin America	30	70	111
Sub-Saharan Africa	47	53	73
North Africa / Middle East	23	77	68

SOURCE: WORLD BANK WORLD DEVELOPMENT REPORT 1994[OF 1992 ZIE REF. LIST]

11.3.3 Genetic Improvements Critical to Climate Adaptation

While yield growth has accounted for over 90 per cent of recent agricultural output growth, scholars credit genetic improvements in crop varieties with half of this yield growth (Duvick, 1992; Byerlee, 1996; Wright 1996). The remainder of the growth is attributed to improved management practices, irrigation, and increased use of fertilisers and other inputs. In the future, genetic improvements are likely to play an even greater role. This is particularly true given other environmental considerations that limit the extent to which higher yields can come from more intensive use of chemical fertilisers and pesticides. The efficient conservation, exchange and use of agricultural genetic resources will be critical for future agricultural technology development and transfer. Despite the impressive achievements of the Green Revolution, nearly 1 billion poor people in developing countries still achieve their sustenance from agriculture using their own traditional plant genetic material (World Bank, 1992).

Because the performance of crop varieties is sensitive to agro-climatic conditions, much of the transfer of improved crop varieties has been North-North between temperate regions and South-South across tropical or sub-tropical regions. The advances of the Green Revolution may be thought of as "North-assisted" South-South technology transfer. The semi-dwarf wheat varieties now widely adopted in India's Punjab were originally developed in Mexico, while Indian rice yields are substantially higher thanks to infusions of germ plasm collected by the International Rice Research Institute (IRRI) from other parts of Asia. In the future, biotechnology may offer significant opportunities to address the need for crop adaptation to changing climate across all countries.

However, the cost of grain increases annually, and funding for plant breeding, especially for developing countries, is now decreasing, breeders must decrease the cost per unit of genetic improvement if gains are to continue. Developing countries must look for more efficient operations and for economies of scale through collaboration with breeding programmes in other countries or the IARCs.

For the past 20 years, great hopes have been placed on the benefit to plant breeding from biotechnology. Biotechnology aids to plant breeding often will be used first in the industrialised nations, but will be available for use in developing countries with very little delay. In some cases the improvements will be publicly available; in other cases, the products will be available on a commercial basis.

Local communities manage an important part of these technologies and hold ownership. This is important in the development of systems for integrated gene management (*e.g.*, the CGIAR's programme) that combine modern and traditional methods of genetic crop improvement, and include the interests of all the stakeholders (including the rights of farmers, local communities, breeders, and biotechnology companies). The funding members of the CGIAR were praised for their readiness to invest and their

unprecedented successfulness, which so far was probably the most productive non-profit international governmental funding ever.

A major constraint to plant breeding for developing countries is the global reduction in allocation of public funds for agricultural research. Such reductions, originating in the developed countries, have especially strong adverse effects on the developing nations.

11.3.4 Current Limitations and Responses

Currently, there are five common limitations for technology transfer in agriculture:

1) Impacts primarily driven by changing patterns of extreme weather events;
2) Opportunities and risks associated with incorporation of climate-change projections in large infrastructure projects that are currently being planned and implemented, and which will still be in place fifty to one hundred years from now;
3) The considerable time it will take to plan and implement a number of adaptation technologies
4) Society's vulnerability to climate change, which largely depends on its economic, technical, institutional and socio-cultural capabilities to cope with adverse effects; and
5) The uncertainty of the impact of future climate change.

This section is concerned with institutional and technical limitations on the global system of agricultural research to develop and supply new technologies. Barriers to farm-level adoption are discussed in section 11.4.1.

Problems of gene bank management

Thus far, the international system of plant genetic resource exchange and research has succeeded at maintaining steady crop yield, while controlling yield variability. There is growing concern that this may not be sustainable in the longer term given current funding levels (United States Congress, Office of Technology Assessment, 1987; NRC, 1993; UN/FAO, 1997). Funding problems arise, in part because individual nations do not capture the full gains from improved crop yields (Frisvold, 1997). This implies that national governments will underfund germ plasm storage (NRC, 1991). The US National Research Council has noted that many public gene banks are not effectively preventing genetic erosion within their collections (NRC, 1993). Public gene banks have even been characterised as "gene morgues" (Goodman, 1990). Multilateral funding of international crop research facilities overcomes this problem partly. Yet, "free-rider" problems (see 9.4.2) imply that funding for international centres will also be difficult. New technologies that are freely available to those who do not pay for their development, may discourage potential funding sources.

A recent comprehensive study shows that:

1. The number of gene banks has increased dramatically since 1970. While much of the emphasis has been placed on collecting materials, less has been given to maintaining the long-term viability of accessions;

2. While representation of many major crops in gene banks is relatively good, coverage of many others (such as root crops, fruits and vegetables) is poor;

3. Only a small fraction of accessions have been characterised; and

4. Many countries have reported that funding has been too unstable and uncertain year to year, hampering investment and planning decisions.

Thus, while plant breeders appear confident that the current germ plasm stock, if properly maintained, is adequate to produce steady yield growth over the next 20-50 years (Knudson, 1999; Frisvold and Condon, 1998), there is widespread concern that this genetic stock is depreciating. Of particular concern is the status of the collections of the Vavilov Institute in Russia, one of the largest collections in the world. It is facing critical financial and structural problems (Zohrabian, 1995).

Limitations of the CGIAR system

As the new seed-fertiliser technology generated at the CGIAR centres, particularly for rice and wheat, began to become available, some donors assumed that the CGIAR centres could bypass the more difficult and often frustrating efforts to strengthen national agricultural research systems. Strong national research centres are essential if the prototype technology that might be developed at the international centres is to be broadly transferred, adopted and made available to producers.

Problems have not only been financial. A number of the CGIAR centres are experiencing the difficulties associated with organisational maturity. There is a natural "life cycle" sequence in the history of research organisations and research programmes (Ruttan, 1992). Certainly, the needs of technology transfer in climate change would encourage the vigor of the system. On the other hand, efforts to strengthen national research institutes have been only partially successful.

Growing role of the private sector

Many studies have considered the public good aspects of genetic resources. Naturally occurring plants are not considered patentable inventions. Genetic resources are easily transported and replicated, making it difficult for a country or individual to exclude others from their use. This discourages private actors from making investments to preserve and collect genetic resources and to screen them for their potential usefulness. Intellectual property protection historically has been weak for biological inventions. While patents on mechanical processes date back hundreds of years, intellectual property rights (IPRs) for commercially developed seed varieties began only this century, and remain considerably weaker than other forms of IPR protection (see also section 3.5 in Chapter 3 on IPRs).

Historically, there have been two major institutional responses to the private sector's inability to gain from and invest in plant breeding. The first, as described above, has been the extensive public funding of an international network of public research facilities and institutions. The second response has been the evolution of increasingly strict IPRs for biological inventions. Both stricter IPRs and advances in hybridisation have stimulated private R&D in plant breeding. The progeny of hybrids have substantially lower yields. This naturally deters purchasers of seed from regenerating new seed for their own use or for resale. The requirement that farmers repurchase seed annually greatly increases returns to private plant (seed) breeders. While public R&D investment has slowed considerably in recent years, private R&D has grown substantially. For example, private plant breeding research in the United States more than quadrupled in real terms between 1970 and 1990.

11.3.5   Likely Conditions for Future Technology Transfer

The uncertainty of climate change, whether a locality will warm a little or a lot or will grow wetter or drier, affects the course of technology transfer greatly. In responding to uncertainty, laying out a portfolio of diverse actions and assuring flexibility to use them may be more appropriate than preparing a specific adaptation or mitigation action.

Promoting sustainable development and protecting the climate system are both goals of the Framework Convention on Climate Change and The Kyoto Protocol. But the trends show that climate change has not been prevented completely. So the transfer of both mitigation and adaptation technology would be considered "no regret". At the same time, the benefits of technology transfer are very unevenly distributed, and are sensitive to the need for adaptation to climate change. Some modest implications for agricultural policy are discussed below.

Market role

The world market allows a country to sell its abundant agricultural production abroad, earning foreign exchange for the nation. The market also puts their consumers in touch with foreign products that are lower priced or more available than domestic products. As the climate changes, the conditions for production will alter and the world market may provide even greater benefits. It can facilitate these changing production patterns by finding new markets for new products and by providing supplies to affected regions. Thus, the flow of trade may relieve food shortages. As the invisible hand that coordinates adaptation, therefore, the world market is a particularly valuable climate change asset. The key question is whether the market will be allowed to operate without distortions.

At the same time, some countries are so dependent on agriculture that vagaries in the weather and economic circumstances make them especially vulnerable and in need of technology transfer. So the special requirements of the countries in technology transfer addressing climate change deserve special consideration. There will remain critical roles for governments in this process.

Growing role of the private sector

An implication of the rise of private sector plant breeding is that new seed varieties so crucial to yield growth across the world will increasingly come from private companies demanding greater levels of IPR protection. Developing countries will have to interact with an increasingly concentrated private agricultural (primarily seed) biotechnology industry. The private sector will thus become a more important vehicle for transferring modern crop varieties in the future.

Growth of transgenic crops

Many of the new innovations in plant breeding are coming in the form of transgenic crops -- crops developed by transferring genes from unrelated species to major food crops. While transgenic crops have already been widely adopted in the United States (*e.g.*, cotton, corn, soybeans, and potatoes), many institutional barriers and controversies may limit their transfer to other countries (industrialised and developing). Chief among these barriers is the international disagreement over the components of a biosafety protocol regulating international field tests of new biotechnology, informed consent, labelling and liability rules governing environmental accidents.

Packages of technologies

There is a need to focus on the potential for packaging technologies to gain greater acceptance. For example, researchers have found that farmers are more willing to adopt a more comprehensive nutrient management plan that included several practices rather than a piecemeal, practice-by-practice approach.

11.4 Programmes, Policies, and other Interventions for Technology Transfer within a Country[2]

11.4.1 Adoption Barriers/Constraints

Generic farm-level adoption constraints.

For farm-level adoption, the barriers include small farm size, credit constraints, risk aversion, lack of access to information, lack of human capital, inappropriate transportation infrastructure, inadequate incentives associated with tenurial arrangements, and unreliable supplies of complementary inputs. Because strategies for new technologies are often imposed from the top down, implementation fails when local people are not consulted or are treated as labourers only, or when local research and extension staff are not sufficiently trained in the specific techniques. Consequently, positive measures to improve soil and water productivity, for example, both through individual and communal action, should receive higher priority for research, extension, and training in the future. Some agricultural measures should also form part of an integrated biotechnical approach that provides appropriate exper-

---

[2] See also Chapter 4 on the role of enabling environments for technology transfer and Chapter 5 on financing and partnerships for technology transfer.

tise and equipment, seeds of improved cultivars, plant nutrients, and pest management, with strong social and economic incentives.

Environmental externalities and technology adoption.

As in the discussion of biotechnology transfer above, cooperation can also be on a commercial basis. Private firms are springing up to do biotechnology for a fee. One commercial organisation, for example, might do molecular marker work on contract. The firm identifies marker linkages with specified traits, or assists in backcrossing marker-linked traits into desired genetic backgrounds. Because of environmental externalities, market prices may not reflect the true social costs and benefits of particular technology adoption choices. Governments can play a role in improving environmental quality, not only by internalising externalities, but also by correcting market failures in the provision of information. Improved production techniques and management practices can improve efficiency and cut both waste and pollution, in effect replacing other, more polluting inputs with clean input information. However, information has certain aspects of a public good, and it is difficult for individual suppliers to restrict its use to only those who have paid for it. Consequently, private markets may undersupply information about environmentally beneficial technologies. Insufficient information can also constrain the adoption of new technologies by farmers. In such cases, the government may be able to improve efficiency by collecting and providing information about resource-conserving practices.

11.4.2 Programmes and Policies to Encourage Technology Transfer

Adaptation

New mechanisms are based on existing technology transfer trends. Adaptation to climate variability and change can be autonomous or planned. Planned adaptation, when it concerns transfer, requires strategic actions, based on an awareness that climate is changing and that action is needed to better respond to such change. In spite of the current uncertainty, a range of adaptation options can be employed to increase the flexibility and adaptability of vulnerable systems, and reverse trends that increase vulnerability. Many technologies that can be used to adapt to climate change are already in use in some places. In order to extend them, technology needs and technology transfer mechanisms should be more fully assessed and reported to increase the role of technology transfer in climate adaptation (IPCC Workshop, 1998).

The transferred technology will focus on key issues. Food security in the 21st century is a major concern of every country. Changes in temperature and precipitation levels may impose a negative impact on that security, especially for the arid regions where water resources are limited and drought is the major risk facing agricultural production. Development, introduction, and adoption of technologies and management systems that enhance water use efficiency represent high priorities. The design of technologies and institutions to complete a successful technology transfer will become increasingly important. During the next century water resources will become an increasingly serious constraint on agricultural production. Irrigation is already playing a critically important role in agricultural production for many countries. See Box 11.1. Weather information, a soft technology, can be very important for managers and producers. In the USA, there are excellent examples of the use of this technology(See Box 11.2).

BOX 11.1	IRRIGATION TECHNOLOGY TRANSFER WITHIN A COUNTRY

The diffusion of irrigation technologies within a country facing water shortage problems can increase that country's adaptability to climate change. The biggest barrier for irrigation technology transfer within a country may be the shortage of financial capital. Usually, the adoption of new technologies imposes additional costs when compared to conventional practices. Technology transfer will be more difficult where the recipients, who are usually farmers, cannot afford the additional costs. Subsidised inputs from governments or international organisations can accelerate technology transfer in these situations.

The second barrier occurs when farmers are not knowledgeable enough to select the most suitable combination of irrigation technology, equipment, crop varieties, and management techniques to obtain the maximum returns in their situation. Technical assistance, such as educational programmes, training, and design aids from the government and other related organisations can help to overcome this obstacle (Ribaudo, 1997).

The third barrier encountered by irrigation technology transfer may result from public attitudes toward the technologies involved. In some regions, the necessity of irrigation to increase agricultural productivity and improve the adaptability of agricultural production to climate change has not been recognised by local governments and farmers. Education, training, demonstration projects and advertising can help create a more positive public attitude toward irrigation, thus eliminating this barrier.

New irrigation technologies require higher levels of management. Limited management knowledge and experience may prevent farmers from realising maximum profits from introduced irrigation technologies. This may slow the adoption of the introduced technologies and require further technology transfer interventions. This barrier can be addressed by training farmers in new irrigation management techniques as an integral part of the planning and installation of new systems, so that proper management is integrated into the initial technology adoption process.

BOX 11.2	FURNISHING WEATHER MEASUREMENTS TO ASSIST IN AGRICULTURAL DECISION MAKING

An automated weather data network (AWDN) was established in Nebraska, USA, in 1981. This network, and associated user interfaces, provides timely access to weather variables known to cause variations in agricultural production (Neue and Boonjawat, 1998).

The role of the government in this case is both to support a weather data collection and dissemination system, and to educate potential climate data users from the various sectors of the economy.

End-user recipients receive information that allows them to reduce their operating costs while maintaining optimal production. This increases net profit while conserving water resources and the energy involved in water delivery to the fields. Technology recipients receive the tools needed to provide timely information that is vital to decision processes in agriculture.

Applications programmes require user friendly interfaces with on-line help. In addition, education is a vital link in the chain from monitoring to application of data. The U.S. system has been replicated in Mexico, Brazil, and India. National and international organisations are recommended to adopt policies that promote near-real time monitoring efforts, and interagency committees should coordinate the needs for near-real time data.

Commodity programmes.

Commodity programmes that are crafted with care can assist the production of crops or crop varieties that have been bred for wider climate adaptation. Introduction of new varieties can be assisted by information provided by commodity programmes.

Trade policies.

Trade is able to play a significant adaptive role in technology transfer, allowing farmers in countries less severely affected by climate change to profit by selling products to consumers in the more severely affected regions. In this way, markets act to pool the risk of locally severe effects. Even with highly uncertain scenarios, regionally differentiated effects are highly likely. Thus, continued strengthening of agriculture with GATT provides important flexibility for adapting to climate change (Reilly, 1995).

Mitigation

Mitigation options exist, but few have been widely adopted or transferred for reasons that are often of a social rather than a technical nature, and it is extremely difficult to gain a sense of their cumulative potential for improving productivity and sustainable natural resource management. For example, the conversion of low intensity agricultural systems to forest has been used as a method for absorbing CO_2, but there are few similar land use change or management practices with a demonstrated impact on CH_4 or N_2O emissions. In the case of N_2O, it is generally accepted that measures which improve the efficiency of nitrogen fertiliser applications may be of value in reducing emissions, but these measures are often difficult to adopt for farmers and their effectiveness is yet to be demonstrated, thus adversely influencing successful technology transfer.

Adoption subsidies.

Although farmers have been able to significantly reduce nitrogen use in certain cases, changes in nitrogen use and profitability of technology adoption are highly location specific. More wide-spread adoption of major changes in management practices may require considerably larger subsidies than are currently available. Adoption subsidies need to be very carefully specified, time limited, targeted and transparent. They need to be monitored and evaluated, and adapted to changing circumstances. See Box 11.3.

Compliance programmes.

Compliance programmes have been successfully implemented in the USA. Such programmes, including transferred mitigation technology, require farmers to adopt certain approved production or land use practices in order to receive income support payments and other USDA programme benefits (Osborn, 1997a). The Conservation Compliance Program encourages the adoption of conservation cropping sequences, crop residue use and conservation tillage on highly erodible land. Similar programmes have been adopted in EU member states in recent years.

Demonstration projects.

Technical opportunities exist to improve animal feed quality and can be applied both on natural pastures and in farming systems. Fortunately, improved feed quality results in the reduction of GHG (methane) emissions from ruminant animals. Box 11.4 describes a demonstration project and its results in technology transfer.

BOX 11.3 ENCOURAGE EFFICIENT USE OF NITROGEN FERTILISERS

The U.S. Department of Agriculture (USDA) administers programmes providing education and technical and financial assistance to encourage more efficient use of nitrogen to control water pollution. Although control of GHG emissions is not an explicit goal, these programmes can reduce N_2O emissions and may be considered to be part of a no-regrets strategy. Stakeholders include farmers as well as federal, state and local environmental and water resource management agencies.

Barriers preventing technology transfer include: lack of producer familiarity with new practices, lack of knowledge on how to integrate new practices into current management systems, need for additional educational support to interpret soil testing data, and lower profitability of some resource-conserving practices. Cost-sharing (adoption subsidies) could be used to overcome adoption barriers. Local environmental conditions play a greater role in the adoption of resource-conserving practices than do the existence of demonstration projects.

Benefits include reduced input costs and a potential for reducing N_2O emission.

An interesting example of improved technology through technology transfer to increase feed supplies is urea treatment, which improves the palatability and digestibility of straw in China, resulting in a reduction of the relative methane emissions from animals. The amount of straw treated has increased from almost 0 tons in 1995 to 4 million tons in 1999, and is projected to reach 30 million tons by 2000 (Hongmin *et al.*, 1996; Tingshang, 1995).

BOX 11.4 THE PROGRAMME OF USING TREATED STRAW AS CATTLE FEED IN CHINA

Straw ammoniation technology can not only increase digestibility of animal feed and feed intake but also reduce 25-75 per cent of methane emissions per unit of animal produce-meat, milk, work, etc. (Sollod and Walters, 1992). This practice, which upgrades straw quality, can increase the digestibility of animal feed and lead to an increase of animal productivity.

Beef and milk production has been increased 230.7% and 38.7% in the past 5 years in China respectively, partly because of the transfer of straw ammoniation technology. The saved feed grain was 19.80 million tons in 1995. It can also increase organic fertiliser used and reduce biomass burning in crop lands.

The China State Council issued a document in 1992 about utilising crop straw to develop livestock in the farming region. There were a series of policies and measures to be adopted by different levels of governments and related departments to support the programme. One hundred and nineteen demonstration counties had been selected for activities that lasted until 1995.

Both the farmers and the government benefited from this programme.

•

The transfer of this technology is limited by the incremental investment (US$ 25 per ton) and by the lack of urea. Subsidisation, credit and assistance may be needed to overcome these financial barriers.

The national network is still weak with respect to animal feeding and nutritional management. Therefore, training should be given to livestock extension agents in the basics of practical ruminant nutrition, in feeding practices, and in the introduction of and on-farm support for key technologies.

This technology was selected to be disseminated nationwide as one of ten key technologies in 1989. The technology of using straw as cattle feed was placed in the agricultural development programme in 1992. The government allocated special funds (US$ 1.2 million) to establish 10 demonstration counties at the beginning, after which the funds increased rapidly to reach US$ 2.5 and 4.3 million in 1993 and 1994, respectively.

Education/technical assistance.

Efforts in the areas of information technology and management are becoming increasingly important in order to establish growth in crop and animal productivity in the face of climate change. Education programmes for youth, technical advisors, and technology recipients are important in improving technology transfer.

During the late 1970s and early 1980s the World Bank devoted very substantial resources to the support of an intensive training and visit (T&V) system of delivering information about practices and technology to farmers. The system involved a highly regimented schedule in which the field level worker is involved one day each week in intensive training about the information that he or she must convey to farmers (Benor and Harrison, 1977). New T&V programmes supported by international organisations for adaptation and mitigation of climate change are desirable.

Communication/outreach activities for technology transfer

Modern modes of extension delivery could greatly reduce traditional face-to-face extension and improve the mobility of extension agents. Radio, television and video display can spread messages suitable for a general audience and leave more time for the agent to concentrate on individual farmers' needs. These methods are common practice in developed countries, and have gained acceptance in developing countries as, for example, agricultural information via state television in India and Brazil; videotaped messages in Brazil, Honduras, Mexico, Paraguay and Peru; and satellite systems for spreading agricultural information over large areas in Indonesia, the Philippines and West Africa. Public information programmes aimed at demonstrating the benefits of new technologies may need to be coupled with cost-sharing to overcome barriers. In some cases, a technology information unit can be very useful to stimulate such public information programmes for technology transfer within a country (see Box 11.5).

Special considerations for agricultural technology transfer

Agriculture, in contrast to many other economic activities, is diffuse across diverse landscapes. Millions of individual decision-makers have to be motivated in order for climate change adaptation or mitigation technology to be transferred and adopted. Technology transfer depends on having good new technology that can provide concrete benefits to farmers and which yields those benefits relatively quickly. The technology also has to fit within the existing social and economic system.

Pilot projects and demonstration projects represent an especially cost-effective approach. However, in order to meet the criteria for technology transfer to farmers, given that the impacts from climate change are slowly cumulative, there will have to be a prompt productivity pay-off to the farmer from the activity. The combination of the gradual accumulation of negative impacts that may come from climate change and the need for almost immediate pay-off required by farmers for technology transfer results in the requirement for subsidies or incentives from the public sector. The public costs of technology transfer to agriculture - especially in mitigation where there may not be clear benefit to the farmer from the new technology or activity - will be high.

11.5 Programmes, Policies, and other Interventions for Technology Transfer between Countries

11.5.1 Barriers to Technology Transfer between Countries

Constraints of supply of new technologies

Shortage of technological information.

Because the developing counties lack access to information, they are not aware of what technologies fit their conditions and where they can find the suitable ones. International technology exchanges are helpful for overcoming this obstacle.

BOX 11.5   CONSERVATION TECHNOLOGY INFORMATION CENTER

The Conservation Technology Information Center (CTIC) was created in the United States by conservation organisations, universities, and private companies in the early 1980s. It was designed to gather and disseminate information to speed the adoption of conservation tillage technologies among American farmers. The Center's goal was to share information, and two early initiatives were launched to establish that capacity: carrying a national survey of tillage practices since 1982, and creating a national network of scientists and scientific information that could serve as a referral service.

Yield reduction and increased risk of failure were the most critical barrier for conserva-

tion tillage. Early field trials were needed on thousands of widely diverse soil and crop combinations to demonstrate that these risks could be managed. The CTIC encouraged conservation districts, USDA-SCS, and universities to conduct such trials, to hold field days for farmers and to communicate the results widely. The benefits to farmers include a reduction in labour, machine time, fuel, and associated costs. Yields are, in general, high enough to make net profits higher due to reduced cost.

Technology transfer that involves a complex system of new crop varieties, new machines, new methods, and unknown risks is a slow process that demands a long-term commitment to research, educa-

tion, and adaptation. Lingering threats of increased federal regulation of farm practices to address water pollution from cropland erosion was a reason for farm businesses and leaders to promote new systems that could achieve water quality goals voluntarily.

The CTIC process can be adopted where both public and private benefits are adequate enough to encourage a long-term commitment. It requires a situation where communication and education hold a major key to technology transfer. It also requires a situation where communications are adequate enough to allow producers to easily reach the information centre.

Shortage of capital.
Due to the long-term aspects of climate change, financial capital may also be a constraint to new technology.

Growth of agricultural research funding is slowing.
Agriculture is heavily dependent on climate in developing countries, so technology transfer is crucial for climate change adaptation and mitigation. The growth of agricultural research funding is slowing down, as Table 11.6 illustrates. This will impede the generation and transfer of technology. Funding trends for international research centres under the auspices of the CGIAR have shown a similar decline. Between 1971 and 1982 real spending for the CGIAR grew 14.3 per cent per year. Growth in real spending decreased to 1.4 per cent per year between 1985 and 1991, and decreased further to 0.5 per cent per year between 1991 and 1996 (Alston *et al.*, 1998). Given the time lags between initial R&D investment and diffusion of technologies, it may be several years before the effect of this slowdown becomes noticeable on the availability of new technologies, but the negative effects seem certain.

Table 11.6.	Growth in Public Agricultural R&D Spending in Developed and Developing Countries, 1971 to 1991	
AVERAGE ANNUAL GROWTH RATE (PER CENT CHANGE)		
	1971 TO 1981	**1981 TO 1991**
Developing countries	6.4	3.9
Sub-Saharan Africa	2.5	0.8
China	7.7	4.7
Other Asia and Pacific	8.7	6.2
Latin America & Caribbean	7.0	-0.5
West Asia and North Africa	4.3	4.1
Developed countries	2.7	1.7
Total	4.3	2.9

SOURCE: ALSTON, ET AL, 1998

Barriers to private sector involvement

The worry about the absence of protection for intellectual property might be the key barrier to more private sector involvement in Technology Transfer. So it is important to adopt stricter IPRs to encourage greater private investment in agricultural R&D, and greater involvement in technology transfer to increase agricultural research funding. Many (particularly developed) countries have adopted stricter intellectual property rights (IPR) regimes for agrochemicals, agricultural machinery and biological innovations. A rationale for adopting stricter IPRs is to increase private appropriability of research benefits and to encourage greater private sector investment in agricultural R&D and greater involvement in technology transfer. Although evidence from the United States suggests that increased plant variety protection has stimulated private R&D and adoption of improved crop varieties, the issue of IPRs for genetic resources remains controversial. Particular areas of controversy are farmer and research exemptions to IPR protection, and whether and to what extent IPRs should be extended to developing countries (Frisvold and Condon, 1995, 1998; Knudson, 1999). Recent theoretical literature suggests that there may be limits to how far IPRs should be extended internationally (Deardorff, 1992, 1993).

Financial barrier for developing countries to access new technology

According to FAO (1990b), technology transfer in developing countries involves some 550,000 staff, most of them in public extension services, and costs about US$4.5 billion annually. Under the influence of structural adjustment and declining public funding, extension services have in recent years tended to shrink. Governments and international organisations have the opportunity to encourage the private sector to promote effective modalities for the access and transfer, in particular to developing countries, of ESTs through grants and concessional loans.

Technology developments respond to local conditions

There is a close relationship between the development of technology and technology transfer. Hayami and Ruttan's theory of induced innovation best describes this process (1985). They link resource scarcity with the economic incentive to overcome that scarcity through the development of new technology (Sanders *et al.*, 1996). Thus, a country like Japan, which has a scarcity of land, will develop an agricultural technology which is land saving. This includes higher yielding varieties, better irrigation, and terracing of land. A country like the United States, which had a scarcity of labour, proceeded in another direction, developing a labour saving technology in agriculture that includes: the mechanical reaper, the steel plow, and the combine harvester. Both in Japan and the United States, and in other countries with specific resource scarcities, institutions were established to develop the technology and assist the technology transfer. Examples would be the Land Grant University in the United States and the itinerant farmer in Japan.

For developing countries, policies to assist technology transfer might include:
- An accelerated international effort to collect germ plasm from species (such as drought resistant crops) whose qualities will be most valuable in the future;
- The expansion of credit and savings schemes, to assist rural people to manage the increased variability in their environment;
- Shifts in the allocation of international agricultural research for the semi-arid tropics towards water-use efficiency, irrigation design, irrigation management and salinity, and the effect of increased CO_2 levels on tropical crops;
- The improvement of food security and disaster early warning systems, through satellite imaging and analysis, national and regional buffer stocks, improved international responses to disasters, and linking disaster food-for-work schemes to adaptation works (*e.g.*, flood barricades);
- The development of institutional linkage between countries with high standards in certain technologies, for example flood control and GPS (Touche Ross, 1991).

Special problems of technology transfer among developing countries

The main flow of technology transfer to deal with climate change is from developed to developing countries, as emphasised by UNFCCC and The Kyoto Protocol. Some cases of the existing technology transfer between developing countries are beneficial to climate. International organisations and relevant developed countries could encourage the existing technology transfer among developing countries.

Institutional capacity on agricultural research is limited in developing countries. A matter of great concern is the state of the national agricultural systems (NARSs) in developing countries, many of which have declined in capacity over the past decade or two. In the past, developed countries provided support to strengthen NARSs and to make them more effective, but since 1985 this support has dwindled hugely and appears poised to disappear entirely. This declining support comes at a time when developing countries are facing problems of competition, trade and economic restructuring. This seems to make the continued linkages with these programmes even more needed.

Operational budgets per researcher in many developing countries have been declining in recent years
Due to the restriction this places on researchers' access to their clients, much national research remains of little practical relevance. Effective links to extension and feedback cannot be properly established. National institutes have often been slow to adopt a client-oriented approach in research programming.

The biggest barrier for technology transfer among developing countries may be the shortage of financial support. Technology recipients need new investments to adopt new technology. The providers need to ensure human and financial resources to transfer the technology. The extent of technology transfer may be limited by the shortage of financial resources. If the technology is to be transferred between developing countries, both the technology provider and the recipient may need new and additional financial resources. Additional financial resources may need to be accompanied by the removal of institutional barriers in order to be effective. Cost-sharing may be an important safeguard.

11.5.2 Transferred Technologies

Adaptation
Adaptation to climate change is likely; the extent depends on the affordability of adaptive measures mentioned in Table 11.4, access to technology and biophysical constrains such as water resource availability, soil characteristics, genetic diversity, crop breeding, and topography. National studies have shown incremental additional costs of agricultural production under climate change which could create a serious burden for some developing countries (IPCC, 1996b).

Nearly all agricultural impact studies conducted over the past 5 years have considered some technological options for adapting to climate change (IPCC, 1996b). Technology transfer and diffusion of new technologies could in particular focus on improvement of irrigation technology and alternative species and varieties.

Improving irrigation technology
Changes in temperature and precipitation levels will alter the hydrological cycle and water supplies. In general, the IPCC (1996b) notes that estimated precipitation increases in high latitude regions may lead to runoff increase; runoff will tend to decrease in lower latitudes due to combined temperature increase and precipitation decrease. Increased rainfall intensity would increase soil erosion, while in other regions agriculture could be affected by drought. Rind *et al.* (1990) use GCM results to calculate that for many mid-latitude locations (*e.g.*, USA) the incidence of severe droughts that currently occur only 5% of the time could rise to a 50% frequency by 2050, based on the difference between precipitation and potential evapotranspiration. Such a change would constitute a severe natural disaster for agricultural production.

About 253 million hectares, 17% of the world's crop land, are irrigated. This land produces more than one-third of the world's food (Geijer *et al.*, 1996). Irrigation is therefore increasingly important for adapting to the effects of climate change on agricultural production. Almost three-quarters of the world's irrigated area is in developing countries. To mitigate the negative effects of climate change on agriculture, developing countries should improve their existing irrigation efficiency through adoption of drip irrigation systems and other water-conserving technologies (FAO 1989, 1990a). An alternative is to import technologies and equipment from developed countries that have advanced irrigation technologies. See Box 11.6 for a discussion of the main barriers involved.

Box 11.6	IRRIGATION TECHNOLOGY TRANSFER BETWEEN COUNTRIES AND ACROSS BARRIERS

The main flow of irrigation technology transfer is from developed to developing countries. The primary barrier encountered by the importing developing countries is that they cannot afford the high cost of patents and equipment. The second barrier is that the importers do not have enough money to build the auxiliary equipment for the introduced technology, because developing countries, in many cases, only buy key equipment due to their limited financial resources. The third barrier for irrigation technology transfer between countries is that the importing developing countries are not clearly aware of what technologies fit their conditions and where they can find the suitable ones. The importers do not always receive satisfactory service when there are problems with their imported equipment or scientific instrument.

New species and varieties adapted to changing climate.
For most major crops, varieties exist with a wide range of maturities and climatic tolerances. For example, Matthews *et al.* (1994) identified wide genetic variability among rice varieties as a reasonably uncomplicated response to spikelet sterility in rice that occurred in simulations for South and Southeast Asia. Studies in Australia showed that responses to climate

BOX 11.7 NEW RICE IN SIERRA LEONE

The development of a new mangrove rice variety in Africa is an important case study of technology development and transfer. Much of the success of this effort hinged on the accident of a critical mass of researchers at the government rice research station in Rokupr Sierra Leone and the interest of the West African Rice Research Development Association (WARDA) in this effort. WARDA provided additional resources to the station at Rokupr to carry out the development of a new rice variety to meet the changed climate conditions, and improve yields above those previously achieved. The Sierra Leone agricultural research establishment was able to demonstrate the value of their rice research effort to the food supply of the nation and WARDA was able to demonstrate to their financial supporters their value in contributing to this new technology and its transfer. There was a German seed distribution project that helped with some seed distribution, but farmers themselves undertook most of the technology transfer to other farmers once the success of the new variety became apparent (Prahah-Asante et al., 1982, 1986; Spencer, 1975; Spencer et al., 1979; Tre et al., 1998; WARDA, 1987; Zinnah, 1992; Zinnah et al., 1993).

change are strongly cultivar dependent (Wang *et al.*, 1992). The genetic base is broad for most crops but limited for some (*e.g.*, kiwi fruit). A study by Easterling *et al.* (1993) explored how hypothetical new varieties would respond to climate change (also reported in McKenney *et al.*, 1992). Heat, drought, and pest resistance, salt tolerance, and general improvements in crop yield and quality would be beneficial for crop adaptation (Smit, 1993). See Box 11.7.

Mitigation

Most developing countries depend heavily on agriculture. Developing countries are barely able to adopt the mitigation technologies mentioned in Table 11.3 to mitigate the GHG emissions in agricultural systems, because of barriers mentioned in Table 11.4. If advanced technologies are transferred to them with demonstration projects, as well as technical assistance and financial support, GHG emissions will be reduced.

Improvement of the efficiency of nitrogen fertiliser.
Nitrogen fertiliser efficiency decreased with increased nitrogen fertiliser input. So farmers need additional information such as soil testing data, as well as educational support to interpret the data. They can also gain experience by participating in demonstration projects. Extension personnel are needed to provide on-farm technical assistance with new practices to increase N efficiency. Availability of application machinery and technologies must be transferred simultaneously to be effective.

Reducing methane emission from rice fields.
Feasible mitigation strategies that have been verified to significantly reduce methane emission from rice fields are temporary midseason aeration of the soil, using fermented instead of fresh organic manure, applying sulfate containing fertiliser, and planting/breeding rice cultivars with low emission capacity.

Reducing methane emission from animal waste.
Biogas digesters can provide clean energy and high quality fertiliser, and can be an important option for reducing methane emission from livestock manure. This technology is widely recognised as an EST and has been widely accepted all over the world, especially in China and India.

11.5.3 Programmes, Policies and other Interventions for Technology Transfer between Countries

Agricultural development has been strongly influenced by technology. New technology has been the most consistent driving force behind agricultural growth. Evenson (1994) estimates that it has contributed from one-half to two-thirds of output gains over recent decades. Thus, strengthening agricultural technology transfer between countries will have significant impact on the improvement of production capability, increased food variety, quality, and security, as well as balanced agricultural development, resource utilisation and conservation.

Technology transfer between developed and developing countries
The flow of technology transfer in the agricultural sector from developed to developing countries dealing with climate change is going to be crucial as the majority of the climate mitigation and adaptation projects are going to be funded by investors, bilateral and multilateral assistance agencies, NGOs and foundations, largely based in developed countries.

Role of Governments
Governments are going to have an important function in the selection and adoption of mitigation and adaptation technologies in the agricultural sector. Governments can promote effective modalities for the access and transfer, in particular to developing countries, of ESTs by means of activities mentioned in Chapter 34 of Agenda 21 and decisions of the CoPs (see also section 3.3.1 in Chapter 3 on Agenda 21). Some of the possible measures for the governments are to: (*i*) increase funding to mitigation and adaptation projects and to programmes on: preventing land degradation, improving water use efficiency, breeding new varieties and manufacturing agrochemicals to mitigate and adapt to climate change; (*ii*) increase funding for improving the capacity to develop and manage EST; (*iii*) increase funding for institution and human capacity building and for improving R&D capabilities in developing countries; (*iv*) facilitate adaptation by farmers by providing incentives, by regulation and by improving existing or setting up new institutions; (*v*) promote research and development activities directed at technological innovation and technology transfer for climate-change adaptation and

mitigation in agriculture; (vi) develop national agricultural information systems to produce reports on state-of-the-art technology, disseminate information on available technologies, their sources, their environmental risk and help users to identify their needs; and, (vii) purchase patent and licenses on commercial terms for their transfer to developing countries on non-commercial terms as part of development cooperation for sustainable development.

Role of International organisations

International organisations could *(i)* increase funding for institutional and human capacity building, and for improving R&D capabilities in developing countries to implement the UNFCCC; *(ii)* provide new and additional grant and concessional funding to meet the agreed incremental costs of projects to achieve agreed global environmental benefits in climate change; *(iii)* increase funding to agricultural sector projects and programmes to mitigate and adapt to climate change; (vi) develop regional and international agricultural information systems to collect and report the state-of-the-art technologies, disseminate the information on available technologies and relevant information and help users to identify their needs; and, (vii) support central laboratories to do analytical work for developing countries, providing uniform methods and equipment and producing comparable results (as originally planned for soil and plant analysis at the IBSRAM in Bangkok).

BOX 11.8   **CAPACITY BUILDING FOR THE IMPLEMENTATION OF UNFCCC**

In the Research Programme on Methane Emissions from Rice Fields funded by GEF in 1993, training has been provided to the country teams working on the ALGAS project. In the ALGAS project, capacity has been raised through training over 160 national technical experts in elements of GHG inventory, mitigation and project identification. These experts are now providing input to the process of national communications and helping to identify other climate change mitigation projects for future development.

Role of Private sector

Technology generation is shared between the public and the private sectors, with the share of the private sector tending to be greater in more developed countries. Private research is attract-

ed to subsectors where markets for research results exist and can be privately appropriated. Therefore, the private sector is going to play a critical role in the agricultural sector to deal with climate change. Some of the measures for the private sector to promote technology transfer could be to: *(i)* provide technical assistance to appropriate users on its new technologies or new varieties; *(ii)* promote cooperation on research and development activities directed at technological innovation and technology transfer for climate-change adaptation and mitigation in agriculture.

Technology transfer among developing countries

Technology transfers among developing countries are limited, because most advanced technologies are developed and owned by industrialised countries. In technology transfer between developing countries, finance can become an issue. For example, the technology transfer of small scale biogas digesters is mainly from developing country to developing country and it is limited by financial problems. Thus, if the technology is transferred from one developing country to other developing countries, both provider and recipient may need new and additional financial resources from international organisations or developed countries. Therefore, international organisations and developed countries could *(i)* increase funding for institutional and human capacity building and for improving R&D capabilities in developing countries; *(ii)* fund both the technology provider and technology recipient to promote a successful technology transfer; *(iii)* facilitate adaptation by farmers by providing incentives, by regulation and by improving existing or setting up new institutions. See Box 11.9.

In order to improve/stimulate technology transfer in agriculture, a multi-level technology transfer system (global, regional, and national) can be established. The global system may be made up of three major players: NARSs of developing countries, IARCs, and advanced laboratories and institutions in developed countries. All of them play common and different roles within government efforts to support technology transfer for climate change.

11.6   Lessons Learned

Technology transfer of agriculture and know-how, similar to other sectors, include financial flows between countries, activities undertaken by governments to facilitate the introduction

---

BOX 11.9   **BIOGAS DIGESTER TECHNOLOGY TRANSFER FROM CHINA TO OTHER DEVELOPING COUNTRIES**

The Asia-Pacific Region Biogas Research and Training Center (BRTC) has been contributing a lot to promote the development of biogas digester technology in developing countries. Since it was established, the BRTC has held 21 training workshops with more than 270 participants coming from 71 countries. Most participants of the programme acquired the skills to construct, operate and maintain small scale biogas digesters in their countries. The center proved to be a valuable tool in demonstrating the usefulness of capacity building in transferring a technology. During the period 1980 to 1990, more than 50 scientists were sent out to assist the construction of over 70 digesters in 22 developing countries. Biogas digester use has been expanded in the developing countries located in Asian, Pacific and African regions. This technology has provided clean and convenient energy for farmers. The diffusion of this technology is limited by financial assistance. New and additional financial resources for both providers and receivers of developing countries are expected to enhance the beneficial practices for abatement of GHG (Yizhang, 1990; Zhao, 1990).

and use of ESTs, private sector banks, small and medium enterprises, and transnational corporations, as well as success stories from different countries. (FCCC/SB/1997/1)

Agriculture will be heavily influenced by climate change. Sustainable agricultural development is an ongoing priority for all countries. Transfer of adaptation and mitigation technologies has significant benefits independent of climate change consideration. This is even more relevant, because now climate change will offer greater challenges and development opportunities for agricultural systems.

Carbon emissions from fossil fuels used in agricultural production comprise 4% of the approximate 3.4Wm-2 radiative forcing of the direct and indirect greenhouse effects of changing gases. Carbon storage in agricultural soil and durable agricultural products can be increased greatly in the future but is still surrounded with uncertainty for the time being. Controlling these emissions or enhancing carbon sinks will be challenging because of technological and other constraints. Available options for GHG reductions, which can be transferred within and between countries are improvements in acquiring new species and varieties, changing tillage systems, treating livestock manure and recovering biogas as energy. However, the specificity in physical, economic, market, historical, and cultural conditions, and also the institutional and technological capacities in different regions and countries create implementing barriers to the diffusion of these options. Exploiting the other benefits of these options such as increasing productivity, environmental externalities and economic gains will improve the likelihood of their implementation.

Adaptation will be important in limiting negative effects and taking advantage of beneficial changes in climate. The extent of adaptability depends on the affordability of such measures, particularly in developing countries, access to know-how and technology, the rate of climate change, and biophysical constraints such as water availability, soil characteristics, and crop genetics. A range of adaptation options can be employed to increase the flexibility and adaptability of vulnerable systems. Many of these options, if they also abate climate change, will be of immediate benefit, and can therefore be considered "no-regret" technologies, giving them the highest priority for technology transfer.

Even though adaptation and mitigation options are clear, integrated options need consideration in technology transfer. These include the following factors: options need to be based on development needs, operate at a desired capacity and technologies need to be adapted to local conditions. For example, technology transfer of fertiliser use, as a main source of GHG, is focused upon and must therefore be balanced by productivity needs and by abatement of GHG emissions.

In the agricultural sector, as most recipients are developing countries with limited financial abilities, the governments of the suppliers' side can play an important role to encourage technology transfer through compensation, tax reduction and so on. They can formulate policies and programmes for the effective transfer of

ESTs that are publicly or privately owned. In the case of privately owned technologies, the adoption of the following measures for technology transfer could be considered:

- Creation and enhancement by developed countries of appropriate measures, fiscal or otherwise, to stimulate the transfer of ESTs and its capacity building, in particular to developing countries;
- Purchase of patents and licenses on commercial terms for their transfer to developing countries on non-commercial terms, as part of development cooperation for sustainable development;
- Implementation of measures to prevent the abuse of intellectual property rights; IPRs should not be a barrier for technology transfer
- Provision of financial resources to acquire ESTs in order to enable developing countries in particular to implement measures to promote sustainable development;
- Development of mechanisms and multi-stakeholder institutions for the access to and transfer of ESTs, in particular to developing countries.

The main flow of technology transfer is from developed to developing countries dealing with climate change, as was emphasised by the UNFCCC and The Kyoto Protocol. Some cases of existing agricultural technology transfer among developing countries, such as CGIAR and other multilateral systems, can be most helpful in assisting countries dealing with climate change if their capacities are strengthened. International organisations and relevant developed countries can make great contributions by encouraging and supporting technology transfer among developing countries.

Climate change requires extra effort to transfer technologies that:

- Increase crop output per liter of irrigation water drawn;
- Increase demand for appropriate technology with incentives if needed, and ensure the provision of reliable supplies and equipment that meet local situations and needs;
- Increase soil carbon and reduce methane emissions;
- Assure that intellectual property rights help and do not hinder technology transfer, especially to small farmers and developing countries; and,
- Provide crops suited to warmer temperatures

At minimum, this extra effort calls for a restored CGIAR and linked NARSs.

References

Adesina, A., and M. Zinnah, 1993: Technology characteristics, farmers' perceptions and adoption decisions: A tobit model application in Sierra Leone. *Agricultural Economics*, **9**, 297-311.

Agyen-Sampong, M., S. Anoop, K. Sandu, M.P. Prakah-Asante, C.A. Jones, S.F. Dixon, W.A.E. Fannah, W.A.E. Cole, and H. M. Bernard, 1986: *A Guide to Better Mangrove Swamp Rice Cultivation in the WARDA Region*. WARDA Regional Mangrove Swamp Rice Research Station, Rokupr, Sierra Leone.

Alston, J., P. Pardey, and V. Smith, 1998: Financing agricultural R&D in rich countries: what's happening and why. *Australian Journal of Agricultural and Resource Economics*, **42**(1), 51-82.

Benor, D., and J.Q. Harrison, 1977: *Agricultural Extension: The Training at Visit System*. World Bank, Washington, DC.

Bouwman, A.F. 1990: Global distribution of the major soils and land cover types. In *Soils and the Greenhouse Effect*. A.F. Bouwman, (ed.), John Wiley and Sons, Chichester, UK, pp. 47-59.

Deardorff, A., 1992: Welfare effects of global patent protection. *Economica*, **59** (233), 35-51.

Deardorff, A., 1993: Should patent protection be extended to all developing countries? In *The Multilateral Trading System: Analysis and Options for Change*. R.M. Stern R.M., (ed.), University of Michigan Press, Ann Arbor, MI, pp. 435-48.

De Janvry, A., 1988: *The Agrarian Question and Reformism in Latin America*. Johns Hopkins University Press, Baltimore, MD.

Duvick, D.N., 1989: Plant Breeding: Past Achievements and Expectations for the Future. *Economic Botany*, **40**, 289-297.

Duxbury, J.M., and A.R. Mosier, 1993: Status and issues concerning agricultural emissions of greenhouse gases. In *Agricultural Dimensions of Global Climate Change*. H.M. Kaiser, T.E. Drennen,(eds.), St. Lucie Press, Delray Beach, FL, pp. 229-258.

Easterling, W.E., P.R. Crosson, N.J.Rosenberg, M.McKenny, L.A.Katz, and K.Lemon, 1993: *Agricultural impacts of and responses to climate change in the Missouri-Iowa-Nebraska-Kansas (MINK) region*, Washington, DC, pp.221-228.

Evenson, 1994: *Science for agriculture: International perspective*. Tale University, New Haven, CT.

FAO, 1989: Guidelines for designing and evaluating surface irrigation systems. Irrigation and Drainage Papers, 45, FAO, Rome.

FAO,1990a: Water harvesting. AGL Miscellaneous Papers, 17, FAO, Rome.

FAO,1990b: *Report of the global consultation in agricultural extension*. Rome.

Fearnside, P., 1987: Deforestation and International Economic Development Projects in Brazilian Amazonia. *Conservation Biology*, **1**, 214-221

Frisvold, G., 1997: Multimarket Effects of Agricultural Research with Technological Spillovers. In *Global Trade Analysis: Modeling and Applications*. T. Hertel, (ed.), Cambridge University Press, Cambridge.

Frisvold, G., and P.T. Condon, 1995: The UN convention on biological diversity: implications for agriculture. *Technological Forecasting and Social Change*, **50**(1), 41-54.

Frisvold, G., and P.T.Condon, 1998: The UN convention on biological diversity and agriculture: implications and unresolved debates. *World Development*, **26** (4 – April), 551-570.

Geijer, J. C. M., A. M. Svendsen, and D. L. Vermillion, 1996: Transferring irrigation management responsibility in Asia: Results of a Workshop. *Short Report Series on Locally Managed Irrigation*, Report No. 13. International Irrigation Management Institute, Food and Agriculture Organization of the United Nations, Colombo, Sri Lanka, 21 pp.

Goodman, M., 1990: Genetic and Germplasm Stocks Worth Preserving. *Journal of Heredity*, **8**, 11-16.

Hayami, Y., and V. Ruttan, 1985: *Agricultural Development*. Johns Hopkins University Press, Baltimore, MD.

Hogan, K.B., 1993: Methane reductions are a cost-effective approach for reducing emissions of greenhouse gases. In *Methane and nitrous oxide: Methods in National Emissions Inventories and Options for Control*. A.R. van Amstel, (ed.), RIVM Report No. 481507003, Bilthoven, The Netherlands, pp.187-201.

Hongmin, D., L.Erda, and L.Yue, 1996: An estimation of methane emission from agricultural activities in China. *AMBIO, A Journal of the Human Environment*, 292-296.

IPCC, 1992: *Climate Change 1992: Supporting Material, Working group III; Response Strategies, Subgroup AFOS*. The IPCC supplementary Report to the IPCC Scientific Assessment, WMO/UNEP, Geneva, pp. 14-22

IPCC 1996a: *Climate Change 1995 – The science of Climate Change: Summary for Policymakers*, in "*Climate Change 1995: Impacts, Adaptations and Mitigation of Climate Change: Scientific-Technical Analyses*." Contribution of Working Group II to the Second Assessment Report of the Intergovernmental Panel to Climate Change, Cambridge University Press, Cambridge and New York , pp. 1-56.

IPCC, 1996b: *Climate Change 1995: Impacts, Adaptations and Mitigation of Climate Change: Scientific-Technical Analyses*. Contribution of Working Group II to the Second Assessment Report of the Intergovernmental Panel to Climate Change, Cambridge University Press, Cambridge and New York, pp. 880.

IPCC Workshop on Adaptation to Climate Variability and Change, 1998: *Summary Report to IPCC*, San Jose, Costa Rica.

Klein, R.J.T., and R.S.J.Tol, 1997: *Adaptation to Climate Change: Options and Technologies*. Institute for Environmental Studies, Vrije Universiteit, Amsterdam, pp.13-14.

Knudson, M., 1999: Agricultural biodiversity: do we have the resources to meet future needs? In *Global Environmental Change and Agriculture: Assessing the Impacts*. G. Frisvold, B. Kuhn, (eds.), Edward Elgar Publishing, Cheltenham, UK/Northampton, MA.

Lal, R., J.M. Kimble, R.F. Follett, and C.V. Cole. 1998: *The Potential of U.S. Cropland to Sequester Carbon and Mitigate the Greenhouse Effect*. Sleeping Bear Press, Inc., Chelsea, MI, p. 61.

Larson, B., 1994: Changing the Economics of Environmental Degradation in Madagascar: Lessons from the National Environmental Action Plan Process. *World Development*; **22**, 671-89.

McKenney, M.S., W.E. Easterling, and N.J. Rosenberg, 1992: Simulation of crop productivity and responses to climate change in the year 2030: The role of future technologies, adjustments and adaptations. *Agricultural and Forest Meteorology*, **59**,103-127.

Mosier, A.R., J.M. Duxbury, J.R. Freney, O. Heinemeyer, K. Minami, and D.E. Johnson, 1998a: Mitigating agricultural emissions of methane. *Climate Change*, **40**, 39-80.

Mosier, A.R., J.M. Duxbury, J.R. Freney, O. Heinemeyer, and K. Minami, 1998b: Assessing and mitigating N_2O emissions from agricultural soils. *Climate Change*, **40**, 7-38.

Mundlak, Y., 1994: Land Expansion, Land Augmentation and Land Saving. Benjamin H. Hibbard Memorial Lecture Series, Department of Agricultural and Applied Economics. University of Wisconsin, Madison, WI.

National Research Council (NRC), 1991: *Managing Global Genetic Resources: The U.S. National Plant Germplasm System*. National Academy Press, Washington, DC.

National Research Council (NRC), 1993: *Crop Diversity: Institutional Responses in Managing Global Genetic Resources: Agricultural Crop Issues and Policies*. National Academy Press, Washington, DC.

Neue, H.U., and J. Boonjawat, 1998: Methane emission from ricefields. In *Asian Change in the Context of Global Change*, IGBP Book Series. J. Galloway, J. Melillo, (eds.), Cambridge University Press, Cambridge, pp.187-207

Oram, P.A., and B. Hojjati, 1995: The Growth Potential of Existing Agricultural Technology. In *Population and Food in the Early Twenty-First Century: Meeting Future Food Demand of an Increasing Population*. Nural Islam,(eds.), IFPRI, 1995, Washington, DC., pp. 167-190.

Osborn, C.T., 1997a: Conservation and environmental programs overview. In *Agricultural Resources and Environmental Indicators,1996-7, Agricultural Handbook Number 712*. M. Anderson, R. Magelby, (eds.), U.S. Department of Agriculture, Economic Research Service, pp. 255-269.

Prahah-Asante, K., M. Samake, and D.S.C. Spencer,1982: Preliminary Results of the WARDA TAT Programme on Floating and Mangrove Swamp Rice, WARDA/82/ARR/4.

Prakah-Asante, K., and D.S.C. Spencer, 1986: *Mangrove swamp rice production in the Great Scarcies Area of Sierra Leone: socio-economic implications*. Socio-Economic Survey Report Part 1, WARDA, Rokupr, Sierra Leone.

Reilly, J., 1995: Climate Change and Agriculture: Research Findings and Policy Consideration. In *Population and Food in the early Twenty-First Century: Meeting Future Food demand of an Increasing Population*. Nural Islam, (eds.), International Food Policy Research Institute (IFPRI).

Ribaudo, M., 1997: Water quality programs. In *Agricultural Resources and Environmental Indicators, 1996-7, Agricultural Handbook Number 712*. M. Anderson, R. Magelby, (eds.), U.S. Department of Agriculture, Economic Research Service, pp. 83-96.

Rind, D., D. Glodberg, J. Hansen, C. Rosenzweig, and R. Ruedy, 1990: Potential evapotranspiration and the likelihood of future drought. *Journal of Geophysical Research*, **95**(D7), 9983-10004.

Ruttan, V.W. (ed.), 1992: *Sustainable Agriculture and the Environment: Perspectives on Growth and Constraints*. Westview Press, Boulder, CO, p. 132.

Ruttan, V.W.(ed.), 1998: Research Systems for Sustainable Agricultural Development. In *Global Environmental Change and Agriculture: Assessing the impacts*. G. Frisvold, B.Kuhn,(eds.), Edward Elgar Publishing, Aldershot, UK.

Safley, L.M., M.E. Casada, J.W.Woodbury, and K.F.Roos, 1992: *Global Methane Emissions from Livestock and Poultry Manure*. USEPA Report 400/1-91-048, Office of Air and Radiation, Washington, DC, 68 pp.

Sanders, J., B. Shapiro, and S. Ramaswamy, 1996: *The Economics of Agricultural Technology in Semiarid Sub-Saharan Africa*, Johns Hopkins University Press, Baltimore, MD.

Smit, B. (ed.), 1993. Adaptation to Climatic Variability and Change. Occasional Paper No. 19, University of Guelph, Guelph, ON, Canada, 53 pp.

Sollod A.E., and M.J. Walters, 1992: Reducing Methane Emission from Ruminant Livestock: China Prefeasibility Study.

Spencer, S.C., 1975: *The Economics of Rice Production in Sierra Leone: Part 2 Mangrove Swamp,* Bulletin No. 2, Dept of Agricultural Economics and Extension, Njala University College, Sierra Leone.

Spencer, S.C., D. Byerlee, and S. Franzel, 1979: Annual Costs, Returns and Seasonal Labor Requirements for Selected Farm and on-Farm Enterprises in Rural Sierra Leone. MSU, African Rural Economy Working Paper, No. 27.

Tingshuang, G., 1995: *Husbandry with Crop Straw*. Agricultural Publishing House, Beijing.

Touche Ross, 1991: *Global Climate Change: The role of Technology Transfer*. Report for the UNCED, financed by the UK Department of Trade and Industry and Overseas Development Administration.

Tre, J.P., M.William, A.Akinwuni, and J. Lowenberg-DeBoer, 1998: The Impact of Regional Research on Micro-Environment Niche Crops: Mangrove Rice in West Africa (unpublished, based J.P. Tre, MS thesis).

UN/ FAO, 1997: *State of the World's Plant Genetic Resources*. Report, FAO, Rome.

United States Congress, Office of Technology Assessment (OTA), 1987: *Technologies to Maintain Biological Diversity. Government Printing Office*, Washington, DC.

Wang, Y.P., Jr. Handoko, and G.M. Rimmington, 1992: Sensitivity of wheat growth to increased air temperature for different scenarios of ambient CO_2 concentration and rainfall in Victoria, Australia - a simulation study. *Climate Research*, **2,** 131-149.

WARDA, 1987: A Decade of Mangrove Swamp Rice Research, Bouake, Cote d'Ivoire.

World Bank, 1992: *Development and the Environment - World Development Report,* Oxford University Press for the World Bank, Oxford.

World Bank 1994: *Infrastructure for Development - World Development Report,* Oxford University Press for the World Bank, Oxford.

Wright, B., 1996: Crop Genetic Resource Policy: Towards a Research Agenda, Environment and Production Technology Division, Discussion Paper No. 19. International Food Policy Research Institute, Washington, DC.

Yizhang, Z.,1990: International exchange on biogas technology during 1980 to 1990. In *China Biogas 1980-1990*. Environmental Protection and Energy Department of Agricultural Ministry of China, (ed.), Chinese Scientific and Technological Publishing House, Beijing, pp.53-57.

Zhao, Y., 1990: International exchange on biogas technology during 1980 to 1990. *In China Biogas 1980-1990*. Environmental Protection and Energy Department of Agricultural Ministry of China (ed.), Chinese Scientific and Technological Publishing House, Beijing, pp. 53-57.

Zinnah, J., L Compton, and A.Adesina, 1993: Research-extension-farmer linkages within the context of the generation, transfer and adoption of improved mangrove swamp rice technology in West Africa. *Quarterly Journal of International Agriculture*, **32**(2).

Zinnah, M.M.,1992: *The Adoption and Impact of Improved Mangrove Swamp Rice Varieties in West Africa: The Case of Guinea and Sierra Leone*. Ph.D. Thesis, University of Wisconsin, Madison, WI.

Zohrabian, A., 1995: Genetic Resources Conservation and Maintenance in the Former Soviet Union. Department of Agricultural and Applied Economics Staff Paper P95-1, University of Minnesota, Minneapolis, MN.

12

Forestry Sector

Coordinating Lead Author:
N. H. RAVINDRANATH (INDIA)

Lead Authors:
Philip M. Fearnside (USA), Willy Makundi (Tanzania), Omar Masera (Mexico), Robert Dixon (USA)

Contributing Authors:
Kenneth Andrasko (USA), Neil Byron (Australia), Antony DiNicola, (USA), Nandita Mongia (India), P. Sudha (India)

Review Editor:
David Hall (United Kingdom)

CONTENTS

EXECUTIVE SUMMARY

In the forestry sector technology transfer has a broad definition, which includes sustainable forest management practices, forest conservation and Protected Area management systems, silvicultural practices for afforestation and reforestation programmes, genetically superior planting material, efficient harvesting, processing, end-use technologies and indigenous knowledge of forest conservation. Some of the potential roles for Governments to promote environmentally sound technology (EST) transfer in the forestry sector are as follows:

1. In the forestry sector, governments could provide targeted financial and technical support through multilateral agencies (such as GEF, World Bank, FAO, UNDP, UNEP), CGIAR institutions (such as CIFOR, ICRAF) and ODA for: capacity building (institutional and human), to adopt sustainable forest management practices, forest certification, reduced impact logging practices, high yielding silvicultural practices, bioenergy technologies, and forest area and status monitoring techniques, and so on.

2. Governments could play a critical role in developing and enforcing the regulations to adopt sustainable forest management, timber certification, recycling, reduced impact logging practices, etc. through financial incentives and regulatory measures.

3. Governments could promote the participation of communities, institutions and NGOs in implementing forestry projects and in enforcing regulations.

4. The forestry sector is highly regulated in many regions and Governments could create conditions to enable participation of industry and farmers -- with adequate guidelines to ensure biodiversity conservation and interests of local communities, and to enable the private sector to have an increasing role in developing and transferring technologies through commercial mechanisms between countries. Relaxing government ownership and control to include private and community ownership and mid-term to long-term leasing may enhance technology transfer.

5. One of the most important approaches that governments along with multilateral agencies can do to promote forestry mitigation technologies and projects is to develop and transfer methodologies for monitoring, measuring and verifying carbon mitigation and forest area changes to enhance the credibility of forestry mitigation projects.

6. The governments and private sector in Annex I and non-Annex I countries, as well as multilateral agencies, have a critical role in establishing and operationalising financial and regulatory mechanisms, monitoring, verification and certification arrangements, and capacity building for technology development, transfer and assimilation. Governments could prepare guidelines and set up institutional mechanisms to process, evaluate, sanction, and monitor forest-sector mitigation and adaptation projects.

7. The role of private sector funding of projects needs to be promoted under the new initiatives, including the proposed flexible mechanisms under the Kyoto Protocol. GEF could fund projects that actively promote technology transfer and capacity building in addition to the mitigation aspects.

Appropriately designed forestry mitigation and adaptation projects contribute to other environmental impacts such as biodiversity conservation, watershed protection, and socio-economic benefits to urban and rural populations through access to forest products and creation of jobs, especially in rural areas, ultimately promoting sustainable development.

12.1 Introduction[1]

Technology transfer in the forestry sector provides a significant opportunity to help mitigate climate change and adapt to potential changes in the climate. Technology transfer strategies in the forestry sector for promoting mitigation options, apart from reducing greenhouse gas (GHG) emissions or enhancing carbon sinks, have the potential to provide other tangible socio-economic and local and global environmental benefits, contributing to sustainable development. Technology is an important component of the forestry-sector mitigation options, and there are barriers to its transfer such as high cost of capital for private sector (particularly developing countries), lack of institutions to enable participation of local communities in forest protection and management (Kadekodi & Ravindranath, 1997), and land tenure policies encouraging deforestation and land degradation in many tropical countries (IPCC, 1996). Existing financial and institutional mechanisms are inadequate, and thus new policies, measures, and institutions are required to promote technology transfer in the forestry sector.

In 1990 forests covered about 3.44 x 109 ha, a quarter of the earth's land surface (FAO, 1995). The total area under tropical forests is estimated to be 1.75 x 109 ha (51% of total global forest area) with a forest biomass roughly equal to 297 X 109 t (67.5 % of total global biomass, FAO, 1995). The world's forests store large quantities of carbon (C) with an estimated 330 GtC in live and dead aboveground biomass vegetation and 660 GtC in soil (Brown *et al.*, 1996). Furthermore, an unknown quantity of C is stored in wood products. High and mid latitude forests are currently estimated to be a net C sink of about 0.5 to 0.9 GtC annually. Low latitude or tropical forests are estimated to be a net C source of 1.1 to 2.1 GtC annually (IPCC, 1995), largely due to clearing and degradation of forests (IPCC, 1996). Forests, in addition to being a source of carbon emissions, are also shown to have a large potential for mitigation, which is estimated to be in the range of 60-87 GtC globally over 50 years (1990 to 2040), through forest conservation (slowing deforestation), forestation, regeneration and agroforestry options. This potential is confirmed by studies conducted at the regional (Sathaye and Ravindranath, 1998; Nabuurs *et al.*, 1998) and national levels (Xu, 1995; Masera 1995; Ravindranath and Somashekhar, 1995). The tropical forest region dominates by accounting for 45-72 GtC of mitigation potential (Brown *et al.*, 1996).

In addition to forest C conservation and sink expansion, the forestry sector offers a large potential for C emission reduction through fossil fuel substitution. For instance, short-rotation woody crops have the potential with advances in energy conversion and yield to reduce global fossil fuel emissions by up to 20 % (Dixon *et al.*, 1994). Further residues from timber logging and processing could also be used as feed stock for energy. A study in India has shown that biomass-based, decentralised, small-scale electricity systems alone could offset nearly a quarter of India's fossil fuel emissions (Ravindranath and Hall, 1995). Several studies have shown that forest-sector mitigation options are cost effective and the unit abatement costs are low, particularly in developing countries (Hourcade *et al.*., 1996; Brown *et al.*, 1996; Sathaye and Ravindranath, 1998; Masera *et al.*, 1997a). Forest-sector mitigation options are also shown to provide multiple, local environmental, as well as socio-economic benefits, apart from C abatement (Kadekodi and Ravindranath, 1997, Masera *et al.*, 1997b). Thus there is a large interest in promoting forest-sector mitigation options. The adoption of forestry mitigation options is subjected to technical, financial and institutional barriers (Sathaye and Ravindranath, 1998). Earlier, the IPCC assessment identified several technical options along with a set of administrative, institutional and policy barriers (IPCC, 1996). Technology, including the "software" (such as methods for monitoring forest area and protected area management practices) as well as the "hardware" (logging or processing equipment), is one of the limiting factors, in addition to lack of financial incentives for large-scale adoption of forest conservation, reforestation, sustainable forest management and fossil fuel substitution options. Article 2 of the Kyoto Protocol recommends promotion of sustainable forest management practices, including afforestation and reforestation to protect and enhance C-sinks (UNFCCC, 1997).

In this chapter, we analyse the existing and emerging technology transfer mechanisms in the forestry sector. This is followed by a consideration of the barriers to transferring ESTs, and an exploration of the institutional, financial and policy measures for promoting technology transfer in the forest-sector mitigation programmes within and between countries. It is important to recognise that technological alternatives can address only a few aspects of the strategies to promote adoption of GHG abatement programmes. For example in Brazil, slowing deforestation represents by far the most attractive option. But the traditional technological solutions can only play a modest role. (Fearnside, 1993).

Forests, in addition to being a source of C emission and having a large potential for mitigation, will be subjected to changing climate. The IPCC (1996) concluded that global models, based on 2xCO2 climate, project that a substantial fraction of the existing forests will experience climatic conditions under which they do not currently exist. Thus, large forested areas are likely to undergo changes from the current forest types to new vegetation types. A number of technological and silvicultural adaptation options have been suggested (Solomon *et al.*, 1996; Ravindranath *et al.*, 1997). Large uncertainties exist in making regional projections of climate change and the potential response of forest ecosystems. And, thus, in this report policies and measures required to promote transfer of adaptation technologies are only very briefly presented.

12.2 Climate Mitigation and Adaptation Technologies

Technologies in the forestry sector need a broader definition than in the other sectors. These technologies could include genetically superior planting material, improved silvicultural practices, sustainable harvest and management practices, protected area management systems, substituting fossil fuels with

---

[1] Relevant cases from the Case Studies Section, Chapter 16, for Chapter 12 are: tree growers cooperative (case 25), reduced impact logging (case 26), and ethnomedicine (case 28).

bioenergy, incorporating indigenous knowledge in forest management, efficient processing and use of forest products, and monitoring of area and vegetation status of forests. These technologies can meet several objectives, including conserving biodiversity and watersheds, enhancing sustainable forest product flows, increasing the efficiency of use of forest products, and maximising the resilience of forest ecosystems to climate change, in addition to enhancing sinks. It must be recognised, however that enhancing sinks will not necessarily or automatically lead to such outcomes. Rather, the drive to enhance sinks must be appropriately harnessed through accounting methodologies, and through rules and guidelines for sink activity, which underpin the achievement of these goals. These could be put in place at the national and sub-national levels. Currently there is a lack of information on the adaptation technologies in the forestry sector.

12.2.1 Features of Forestry Mitigation Technologies

Long gestation period. Decisions on transfer and adoption of technologies are going to be determined by the anticipated carbon abatement and long-term biological (such as biodiversity) and socio-economic factors. Forestry projects (such as hard wood plantations) could take 50–100 years to provide carbon mitigation benefits. The long gestation period leads to uncertainties regarding carbon abatement and socio-economic impacts.

Linked to subsistence economy. In tropical countries millions of indigenous and rural households depend on forests for their livelihood, whereas in temperate countries dependence on forestry is largely commercial, although protective functions of the forests are important too. Forestry projects could impact on the livelihood and local economies in developing countries.

Subject to natural calamities. Forests and plantations are subjected to fire, drought, pests and diseases affecting the C stocks and flows. Thus, any intervention will have to be carefully evaluated.

Climate and location specificity of technologies. Forestry technologies vary among tropical, temperate and boreal regions, as well as with varying forest and plantation types, precipitation regions and socio-economic pressures.

State control of forests. In most countries, and particularly in tropical countries, forests are largely controlled and managed by the state forest departments.

Mitigation options involving minimal technology transfer. Some of the mitigation options require only policy initiatives, funding, regulations and control (*e.g.*, ban on forest conversion or extraction).

Links to local and global environmental factors. Decisions on forestry will affect biodiversity and other ecological benefits such as watershed protection, soil erosion protection, resilience to climate change and prevention of desertification. Sustainable forest management practices could be beneficial to timber logging countries for conserving biodiversity and watersheds. Most of the local and global environmental benefits have not been well quantified in monetary values.

Low economic return: Forestry technologies generally have low economic return. This impedes the investments from private (commercial) setors.

Long term sustainable approach. Requires adoption of long term sustainable management practices.

Participation of local communities. Local participation is required for implementing mitigation projects where local communities currently reside in or depend on the forests.

12.2.2 Climate Change Mitigation Technologies

Forest-sector mitigation technologies could be grouped into four general categories according to their main expected impact on C emissions and removals:
  a) C conservation (avoided emissions by halting deforestation, forest protection, etc);
  b) C sequestration;
  c) C offsets (substitution of fossil fuels or non-sustainable timber extraction);
  d) C offsets from wood products.
The technologies are listed in Table 12.1.

12.3 Current and Emerging Pathways in Technology Transfer

12.3.1 Current Pathways

Global experience in technology generation and transfer in the forestry sector is limited, compared to the agricultural and energy sectors. The pathways are complex and country-specific. The three major pathways are: government, private sector, and community-initiated. The existing institutions involved in technology transfer along with barriers to large-scale technology transfer are listed in Table 12.2.

Government Initiated. One of the features of technology transfer in the forestry sector is the non-commercial nature of the transfer of some technologies as well as low levels of involvement of commercial institutions. Furthermore, the forests, forestry research and extension services are largely controlled by the governments, particularly in tropical countries. Currently technology transfer is largely from the government-controlled universities and research institutions to forest departments and farmers. Forestry research and development is largely restricted to university and research institutions. These are largely funded by the national governments and external development assistance

Table 12.1	Climate Change Mitigation Technologies, Benefits, and Impacts			
TYPES OF MITIGATION PROJECTS	**TECHNOLOGIES, PRACTICES AND SYSTEMS**	**CARBON SEQUESTRATION OR EMISSION REDUCTION POTENTIAL**	**BIO-DIVERSITY CONSERVATION**	**SOCIO-ECONOMIC BENEFITS**
Carbon Conservation	Deforestation reduction through policy changes	+++	+++	++
	Formation of Protected Areas	+++	+++	++
	Monitoring forest; area and vegetation status	++	++	
	Recreational reserves	++	++	++
	Sustainable Forest Management	+++	++	+++
	Fire protection techniques	++	++	
	Reduced impact logging	++	++	+
Carbon sequestration	Afforestation	++	++	+++
	Reforestation	+++	++	+++
	Industrial Plantations	++		++
	Agroforestry	++	+	++
	Urban forestry	++	+	+++
Carbon Offsets (substitution for fossil fuels and unsustainably harvested wood)	Short rotation forestry for biofuels	+++	+	+++
	Sustainable biomass plantation	+++		+++
	Waste use for energy	++	+	
	Efficient processing technologies	++		+++
	Recycling of forest products	++	+	++
	Bio-energy (bio-electricity through gasification of biomass or combustion)	+++		++
	Fuel efficient stoves	+	+	+++
	Biogas for cooking	+++	+	+++
	Efficient charcoal kilns	++		++
Carbon offsets from wood products	Recycling forest products	++	+	+
	Substitution of fossil-fuel intensive products with wood products (e.g., steel with wood for construction)	++	+	+
	Storage in long-term wood products	++	+	

+++ HIGH POSITIVE IMPACT
+ LOW POSITIVE IMPACT

agencies. Yet these developing country research institutions are inadequately funded. Nevertheless, in Annex I countries, significant research and development (R&D) and technology transfer occurs in the private sector.

An analysis of forest research institutions and universities throughout the world and their staffing levels clearly shows that very few developing countries have capacity in forest science (in terms of numbers of institutes, staff or budgets). The effectiveness of the few scientists working on forestry issues in developing countries is frequently constrained by shortages of equipment and operating budgets, by limited access to scientific advances elsewhere and by local institutional issues (FAO, 1997). There are several international institutions involved in technology (including information) generation, dissemination and capacity building. Some examples are the Food and Agriculture Organization (FAO), "Centro Agronómico Tropical para Investigación y Enseñanza"(Tropical Agriculture Centre for Research and

Education) (CATIE), Centre for International Forestry Research (CIFOR), (IUFRO), International Centre for Research in Agroforestry (ICRAF), (IFGRI) and World Conservation and Monitoring Centre (WCMC). A new global framework might, for example, build upon the extensive coverage of IUFRO, the interdisciplinary and collaborative mechanisms of CIFOR, and the practical field presence and global data sources of the FAO.

Currently, almost all of the investment projects funded by multilateral agencies (World Bank, 1993) and bilateral agencies contain a technical assistance component. When the World Bank, the largest multilateral funding agency is considered, only US$5.4 billion out of US$396 billion was allocated for environmental projects, where forestry is only one of the components and, furthermore, only a marginal part of it was devoted to R&D (World Bank, 1997).

Private Sector Dominated. Private sector timber companies and industries (paper mills) are increasingly participating in R&D and

Table 12.2	Barriers to Technology Transfer in the forestry sector		
INSTITUTIONS AND MECHANISMS	**EXAMPLES**	**FEATURES**	**BARRIERS**
Multilateral	World Bank, ADB, GEF, FAO, IPF, IFF, IUFRO, CIFOR, UNDP, Technical cooperation	Commercial and non-commercial lending agencies	• Currently limited role • Absence of local institutions to assimilate and adopt technologies in developing countries • Limited funding to forestry (World Bank, GEF) • Narrowly defined economic criteria
Bilateral	Development assistance programmes of OECD countries	• Aid agencies • Intergovern-mental transfers	• Project or location specific • Limited experience • Largely limited to reforestation and Protected Area formations • Influenced by donor's foreign policy objectives • Diversion by recipients to non-target programmes
Forest department	• Wildlife divisions • Social Forestry Dept. • Extension service • National Forest Service (USFS)	• Part of government bureaucracy • Multi-objective outlook	• Inadequate research and extension capability • Inadequate funding • Focussed more on forest protection, afforestation, enforcing regulations • Often exclude local communities • Subject to political strings
Research institutions	• International • FAO • CIFOR • IUFRO • National institutes • Universities • Industry or timber companies	• Generate and assimilate technologies • Monitor forest area, status	• Limited role due to inadequate funding or national policies or regulations • Industry research focussed only on certain commercial aspects of forestry
Industry/ private sector.	• Timber plantation companies • Industries (paper) • Weyerhauser, Georgia Pacific International Paper, Jaakko -Poyry, etc.	• Commercial transfer • Profit motive	• Focus limited to commercial aspects of forestry only • Inadequate foreign investment policies • Protection measures and economic embargoes • Large uncertainties
Joint ventures	AIJ / USIJI	• No C-crediting • Private utility in OECD and developing country agency	• Many uncertainties on the mechanisms and C-abatement achieved • Absence of C-credits
NGOs	• International • WWF • Nature Conservancy • WRI • CARE • National and local NGOs • Community based organisations	• Activist in origin • Main base in tech-source countries • Mostly not-for profit	• Focus on technology inadequate, narrow focus on issues • Inadequate arrangements to protect the interests of forest dwellers and promotion of participatory management • Limited technical capacity • Little or no weight on economic efficiency
Community organisations	• Grameen Bank funded organisations • Joint forest management committees	• Grassroot affiliations • Organised on special interests e.g.: Women, artisans, etc.	• Mostly loose structure • Funding and accountability usually weak. • No legal status, inadequate powers.

technology transfer in the forestry sector. Technologies and machinery for recycling, harvesting, and processing are well developed and largely originate from industrialised countries. The private sector is likely to play a larger role in the future, even in developing countries (refer to Tables 12.2 and 12.3).

Community Initiated. A substantial part of technology transfer in the forestry sector within developing countries is driven by local communities and NGOs. This is particularly true for forest conservation practices, agroforestry systems, and systems for harvesting of non-timber and other subsistence products. Much of the transfer also takes place in the form of "software", *i.e.*, training and capacity building. Local institutions, such as NGOs and grassroots organisations, are increasingly participating in these technology transfer programmes. In some countries, such as Mexico, communities with commercial forest resources have organised nation-wide organisations of social forestry enterprises. Through these organisations villages receive administrative and technical training and financial resources.

The existing institutions are currently playing a limited role compared to the demands. Currently, there is a marginal role for private sector or industry participation in technology transfer. Forest departments, research institutions and NGOs are even more constrained by limited financial support and technical capability. The existing institutional mechanisms have several limitations to promote climate mitigation technologies, namely, limited resources and absence of policies and institutions to process, evaluate and approve mitigation projects for implementation (Sathaye *et al.* 1999).

The existing institutions and arrangements are inadequate to meet the emerging challenges of promoting forestry-based mitigation projects. In tropical countries, the state forest departments play a predominant role in all aspects of forest protection, regeneration and management. There are uncertainties regarding the areas under tropical forests, rates of deforestation, causes of deforestation and the C-densities in vegetation and soil of different forest categories. Such information is crucial for developing strategies, projects and technologies to reduce deforestation. Currently, the lack of funding and technical capabilities in tropical countries limit the generation of information on all of the above aspects.

12.3.2 Emerging Pathways

In the face of limitations of existing institutions, new mechanisms are emerging. Recent developments indicate an increased importance ascribed to environmental functions of forests and their integral role in sustainable forest management. These efforts include attempts to manage forests as ecological systems (taking into consideration forests' protective functions and their role in the conservation of biological diversity); adoption of reduced-impact logging systems and development of codes of harvesting practice; and restrictions placed on timber harvesting in forests in North America and some tropical Asian and Pacific countries.

Environmental concerns have also led to certification schemes and export controls for forest products. The trend towards increased involvement of nearby communities in forest management, particularly in developing countries, allows for greater consideration to be given to local environmental concerns and to the social benefits derived locally from forests (FAO, 1997). Many Governments have regulations to promote adoption of sustainable forest management practices, and prevention of damage to forests for short-term benefits to timber companies.

A majority of these efforts have emerged independently of the climate change-related debates. Many efforts, governmental and non-governmental, national, regional and international, have been made to promote sustainable forest management. Major international initiatives include: the International Tropical Timber Organization's Year 2000 Objective, in which producer member countries have committed themselves to having all their internationally-traded tropical timber come from sustainably-managed forests by the year 2000; and national and regional efforts to define criteria and indicators for sustainable forest management, and to determine means of assessing progress towards achieving it. The latter involves a number of regional initiatives, most of which have been launched since 1995, focusing on: humid tropical forests in ITTO producer countries; boreal, temperate and Mediterranean forests in Europe (the Helsinki Process); temperate and boreal forests outside Europe (Montreal Process); Amazon basin forests (the Tarapoto Proposal); and forests in dry zone sub-Saharan Africa (the UNEP/FAO Dry-Zone Africa initiative); in the Near East region (FAO/UNEP Expert Meeting for the Near East); and in Central America (FAO/CCAD Expert Meeting on Criteria and Indicators for Sustainable Forest Management in Central America) and "towards a sustainable paper cycle" (World Business Council for Sustainable Development).

Some of the non-climate-related trends and mechanisms, which have implications for climate mitigation programmes, are:
- Timber Certification. This activity promotes the sustainable management of forests by issuing "green certificates" according to a set of standard social, economic and environmental criteria. There are several certification schemes being developed at the regional and international levels. The most important of them are FSC-certification and the ISO-certification. Currently there are a little more than 10.3 x 106 ha in 1997 (which has increased to 16.2 x 106 ha) of forests certified, through entities accredited, by the Forest Stewardship Council worldwide (FSC, 1997). The slow growth of markets for certified timber is still one of the main barriers for a wider and faster dissemination of this particular form of technology transfer.
- Increased forest product recycling.
- Adoption of reduced impact-logging practices.
- Substitution of fossil fuels by biofuels.

12.3.3 Climate Mitigation Related Pathways

A number of new mechanisms have emerged in the forestry sector for technology transfer, including the concept of joint implementation and emissions trading. With the advent of the Kyoto Protocol, the Clean Development Mechanism has also emerged as a multilateral entity which will significantly influence forestry sector technology transfer and finance (Bolin, 1998). New brokerage, as well as monitoring and verification institutions are also emerging associated to climate change institutions and mechanisms.

GEF. The Global Environmental Facility (GEF) is a financial mechanism that promotes international cooperation and fosters actions to protect the global environment. The grants and concessional funds disbursed, complement traditional development assistance by covering the additional costs (also known as 'agreed incremental costs') incurred when a national, regional, or global development project also targets global environmental objectives. Forestry does not appear among the operational programmes in the current phase of GEF, though it appears under the short-term measures. However, fossil fuel substitution through bioenergy, fuelwood conservation and bioelectricity systems for decentralised application are included in the operational programmes. In addition, the majority of biodiversity conservation projects also contribute to C abatement. GEF operational programme-12 on carbon sequestration is being currently formulated. Under this programme, GEF can facilitate transfer of technology to near commercial forestry sector projects that face incremental risk, and hence justify funding the incremental cost of undertaking a forestry sector initiative over its relevant growth cycle. Furthermore, this GEF programme can address the barriers and impediments to widespread implementation of viable carbon sequestration opportunities, addressing risks and risk financing. GEF initiatives can prepare an environment for replication and sustainable continuation of activities. By breaking a first time barrier, it can pave the way for the flow of private sector resources that can be coupled with international initiatives.

AIJ/CDM/JI. At the first UN-FCCC Conference of the Parties, Berlin, 1995, the Parties established a pilot technology transfer programme, termed the 'Activities Implemented Jointly (AIJ)'. To date, over 70 countries have established government institutions to develop and monitor AIJ projects. Over 50 projects have been registered with the UN-FCCC Secretariat. The US Initiative on Joint Implementation (US-IJI) is the largest AIJ pilot programme, with 32 projects in 12 countries. US-IJI has attracted over US$160 million in private sector finance. Approximately one-half of the US projects are in the forestry sector. For example, private sector partners from the US and Bolivia are working under the AIJ pilot phase to establish the largest private sector national park in the world. However, there are uncertainties regarding the inclusion of land use change and forestry activities under CDM. Market-based trading of carbon credit could enhance the flow of resources to forestry projects worldwide. It is also not clear how this would affect other aspects like biodiversity and local needs, and there are many other uncertainties, like those associated with carbon-accounting and the life-time of the carbon-stock. To clarify these aspects and uncertainties the IPCC will issue its Special Report on Land-Use, Land-Use Change and Forestry in 2000 (sections 3.4 and 3.6 in Chapter 3 provide more information on technology transfer in the Kyoto Protocol and other UNFCCC agreements).

12.3.4   *Generic Barriers to Technology Transfer*

The current and emerging pathways and mechanisms have several limitations to promote climate mitigation technologies, namely, *(i)* limited financial resources, *(ii)* inadequate information on the costs and potential benefits, *(iii)* limited technical capacity, *(iii)* absence of policies and institutions to process, evaluate and clear mitigation projects, *(iv)* uncertainty regarding quantity of carbon abated and its permanence, *(v)* a longer period to realise carbon benefits (eg. hardwood timber plantations), (vi) low economic returns for some technologies, and (vii) absence of consideration of the economic value of environmental benefits. In addition, the forestry sector faces land use regulation and other macroeconomic policies that usually favour conversion to other land uses such as agriculture and cattle ranching. Insecure land tenure regimes and tenure rights, and subsidies favouring agriculture or livestock are among the most important barriers for ensuring sustainable management of forests as well as sustainability of C abatement.

In many tropical countries, the national and state forest departments play a predominant role in all aspects of forest protection, regeneration and management. Currently, lack of funding and technical capabilities in most tropical countries limit generation of information required for planning and implementation of forestry mitigation projects. Apart from a few exceptions, developing countries do not have adequate capacity to participate in international research projects, and to adapt and transfer results of the research to the local level. Research on forests has not only suffered from a lack of resources; it has not been sufficiently inter-disciplinary to provide an integrated view of forestry (FAO, 1997). However, the majority of the forestry research institutions in developing countries do not function as R&D laboratories on the patterns of research institutions in industrialised countries. Unlike in the energy or transportation sectors, the technologies or even the management systems are going to be forest type or country specific. Complimentary to R&D activities of research institutes, special information transfer and extension services are desirable.

Forestry-sector GHG mitigation activities and joint implementation projects generally face a wide range of technical issues that challenge their credibility. The twin objectives of using forestry to mitigate climate change and managing forests sustainably do pose a challenge in monitoring and verifying benefits from carbon offset projects in the sector (Andrasko, 1997). The emergence of improved monitoring methods could force

reappraisal of the relative credibility of activities to manage carbon sinks. Monitoring and verification are key elements in gaining the credibility needed to capture the potential benefits of forestry sector response options, particularly in reducing deforestation (Fearnside, 1997). While this is a generic barrier to deforestation reduction initiatives, it also represents an opportunity for transferring the technologies needed to monitor land-use change, and carbon stocks and flows. Among the mitigation options, there is quite a high degree of certainty about reforestation/afforestation, less on forest management and even less on forest conservation.

12.4   Technology Transfer within Countries

Climate mitigation projects under the bilateral and multilateral arrangements in the forestry sector could in some cases involve only transfer of funds from donor to host countries and agencies, without any external transfer of technology. Thus, diffusion of technology within countries becomes important, particularly in countries with large mitigation potential such as China, India, Indonesia (Sathaye and Ravindranath, 1998), Brazil (Fearnside, 1999) and Russia. The existing and emerging arrangements within countries need to be strengthened and reoriented. The current level of technology transfer in the forestry sector, particularly in developing countries, is marginal due to limitations of infra-structure and barriers in dissemination of R&D outputs. This has not led to enhanced productivity in either the state-managed or farmer-managed plantation forestry systems (Ravindranath and Hall, 1995). Forestry needs to be more productive and profitable to be attractive to private investors under some of the climate change mitigation programmes. Technology transfer within countries is very crucial in the forestry sector. The sources of technology or management practices for in-country technology transfer could be:

- Forest departments;
- Research institutions and university laboratories;
- Industry; paper and pulp, timber logging, and plantation companies;
- Indigenous communities.

The main drivers or incentives for diffusing technologies, the key stakeholders, the potential barriers, and measures to overcome the barriers vary with technologies and some examples are listed in Table 12.3.

The categories of technologies likely to be involved in technology transfer within countries are: a) silvicultural practices for high yields, b) improved genetic stock for planting, c) practices for sustainable forest management and Protected Area Management, d) monitoring and verification of C flows in forestry projects, e) efficiency improvements, f) utilisation and management of secondary forests, and g) traditional forest management practices adopted by indigenous communities. The policies and measures required for promoting technology transfer within Annex I and non-Annex I countries are given in Tables 12.3. and 12.4.

In many developing countries, logging of natural forests constitutes an important economic activity within the forestry sector (Masera, 1995). There is significant transfer of "hardware" at the timber processing stages, usually from industrialised countries.

Table 12.3	Potential Drivers, Stakeholders, Pathways for Technology Transfer, Barriers to Transfer and Policies and Measures for Promoting Technology Transfer
CARBON CONSERVATION MEASURES	
Drivers or incentives	Biodiversity conservation, watershed protection (national) Access to forest products, eco-tourism, rural job (individual) Carbon conservation, credits and financial rewards (global)
Stakeholders	Indigenous communities National governments Conservation groups
Barriers	Inadequate understanding of deforestation causes to propose effective counter measures. Potential loss of revenue for the government and timber logging companies. Lack of institutions for enabling community participation. Inadequate skills for managing Protected Area and SFM practices. Inadequate financial incentives for forest conservation
Policies and measures	Financial support for compensating loss to the government, particularly to local communities. Institutions for implementing policies for decreasing deforestation. Institutions to facilitate community participation and management. Training and capacity building for forest management, sustainable logging techniques, Protected Area management Education and awareness
CARBON SEQUESTRATION	
Incentives and driver	Income generation from agroforestry and reforestation (non timber products). Watershed protection – if degraded lands used. Potential financial rewards for carbon credits
Stakeholders	Governments, Farmers and Companies, local communities.
Barriers	Lack of funding and high cost of credit. Regulations on land use. Lack of technology for high growth rate. Lack of policies to ensure sustainability of C sequestered Opportunity cost of land and current product flows
Policies and measures	Financial support and mechanisms to compensate for opportunity lost. Financial rewards for carbon sink creation. Land tenure policies for sustaining carbon sink in selected categories of land.
SUBSTITUTION OF FOSSIL FUELS AND NON-SUSTAINABLY EXTRACTED TIMBER	
Incentives and drivers	Profit motive from sale of wood to bioenergy utility and timber companies. Potential financial rewards for pollution abatement and carbon credits. Reclamation of degraded lands and meeting biomass needs.
Stakeholders	Farmers, bioenergy utility, plantation owners, paper mills and NGOs.
Barriers	Absence of policies to promote sustainable bioenergy or timber plantations. Subsidies to companies to extract timber from natural forests and to use fossil fuel energy. Lack of finance and financing institutions. Lack of access to technology; bioenergy and high yielding silvicultural practices and quality seedling.
Policies and measures	Financial incentives to sustainable timber producers and bioenergy utilities. Sale of technology. Level playing field for bioenergy and sustainable timber.

Barriers for sustainable logging include:

- Regulations that encourage short-term profit maximisation without taking into consideration long-term effects;
- Lack of appropriate forest management systems, which cause unwanted changes in species composition and a reduction in the inventories of commercial species;

Table 12.4	Technology Transfer Within Countries: Pathways, Policies, Programmes, and Measures
POLICIES AND MEASURES	
GOVERNMENT INITIATED	
Financial incentives for companies importing sustainable timber. Financial incentives and tax rebates to promote recycling. Regulations restricting deforestation and policy changes reducing motivation for deforestation. Regulations on timber extraction companies for adopting sustainable logging practices. Awareness programmes regarding forest conservation. Financial incentives for adopting sustainable forest management and reduced impact logging practices. Capacity-building programmes for monitoring of carbon flows. Funding capacity building in R&D institutions. Promoting research to understand causes of deforestation. Training in sustainable logging and management practices for forest department and timber logging companies. Strengthening forest extension service. Formation of Protected Areas. Framework for policies on land and product tenures to promote community participation. Linkage between research institutions and forest departments.	
PRIVATE SECTOR INITIATED	
Timber certification. Investment in forestry R&D. Joint ventures between industry and forest departments for technology transfer Farmer and industry partnership. Financial incentives; tax incentives, low cost credit to farmers for raising commercial wood in low carbon density lands. Industry providing technology as a package for farmers will ensure flow of modern technology.	
COMMUNITY INITIATED	
Consumer interest groups/NGOs to ensure enforcement of regulations; paper recycling, marketing of sustainable timber. Public awareness to promote use of sustainably logged timber. NGOs to promote community participation for adopting forest conservation measures. Community awareness on forest conservation.	

- Inadequate institutions and lack of funding for forestry services;
- Excessive regulation of local communities without taking into account the needs and traditional rights of local communities that leads to clandestine logging,
- Poor training of forest owners in technical aspects, and lack of financial resources to invest in basic infrastructure and detailed forest inventories, and
- Higher opportunity costs of alternative uses of the land (Klooster, 1996; Merino, 1996; Masera *et al.*, 1997a).

Policies and measures include increased access to funding for investment in basic infrastructure, technical and administrative training of local communities, and internalisation of "environmental services" to compete with alternative land uses.

12.4.1   *Technology Transfer within Annex I Countries*

The policies in timber importing countries are critical to ensure environmentally sound technology transfer within timber exporting countries. Annex I countries are the dominant timber importing countries both from non-Annex I as well as from other Annex I countries. Thus, it is very important for Annex I countries to promote the diffusion of sustainable forest management (SFM) practices. SFM practices are also supposed to be beneficial to timber logging countries for conserving biodiversity and watersheds.

Government initiated measures. Some of the measures to promote technology transfer on SFM practices and Protected Area management are as follows:
- Application of certification and labelling schemes in a non-discriminatory and correct manner;
- Regulations to control the import of uncertified forest products within the framework of international trade agreements;
- Development of criteria and indicators for SFM;
- Provision of SFM principles and practices at national level;
- Preference for SFM-certified wood for public construction;
- Investment in forestry R&D;
- Recycling of paper through regulations, incentives and agreements;
- Promotion of public awareness;
- Economic valuation of forest functions.

Community-initiated measures. NGOs could contribute to creating awareness in the consumers to promote the use of certified timber and recycled paper. Consumer interest groups/NGOs could ensure enforcement of regulations; paper recycling, and marketing of sustainable timber.

Private sector. The private sector in Annex I countries will play a role in the development of criteria and indicators for SFM, development of certification schemes and their adaptation to the national/regional level, ensuring the adoption of SFM practices in timber logging, marketing sustainably logged timber, and in generating forestry technologies.

12.4.2   *Technology Transfer within non-Annex I Countries*

In the forestry sector, technology transfer within non-Annex I countries (largely referring to tropical countries) is crucial (Table 12.4). Forests in most countries are largely controlled and managed by forest departments. Thus, government will be the dominant actor for promoting technology transfer.

Government initiated mechanisms and measures:
- Regulations on timber extraction companies for practicing sustainable logging
- Enforcing forest conservation regulations and adoption of

effective monitoring techniques
- Removing subsidies for deforestation
- Financial incentives for adopting SFM and particularly for reduced impact logging practices
- Funding capacity building in R&D institutions for technology generation as well as assimilation of imported technology
- Training in sustainable logging and management practices for forest department and timber logging companies
- Strengthening forest extension services
- Formation of Protected Areas
- Framing policies on land and product tenures to promote community participation
- Establishing linkage between research institutions and forest department
- Creating awareness regarding the potential for mitigation projects, costs, and benefits (private, local and global), sources of funding and technology
- Understanding, recording, and adopting indigenous forest conservation and management practices
- Increasing financial allocation to environmental education programmes
- Setting up agencies to assist local groups (NGOs, farmers' organisations, local forest departments and even industries) to generate forest-sector mitigation project proposals
- Creating technical capacity for monitoring of forest areas and status for assessing impacts of removal of subsidies to deforestation, on causes of deforestation and the likely effects of different measures intended to reduce it
- Assigning economic valuation to forest services, such as biodiversity, soil and water conservation, etc., and collecting carbon tax from potential beneficiaries such as farmers and industries.

Funding for forestry research and development needs to be increased substantially, as usually the forestry sector competes poorly with agriculture; for example, in Mexico, 98% of subsidies still go to promote agriculture rather than forestry (SEMARNAP, 1996). Licensing of logging companies or timber export organisations could be linked to adoption of reduced impact logging techniques. Strong forest conservation policies are required. India passed a Forest Conservation Act in 1980, under which conversion of forest land is highly regulated. The area under forest has begun to stabilise in India, as deforestation rates have declined and conversion of forest land to agriculture has nearly stopped (Ravindranath and Hall, 1994). The result is that there is a shift in sourcing industrial wood from forests to private farm based plantations. Protected Areas account for 15% of forests globally (FAO, 1995). There is a large potential to increase the protected area coverage. Appropriate legislation is required to demarcate new Protected Areas. This would necessitate adoption of protected-area management practices based on experience from other protected areas. It may be possible to legislate to ensure that a certain percentage of newsprint is recycled even in developing countries, to conserve forests and in turn the C sinks. Thus, regulations on forest conversion, harvest of forest products and its processing could automatically ensure diffusion of forest conservation, sustainable forest management and processing technologies.

Diffusion of technology from R&D institutions to forest departments. The focus of forest departments in most countries, particularly tropical countries, is on forest protection, afforestation and enforcing regulations. They will also continue to implement mitigation projects. R&D, however, is a low-priority area in these organisations. Forest departments in most tropical countries, being the dominant agency involved in forest protection and afforestation, will require input of technology. These departments are unlikely to be driven by profit motive. However, to maximise returns (monetary, as well as environmental, such as biodiversity conservation) forest departments could adopt efficient management say for Protected Areas, silvicultural for afforestation and (reduced impact) logging practices. The national governments, particularly ministries in charge of the forest department, could evolve national mechanisms to promote closer interaction between R&D institutions (government, university and industry) and the local forest departments to promote technology transfer.

Private sector. Linkages between industry and farmers is one of the most important emerging pathways for future technology transfer. The degraded farmland in many countries has a large potential for forestry mitigation projects (Ravindranath and Hall, 1995; Sathaye and Ravindranath, 1998). Both industry and farmers are driven by the profit motive. The goals for promoting "farm forestry" and agroforestry could be: firstly, to grow biomass for meeting rural, urban, industrial and export needs to reduce pressure on forests (ultimately conserving C sinks); secondly, to create new sinks of C particularly in sawn timber from trees on farms and fruit trees on farm lands and; thirdly, to grow biomass feedstock for bioenergy utilities. Industry needs to have in-house R&D or could access technology from the research institutions within the countries. The role of industry is crucial in facilitating technology transfer to a large number of small and dispersed farmers. Technology transfer could be facilitated as part of a package from industry to farmers including; credit, technology and marketing arrangements.

Community initiated. NGOs in developing countries could create public awareness regarding forest conservation, SFM practices, recycling and their local and global implications, to ensure compliance with legislation and policies by the government departments, industries, timber logging and exporting companies. Forestry extension service is very poorly developed compared to agricultural extension service, thus NGOs could play a major role. An example of an NGO driven large scale revegetation programme, involving Tree Growers' Cooperatives, is given in Box 12.1 (see also Case Study 25).

Technology transfer with multiple actors. The foregoing discussion covers the role of different technology transfer mechanisms separately. However, in practice pathways are often composed of complementary links among the actors (government, private sector, NGOs). An example is the social forestry programmes in India. During the 1980s, these programmes were exclusively promoted by the government forest department. From the early 1990s the programmes were promoted on village communal land by NGOs, such as the National Tree Growers' Cooperative Federation (NTGCF), and on private farms by industries, mainly for pulp (Ravindranath and Hall, 1995).

BOX 12.1   TREE GROWERS' COOPERATIVES; A PARTICIPATORY APPROACH TO RECLAIM DEGRADED LANDS

National Tree Growers' Cooperative Federation Limited (NTGCF) is an NGO based in India. It was established in 1988 with the main objective to restore the degraded and marginalised village lands. The NTGCF has been operating in six states of India since 1986. Professionals from forestry, social and economic sectors work together to achieve this goal. A participatory approach is adopted by the NTGCF to create self-sustaining village cooperatives called the Tree Growers' Cooperative Society (TGCS). The TGCS works towards restoring the biological productivity of marginally productive and unproductive degraded lands, establish ecologically self-sustained fuelwood and fodder plantations to meet the essential needs of the villagers, and also cater to the urban demand for fuelwood, timber and tree based products through ecologically sustainable modes. The NTGCF provides financial assistance and organises training, extension and orientation programmes for the rural communities to revegetate the degraded village lands. As of March 1998, the NTGCF has revegetated 8911 hectares of land with a carbon sequestration potential of about 14258 tC. If TGCS' concept is extended to cover all the degraded village commons of 11.8 x 106 ha, the annual C sequestered could be 19 MtC. Revegetation under the growers cooperative concept will ultimately lead to forest C sink conservation.

12.5   Technology Transfer between Countries

12.5.1   *Technologies to be transferred*

The categories of technologies likely to be involved in technology transfer between countries are: a) silvicultural practices for high yields, b) genetic stock for planting, c) practices for SFM and Protected-Area Management, d) monitoring and verification of C flows in forestry projects, e) efficiency improvements, f) modelling for projecting changes in carbon stock and forest area, g) fossil fuel substitution techniques, h) agroforestry and i) industrial forest processing. The flows of technologies among non-Annex I countries, as well as between Annex I and non-Annex I countries are equally important. Technology transfer between Annex I and tropical countries could include transfer of R&D infrastructure and capacity, monitoring and verification systems, SFM principles and practices, and efficient harvesting and processing technologies. Technology transfer between the non-Annex I countries (particularly tropical countries) is crucial as there are many similarities in the their ecological and socio-economic conditions. However, what is critical is that whenever technology is transferred between two countries (often even within a country), it is very important to modify and adopt the technology or the practice to the local conditions.

12.5.2 *Barriers of Technical Capacity*

The majority of developing countries do not have adequate technical capacity to participate in international research projects and to adopt and transfer results of the research to the local level. Research on forests has not only suffered from a lack of resources; it has also not been sufficiently interdisciplinary to provide an integrated view of forestry (FAO, 1997). However, the majority of forestry research institutions in developing countries do not function as R&D laboratories, as they do in industrialised countries, and their main focus is on research and not technology development and dissemination. Unlike in the energy or transportation sectors, the technologies or even the management systems are going to be forest type- or country-specific. Additionally, there is lack of training structures to promote technology dissemination.

Under the GEF-UNDP sponsored Asia Least-Cost Greenhouse Gas Abatement Strategy (ALGAS), the US Country Studies Program, and other forestry sector capacity building and analytical activities have identified mitigation options and technologies. Furthermore, the policies to promote technology transfer have been identified (*e.g.*, regulations, financial incentives) and sometimes implemented (*e.g.*, Mexico, Bolivia). Under the UN-FCCC, each party is required to communicate a national inventory of GHG emissions by sources and sinks. A large portion of the parties have completed this task and are trying to understand forestry sector emissions and removals by sinks, which has improved dramatically. Many parties are taking steps to manage forest systems as C reservoirs (Kokorin, 1997).

As a result of the UN-FCCC and Kyoto Protocol, many developing and transition countries are developing National Climate Change Action Plans which incorporate forestry-sector mitigation and adaptation options (Benioff *et al.*, 1997). "No regrets" adaptation and mitigation options have been identified that are consistent with national sustainable development goals. Bulgaria, China, Hungary, Russia, Ukraine, Mexico, Nigeria, and Venezuela all have developed very specific forestry sector climate action plans. The Russian Federation has a very progressive forestry sector climate change action plan (Kokorin, 1997). Based on current economic and climate change scenarios several mitigation and adaptation strategies have emerged: (1) creating economic mechanisms to increase forestry sector effectiveness and efficiency in logged (removal) areas, (2) provide assistance for afforestation in Europe's Ural region, (3) promote fire management and protection for central and north-eastern Siberia, and (4) limit clear-cut logging in southern Siberia. These steps are significant since Russia contains approximately 22% of the world's coniferous forests.

12.5.3 *Barriers of Credibility and Uncertainties in Mitigation in the Forestry Sector*

Forestry mitigation projects are likely to be largely funded by Annex I countries and implemented in non-Annex I countries and countries with economies in transition (CEITs). Technology, including management systems, is an integral part of all projects funded by bilateral or multilateral or commercial agencies. Thus, promotion of mitigation projects also automatically promotes flow of technology from donor agencies or countries to host agencies or countries. Thus, technology transfer is already happening. Forestry sector options are of relatively low cost compared to those in the energy sector (Sathaye and Ravindranath, 1998). But there are some problems and uncertainties regarding the incremental C abated, its sustainability, its measurement, verification and certification. All forestry sector GHG mitigation projects must ensure that they meet accepted standards for sustainable forest management (Sathaye *et al.*, 1997). An independent international verification and certification of carbon abatement is fundamental to ensure the flow of funding to forestry-sector mitigation projects. Some of the policies and measures for promoting technology transfer according to different pathways are listed in Table 12.5.

12.5.4 *Technology Transfer between Annex I and non-Annex I Countries*

Technology transfer between Annex I and non-Annex I countries is going to be crucial for climate mitigation projects, which are going to be funded by investors (including equipment suppliers), bilateral and multilateral assistance agencies, CGIAR institutions, NGOs and foundations, largely based in Annex I countries.

Government Initiated. Governments are going to play a critical role in promoting mitigation projects and the accompanying technology transfer between Annex I and developing countries. Multilateral agencies as well as bilateral development assistance are controlled by governments. Some of the measures for the multilateral and bilateral agencies are to: i) support funding to forestry-sector mitigation projects and programmes through grants and low interest loans for SFM practices, industrial plantations, Protected Areas, and forest restoration programmes, ii) strengthen forestry certification programmes, iii) set up forestry monitoring and verification programmes in non-Annex I countries, iv) link technology transfer to grants and concessional loans, and v) provide funding for institution and human capacity building and for improving R&D capabilities in non-Annex I countries. The GEF could include forestry mitigation projects in its operational programmes.

Private sector. If CDMs would be operationalised and include forestry projects, they would offer potential for facilitating technology transfer in forestry mitigation and adaptation projects. Brokerage institutions could assist developing countries in preparation of proposals and in accessing funds under climate mitigation programmes. These institutions could facilitate agreements between industries in Annex I countries with timber industries and farmers' cooperatives in non-Annex I countries. Timber companies could import technologies and transfer them to farmers, cooperatives and forestry departments, if required. Two private sector-driven examples, namely, Reduced-Impact Logging (Malaysia) and the Rio Bravo Carbon Sequestration Project (Belize) are presented in Boxes 12.2 and 12.3 respectively.

The private sector role in movement of technologies is playing an increasingly prominent role. For example, many European, Japanese, Korean, and US private forest products companies are introducing more efficient sawtimber and plywood mill technologies to Siberia, Southeast Asia, West Africa, and Latin America. Other technologies being widely improved and exported include seedling nursery practices, alternative logging techniques (like reduced impact logging to Malaysia, Indonesia, and Latin America), software for forest management and planning, harvest and processing equipment, operations monitoring systems, and fire management (recently, these technologies were transferred from the EU, Canada, and the US to Indonesia, Malaysia, Mexico, and the Russian Federation).

Community initiated. International NGOs could play a role in capacity building among NGOs in developing countries. NGOs could contribute to international verification and monitoring programmes on sustainable logging practices and for monitoring

carbon flows in projects. The sourcing of technologies and management practices for tropical countries could be largely from other tropical countries.

Technology transfer with multiple actors. The foregoing section has discussed the role of each actor separately, however, in reality technology transfer mechanisms are often composed of complementary links among several actors. In some cases, a project incubation is done by the government, and then passed on to a community organisation and/or the private sector. In other cases a technology transfer initiated by an NGO or a private company can be adopted by the government. An example of a cross pathway interaction is a demonstration of Reduced-Impact Logging (RIL) in 1992 in Sabah, Malaysia (see Case Study 26), funded by a private Annex I company (See Box 12.2).

12.5.5   Technology Transfer among Annex I Countries

Technology transfer among the Annex I countries is largely private sector driven. There are strong R&D institutions, and interfaces exist between R&D institutions, forest departments and private plantation companies. There is a significant quantity of timber export among Annex I countries. Thus mechanisms such as timber certification and financial incentives for sustainably logged timber could facilitate adoption of sustainable practices even in Annex I countries. NGOs could play a significant role in educating the consumers about import and certification of sustainable timber coming from other Annex I countries and recycling of forest products.

12.5.6   Technology Transfer among non-Annex I Countries

Technology transfer among tropical countries, and in general among developing countries is crucial, given the similarities of forest types, silvicultural practices, socio-economic and resource conditions. There is a relatively less prominent but increasing technology transfer among non-Annex I countries. The government-initiated programmes are either on a bilateral basis or through regional cooperation arrangements such as the Association of South East Asian Nations (ASEAN) or Southern African Development Community (SADC) with collaboration in areas likely to enhance technology transfer, including natural resource management, trade and investment. For example, during the 1997 forest fires in Indonesia, there was an extensive regional collaboration and exchange of techniques on fire suppression and subsequent fire prevention measures. Regional community networks, involved in conservation and climate-friendly activities, have been used to exchange techniques among developing countries. Private sector driven technology transfer has not been as prominent, but is likely to increase due to economic liberalisation and globalisation in non-Annex I countries. Technology transfer among developing countries in the forestry sector has mainly been through capacity building and exchange of key needs such as genetically improved species (*e.g.*, Mexico and Brazil) and procurement of provenance-relevant seeds. For example, in 1998 the political party subsequently elected to the state government in Acre, Brazil, sent a resource manager to Costa Rica to learn from the experiences of that country about applicable conservation techniques. Inadequate technology transfer among non-Annex I countries is primarily due to lack of targeted funding. This could be addressed by the multilateral and bilateral mechanisms under the convention.

South-South technology transfer is increasing in scale and has potential for being highly cost-effective. Examples in the climate change context include the Costa Rican Joint Implementation Office assistance to other Latin countries in establishing programmes to evaluate JI projects; sharing of carbon accounting and monitoring experience and tools from ECOSUR in Chiapas, Mexico, with other institutes and NGOs throughout Mexico and other Latin countries, including Brazil; and Malaysian reduced impact logging technicians training Indonesians (Tipper and de Jong, 1998; Jones, 1996).

12.5.7 *Institutional Mechanisms*

A large number of institutions will be involved in technology transfer between countries. They are multilateral agencies, bilateral agencies, commercial financing agencies, international R&D institutions, national and international NGOs, national Ministries of Environment, forest departments, timber companies, brokerage agencies, and national research institutions. The governments have to play a key facilitative role. There is an increased need and scope for private sector participation particularly in the forestry sector projects, currently dominated by the state forest departments. Brokerage institutions could play a key role in providing information on potential sources of funding, technological alternatives and sources, and in assisting local institutions in developing countries in preparing project proposals and in negotiating terms and conditions. Similarly, local, national and international NGOs could ensure appropriate technologies are transferred and interests of forest-dependent communities are protected. Institutions such as FAO, CIFOR, and ICRAF could promote international scientific collaborations and capacity building in developing countries.

The current low level of technology transfer in the forestry sector can partly be attributed to institutional inadequacies, some of which are listed in Table 12.2, as institutional barriers. The

Table 12.5	Policies and Measures to Promote Technology Transfer Between Countries	
DOMINANT PATHWAY	**POLICIES, MEASURES AND PROGRAMMES**
GOVERNMENT DRIVEN	
Multilateral Agencies (World Bank, FAO, ADB, GEF, etc)	• Increase funding to forestry sector projects and programmes. • Funding for research on causes of deforestation and impacts of policies. • Facilitate grants and low interest loans for sustainable forest harvesting, industrial plantations, Protected Areas, forest restoration. • Strengthen forest certification programmes. • Tax or import Duty on timber from non sustainable logged forests. • Setting up forest monitoring and verification programmes in non-Annex I countries. • Funding for institution and human capacity building and for improving R&D capabilities in non-Annex I countries. • GEF should include forestry in mitigation projects in its operational programmes.
Bilateral development assistance	• Increase funding to forestry sector projects and programmes. • Link technology transfer to grants and concessional loans. • Facilitate grants and low interest loans for sustainable forest harvesting, industrial plantations, Protected Areas, forest restoration. • Funding for institution and human capacity building and for improving R&D capabilities in non-Annex I countries.
PRIVATE SECTOR DRIVEN	
Climate Change Related Mechanisms (JI/CDM)	• CDM potentially offers opportunity for facilitating technology transfer in forestry mitigation and adaptation projects. • Set up and expedite monitoring and verification procedures. • Brokerage institutions to assist in preparation of proposals.
Timber Industries	• Financial incentives (such as tax concessions) to sustainable harvesting of natural forests and forest plantations. • Facilitate agreements between industries in Annex I countries with timber industries and farmer cooperatives in non-Annex I countries.
COMMUNITY DRIVEN	
International NGOs	• NGOs could monitor timber certification and import of sustainable timber. • Alliance for conservation and sustainable use of forests. • International verification and monitoring on logging practices and methods. • Public awareness campaigns.

majority of the institutions have to be set up in non-Annex I countries and financial support for capacity building and providing incentives to local communities, timber logging companies may have to come from Annex I countries. However, timber certification agencies may have to be set up in Annex I countries. These certification agencies could be internationally approved by credible agencies to ensure an unbiased evaluation.

Adoption of mechanisms to promote technology transfer in the forestry sector requires the existing institutions as well as some new institutions. The existing bilateral and multilateral institutions will continue to play a dominant role in funding forestry-sector mitigation projects. Once the new mechanisms such as JI and CDM are approved, there will be an increased interest in funding forestry mitigation projects, provided that forestry projects are permitted. One of the key barriers to increased participation of bilateral, multilateral and private sector institutions is the uncertainty regarding the sustainability, measurement and verification of the C abated. Thus, it is very important to establish internationally acceptable monitoring and verification procedures and institutions. International institutions such as FAO, NGOs and private firms pre-approved by the FCCC Secretariat for monitoring and verification are required (Sathaye *et al.*, 1997). These institutions could also play a role in the timber certification process.

Currently, many of the institutions, particularly in developing countries, are not aware of the opportunities provided under new and emerging mechanisms. In addition, these institutions in developing countries may not have access to all the information required and the procedures for preparing proposals and for negotiating terms and conditions. Thus, there is a need for setting up technology transfer mechanisms or brokerage institutions to assist multilateral and bilateral institutions, industries and utilities in Annex I countries as well as the governments, farmers' organisations, NGOs and industries in developing countries. Even agencies such as the World Bank are already acting as intermediaries between the utilities in OECD countries and recipient agencies in developing countries (*e.g.*, carbon fund). Promotion of forestry-sector mitigation projects and the accompanying technology component would require careful attention as its adoption could impact biodiversity and the watershed role of forests, and further affect the poorest and indigenous communities. Due to the potential of forestry mitigation programmes for biodiversity conservation, watershed protection, and provision of socio-economic benefits, it may be necessary to provide additional incentives.

12.5.8 Financial Mechanisms[2]

 The existing and emerging financial mechanisms may have to be strengthened and reoriented to promote forestry mitigation projects. The traditional funding sources for forestry are domestic official, multilateral, external developmental assistance and commercial (private). In developing countries, large domestic financial gaps will make increasing internal funding for the forestry sector difficult, and alternate sources will be required (FAO, 1997). Post-UNCED international opinion is that forests warrant preferential allocation of funds. Estimates suggest that forestry accounts for only 1.6% of combined 1993 official and private funding transfers to developing countries. In 1993 bilateral aid accounted for 60% of developmental assistance flows to the forestry sector followed by multilateral banks (27%). An FAO survey revealed that 60% of responding countries relied on foreign sources for the greater part of their forestry sector funding (FAO, 1997). Thus, external funding is critical to the forestry sector. The private sector is increasingly becoming an important source of such funding. There are a number of domestic sources of funding to the forestry sector, particularly industries and farmers. Multilateral and private agencies may prefer funding private sector institutions in developing countries. However, bilateral developmental assistance may largely concentrate on forest conservation and protected areas that come under the control of state forest departments. Significant additional investment is required for R&D institutions in developing countries as currently, only 5% of developmental assistance is channelled to forestry research compared with about 10% to agriculture.

GEF. GEF is an institutional arrangement for supporting climate change mitigation programmes through providing incremental costs and supporting capacity building programmes. GEF needs to re-orient its operational programmes to incorporate forestry mitigation, which is being currently attempted through the new operational programme-12 and adaptation programmes. GEF could reorient and intensify support to programmes for capacity building (as suggested by FCCC/CP/1998/ADD1) for:
- Implementing adaptation measures through environmentally sound technologies
- Assessing technology needs and identifying the sources and suppliers of technologies and the modalities for the acquisition and absorption in developing countries
- Designing, formulating, evaluating, implementing and managing mitigation and adaptation projects
- Facilitating national/regional access to information generated by international centres and networks and for disseminating information services and transfer of environmentally sound technologies.

The new operational programme, which is currently being finalised, is likely to facilitate a systematic and widespread implementation of viable carbon sequestration opportunities. It may finance information and advisory and capacity building services to public or private decision-makers, provide exposure to policy and business concepts, and provide access to mainstream sources of financing. Furthermore, GEF may consider investment financing and risk management services through various forms of innovative financing including contingent financing or risk guarantee schemes. Investment in the majority of climate mitigation projects in the forestry sector also leads to conservation of biodiversity, a major area of concern to GEF.

Flexibility mechanisms. CDM, if approved by the UNFCCC inclusive of forestry activities, along with JI, could be viewed as potential investment opportunities when they become operational. The Kyoto Protocol restricts JI to activities between Annex B countries. Some forestry sector JI projects may be undertaken between the present members of Annex B, such as two projects now underway between the US and Russia, but the potential might increase substantially if some developing countries in tropical forest areas join Annex B in the future. The lessons from JI-like projects begun in tropical countries under such programmes as the US initiative on Joint Implementation (US-IJI) could be valuable both in the context of possible future JI activities if more countries join Annex B, and, potentially, for similar activities under the CDM. If some tropical countries join Annex B, the potential for emissions trading under Article 17 of the Kyoto Protocol would offer substantial opportunities for these countries to capture carbon benefits by reducing their emissions from deforestation. Because such trading is based on national-level carbon accounts, these benefits are not affected by problems of leakage and boundary definition that are common in project-

---

[2] See also Chapter 5 on financing and partnerships for technology transfer.

based initiatives such as those under JI and the CDM. Decisions by the governments of these countries on policies that affect deforestation would require reliable scenarios to predict the results of possible policy changes on deforestation and other measures. This could often benefit from technology transfer of analytical tools and experience from elsewhere. A good example of using a mix of financial incentives for forest conservation and reforestation programmes is Costa Rica (Castro and Arias, 1998).

12.6 Examples of Conditions and Policies that Facilitate Technology Transfer

What policies, measures and other preconditions help facilitate technology transfer in the forestry sector with climate change benefits? A brief review of the limited experience suggests that some such facilitating conditions can be identified. Table 12.6 presents a few examples of technology transfer and identifies policy, socio-economic, or technological conditions that helped facilitate the successful transfer of new practices.

The patterns and relative ease or difficulty of transfer of forestry technologies with climate change mitigation benefits largely have been guided by three major factors. The first is a national government's position and policies on climate change, emerging in the context of the UNFCCC Convention and the proposed Kyoto Protocol. These have provided signals, if not clear economic or policy drivers, of potential future incentive, regulatory, tax, or other policies to reduce GHG emissions in Annex I countries, and to encourage or inhibit GHG mitigation activities in CEITs and developing countries. The second factor is national economic and forest policies, especially relating to forest concessions, timber production, and exports. The third factor is competitiveness of new technologies, in terms of offering enhanced efficiency and cost savings.

The Energy Policy Act of 1992 in the U.S. established a voluntary programme for companies to report GHG emissions reduction activities. U.S. electric utility companies began to contract with private and NGO partners like The Nature Conservancy and World Resources Institute, to identify and fund pilot carbon offset projects (Dixon *et al.*, 1994; Panayotou *et al.*, 1994). It added capability to assess natural forest management and preserved areas, and led to the implementation of training and methodology development programmes in Belize (Belize, 1997) and Bolivia, where informal climate policies were receptive to cooperation. Private utilities seek cost-effective and reduced-risk delivery of tonnes of carbon. Dual camera aerial videography is currently being field tested or deployed in forest management or carbon offset projects in Ohio (USA), Indonesia, and Bolivia, as a means of improving the accuracy, and decreasing the costs of carbon estimates and monitoring.

Examples of enabling conditions for a few selected technologies are presented in Table 12.7 (chapter 4 provides an general overview on the role of enabling environments for technology transfer). The selection of the mitigation technology to be utilised

Table 12.6	Examples of Policies, Measures and Conditions That Facilitate Forestry Technology Transfer with Climate Change Benefits.	
TECHNOLOGY TRANSFER EXAMPLE	**FACILITATING POLICIES, MEASURES OR CONDITIONS IN PLACE**	**CONTRIBUTION TO OVERCOMING BARRIERS AND IMPLEMENTING THE TRANSFER**
GOVERNMENT INITIATED		
FACE Foundation (of Dutch Electricity Generating Board) for the development of efficient propogation of native dipterocarp high-value timber from cuttings, not seedlings, in Sabah, Malaysia.	FACE has high GHG emissions and expects high energy production GHG mitigation costs, and high carbon taxes. Netherlands has high vulnerability to sea level rise.	FACE started a carbon forestry programme in 1990, planned to undertake forest plantings on 150,000 ha of new forest to absorb the Generating Board's GHG emissions. FACE contracted Innoprise Corp. of the Sabah Foundation to establish 5,000 ha of dipterocarps. Propagation was limited by supply of seedlings that flowered only every few years (Jones, 1996).
PRIVATE SECTOR INITIATED		
Reduced Impact Logging (RIL) project, Innoprise Corp., Sabah, Malaysia. (Pinard and Putz, 1996; Putz and Pinard, 1993; Jones, 1996). RIL techniques developed in Australia; transferred to Malaysia by US experts.	US Energy Policy Act of 1992 led US utility NEEP to start project in 1992. Sabah has extremely high timber royalties, but high rates of residual stand damage during harvest (Miranda et al., 1992).	US utility funded improved mapping retraining of loggers in directional felling techniques and skid road planning, carbon data collection, and training and tools for carbon benefit analysis. US utilities began to seek low-cost carbon offsets after 1992. Sabah forestry regulations, and very high rent capture in forest concessions, helped Innoprise open to innovations to reduce residual tree damage, raising profits in later harvests. Sabah State government includes RIL techniques in new logging regulations, 1998.
COMMUNITY INITIATED		
Deforestation reductionin indigenous villages, Chiapas, Mexico. Scolel Te project (de Jong, et al., 1997; Tipper and de Jong, 1998)	SEMARNAP federal agency identified rural land tenure stabilisation, and carbon forestry projects, as objectives for Chiapas, due to insurgency movement	Academic researchers from UK brought computer models of forest growth and carbon benefits, monitoring methods, funding for remote sensing analysis of deforestation trends, and training to regional institute ECOSUR. Analysis of carbon sequestration potential practices performed after farmers identified candidate practices.

is likely to require the presence of specific types of policies, measures, and economic considerations. Similarly, the type of mitigation activity will largely determine the kind of technologies likely to be transferred by private, public, or multilateral entities participating in the activity.

Costs involved in Technology Transfer: The cost and funding of technology transfer varies by technology and the biophysical, economic and policy setting in which it is deployed. Generally, technologies are more likely to be transferred if they are cost-effective relative to existing practices, if the cost savings benefit the technology provider directly or indirectly, and if the relative ease of transferal is high (*e.g.*, Panayotou *et al.*, 1994). The incentives for investor companies or governments are high whenever the forest concession or timber production policies, or rent capture demands of alternative forestry practices are likely to confer benefits (public, social welfare, environmental or private sector profit benefits) (Gillis,1992; Vincent and Binkley, 1992). Financial conditions that favour the transfer of new technologies might

Table 12.7	Selected Climate Change Mitigation Technologies: Examples of Candidate Technologies for Transferal, and Facilitating Policy or Conditions	
TECHNOLOGY PRACTICE OR SYSTEM	**CANDIDATE TECHNOLOGIES AND TOOLS FOR TRANSFERAL (EXAMPLES)**	**POTENTIAL FACILATATING POLICIES OR CONDITIONS (EXAMPLES)**
Deforestation reduction via policy changes	1) remote sensing imagery and analysis capability; 2) forest and agricultural sector economic models and analytic training; 3) database management and compliance enforcement training	1) removal of government subsidies for forest conversion; 2) tax or other incentives for afforestation; 3) capability to enforce legislation or rules on forest conversion
Formation of protected areas	1) remote sensing; 2) natural resource and biodiversity assessment techniques; 3) new economic opportunities, e.g., ecotourism, butterfly farming	1) policy expanding protected areas; 2) commitment to sustainable development; 3) community support for resource management
Reforestation	1) advanced silvicultural and genetic stock techniques; 2) seedling production methods and facilities	1) incentives for reforestation; 2) incentives for maintenance of reforested areas; 3) community support and land tenure
Short rotation forestry for biofuels	1) advanced silvicultural and genetic stock techniques; 2) harvesting equipment; 3) biomass combustion or gasification technology	1) electric power purchase agreements, and grid access, available to small producers; 2) removal of subsidies for rural fossil fuel use
Reduced impact logging	1) forest management and carbon estimation software; 2) low-impact logging equipment; 3) directional felling and skid road design training	1) forest concessions granted over long timeframes, with environmental regulations; 2) forest polices to increase rent capture from concessions

include both low initial costs but demonstrated benefits (*e.g.*, reduced impact logging, if existing harvest equipment can be used; Putz and Pinard, 1993); and conditions where low to high initial costs are offset by high investor confidence in recovery of investment (*e.g.*, more efficient milling equipment in a mill owned by a private investor with a secure, long-term concession or lease; Vincent and Binkley, 1992). Technologies that are developed by public or nonprofit institutions, especially analytic and resource management innovations like carbon estimation software or methods, incorporating high intellectual but low capital inputs, tend to be transferred readily and early (Brown *et al.*, 1997).

12.7 Adaptation

Article 12 of the Kyoto Protocol to the UNFCCC provides for the CDM to channel part of funding to assist with the cost of adapting to the adverse effects of climate change. For tropical countries, deforestation and land use change driven by socio-economic factors are going to be the dominating factors affecting the forest area and vegetation status (Solomon *et al.*, 1996). However, the projected climate change under the $2xCO_2$ scenario could be at a rate higher than the capacity of the ecosystems and plant species to adapt. The projected climate change will be an additional stress on the forests affecting the vegetation status (Ravindranath *et al.*, 1997). In the event of the projected climate change, increasing forest resilience and adaptation to climate change will be as much or more important than implementation

of mitigation programmes for some of the tropical countries. Studies have shown that, in the event of an adverse impact on vegetation due to climate change, the forest dependent communities will be adversely affected through loss or change in forest area and diversity, and through forest dieback. The countries where forest vegetation is likely to be adversely affected by climate change may have to set up institutions to assess the impacts of projected climate change in their region and to develop adaptation strategies. A number of technological and silvicultural adaptation practices have been suggested in the literature (Solomon *et al.*, 1996; Ravindranath *et al*, 1997). Practices that promote biodiversity conservation or reduce current pressures on forests are also likely to enhance forest resilience under climate stressed situations (Secrett, 1996). There is likely to be a need for intact natural vegetation corridors to assist species migration to enhance habitat resilience on a regional/continental scale. Planning on this scale could be supported by the spread of remote sensing and geographical information system technologies. There are large uncertainties in projecting climate change as well as in the responses of vegetation at the regional level.

12.8 Lessons Learned

In the forestry sector, technology transfer has a broad definition, which includes sustainable forest management practices, forest conservation and Protected Area management systems, silvicultural practices for afforestation and reforestation programs, genetically superior planting material, efficient harvesting, processing, end-use technologies and indigenous knowledge of forest conservation. In the forestry sector, technology transfer could take place between Annex I and non-Annex I countries; among Annex I and non-Annex I countries; and within countries of Annex I and non-Annex I groups. The technology flow among the non-Annex I countries (tropical countries) is likely to be very important due to similarities in ecological and socio-economic conditions. The governments and private sector in Annex I and non Annex I countries, as well as multilateral agencies, have a critical role in establishing and operationalising financial and regulatory mechanisms, monitoring, verification and certification arrangements, and capacity building for technology development, diffusion and assimilation. Governments, particularly in non-Annex I countries, could prepare guidelines and set up institutional mechanisms to process, evaluate, sanction, and monitor forestry-sector mitigation and adaptation projects.

Financial. The bilateral and multilateral agencies could consider how to attract additional financial support and how they might best enhance their contribution to forestry sector mitigation projects, where technology is an integral component. The role of private sector funding of projects needs to be promoted under new initiatives, including the emerging mechanisms such as JI and the CDM, if operationalised under the Kyoto Protocol, and forestry projects are included. The role of GEF could be crucial assuming that it will reorient its operational programmes to include forestry sector carbon abatement projects. Adoption of sustainable logging practices, efficient processing and recycling technologies could be promoted by providing financial incentives such as preferential market access, lower taxes or duty and low-cost credit to companies adopting such technologies.

Regulatory Measures. Governments in Annex I and non-Annex I countries could benefit by adopting regulations to safeguard sustainable management and the use of forest resources. Governments could also pass regulations to ensure sustainable logging, efficient wood processing, recycling of forest products, timber certification, and regulating access to industry for industrial wood from natural forests. This could improve transfer of technologies for sustainable logging, high yielding plantation forestry and efficient processing technologies. Regulations to enhance the coverage of Protected Areas will ensure transfer and adoption of Protected Area management practices. Regulations affecting deforestation and its underlying causes are particularly important, and require a sound analysis and understanding of the actors and their motivations in the deforestation process.

Methods for monitoring, verification and certification. One of the most important approaches to enhance the credibility of forestry mitigation technologies and projects is to develop and transfer methodologies for monitoring, measurement and verification of carbon abatement in the mitigation projects as well as the changes in forest area. It may be necessary to develop internationally credible institutional arrangements for monitoring and verification in forestry mitigation projects.

Capacity building for environmentally sound technology development, transfer and assimilation. Non-Annex I countries often have inadequate institutional capabilities to develop technologies or to assimilate the transferred technologies, as well as to develop an understanding of the effective measures to reduce deforestation. Annex I countries could increase financial and institutional support and training to non-Annex I countries to enable them to develop, evaluate and assimilate climate-friendly forestry technologies and practices.

Awareness and education. To ensure compliance with regulations reducing deforestation and logging, adoption of sustainable forest management practices, efficient processing and recycling of forest products, it is necessary to create awareness among local communities, NGOs and the general public. For example, consumers could insist on certified timber coming from sustainably logged forests or recycled forest products such as paper.

Appropriately designed forestry mitigation and adaptation projects contribute to other environmental impacts such as biodiversity conservation, watershed protection, and socio-economic benefits to urban and rural populations through access to forest products and creation of jobs, ultimately promoting sustainable development and amelioration of the process of land degradation and desertification. Forest conservation, reforestation and sustainable forest management practices for carbon sink conservation or enhancement will particularly benefit the forest dwellers and rural communities by providing forest products and livelihood.

References

Andrasko, K., 1997: Forest management for greenhouse gas benefits: Resolving monitoring issues across project and national boundaries. *Mitigation and Adaptation Strategies for Global Change*, **2**, 117-132.

Benioff, R., E. Ness, and J. Hirst, 1997: *National Climate Change Action Plans: Interim Report for Developing and Transition Countries*. U.S. Country Studies Program, Washington, DC, 156p

Belize (Programme for), 1997. Rio Bravo Carbon Sequestration Pilot Project: Operational protocol 3. Belize City, June.

Bolin, B., 1998: The Kyoto Negotiations on Climate Change: A Science Perspective. *Science*, **279**, 330-331.

Brown, S., A.Z. Shvidenko, W. Galinski, R.A. Houghton, E.S. Kasischke, P. Kauppi, W.A. Kerz, I.A. Nalder, and V.A. Rojkov, 1996: In *The role of forest Ecosystems and Forest Management in the Global Carbon cycle*. M. Apps, D. Price, (eds.), NATO ARW Series, Springer-Verlag, New York, NY.

Brown, P., B. Cabarle, and R. Livernash, 1997: *Carbon counts: estimating climate change mitigation in forestry projects*. World Resources Institute, Washington, DC, 25 pp.

Castro, R., and G. Arias, 1998: *Costa Rica: Toward the Sustainability of its Forest Resources*. Report by the Ministerio de Ambiente y Energía-Fondo Nacional de Financiamiento Forestal, Gobierno de Costa Rica, San José.

Dixon, R.K., S. Brown, R.A. Houghton, A.M. Solomon, M.C. Trexler, and J. Wisniewski, 1994: Carbon and flux of global forest ecosystems. *Science*, **263**, 185-90.

Fearnside, P.M., 1993: Deforestation in Brazilian Amazonia: The effect of population and land tenure. *Ambio*, **22**(8), 537-545.

Fearnside, P.M., 1997: Monitoring needs to transform Amazonian forest maintenance into a global warming mitigation option. *Mitigation and Adaptation Strategies for Global Change*, **2**(2-3), 285-302.

Fearnside, P.M.,1999: Forests and global warming mitigation in Brazil: Opportunities in the Brazilian forest sector for responses to global warming under the "Clean Development Mechanism." *Biomass and Bioenergy*, **16**(3), 171-189.

Food and Agriculture Organization of the United Nations (FAO): 1995. Forest Resources Assessment 1990: Global Synthesis. FAO Forestry Paper 124. FAO, Rome. 90 pp.

Food and Agriculture Organization (FAO), 1997: Environmentally Sound Forest Harvesting: Testing the Applicability of the FAO Model Code in the Amazon in Brazil. FAO Forest Harvesting Case Study 8, Food and Agriculture Organization of the United Nations (FAO), Rome, 78 pp.

Forest Stewardship Council (FSC), 1997. Forests Certified by FSC-accredited Certification Bodies. FSC WWW site: http://antequera.antequera.com/FSC/

Gillis, M., 1992: Forest concession management and revenue policies. Forest-based industrialization: a dynamic perspective. In *Managing the world's forests*. N.Sharma, World Bank, Kendall/Hunt Publ. Co., Dubuque, IA, pp. 139-176.

Hourcade, J.C., Halsnaes, K., Jaccard, M., Montgomery, W.D., Richels, R., Robinson, J., Shukla, P.R., and P. Sturm, P. 1996. A Review of Mitigation Cost Studies. In *Climate Change 1995, Economic and Social Dimensions of Climate Change*. R.T. Warson, M. C. Zinyomera, R.H. Moss). Contribution of Working Group III to the second assessment report of the IPCC, Cambridge University Press, NY.

IPCC, 1995. *Climate change 1994: Radiative forcing of climate change and an evaluation of the IPCC IS92 emission scenarios*, J.T. Houghton, L.G. Meira Filho, J. Bruce, Hoesung Lee, B.A. Callander, E.Haites, N. Harris, and K. Maskell, (eds.), Cambridge University Press, Cambridge.

IPCC, 1996. *Climate Change 1995: Impacts, Adaptations and Mitigation of Climate Change*: Scientific-Technical Analyses. IPCC, Cambridge University Press, Australia.

Jones, D.J., 1996: Criteria for AIJ Project Design in Forestry: The Private Sector Perspective. In Regional Workshop on Activities Implemented Jointly. Proceedings of a workshop in Jakarta, Indonesia, June 25-27, 1996, US Initiative on Joint Implementation, Washington, DC, pp. 34-37.

Jong, de., B.H.R Tipper and J. Taylor, 1997: A Framework for Monitoring and Evaluation of Carbon Mitigation by Farm Forestry Projects - Example of a Demonstration Project in Chiapas, Mexico, *Mitigation and Adaptation Strategies for Global Change*, 2: 231-246.

Kadekodi, G.K., and Ravindranath, N.H. 1997: Marco-economic analysis of forestry options on carbon sequestration in India. *Ecological Economics*, 23, 201-223.

Klooster, D. 1996: Como Conservar el Bosque: La Marginación del Campesino en la Historia Forestal Mexicana. In *Bosques y Plantaciones Forestales*. L. Paré, Y S. Madrid, (eds)., Cuadernos Agrarios 14 Nueva Epoca. Edición Federación Editorial Mexicana, S.A. de C.V. Tlalpan, D.F. México, pp 144-156.

Kokorin, A., 1997: Forest carbon sequestration scenarios and priorities for the Russian Federation Action Plan. *Energy Policy*, **56**, 407-421.

Makundi, W., W. Razali, D.J. Jones, and C. Pinso, 1998: Tropical Forests in the Kyoto Protocol. *Tropical Forest Update*, **8**(4), 5-8.

Masera, O.R., 1995: Carbon Mitigation Scenarios for Mexican Forests: Methodological Considerations and Results. *Interciencia*, **20** , 388-395.

Masera, O.R., M. Bellon, and G. Segura, 1995: Forest Management Options for Sequestering Carbon in Mexico. *Biomass and Bioenergy*, **8**(5), 357-367.

Masera, O. R., M.R. Bellon, and G. Segura, 1997a: Forestry Options for Sequestering Carbon in Mexico: Comparative Economic Analysis of Three Case Studies. *Critical Reviews in Environmental Science and Technology*, **27** (Special), S227-S244.

Masera, O.R., M.J. Ordoñez, and R. Dirzo, 1997b: Carbon emissions from Mexican Forests: Current Situation and Long-term Scenarios. *Climatic Change*, **35**,. 265-295.

Merino, L., 1996: Los Bosques de México, una Perspectiva General. In *Bosques y Plantaciones Forestales*. Cuadernos Agrarios 14 Nueva Epoca. L. Paré, Y S. Madrid, (Coords.), Edición Federación Editorial Mexicana, S.A. de C.V. Tlalpan, D.F. México, pp 157-162.

Miranda, M.L., O.M. Corrales, M. Regan, and W. Ascher, 1992: Forestry institutions. In *Managing the World's Forests*. N. Sharma, Kendall/Hunt Publishing Co., Dubuque, IO for the World Bank, pp. 269-300.

Nabuurs, G.J., R. Paivinen, R. Sikkema, and G.M. Mohren, 1998: The Role of European Forests in the Global Carbon Cycle- A Review. *Biomass and Bioenergy*, **13**(6), 345-358.

Owang, J.B., and P. Karani, 1994: *The Climate Convention: Joint Implementation of Greenhouse Gas Abatement Commitments*. Africa Center for Technology Studies, Nairobi.

Panayotou, T., A. Rosenfeld, and L. Kouju, 1994: To offset or not to offset: US Power utility offsets CO_2 emissions by financing reduced impact logging in Malaysia. Case study for Harvard Institute for International Development, Cambridge, MA., 22 pp.

Pinard, M.A., and F.E. Putz, 1996. Retaining forest biomass by reduced impact logging damage. *Biotropica*, **28**, 278-295.

Putz, F.E., and M.A. Pinard, 1993: Reduced-impact logging as a carbon-offset method. *Conservation Biology*, **7**(4), 755-759.

Ravindranath, N.H., and D.O. Hall, 1994: Indian forest conservation and tropical deforestation. Ambio, **23** (8), 521-523.

Ravindranath, N.H., and D.O. Hall, 1995: *Biomass energy and environment: A developing country perspective from India*. Oxford University Press, Oxford.

Ravindranath, N.H., and B.S. Somashekar, 1995. Potential and economics of forestry options for carbon sequestration in India. *Biomass and Bioenergy*, **8**(5), 323-336.

Ravindranath, N.H.,Sukumar, R and Deshingkar, P. 1997. *Climate Change and Forests: Impacts and Adaptation, A regional assessment for the Western Ghats, India*. Stockholm Environment Institute. Stockholm.

Rose, A., and T.Titetenberg, 1994: An International System of Tradable CO_2 Entitlement: Implication of Economic Development. *Journal of Environment and Development*, **2**(1),1-33.

Sathaye, J., and N.H. Ravindranath,1998: Climate change mitigation in the energy and forestry sectors of developing countries. *Ann. Rev Energy Environment*, **23**, 387-437.

Sathaye, J., W. Makundi, B.Goldberg, M. Pinard, and C. Jepma, 1997: International workshop on sustainable forestry management: monitoring and verification of greenhouse gases. Summary report. *Mitigation and Adaptation Strategies for Global Change,* **2**, 91-99

Sathaye, J., and K. Andrasko, W. Makundi, E.L.La Rovere, N.H.Ravindranath, A. Meili, A. Rangachari, M. Imaz, C. Gay, R. Friedmannn, B. Goldberg, C. van Horen, G. Simmonds, and G. Parker, 1999: Concerns about climate change mitigation projects; summary of findings from case studies in Brazil, India, Mexico and South Africa. *Environmental Science and Policy,* **2**(2) 187-198.

Secretaría de Medio Ambiente, 1996: Recursos Naturales y Pesca (SEMAR-NAP), Programa Forestal y de Suelo 1995-2000. Poder ejecutivo Federal, SEMARNAP, México City. 79 pp.

Secrett, C.M., 1996: *Adapting to climate change in forest-based land-use systems: a guide to strategy.* Stockholm Environment Institute, Stockholm.

Solomon, A.M., N.H. Ravindranath, R.B. Stewart, M.Weber, and S.Nilsson, 1996: Wood production under changing climate and land use. *In Climate Change 1995, Impacts, Adaptation and Mitigation of climate change.* Scientific-Technical Analyses, Cambridge University Press, Cambridge.

Stowall, D., 1996: US Agreements on Joint Implementation . Joint Implementation Online, January.

Sughandy, A., 1996: An Interview: AIJ to Support Indonesia's Climate Change Policies. *Joint Implementation Quarterly,* **2**(1), 3-4.

Tipper, R., and B.H. de Jong, 1998: Quantification and regulation of carbon offsets from forestry: a comparison of alternative methodologies, with special reference to Chiapas, Mexico. *Commonwealth Forestry Review,* **77**, 219-228.

UNFCCC, 1997: Kyoto Protocol to the United Nations Framework Convention on Climate Change. FCCC/CP/1997/L7/Add.1.Dec 10.

Vincent, J. R., and C.S.Binkley, 1992: Forest-based industrialization: a dynamic perspective. In *Managing the world's forests.* N. Sharma, Kendall/Hunt Publishing Co., Dubuque, Iowa, for the World Bank, pp. 93-138.

World Bank, 1993: China Energy Conservation Study. Rep. No. 10813-CHA. World Bank, New York, NY.

World Bank, 1997. *Annual Report.* World Bank, Washington, DC.

Xu, Deying, 1995: The Potential for Reducing Atmospheric Carbon by Large-scale Afforestation in China and related Cost/Benefit Analysis. *Biomass and Bioenergy,* **8**, 337-344.

13

Solid Waste Management and Wastewater Treatment

Coordinating Lead Author:
DINA KRUGER (USA)

Lead Authors:
Tom Beer (Australia), Ron Wainberg (Australia), Xu Huaqing (China)

Review Editor:
Carlos Pereyra (Argentina)

CONTENTS

EXECUTIVE SUMMARY

Methane is generated from solid waste and wastewater through anaerobic decomposition. Together, solid waste and wastewater disposal and treatment represent about 20 per cent of human-induced methane emissions. Emissions are expected to grow in the future, with the largest increases coming from developing countries.

Methane emissions can be reduced in many ways, including reducing waste generation (source reduction), diverting waste away from disposal sites (*i.e.*, through composting, recycling, or incineration), recovering methane generated from the waste, or ensuring that waste does not decompose in an anaerobic environment. In general, any technique or technology that reduces methane generation or converts methane into carbon dioxide through combustion will reduce greenhouse gases (GHGs). The most effective mitigation approaches are those that either reduce overall methane generation (because methane collection efficiencies rarely approach 100%) or ensure that the combusted methane is substituted for fossil-based energy.

Extensive technology transfer aimed at improving waste management is underway both within and between countries, although most activities have been, and will likely continue to be, domestic in nature. In many regions, large investments are still required to provide adequate waste management services. In the past, the climate-related impacts of waste management choices were not routinely considered. Mitigation technologies can be readily deployed in this sector, however, and provide benefits beyond the reduction of GHG emissions, such as reduced landfill space requirements or additional energy generation through methane recovery.

Technology transfer in the waste management sector occurs predominately along government driven pathways, with several levels of government (from the national to the municipal level) participating. Key government priorities are establishing appropriate policy/regulatory frameworks, supporting the expansion of private sector participation, participating in technical assistance and capacity building activities, particularly with community groups, and in some cases providing incentives to catalyze desirable actions. This is discussed in more detail in Section 5.2.

Historically, the private sector (including both domestic and multinational companies, as well as more informal local enterprises) and community-based organisations have been somewhat limited participants in government-driven technology transfer. The private sector has an increasingly important role, however, because meeting future waste management needs depends on expanded private investment. Private sector driven pathways are already used routinely for some types of investments (such as methane recovery at landfills), and efforts are underway to expand private sector participation across the full range of waste management services and technologies. The involvement of community organisations is also increasing as the link between community support and project sustainability has become clear. Soliciting local input and providing local training are two ways of ensuring sustainability. In many areas, locally developed and implemented projects are also being used to quickly address serious local concerns.

This review of the waste management sector reveals several key findings. This sector can contribute to greenhouse gas mitigation in ways that are economically viable and meet many social priorities. Already, extensive technology transfer is underway, and it will continue due to the continuing need to provide and improve waste management services for the world's population. In the past, the government driven pathway has dominated this sector, and it will likely dominate in the future as well. However, additional levels of government are becoming involved, as national government agencies devolve responsibilities for waste management to regional and municipal agencies. Private sector and community driven pathways are also becoming more important in this sector. Regardless of the pathway, it is important that projects emphasise the deployment of locally appropriate technologies, and minimise the development of conventional, large, integrated waste management systems (with their attendant financial, institutional and technical requirements) in situations where lower cost, simpler alternative waste management technologies can be used.

13.1 Introduction[1]

Methane is generated from solid waste and wastewater through the anaerobic decomposition of the organic component of the waste stream. Each of these sources is believed to account for about 10% of human-induced methane emissions, although there are large uncertainties in the precise emission levels (IPCC 1996a, Section 22.4.4.1). In the future, methane emissions from these sources are expected to grow, with the largest increases expected in developing countries. Population and economic growth are the major drivers behind the rising emission levels of developing countries. In addition, current efforts to improve solid waste and wastewater management will affect future emission levels. If developing countries pursue solid waste disposal in large landfills and anaerobic wastewater treatment without including methane recovery technologies, emissions could increase significantly. There are a variety of technologies and approaches available that require less capital, use less complicated technologies and avoid GHG emissions. These technologies may be particularly appropriate in developing countries, but could also be more widely used in industrialised countries. The key challenge will be ensuring that the opportunities are identified and implemented.

Solid waste disposal sites (landfills and open dumps) are reported to emit about 20-40 Mt of methane (or 110-230 Mt Carbon-equivalent, based on a 100-year GWP of 21), with the largest share of emissions coming from Annex I countries. Emissions from most Annex I countries are expected to remain stable or decline over the next 10-20 years (IEA, 1996). A recent review of national communications and the Second Compilation and Synthesis report, for example, estimated that overall Annex I emissions from solid waste disposal sites were 24 Mt (138 MtCe) in 1990 and are estimated to fall slightly to 23 Mt (132 MtCe) in 2010 (USEPA, 1999a). The principal reasons cited are increasing attention on reducing the amount of organic material disposed of in landfills and the expanded use of methane collection systems. In contrast, methane emissions from developing countries are expected to increase in the future (IEA, 1996). Key factors are population growth, additional waste generation associated with economic development, and the continuing priority of many developing countries to reduce unmanaged dumping and develop larger, solid waste disposal sites, which typically have higher methane emissions. If, however, the use of locally appropriate technologies that avoid methane generation is expanded, and those solid waste landfills that remain essential are modified to include methane recovery systems, then methane emissions could be significantly reduced.

Methane emissions from domestic and industrial wastewater disposal are estimated to be about 30-40 Mt (170-230 Mt Ce) annually (IPCC 1996a, Section 22.4.4.1). Industrial processes, principally food processing and pulp and paper, are the major contributor, accounting for up to 95% of total emissions from this source. Domestic and commercial wastewater emissions are believed to emit only about 2 Mt of methane annually (USEPA, 1997a). Most countries only report domestic wastewater emissions in their National Communications; these emissions are expected to decline slightly in Annex I countries between 1990 and 2010. Emissions are believed to be significantly higher in developing countries, where domestic sewage and industrial waste streams are often unmanaged or maintained under anaerobic conditions without methane control. Future emissions are expected to grow, primarily driven by population and industrial growth. Emission levels are highly uncertain, however, and depend heavily on the rate at which wastewater infrastructure is developed. To the extent that developing countries successfully expand wastewater management services and reduce the proportion of their wastewater managed anaerobically without methane control, emissions would be lower than under a "business as usual" case.

13.2 Climate Mitigation Technologies

A variety of technologies and approaches are available to reduce the methane emissions associated with solid waste disposal and wastewater treatment. In the area of solid waste disposal, options include source reduction, methane recovery from disposal sites, and in some cases aerobic treatment of solid waste, through composting or other means. Similarly, methane emissions from wastewater treatment can be reduced through methane recovery or use of aerobic treatment facilities that do not generate methane. The available approaches are discussed in detail in several references (IPCC 1996a, Section 22.4.4.2; IPCC, 1996b, Thorneloe *et al.*, 1993). The principal approaches are described briefly below and summarised in Table 13.1.

There are some important points to note:
- Any technique, such as flaring, which converts methane (which has a global warming potential of 21) to carbon dioxide (which has a global warming potential of 1) is climate friendly;
- All technologies mentioned are considered environmentally sound technologies (ESTs), according to the Glossary definition;
- If the methane conversion also involves energy substitution then it is even more climate friendly;
- The complete life-cycle of waste products needs to be considered. Thus, at first sight, the use of household anaerobic compost systems for organic waste would not appear to be a mitigation technology. Such compost bins generate methane, although most compost bins maintain aerobic conditions through frequent turning. However, if the compost from these bins is used instead of inorganic fertilisers or is used to fertilise growing plants which act as a carbon sink, then it can be argued that the technology is a mitigation technology, because it either replaces a source of carbon emission (manufacture of inorganic fertiliser), or enhances a carbon sink. The alternative use of the household organic waste -- namely disposal to landfill -- would not do this, especially if the landfill lacks a methane recovery system.

---

[1] Relevant case from Case Studies Section, Chapter 16, for Chapter 13: biogas digester (case 19).

Table 13.1:	Comparison of Mitigation Technologies in the Waste Management Sector			
MITIGATION OPTIONS	EFFECTIVENESS	TECHNICAL REQUIREMENTS	APPLICABILITY	COST
SOLID WASTE DISPOSAL				
• Waste Reduction	High	Low-High (depending on site)	High	Low-Moderate
• Waste Diversion				
1) Recycling	High (if focused on organic wate)	Low to Moderate	High	Low-Moderate
2) Composting	High (if well managed)	Low	High	Low
3) Incineration	High	High	Low-Moderate (less applicable in developing countries)	High
• Methane Recovery	Moderate-High (50-75% of methane recoverable: most applicable at large sites)	Moderate	High (especiallyin the near term)	Low-Moderate (depending on site)
WASTEWATER TREATMENT				
• Waste reduction	High	Low-High (depending on site)	High	Low
• Waste Diversion	High	Low	High	Low
• Aerobic Treatment	High	Moderate-High	Low-Moderate	Moderate-High
• Methane Recovery	Moderate-High	Moderate	High (especially in near term)	Low-Moderate (depending on site)

13.2.1   Solid Waste Disposal

The amount of methane emitted to the atmosphere from solid waste disposal depends on the amount of waste that is disposed, its composition, and the nature of the disposal mechanism. Mitigation technologies involve management techniques that can be incorporated at any one, or a combination of these points. Furthermore, greenhouse gases emissions can also be reduced by collecting and using the methane generated.

Amount

Developed countries seek to reduce solid waste disposal by management techniques that concentrate on waste minimisation and recycling (CEPA, 1992). Options to reduce the amount of waste, in order of preference, are:

> AVOID producing the waste (become more efficient).
> REDUCE the amount of waste produced (substitute materials if possible).
> RE-USE the materials instead of turning them into waste (internal recycling by households and industry).
> REPROCESS the materials during manufacture, instead of using new materials (community recycling)

Solid waste generated in less developed countries is considerably less, per capita, than that in developed countries, because there is a relatively high recycling rate through an active informal sector (Yhdego, 1995).

Composition

Solid waste composition can change as a result of increasing material consumption. Active recycling techniques, such as composting, can be used to minimise the unnecessary disposal of organic waste in landfills.

Mechanism

Solid waste disposal in less developed countries is often in tips or open dumps. Such aerobic disposal tends to generate less methane than landfills (where operating practices generally require that the waste be covered with dirt for odor control and other environmental reasons). However, such tips can become anaerobic and generate methane under certain conditions. Countries, such as Austria and Japan, where the available land area is limited, use waste incineration for solid waste disposal. This reduces methane emissions, but increases carbon dioxide (and other air pollutant) emissions, and produces ash which must be disposed of. Cogeneration, in which the heat from an incinerator is used for energy production or district heating, further reduces GHGs (Fernwaerme Wien, 1994). Incineration may be more widely applied in the future, although care must be taken to ensure that waste composition and other factors are appropriate for effective combustion, and attention should be given to improving combustion efficiency when possible. Appropriate control of air emissions from a variable composition waste stream is also of concern.

Methane Recovery

Because organic matter can generate methane over 10-30 years or more, methane recovery programmes may be particularly appropriate to address the near and mid-term GHG emissions in regions where large amounts of organic waste have already been or are presently being landfilled. Depending on the site and the type of gas collection system installed, upward of 50% of the emitted methane can be recovered and used. Numerous projects have been developed throughout the world, which use the gas for heat, electricity generation, vehicle fuel, and purification and injection into gas systems.

13.2.2   Wastewater Treatment

There are three basic approaches toward reducing methane emissions from wastewater treatment systems. If wastewater and sludge are stored and treated under aerobic conditions, methane emissions can be virtually eliminated. Available technical options include aerobic primary and secondary treatment, and land treatment. However, it is to be noted that aerobic treatment of wastewater requires energy input for aeration, which may result in off-site increases in greenhouse gas emissions. Alternatively, wastewater can be treated under anaerobic conditions and the generated methane can be captured and used as an energy source. Depending on the amount of methane available, it can be used to heat the digestion tank, as a fuel for other processes, or to generate electricity. Finally, in many countries, attention is focused on reducing industrial waste generation, thereby minimising the amount of wastewater requiring treatment.

13.3 Magnitude of Current and Future Technology Transfer

Currently, technology transfer activities in the waste sector focus primarily on the development of basic urban infrastructure. In many regions, particularly in developing countries, urban populations lack basic solid waste disposal and wastewater treatment services. As a result, these populations suffer from a host of public health and environmental problems, including increased disease and mortality. The need for expanded services is well recognised, and substantial investments have been made in the past 10-20 years to develop infrastructure. In spite of large investments, however, the task of providing adequate waste services to the world's population is far from complete. Substantial additional investments are needed in many countries to keep pace with rapid urbanisation, economic development and population growth.

There is limited information of a comprehensive nature on current and future capital flows in this sector. A recent study (Bartone, 1997) reported that between 1988 and 1997, the World Bank lent over US$1 billion to 70 municipal solid waste projects. Most of these projects addressed solid waste management as part of larger municipal improvement initiatives; only three were dedicated to the solid waste management sector. In addition, this report noted that solid waste management consumes between 10 and 40 per cent of many municipal operating budgets. Providing water services (including both supply and wastewater treatment) is similarly expensive, reaching millions of dollars at the municipal level and billions at the national level (Gentry, 1997).

Between 1960 and the 1980s, public investment in water and sanitation in developing countries was roughly 0.4 per cent of gross domestic product (Serageldin, 1994). More recently, interest in private investment has grown. According to the United Nations, for example, about 250 infrastructure projects were being considered as of 1996 for private finance. Only 5 per cent of financing for water projects currently comes from the private sector, however (Gentry and Fernander, 1997).

An examination of the remaining needs for wastewater and solid waste management services indicates that substantial investments will be required in the future. Thus, identifying funding sources and encouraging private sector investments will be increasingly important.

13.4 Enhanced Technology Transfer within Countries

Extensive technology transfer in the waste management sector is underway. In developing countries, the principal activities focus on expanding waste management services and upgrading existing systems. Countries with economies in transition (CEITs) tend to have a more extensive infrastructure in place, but require substantial investments in upgrading and maintenance. Even in developed countries, many waste management systems are aging and must be upgraded to meet community demands or increasing regulatory requirements.

Expanded investment will provide numerous opportunities to incorporate mitigation technologies, which can be readily integrated into many projects. To date, however, most projects have not specifically considered the climate-related impacts of different waste management systems. When these impacts are considered, it is usually late in the process, and the focus is on possible "add-on technologies" such as methane recovery systems. In part this is a function of timing; in countries with an existing waste management infrastructure, addressing climate issues in the near-term will likely involve significant deployment of add-on technologies. Where there is a long history of waste disposal in landfills, for example, add-on methane recovery will be an important element of mitigation strategies. There are also design and operating procedures that help increase the amount of methane recovered for existing control systems. Both developing and industrialised countries can significantly expand investments in alternative waste management technologies that avoid GHG emissions. In particular, local composting and recycling programmes can reduce the need for large centralised landfills, with their GHG emissions and other environmental implications. The potential of alternative technologies is largest in developing countries, where the choice of lower cost, less complex waste management systems can be more acceptable to local communities and impose fewer institutional, technical and financial burdens on governments than conventional integrated waste management systems. Industrialised countries are also focusing increasingly on alternative waste management systems, however, and are promoting composting and recycling projects as an alternative to continued reliance on large centralised landfills.

13.4.1 Barriers

In recent years, attention has focused on the limitations of past waste management investments and the identification of more effective means of technology transfer. The principal barriers to technology transfer in the waste management sector have been discussed more fully elsewhere (see, for example, IPCC, 1996b; Bartone, 1997; Serageldin, 1994). Briefly, the principal barriers within countries include:

- Limited financing: Developing basic infrastructure to collect and treat solid waste and wastewater can be extremely expensive. Frequently, those regions where the lack of waste management is felt most severely are also some of the poorest and fastest growing. Local governments often find that they cannot generate the investment required, and availability of private financing for these types of projects can be limited, particularly if the recipient governments are not considered "credit-worthy." Case Study 19, in Section III, gives an example of a financial barrier.

- Limited Institutional Capabilities: Waste management systems require well developed institutional frameworks to ensure that waste is collected as expected, disposal and treatment facilities are operated and maintained effectively, and revenues collected.

- Jurisdictional complexity: Effective waste management involves different levels of government (local, state or provincial, and national), as well as different departments within a jurisdiction. Conflicting and competing priorities can impede the efficient development and implementation of systems.

- Need for Community Involvement: Ultimately, the success of a waste management system depends upon the willingness of the public to use it. Unfortunately, many waste management projects have historically focused on large, centralised infrastructure investments and have either ignored the role of the community or included it too late in the process. The results of such investments have been poorly designed or ill-suited projects lacking public support. Indeed, reviews of waste management projects have indicated that sustainability and performance improve to the degree that end-users are involved in the design and financing of the project (Serageldin, 1994).

Mitigation projects can be successfully integrated into larger waste management efforts provided they are able to meet the needs and priorities of end-users, decision-makers, and financial supporters. However, mitigation projects may confront additional barriers, including:

- Lack of familiarity with the potential to reduce methane generation or capture the methane emissions associated with waste management;

- Unwillingness or inability to commit additional human or financial resources to investigating and addressing the climactic implications of the waste management project; and

- Additional institutional complexity when new groups, representing issues such as energy generation or byproduct marketing, are incorporated into the project.

13.4.2 Encouraging Technology Transfer

Many of the actions that promote technology transfer within a country should involve government agencies, private entities and local community organisations to be fully successful (see

chapter 4 on the role of enabling environments for technology transfer). In general, the government-driven pathway tends to dominate in this sector. National governments typically set standards and priorities for waste management through legislation or regulation. National governments also historically have provided funding for major investments, either from internal sources or by obtaining loans from multilateral development banks. However, all levels of government are involved in waste management and this pathway has increasingly expanded to include regional and municipal governments. Activities along the government-driven pathway are also expanding to include additional actors, especially the private sector and community groups.

In addition, private sector and community-driven pathways are playing more important roles in technology transfer within the waste sector. Private entities ranging from individual entrepreneurial garbage collectors to large domestic or multinational enterprises are participating in waste management operations (Mangal, 1996). While the private sector's participation in waste management projects is often in government-driven projects, private enterprises are developing independent projects as well. Similarly, community groups are beginning to develop projects outside of government pathways that support the provision of waste management services to local populations. These informal activities are more common in developing countries and CEITs, and hold significant promise for the expansion of waste management services in the future. As the importance of community involvement for successful waste management has become evident, moreover, community groups have become more important actors in the government and private sector pathways too.

The discussion that follows describes five broad types of actions to encourage technology transfer in the waste management sector, and particularly to ensure deployment of mitigation technologies. Many of these actions will be most frequently employed along the dominant government-driven pathway. However, depending upon the specific circumstances of countries, different pathways and actors may be involved.

Policy and Regulatory Development: A framework for waste management is essential to promote technology transfer in this sector. Activities involving policies and regulations are typically government driven, with various government agencies responsible for development and implementation. Most often, this framework is defined by the national government, with various implementation requirements borne by local government agencies (or in some cases, private companies) that operate the waste management system.

Extensive technology transfer activities are already underway within countries as governments strive to improve their policy and regulatory frameworks for waste management. Consideration of policy and regulatory issues can be beneficial in all countries. Challenges within regulatory and policy frameworks may be particularly acute in developing countries and CEITs, and even more so where government agencies and institutional frameworks may be weak. In particular, enforcement capabilities may be limited in many developing countries and CEITs. Thus, new regulatory frameworks frequently may need to be coupled with capacity building programmes aimed at facilitating the effective implementation of policies and regulations.

Based on past experience, the most successful frameworks will be those that were developed through the participation of many parties, especially those directly affected by changes in waste management practices and those directly responsible for maintaining the system. If regulatory and policy frameworks for waste management are impractical or unachievable, they will not work. Some of the most effective frameworks are those that develop integrated approaches toward waste management, encouraging a mix of waste management approaches (such as source reduction, recycling and appropriate disposal and treatment) that are specifically adapted to local conditions (Bartone, 1997). Careful attention to the needs and capabilities of those that will implement the policies and regulations is also crucial, in order to ensure that the new frameworks are practical.

The involvement of community groups is necessary if the continuing challenges of the waste management are to be addressed. Ultimately, the success of any waste management system will depend upon the acceptance and participation of local populations. Thus, a country's policy and regulatory framework will likely be most effective if it reflects the needs and priorities of local populations. While the importance of local support might seem most obvious in developing countries and CEITs, where significant changes in waste disposal practices may be required, it is also important for developed countries (Serageldin, 1994).

The involvement of the private sector is also important. In most countries, the private sector has not traditionally played a major role in the design of regulatory or policy frameworks in the waste management sector. If enhanced private sector participation is desired, however, it is essential that appropriate legal and regulatory frameworks are put in place (ADB, 1996). Private sector feedback to government agencies can greatly improve the effectiveness of policies by identifying barriers constraining private investment. Lack of clarity in requirements, unclear procedures and timelines for obtaining permits or approvals, and overlapping jurisdictions are examples of barriers that can be addressed.

Emphasis on public-private exchanges on regulatory and policy issues can be beneficial in all countries. They may be particularly valuable in CEITs and developing countries, however, where the private sector may be weak, the judicial system less developed, and relations between the private sector and the government more tenuous. CEITs, especially, may confront special challenges due to the lack of a strong private sector and the suspicion government agencies frequently have for private companies. In developing countries, another challenge will likely be how to involve the variety of private entities (including micro-enterprises and informal groups of individuals) that play essential roles in the waste management sector.

Mitigation technologies can be encouraged in many ways through policies and regulations. First, governments can directly mandate certain approaches as a means of minimising GHG emissions or achieving other environmental benefits. In the United States, for example, methane emissions from landfills will be reduced by almost 50% by 2000 as a result of a regulatory action taken to reduce VOC emissions (USEPA, 1999b). Similarly, methane generation can be reduced through government policies or regulations for waste separation or recycling. To date, such policies have been motivated by the scarcity of suitable landfill capacity, and have had the secondary benefit of reducing GHG emissions. In 1992, for example, Australia adopted a National Waste Minimisation and Recycling Strategy, the goal of which was to reduce landfilled waste by 50 per cent between 1990 and 2000. Governments can also modify regulations in other sectors to reduce barriers to the use of mitigation technologies in the waste sector. In Japan, for example, the national government revised existing regulations in the power sector to create a market for energy from waste incineration plants. This action facilitated rapid expansion of waste incineration, which has contributed to lower overall GHG emissions.

Innovative Financing Approaches[2]
Waste management projects, especially large centralised landfills and wastewater treatment plants, can require large investments that traditionally have been provided by governments. This approach has limitations, however, particularly when financial requirements exceed a government's capabilities. Limited financing capacities are most severe in developing countries and CEITs, where local populations may lack adequate waste management services and existing infrastructure is deteriorating from lack of maintenance. Even in developed countries, however, the costs of maintaining and upgrading waste management systems can be high (Serageldin, 1994).

Both national and local governments are turning to the private sector in this era of financial shortages. Some governments are leasing concessions for solid waste management services, others are encouraging project finance using end-user fees to recoup investment costs, and still others are privatising waste disposal and sanitation services (World Bank, 1996). National governments may be able to assist municipalities in the development of innovative approaches for attracting private sector participation. Attracting private capital can be difficult, however, unless the relevant agencies have strong financial management and accounting practices, transparent procedures, competitive procurements, and reliable supervisory and monitoring capabilities (World Bank, 1996). Where financial management is a barrier, the national government may be able to assist municipalities in addressing financial weaknesses and attracting private investment.

The municipalities that have dealt most successfully with their increased role in waste management have tended to be those that emphasise locally appropriate waste management approach-

---

[2] See also chapter 5 on financing and partnerships for technology transfer.

es with lower financial requirements. In Badong, Indonesia, for instance, a sustainable recycling project employing people who had previously scavenged at open dumps was developed as an alternative to the more conventional centralised landfill model. By creating a strong recycling industry, needs for an expensive landfill were reduced and a variety of secondary markets in recycled materials were created (ICLEI, 1997). Similarly, a recent project in Cairo, Egypt, focused on no- and low-cost measures for wastewater treatment including waste minimisation, and wastewater treatment facilities were installed only where necessary. In addition, this project emphasised local design and manufacturing of facilities to ensure sustainability (Myllyla, 1995).

If innovative financing approaches are in use for basic waste management services within a country, financing should be available for mitigation technologies. In many developed countries, mitigation technologies are frequently developed by the private sector, with limited direct involvement of government agencies. Examples include some recycling programmes and many projects that recover methane from landfills. Similar approaches may be replicable in developing countries and CEITs, particularly in middle-income cities with strong institutions. Moreover, successful technology transfers will likely increase emphasis on alternative waste management approaches, which are often very environmentally beneficial. It is likely that central governments will continue to underwrite some investments in the sector through grants, subsidies, and revenue sharing (World Bank, 1996). If appropriate, these government agencies could require that available mitigation approaches be employed as a condition of support.

Capacity Building: Efforts to improve the institutional capacity for waste management are critical. Many of these activities will be initiated by government agencies, but in many cases the target audience may include the private sector and community groups. Appropriate capacity building activities are varied, but could include:

- **Training:** Training needs are likely to be extremely varied depending on national circumstances. For government staff, training aimed at improving management and administrative skills may be important, particularly where responsibility for waste management programmes is devolving from national agencies to lower levels of government. Technology transfer regarding waste management options, project finance, and regulatory development and enforcement are other examples of possible training needs. Local private companies may need management and technical training to assist them in establishing viable small enterprises that can implement locally based projects, many of which are very environmentally desirable. This type of training may be particularly valuable in CEITs and developing countries. It may also be desirable to train local community organisations and decision makers in both general waste management issues and those issues specific to the project under consideration. Such training can ensure that the community understands the

project context, is committed to the project goals, and has the ability to maintain it.

- **Local Assessments:** Cooperative assessment activities involving government agencies, the private sector, and community representatives should be undertaken to determine waste management priorities. Local representatives should participate in the assessment and prioritisation of waste management issues, so as to ensure that local needs and priorities are incorporated into project design and also obtain the support of local populations. In addition, these representatives can ensure that the cost of services is in line with the benefits to the population. The differences in cost between basic and full waste management services can be substantial, for example, and it is important to determine the level of service communities want and are willing to pay for before developing the system (Serageldin, 1994).

- **Information Exchanges:** Where waste management systems are changing rapidly or new approaches are being piloted within a country, information exchange and discussion among a wide range of parties can be valuable. In addition to national or regional experts, the involvement of local municipal decision makers and administrative staff, NGOs, and community leaders can be important. Information exchange can occur through clearinghouses, peer networks, government outreach or education programmes, or other means, depending on the type of information being shared and the participants.

Capacity building activities are necessary for all types of waste management projects, including those that include mitigation technologies. Depending on the extent to which mitigation strategies encourage waste management approaches that are more innovative, use newer technologies, or involve more groups, however, expanded technical and institutional assistance may be needed.

Incentives: Incentives can encourage specific activities or technologies, and can be employed by government to catalyse development of mitigation technologies. Incentives can take many forms, including tax breaks, subsidies, preferential financing, or expedited regulatory approvals. Incentives can be designed to draw attention to desired project types, and to address barriers to project implementation. Many incentives affecting mitigation technologies have been implemented, although to date few have been explicitly motivated by the desire to reduce GHG emissions. In the U.S., for example, methane recovery projects at landfills have, at various times, been eligible for tax incentives provided by both Federal and state governments, as well as preferential financing and streamlined regulatory approvals (USEPA, 1997b). For the most part, these policies have been designed to promote renewable energy or to diversify energy supply, as opposed to explicitly intending to reduce methane emissions. Similarly, the national government of Japan has provided subsidies to local governments for construction of waste incineration plants (Tanaka and Ikeguchi, 1998). The policy, which was motivated by extremely limited landfill capacity in Japan, has resulted in construction of almost 2,000 incinerators.

Voluntary Programmes: Many of the activities described above can be incorporated into voluntary programmes, which can be used to encourage wider participation in project development, to foster better coordination among agencies, or to encourage the consideration and implementation of desired technologies. Government agencies at all levels may undertake voluntary programmes, depending on the particular barriers to be addressed. Voluntary programmes promoting recycling, composting, and waste minimisation have also been implemented in many countries with much success. The U.S. EPA also uses a voluntary programme to promote methane recovery from landfills (USEPA, 1997b).

13.5 Enhanced Technology Transfer between Countries

Expanding technology transfer between countries will require addressing many of the same barriers and undertaking activities similar to those discussed in the previous section. Many of these activities will occur between governments, although involving the private sector and community groups is important for success. In the waste management sector, there are opportunities for North-North, North-South and South-South bilateral cooperation. In fact, government-to-government South-South exchanges are likely to be particularly valuable, because these countries confront different waste management issues than developed countries and may gain valuable insights from working together. Moreover, information should also flow from the South to the North, since many developing countries are implementing alternative waste management strategies with untapped potential in developed countries.

Traditionally, national government agencies have been the primary actors for donor countries, although some municipalities have developed partnerships (such as sister city or "twinning" relationships). Some of these community-based programmes directly promote mitigation technologies, such as the Cities for Climate Protection Campaign. To date, more than 180 cities worldwide have committed to reduce their GHG impacts as part of the programme, and almost one-quarter of the proposed activities will address GHG emissions from waste management (ICLEI, 1997). Where national governments take the lead in organising or funding bilateral activities, municipal governments on both sides should be involved since agencies at these levels are frequently responsible for basic waste management services. Efforts should also be made to involve the community that will be directly affected by the project from the earliest stages of its development.

13.5.1 Barriers

Technology transfer between countries confronts many of the barriers already discussed in section 13.4.1, although there are some additional barriers:

- **Lack of clear regulatory and investment frameworks:** Unclear, outdated or impractical regulatory frameworks can pose significant challenges for project development, and the problems may be particularly acute where international ventures are concerned. International projects require not only clear regulations and policies for the waste management sector, but also strong legal institutions and clear investment and tax policies, as discussed in Section I.

- **Limited financing for South-South activities:** Most technology transfer to date has been along a North-South axis, and given financial constraints in many developing countries and CEITs , this situation is likely to continue. Creative means of using developed country bilateral aid to provide opportunities for South-South transfers should be emphasised, however, since these countries may confront challenges that are unlike those found in developed countries. In particular, these countries may find that particular waste management approaches that are not used or not workable in developed countries are highly effective and replicable in their situations.

- **An overemphasis on projects, at the expense of capacity building activities:** Some of the most significant technology transfer activities in this sector will be those that develop the capacities of local governments, community groups and small private enterprises to deliver appropriate services to local populations. Such projects typically do not require large capital investments, but instead rely on the capacities of local groups to work together and marshal their own resources to implement effective projects. Too often, however, technology transfer activities focus on the development of large capital-intensive projects that may not be appropriate or sustainable. The provision of capacity building services requires significant commitments of time and personnel, and may lead to the ultimate choice of technologies that are not being promoted by the funding country (USAID, 1997). These elements may pose barriers to donor countries attempting to develop effective capacity building programmes.

13.5.2 Encouraging Technology Transfer

Many of the technology transfer approaches described in section 13.4.2 can also be undertaken between countries. Some of the key approaches for consideration are described below.

Policy and Regulatory Development: Bilateral or multilateral regulatory or policy development assistance can be very useful to countries seeking to develop an appropriate framework for waste management. Donor government agencies of all levels can provide information about their domestic policies and regulations, enabling the host country to identify useful elements for its own situation. Regulatory training programmes can also be used to outline how to analyse, draft, implement, and enforce new regulatory or policy systems. Given the importance of the regulatory/policy framework for international private sector investment, private firms should be encouraged to participate in such bilateral activities. Private firms may also partner with domestic firms and work together to focus attention on key regulatory or policy issues.

When considering technology transfer between countries aimed at improving regulatory or policy frameworks, it is important to recognise differences between host and donor countries and to ensure that the proper solutions are developed. This may be particularly important in the waste management sector, where the social aspects of waste management and the different technologies that will likely be most sustainable can differ greatly between countries. This issue may be of particular concern between where North-South transfers are concerned.

Programmes to improve and develop regulations and policies for the waste management sector can reduce GHG emissions to the extent that they promote the full range of waste management alternatives (not just centralised landfilling). However, more directed activities to encourage the deployment of mitigation technologies can also be pursued. Possible examples could include cooperative activities to develop specific regulatory requirements for mitigation technologies, or consideration of how waste management technologies could be promoted under various articles of the Kyoto Protocol (such as the Clean Development Mechanism or emissions trading).

Innovative Financing Approaches: The private sector can participate in technology transfer in the waste management sector by serving as the developer of bilateral or multilateral funded projects or through direct investment. Traditionally, private sector participation has been through government driven activities, given the dominance of this pathway, the magnitude of conventional waste management investments, and the difficulty of investing directly in the waste management sector in many countries.

Private sector driven pathways for direct investment are emerging, however, and many countries are increasingly seeking private participation in waste projects. Section 13.4.2 discusses several ways that governments are seeking to involve private firms in waste management projects. International as well as domestic firms may find attractive opportunities. International companies will probably be most attracted to the larger, centralised projects that would typically be developed in medium to large cities. International companies will also be most attracted to countries with clear investment frameworks, both generally and in the waste management sector.

Technology transfer aimed at assisting government agencies, particularly at the municipal level, to privatise or otherwise encourage private participation may be useful. The World Bank and other multilateral agencies have emphasised private sector solutions to waste management challenges in recent years as a means of improving the scope and effectiveness of services (Serageldin, 1994). Especially in developing countries, however, the structure and function of the existing waste management system is likely to be very different from the norm in developed countries. If these differences are not recognised and addressed, attempts to emulate developed country models can fail. Thus, as countries develop approaches to encourage private financing of waste management services, they should consider specific local conditions and develop locally appropriate strategies (Ali, 1997).

Capacity Building: Many bilateral technical assistance and capacity building activities are already underway in the waste management sector. Technical experts in many national and local government agencies participate in bilateral activities designed to transfer information about available waste management technologies and techniques and enhance the capacity of host country experts to assess their options and identify those most suited to local conditions.

Given the magnitude of the needs in the waste management sector, current efforts are not adequate to increase the capacity to improve waste management in all areas. In addition, typical capacity building programmes involve developed countries providing training to developing countries and CEITs. Such programmes can be of limited usefulness if they are not customised to the particular needs of the recipient country, which potentially means that the donor country would need to promote approaches that are not widely used in their own country.

Section 13.4.2 discusses several types of technology transfer for capacity building within countries, and many of the same approaches can be used between countries. All types of training, as well as facilitating local assessments and information exchange can be undertaken between countries. Some areas of capacity building may be particularly appropriate for bilateral activities, including:

- **Training to facilitate public participation:** In cases where national or local governments lack an understanding or tradition of public participation, bilateral capacity building activities may be very useful. Developed countries may be able to provide such training to developing countries or CEITs; in addition, sharing success stories between developing countries and CEITs could also be useful. Only when training is integrated into a more attractive process, are local people more likely to apply what they have learned in the trainingabout public participation.

- **Training in financial management:** Another important area for bilateral government training is in the area of improved financial management. Such assistance may be especially useful if provided to municipal governments in developing countries and CEITs, whose role in waste management is increasing, but who have historically had less information and training in financial issues.

- **Training in alternative technologies:** Training in alternative technologies, particularly those that are low technology, small scale and less capital intensive, can be useful. Since many developed countries do not use such technologies widely, CEITs and developing countries may find it useful to exchange such information between themselves, as well as to transfer information on their experiences to developed countries.

- **Training in formulating business plans:** Developing countries often lack the capacity to formulate good project frameworks and business plans for urban waste management services. This turns out to be a substantial barrier to development of urban environmental services in developing countries and CEITs. Bilateral capacity building activities can be of great use in overcoming this obstacle in this regard.

Existing and future capacity building activities in the waste management area can be readily expanded to facilitate the transfer of mitigation technologies. Activities could include increased emphasis on the climate impacts of various technologies as part of basic technical assistance programmes, developing specific decision tools and information for assessing climate impacts, and working with government counterparts on how to encourage mitigation technologies using a menu of voluntary, regulatory and incentive-based programmes. Expanded involvement of local and regional government officials from both donor and host countries would be beneficial, given their role in project development and implementation. Given the still emerging understanding of these issues, technical assistance activities in all directions (North-North, North-South, South-North, and South-South) are likely to be useful.

Incentives: The Kyoto Protocol will likely provide important incentives for expanded activities between governments to develop mitigation projects. Provisions related to both the Clean Development Mechanism and emissions trading could encourage expanded bilateral investments in projects in exchange for emission reduction units. In the near-term, projects to reduce methane emissions from landfills and wastewater treatment facilities may be particularly attractive, given methane's potency as a greenhouse gas, the favourable economics of many such projects, and the ability to monitor methane capture (see also section 3.4 and 3.6 in Chapter 3 on the Kyoto Protocol and technology transfer for further reference).

13.6 Lessons Learned

The waste management sector is an important one because of its contribution to global warming, and because it provides attractive opportunities to deploy mitigation technologies. Numerous alternative waste management strategies are available, many of which offer significant climate benefits. Available technologies can be adapted to a variety of local conditions to recover methane already generated and use it for energy or to avoid methane generation through recycling, composting, incineration, or aerobic waste decomposition processes. Moreover, the opportunities to reduce emissions are global. Although the waste management sector has different characteristics in developed, developing countries and CEITs, all regions have the potential to deploy mitigation technologies.

Many actors have important roles in encouraging the deployment of these technologies. Traditionally, large public agencies and public sector investment have dominated the sector. This situation is changing rapidly, however, and national governments need to find new roles as facilitators of municipal, private sector and community-based initiatives. In this changing environment, key roles for the national government will be in establishing workable policy and regulatory frameworks, strengthening financial management to facilitate private investment, providing technical assistance to improve technical and managerial effectiveness, and (where appropriate) providing incentives to encourage specific actions or technologies. In accomplishing these activities, national governments should seek opportunities to involve and partner with municipal and local government agencies, the private sector, and community organisations. Where national governments lack the resources or capacity to undertake these activities independently, they should seek foreign assistance.

Like government, the role of the private sector is also changing. Opportunities to participate in waste management projects are growing, and an increasing emphasis on deploying mitigation technologies will provide further opportunities. The role of the private sector will vary between regions, based on factors such as investment conditions and the readiness of various waste management systems for private sector investment. In the near-term, traditional types of private sector participation will likely expand most rapidly in developed countries. Developing countries and CEITs have the opportunity to encourage more innovative, untraditional private sector activities because of the myriad of small, informal private sector actors involved in waste management in these areas. These countries may need bilateral or multilateral assistance to assess the best ways of engaging this informal private sector, creating optimal frameworks for investment and developing strong, local private companies. The key challenge in these countries is to fully assess the range of alternative waste management approaches and project structures, instead of automatically pursuing the conventional integrated waste management systems of developed countries.

Perhaps one of the most significant developments in the waste management sector has been the recognition of the key role of community organisations. The importance of obtaining community input from the beginning, of identifying and evaluating community priorities, and of designing innovative systems that provide an appropriate level of service at reasonable cost has been well documented. Projects of all types should include community participation, regardless of whether they are undertaken by government agencies or the private sector. Community organisations are increasingly initiating their own projects, moreover, and seeking out private or government partners. This new role emphasises the importance of making climate protection a community activity, and also bodes well for the deployment of mitigation technologies. Not only are extensive networks of communities organising around climate protection and sustainable development issues, but many of the waste management alternatives pursued by communities (such as composting, recycling and methane recovery) are environmentally beneficial.

Technology transfer in the waste management sector has a long history, and has been extremely important in improving quality of life by providing waste management services to an increasing share of the world's population. The challenge is far from over, however, and expanded investments will be necessary just to keep pace with population and economic growth and urbanisation. Fortunately, a future in which the world's waste management needs are met with increased use of mitigation technologies is achievable. Meeting this goal requires the willingness of government agencies, private companies, and local organisations to expand their partnerships, consider the climate implications of waste management choices, and learn from the innovative project structures and alternative waste management approaches already being demonstrated around the world.

References

ADB,1996: *The Development and Management of Asian Megacities.* Asian Development Bank, Manila.

Ali, 1997: *Micro-enterprise Development for Primary Collection of Solid Waste.* Loughborough University, Leics, UK, 14 pp.

Bartone, C.R., 1997: *Strategies for Improving Urban Waste Management: Lessons from a Decade of World Bank Lending.* In Proceedings of the workshop in Washington, DC, 2-4, December 1997, HazWaste World/Superfund XVII Conference.

Commonwealth Environment Protection Agency (CEPA), 1992: National Waste Minimisation and Recycling Strategy, Department of the Arts, Sport, Environment and Territories, Canberra, Australia.

Fernwaerme Wien, 1994. Ges.m.b.H. Spittelau District Heating Plant, Fernwaerme Wien, Vienna.

Gentry, B.S., and L.O. Fernandez, 1997: Evolving Public-Private Partnerships: General Themes and Urban Water Examples. In *Globalization and the Environment: New Challenges for the Public and Private Sectors,* Proceedings of the workshop in Paris, 13-14, November 1997, Organization of Economic Cooperation and Development.

ICLEI, 1997: Cities Report Success in Reducing CO_2 Emissions. In *ICLEI Initiatives,* No. 17 (November).

IEA, 1996: *Methane Emissions from Land Disposal of Solid Waste.* IEA Greenhouse Gas R&D Programme, Stoke Orchard, Cheltenham, UK.

IPCC, 1996a: *Climate Change 1995: Impacts, Adaptations, and Mitigation of Climate Change: Scientific and Technical Analyses.* Contribution of Working Group II to the Second Assessment Report of the Intergovernmental Panel on Climate Change, R.T. Watson, M.C. Zinyowera, R.H. Moss, (eds.), Cambridge University Press, Cambridge, MA/New York, NY, 880 pp.
   - Levine, M, *et al,* Chapter 22, *Mitigation Options for Human Settlements*

IPCC, 1996b: Technologies, *Policies and Measures for Mitigating Climate Change.* IPCC Technical Paper 1, R.T. Watson, M.C. Zinyowera, R.H. Moss, (eds.), Cambridge University Press, Cambridge, MA/New York, NY.
   - Kruger, D.W. and G. Dutt, Chapter 8, *Solid Waste and Wastewater Disposal,* pp. 63 – 67.

Mangal, S., 1996: Participation and the Public-Private Interface in Urban Environmental Management. Background Paper for UNDP's Internet Conference on Public Private Partnerships for the Urban Environment, United Nations Development Programme, www.undp.org/ppp, 12 pp.

Myllyla, S., 1995: *Cairo – A Mega-City and its Water Resources.* In Proceedings of the third Nordic conference on Middle Eastern Studies: Ethnic Encounter and Culture Change, Joensuu, Finland, 19-22 June 1995.

Serageldin, I. 1994: *Water Supply, Sanitation, and Environmental Sustainability: the Financing Challenge.* Directions in Development, The World Bank, Washington, DC, 35 pp.

Tanaka, and Ikeguchi, 1998: Power Generation through Waste Incineration. National Institute of Public Health, Japan.

Thorneloe, S.A. et. al., 1993: Global Methane Emissions from Waste Management. In *Atmospheric Methane: Sources, Sinks and Role in Global Change,* NATO ASI Series I, Environmental Change, **I** (13).

USAID, 1997: Participatory Practices: *Learning from Experience: Partnership Among Government Officials and Local Communities: Community Involvement in Management of Environmental Pollution (CIMEP) in Tunisia.* U.S. Agency for International Development, Washington, DC. 5 pp.

USEPA, 1997a: *Estimates of Global Greenhouse Gas Emissions from Industrial and Domestic Wastewater Treatment.* U.S. Environmental Protection Agency, Office of Research and Development, EPA 600-R-97-091, Washington, DC.

USEPA, 1997b: *Turning a Liability into an Asset: A Handbook for Landfill Project Developers.* U.S. Environmental Protection Agency, Office of Air and Radiation, Washington, DC.

USEPA, 1999a: *Emissions and Projections of Non-CO_2 Gases for Developed Countries: 1990 – 2010* (Draft Report). U.S. Environmental Protection Agency, Office of Air and Radiation, Washington, DC.

USEPA, 1999b *U.S. Methane Emissions 1990 – 2020: Inventories, Projections and Opportunities for Reductions.* U.S. Environmental Protection Agency, Office of Air and Radiation, EPA 430-R-99-013, Washington, DC.

World Bank, 1996: *Livable Cities for the 21st Century. Directions in Development.* The International Bank for Reconstruction and Development, The World Bank, Washington, DC. 47 pp.

Yhdego, M., 1995: Environmental Pollution Management for Tanzania: Towards Pollution Prevention. Journal of Cleaner Production, **33**, 143–151.

14

Human Health

Coordinating Lead Author:
ANTHONY MCMICHAEL (UNITED KINGDOM)

Lead Authors:
Ulisses Confalonieri (Brazil), Andrew Githeko (Kenya), Pim Martens (Netherlands), Sari Kovats (United Kingdom), Jonathan Patz (USA), Alistair Woodward (New Zealand), Andrew Haines (United Kingdom), Akihiko Sasaki (Japan)

Contributing Authors:
Gregg Greenough (USA), Simon Hales (New Zealand), Larry Kalkstein (USA), Pete Kolsky (United Kingdom), Len Lerer (USA), Rudi Slooff (The Netherlands), Kirk Smith (USA)

Review Editor:
Tord Kjellstrom (New Zealand)

CONTENTS

EXECUTIVE SUMMARY

Climate change is anticipated to have wide-ranging consequences for human population health. Public health depends particularly on sufficiency of food, safe drinking water, secure community settlement and family shelter, and the environmental and social control of various infectious diseases. These health determinants can be affected by climate. Technological actions taken to reduce greenhouse gas (GHG) emissions or to reduce the health (or other) impacts of climate change can themselves affect population health. Therefore, the evaluation and full-cost accounting of technologies introduced to mitigate climate change must include an assessment of the health impacts of those technologies.

The uncertainty about future local and regional health impacts of climate change means that intervention strategies that also produce current public health benefit are more acceptable. For example, reductions in fossil fuel combustion and changes in transport policy have the potential to achieve prompt reductions in mortality and morbidity. Since these "secondary" health benefits of mitigation occur primarily within local populations, Clean Development Mechanism projects that entail bilateral investment in GHG mitigation strategies can achieve immediate and substantial health gains in low-income countries.

A population's vulnerability to adverse health impacts of climate change is usually amplified by socio-economic deprivation. Social policies that reduce socio-economic and environmental vulnerability will therefore mitigate the effects of climate change. Maintaining and strengthening national public health infrastructure is therefore fundamental to effective adaptation to climate change. Adaptation strategies should be undertaken principally via public agencies. To facilitate the process, the public policies and institutional arrangements that currently impede adaptation to environmental conditions should be identified.

Both the existing and the future-potential environmental health problems share many underlying causes, relating to poverty, inequality and social and economic practices. Reducing levels of resource consumption in Annex I countries and slowing down population growth in other countries will help to mitigate greenhouse gas emissions and their consequent climate-mediated health impacts, while also bringing early public health benefits. The protection and improvement of population health must be recognised as a central goal of environmentally sustainable development.

14.1 Introduction[1]

The sectors treated in the preceding seven chapters refer to technology options for mitigating climate change and for adapting to the impacts of climate change. The consideration of human health is largely limited to health impacts and benefits that arise from:*(i)* climate change, *(ii)* mitigation, and *(iii)* adaptation. The formal health sector is not substantively involved in the reduction of greenhouse gas emissions – other than incidentally via participation in society-wide improved energy efficiency (hospital building design, institutional energy-use policies, etc.), and by promoting alternative energy-saving systems of transport and mobility to increase physical activity levels. Nevertheless, by providing reproductive health and family planning services (in concert with education and social liberalisation), the health sector contributes to climate change mitigation via reductions in birth rates. Human population size is one of the prime determinants of total greenhouse gas emissions (Engelman, 1998).

Long-term changes in world climate would affect the foundations of public health — sufficient food, safe and adequate drinking water, secure community settlement and family shelter, and the environmental and social control of various infectious diseases. Currently, malnutrition is a major factor in 11.7% of deaths worldwide (*i.e.*, 5.8 million deaths every year) and accounts for an estimated 15.9% of total DALYs lost (Disability-Adjusted Life Years — a composite measure of the burden of ill health) (Murray and Lopez, 1996). Climate change is likely to make food and water supplies more uncertain in various regions, thereby increasing the risk to health for hundreds of millions of vulnerable persons, particularly in sub-Saharan Africa.

Despite the fundamental long-term dependence of human health upon the sustainability of physical and ecological systems, this dependency is not yet well appreciated by the general public and policymakers. Ecosystem services (i.e., processes maintained by the natural ecosystems) are essential for the support of human health and well-being. Natural processes that are maintained at levels adequate for human health include: the chemical composition of the atmosphere; decontamination of rivers and oceans; fertility of soils; and the availability of genetic resources. Climate change, by contributing either directly or indirectly to the impairment of natural unmanaged ecosystems can affect human health. This damage can occur via changes in species composition occurring in response to new climatic conditions. Climate change may also cause an increase in forest fires due to drought; salinisation of water sources; increased soil erosion due to heavy rainfall; and the reduction of GHG sinks due to reductions in vegetation cover.

Climate change, by stretching limited social resources across a broad range of additional health and other problems, could also impair existing public health programmes. Enhancement of the public health and nutrition interventions that are cornerstones of development (e.g., food and nutrition policies, environmental management, disease surveillance, and access to high quality health services) is an important way to reduce vulnerability to the health impacts of climate change. Whatever the economic or developmental level of the population in question, adaptations that also meet more general policy needs and specific existing local needs should generally be favoured, for reasons of cost–effectiveness, equity (especially between present and future generations), and political acceptability (McMichael and Hales, 1997).

14.2 Categories of Health Impacts

The "costs" and "benefits" of alternative technologies and policies must include an evaluation of health impacts. Public health professionals should be consulted about how to reduce adverse health impacts and promote positive health outcomes. Uncertainty about the probabilities and timing of local and regional impacts, and particularly health impacts, means that policymakers may be unwilling to act. One barrier to the understanding of health impacts assessments by policymakers is the lack of a common currency for the quantification of diverse types of health outcomes (Robine, 1998). Methods such as DALYs (Murray and Lopez, 1997), which quantify morbidity impacts in a single unit of disability-adjusted years of life lost, may make health impacts assessment more acceptable to policymakers. Furthermore, many policymakers are preoccupied with current pressing health problems. Technologies that address climate change will therefore be more acceptable if they improve public health in the near-term (Adger and Kelly, 1999).

14.2.1 Potential Health Consequences of Climate Change

The potential health impacts of climate change would occur via pathways of varying complexity, scale, and directness (Figure 14.1) (see McMichael *et al.*, 1996 and IPCC, 1996 WGII [Chapter 18]). The limited capacity for social adaptation in some populations and the unpredictability of social, economic, demographic change make forecasts of the extent and timing of health impacts difficult.

The direct adverse impacts on human health are likely to be due to changes in exposure to temperature extremes (heatwaves) and increases in the frequency or intensity of other extreme events (floods, cyclones, droughts, etc.). The extent to which the frequency of extreme weather events will be altered by climate change remains uncertain, although a changed pattern of floods and droughts is anticipated. In addition, an increase in winter temperatures may lead to a decrease in winter mortality (Langford and Bentham, 1995; Martens, 1998), since many temperate-zone countries experience significantly higher death rates in winter than in other seasons (Curwen, 1991).

Less direct impacts include net increases or decreases in the geographic distribution of vector organisms such as malarial mosquitoes. The distribution and abundance of vectors and interme-

---

[1] Relevant case from the Case Studies Section, Chapter 16, for Chapter 14: Ethnomedicine (case 28).

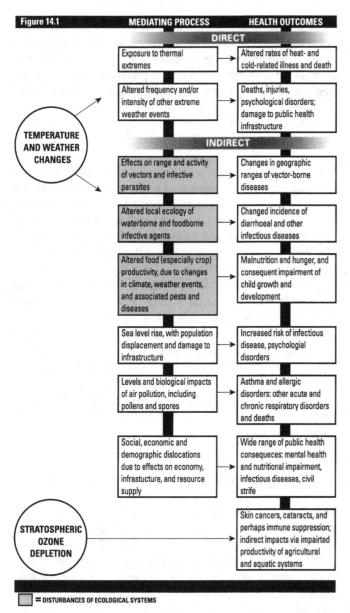

Figure 14.1

MEDIATING PROCESS | HEALTH OUTCOMES

DIRECT

TEMPERATURE AND WEATHER CHANGES

| Exposure to thermal extremes | → | Altered rates of heat- and cold-related illness and death |
| Altered frequency and/or intensity of other extreme weather events | → | Deaths, injuries, psychological disorders; damage to public health infrastructure |

INDIRECT

Effects on range and activity of vectors and infective parasites	→	Changes in geographic ranges of vector-borne diseases
Altered local ecology of waterborne and foodborne infective agents	→	Changed incidence of diarrhoeal and other infectious diseases
Altered food (especially crop) productivity, due to changes in climate, weather events, and associated pests and diseases	→	Malnutrition and hunger, and consequent impairment of child growth and development
Sea level rise, with population displacement and damage to infrastructure	→	Increased risk of infectious disease, psychologial disorders
Levels and biological impacts of air pollution, including pollens and spores	→	Asthma and allergic disorders: other acute and chronic respiratory disorders and deaths
Social, economic and demographic dislocations due to effects on economy, infrastucture, and resource supply	→	Wide range of public health consequeces: mental health and nutritional impairment, infectious diseases, civil strife

STRATOSPHERIC OZONE DEPLETION → | Skin cancers, cataracts, and perhaps immune suppression; indirect impacts via impairted productivity of agricultural and aquatic systems |

☐ = DISTURBANCES OF ECOLOGICAL SYSTEMS

diate hosts are determined by meteorological factors as well as by human interventions. Temperature-related changes in the life-cycle dynamics of vector species and pathogens (helminths, protozoa, bacteria and viruses) may change the transmission of certain infectious diseases such as malaria, dengue and leishmaniasis (Patz *et al.*, 1996; Martens, 1998). While most anticipated change would increase the transmission, decreases would occur in some locations (Martens, 1998; Faye *et al.*, 1995). Climate change will require some changes in crops and food production systems. Some regions are likely to benefit from increased agricultural productivity while others may lose out, according to their location and dependence on the agricultural sector.

14.2.2 Health Impacts of Mitigation Technologies

There is an opportunity to achieve near-term gains in population health through steps taken to reduce GHG emissions. These benefits are concrete examples of "no-regrets" or "win-win" policies

While, reductions in GHG emissions are directed principally at achieving long-term global benefit by mitigating climate change, secondary health effects of mitigation technologies are likely to occur at a local level and more immediately. This has important implications for "joint implementation" and "clean development mechanisms". Governments can thus act to optimise health as well as GHG emissions reduction in their populations.

Through the Clean Development Mechanism, investment in GHG mitigation strategies that promote more efficient or low-carbon energy generation can improve health in less developed countries. Fossil fuel combustion produces air pollutants that have both short- and long-term impacts on mortality and morbidity rates (Katsouyanni *et al.*, 1997). The secondary health benefit of reducing air pollutant concentrations can be substantial, particularly for the impacts of particulates, nitrogen oxides and sulphur dioxide. For example, the Working Group on Public Health and Fossil Fuel Combustion (1997) estimated the global health benefit of reduced outdoor exposure to particulates as 700,000 fewer premature deaths per year by 2020 under a Kyoto-like mitigation scenario compared to a business-as-usual scenario. The authors emphasised that simplifying assumptions in the model precluded precise predictions of the number of avoidable deaths and that the estimate of avoided deaths is merely indicative of the approximate magnitude of the likely health benefits of the climate policy scenario. Moreover, comparisons of premature mortality are difficult to interpret across differing populations.

Some country-specific estimates of air pollution-related secondary benefits have also been undertaken. China is an important source of GHG emissions and already suffers a high burden of ill health due to air pollution (WRI, 1998). Reductions in GHGs emissions would have large benefits for the Chinese population through reductions in indoor air pollution (Wang and Smith, 1999a,b). Several studies have evaluated other secondary health benefits associated with air pollution reduction — such as the direct costs of health services used (e.g., Aaheim *et al.*, 1997); or costing lives lost or years-of-life lost (e.g., Ontario Medical Association, 1998; see also IPCC TAR Working Group III (Chapter 9), forthcoming).

The degree of health benefit depends markedly on the particular mitigation scenario that is used in the assessment (Wang and Smith, 1999a). Furthermore, in countries with high levels of air pollution, mitigation strategies will have a greater health benefit per unit GHG emission reduction than in those countries with low levels of air pollution. The degree of health benefit also depends on the current source of energy and the proposed alternative. Switching from natural gas power plants to wind or solar power sources has little near-term health benefit because gas burns relatively cleanly. Reductions in sectors where emissions occur near human activities, e.g., in transport and household/domestic sectors, will have more near-term health benefit per unit GHG reduction than in other sectors (see Box 14.1).

The adverse impacts on health of mitigation technologies must also be considered. An increased demand for hydropower may increase

Table 14.1	Potential impact of GHG reduction technologies and policies on human health. *Note: the magnitude of the benefits and negative impacts will depend on the particular technologies used.*	
SECTOR / TECHNOLOGY	**HEALTH BENEFITS**	**NEGATIVE HEALTH IMPACTS**
BUILDINGS SECTOR		
Improved efficiency and low emissions cookstoves	Reduced air pollution exposure; decreased burn hazard; lowered physical burden of fuel gathering	
Improved building design (including insulation)	Reductions in heat/cold related mortality	Increase in indoor air pollution from lower ventilation rates. Increases in humidity and asthma
TRANSPORTATION		
High-efficiency, low-emissions vehicles, e.g., hybrid, electric.	Reduced air pollution and associated reductions in mortality and morbidity.	Increased accident risk if vehicles become smaller; hazardous material in batteries. [See biomass under energy supply]
Alternate fuels, e.g., ethanol from biomass	Reduced air pollution and associated reductions in mortality and morbidity.	
Increased cycling and walking.	Increased fitness and well-being.	
Better land-use planning and public transport.	Lower air pollution; reduced accidents and congestion	
ENERGY SUPPLY		
Photovoltaic systems	Low pollution; availability of cold storage for medications in remote areas.	
Wind energy	No air pollution	
Nuclear power	Low air and water pollution during routine operation.	Risk of large accidents; increase of nuclear proliferation risks. Risks from unsafe waste storage
Hydropower (Micro-hydropower may not have so many potential negative effects)	Increased water supplies Improved flood control Low air pollution	Increased incidence of certain disease, e.g., schistosomiasis, malaria, filariasis. Displacement of populations. Risk of large accidents.
Biomass fuel, e.g. wood	Lower air pollution if an efficient form of combustion is used, e.g., fuel-efficient cookstoves.	Indoor air pollution with non-fuel efficient forms of combustion.
Natural Gas	Lower air pollution.	Large-accident risks; energy security risks.
High-Efficiency Clean Coal	Lower air pollution.	Water pollution and occupational hazards.
SOLID WASTE / WASTE WATER		
Constructed wetlands		Increased vector breeding sites and increased risk of disease transmission.
Advanced treatment to reduce methane emissions	Lower air and water pollution	

*Health benefits of mitigating technology are compared to existing technology.

Note: Health benefits can be highly variable depending on the type of industry. Whenever there is a change in
industry there is a change to occupational exposures and health risks, therefore,
reference to specific occupational hazards are not addressed in this table.

the building of large dams; yet, there has been growing concern about the social and health impacts of large hydropower projects (Goodland, 1997). In 1998, a World Commission on Dams was set up by the World Bank and the World Conservation Union to set new international guidelines. Large water projects in tropical and sub-tropical countries have resulted in increases in the prevalence of schistosomiasis and other diseases, loss of food security and social problems that negatively influence health (Oomen *et al.*, 1994; Brantly and Ramsey, 1998; Lerer and Scudder, 1999). Social impacts include the dislocation of the rural population living in the area to be inundated; over 40 million people are estimated to have been displaced by dam projects over the past 10 years (Cernea, 1996). Resettled families lose homes, land, food sources and employment. Communities that host the resettlers face increased population densities, which places severe pressure on natural resources and water and sanitation infrastructure. Reduction of fish populations downstream has affected indigenous populations that rely on fish as their main source of animal protein. For example, downstream of the Tucurui dam, Brazil, affected communities along the Tocantins River complained that seven fish species have almost disappeared (Confalonieri, personal communication, 1998). The health impact of dams may also include increases in the transmission of vector-borne diseases. For example, an increase in the population of mosquito vectors of malaria due to the availability of more breeding sites has been observed at the Tucurui dam (Tadei, 1993).

Small-scale hydroelectric power generation schemes rarely use dams, but instead collect water from smaller structure such as weirs. Such schemes may also have undesirable local effects, such as providing vector breeding sites (Ghebreyesus *et al.*, 1999), although small-scale dams are generally less environmentally damaging than are large-scale projects. The most promising application of small-scale hydropower appears to be in isolated communities, to provide electricity for limited uses (e.g., lighting, communications, refrigeration) when there is no other feasible means of providing a continuous supply. In these conditions hydro schemes may have positive effects (including benefits for public health such as better vaccine storage, telemedicine, education) that outweigh relatively minor, local environmental and health impacts (EECA, 1996).

The evaluation and full-cost accounting of GHG mitigation technologies should include an assessment of the *health* impacts of these technologies. Regrettably, although environmental health impact assessments are an important part of environmental impact assessment, they are often omitted and are all too rarely a prime deter-

BOX 14.1   COOKSTOVES AND INDOOR AIR POLLUTION (SEE CASE STUDY 1, CHAPTER 16)

Old technologies using traditional non-fossil fuels produce large health-damaging exposures and significant greenhouse-gas emissions. Simple household stoves, which burn mainly biomass fuels (wood, dung, crop residues), provide cooking and heating needs for nearly half the world's households. A large fraction of the carbon in the fuel is diverted into airborne products of incomplete combustion, e.g., particulates, CH_4, CO, and hundreds of organic compounds. More than two million premature deaths per year could be attributed globally to the indoor air pollution caused by household solid fuels (WHO, 1997b). Although the total health-damaging emissions of such stoves are less than the emissions in cities, the exposures are much higher, because the pollutants are released indoors at the times and places where people are (Smith *et al.*, 1999). The fraction of HDP that reaches people's breathing zones can vary by 2–3 orders of magnitude depending on where emissions occur (Smith, 1995).

minant of the ultimate policy decision. Lack of awareness of long-term objectives and of a political will to value human well-being and health over material gain contribute to this problem (Last, 1997).

14.2.3 Health Impacts of Adaptation Technologies

The application of technologies to adapt to climate change should have net health benefits. Table 14.2 lists some of the health implications of adaptation technologies. For example, measures applied to control climate-sensitive invasive species (e.g., agricultural pests) may cause new human health problems. The occurrence of secondary health impacts is well illustrated by water development projects that have had significant effects on the local transmission of parasitic diseases, including malaria, lymphatic filariasis and schistosomiasis (Hunter *et al.*, 1993). For example, improved water supply in some rural areas of India has resulted in an increase in *Aedes aegypti* mosquito breeding sites and, consequently, may have contributed to outbreaks of dengue. More often, however, reliable piped water systems result in reduced *Ae. aegypti* densities and lower risk of epidemics (Gubler, 1994). Large-scale irrigation schemes for cultivating crops can increase malaria transmission, as illustrated by a malaria epidemic in Burundi, which was linked to the expansion of local rice fields (Marimbu *et al.*, 1993). The impacts of water management projects can be minimised by appropriate design measures such as flushing or varying water levels to kill snails and mosquito larvae.

Table 14.2	Potential health impacts of various adaptation responses *Note: magnitude of impact will depend on details of the particular responses taken*		
GLOBAL CHANGE FACTOR	**SECTOR**	**ADAPTIVE RESPONSE**	**HEALTH EFFECT**
Change in local temperature	Buildings	Increased Cooling*	Increased energy demand leading to air pollution and other hazards from energy supply
	Transport	Increased Cooling*	As above
	Energy Supply	Increased energy demand due to lowered efficiency of thermal conversion devices, e.g., power plants*	As above
Change in local precipitation	Water Supply	Build large hydro schemes to transport water.	Vector-borne and parasitic disease, accident, and population displacement risks.
	Land-use	Shift populations.	Impacts of social and economic disruption.
Change in sea level	Land-use	Shift populations.	Impacts of social and economic disruption.

* Technological adaptations with potential positive feedback, i.e., leading to even larger GHG emissions.

14.3 Current Social, Economic and Technological Influences on Population Vulnerability to Adverse Health Impacts

Despite rapid economic growth over the past few decades, inequalities in income have generally increased both within and between countries, in association with an increasingly market-dri-ven global economy. Between 1960 and 1993, the difference between average per capita incomes in industrialised and in developing countries tripled (UNDP, 1996). Inequalities in health have also increased because, although the health of most populations has improved over the past half-century, these gains have not been shared equally. International reviews indicate that present trends in population growth, consumption, and economic inequality are likely to lead to ecologically unsustainable patterns of development that would undermine the basic prerequisites for good health (UNDP, 1996, 1998; UNEP, 1999).

There is a large and growing population that suffers the double burden of poverty and disease (WHO, 1995). This population, especially in developing countries and countries with economies in transition (CEIT), is also increasingly having a double disease burden, with the highest incidence of both infectious endemic diseases (tuberculosis, dengue, cholera, etc.) and the degenerative diseases typical of urban-industrial societies. The "technology" exists to prevent or control most of the world's biggest killers, because they are linked to factors such as inadequate: food supply, mosquito control, safe water, secure shelter, access to education and healthcare (WHO, 1992). The barriers that exist to improved public health are to a large extent the same social and political barriers to the elimination of poverty.

Assessment of the potential for adaptation is dependent on both knowledge of potential impacts and on the vulnerability. Vulnerability is a function of the extent to which a health outcome is sensitive to climate change, and on the capacity of the population to adapt to new climate conditions (Parry and Carter, 1998). Thus, technologies that strengthen adaptive capacity are technologies that reduce vulnerability and vice versa.

14.3.1 Population Vulnerability: Social

Measures of the vulnerability of human populations to disease and injury are not well developed. Vulnerability is intrinsically a multi-faceted characteristic. Simple "indicators" of vulnerability include average income, existing health status (as reflected in mortality rates) and proximity to potential hazards. Little research has been undertaken on population vulnerability specifically in relation to climate change and health. Progress has been made defining and forecasting areas with food insecurity (e.g., Bohle *et al.*, 1994; Adger and Kelly, 1999; Vogel, 1998).

Woodward *et al.* (1998) have identified five causes of population vulnerability to ill health in the face of environmental stress: destructive economic growth that depletes capital stocks, poverty, political rigidity, dependency, and geographic isolation. With the partial exception of the first two, each of these act principally by reducing the population's capacity for adaptive response. The attainment of good public health depends on there being a responsive social order. The lack of a flexible and responsive political system and public institutions may have contributed to recent, severe impacts of climate-related disasters in parts of

Box 14.2	VULNERABILITY TO CLIMATE VARIABILITY: EXAMPLES FROM BRAZIL.

Landslides during torrential rains in vulnerable urban areas kill many people each year. Squatter settlements are constructed on denuded hills as a consequence of the structure of urban land ownership and rent, and the need of the poor people to get closer to economic opportunities. For example, in February 1988 in Rio de Janeiro, 277 people were killed, 735 injured and more than 22,000 displaced due to rains, floods and landslides (Munasinghe *et al.*, 1991)

Floods following storms in urban areas have caused epidemics of leptospirosis (Weil's disease). The vulnerability to the disease is caused by improper housing in flood-prone areas, poor drainage of rainwater and inadequate garbage disposal, which favours an increase in the population of sewer rats, the animal reservoir for the disease. This has been repeatedly reported in summertime in several coastal cities in Brazil. In some cities, such as Recife, floods could become worse with sea level rise (Neves and Muehe, 1995). Severe storms causing extensive flooding in the city of Rio de Janeiro have increased the annual incidence of leptospirosis by 10–30 fold, as has happened in the years 1988 and 1996 (Confalonieri, unpublished data)

Migration of the rural population affected by drought in the northeastern region of Brazil has lead to the spread of endemic diseases to the periphery of urban areas. This has happened with visceral leishmaniasis, which was re-introduced in the cities of Teresina and Sao Luis (Silva *et al.*, 1997; Costa, 1993) following migrant waves of peasants affected by severe droughts in El Niño years, in the late 1980s and early 1990s.

Asia, such as widespread forest fires in Indonesia and famine following flooding in North Korea. Dependency (such as reliance on others for information, resources and expertise) is a cause of vulnerability, because support is not always provided when it is needed. In the past, geographic isolation frequently protected populations from the introduction of infectious organisms and agricultural pests. Today, all countries are tied in to the global economy by the ease with which people and goods move around the globe. Unless new methods of protection are developed in isolated countries, remoteness may become a liability.

Vulnerability to climate variability provides a good framework to approach social vulnerability to climate change (Adger, 1996). Blaikie *et al.* (1994) defined vulnerability as the capacity of a population to anticipate, cope with, resist and recover from the impacts of a natural hazard. At the *individual* level, a complex mix of factors determines the degree to which someone's life and livelihood is put at risk by an event. These factors include: age, gender, disability, information, social engagement, income, cultural knowledge, legal rights, political power, built and natural environment, and physical resources, etc. Many social factors generate vulnerability to natural hazards at the *community* level; Box 14.2 illustrates some of these factors using examples from Brazil.

14.3.2 Population Vulnerability: Economic

Socio-economic vulnerability means that low-income countries are relatively less able to purchase and maintain resources and technology to protect and promote population health. Increases in information, education, transportation and social services can promote good health and reduce the potential health impacts of climate change. There is a strong positive relationship between absolute poverty and ill health (WHO, 1992). Deprived communities, which lack wealth, social institutions, environmental security and robust health, are likely to be at the greatest risk of adverse health from environmental change.

This constitutes a powerful argument for policies to reduce poverty. Inequalities in income both within and between countries may be an important cause of vulnerability to the health impacts of climate change. Within countries, this can only be achieved by income redistribution, full employment, better housing and improved public health infrastructure. In a world that is currently dominated by market-driven economics, and by associated evidence of a widening gap between rich and poor, this presents a major challenge to national governments and international agencies. Macroeconomic policies have major influences on population health in all countries, yet are usually established with little or no consideration of health impacts (WHO, 1992). For example, income tax reductions measures taken under regimes of structural adjustment policies in some countries have led to cutbacks in health services and increased the disease burden. Improvement in population health requires increased consideration of equity, and how risks are distributed among populations. Both existing and future-potential environmental health problems share many of the same underlying causes related to poverty, inequality, excessive consumption in affluent groups and population growth.

14.3.3 Population Vulnerability: Technological

A population's technological vulnerability necessarily reflects its economic conditions and cultural values. However, it refers particularly to the extent and quality of those of its technologies that would, intentionally or unintentionally, affect the health impacts of climate change. Some technologies affect basic community vulnerability; other technologies affect the capacity to prevent or treat specific injuries and diseases. The former category includes such things as: urban design (influencing the heat island effect); types of air-conditioning (domestic evaporative coolers in New Delhi, India, largely accounted for an outbreak of dengue in 1997); capacity and integrity of sewer systems (in various rapidly-expanding developing country cities, systems are seriously overloaded and would be unable to cope with an increase in precipitation intensity); wholesale and retail food storage and safety inspection facilities (could be a source of summer food poisoning outbreaks in warmer climate); housing design (control of infectious disease vectors); and coastal barriers commensurate with projected sea level rise and storm surges. The latter category includes such things as: public education systems (especially via popular media); vaccination programmes; health-care system capacity, accessibility and organisation; disaster response capacity; and food reserves and food distribution system. Current barriers to the implementation of public health "technologies" are listed in Table 14.3.

Table 14.3	Current barriers to the implementation of public health strategies and application of preventive technologies	
BARRIER	**MANIFESTATION**	**EXAMPLE**
Low income	Some people unable to purchase the means to improve health - either directly from health care provider or via other means.	During cholera epidemic in Brazil, very poor had difficulties in purchasing filtering devices or rain collecting devices.
Institutional decentralisation.	Political-administrative decentralisation of the health systems and disease control programmes.	Local governments are not prepared to take responsibility for health services, such as malaria control, sanitation, infrastructure building, etc.
Lack of funding from central government	Sometimes linked to regional or global economic crises; often linked to high levels of military spending or to debt repayment programmes from World Bank and IMF.	Susceptibility of populations in Tajikistan, Somalia and northern Kenya to mosquito-borne infectious diseases, in wake of political crises, military action and lack of centrally-provided funds.
Lack of technological capacity	Lack of investment in new/up t o-date equipment, trained personnel, transport, etc.	Lack of refrigerators/cold chain for vaccines, etc. No meteorological radar for weather early warning systems Lack of aeroplanes for the control of forest fires.
Poor communication between institutions	Inefficient use of limited resources.	Poor inter-institutional co-ordination between disaster relief and public health agencies
Lack of under-standing of the underlying issues.	Lack of education and under-standing on the links between environment, ecology and human health.	Ignorance in some traditional communities about the vector-borne basis of certain infectious diseases. Low appreciation of the physiological stressfulness of prolonged thermal extremes.
Poor policy decisions	Indiscriminate use of pesticides and anti-malarial drugs.	Rise of resistant mosquitoes and parasite, leading to increased disease transmission.
Discrimination	Vulnerable populations (e.g., poor, illiterate, powerless) may not be seen as a priority by the politicians and decision makers.	Refugee or ethnic minority populations often do not receive equal access to health services.

14.3.4　Population Vulnerability: Demographic

Increases in total population when resources are scarce increase population vulnerability. Assuming a continuation of recent trends in nearly all countries, we will see a demographic "bulge" in the population moving into older age groups and increasing life expectancies. Consequently, the proportion of elderly persons within most populations will increase. The increased longevity of the elderly population entails an increase in the burden of disabilities (Crimmins *et al.*, 1997). In Australia, elderly populations have increased disproportionately in coastal areas where the risk of flooding may increase (Curson, 1995).

14.4　Adaptation Options to Reduce Health Impacts

Adaptation refers to actions taken to lessen the impact of the (anticipated) change in climate. Adaptive actions to reduce health impacts can be thought of in terms of the classical categorisation of preventive measures in public health (Last, 1995):

1. *Primary prevention*: actions taken to prevent the onset of disease due to environmental disturbances, in an otherwise unaffected population (e.g., supply of bed nets to all members of a population at risk of exposure to malaria).
2. *Secondary prevention*: preventive actions taken in response to early evidence of health impacts (e.g., strengthening disease surveillance and responding adequately to disease outbreaks).
3. *Tertiary prevention*: health-care actions taken to lessen the morbidity or mortality caused by the disease (e.g., improved diagnosis and treatment of cases of malaria).

Primary prevention can be addressed at many levels – including, most radically, the mitigation of the climate change process. In general, secondary and tertiary prevention are less effective than primary prevention, although they are often practically and politically easier to implement. In addition, there are both ethical and social reasons to prefer primary preventive action wherever it is feasible. In the long term, secondary and tertiary prevention measures usually turn out to be more expensive than primary prevention.

A three-fold categorisation of strategies to protect population health is:*(i)* administrative or legislative, *(ii)* technical/engineering, and *(iii)* personal (behavioural) (Patz, 1996). Legislative or regulatory action can be taken by government, requiring compliance by all, or by designated classes of, persons. Alternatively, an adaptive action may be encouraged on a voluntary basis, via advocacy, education or economic incentives. The former type of action would normally be taken at international, national or community level; the latter would range from international to individual levels. Adaptation options can operate at different spatial levels, from local to global. Some of the options are of a structural and general kind, facilitating and maximising preventive impacts. Other options are of a more specific kind, entailing procedures, technologies or behavioural changes. Table 14.4. shows the major adaptation options for reducing the health impacts of climate change, and the major considerations that bear upon their effectiveness.

Actions to reduce the health impacts of climate change are basically public policy response options and are, therefore, in the latter system. The ultimate goal of these interventions is the reduction, with the least cost, of diseases, injuries, disabilities, suffering and death. There is little quantitative information about how humans adapt either biologically or culturally to climate change. Most assessments of the health impacts of climate change have therefore not addressed adaptation explicitly and quantitatively. However, some assessments of the impacts of thermal stress have modelled, by extrapolation from short-term observations, the effect of longer-term acclimatisation at the population level (e.g., Kalkstein and Greene, 1997). The most effective way to reduce potential health impacts will be through adaptation technologies that reduce the overall level of population vulnerability. The potential impacts of climate change on food and water are not seen as the responsibility of public health agencies.

TABLE 14.4 ADAPTATION OPTIONS TO REDUCE THE POTENTIAL HEALTH IMPACTS OF CLIMATE CHANGE					
ADAPTATION OPTION	LEVEL	NO. OF PEOPLE THAT BENEFIT	FEASIBILITY	BARRIERS	COST
Interagency cooperation	G,R,N	+++	++	++	+
Reduction of social vulnerability	N,L	+++	+	++	+
Improvements of public health infrastructure	N,L	++	+	+	++
Early warning and epidemic forecasting	L	++	++	+	+
Support for infectious disease control	N,L	++	+++	+	+
Monitoring and surveillance of environmental, biological and health status		++	+++	+	++
Integrated environmental management	L	+	++	+	++
Urban design (including transport systems)	L	+	+	++	++
Housing, sanitation, water quality	L	+	+	+	+
Specific technologies (e.g., air conditioning)	L	+	+++	+	+
Public education	L	++	+++	+	+

*G = GLOBAL, N = NATIONAL, L = LOCAL.

14.4.1   Health Sector

Public health infrastructure is a fundamental resource. The last decade has witnessed the resurgence of several major diseases that were once thought to have been controlled. The resurgence of tuberculosis — which is both cost-effective to treat and curable in virtually all cases — has been caused by persistent, and in some cases increasing, poverty and a lack of political will to develop and sustain effective control programmes (WHO, 1998).

The recent resurgence of malaria in areas where it had previously been eradicated (Azerbaijan, Tajikistan) or under control (Iraq, Turkey) are the consequences of deteriorating malaria prevention and vector control programmes due to conflict and economic crises (WHO, 1997a). In the 1950s, vector control programmes in Madagascar led to the eradication of the main vector in the central highland plateau and almost total eradication of malaria (Lepers *et al.*, 1988). Since then there has been a progressive increase in malaria due to the collapse of the spraying programme and population movements (Fontenille *et al.*, 1990). In Ethiopia, indoor spraying campaigns with DDT were shown to be effective at reducing malaria (Fontaine *et al.*, 1961), but over the last 20 years there has been an increase in cases, partly due to a breakdown in the health service due to civil war and forced movement of people. Similarly, towns in the highlands of Zambia where malaria was once rare now experience a substantial number of cases as a result of the cessation of vector control activities (Fisher, 1985). It is now also recognised that effective control of infectious disease cannot occur without active community support (WHO, 1997a). In the past, most disease control programmes were vertically structured, lacking robust, horizontal community-based support, but this has proven to be non-sustainable (Gubler, 1989).
The economic crises of the 1980s, in addition to poor policy decisions in the late 1960s and 1970s, have led to cuts in both government and household expenditure on health in many developing countries and CEIT (Evlo and Carrin, 1992). Cost-sharing policies recently implemented in Africa have resulted in people delaying treatment and disease progression to more life threatening forms (WHO, 1997a). In several countries, declines in health facility use of over 30% have been recorded following the policy of cost sharing (Waddington *et al.*, 1989).

Decision-making whether at the policy, implementation or at the health-seeking (individual) level depends on availability of relevant, accurate, and useful information (Emmanuel, 1998; Sayer, 1998). The cost of data collection and analysis is often beyond the resources of developing countries and CEIT. Thus, decision-making is often delayed and this introduces uncertainty in the choice of policy and interventions. In Bolivia and Zimbabwe, Nugroho *et al.* (1997) have observed that malnutrition and other health problems in underprivileged communities cannot be tackled effectively unless attention is paid to family income, housing, water supply, sanitation, food and environmental safety. Communities with so many needs may downgrade the importance of some diseases. Thus, the additional hazards of climate change have to compete with existing community needs for a local decision-maker.

Human behaviour sometimes changes dramatically following well-targeted, culturally-sensitive dissemination of health information, especially when a change in attitude is first induced (as has happened with respect to exposure to passive smoke). Some issues relating to climate change could form the subject of effective health education programmes, for example to encourage the elimination of human-made vector breeding sites, and promote the use of mosquito nets impregnated with pyrethroid compounds to reduce malaria transmission, particularly among children and pregnant women.

Monitoring and Surveillance
The most elementary form of adaptation is to launch or improve health monitoring and surveillance systems (McMichael *et al.*, 1996; WHO/MRC/UNEP, 1998; Stanwell-Smith, 1998). Table 14.5 summarises the mechanisms for a comprehensive monitoring scheme for the types of potential health impact of climate change (Haines and McMichael, 1997).

In the health sector, only the basic measures of public health status (e.g., infant mortality) can be measured simply and uniformly around the world because births and deaths are monitored in most countries. However, disease (morbidity) surveillance varies widely depending on the locality, the country and the disease. Most of the least developed countries have poorly developed sur-

Table 14.5	Summary of methods needed to monitor the potential impacts of climate change and climate variability on human health.		
WHAT	**WHERE**	**HOW**	
Heat stress.	Urban centres in developed and developing countries.	Daily mortality and morbidity data.	
Changes in seasonal. patterns of disease (e.g., asthma, allergies)	"Sentinel" populations at different levels.	Primary health care morbidity data, hospital admissions, emergency room attendance.	
Vector-borne diseases.	Margins of distribution (latitude and altitude). Areas with seasonal and sporadic incidence.	Primary health care data; local field surveys, communicable disease surveillance centres; remote sensing data. Surveillance of infectious disease must be active and laboratory-based.	
Marine ecosystems.	Coastal populations, coastal zones.	Sampling of phytoplankton for biotoxins, pathogens. Remote sensing of algal blooms. Epidemiology of cholera, other Vibrios and shellfish poisoning.	
Natural disasters	All regions.	Mortality and morbidity data.	
Effects on health of sea level rise.	Low-lying regions.	Local population surveillance.	
Freshwater supply.	Critical regions especially in the interior of continents.	Measures of runoff; irrigation patterns; pollutant concentrations.	
Food supply.	Critical regions.	Remote sensing; measures of crop yield; food access and nutrition from local surveys. Agricultural pest and disease surveillance.	
Emerging diseases.	Areas of population movement or ecological change.	Identification of "new" syndrome or disease outbreak; population-based time series; laboratory characterisation.	

SOURCE: HAINES AND MCMICHAEL, 1997.

veillance systems. Many developing country governments lack the resources and expertise for collecting appropriate data for effective monitoring of the impacts of climate change. Data sharing and capacity strengthening for local data collection and development of integrated early warning systems are very important. A strong public health infrastructure – international, national and local – along with active local community involvement is necessary to achieve effective response to information provided by the surveillance of infectious diseases.

Reliable, continuous monitoring of cause-specific mortality in vulnerable populations would be invaluable. Effective infectious disease surveillance requires good laboratory support. In addition, low-cost data from primary care facilities could be collected in sentinel populations in vulnerable zones. The use of animal sentinel populations (including the investigation of outbreaks of diseases in animals) can be used to detect early changes in patterns of human disease as part of a comprehensive surveillance programme. For example, sentinel caged chicken flocks are

used to monitor encephalitis virus in the US (Tsai and Mitchell, 1989). Animal reservoirs are also used to monitor leishmaniasis (Semiao Santos *et al.*, 1996; Mancianti *et al.*, 1994). Effective surveillance demands global cooperation and exchange of information, as well as the modernising of monitoring and surveillance systems. Such initiatives should build on current successes like ProMED — a network for the exchange of information on outbreaks of new and resurgent infectious diseases (Morse, 1995).

Mosquito vectors of malaria are expected to increase their altitudinal range as the world warms, and the incidence of malaria may increase in certain highland areas in the tropics and subtropics. There is some indication that increases may already be occurring (Loevinsohn, 1994; Epstein *et al.*, 1998) although this remains contentious (Reiter, 1998). To track these possible changes, new surveillance measures must be initiated to monitor vector populations and disease incidence in many highland areas that are not well served by clinical health services (Le Sueur *et al.*, 1997). There is also a need for additional data that would enable researchers to distinguish the effect of climate change from other environmental factors which affect malaria distribution, e.g., deforestation.

Control of Vector-borne and Water-borne Diseases

In addition to the vector control and surveillance strategies discussed above, populations can be protected from vector-borne diseases by immunisation campaigns when a suitable vaccine exists. The coverage of existing vaccination programmes aimed at elimination of diseases such as yellow fever should be expanded. Unfortunately, no vaccines yet exist for some of the diseases most sensitive to climate change, e.g., malaria, dengue, schistosomiasis, nor for many newly emerging infections. While there is no vaccine for dengue currently approved for general use, there are vaccines at an early stage of development. Other strategies are important to combat diseases like malaria. For example, periodic checks may be carried out on parasite sensitivity to the commonly used antimalarial drugs. The use of insecticide-impregnated bed nets has been successful in reducing malaria transmission in endemic areas. However, there have been economic barriers and difficulties in obtaining the appropriate bed nets because of distribution problems.

The control of some epidemic diseases, such as malaria, could benefit from the application of new technology (*e.g.*, geographical information systems (GIS) and remote sensing technologies) to forecast outbreaks using meteorological data (*e.g.*, Snow *et al.*, 1996; Le Sueur *et al.*, 1997; GCTE, 1998). For example,

Table 14.6	Types of adaptive strategies, illustrated with malaria as an example					
LEVEL	**VECTOR CONTROL**	**VACCINE DEVELOPMENT**	**ACCESS TO ANTI MALARIAL DRUGS**	**HOUSING DESIGN**	**EPIDEMIC FORECASTING**	**ENVIRONMENTAL MANAGEMENT**
International	++	+	+++	+	-	-
Regional or Federal	++	-	++	+	-	-
National or State	+++	-	+++	+	+++	+
Local or community	++	-	+	++	++	+++
Individual	+++	-	+	+++	+	++

a prediction system for malaria outbreaks in the western Kenyan highlands is being developed (Githeko, 1998). Initial investment in such predictive modelling is relatively high. Once established, however, these systems become cost-effective.

Table 14.6. highlights malaria to indicate some types of adaptive strategies, and the level at which they operate. For example, specific products for vector control, malaria vaccines, and drugs are developed, around the world, under the guidelines of WHO. However, the local use of the products depends upon national policies and demand by users.

Populations that are vulnerable to water-borne diseases should have access to technology for safe drinking water. *Cryptosporidium* oocysts are resistant to chlorine and other disinfectants (Venczel *et al.*, 1997) and have a very low sedimentation rate (Medema *et al.*, 1998). Consequently boiling may be the most appropriate method of disinfecting water where risks of infection exist (Willcocks *et al.*, 1998). The use of submicron point-of-use filters may reduce the risk of waterborne cryptosporidiosis (Addiss *et al.*, 1996). In addition, a number of simple and cheap techniques have been found to be effective in reducing the risk of infection with cholera from contaminated water. A simple filtration procedure involving the use of domestic sari material can reduce the number of *Vibrios* attached to plankton in raw water (Huo *et al.*, 1996). In Bolivia, the use of 5% calcium hypochlorite to disinfect water, and the subsequent storage of the treated water in a narrow-mouthed jar produced drinking water from non-potable sources that met the WHO standards for microbiologic quality (Quick *et al.*, 1996). These examples of low cost technologies should become widely available to populations that are likely to be affected by contaminated water supplies, for example, following flooding.

14.4.2 Intersectoral Adaptation Options

Intersectoral technologies are those that involve collaboration between health and other sectors. Recent experiences with the control of emerging and resurging diseases, e.g., dengue, have shown that emergency responses by the health sector have limited success — usually because they are begun too late to have any effect. There is a need to think more broadly than the health sector for effective interventions for infectious disease threats (Pinheiro and Corber, 1997; Gubler, 1989).

Collaboration across research disciplines is essential. For example, the use of a hydrological soil moisture model to forecast *Anopheles* mosquito biting rates in Kenya was only possible because of collaboration between epidemiologists, entomologists and hydrologists (Patz *et al.*, 1998). Barriers to global change research include: the lack of national strategic research plans; lack of communication between disciplines; single-discipline funding agencies that do not fund interdisciplinary research; the pressure of other existing health needs; and the lack of public concern, which would increase priority with decision-makers (CGCP, 1995).

Integrated Environmental Management

The environmental management of ecosystems upon which health depends (i.e., freshwater resources, agricultural areas) should be improved. The incidence of certain water-borne and vector-borne infections can be reduced by several environmental measures. Experience with the WHO/FAO/UNEP/UNCHS Panel of Experts on Environmental Management (PEEM) has shown that early consultations between health and agricultural sectors can greatly reduce the burden of vector-borne diseases (e.g., malaria, schistosomiasis) in large-scale irrigation projects by appropriate adaptations in the management of irrigation water (FAO, 1987). Likewise, collaboration between oceanographers, marine biologists and fisheries experts should lead to the delineation of non-fishing zones on coral reefs, to maintain fish breeding sites and to provide sentinel sites for the undistorted monitoring of ocean warming effects.

Traditional public health interventions for water- and food-borne diseases, which focus on personal hygiene and food safety, have limited effectiveness. A broader approach would consider the interactions between climate, vegetation, agricultural practices and human activity. For example, some farming practices remove vegetation from hillsides and river banks, increasing run-off, leading in turn to more erosion and sedimentation, and the retention of encysted pathogens from livestock excrement. Ecological analysis can result in recommendations for the type, time and place of public health interventions, such as changes in management of water catchment areas, or targeting water treatment to cover high risk periods.

Strategies to control the invasion of climate-sensitive disease vector and pest species require intersectoral collaboration between the health, forestry, environment, and conservation sectors. Climate change is likely to amplify the challenge of pest control, as new ecological niches appear that may sustain exotic pathogens and disease vectors. The recent establishment of the Environmental Risk Management Authority (ERMA) in New Zealand is an example of such a strategy (Anon, 1996). As an evolutionarily-isolated island ecosystem New Zealand is particularly vulnerable to invading species (Leakey and Lewin, 1996). ERMA provides an integrated approach with a wide-ranging brief that includes regulation of importation, investigation of incidents and emergencies, and review of existing hazards. The Authority has formal links with many sectors, must consider public input from diverse interest groups, and reports directly to a senior Minister who holds both the Environment and Biosecurity portfolios.

Urban Planning

Within the next decade, more than half the world's population (more than 3.3 billion people) will be living in cities (WRI, 1997). Rapid urbanisation has many adverse environmental and health consequences, and health inequalities between rich and poor are widest in cities of developing countries. Current strategies to reduce poverty and vulnerability to climate variability in urban environments will serve to enhance adaptation to the health impacts of climate change. Some examples are described below.

Natural Disasters: Land use planning can reduce vulnerability to weather disasters by maintaining strict control over any development in high-risk zones and to discourage human settlement in such zones. In addition, governments must ensure that buildings and the urban infrastructure are built to withstand extreme events. Populations in shanty settlements in developing countries are very vulnerable to climate variability, as structures are often flimsy and located on land subject to frequent flooding. Future projections of land-use changes indicate that in many countries the poorest population sectors will congregate increasingly on the land that is most vulnerable to the impacts of natural disasters. Therefore, these countries need to formulate national disaster-response strategies as part of their sustainable development planning.

Urban micro-climates: At the local scale, human activities alter the climate in cities principally by the "heat island" effect that increases temperatures and decreases precipitation. The heat island effect can be reduced by appropriate urban planning (Oke, 1997), including improved insulation of buildings and other design features that reduce heat load, planting trees, and selection of materials with high albedo for roads, parking lots and roofs.

Water quality: A large urban population in developing countries currently does not have access to a safe drinking water source (standpipe or borehole) or to sanitation services (sewers, septic tanks or wet latrines) (WRI, 1997). Flooding may become more frequent with climate change and can affect health through the spread of disease. The only way to reduce vulnerability is to build the infrastructure to remove solid waste and waste water, and to supply potable water. No sanitation technology is "safe" when covered by flood waters, as faecal matter mixes with flood waters and is spread wherever the flood waters go. The usual focus of drainage design on the performance of the minor drainage system (pipes, channels, and overflows) neglects the performance of the major drainage system, including flows over land and in the streets, which occurs during major floods. As the risk of flooding increases with climate change, so does the importance of the major drainage system. New design approaches, which explicitly design roads to act as drains, can radically reduce the duration of flooding. Litter management is critical to the management of urban drainage systems; often the best investment in drainage is better handling of solid waste to prevent systems becoming rapidly blocked with debris.

Early Warning Systems
Weather and climate forecasts may be used, where appropriate, in preventing deaths, injuries and in disease prevention and control. The five-day weather forecasts, achieved over the last decade and now routine in many countries, have saved millions of lives through warnings of hurricanes, floods and other severe weather (Noji, 1997). For example, deaths due to tropical cyclones in South Asia have been dramatically reduced due to meteorological forecasts, systems to disseminate forecasts and the building of storm shelters. The use of historical climate data for planning may reduce food insecurity in many countries. However, meteorological and related services are not used to their full potential in mitigating the adverse health impacts of climate variability.

Illness and deaths due to heatwaves can be prevented by hot weather watch/warning systems, which alert people to impending dangerous weather and provide information on how to avoid illness during weather extremes (Kalkstein *et al.*, 1996). Sentinel surveillance in emergency departments can also be used to monitor increases in heat-related cases (Kellermann and Todd, 1996).

Seasonal regional climate forecasts have proved useful in guiding farmers in Brazil, Zimbabwe, Peru, and other countries in the planting of drought resistant crops and management of fisheries during El Niño years (Glantz, 1996). Seasonal forecasts have been included into many local and regional famine and drought early warning systems (Glantz, 1994). Associations between El Niño and malaria epidemics in Asia and South America have also been described (Bouma and Dye, 1997; Bouma and van der Kaay, 1996; Kovats *et al.*, 1999). The use of climate forecasts for epidemic forecasting needs to be incorporated with early warning systems based on known epidemic risk factors (e.g., heavy rainfall). Initiatives that are underway in East and Southern Africa and India to develop these new approaches for the surveillance and control of epidemic malaria should be supported.

14.4.3 Cross-sectoral Adaptation to Climate Change

Trends in inequality, resource consumption and depletion, environmental degradation, population growth and ill-health are closely interrelated (McMichael, 1995; Dasgupta, 1995; see Section 14.3) and will strongly interact with potential climate change impacts (Petersen *et al.*, 1998). Such problems cannot be effectively addressed solely by implementing improved intersectoral (energy, agriculture) or public health technologies. Cross-sectoral policies that promote ecologically sustainable development and address underlying driving forces will be essential.

Agenda 21 and the Rio Declaration on Environment and Development describe a comprehensive approach to ecologically sustainable development incorporating cross-sectoral policies. Economic sectors, such as industry, agriculture, energy, transport and tourism, must take responsibility for the impact of their activities on social and ecological systems. National and local strategies for sustainable development should be completed in all countries, in accordance with Local Agenda 21 principles. The empowerment and the full and equal participation of women in all spheres of society, including the decision-making process, are necessary (UN, 1996). Rich countries should fulfil their commitments to reach the United Nations target of 0.7 per cent of gross national product devoted to international development assistance as soon as possible (UN, 1997).

14.5 Adaptation: Policy and Institutional Context

The Universal Declaration of Human Rights (UN, 1948) affirms

> "the right to a standard of living adequate for the health
> and well-being — including food, clothing, housing and
> medical care and necessary social services"

Collective international action is required to ensure that dangerous anthropogenic interference with global climate does not worsen the plight of vulnerable populations. Many of the world's poorest countries are currently obliged to divert resources from essential social services to debt repayments. Reducing those international debt repayments is likely to increase health spending in some countries, and initiatives like the IMF/World Bank Heavily Indebted Poor Countries Initiative could be extended (Oxfam, 1998). The adverse public health impacts of climate change are likely to be worsened by Structural Adjustment Policies, which should now be reviewed in the light of FCCC commitments.

The vulnerability of certain groups may be a direct threat to the well-being of more advantaged members of the same population. An example of such a "spillover" effect is the spread of infectious diseases from primary foci in poor populations. What applies within populations may apply also between countries. The vulnerability of poorer countries may jeopardise the security of others, for example, by triggering population movements.

14.5.1 Role of International Agencies

Awareness of the multisectoral interactions that influence human health has recently induced many international agencies to address health issues, in addition to the World Health Organization, UNICEF and other UN agencies that have an explicit health mandate. The World Bank has taken a leading role in international health policy through providing loans for health sector reform in developing countries and CEIT (Walt, 1994, 1998; World Bank, 1993). There are also many health and welfare-related NGOs that operate at the international level. Many international agencies play an important role in disaster and humanitarian relief, such as the International Federation of Red Cross and Red Crescent Societies, the UN High Commission for Refugees, WHO, and international disaster relief agencies. UNESCO supports the education sector in member states and has an important role in supporting climate change mitigation and adaptation.

The specialised agencies of the United Nations are *primarily* mandated to support government planning and management in Member States, by strengthening sectors such as industry, health and agriculture. These agencies also play an important role in assisting countries to apply internationally-agreed rules and regulations, and quality standards. Accordingly, these agencies consider that technology transfer, as it is broadly defined, should principally serve the needs of Member States regarding selection and implementation of mitigation and adaptive measures for the impacts of global climate change.

A collaborative programme on climate and human health, tentatively defined by WHO, WMO and UNEP, has been approved by the Interagency Committee for the Climate Agenda (IACCA). Execution of this programme is entrusted to WHO as part of its newly defined programme on Sustainable Development and Healthy Environments (SDE). Depending on the availability of resources, it is foreseen that WHO in collaboration with UNEP and WMO will:

1. assist capacity building in Member States, in order to promote vulnerability assessment, adaptation strategies, and the adoption of technologies to promote health and reduce GHG emissions;
2. exchange and provide information on the health impacts of climate change and variability and of mitigation strategies, as well as effective approaches to adaptation;
3. promote research on the above topics.

Nations have both abilities and responsibilities, resources and needs. There will be benefit in sharing and networking, and in working within the framework provided by WHO and other international agencies.

14.5.2 Promoting Public Health Trans-sectorally as a Policy Criterion

Greater emphasis must now be given to helping countries where there is a particular need to improve vulnerability assessment, to identify climate change health risks, and to define needs and resources for mitigation and adaptation programmes. The strengthening of intersectoral efforts at the international level (for example, between the health, meteorology, agriculture, and fisheries sectors) could be a longer-term goal. WHO collaborates with other UN agencies on a number of monitoring programmes, such as GEMS (Global Environmental Monitoring System) that provides monitoring and assessment of air and water quality. International monitoring programmes could be extended to include exposures to the direct and indirect health hazards associated with climate change and sea level rise (McMichael *et al.*, 1996; WHO/MRC/UNEP, 1998). Linkages between health-related early warning systems that are already in existence, such as Global Information and Early Warning System (GIEWS) and ProMED, will need to be strengthened.

Research requires the availability of quality data. Many international agencies have a responsibility to collect and disseminate data; e.g., WHO supports a major international database focused on the incidence of infectious diseases worldwide. They also have a responsibility to ensure that financial considerations concerning access to data do not impede research (Colwell and Patz, 1998). International agencies could support the formation of a central clearinghouse for data on climate-health linkages in order to facilitate the gathering of data for research, especially for developing countries and particularly for data generated by global monitoring systems.

Food security in some countries may be worsened by climate change while it may be improved in others (IPCC, 1996 WGII [Chapter 13]). In the former countries, personal, community and national level adaptation options are likely to be severely limited. The most effective strategies for adaptation would entail changes in the international mechanisms of agriculture, trade and finance. Global organisations such as the World Trade Organisation may be best placed to implement the required changes of policy — provided there is the political will to address the issue seriously.

In recent decades, there has been a shift of economic power away from national governments to various private and public organisations (e.g., multi-national corporations, the World Trade Organization, and regional free trade associations) which are not explicitly accountable for, nor concerned by, the social, health and environmental effects of their actions. The "globalisation" and liberalisation of international systems of trade and finance has facilitated the exploitation of poorly protected environmental resources in the short term (McMichael, 1995). Although responsibility for population health remains primarily national, there is need for new mechanisms for international collective action in favour of forms of social and economic development that are compatible with sustained good health (Jamison *et al.*, 1998; McMichael *et al.*, 1999).

14.6   Lessons learned

The reduction of population vulnerability is, in general, the most cost-effective strategy to deal with the health impacts of climate change. In the lowest-income countries, where impacts are likely to be greatest, strengthening the public health infrastructure should be a priority. There has been a widespread decline in public health training, facilities, and programmes over the past decade (McCally *et al.*, 1998). Equally important, the health of communities will be improved by social-economic development that promotes income redistribution and the reduction in health inequalities. Impoverished populations are at greater risk of adverse health outcomes (such as those due to exposure to infectious agents, vulnerability to thermal extremes, and marginal nutritional status) because they have fewer choices. For populations whose health status is already compromised, the consequences of climate change will often be a critical further stress upon health.

Population health vulnerability would also be reduced by taking specific steps to lessen the health impacts of climate change. Examples include improved urban and housing design, vaccination programmes, upgraded wastewater and refuse disposal systems, and public education and early-warning programmes. Improved understanding by the public and policymakers of the potential health impacts of climate change is a prerequisite to reducing the impacts. This can be achieved by:

- Encouragement of research institutions to pursue long-term, multidisciplinary research.

- Dissemination and public discussion of research results and risk assessments.
- Monitoring of, and quantification of, any early health outcomes.

With respect to the mitigation of climate change, awareness of potential health impacts should be a central consideration in the public discourse. The social incentives for GHG mitigation would be further enhanced by market mechanisms, particularly pricing, that take account of human health costs.

With respect to the capacity of populations to adapt to climate change, public policies to reduce socio-economic and physical-environmental vulnerability are necessary. Health sector institutions and processes, however, are not able to deal effectively with health problems on their own. The task is multi-sectoral. A structured approach for cross-sectoral strategies and international cooperation is therefore required, entailing explicit reference to environmental health impact assessment of all policies. Low-income countries are unable to afford many of the basic health protection measures without international assistance. Overall, intergovernmental agreements are needed for national policy making and funding, and for coordinated research and monitoring to reduce the health impacts of climate change.

References

Aaheim, H.A., K. Aunan, and H.M. Seip, 1997: *Social Benefits of Energy Conservation in Hungary. An examination of alternative methods of evaluation.* Working Paper 1997:10. Center for International Climate and Environmental Research – Oslo (CICERO), Oslo.

Addiss, D.G., R.S. Pond, M. Remshak, D.D. Juranek, S. Stokes, and J.P. Davis, 1996: Reduction of risk of watery diarrhea with point-of-use water filters during a massive outbreak of water-borne Cryptosporidium infection Milwaukee, Wisconsin, 1993. *American Journal of Tropical Medicine and Hygiene,* **54**, 549-553.

Adger, W.N., 1996: *Approaches to Vulnerability to Climate Change.* CSERGE Working Paper GEC 96–05. Centre for Social and Economic Research on the Global Environment, University of East Anglia, Norwich.

Adger, W.N., and M.P. Kelly, 1999: Social vulnerability to climate change and the architecture of entitlements. *Mitigation and Adaptation Strategies for Global Change,* in press.

Anon, 1996: *New Zealand Hazardous Substances and New Organisms* Act. Bennetts, Wellington, New Zealand.

Blaikie, P., T. Cannon, and B. Wisner, 1994: *At Risk, Natural Hazards, People's Vulnerability and Disasters.* Routledge, London.

Bohle, H.G., T.E. Downing, and M.J. Watts, 1994: Climate change and social vulnerability: towards a sociology and geography of food insecurity. *Global Environmental Change,* **4**, 37–48.

Bouma, M.J., and C. Dye, 1997: Cycles of malaria associated with El Niño in Venezuela. *Journal of American Medical Association,* **278**, 1772–1774.

Bouma, M.J., and H.J. van der Kaay, 1996: The El Niño Southern Oscillation and the historic malaria epidemics on the Indian subcontinent and Sri Lanka: an early warning system for future epidemics? *Tropical Medicine and International Health,* **1**, 86–96.

Brantly, E.P., and K.E. Ramsey, 1998: Damming the Senegal River. *In World Resources 1998–9: Environmental Change and Human Health.* Joint Publication: World Resources Institute/UNEP/UNDP/World Bank., Oxford University Press, New York, pp. 108–114.

Cernea, M.M., 1996: Public policy responses to development-induced population displacements. *Economic and Political Weekly,* **31**, 1515-1523.

CGCP (Canadian Global Change Program), 1995: *Implications of global change and human health. Final report of the Health Issues Panel to the Canadian Global Change Program,* The Royal Society of Canada (CGCP Technical Report Series), Ottawa.

Colwell, R.R. and J.A. Patz (eds.), 1998: *Climate, Infectious Disease and Human Health: An Interdisciplinary Perspective.* American Academy of Microbiology, Washington DC.

Costa, C.H., 1993: Urbanization and kala-azar in Brazil: Kala-azar in Teresina. *In Research and Control of Leishmaniasis in Brazil.* Proceedings of a Workshop, S.P. Brandao Filho, (ed.), CPAG, Fiocruz, Recife.

Crimmins, E.M., Y. Saito, and D. Ingegneri, 1997: Trends in disability-free life expectancy in the United States, 1970–90. *Population Development Review,* **23**, 555–570.

Curson, P.,1995: Human Health, Climate and Climate Change - An Australian perspective. Climate Change, People and Policy: Developing Southern Hemisphere Perspectives. A, Henderson-Sellers, T. Giambelluca, (eds.), Wiley & Sons, Chichester, UK, pp. 319-348

Curwen, M., 1991: Excess winter mortality: a British phenomenon? *Health Trends,* **22**, 169-175.

Dasgupta, P., 1995: Population, poverty and the local environment. *Scientific American,* **272**, 26–31.

EECA (Energy Efficiency and Conservation Authority), 1996: *Small hydro: Planning guidelines for renewable energy developments.* EECA, Wellington, New Zealand

Emmanuel, S.C., 1998: Information and research for decision makers. *World Health Forum WHO,* **19**, 12–14.

Engelman, R., 1998: *Profiles in carbon: an update on population, consumption and carbon dioxide emissions.* Population Action International, Washington, DC.

Epstein, P.R., H.F. Diaz, S.A. Elias, G. Grabherr, N.E. Graham, W.J.M. Martens, E. Mosley-Thompson, and J. Susskind, 1998: Biological and physical signs of climate change: focus on mosquito-borne diseases. *Bulletin of the American Meteorological Society* **78**, 409–417.

Evlo, K., and G. Carrin, 1992: Finance for health care: part of a broad canvas. *World Health Forum WHO,* **13**, 165–170.

FAO, 1987: *Effects of Agricultural Development on Vector-Borne Diseases.* Edited Versions of Working Papers presented to the 7th Annual Meeting of the Joint WHO/FAO/UNEP Panel of Experts on Environmental Management for Vector Control (PEEM), 7-11 September 1987, Rome. Document AGL/MISC/12/87, FAO, Rome.

Faye, O., O. Gaye, D. Fontenille, L. Konate, J.P.Herve, Y.Toure, S.Diallo, J.F. Molez J. *et al.,* 1995: Drought and malaria decrease in the Niayes area of Senegal. *Sante,***5**, 199-305.

Fisher, M., 1985: Malaria at high altitudes in Africa. *British Medical Journal,* **291**, 56.

Fontaine, R.E., A.E. Najar, and J.S.Prince, 1961: The 1958 malaria epidemic in Ethiopia. *American Journal of Tropical Medicine and Hygiene,* **10**, 795–803.

Fontenille, D., J.P. Lepers, G.H Campbell, M. Coluzzi, I. Rakotoarivony and P. Coulanges, 1990: Malaria transmission and vector biology in Manarintsoa, high plateaux of Madagascar. *American Journal of Tropical Medicine and Hygiene,* **43**, 107–115.

GCTE, 1998: *Global Change Impact Assessment and Approaches for Vectors and Vector-borne Diseases. Report of an international workshop, 3-6 September 1997, Nairobi.* R.W. Sutherst, J.S. Ingram, T. Yonow, H. Scherm, K. Sutton, (eds.), GCTE Working Document No.27.

Ghebreyesus, M. Haile, K.H. Witten, A. Getachew, A.M. Yohannes, M. Yohannes, Teklehaimnot, S.W. Lindsay, and P. Byass, 1999: Incidence of malaria among children living near dames in northern Ethiopia: a community based incidence survey. *British Medical Journal,* **319**, 663-666.

Githeko, A. K., 1998: Climate change and human health in East Africa. Unpublished report. Centre for Vector Biology and Control Research, Kenya Medical Research Institute, Kisumu, Kenya.

Glantz, M.H. (ed.), 1994: *Usable Science: Food security, early warning, and El Niño.* Proceedings of the Workshop held 31 October – 3 November 1994 in Boulder, Colorado, National Center for Atmospheric Research, Boulder, CO.

Glantz, M.H., 1996: *Currents of change: El Niño's impact on climate and society.* Cambridge University Press. Cambridge.

Goodland, R., 1997: Environmental sustainability in the hydro industry: Disaggregating the debate. *In. Large dams: learning from the past, looking at the future. Workshop Proceedings,* IUCN, The World Conservation Union and The World Bank Group IUCN, Gland, Switzerland and Cambridge, UK/ World Bank Group, Washington, DC, July, pp. 69–102.

Gubler, D.J, 1989: Aedes aegypti and Aedes aegypti-borne disease control in the 1990s: Top down or bottom up. *American Journal of Tropical Medicine and Hygiene,* **40**, 571–578.

Gubler, D.J., 1994: Perspectives on the prevention and control of dengue haemorrhagic fever. *Kaohsiung Journal of Medical Science,* 10, S15-S18.

Haines, A., and A.J. McMichael, 1997: Climate change and health: implications for research, monitoring and policy. *British Medical Journal,* **315**, 870–874.

Hunter, J.M., L.Rey, K.Y.Chu, E.O.Adekolu, and K.E. Mott, 1993: *Parasitic Diseases in Water Resource Development: The Need for Intersectoral Negotiation,* World Health Organization, Geneva.

Huo, A., B. Xu, M.A. Chowdhury, M.S. Islam, R. Montilla, and R.R. Colwell, 1996: A simple filtration method to remove plankton-associated *Vibrio cholerae* in raw water supplies in developing countries. *Applied Evironmental Microbiology,* **62**, 2508-2512.

Jamison, D.T., J. Frenk, J., and F. Knaul, 1998: International collective action in health: objectives, functions, and rationale. *Lancet,* **352**, 514-517.

Kalkstein, L.S., and J.S. Greene, 1997: An evaluation of climate/mortality relationships in large US cities and the possible impacts of climate change. *Environmental Health Perspectives,* **105**, 84–93.

Kalkstein, L.S., P.F. Jamason, J.S. Greene, J. Libby, and L. Robinson, 1996: The Philadelphia hot weather–health watch warning system: development and application, Summer 1995. *Bulletin of the American Meteorological Society,* **77**, 1519–1528.

Katsouyanni, K., Touloumi, G., Spix, C.M., Schwartz, J. *et al.*, 1997: Short term effects of ambient sulphur dioxide and particulate matter on mortality in 12 European cities: results from time series data from the APHEA project. *British Medical Journal*, **314**, 1658–1663.

Kellermann, A.L., and K.H. Todd, 1996: Killing heat. *New England Journal of Medicine*, **335**, 126–127.

Kovats, R.S., M.J. Bouma, and A. Haines, 1999: *El Nino and Health*. Technical Report. WHO, Geneva.

Langford, I.H., and G. Bentham, 1995: The potential effects of climate change on winter mortality in England and Wales. *International Journal of Biometeorology*, **38**, 141–147.

Last, J.M., 1995: *A Dictionary of Epidemiology*, 3rd edition. Oxford University Press, Oxford.

Last, J.M., 1997: *Public Health and Human Ecology*, 2nd edition. Appleton and Lange, Stamford CT.

Leakey, R,. and R. Lewin, 1996: *The Sixth Extinction: Biodiversity and Its Survival*. Doubleday, New York.

Lerer, L.B., and T. Scudder, 1999: Health impacts of large dams. *Environmental Impact Assessment Reviews*, **19**, 113-123.

Le Sueur, D., F. Binka, C. Lenegeler, D. de Savigny, R. Snow, T. Teuscher, and Y. Toure, 1997: The Mapping of Malaria Risk in Africa (MARA)/Atlas due Risque de la Malaria en Afrique (ARMA) Initiative. *Africa Health*, **19**, 23-4.

Lepers, J.P., Deloron, P., Fontenille, D., and Coulanges, P., 1988: Reappearance of falciparum malaria in central highland plateaux of Madagascar [letter]. *Lancet*, **I**, 586.

Loevinsohn, M.E., 1994: Climatic warming and increased malaria incidence in Rwanda. *Lancet*, **343**, 714-8.

Mancianti. F., W. Mignone, and F. Galastri, 1994: Serologic survey for leishmaniasis in free-living red foxes (Vulpes vulpes). *Journal of Wildlife Diseases*, **30**, 454-456.

Marimbu, J., A. Ndayiragije, M. le Bras, and J. Chaperon, 1993: A propos d'une épidèmie de paludisme dans une règion montagneuse non endèmique. *Bulletin de la Societè Pathologie Exotique*, **86**, 399–401.

Martens, W.J.M., 1998: *Health and Climate Change: Modelling the Impacts of Global Warming and Ozone Depletion*. EarthScan, London.

McCally, M., A. Haines, O. Fein, W. Addington, R.S. Lawrence, and C.K. Cassel, 1998: Poverty and ill health: physicians can and should make a difference. *Annals of Internal Medicine*, **129**, 726-733.

McMichael, A.J., 1995: *Nexus between population, demographic change, poverty and environmentally sustainable development*. Paper for the Third Annual World Bank Conference on Environmentally Sustainable Development, London.

McMichael, A.J., and S. Hales, 1997: Global health promotion: looking back to the future. *Australian and New Zealand Journal of Public Health* **21**, special issue, 425–428.

McMichael, A.J., A. Haines, R. Slooff, and S. Kovats (eds.), 1996: *Climate Change and Human Health. An assessment prepared by a Task Group on behalf of the World Health Organization, the World Meteorological Organization, and the United Nations Environment Programme*. WHO, Geneva.

McMichael, A.J., B. Bolin, R. Constanza, G.C. Daily, C. Folke, K. Lindahl-Kiessling, E. Lindgren, and B. Niklasson, 1999: Globalization and the sustainability of human health: towards an ecological perspective. *BioScience*, **49**, 205-210.

Medema, G.J., F.M. Schets, P.F. Teunis, and M. Havelaar. 1998: Sedimentation of free and attached *Cryptosporidium oocyst and Giardia cysts in water. Applied Environmental Microbiology*, **64**, 4460-4466.

Morse, S., 1995: Factors in the emergence of infectious diseases. *Emerging Infectious Diseases*, **1**, 7–15.

Munasinghe, M., *et al.*, 1991: Case Study: Rio flood reconstruction and prevention project. In *Managing Natural Disasters and the Environment*. A. Kreimer, M. Munasinghe, (eds.), World Bank, Washington, DC, pp 28–31.

Murray, C.J.L., and A.D. Lopez, 1996: Evidence-based health policy: lessons from the Global Burden of Disease Study. *Science*, **274**, 740–743.

Murray, C.J.L., and A.D. Lopez, 1997: Global mortality, disability, and the contribution of risk factors: global burnden of disease study. *Lancet*, **349**, 1436–1442.

Neves, C.F., and D. Muehe, 1995: Potential impacts of sea-level rise on the metropolitan region of Recife, Brasil. *Journal of Coastal Research*, **14** (Special issue).

Noji, E. (ed), 1997: *The Public Health Consequences of Disasters*. Oxford University Press, New York.

Nugroho, G., Macagba, R.L., Dorros, G.L., and Weinstock A., 1997: Challenges in Health Development. *WHO World Health Forum*, **18**, 44-47.

Oke, T.R., 1997: Urban climates and global environmental change. In *Applied Climatology: Principles and Practice*. R.D. Thompson, A. Perry, (eds.), Routledge, London, pp. 273–287.

Ontario Medical Association, 1998: *Health Effects of Ground Level Ozone, Acid Aerosols and Particulate Matter*. OMA Position Paper, Ontario Medical Association,Toronto.

Oomen, J. J. de Wolf, and W. Jobin, 1994: *Health and Irrigation. Vols 1 and 2*, ILRI Publication No.45. International Land Reclamation and Improvement, Wageningen, The Netherlands.

Oxfam, 1998: *Making Debt Relief Work: A Test of Political Will*. Oxfam International, Oxford..

Parry, M.L., and T. Carter, 1998: *Climate Impact and Adaptation Assessment*. EarthScan, London.

Patz, J.A., 1996: Health adaptations to climate change: Need for farsighted, integrated approaches. *In Adapting to Climate Change: An International Perspective*. J.B. Smith *et al.*, (eds),. Springer, New York, pp. 440–464.

Patz, J.A., P.R. Epstein, T.A.Burke, and J.M. Balbus, 1996: Global climate change and emerging infectious diseases. *Journal of the American Medical Association*, 275, **217–223**.

Patz, J.A., K. Strzepek, S. Lele *et al.*, 1998: Predicting key malaria transmission factors, biting and entomological rates, using modelled soil moisture in Kenya. *Tropical Medicine Hygiene and International Health*, **3**, 818-827.

Petersen, G., G. de Leo, J. Hellmann, M. Janssen, A. Kinzig, J. Malcolm, K. O'Brien, S. Pope, D. Rothman, E. Sheviliakova, and R. Tinch, 1998: Uncertainty, climate change and adaptive management. *Conservation Ecology*, **1**, 4.

Pinheiro F., and P. Corber, 1997: Global situation of dengue and dengue haemorrhagic fever, and its emergence in the Americas. *World Health Statistics Quarterly*, **50**, 161–169.

Quick, R.E., L.V. Venzel, O. Gonzalez, E.D. Mintz, A.K. Highsmith, A. Espanda, N. Bean, E.H. Hannover, and R.V. Tauxe, 1996: Narrow-mouthed water storage vessels and in situ chlorination in a Bolivian community: a simple method to improve drinking water quality. *American Journal of Tropical Medicine and Hygiene*, **54**, 511-516.

Reiter, P., 1998: Global warming and vector–borne disease in temperate regions and at high altitude [letter]. *Lancet*, **351**, 839–840.

Robine, J-M., 1998: Measuring the burden of disease. *Lancet*, **352**, 757–758.

Sayer, B., 1998: Knowledge based technology in the service of health. *World Health Forum WHO*, **19**, 15–20.

Semiao Santos, S.J., P. Abranches, M.C. Silva Pereira, G.M. Santos Gomes, J.P. Fernandes, and J.C. Vetter, 1996: Reliability of serological methods for detection of leishmaniasis in Portuguese domestic and wild reservoirs. *Memorias do Instituto Oswaldo Cruz*, **91**, 747-750

Silva, A.R. *et al.*, 1997: Leishmaniose visceral (Calazar) na Ilha de Sao Luis, Maranhao, Brasil: evolucao e perspectivas. *Revista da Sociedade Brasileira de Medicina Tropical*, **30**, 359–368.

Smith, K.R., 1995: Health, Energy, and Greenhouse-Gas Impacts of Biomass Combustion. *Energy for Sustainable Development*, **1**, 23–29.

Smith, K.R. *et al.*, 1999: *Greenhouse Gases from Small-Scale Combustion Devices: Household Stoves in India and China*. Air Pollution Prevention and Control Division, USEPA, Research Triangle Park, NC. (forthcoming)

Snow, R.W., K. Marsh, and D. Le Sueur, 1996: The need for maps of transmission intensity to guide malaria control in Africa. *Parasitology Today*, **12**, 455–456.

Stanwell-Smith, R., 1998: European infection surveillance prepares for climate change. *EuroSurveillance Weekly*, Nr. 23.

Tadei, W.P., 1993: Incidencia, distributio e aspectos ecologicos de especies de Anopheles (Diptera: Culicidade) em regios naturais e sob impacto ambiental da Amazonia Brasileira. *In Bases Empiricas para Estrategias de Desenvolvimento e Preservatao da Amazonia, Vol 2*. F.J.G. Ferreira *et al.*, (eds.), INPA, Manaus, pp. 167–196.

Tsai, T.F., and C.J. Mitchell, 1989: St Louis encephalitis. In *The arboviruses: epidemiology and ecology*, Vol.4. T.P. Monath, (ed.), CRC Press, Boca Raton, pp. 113-43.

UN, 1948: *Universal Declaration of Human Rights*, Resolution 217 A (III). United Nations General Assembly, Geneve.

UN, 1996: *The Beijing Declaration and the platform for action. Fourth World Conference on Women, Beijing, China, 4-15 September 1995*. UN Department of Public Information, New York.

UN, 1997: Programme for the Further Implementation of Agenda 21. United Nations General Assembly, A/RES/S–19/2 19. Geneva.

UNDP, 1996: *Human Development Report 1996*. Oxford University Press, Oxford.

UNEP, 1999: *Global Environment Outlook 2000*. United Nations Environment Programme, Geneva.

Venczel, V.L., M. Arrowood, M. Hurd, and M.D. Sobsey, 1997: Inactivation of *Cryptosporidium parvum oocyst* and *Clostridum perfrigens spores by a mixed-oxidant disinfectant and by free chlorine. Applied Environmental Microbiology*, **63**, 1598-1601.

Vogel, C. 1998: Vulnerability and global environmental change. *LUCC Newsletter*, **3**, 15-19.

Waddington E.G., J. Catriona, and K.A. Enmayew,1989: A price to pay: the impact of user charges in Ashanti-Akim District, Ghana. *International Journal of Health Planning and Management*, **4**, 17–47.

Walt, G., 1994: *Health Policy: An Introduction to Process and Power*. Zed Books. London.

Walt, G., 1998: The globalisation of international health. *Lancet*, **351**, 434–438.

Wang, X., and K.R. Smith, 1999a: *Near-term health benefits of greenhouse gas reductions: a proposed assessment method and application in two energy sectors in China*, WHO/SDE/PHE/99.1,) WHO, Geneva,

Wang, X., and K.R. Smith, 1999b: Secondary benefits of greenhouse gas control: health impacts in China. *Environmental Science and Technology*, in press.

WHO, 1992: *Our Planet, Our Health*. Report of the WHO Commission on Health and the Environment (WCHE). World Health Organization, Geneva.

WHO, 1995: *World Health Report 1995: Bridging the Gaps*. World Health Organization, Geneva.

WHO, 1997a: *Division of Control of Tropical Diseases (CTD) Progress Report 1997*, CTD/PR/x, World Health Organization, Geneva.

WHO, 1997b: *Health and Environment in Sustainable Development*. WHO, Geneva.

WHO, 1998: WHO *Committee Identifies Constraints that Stall Progress against TB*. Statement by the Ad Hoc Committee on the Global TB Epidemic. World Health Organization, Geneva.

WHO/MRC/UNEP, 1998: *First InterAgency Workshop on Monitoring the Health Impacts of Climate Change*. Medical Research Council, London.

Willocks L., A. Crampin, L. Milne, C. Seng, M. Susman, R. Gair, M. Moulsdale, S. Shafi, R. Wall, R. Wiggins, and N. Lightfoot, 1998: A large outbreak of cryptosporidiosis with a public water supply from a deep chalk borehole. Outbreak Investigation team. *Commun. Dis. Public Health*, **1**, 239-243.

Woodward, A., S. Hales, and P. Weinstein, 1998: Climate change and human health in the Asia Pacific region: who will be the most vulnerable? *Climate Research*, **11**, 31-38.

Working Group on Public Health and Fossil-Fuel Combustion, 1997: Short-term improvements in public health from global-climate policies on fossil-fuel combustion: an interim report. *Lancet*, **350**, 1341–1349.

World Bank, 1993: *World Development Report 1993: Investing in Health*. Oxford University Press, New York.

WRI (World Resources Institute), 1997: *World Resources 1997–8: The Urban Environment*. Joint Publication: World Resources Institute/UNEP/UNDP/World Bank/Oxford University Press, New York.

WRI, 1998: *World Resources 1998–9: Environmental Change and Human Health*. Joint Publication: World Resources Institute/UNEP/UNDP/World Bank/ Oxford University Press, New York.

15

Coastal Adaptation

Coordinating Lead Author:
RICHARD J.T. KLEIN (THE NETHERLANDS / GERMANY)

Lead Authors:
Earle N. Buckley (USA), Robert J. Nicholls (United Kingdom), Sachooda Ragoonaden (Mauritius), James Aston (Western Samoa), Michele Capobianco (Italy), Norimi Mizutani (Japan), Patrick D. Nunn (Fiji)

Contributing Authors:
Darius J. Bartlett (Ireland), James G. Boyd (USA), Eugene L. Lecomte (USA), Xenia I. Loizidou (Cyprus), Claudio R. Volonté (Uruguay)

Review Editor:
Isabelle Niang-Diop (Senegal)

CONTENTS

EXECUTIVE SUMMARY

This chapter assesses the current state of knowledge on technology transfer for coastal adaptation to climate change. It aims to inform coastal managers, planners, scientists and other interested actors of the process of technology transfer in coastal zones and of its importance when seeking to reduce coastal vulnerability to climate change. It identifies barriers to technology transfer as well as enabling policies, programmes and measures to overcome these barriers.

The five key messages of this chapter are as follows:

- **Adaptation to climate change in coastal zones is becoming increasingly important and many proven technologies are available for coastal adaptation.** Existing coastal technologies that have been used to deal with climate variability in coastal zones can also be applied to adapt to climate change. A range of opportunities exists for the application of both hard and soft technologies to complement economic, legal and institutional options. Available and effective technologies include traditional, indigenous, non-western technologies. They also include technologies to develop and exchange knowledge and information. Given that many technologies are available, extra efforts in technology transfer should focus on promoting and adjusting existing technologies, rather than on the development of new technologies.

- **Effective adaptation to climate change needs to consider the numerous non-climate stresses in coastal zones and be consistent with existing policy criteria and development objectives.** Adaptation in coastal zones must strike a balance between current pressures resulting from climate variability and unsustainable development, and anticipated impacts of climate change and associated sea-level rise. Adaptation technologies are best implemented as part of a broader, integrated coastal-management framework that recognises immediate and longer-term sectoral needs. Win-win situations could be established when coastal-adaptation technologies also provide benefits unrelated to climate change.

- **Successful coastal adaptation depends on many local factors and cannot be simply transferred to other vulnerable areas.** The purpose of technology application in coastal zones is to reduce risks and to increase adaptive capacity. The effectiveness of a particular technology depends on local circumstances, including the biophysical setting and economic, institutional, legal and socio-cultural conditions. Protection is not a feasible option for all coastlines. In many, particularly lesser developed places, retreat and accommodate strategies will be most effective. Local expertise is essential to identify and design appropriate coastal-adaptation technologies, as well as to implement, operate and maintain these.

- **Pathways of technology transfer in coastal zones are predominantly driven by government interests.** The strongest and most direct incentives to adapt to climate change in coastal zones are with the public sector, although particularly tourism and marine transportation represent important private-sector interests. The private sector is typically not the stakeholder that drives technology transfer for coastal adaptation, because benefits are small or uncertain, and action is expected from the government to protect private-sector interests. In developing countries, the private sector is generally a less significant economic force, so again governments are expected to lead the way in coastal adaptation. In government-initiated coastal technology transfer, the private sector is often involved in the planning, design and implementation of adaptation technologies.

- **Technology transfer in coastal zones faces a number of important barriers.** A refocus of institutions and funding priorities may be required to overcome these barriers. Many barriers to effective technology transfer are site-specific and require site-specific solutions. Four major general barriers exist: *(i)* lack of data, information and knowledge to identify adaptation needs and appropriate technologies, *(ii)* lack of local capacity and consequent dependence of customers on suppliers of technology for operation, maintenance and duplication, *(iii)* disconnected organisational and institutional relationships between relevant actors and *(iv)* access to financial means. Overcoming these barriers does not require setting up new institutions. Instead, existing activities and institutions need to be refocused to improve the efficiency and effectiveness of coastal technology transfer. Additionally, regional collaboration and a redirection of funds to support appropriate coastal adaptation to climate change are required.

15.1 Introduction

A variety of coastal systems produce a large number of goods and services that are valuable to society. This has attracted many people and major investments to coastal zones, even to places that are susceptible to hazards such as storm surges and coastal erosion. In many places, technology has been instrumental in reducing society's vulnerability to coastal hazards, in three basic ways:

- *Protect* — reduce the risk of the event by decreasing its probability of occurrence;
- *Retreat* — reduce the risk of the event by limiting its potential effects;
- *Accommodate* — increase society's ability to cope with the effects of the event.

Extensive research has shown that climate change will increase the hazard potential for many coastal zones. The same three strategies of protect, retreat and accommodate can be followed to reduce vulnerability to climate change, including application of the same range of hard and soft technologies as are used today. However, many of the world's vulnerable coastal countries currently do not have access to appropriate adaptation technologies, nor to the knowledge or resources that are required to develop or implement these. Effective coastal adaptation by these countries could benefit from increasing current efforts of technology transfer.

This chapter uses the term "technology" in its broadest possible sense, including knowledge. A technology is considered appropriate when it is environmentally sound, economically viable and culturally and socially acceptable, amongst other criteria. The extent to which a technology meets these criteria differs from one location to another, suggesting that the appropriateness of a technology is not universal but—at least in part—determined by local factors. Technology transfer, as explained in Chapter 1, is the broad set of processes covering the exchange of knowledge, money and goods amongst different stakeholders that lead to the spreading of technology for adapting to or mitigating climate change.

Coastal adaptation to climate change must be seen as part of a broader coastal-management policy, which includes consideration of numerous non-climatic issues (Harvey *et al.*, 1999). It typically follows a continuous and iterative cycle involving four main steps: *(i)* information development and awareness raising, *(ii)* planning and design, *(iii)* implementation and *(iv)* monitoring and evaluation (Klein *et al.*, 1999). To date, technology transfer for coastal adaptation has focused primarily on the implementation stage: the actual hardware that can be employed to protect or, to a lesser extent, retreat or accommodate. As argued by Klein *et al.* (1999), coastal adaptation should also aim at increasing the extent to which mechanisms are in place and technologies, expertise and other resources are available to assist the other three steps.

Section 15.3 therefore identifies available technologies for each of the four steps of adaptation, with Section 15.4 assessing cur-

rent and future trends in the transfer of these technologies. Sections 15.5 and 15.6 analyses important barriers and opportunities for coastal-technology transfer in light of economic, institutional, legal and socio-cultural aspects, within and between countries respectively. Section 15.7 summarises the lessons learnt and suggests future actions. First, however, Section 15.2 gives an overview of anticipated coastal impacts of climate change and concurrent non-climate stresses affecting coastal-adaptation needs.

15.2 Current and Future Adaptation Needs in Coastal Zones

In the 21st century and beyond, climate change can have important impacts on coastal zones. The magnitude of these impacts and hence the need for adaptation and technology transfer will depend on the magnitude of climate change and the interaction of climate change with other stresses. This section discusses both factors.

15.2.1 Coastal Vulnerability to Climate Change

Sea-level rise and changes in the frequency and intensity of extreme events, such as cyclones, storm surges and high river flows, can cause:

- Inundation and displacement of wetlands and lowlands;
- Erosion and degradation of shorelines and coral reefs;
- Increased coastal flooding;
- Salinisation of estuaries and freshwater aquifers.

Bijlsma *et al.* (1996) concluded that most coastal areas are vulnerable to such impacts to some degree and some adaptation will be necessary. However, certain settings are more vulnerable than other ones. Deltaic areas, small islands—especially coral atolls—and coastal wetlands appear particularly vulnerable to climate change. In addition, developed sandy shores could be vulnerable because of the large investment and significant sand resources required to maintain beaches and adjoining infrastructure. Taking a regional perspective, IPCC (1998) concluded that the threat of increased coastal flooding is most severe for South and Southeast Asia, Africa, the southern Mediterranean coasts, the Caribbean and most islands in the Indian and Pacific Oceans.

15.2.2 Concurrent Stresses in Coastal Zones

Bijlsma *et al.* (1996) noted that climate-related change in coastal zones "represents potential additional stresses on systems that are already under intense and growing pressure". Climate change is one factor amongst many that affect coastal ecological systems and societies. Other factors that interact with climate change include overexploitation of resources, pollution, increasing nutrient fluxes, decreasing fresh-water availability, sediment starvation and urbanisation (Goldberg, 1994; Viles and Spencer, 1995).

These non-climate stresses decrease the resilience of coastal systems to cope with natural climate variability and anticipated climate change (Nicholls and Branson, 1998; Klein and Nicholls, 1999). Bijlsma *et al.* (1996) concluded that "although the potential impacts of climate change by itself may not always be the largest threat to natural coastal systems, in conjunction with other stresses they can become a serious issue for coastal societies, particularly in those places where the resilience of natural coastal systems has been reduced."

Policies and practices that are unrelated to climate but which do increase a system's vulnerability to climate change are termed "maladaptation" (Burton, 1996, 1997). Examples of maladaptation in coastal zones include investments in hazardous zones, inappropriate coastal-defence schemes, sand or coral mining and coastal-habitat conversions. A common cause of maladaptation is a lack of information on the potential external effects of proposed developments on other sectors, or a lack of consideration thereof. More proactive and integrated planning and management of coastal zones is widely suggested as an effective mechanism for strengthening sustainable development (*e.g.*, Cicin-Sain, 1993; Ehler *et al.*, 1997; Cicin-Sain and Knecht, 1998) and can be both environmentally sound and economically efficient (Tol *et al.*, 1996).

To identify the most appropriate coastal-adaptation strategy, one must consider the full context in which impacts of climate change arise, and realise that the three earlier-mentioned strategies—protect, retreat, accommodate—happen within a broader policy process. Within this process, increasing resilience by reversing maladaptive trends could be an important option to reduce coastal vulnerability to climate variability and change. This approach will often address more than climate issues alone and generally involve a change in adaptation strategy, for example, nourishing beaches instead of constructing seawalls, or introducing a building setback instead of allowing construction next to the coast.

15.3   Available Technologies for Coastal Adaptation

As stated in Section 15.1, the process of coastal adaptation to climate change comprises four steps. This section briefly describes these steps and provides important examples of technologies that can be employed to accomplish them.

15.3.1   *Information Development and Awareness Raising*

Data collection and information development are essential prerequisites for coastal adaptation. The more relevant, accurate and up-to-date the data and information available, the more targeted and effective adaptation can be. Coastal adaptation requires data and information on coastal characteristics and dynamics, patterns of human behaviour, as well as an understanding of the potential consequences of climate change. It is also essential that there is a general awareness amongst the public and coastal planners and managers of these consequences and of the possi-

ble need to act. In countries where the central government has neither the means nor the expertise to address problems in every part of the coast, the information is used most effectively when targeted at the most influential people in the community.

Large-scale global and regional databases exist for a great number of climatic and socio-economic variables relevant to coastal zones. These databases may be accessed and downloaded from the Internet. However, coastal adaptation to climate change cannot rely exclusively on these readily available databases. Table 15.1 lists a number of relevant technologies that can serve to increase the understanding of the coastal system (which involves data collection and analysis), to conduct climate-impact assessment in coastal zones and to raise public awareness. Where appropriate, reference is made to publications that either describe the technology in detail or provide examples of its application. Further information on a broad range of technologies for coastal-system description can be found in Morang *et al.* (1997a), Larson *et al.* (1997), Morang *et al.* (1997b) and Gorman *et al.* (1998). Bush *et al.* (1999) described the use of geoindicators for rapid coastal risk assessment, while Capobianco (1999) discussed technologies in relation to integrated coastal zone management.

15.3.2   *Planning and Design*

When the available data and information point towards a potential problem justifying action, the next stage is to decide *which action* to take and *where* and *when* to take it. The answers to these questions depend on prevailing criteria that guide local, national or regional policy preparation, as well as on existing coastal-development and management plans that form the broader context for any adaptation initiative. Important policy criteria that could influence adaptation decisions include cost-effectiveness, environmental sustainability, cultural compatibility and social acceptability. In addition, countries may choose to take a precautionary approach when postponing action would involve substantial risks, even though uncertainty may still be considerable (CEC, 1999).

Coastal planners will always face a certain degree of uncertainty, not only because the future is by definition uncertain, but also because knowledge of natural and socio-economic coastal processes is and always will remain incomplete. Limits to predictability require planners to assess the environmental and societal risks of climate change with and without adaptation (Carter *et al.*, 1994). The information thus obtained can help to determine the optimal adaptation strategy *(which action?)* and timing of implementation *(when?)* (*e.g.*, Chao and Hobbs, 1997; Yohe and Neumann, 1997). There are a number of decision tools available to assist in this process. Examples of these tools include cost-benefit analysis, cost-effectiveness analysis, risk-effectiveness analysis and multi-criteria analysis (Turner and Adger, 1996). The latter technique is particularly relevant when great significance is attached to values that cannot be easily expressed in monetary terms.

Table 15.1 Examples (*i.e.*, not an exhaustive list) of important technologies to collect data, provide information and increase awareness for coastal adaptation to climate change

APPLICATION	TECHNOLOGY	ADDITIONAL INFORMATION
COASTAL-SYSTEM DESCRIPTION		
• Coastal topography and bathymetry	• Mapping and surveying	• Birkemeier *et al.* (1985)
	• Videography	• Debusschere *et al.* (1991); Holman *et al.* (1994); Plant and Holman (1997)
	• Airborne laser-scanning (lidar)	• Lillycrop and Estep (1995); Sallenger *et al.* (1999)
	• Satellite remote sensing	• Leu *et al.* (1999)
• Wind and wave regime	• Waverider buoys	• Morang *et al.* (1997a)
	• Satellite remote sensing	• Martinez-Diaz-De-Leon *et al.* (1999)
• Tidal and surge regime	• Tide gauges	• Pugh (1987)
• Relative sea level	• Tide gauges	• Emery and Aubrey (1991); Woodworth (1991); Gröger and Plag (1993); NOAA (1998)
	• Historical or geological methods	• Van de Plassche (1986)
• Absolute sea level	• Satellite remote sensing	• Nerem (1995); Fu *et al.* (1996); Nerem et al. (1997); Cazenave *et al.* (1998)
	• Tide gauges, satellite altimetry and global positioning systems	• Baker (1993); Miller *et al.* (1993); Zerbini *et al.* (1996); Neilan *et al.* (1997)
• Land use	• Airborne and satellite remote sensing	• Redfern and Williams (1996); Clark *et al.* (1997); Henderson *et al.* (1999)
• Natural values	• Resource surveys	• Lipton and Wellman (1995); Turner and Adger (1996)
• Socio-economic aspects	• Mapping and surveying	
• Legal and institutional arrangements	• Interviews, questionnaires	
• Socio-cultural factors	• Interviews, questionnaires	
CLIMATE-IMPACT ASSESSMENT		
• Index-based methods	• Coastal vulnerability index	• Hughes and Brundrit (1992); Gornitz *et al.* (1994); Shaw *et al.* (1998)
	• Sustainable capacity index	• Kay and Hay (1993); Yamada *et al.* (1995); Nunn *et al.* (1994a,b)
• (Semi-) quantitative methods	• IPCC common methodology	• IPCC CZMS (1992); Bijlsma *et al.* (1996)
	• Aerial-videotape assisted vulnerability assessment	• Leatherman *et al.* (1995); Nicholls and Leatherman (1995)
	• UNEP impact and adaptation assessment	• Klein and Nicholls (1998, 1999)
• Integrated assessment	• Coupled models	• Engelen *et al.* (1993); Ruth and Pieper (1994); West and Dowlatabadi (1999)
AWARENESS RAISING		
• Printed information	• Brochures, leaflets, newsletters	
• Audio-visual media	• Newspapers, radio, television, cinema	
• Interactive tools	• Board-games	
	• Internet, worldwide web	
	• Computerised simulation models	

Geographical information systems (GIS) are an important technology for spatial planning *(where?)*, while they can also contribute to the other adaptation steps described in this section and indeed to all aspects of coastal management (Box 15.1).

The modelling of potential futures based on plausible scenarios is particularly pertinent for the planning and design of adaptation technologies when relevant impacts are quantified, alternative adaptation options are evaluated and one course of action is selected. Modelling capabilities are increasing rapidly, leading, for example, to better morphodynamic models (De Vriend *et al.*, 1993; Stive and De Vriend, 1995), dynamic biogeophysical models (*e.g.*, Capobianco *et al.*, 1999) and integrated models (*e.g.*, Engelen *et al.*, 1993). The rapid developments in information technology are facilitating the transfer of these tools. However, the limitations inherent in all models (*i.e.*, they are representations of a part of reality for a specific purpose) must not be overlooked. Human expertise remains essential for the intelligent use of models.

The quality and effectiveness of the planning and design process is affected by the context in which the decision is made. Coastal management in many countries used to be top-down by nature, but as public interest and involvement in coastal issues has grown, so has resistance to top-down decision-making (*e.g.*, Taiepa *et al.*, 1997). The successful implementation of many coastal policies, including adaptation to climate change, is now increasingly dependent on public acceptance at the community level (King, 1999). Hence, in addition to informing the public to raise their awareness of the issues at stake (Section 15.3.1), it is also important to involve them throughout the planning process to inform decision makers (CEC, 1999). Gaining public acceptance, for example by two-way interaction and partnerships, is an important prerequisite for finding and transferring appropriate adaptation technologies. Furthermore, local expertise will be required for technology implementation, application, maintenance and enforcement.

In some settings, however, public involvement can be difficult to accomplish. In situations where there is little truly private land, coastal inhabitants may have little long-term stake and therefore interest in the land they occupy (*e.g.*, in parts of Tonga; Nunn and Waddell, 1992). Moreover, governments may have neither the resources to address country-wide coastal management (particularly in archipelagic nations) nor, compared to long-resident inhabitants, the local knowledge or experience that are essential for effective management (*e.g.*, in parts of Fiji; Nunn *et al.*, 1994a).

15.3.3 Implementation

Once all options for coastal adaptation have been considered and the optimal strategy has been selected and designed, implementation is the next stage. As stated before, an adaptation strategy to sea-level rise can comprise one or more options that fall under the three broad categories protect, retreat and accommodate. Table 15.2 provides an overview of these options and technologies. In addition to the subdivision between protect, retreat and accommodate, there are various other ways to classify or distinguish between adaptation strategies, both in generic terms (*e.g.*, Smit, 1993; Burton, 1997; Klein and Tol, 1997; Smit *et al.*, 2000) and for coastal zones (*e.g.*, Kay *et al.*, 1996; Pope, 1997). To date, the assessment of possible response strategies has

| Box 15.1 | THE ROLE OF GEOGRAPHICAL INFORMATION SYSTEMS (GIS) IN COASTAL ADAPTATION AND MANAGEMENT |

GIS combines computer mapping and visualisation techniques with spatial databases and statistical, modelling and analytical tools. It offers powerful methods to collect, manage, retrieve, integrate, manipulate, combine, visualise and analyse spatial data and to derive information from these data (Burrough and McDonnell, 1998; Longley *et al.*, 1999; Wright and Bartlett, 1999). One simple, first-order application of GIS in coastal adaptation would be overlaying scenarios of sea-level rise with elevation and coastal-development data to define impact zones. More sophisticated applications may include morphodynamic modelling (*e.g.*, Capobianco *et al.*, 1999). GIS technology is evolving rapidly and is increasingly used for sophisticated modelling. Hence, GIS can provide excellent support to coastal managers for making decisions about adaptation.

GIS can contribute to each of the four adaptation steps. Collected data can be stored in a GIS, combined to develop new insights and information, and visualised for interpretation and educational purposes. In combination with scenarios of relevant developments and models to assess and evaluate changes in important natural and socio-

economic variables, GIS can assist planners to identify appropriate adaptation technologies as well as their optimal locations for implementation. It allows for the non-invasive, reversible and refinable testing of specific adaptation technologies before these are implemented in the real world. After implementation, newly acquired data can be analysed to evaluate technology performance. Once created, a GIS database will have further utility in other aspects of coastal management (Jones, 1995; O'Regan, 1996).

In spite of its clear utility, GIS cannot substitute for fieldwork or common sense (Crawford, 1993). It will never eclipse the importance of economic, institutional, legal and socio-cultural factors in coastal management. In addition, true three-dimensional modelling in GIS (*e.g.*, for sediment budgets) remains problematic. Finally, some commentators have questioned whether GIS can always be used effectively in developing countries. Specific issues in this regard include:

- The costs of computer hardware and most GIS software;

- The lack of raw data to input to the system;
- The lack of consistency between data sets;
- Restrictions on free access to information for strategic, political, economic or other reasons;
- Limited salaries and career opportunities for GIS-literate operators compared to the industrialised world;
- The prevailing Western conceptual model of geographical space, which may be different from local ways of perceiving and interpreting spatial relationships;
- The fear that the introduction of GIS could lead to or facilitate oppressive government, misuse of power, civil unrest or other non-democratic activities.

The rapid ongoing developments of all aspects of GIS may remove some of these concerns. There is no doubt that GIS presents great potential for societies wishing to anticipate and understand the consequences of climate change and develop adaptation strategies to cope with the potential impacts.

focused mainly on protection. Bijlsma *et al.* (1996) noted the need to identify and evaluate the full range of options listed in Table 15.2. The range of appropriate options will vary amongst and within countries, and different socio-economic sectors may prefer conflicting adaptation options for the same area. This is one of the reasons why adaptation to climate change would best take place within the framework of integrated coastal zone management (Section 15.2.2). Another important issue is that successful adaptation involves more than just technological options. Technological options can only be implemented effectively in an appropriate economic, institutional, legal and socio-cultural context (Klein and Tol, 1997).

15.3.4 Monitoring and Evaluation

It is common practice in any field of policy that the performance of implemented measures is periodically or continuously evaluated in terms of the original objectives. Such evaluation can yield new insights and information, which could give rise to adjusting the strategy as appropriate (NRC, 1995). This post-implementation evaluation must be distinguished from the evaluation exercise that is done to identify the most appropriate technology. The lat-

ter can be considered pre-implementation evaluation and is part of the planning and design phase (Section 15.3.2).

Effective evaluation requires a reliable set of data or indicators, to be collected at some regular interval using an appropriate monitoring system. Evaluation will often be necessary for decades and the monitoring should be planned accordingly. There is limited experience of such long-term monitoring, so in many situations it is unclear which are the most appropriate data or indicators (Basher, 1999). For physical systems, experience can be drawn from countries where the coast has been monitored for long periods. In The Netherlands, for example, the position of high water has been collected annually for nearly a century and cross-shore profiles have been measured annually since 1963 (Verhagen, 1989; Wijnberg and Terwindt, 1995; Hinton *et al.*, 1999). Observations of the "natural" evolution of the coast allow trends to be reliably estimated and hence the impact of human interventions on the coast (breakwaters, nourishment, etc.) to be evaluated.

In general, monitoring technologies are the same as those used for initial description of the coastal system. They are listed in the upper part of Table 15.1 and discussed by Morang *et al.* (1997a), Larson *et al.* (1997), Morang *et al.* (1997b) and Gorman *et al.*

(1998). After a number of years, a new climate-impact assessment, using updated climatic and socio-economic scenarios and including consideration of adaptive capacity and non-climate stresses, could serve to see if overall coastal vulnerability to climate variability and change has been successfully reduced.

15.4 Current and Future Trends in Coastal-Adaptation Technology Transfer

Chapter 1 identified three fundamentally different pathways of technology transfer. These pathways are based on the stakeholder as the primary driving force of the technology transfer: the government, the private sector or the community. Most of the literature describing and analysing climate-relevant technology transfer deals with mitigation technologies. In technology transfer for mitigation, the private sector usually plays a crucial part throughout the entire pathway from development to diffusion. Market opportunities, investment procedures and profitability criteria are key words to discuss the incentives and behaviour of both the mitigation-technology provider and recipient.

Most coastal impacts of climate change will impinge on collective goods and systems, such as food and water security, biodiversity and human health and safety. These impacts could affect commercial interests indirectly, but usually the strongest and most direct incentives to adapt are with the public sector. Coastal management is therefore usually a public-sector responsibility, and the planning and design of coastal adaptation to climate change needs to be tuned to existing policy criteria and development objectives (Klein *et al.*, 1999). Only in cases where a particular stretch of coastline provides direct financial benefits is the private sector likely to invest in coastal management. Prime examples of this case are coastal tourist resorts, for which beach erosion represents a direct threat to their profitability, and ports and harbours, which will have to raise their infrastructure as sea level rises. In most industrialised countries, however, the private sector is typically not the stakeholder that drives technology transfer for coastal adaptation, because benefits are small or uncertain and action is expected from the government to protect private-sector interests. In developing countries, the private sector is generally a less significant economic force so again governments are expected to lead the way in climate-change adaptation strategies.

Coastal-adaptation technology transfer is therefore predominantly government-driven (or donor-driven). Case Studies 16, 20 and 21 in Chapter 16 provide illustrations of this. Community-driven pathways may be found in places where a local need for adaptation is recognised but no government or private-sector interest is anticipated. Case Study 30 presents an example of collective action and joint management to combat erosion in Tuvalu. On Viti Levu (Fiji), a traditional village community has been actively involved in a mangrove rehabilitation project. This donor-funded project has been strongly driven by local concerns, taking into account the particular cultural and political settings (Nunn, 1999).

Table 15.2 Examples (*i.e.*, not an exhaustive list) of important technologies to protect against, retreat from or accommodate sea-level rise and other coastal impacts of climate change (see also NRC, 1987; IPCC CZMS, 1990; Bijlsma *et al.*, 1996)

APPLICATION	TECHNOLOGY	ADDITIONAL INFORMATION
PROTECT		
• Hard structural options	• Dikes, levees, floodwalls	• Pilarczyk (1990); Silvester and Hsu (1993)
	• Seawalls, revetments, bulkheads	
	• Groynes	
	• Detached breakwaters	
	• Floodgates and tidal barrier	• Gilbert and Horner (1984); Penning-Rowsell *et al.* (1998)
	• Saltwater-intrusion barriers	• Sorensen *et al.* (1984)
• Soft structural options	• Periodic beach nourishment	• Delft Hydraulics and Rijkswaterstaat (1987); Davison *et al.* (1992); Stauble and Kraus (1993)
	• Dune restoration and creation	• Vellinga (1986); Hallermeier and Rhodes (1988)
	• Wetland restoration and creation	• NRC (1992; 1994); Boesch *et al.* (1994); Tri *et al.* (1998)
• Indigenous options	• Afforestation	• McLean *et al.* (1998); Mimura and Nunn (1998)
	• Coconut-leaf walls	
	• Coconut-fibre stone units	
	• Wooden walls	
	• Stone walls	
(MANAGED) RETREAT		
• Increasing or establishing set-back zones	• Limited technology required	• NRC (1990); Kay (1990); Caton and Eliot (1993); OTA (1993)
• Relocating threatened buildings	• Various technologies	• Rogers (1993)
• Phased-out or no development in susceptible areas	• Limited technology required	• OTA (1993)
• Presumed mobility, rolling easements	• Limited technology required	• Titus (1991, 1998)
• Managed realignment	• Various technologies depending on location	• Burd (1995); English Nature (1997); French (1997; 1999)
• Creating upland buffers	• Limited technology required	• Kaly and Jones (1998)
ACCOMODATE		
• Emergency planning	• Early-warning systems	• Burkhart (1991); Haque (1995, 1997); Rosenthal and 't Hart (1998)
	• Evacuation systems	• Parker and Handmer (1997); Rosenthal and 't Hart (1998)
• Hazard insurance	• Limited technology required	• Davison (1993); OTA (1993); Crichton and Mounsy (1997); Clark (1998)
• Modification of land use and agricultural practice	• Various technologies (e.g., aquaculture, saline-resistant crops), depending on location and purpose	
• Modification of building styles and codes	• Various technologies	• FEMA (1986, 1994, 1997)
• Strict regulation of hazard zones	• Limited technology required	• May *et al.* (1996)
• Improved drainage	• Increased diameter of pipes	• Titus *et al.* (1987)
	• Increased pump capacity	• Titus *et al.* (1987)
• Desalination	• Desalination plants	• Ribeiro (1996)

Many of the technologies listed in Section 15.3 have been applied to adapt to the effects of climate variability in coastal zones. The emphasis has traditionally been on protecting developed areas using hard structures. The need for technology transfer to plan, design and build these structures depends on their required scale and level of sophistication. At a small scale, local communities can use readily available materials to build protective structures (Mimura and Nunn, 1998). However, these communities often lack the information to know whether or not these structures are appropriate and whether or not their design standards are acceptable. For larger-scale, more sophisticated structures, technical advice is required, as well as a contracting company to build the structure. Developing countries may receive bilateral or multilateral funding to meet some or all costs involved.

Until recently, it was rarely questioned whether a country's entire coastline could be protected effectively if optimal management conditions prevail. It has become clear, however, that even with massive amounts of external funding, coastlines in the developing world (particularly of archipelagic countries) cannot be effectively protected by hard structures. In addition, increasing awareness of unwanted effects of hard structures on erosion and sedimentation patterns has led to growing recognition of the benefits of "soft" protection (*e.g.*, beach nourishment, wetland restoration and creation) and of the adaptation strategies retreat and accommodate (Capobianco and Stive, 1997). An increasing number of private companies are now discovering market opportunities for implementing soft-protection options. Interest in the retreat and accommodate strategies is also growing, but markets for these are as yet less developed. In spite of this trend to consider adaptation options other than hard protection, many structures are still being built without a full evaluation of the alternatives.

A second trend in coastal adaptation is an increasing reliance on technologies to develop and manage information (Wright and Bartlett, 1999). This trend stems from the recognition that designing an appropriate technology to protect, retreat or accommodate requires a considerable amount of data on a range of coastal parameters, as well as a good understanding of the uncertainties involved in the impacts to be addressed (Capobianco, 1999). National, regional and global monitoring networks are being set up to help to assess adaptation needs and opportunities. In the Caribbean, for example, developing information has been presented as the first phase of a regional adaptation process and as such has been found eligible for funding from the Global Environment Facility (GEF) (see Case Study 20).

Thirdly, many efforts are now initiated to enhance awareness of the need for appropriate coastal adaptation, often as maladaptive practices are becoming apparent. For example, before a new hospital was built in Kiribati in 1992, a substantial site-selection document had been prepared, examining numerous aspects of three alternative sites but without consideration of coastal processes. A serious shoreline erosion problem advancing rapidly to within eight metres of the hospital was discovered by 1995 (Forbes and Hosoi, 1995). Efforts to enhance awareness include national and international workshops and conferences, training programmes, on-line courses and technical assistance and capacity building as part of bilateral or multilateral projects. In view of the many sectoral interests in coastal zones it will become increasingly important to involve decision-makers

without direct responsibility for coastal issues and other stakeholders in this ongoing learning process (Humphrey and Burbridge, 1999; King, 1999).

15.5   Transfer of Coastal-Adaptation Technologies within Countries

The predominant nature and goal of coastal adaptation require a form of in-country technology transfer that differs from current explanations of transfer and the conditions under which it takes place (*cf.* BCSD, 1992). Common definitions of technology transfer describe the process as a company-to-company transaction, but coastal-adaptation technologies—with few exceptions—are not developed and owned by business and industry. Economic considerations are a major force in driving technology transfer for coastal adaptation, but objectives are less focused on commercial terms. Rather, considerations of public well being are essential, such as the reduction of loss of property and lives and the protection of essential coastal habitats. Additionally, many coastal-technology transactions involve the exchange of information, knowledge and even wisdom. Knowledge transactions have characteristics that make them quite different from trade transactions (STAN, 1998). Incentives, pathways, the role of stakeholders and barriers reflect these distinctions.

15.5.1   Incentives for In-Country Technology Transfer

The use and transfer of coastal-adaptation technologies worldwide has occurred mostly because of societal interventions and not in response to market forces. Examples of such interventions include direct governmental expenditures, regulations and policies and public choices. Each stakeholder—governments, universities and government-sponsored laboratories, the private sector, non-governmental organisations and local communities and landowners—has its specific interests, directions and objectives. Incentives for technology transfer vary according to each of these agendas, independent of the technology considered (Kuhn, 1990).

National governments are responsible for national security in the broadest sense, embedding issues of international economic competitiveness and national prosperity, national defence, environmental stewardship and improving the standard of living. Public-policy formulation and implementation of coastal adaptation are motivated principally by the need to reduce public costs associated with loss of property and life and preserving common goods. Likewise, economic considerations are a driving force for research and development (R&D) on coastal-adaptation technologies. Excluding information technologies such as GIS, most development and innovation of coastal-adaptation technologies occur in R&D programmes sponsored by national governments, either directly by government-owned or -controlled laboratories or indirectly by universities, R&D consortia and international research programmes (Capobianco, 1999). The private sector then refines and applies these technologies, usually in a government-sponsored project. Only rarely have government institutions themselves pursued opportunities for commercial exploitation of research results that lead to the recovery of royalties (Case Study 21).

Traditionally, the motivation for university-based research was scholarly reputation; user need and involvement and technology transfer was not considered important. Value was based on authorship of new knowledge and peer recognition. However, a growing number of universities are increasing their focus on applied R&D and more extensive connections are established with potential users of tangible results. For some universities, it is the desire to increase support for academic research. For other institutions, such as sea-grant colleges in the United States, the impetus is an expanded concept of their mandate to serve the coastal community.

The private sector seeks to generate profits. The community of coastal-technology private-sector companies is predominantly made up of small companies providing engineering and consulting services, with a small number of global players (Funnell, 1998; UNFCCC, 1999). Few are actively pursuing technology transfer (Stockdale, 1996). For most technological developments, the market is too small or diffuse to justify private-sector investment. The sector may generate appropriate technologies, but then relies on government funding to see it applied, usually on a project-specific basis. Motivations for technology transfer include leveraging limited resources with public financing for R&D, reducing commercial risk and increasing competitive advantage and market share for products and services (Stockdale, 1996).

15.5.2 Pathways of In-Country Technology Transfer

Chapter 1 identified five stages within the process of technology transfer: *(i)* assessment, *(ii)* agreement, *(iii)* implementation, *(iv)* evaluation and adjustment and *(v)* replication. These stages should not be confused with the four steps outlined in Section 15.1. These latter four steps describe the process of coastal adaptation, while the above five stages define the process of technology transfer. R&D, aimed at technology development, innovation and acceptance, is a crucial stage before the actual technology transfer (*cf.* Gibson and Rogers, 1994).

As noted above, national governments, directly or indirectly, are the primary developer of coastal-adaptation technologies and governments at all levels are the principal users. Transfer mechanisms relevant to R&D include *(i)* sharing information via workshops, briefings and visits, *(ii)* publications in professional journals and presentations at scientific and technical conferences, and *(iii)* where appropriate, patents. At this stage, research quality and strength and scholarly reputation are most significant, while technology-transfer plans and processes are considered less important.

Technology *assessment* and *agreement* call for shared responsibility between technology developers and users. Success occurs when a technology is transferred across personal, functional or organisational boundaries and is accepted and understood by designated users. Implicit in these stages is the belief that successful technology transfer is simply a matter of getting the right information to the right people at the right time. Typical mechanisms include technical consulting, good-practice schemes and manuals, exchange programmes and various grants and cooperative agreements in which work is undertaken to benefit both parties.

The success of technology *implementation* is marked by the timely and efficient employment of the technology. Users should have the knowledge and resources to implement the technology. Following *evaluation* and *adjustment* of the technology, replication is directed at the fully integrated use of the technology by the user community and its further dissemination. Technology *replication* builds on the successes achieved in obtaining the objectives of the previous stages. Where hardware is concerned, such as structural technologies and monitoring equipment, successful replication may require an industrial provider. In these cases, market strength is required.

Moving from R&D to technology replication is not a linear process. As illustrated by the new breakwater technology CoreLoc (Case Study 21), complexity increased significantly as the transfer process moved from intra-agency technology, via the presentation of experimental findings at other government facilities, to implementation in partnership with private-sector entities. Box 15.2 illustrates the obstacles encountered between the development and application of an important concept for shoreline management in England and Wales, which eventually resulted in a diffusion period of 30 years. An important lesson from these two examples is that feedback from the technology users drives the transfer process.

15.5.3 The Role of Stakeholders

By means of their regulatory and decision-making apparatus, national governments can stimulate the implementation of coastal-adaptation technologies by providing incentives to lower levels of government, as demonstrated in the case of the application of littoral cell-based management in England and Wales (Box 15.2). Thus, a government's operational responsibilities for coastal adaptation make it a developer and customer of adaptation technologies at the same time. As such, national governments need to balance investment in the development of new technology with support for information infrastructure and education and training. For example, governments can create mechanisms to bring together the fragmented coastal-technology private sector and foster partnerships and alliances that improve access to government-owned innovations, leverage talent and capital and share risks. National governments of many small and/or developing countries have neither the financial nor the human resources to facilitate development of technologies for coastal adaptation. The dilemma posed consequently is that such countries must depend either on other countries to produce appropriate technologies or upon developed countries and international agencies to sponsor R&D.

In countries with several levels of government, governments below the national level play a principal part in managing coastal resources and development. It is at this level that most coastal-adaptation technologies are implemented. Subnational governments do not account for a very large share of total investment in R&D for coastal-adaptation technologies. However, with their business and academic partners, they have knowledge of local and regional conditions that national governments cannot equal (Carnegie Commission, 1992; Moser, 1998). Subnational governments are most effective at timely communicating new ideas and priorities to the public, and putting the public's needs forward in national science and technology forums.

The traditional paradigm for science and technology innovation is that government laboratories concentrate their R&D on mission-oriented projects, universities confine themselves to basic research, and the private sector concentrates on shorter-term, more profit-oriented developments (Kozmetsky, 1990). National governments have the primary responsibility for creating conditions for successful coastal-adaptation technology transfer (Capobianco, 1999). This role includes nurturing and optimising a synergetic and symbiotic interrelationship between stakeholders in industry, academia and government (Kuhn, 1990). Accessible technologies, high-quality human resources, adequate physical infrastructure and a favourable public-policy environment are especially important.

The function of universities in the national research system is the creation of public knowledge. The primary role of academic research in coastal adaptation is the development and testing of generic design methodologies and tools, and the presentation of this emerging knowledge for effective exchange. When academic research is applied to a specific problem-solving situation, it may serve as a stimulus for new areas of research, resulting from reformulation of problems uncovered in technological development or assessment (Box 15.2). The research thus seeks a more in-depth understanding than required by the immediate needs of the original problem, ideally giving rise to new questions of conceptual importance (Brooks, 1993). Academic institutions also play a critical part in the dissemination of knowledge to numerous audiences. Knowledge dissemination serves to increase awareness amongst the public, and contributes to basic education and training in technical and problem-solving skills that are needed for effective technology transfer.

As opposed to companies focusing on climate mitigation, only a few of the larger coastal-technology companies are strongly involved in technology transfer (Stockdale, 1996). Where the

---

Box 15.2	SHORELINE MANAGEMENT PLANS

Littoral or sediment cells are self-contained systems that contain all the sources, pathways, stores and sinks of beach sediment. They were first recognised over 30 years ago (Komar, 1998), and define natural divisions of the coastline as opposed to the arbitrary geopolitical divisions that are normally used to break up the coast into management units. Academics quickly recognised that littoral cells provided a more effective basis for shoreline management (e.g., Carter, 1988), but until recently there has been limited transfer of this approach to real-world application (see also Case Study 16).

England and Wales have a long tradition of using rigid defences against flooding and erosion in many areas (Klein et al., 1999). Large parts of eastern England would be flooded on every high tide without such defences. These defences were often planned by relatively small geopolitical units (e.g., maritime district councils or local drainage boards), although most funding came from central government. While long-term rates of sea-level rise were often considered in design (e.g., Gilbert and Horner, 1984), the long-term consequences

of defences in terms of sediment supply to the littoral cell were not considered. Therefore, protection at one site often led to the need to protect adjoining sites, and the slow but progressive expansion in the length of defended coastline has starved much of the coast of new beach sediment.

These problems have led to a number of policy changes, such as a move towards softer approaches to protection. This has included one of the first transfers of the cell concept to shoreline management. The coast of England and Wales has been divided into 11 littoral cells and about 50 littoral subcells (MAFF et al., 1995). About 40 shoreline management plans (SMPs) have been defined and initiated based on one or more subcells. Most subcells are larger than the traditional geopolitical units involved in shoreline planning, necessitating partnerships to produce the SMPs. Each SMP takes a strategic view of future shoreline defence and after dividing the coastline into management units, four possible policies are evaluated: (i) do nothing, (ii) advance the line, (iii) hold the line and (iv) retreat the line. Importantly, SMPs are "living" documents, and regular revisions

are expected to reflect changing policy, improved understanding and opportunities for innovation (Leafe et al., 1998).

The slow diffusion of the SMP approach to practice around the world reflects two key factors. First, the cell concept was developed by coastal scientists, while coastal managers were largely trained in more technical disciplines, hindering exchange. Second, there was no immediate market for the approach as the problems of existing management approaches do not become manifest for decades or longer. In England and Wales, diffusion was facilitated by the need to (i) improve understanding of coastal processes, (ii) predict likely future coastal evolution, (iii) identify assets likely to be affected by coastal change, (iv) identify need for regional or site-specific research and (v) facilitate consultation (MAFF et al., 1995). The threat of sea-level rise helped to raise these concerns and needs, while economic appraisal has helped to distinguish more efficient approaches to shoreline management (e.g., Penning-Rowsell et al., 1992; Turner et al., 1995).

manufacturing of hardware is involved, successful technology transfer requires an industrial provider. Furthermore, government and university laboratories generally do not have the mandate or resources for commercial-scale testing of new innovations, so leveraging resources by means of public-private sector partnerships are vital to widespread technology transfer and continuous improvement. Organisations such as WL | Delft Hydraulics, HR Wallingford and the Danish Hydraulic Institute combine original research with facilities capable of prototype testing. A significant portion of each organisation's R&D is done in collaboration with government partners (see also UNFCCC, 1999).

Non-governmental organisations (NGOs) increasingly offer a new channel for accomplishing technology transfer in ways that governments and the private sector cannot, should not or will not act. The advantage of NGOs is that they can serve as catalysts to spark action and create diffusion networks at the grassroots level. NGOs have the unique capability of reaching isolated communities and stakeholders to provide the proper cultural and socio-economic contexts required for successful technology transfer (see also section 4.4).

15.5.4 Barriers to In-Country Technology Transfer

Government policies help to define and shape markets for coastal-adaptation technologies, but regulatory frameworks often lack incentives for innovative technology application and may include disincentives to the development and adoption of new technologies (NSTC, 1995). For example, certain regulations may prescribe the use of specific technologies, while approval of new, more efficient or effective technologies must go through lengthy legal processes. In addition, relevant government agencies may be fragmented, which impedes decision-making on all aspects of technology transfer (Chapters 4 and 5 provide an extensive treatment of barriers that impede technology transfer).

Effective technology transfer depends on the ability of the technology supplier to deliver the desired technology and on the capabilities of the technology recipient to employ it. Scientists in government and university laboratories may be too removed from the marketplace to facilitate rapid application of their technology innovations. Stockdale (1996) attributed the lack of participation of the private sector in partnerships with national laboratories to the lack of a "road-map" for locating the appropriate laboratory to cooperate. When collaboration with private industry occurs, conflicts may arise between the desire of government and university-based scientists to conduct independent, open research and the needs of industry to keep new knowledge proprietary in order to benefit financially. The lack of collaboration within and between stakeholders significantly impedes the deployment of coastal-adaptation technologies.

There is an acute shortage of adequate and accurate information on the impacts of climate change on coastal systems and on adaptation possibilities. This information is required to underpin public and private spending on coastal-adaptation technologies.

Availability of data does not automatically result in increased information use for decision-making. Coastal and climate information often involves huge sets of unprocessed data. Unless a bridging organisation exists to "translate" data into usable information products—such as GIS analyses or assessments—and present it to the user community (*e.g.*, via training programmes), these data sets go unused. For example, the UNEP Infoterra programme in the Caribbean provides large amounts of environmental data to national focal points via the Internet. However, few of these data are used in environmental management or local development planning, partly because the focal points have neither the tools nor training to convert the data into forms required to support the decision process (Potter, 1995).

Furthermore, as the means for developing information become more sophisticated, the transfer process is delayed by a lack of technology-verification to inform public officials of technology performance and by a lack of training to maintain a workforce technically skilled in application of the new technologies. Conveying information to the public can be problematic, because of limited reception of messages that are not readily related to immediate concerns.

15.5.5 Enabling Policies, Programmes and Measures

Acceleration of coastal-adaptation technology transfer across organisational boundaries requires actions at all levels of the process by all stakeholders. Of the newer institutional developments evolving, the best will be those that are effective in fostering corporate and community collaborative efforts, while nurturing positive government-academic-business relationships (Kozmetsky, 1990; see also Chapter 4). Common elements of these new policies, programmes and measures include:

- New governmental technology policies that provide increased support for helping users to obtain adjust and use technology. This includes major investments in information infrastructure, in training and education of the workforce and in outreach and extension services (Brooks, 1993). For example, the U.S. NOAA Coastal Services Center produces environmental data sets and information products for coastal managers, provides access to information clearinghouses for coastal scientists, managers and the public, and conducts technical training for coastal planners and managers.
- Decisions on technology activities at the national level that are increasingly made in collaboration with other levels of government and private-sector users. This may be accomplished by government-sponsored needs assessments (*e.g.*, via surveys, stakeholder meetings and workshops). However, it implies that focal points must exist and be well sited within the national government to be able to input user needs and capabilities into the decision process.
- Research universities that are becoming more deeply engaged in applied R&D, and are expanding ways for

accelerating technology diffusion to users at all levels of government and the private sector. Mechanisms that have proven to be effective include extension services, such as training and guidance provided by sea-grant colleges in the United States and joint university-business research parks to promote collective R&D.

- New business alliances with government collaborators to leverage resources and capabilities. These could include joint projects with consortia of business, such as those fostered by the Danish Hydraulic Institute.
- Increased involvement of NGOs and trainers who have a sound grasp of social methodology, understand the concerns of target groups and are able to devise the most appropriate means for conveying information at the grassroots level.

Other powerful mechanisms to facilitate successful coastal adaptation and technology transfer are regulation, standards and insurance. For example, the Coastal Barrier Resources Act in the United States provides protection to coastal barriers by prohibiting most expenditures of federal funds in these areas, thus strongly discouraging development. Insurance can have a similar role, although it may also stimulate maladaptive practices (Box 15.3). Building codes and standards are covered in Chapter 7, while international performance standards are discussed in Section 15.6.5.

15.6 Transfer of Coastal-Adaptation Technologies between Countries

Coastal-adaptation technology transfer in an international context has many of the same characteristics as that within countries. It consists primarily of knowledge transactions, innovations predominantly emerge from government laboratories and universities, stakeholders' incentives are not purely commercial, and pathways and barriers are comparable, albeit more complex and far-reaching. However, effective globalisation of coastal-adaptation technologies requires creativity in two areas not necessarily required for technology transfer within countries.

First, governments need to develop policies that go beyond meeting predetermined goals or maintaining current standards, while all stakeholders must recognise that the traditional paradigm that technology can be transferred full-blown from one cultural context to another is flawed. Technologies must be adjusted, oriented and made appropriate for local conditions in the host country, possibly in the context of integrated coastal zone management. Second, the importance of global networks to improve and accelerate coastal-adaptation technology transfer should not be underestimated. Such networks provide access to up-to-date information and real-time tracking of global trends, accelerate the formation of joint ventures and permit direct participation in strategic locations around the world. The process for building these networks must include not only personal links but also institutional and functional linkages (Kozmetsky, 1990).

Box 15.3   THE ROLE OF INSURANCE

In countries with climate-related insurance markets, insurance can have a positive or a negative role in promoting adaptation to climate change and any associated technology transfer (Clark, 1998). This may happen directly via contacts with customers or indirectly as the insurance industry lobbies institutions. Technology underpins this interaction as improving data management and modelling capability give the insurance industry more detailed information of both the risks and opportunities that climate variability and change present (Crichton and Mounsey, 1997). However, as Clark (1998) argued, more knowledge may benefit the insurance industry, but it does not necessarily lead to overall societal benefits. Determining if insurance is acting in a negative or positive role will depend on the assessment criteria and is linked with the issue of distinguishing adaptation from maladaptation. Clark (1998) argued that a government-insurance industry partnership can benefit both the industry and wider society in terms of reduced exposure to both climate variability and climate change and maintain the long-term health of the insurance industry.

At the one extreme, the insurance industry might be involved in the planning process by using its knowledge to discourage development in hazardous locations. Information technology is critical for this type of role. At the other extreme, the insurance industry might simply ignore the problem (*i.e.*, use no technology) and promote maladaptation by repeatedly providing the resources for redevelopment. In the short term, this may seem sustainable, but a large natural disaster may promote sudden withdrawal of insurance and have serious consequences for coastal property values and general development.

An intermediate approach is to link the availability of insurance to appropriate regulations on land use in hazardous areas. The U.S. National Flood Insurance Program (NFIP) offers an example of this approach. It requires information technology to define the hazardous zone, appropriate building technology and continuous monitoring of performance (Davison, 1993; OTA, 1993). Training and technology transfer are an integral part of the NFIP. In return for obeying the codes, all participating communities are eligible for flood insurance. Opinions on the success or failure of the NFIP differ. An NFIP perspective is that as the property stock is progressively replaced and buildings built to code increase as a proportion of the vulnerable stock, insured losses will decline significantly relative to a baseline without the NFIP. A first-order assessment of the impacts of accelerated sea-level rise suggests that the NFIP can be adapted to this change (FEMA, 1991). The major deficiency, however, is that erosion is ignored (NRC, 1990; Davison, 1993). An alternative view is that the NFIP is a classic example of maladaptation. Whichever perspective is correct, the NFIP demonstrates that incentives and regulations enforced via insurance can promote technology transfer with large potential benefits.

While the above examples are mainly focused on climate variability, with appropriate anticipatory measures insurance could be used to encourage adaptation to climate change, including the necessary technology transfer (*e.g.*, OTA, 1993; Berz, 1998). The insurance industry is increasingly assessing the potential implications of climate change and where they have a market presence, their opinion will be an important influence on the timing and nature of adaptation.

15.6.1 Incentives for Transnational Technology Transfer

Similar to in-country technology transfer, the incentives for the international transfer and diffusion of coastal-adaptation technologies are not predominantly commercial. Amongst the stakeholders, the goals of the international business community are increased global market share for their products and hence, increased profits. However, as stated before, the coastal-technology sector is fragmented and consists mostly of small companies that lack the resources to initiate technology transfer. Scientists need to be aware of new scientific advances so that their own accomplishments do not re-invent already published work, and that their research both builds on complimentary work and identifies other leading researchers. To fulfil their missions, government organisations must have access to the best available technologies and latest information. Therefore, national governments are the primary investors in and sources of coastal-adaptation technologies, and governments at all levels comprise the predominant market.

Coastal adaptation to climate change is a transnational issue - it cannot be addressed within the borders of one country, no matter how effective and creative the decision-makers, innovative the academic relationships and dynamic the private sector are in advancing the deployment of appropriate technologies. Alliances for technology transfer between countries are a way to leverage shortages or complement the skills of local scientists and engineers, share financial and other resources and develop and extend access of technical assistance and capacity building to strategic locations around the world (Kozmetsky, 1990). An example is the EU Marine and Science Technology (MAST) programme, which requires partnerships of at least two non-affiliated partners from two member states or from one member state and an associated state. The EU International Cooperation with Developing Countries (INCO-DC) and ASEAN Environment Programme (ASEP) are two examples of many programmes supporting scientific cooperation between industrialised and developing countries to stimulate collaborative research and broaden knowledge of all partners.

An example of effective multinational, cross-institutional cooperation is the development of airborne lidar (light detection and ranging) technology for shoreline topography and bathymetry surveying (Lillycrop and Estep, 1995). The technology offers a significant advance in capabilities for collecting, analysing, using and retaining critical data for coastal management. One airborne lidar system for bathymetry surveying was developed by a small Canadian company, with financial support from the Canadian government. In 1988, the United States' Army Corps of Engineers and the Canadian government began a shared-cost programme for the design, construction and field verification of the lidar technology. The original Canadian private-sector developer provided technical-design support, an American surveying company provided evaluation services, and both governments provided financial and operational support. Lidar is now operational and has been applied in a number of countries, including the United States, the United Kingdom, The Netherlands and Australia.

15.6.2 Pathways of Transnational Technology Transfer

The five stages of technology transfer (Section 15.5.2) are also applicable to international pathways of coastal-adaptation technology transfer. Similar to in-country transfer, advancing from R&D to replication is not a linear process. However, the complexity of technology transfer increases not only as the process proceeds, but also within stages as differences in the economic, institutional, legal and socio-cultural contexts of the technology transfer broaden. In the case of CoreLoc, for example, cooperative agreements with companies in the host countries were delayed because of the home-government researchers' inexperience in obtaining international patents (see Case Study 21).

As mentioned earlier, the use and transfer of most coastal-adaptation technologies has been the result of societal intervention, rather than market forces. Therefore, official development assistance (ODA; see also Chapter 5) will remain crucial for vulnerable developing countries to obtain access to appropriate technologies. Currently, the majority of ODA-funded coastal projects are carried out for economic purposes, such as fisheries, tourism and port development. In this context, technology is often equated with hardware, transferable in single, point-in-time transactions. Such technology implementation has proven maladaptive in many occasions (WCC'93, 1994). Successful coastal adaptation to sea-level rise also requires the transfer of soft technologies, which enhance human skills and capacity needed to adopt and adjust new approaches to coastal management. Long-term relationships involving technical assistance and in-situ training are important elements of effective pathways.

15.6.3 The Role of Stakeholders

Currently, the international flow of coastal-adaptation technologies is mostly government-driven. As in any technology transfer, governments create and sustain the conditions for the successful development and delivery of coastal-adaptation technologies. As the primary technology source, they must support continuous and targeted innovation by means of investments in education, R&D and scholarly and technical exchange programmes. Governments can take a proactive role in technology transfer by encouraging collaborative relations of national laboratories and universities with foreign-affiliated businesses, as well as the involvement of foreign nationals in publicly supported research. Claims that these policies result in the loss of knowledge assets paid for with public funds without fair compensation are mostly unfounded. Instead, the driving force of coastal technology transfer is often information sharing and, in an economic context, both parties benefit only when new knowledge, rather than currency or goods, is exchanged equally (Rollwagon, 1990).

For their part, universities must continue to invest and reward the conduct of basic research, since this is the "seed corn" for innovation. However, under the paradigm of international technology transfer they also have a responsibility to present new knowledge in forms that can be readily utilised by diverse audiences and to provide technical assistance. Since these activities do not generally lead to peer-reviewed publications, universities may wish to restructure reward systems to meet this new global social contract. Moreover, interdisciplinary approaches to understand systems rather than processes may benefit coastal adaptation and related technology transfer.

Although the companies that make up the coastal-technology sector generally have limited resources, many of the service providers in this sector can adopt practices that call for active participation and training of host-country users and joint project management. This leads to long-term relationships and potentially new business opportunities.

NGOs play a vital part in international technology transfer. They often serve as "knowledge translators" to bridge the gap between technology acceptance and application, and create and promote adaptive capabilities within the receiving country to sustain technology operation and maintenance, as well as enable replication (Box 15.4; see also section 4.4).

15.6.4 Barriers to Transnational Technology Transfer

Traditional barriers to technology transfer between countries, such as concerns by national governments about issues of national defence or economic competition, are generally not applicable to the transfer of coastal-adaptation technologies. However, government investment in R&D is primarily mission-oriented and focuses on needs and capacities within the home country. Consequently, many technology transactions across countries tend to be driven by (the desire to sell) technology rather than by the needs and special requirements of the host country. On the other hand, host countries often lack the necessary infrastructure for developing a market for the technology. In addition, government officials may lack technology awareness, while the workforce may lack technical skills and expertise (BCSD, 1992).

As noted before, ODA-funded projects have traditionally been directed towards costly infrastructure projects, which may be maladaptive. This suggests that the principal financial barrier to technology transfer for coastal adaptation is not a shortage of funds, but the preference of donors as well as recipients to underwrite projects involving hard technologies developed in industrialised countries. Financial means for coastal adaptation to climate change have been limited thus far. For example, the GEF has only funded activities to plan for adaptation (*i.e.*, information development), rather than supporting actual adaptation activities. Two successful GEF-funded projects to help Pacific and Caribbean countries to plan adaptation to global climate change are the Pacific Islands Climate Change Assistance Programme (PICCAP) and the Caribbean Planning for Adaptation to Climate Change (CPACC). Case Study 20 provides a summary of relevant technology-transfer issues in CPACC.

Box 15.4   The role of NGOs

Non-governmental organisations can be used to promote new coastal-adaptation technologies, because they typically have close links to communities and individuals with a direct interest in the coast. NGO activities in the coastal zone are very diverse and range from environmental education, consultation, lobbying and campaigning to research into alternative coastal-management strategies (De Waal, 1994).

Much of the coastal-adaptation technology transfer can be developed at the grassroots level, based on the needs as identified by communities and individuals. The advantage of NGOs is their ability to reach into isolated communities, which often have little or no experience in communicating problems and concerns to government representatives, to manage the important community-development components of technology transfer. NGOs may have their headquarters in metro-politan countries as well as country field offices to facilitate grassroots networks. These country field offices can facilitate country-specific bilateral and non-government funding and technical assistance. On the other hand, regional programmes also allow for cost-effective provision of technical assistance, training and management of project funds.

The commonality of NGOs can be seen in terms of their wishing to retain maximum involvement of individuals in projects even where those projects involve institutional development for government. This approach is ideal for testing and introducing a range of adaptation technologies, especially simple ones, on a cost-effective basis. Cross-training and sharing of experiences, successes and lessons learnt between projects are essential features of the way NGOs operate. Quality control may also be enhanced for project

---

Barriers of distance and culture must be overcome in all international transactions. These difficulties are multiplied if markets, information sources and the means of matching potential partners are poorly developed.

15.6.5 Enabling Policies, Programmes and Measures

Several critical factors, which may be part of pathways within countries to varying degrees, are suggested to be integral components of any international coastal-adaptation technology transfer (BCSD, 1992):

- Technologies must be adjusted, oriented and made appropriate for local conditions in the host country. For example, the Japanese government designed and constructed a seawall to protect the shoreline fronting Nuku'alofa, the capital of Tonga, against erosion (Mimura and Pelesikoti, 1997). Despite the superior design and effectiveness of the structure, the project was not extended and duplicated because the host government had neither the funds, the physical capacity to produce the necessary raw materials nor the technological expertise to build another such sea-

onitoring and implementation y sharing tools and expertise. he success of this approach can e attributed to, amongst other iings, the integration of individ- al needs with government pro- rammes and cultural sensitivity.

lore visible and controversial chnology-transfer initiatives ould engage NGOs (UNEP, 1996), ut only where there is a high egree of mutual trust and two- ay learning, not merely the ansfer of money (Earle, 1997). xperience and case studies (*cf.* e Waal, 1994) show that of the arious reactionary strategies vailable to NGOs, extensive ublic awareness and education ampaigns, dialogue with all terest groups and extensive edia attention are amongst the ost effective.

GOs may or may not have core- inding or an endowment source f funds to finance their opera- ons. However, NGOs are often in

a unique position to harness for- eign funds for assistance and in administering development pro- jects. Some NGO projects are implemented independently, other ones in cooperation with other NGOs, national institutions, government or grassroots organi- sations.

The UK-based NGO Oxfam pro- vides a good example of technol- ogy transfer relying on generat- ing synergies. Oxfam supports grassroots communities in coastal-management activities such as replanting mangroves and developing artificial reefs, while concurrently assisting national fishery organisations that provide support to those communities. Links are encour- aged between the two organisa- tions and more specialised lobby- ing organisations in order to change national legislation (Earle, 1997).

wall. Another example involves the copying of a Northern European harbour design, including a device to reduce ice pressure, which was installed in its entirety in a tropical country (UNFCCC, 1999).

* Technology transfer must involve long-term partnerships and be linked to training and human-resource develop- ment. In the case of the first application of CoreLoc (Case Study 21), the researchers were not involved in planning and construction of the breakwater units. Results were more satisfactory in subsequent deployments when the researchers were able to work with local operators dur- ing project implementation. Furthermore, engaging local operators in the development of technology generates skills necessary to provide feedback required to contin- ue the innovation process.

* Institutional regional or global networks are powerful mech- anisms for fostering all aspects of coastal-adaptation tech- nology transfer, including assessment of needs, information exchange, training and technical assistance, capacity build- ing and technology development. Examples are the above- mentioned GEF-funded projects (PICCAP and CPACC) as well as the training and outreach activities of the Coastal Resources Center of the University of Rhode Island (USA).

In addition, Kozmetsky (1990) and Heaton *et al.* (1994) have identified a number of mechanisms that would be particularly important in the development and implementation of science and technology policies to be responsive to global economic trends. All of these are applicable to enabling and accelerating coastal-adaptation technology transfer:

* the creation of strategic alliances;
* the development of innovative financial arrangements;
* leveraging international standards on environmental man- agement;
* establishing technology "intermediaries".

The creation of new strategic alliances is a way to leverage sci- entific and technical talent, share financial resources and risks and extend access and delivery of new knowledge and technologies to other countries. These alliances may be intra- or intersectoral, but work best when participants share similar strategic goals, mutually understand and respect their respective cultures, and understand short- and long-term expectations. The PICCAP approach, for example, involves the appointment of a national coordinator and country teams in each of the ten participating countries. This strategy helps to ensure that technical support is provided in a socially and culturally sensitive manner, that pro- ject outputs are compatible with regional environmental and development strategies and priorities and that regional institutions are strengthened for future activities (Sem, 1998).

Development of innovative financial mechanisms may include linking ODA to foreign direct investment (BCSD, 1992), tax incentives for joint ventures in technology development and tax exemptions of income derived from technology transfer (Midlock, 1990). Chapter 5 presents a comprehensive overview of oppor- tunities to overcome financial barriers to climate-related tech- nology transfer.

Private-sector demand for adaptation technologies could be enhanced and refocused by international standards on coastal management. International standards allow companies to mea- sure environmental performance, establish best practices across industries and establish a degree of accountability through cer- tification, thus improving a company's worldwide competi- tive advantage (Heaton *et al.*, 1994). For example, the ISO 14000 Series Standards Environmental Management System (EMS) provides a framework that responds to the short-term changes in coastal environments brought about by develop- ment pressures as well as the long-term changes resulting from natural forces. Adoption by countries could provide mecha- nisms necessary to integrate vulnerability assessment with existing coastal-adaptation practices.

As stated, NGOs can play an important part as intermediaries and knowledge translators in the technology-transfer process. As intermediaries, they can identify sources of currently available and emerging technologies, facilitate investment arrangements, and provide management, technical and other assistance to devel- oping countries. As knowledge translators, NGOs can ensure that technology transfer is designed to create adaptive capabilities within the receiving country to adapt technology rather than simply to encourage its passive acceptance after transfer. NGOs can provide the foundation for the long-term relationship needed to replace casual or short-term connections between technology providers and users. Finally, they are particularly suited to link technology transfer to training and human-resource development and to public awareness raising.

15.7   Lessons Learned and Future Work

Climate-related coastal-adaptation technology transfer is a relatively new phenomenon for coastal scientists, planners and managers. This is illustrated by the limited number of publications that deal explicitly with the subject. This does not suggest, however, that technology transfer for coastal adaptation represents fundamentally new scientific, planning or management concepts. Coastal adaptation to climate change is a logical extension of ongoing coastal-management activities, for which the transfer of technologies has always been important. However, many of the stakeholders involved in technology transfer have yet to recognise the relevance of these activities to climate adaptation.

This chapter concludes that adaptation technologies can play an important part in reducing coastal vulnerability to climate change. However, empirical information on coastal adaptation to climate change is still scarce and past and current efforts of technology transfer in coastal zones have been poorly documented, so uncertainty about the appropriateness and generic applicability of adaptation technologies remains considerable. Moreover, the absence of any criteria to measure a project's outcome rather than its output has thus far impeded any realistic evaluation of the effectiveness of technology transfer in terms of reducing risks of coastal hazards and increasing adaptive capacity. In other words, given the current state of knowledge, it is difficult to say whether technology transfer in coastal zones to date has been a success or a failure.

15.7.1   Evaluation of Current Status

Ever since humans have lived near the sea they have developed and applied technologies to reduce their vulnerability to coastal hazards. The same technologies can be applied to adapt to anticipated impacts of climate change. However, access to these technologies in vulnerable areas can be a problem without effective technology transfer. Improving and facilitating the process of technology transfer are key challenges to reduce coastal vulnerability worldwide.

This chapter has shown that effective coastal adaptation and associated technology transfer may have to overcome a large number of very diverse obstacles. Many of these are site-specific and require site-specific solutions. The four major general barriers to coastal-adaptation technology transfer are the following:

- lack of data, information and knowledge to identify adaptation needs and appropriate technologies;
- lack of local capacity and consequent dependence of customers on suppliers of technology for operation, maintenance and duplication;
- disconnected organisational and institutional relationships between relevant actors;
- access to financial means.

Additionally, coastal adaptation itself is often hampered by a number of factors (see also Chapter 4):

- the uncertainty about the location, rate and magnitude of climate-change impacts;
- the local nature of coastal-adaptation requirements;
- the absence of global benefits of coastal adaptation, which constrains its international financing;
- the fact that adaptation is often not considered a development objective.

15.7.2   Possible Approaches to Promote Technology Transfer

The barriers identified above will not be overcome overnight. Many relate to fundamental and intrinsic characteristics and principles of today's society. Adjusting the process of technology transfer to accommodate societal imperfections will be easier to accomplish than the reverse. There are a number of important opportunities to promote technology transfer for coastal adaptation. All relevant stakeholders identified in this chapter have important responsibilities to seize these opportunities. Each of them has specific roles in technology transfer, but generally, neither of them can trigger all steps involved. It is therefore essential that there is true collaboration and interaction amongst all stakeholders. All stakeholders' interests must be known and considered to prevent the transfer of inappropriate technology, including technology that is culturally or socially unacceptable or cannot be operated or maintained using local expertise. Already in the project-formulation stage, it is crucial to consult and involve local experts.

Box 1.1 in Chapter 1 provided a list of questions that were raised by the fourth Conference of the Parties to the United Nations Framework Convention on Climate Change (UNFCCC CoP-4) in the Annex to its Decision 4/CP.4 (see box 1.1 in Chapter 1 for all questions). These questions can be summarised and reinterpreted as:

- How can a receptive environment for technology transfer be created?
- What additional bilateral and multilateral efforts are needed, if any?
- What steps can be taken by Annex II Parties?
- What additional role can the private sector play?

The number and magnitude of ongoing coastal infrastructural projects suggest that coastal planners and managers are already receptive to technology transfer of a particular kind.

However, there is a need to proceed from single transactions involving only hardware to long-term partnerships that also concentrate on enhancing human capacity by providing technical assistance and in-situ training. Moreover, awareness that protection is often not the most appropriate strategy to adapt to climate change is growing, but it remains difficult for coastal planners to oversee the entire spectrum of available options. The establishment of a clearinghouse for coastal-adaptation technologies could facilitate the task of technology selection and evaluation. Such a clearinghouse would develop and hold an extensive catalogue of available technologies, including information on their costs, performance, owner (if not publicly owned), availability, implementation requirements and other relevant issues.

Addressing the prevalent barriers does not require setting up new bilateral and multilateral institutions or mechanisms. Instead, existing activities and institutions need to be refocused to improve the efficiency and effectiveness of coastal technology transfer, taking climate change into account. Technology transfer will be served by regional collaboration, especially when it concerns countries that do not have the necessary human or financial resources to develop or manage large projects independently. Examples of successful projects involving technology transfer that are the result of regional collaborations are PICCAP and CPACC. Primarily aimed at information development, these projects recognise that effective coastal adaptation and related technology transfer require a better understanding of local adaptation needs.

Annex II Parties have an important responsibility to facilitate coastal adaptation in vulnerable countries. Many donors have traditionally attached a higher priority to projects aimed at mitigating climate change than to adaptation projects. The UNFCCC CoP-4, however, decided that activities directed at the preparation of adaptation activities are now eligible for funding by the GEF. This decision, along with the pressing need for adaptation, may stimulate other donors to revisit their priorities. For most locations, there is no need to invest in the development of new technologies. Rather, the evaluation, adjustment and replication of existing technologies need to be stimulated. Annex II Parties may also promote the establishment of collaborative mechanisms for technology transfer.

The private sector may be able to extend its role in technology transfer when provided with the right incentives. These incentives would tend to increase the profitability of socially desirable technology-transfer projects and could include subsidies for investment and tax exemptions of income. The role of the private sector may also be extended by regulation. In transnational technology transfer, for example, the home company could be required to involve a partner company in the host country. Finally, professional organisations can stimulate private-sector involvement by lobbying and engaging in relevant networks.

Finally, in a Special Report that discusses the methodological and technological issues involved in technology transfer one might lose sight of the fact that technology by itself is not a panacea.

Coastal-adaptation technologies can provide an important contribution to the sustainable development in coastal zones, but their effectiveness depends strongly on the economic, institutional, legal and socio-cultural contexts in which they are implemented. Furthermore, climate change is but one of the many interacting stresses in coastal zones. The importance of controlling non-climate stresses in the quest to reduce coastal vulnerability to climate change must not be underestimated.

References

Baker, T.F., 1993: Absolute sea level measurements, climate change and vertical crustal movements. *Global and Planetary Change*, **8**(3), 149–159.

Basher, R.E., 1999: Data requirements for developing adaptations to climate variability and change. *Mitigation and Adaptation Strategies for Global Change*, **4**(3–4), 227–237.

(BCSD) Business Council for Sustainable Development, 1992: *Report on Technology Cooperation*. Technology Advisory Group Working Paper, Business Council for Sustainable Development, Geneva, iii+49 pp.

Berz, G.A., 1998: Global warming and the insurance industry. In *Cost-Benefit Analyses of Climate Change—The Broader Perspectives*. F.L. Tóth, (ed.), Birkhäuser Verlag, Basel, pp. 41–56.

Bijlsma, L., C.N. Ehler, R.J.T. Klein, S.M. Kulshrestha, R.F. McLean, N. Mimura, R.J. Nicholls, L.A. Nurse, H. Pérez Nieto, E.Z. Stakhiv, R.K. Turner, and R.A. Warrick, 1996: Coastal zones and small islands. In *Climate Change 1995—Impacts, Adaptations and Mitigation of Climate Change: Scientific-Technical Analyses*. Contribution of Working Group II to the Second Assessment Report of the Intergovernmental Panel on Climate Change, R.T.Watson, M.C. Zinyowera, R.H. Moss, (eds.), Cambridge University Press, Cambridge, UK, pp. 289–324.

Birkemeier, W.A., H.C. Miller, S.D. Wilhelm, A.E. DeWall, and C.S. Gorbics, 1985: *A User's Guide to the Coastal Engineering Research Center's (CERC's) Field Research Facility*. Instruction report CERC-85-1. U.S. Army Corps of Engineers, Waterways Experiment Station, Vicksburgh, MI, 121 pp.

Boesch, D.F., M.N. Josselyn, A.J. Mehta, J.T. Morris, W.K. Nuttle, C.A. Simenstad, and D.J.P. Swift, 1994: Scientific assessment of coastal wetland loss, restoration and management in Louisiana. *Journal of Coastal Research*, **20**, Special Issue, v+103 pp.

Brooks, H., 1993: Research universities and the social contract for science. In *Empowering Technology: Implementing a U.S. Strategy*. L.M. Branscomb, (ed.), MIT Press, Cambridge, MA, pp. 202–234.

Burd, F., 1995: *Managed Retreat—A Practical Guide*. English Nature, Peterborough, UK, 27 pp.

Burkhart, F.N., 1991: *Media, Emergency Warnings and Citizen Response*. Westview Press, Boulder, CO.

Burrough, P.A., and R.A. McDonnell, 1998: *Principles of Geographical Information Systems*. Oxford University Press, Oxford, 333 pp.

Burton, I., 1996: The growth of adaptation capacity: practice and policy. In *Adapting to Climate Change: An International Perspective*. J.B. Smith, N. Bhatti, G.V. Menzhulin, R. Benioff, M. Campos, B. Jallow, F. Rijsberman, M.I. Budyko,R.K. Dixon, (eds.), Springer-Verlag, New York, NY, pp. 55–67.

Burton, I., 1997: Vulnerability and adaptive response in the context of climate and climate change. *Climatic Change*, **36** (1-2), 185–196.

Bush, D.M., W.J. Neal, R.S. Young, and O.H. Pilkey, 1999: Utilization of geoindicators for rapid assessment of coastal-hazard risk and mitigation. *Ocean & Coastal Management*, **42**(8), 647–670.

Capobianco, M., 1999: *Role and Use of Technologies in Relation to Integrated Coastal Zone Management*. Report to the European Union Demonstration Programme on Integrated Coastal Zone Management, Tecnomare S.p.A., Venice, x+106 pp + apps.

Capobianco, M.., and M.J.F. Stive, 1997: Soft protection technologies as a tool for integrated coastal zone management. In *MEDCOAST'97*. Proceedings of the third international conference on the Mediterranean coastal environment, E. Orhan, (ed.), Valletta, Malta, 11–14 November 1997, Middle East Technical University, Ankara, pp. 469–484.

Capobianco, M., H.J. de Vriend, R.J. Nicholls, and M.J.F. Stive, 1999: Coastal area impact and vulnerability assessment: the point of view of a morphodynamic modeller. *Journal of Coastal Research*, **15**(3), 701–716.

Carnegie Commission, 1992: *Science, Technology, and the States in America's Third Century*. Carnegie Commission on Science, Technology, and Government, New York, NY, 73 pp.

Carter, R.W.G., 1988: *Coastal Environments—An Introduction to the Physical, Ecological and Cultural Systems of Coastlines*. Academic Press, London, 617 pp.

Carter, T.R., M.L. Parry, S. Nishioka, and H. Harasawa (eds.), 1994: *Technical Guidelines for Assessing Climate Change Impacts and Adaptations*. Report of Working Group II of the Intergovernmental Panel on Climate Change, University College, London and Centre for Global Environmental Research, London, and Tsukuba, Japan,x+59 pp.

Caton, B., and I. Eliot, 1993: Coastal hazard policy development and the Australian federal system. In *Vulnerability Assessment to Sea-Level Rise and Coastal Zone Management*. Proceedings of the IPCC/WCC'93 Eastern Hemisphere workshop, Tsukuba, Japan, 3–6 August 1993, R.F. McLean, N. Mimura, (eds.), Department of Environment, Sport and Territories, Canberra, Australia, pp. 417–427.

Cazenave, A., K. Dominh, M.C. Gennero, and B. Ferret, 1998: Global mean sea level changes observed by TOPEX-POSEIDON and ERS-1. *Physics and Chemistry of the Earth*, **23**(9–10), 1069–1075.

(CEC) Commission of the European Communities, 1999: *Towards a European Integrated Coastal Zone Management (ICZM) Strategy: General Principles and Policy Options—A Reflection Paper*. Directorates-General Environment, Nuclear Safety and Civil Protection, Fisheries and Regional Policies and Cohesion, Office for Official Publications of the European Communities, Luxembourg, 31 pp. + maps.

Chao, P.T., and B.F. Hobbs, 1997: Decision analysis of shoreline protection under climate change uncertainty. *Water Resources Research*, **33**(4), 817–829.

Cicin-Sain, B., 1993: Sustainable development and integrated coastal management. *Ocean & Coastal Management*, **21**(1-3), 11–43.

Cicin-Sain, B., and R.W. Knecht, 1998: *Integrated Coastal and Ocean Management: Concepts and Practices*. Island Press, Washington, DC, 499 pp.

Clark, C.D., H.T. Ripley, E.P. Green, A.J. Edwards, and P.J. Mumby, 1997: Mapping and measurement of tropical coastal environments with hyperspectral and high spatial resolution data. *International Journal of Remote Sensing*, **18**(2), 237–242.

Clark, M., 1998: Flood insurance as a management strategy for UK coastal resilience. *The Geographical Journal*, **164**(3), 333–343.

Crawford, M.J., 1993: Coastal management in island states: potential uses of satellite imagery, aerial photography and geographic information systems. In *Vulnerability Assessment to Sea-Level Rise and Coastal Zone Management*. Proceedings of the IPCC/WCC'93 Eastern Hemisphere workshop, Tsukuba, Japan, 3–6 August 1993, R.F. McLean, N. Mimura, (eds.), Department of Environment, Sport and Territories, Canberra, Australia, pp. 269–272.

Crichton, D., and C. Mounsey, 1997: *How the insurance market will use its flood research*. In Proceedings of the 32nd MAFF Conference on Coastal and River Engineering, Keele, UK, 2–4 July 1997, Ministry of Agriculture, Fisheries and Food, London, pp. J31–J34.

Davison, A.T., 1993: The national flood insurance program and coastal hazards. In *Coastal Zone '93*. Proceedings of the eighth symposium on coastal and ocean management, O.T. Magoon, W.S. Wilson, H. Converse, L.T. Tobin, (eds.), New Orleans, LA, 19–23 July 1993, American Society of Civil Engineers, New York, NY, pp. 1377–1391.

Davison, A.T., R.J. Nicholls, and S.P. Leatherman, 1992: Beach nourishment as a coastal management tool: an annotated bibliography on developments associated with the artificial nourishment of beaches. *Journal of Coastal Research*, **8**(4), 984–1022.

Debusschere, K., S. Penland, K.A. Westphal, P.D. Reimer, and R.A. McBride, 1991: Aerial videotape mapping of coastal geomorphic changes. In *Coastal Zone '91*. Proceedings of the seventh symposium on coastal and ocean management, Long Beach, CA, 8–12 July 1991, O.T. Magoon, H. Converse, V. Tippie, L.T. Tobin, D. Clark, (eds.), American Society of Civil Engineers, New York, NY, pp. 370–390.

Delft Hydraulics and Rijkswaterstaat, 1987: *Manual on Artificial Beach Nourishment*. CUR Report No. 130. Centre for Civil Engineering Research, Codes and Specifications, Gouda, The Netherlands, 195 pp.

De Vriend, H.J., M. Capobianco, T. Chesher, H.E. de Swart, B. Latteux, and M.J.F. Stive, 1993: Approaches to long-term modelling of coastal morphology: a review. *Coastal Engineering*, **21**(1–3), 225–269.

De Waal, J.C., 1994: *Where Water Meets Land: NGO Experiences in Coastal Management*. Both Ends, Amsterdam, 65 pp.

Doody, P. (ed.), 1985: *Sand Dunes and their management*, Nature Conservancy Council Peterborough, UK, 262 pp.

Earle, D., 1997: *Capacity Building: An Approach to People Centred Development*. Oxfam, Oxford, 226 pp.

Ehler, C.N., B. Cicin-Sain, R. Knecht, R. South, and R. Weiher, 1997: Guidelines to assist policy makers and managers of coastal areas in the integration of coastal management programs and national climate-change action plans. *Ocean & Coastal Management*, **37**(1), 7–27.

Emery, K.O., and D.G. Aubrey, 1991: *Sea Levels, Land Levels, and Tide Gauges.* Springer-Verlag, New York, NY, 237 pp.

Engelen, G., R. White, and I. Uljee, 1993: Exploratory modelling of socio-economic impacts of climate change. In *Climatic Change in the Intra-Americas Sea—Implications of Future Climate on the Ecosystems and Socio-Economic Structure in the Marine and Coastal Regions of the Caribbean Sea, Gulf of Mexico, Bahamas, and the Northeast Coast of South America.* G.A. Maul, (ed.), Edward Arnold, London, pp. 350–368.

English Nature, 1997: *Coastal Zone Conservation.* English Nature, Peterborough, 20 pp.

(FEMA) Federal Emergency Management Agency, 1986: *Coastal Construction Manual.* FEMA-55, Federal Emergency Management Agency, Washington, DC, xxi+114 pp.+apps.

(FEMA) Federal Emergency Management Agency, 1991: *Projected Impact of Relative Sea-Level Rise on the National Flood Insurance Program.* Unpublished report, Federal Emergency Management Agency, Washington, DC, vii+72 pp.

(FEMA) Federal Emergency Management Agency, 1994: *Mitigation of Flood and Erosion Damage to Residential Buildings in Coastal Areas.* FEMA-257, Mitigation Directorate, Federal Emergency Management Agency, Washington, DC, ii+34 pp..

(FEMA) Federal Emergency Management Agency, 1997: *Building Performance Assessment: Hurricane Fran in North Carolina— Observations, Recommendations and Technical Guidance.* FEMA-290, Mitigation Directorate, Federal Emergency Management Agency, Washington, DC, viii+60 pp.+apps.

Forbes, D.L., and Y. Hosoi, 1995: *Coastal Erosion in South Tarawa, Kiribati.* Technical Report No. 225, South Pacific Applied Geoscience Commission, Suva, Fiji, 77 pp. + apps.

French, P.W., 1997: *Coastal and Estuarine Management.* Routledge, London, xv+251 pp.

French, P.W., 1999: Managed retreat: a natural analogue from the Medway estuary, UK. *Ocean & Coastal Management*, **42**(1), 49–62.

Fu, L.-L., C.J. Koblinsky, J.-F. Minster, and J. Picaut, 1996: Reflecting on the first three years of TOPEX/POSEIDON. *Eos Transactions*, **77**(12), 109,111,117.

Funnell, C., 1998: Europe's ocean industry must be ready for the EC's FP5. *International Ocean Systems Design*, **2**(1), 15–18.

Gibson, D., and E.M. Rogers, 1994: *R&D Collaboration on Trial.* Harvard Business School Press, Boston, MA, xvii+607 pp.

Gilbert, S., and R. Horner, 1984: *The Thames Barrier.* Thomas Telford, London, 182 pp.

Goldberg, E.D., 1994: *Coastal Zone Space—Prelude to Conflict?* IOC Ocean Forum I, UNESCO Publishing, Paris, 138 pp.

Gorman, L., A. Morang, and R. Larson, 1998: Monitoring the coastal environment; part IV: mapping, shoreline changes, and bathymetric analysis. *Journal of Coastal Research*, **14**(1), 61–92.

Gornitz, V.M., R.C. Daniels, T.W. White, and K.R. Birdwell, 1994: The development of a coastal risk database: vulnerability to sea-level rise in the U.S. Southeast. *Journal of Coastal Research*, **12**, Special Issue, 327–338.

Gröger, M., and H.-P. Plag, 1993: Estimations of a global sea level trend: limitations from the structure of the PSMSL global sea level data set. *Global and Planetary Change*, **8**(3), 161–179.

Haque, C.E., 1995: Climatic hazards warning process in Bangladesh: experience of and lessons from the 1991 April cyclone. *Environmental Management*, **19**(5), 719–734.

Haque, C.E., 1997: Atmospheric hazards preparedness in Bangladesh—a study of warning, adjustments and recovery from the April 1991 cyclone. *Natural Hazards*, **16**(2–3), 181–202.

Harvey, N., B. Clouston, and P. Carvalho, 1999: Improving coastal vulnerability assessment methodologies for integrated coastal zone management: an approach from south Australia. *Australian Geographical Studies*, **37**(1), 50–69.

Heaton, G.R., R.D. Banks, and D.W. Ditz, 1994: *Missing Links: Technology and Environmental Improvement in the Industrializing World.* World Resources Institute, Washington, DC, 52 pp.

Henderson, F.M., T.F. Hart, Jr., B.P. Heaton, and J.E. Portolese, 1999: Mapping coastal ecosystems over a steep development gradient using C-CAP protocols. *International Journal of Remote Sensing*, **20**(4), 727–744.

Hinton, C., R. Nicholls, and D. Dunsbergen, 1999: *Profile reopening on the Dutch shoreface.* In Proceedings of Coastal Sediments 99, N. Kraus, W. McDougal, (eds.), Hauppage, Long Island, NY, 21–26 June 1999, American Society of Civil Engineers, New York, NY, pp. 535–550.

Holman, R.A., A.H. Sallenger Jr., T.C. Lippmann, and J.W. Haines, 1994: The application of video image processing to the study of nearshore processes. *Oceanography*, **6**(3), 78–85.

Hughes, P. and G.B. Brundrit, 1992: An index to assess South Africa's vulnerability to sea-level rise. *South African Journal of Science*, **88**(6), 308–311.

Humphrey, S., and P. Burbridge, 1999: *Planning and Management Processes: Sectoral and Territorial Cooperation.* Report to the European Union Demonstration Programme on Integrated Coastal Zone Management, Department of Marine Sciences and Coastal Management, University of Newcastle, Newcastle upon Tyne, xi+65 pp.+apps.

IPCC CZMS, 1990: *Strategies for Adaptation to Sea Level Rise.* Report of the Coastal Zone Management Subgroup, Response Strategies Working Group of the Intergovernmental Panel on Climate Change, Ministry of Transport, Public Works and Water Management, The Hague, x+122 pp.

IPCC CZMS, 1992: A common methodology for assessing vulnerability to sea-level rise—second revision. In *Global Climate Change and the Rising Challenge of the Sea.* Report of the Coastal Zone Management Subgroup, Response Strategies Working Group of the Intergovernmental Panel on Climate Change, Ministry of Transport, Public Works and Water Management, The Hague, Appendix C.

IPCC 1998: *The Regional Impacts of Climate Change—An Assessment of Vulnerability.* Special Report of Working Group II, Watson, R.T., M.C. Zinyowera, and R.H. Moss (eds.),Cambridge University Press, Cambridge, UK, x+517 pp.

Jones, A.R., 1995: GIS in coastal management: a progress review. In *CoastGIS '95,* Proceedings of the International Symposium on GIS and Computer Mapping for Coastal Zone Management, R. Furness, (ed.),Cork, Ireland, 3–5 February 1995, University College Cork, Cork, Ireland, pp. 165–178.

Kaly, U.L., and G.P. Jones, 1998: Mangrove restoration: a potential tool for coastal management in tropical developing countries. *Ambio*, **27**(8), 656–661.

Kay, R.C., 1990: Development controls on eroding coastlines—reducing the future impact of greenhouse-induced sea level rise. *Land Use Policy*, **7**(4), 169–172.

Kay, R.C., and J.E. Hay, 1993: A decision support approach to coastal vulnerability and resilience assessment: a tool for integrated coastal zone management. In *Vulnerability Assessment to Sea-Level Rise and Coastal Zone Management.* Proceedings of the IPCC/WCC'93 Eastern Hemisphere workshop, R.F.McLean, N. Mimura, (eds.)Tsukuba, Japan, 3–6 August 1993, Department of Environment, Sport and Territories, Canberra, Australia, pp. 213–225.

Kay, R.C., A. Kirkland, and I. Stewart, 1996: Planning for future climate change and sea-level rise induced coastal change in Australia and New Zealand. In *Greenhouse—Coping with Climate Change.* W.J. Bouma, G.I. Pearman, M.R. Manning, (eds.), CSIRO Publishing, Wellington, New Zealand, pp. 377–398.

King, G., 1999: *Participation in the ICZM Processes: Mechanisms and Procedures Needed.* Report to the European Union Demonstration Programme on Integrated Coastal Zone Management, Hyder Consulting, Swansea, UK, vi+114pp.

Klein, R.J.T., and R.S.J. Tol, 1997: *Adaptation to Climate Change: Options and Technologies—An Overview Paper.* Technical Paper FCCC/TP/1997/3, United Nations Framework Convention on Climate Change Secretariat, Bonn, iii+33 pp.

Klein, R.J.T., and R.J. Nicholls, 1998: Coastal zones. In *Handbook on Climate Change Impact Assessment and Adaptation Strategies, Version 2.0.* J.F.Feenstra, I. Burton, J.B. Smith, R.S.J. Tol, (eds.), United Nations Environment Programme and Institute for Environmental Studies, Vrije Universiteit, Amsterdam and Nairobi, pp. 7.1–7.35.

Klein, R.J.T., and R.J. Nicholls, 1999: Assessment of coastal vulnerability to climate change. *Ambio,* **28**(2), 182–187.

Klein, R.J.T., R.J. Nicholls, and N. Mimura, 1999: Coastal adaptation to climate change: can the IPCC Technical Guidelines be applied? *Mitigation and Adaptation Strategies for Global Change,* **4**(3–4), 239-252.

Komar, P.D., 1998: *Beach Processes and Sedimentation,* Second edition. Prentice Hall, Upper Saddle River, NJ, 544 pp.

Kozmetsky, G., 1990: Shaping science and policy in a globally competitive marketplace. In *Technology Commercialization and Competitiveness.* IC2 Institute, University of Texas, Austin, TX, pp. 11–22.

Kuhn, R.L., 1990: Joining industry, university, and government: strategy, finance, commitment. In *Technology Commercialization and Competitiveness.* IC2 Institute, University of Texas, Austin, TX, pp. 223–233.

Larson, R., A. Morang, and L. Gorman, 1997: Monitoring the coastal environment; part II: sediment sampling and geotechnical methods. *Journal of Coastal Research,* **13**(2), 308–330.

Leafe, R., J. Pethick, and I. Townend, 1998: Realising the benefits of shoreline management. *The Geographical Journal,* **164**(3), 282–290.

Leatherman, S.P., R.J. Nicholls, and K.C. Dennis, 1995: Aerial videotape-assisted vulnerability analysis: a cost-effective approach to assess sea-level rise impacts. *Journal of Coastal Research,* 14, Special Issue, 15–25.

Leu, L.-G., Y.-Y. Kuo, and C.-T. Liu, 1999: Coastal bathymetry from the wave spectrum of SPOT images. *Coastal Engineering Journal,* **41**(1), 21–41.

Lillycrop, W.J., and L.L. Estep, 1995: Generational advancements in coastal surveying, mapping. *Sea Technology,* **36**(6), 10–16.

Lipton, D.W., and K. Wellman, 1995: *Economic Valuation of Natural Resources—A Handbook for Coastal Resource Policymakers.* NOAA Coastal Ocean Program Decision Analysis Series No. 5, NOAA Coastal Ocean Office, Silver Spring, MD, viii+131 pp.

Longley, P., M. Goodchild, D. Maguire, and D. Rhind (eds.), 1999: *Geographical Information Systems: Principles, Techniques, Applications and Management, Second edition.* Wiley, London (2 volumes).

(MAFF) Ministry of Agriculture, Fisheries and Food (Welsh Office, Association of District Councils, English Nature, and National Rivers Authority) 1995: *Shoreline Management Plans—A Guide for Coastal Defence Authorities.* Her Majesty's Stationery Office, 24 pp.

Martinez-Diaz-De-Leon, A., I.S. Robinson, D. Ballestero, and E. Coen, 1999: Wind driven ocean circulation features in the Gulf of Tehuantepec, Mexico, revealed by combined SAR and SST satellite sensor data. *International Journal of Remote Sensing,* **20**(8), 1661–1668.

May, P., R.J. Burby, N.J. Ericksen, J.W. Handmer, J.E. Dixon, S. Michaels, and D.I. Smith, 1996: *Environmental Management and Governance: Intergovernmental Approaches to Hazards and Sustainability.* Routledge, London, xvii+254 pp.

McLean, R., L. Rose, C. Kaluwin, and J. Aston (eds.), 1998: *The Australia/SPREP Coastal Vulnerability Initiative for Atoll States.* Workshop Report, Tarawa, Kiribati, 10–13 February 1997, Environment Australia, Canberra, 26 pp.

Midlock, E., 1990: Fiscal incentives available for international technology ventures. In *Technology Commercialization and Competitiveness.* IC2 Institute, University of Texas, Austin, TX, pp. 125-142.

Miller, L., R. Cheney, and J. Lillibridge, 1993: Blending ERS-1 altimetry and tide-gauge data. *Eos Transactions,* **74**(16), 185,197.

Mimura, N., and N. Pelesikoti, 1997: Vulnerability of Tonga to sea-level rise. *Journal of Coastal Research,* special issue **24**, 117–151.

Mimura, N., and P.D. Nunn, 1998: Trends of beach erosion and shoreline protection in rural Fiji. *Journal of Coastal Research,* **14**(1), 37–46.

Morang, A., R. Larson, and L. Gorman, 1997a: Monitoring the coastal environment; part I: waves and currents. *Journal of Coastal Research,* **13**(1), 111–133.

Morang, A., R. Larson, and L. Gorman, 1997b: Monitoring the coastal environment; part III: geophysical and research methods. *Journal of Coastal Research,* **13**(4), 1064–1085.

Moser, S.C., 1998: *Talk Globally, Walk Locally—The Cross-Scale Influence of Global Change Information on Coastal Zone Management in Maine and Hawai'i.* John F. Kennedy School of Government, Harvard University, Cambridge, MA, 114 pp.

Neilan, R.E., P.A. Van Scoy, and P.L. Woodworth (eds.), 1997: *Workshop on Methods for Monitoring Sea Level—GPS and Tide Gauge Benchmark Monitoring and GPS Altimeter Calibration.* Proceedings of an IGS/PSMSL workshop, Pasadena, CA, 17–18 March 1997, Jet Propulsion Laboratory, California Institute of Technology, Pasadena, CA,xvii+202 pp.

Nerem, R.S., 1995: Global mean sea level variation from TOPEX/POSEIDON altimeter data. *Science,* **268**, 708–710.

Nerem, R.S., B.J. Haines, J. Hendricks, J.F. Minster, G.T. Mitchum, and W.B. White, 1997: Improved determination of global mean sea level variations using TOPEX/POSEIDON altimeter data. *Geophysical Research Letters,* **24**(11), 1331–1334.

Nicholls, R.J., and S.P. Leatherman (eds.), 1995: The potential impact of accelerated sea-level rise on developing countries. *Journal of Coastal Research,* Special Issue **14**, 1–324.

Nicholls, R.J., and J. Branson (eds.), 1998: Enhancing coastal resilience - planning for an uncertain future. *The Geographical Journal,* **164**(3), 255–343.

(NOAA) National Oceanic and Atmospheric Administration, 1998: *International Sea Level Workshop.* Workshop Report, Honolulu, HI, 10–11 June 1997, Global Climate Observing System Publ. No. 43, Global Ocean Observing System Publ. No. 55, International CLIVAR Project Office Publ. No. 16, Silver Spring, MD, iv+133 pp.

Nordstrom, K.F. and S.F. Arens, 1998: The role of human actions in evolution and management of foredunes in The Netherlands and New Jersey. *Journal of Coastal Conservation,* **4**, 169-180.

Nordstrom, K.F., R. Lampe, and L.M. Vandemark, 1998: Reestablishing naturally functioning dunes on developed coasts. *Environmental Management,* **25**(1), 37-51.

(NRC) National Research Council, 1987: *Responding to Changes in Sea Level: Engineering Implications.* National Academy Press, Washington, DC, 148 pp.

(NRC) National Research Council, 1990: *Managing Coastal Erosion.* National Academy Press, Washington, DC, 522 pp.

(NRC) National Research Council, 1992: *Restoration of Aquatic Ecosystems.* National Academy Press, Washington, DC, 522 pp.

(NRC) National Research Council, 1994: *Restoring and Protecting Marine Habitat: The Role of Engineering and Technology.* National Academy Press, Washington, DC, 193 pp.

(NRC) National Research Council, 1995: *Science, Policy and the Coast: Improving Decisionmaking.* National Academy Press, Washington, DC, 85 pp.

(NSTC) National Science and Technology Council, 1995: *Bridge to a Sustainable Future.* Interagency Environmental Technologies Office, Washington, DC, vi+87 pp.

Nunn, P.D., 1999: Pacific island beaches—a diminishing resource? *Asia-Pacific Network for Global Change Research Newsletter,* **5**, 1–3.

Nunn, P.D., and E. Waddell, 1992: *Implications of Climate Change and Sea-Level Rise for the Kingdom of Tonga.* South Pacific Regional Environment Programme Reports and Studies No. 58, Apia, Samoa, 39 pp.

Nunn, P.D., A.D. Ravuvu, W. Aalbersberg, N. Mimura, and K. Yamada, 1994a: *Assessment of Coastal Vulnerability and Resilience to Sea-Level Rise and Climate Change, Case Study: Yasawa Islands, Fiji—Phase II: Development of Methodology.* Environment Agency Japan, Overseas Environment Cooperation Centre Japan, South Pacific Regional Environment Programme, xiv+118 pp.

Nunn, P.D., A.D. Ravuvu, E. Balogh, N. Mimura, and K. Yamada, 1994b: *Assessment of Coastal Vulnerability and Resilience to Sea-Level Rise and Climate Change, Case Study: Savai'i Island, Western Samoa—Phase II: Development of Methodology.* Environment Agency Japan, Overseas Environment Cooperation Centre Japan, South Pacific Regional Environment Programme, xiv+109 pp.

O'Regan, P.R., 1996: The use of contemporary information technologies for coastal research and management—a review. *Journal of Coastal Research,* **12**(1), 192–204.

(OTA) Office of Technology Assessment, 1993: *Preparing for an Uncertain Climate, Volume 1*.U.S. Congress Publ. No. OTA-O-567, U.S. Government Printing Office, Washington, DC, 359 pp.

Parker, D.J., and J.W. Handmer, 1997: The role of unofficial flood warning systems. *Journal of Contingencies and Crisis Management*, **6**(1), 45–60.

Penning-Rowsell, E., C.H. Green, P.M. Thompson, A.M. Coker, S.M. Tunstall, C. Richards, and D.J. Parker, 1992: *The Economics of Coastal Management—A Manual of Benefit Assessment Techniques*. Belhaven Press, London, xviii+380 pp.

Penning-Rowsell, E., P. Winchester, and G. Gardiner, 1998: New approaches to sustainable hazard management for Venice. *The Geographical Journal*, **164**(1), 1–18.

Pilarczyk, K.W. (ed.), 1990: *Coastal Protection*. Proceedings of a short course on coastal protection, Delft, 30 June – 1 July 1990, Balkema, Rotterdam, 500 pp.

Plant, N.G., and R.A. Holman, 1997: Intertidal beach profile estimation using video images. *Marine Geology*, **140**(1–2), 1–24.

Pope, J., 1997: Responding to coastal erosion and flooding damages. *Journal of Coastal Research*, **13**(3), 704–710.

Potter, B., 1995: *Strengthening Caribbean Environmental Information Services*. Reflections on a Conference Sponsored by UNEP/INFOTERRA, the Caribbean Conservation Association and the European Centre for Development Policy Management. http://www.irf.org/irenvinf.html.

Pugh, D.T., 1987: *Tides, Surges and Mean Sea-Level: A Handbook for Engineers and Scientists*. Wiley, Chichester, UK, 472 pp.

Redfern, H., and R.G. Williams, 1996: Remote sensing—latest developments and uses. *Journal of the Institution of Water and Environmental Management*, **10**(6), 423–428.

Ribeiro, J.F., 1996: *Desalination Technology—Survey and Prospects*. European Commission Joint Research Centre Publ. No. EUR 16434 EN, Institute for Prospective Technological Studies, Seville, vii+56 pp.

Rogers, S.M., 1993: Relocating erosion-threatened buildings: a study of North Carolina housemoving. In *Coastal Zone '93*. Proceedings of the eighth symposium on coastal and ocean management, O.T. Magoon, W.S. Wilson, H. Converse, L.T. Tobin, (eds.), New Orleans, LA, 19–23 July 1993, American Society of Civil Engineers, New York, NY, pp. 1392–1405.

Rollwagon, J., 1990: Can U.S. high-technology companies compete internationally? In *Technology Commercialization and Competitiveness*. IC2 Institute, University of Texas, Austin, TX, pp. 1–9.

Rosenthal, U., and P. 't Hart (eds.), 1998: *Flood Response and Crisis Management in Western Europe: A Comparative Analysis*. Springer-Verlag, Berlin, 236 pp.

Ruth, M., and F. Pieper, 1994: Modeling spatial dynamics of sea-level rise in a coastal area. *System Dynamics Review*, **10**(4), 375–389.

Sallenger, A.H., Jr., W. Krabill, J. Brock, R. Swift, M. Jansen, S. Manizade, B. Richmond, M. Hampton, and D. Eslinger, 1999: Airborne laser study quantifies El Niño-induced coastal change. *Eos Transactions*, **80**(8), 89,92–93.

Sem, G., 1998: *Pacific Islands Climate Change Assistance Programme (PICCAP): a programme to assist Pacific Island countries implement the UNFCCC*. In Proceedings of The Eighth Asia-Pacific Seminar On Climate Change. Phuket, Thailand, 22–25 June 1998, Global Environment Department, Environment Agency of Japan, Tokyo, pp. 133–141.

Shaw, J., R.B. Taylor, D.L. Forbes, M.-H. Ruz, and S. Solomon, 1998: *Sensitivity of the Coasts of Canada to Sea-Level Rise*. Bulletin 505, Geological Survey of Canada, Ottawa, Canada, 79 pp.

Silvester, R., and J.R.C. Hsu, 1993: *Coastal Stabilization—Innovative Concepts*. Prentice Hall, Englewood Cliffs, NJ, 578 pp.

Smit, B. (ed.), 1993: *Adaptation to Climatic Variability and Change*. Report of the Task Force on Climate Adaptation, Occasional Paper No. 19, University of Guelph, Guelph, Canada, 53 pp. (see viii).

Smit, B., I. Burton, R.J.T. Klein, and J. Wandel: The anatomy of adaptation to climate change and variability. *Climatic Change*, in press.

Sorensen, R.M., R.N. Weisman, and G.P. Lennon, 1984: Control of erosion, inundation and salinity intrusion caused by sea-level rise. In *Greenhouse Effect and Sea Level Rise*. M.C. Barth, J.G. Titus, (eds.), Van Nostrand Reinhold, New York, NY, pp. 179–214.

(STAN) Science and Technology Awareness Network, 1998: *Planning and Implementing Science Awareness Activities*. Centre for Marine Geology, Dalhousie University, Halifax, Canada, 90 pp.

Stauble, D.K., and N.C. Kraus, 1993: *Beach Nourishment Engineering and Management Considerations*. American Society of Civil Engineers, New York, NY, 245 pp.

Stive, M.J.F., and H.J. de Vriend, 1995: Modelling shoreface evolution. *Marine Geology*, **126**(1–4), 235–248.

Stockdale, G., 1996: Federal technology transfer—oceans of opportunity for industry. *MTS Journal*, **30**(3), 40–42.

Taiepa, T., P. Lyver, P. Horsley, J. Davis, M. Bragg, and H. Moller, 1997: Co-management of New Zealand's conservation estate by Maori and Pakeha—a review. *Environmental Conservation*, **24**(3), 236–250.

Titus, J.G., 1991: Greenhouse effect and coastal wetland policy: how Americans could abandon an area the size of Massachusetts at minimum cost. *Environmental Management*, **15**(1), 39–58.

Titus, J.G., 1998: Rising seas, coastal erosion and the takings clause: how to save wetlands and beaches without hurting property owners. *Maryland Law Review*, **57**(4), 1279–1399.

Titus, J.G., C.Y. Kuo, M.J. Gibbs, T.B. LaRoche, M.K. Webb, and J.O. Waddell, 1987: Greenhouse effect, sea level rise and coastal drainage systems. *Journal of Water Resources Planning and Management*, **113**(2), 216–227.

Tol, R.S.J., R.J.T. Klein, H.M.A. Jansen, and H. Verbruggen, 1996: Some economic considerations on the importance of proactive integrated coastal zone management. *Ocean & Coastal Management*, **32**(1), 39–55.

Tri, N.H., W.N. Adger, and P.M. Kelly, 1998: Natural resource management in mitigating climate impacts: the example of mangrove restoration in Vietnam. *Global Environmental Change*, **8**(1), 49–61.

Turner, R.K., and W.N. Adger, 1996: *Coastal Zone Resources Assessment Guidelines*. Land-Ocean Interactions in the Coastal Zone Reports and Studies No. 4, IGBP/LOICZ, Texel, The Netherlands, iv+101 pp.

Turner, R.K., W.N. Adger, and P. Doktor, 1995: Assessing the economic costs of sea level rise. *Environment and Planning A*, **27**(11), 1777–1796.

(UNEP) United Nations Environment Programme, 1996: *Guidelines for Integrated Planning and Management of Coastal and Marine Areas in the Wider Caribbean Region*. UNEP Caribbean Environment Programme, Kingston, Jamaica, 141 pp.

(UNFCCC) United Nations Framework Convention on Climate Change, 1999: *Coastal Adaptation Technologies*. Technical Paper FCCC/TP/1999/1, United Nations Framework Convention on Climate Change Secretariat, Bonn, 49 pp.

Van de Plassche, O. (ed.), 1986: *Sea Level Research—A Manual for the Collection and Evaluation of Data*. Geo Books, Norwich, UK, xxv+618 pp.

Vellinga, P., 1986: *Beach and Dune Erosion during Storm Surges*. PhD thesis, Delft University of Technology, Delft, 169 pp.

Verhagen, H.J., 1989: Sand waves along the Dutch coast. *Coastal Engineering*, **13**(2), 129–147.

Viles, H., and T. Spencer, 1995: *Coastal Problems—Geomorphology, Ecology and Society at the Coast*. Edward Arnold, London, x+350 pp.

WCC'93, 1994: *Preparing to Meet the Coastal Challenges of the 21st Century*. Report of the World Coast Conference, Noordwijk, The Netherlands, 1–5 November 1993, Ministry of Transport, Public Works and Water Management, The Hague, 49 pp.+ apps.

West, J.J., and H. Dowlatabadi, 1999: On assessing the economic impacts of sea-level rise on developed coasts. In *Climate, Change and Risk*. T.E. Downing, A.A. Olsthoorn, R.S.J. Tol, (eds.), Routledge, London, pp. 205–220.

Wijnberg, K.M., and J.H.J. Terwindt, 1995: Extracting decadal morphological behaviour from high-resolution, long-term bathymetric surveys along the Holland coast using eigenfunction analysis. *Marine Geology*, **126**(1–4), 301–330.

Woodworth, P.L., 1991: The Permanent Service for Mean Sea Level and the Global Sea Level Observing System. *Journal of Coastal Research*, **7**(3), 699–710.

Wright, D.J., and D.J. Bartlett, (eds.), 1999: *Marine and Coastal Geographical Information Systems*. Taylor and Francis, London, 348 pp.

Yamada, K., P.D. Nunn, N. Mimura, S. Machida, and M. Yamamoto, 1995: Methodology for the assessment of vulnerability of South Pacific island countries to sea-level rise and climate change. *Journal of Global Environment Engineering*, **1**(1), 101–125.

Yohe, G.W., and J. Neumann, 1997: Planning for sea level rise and shore protection under climate uncertainty. *Climatic Change*, **37**(1), 243–270.

Zerbini, S., H.-P. Plag, T. Baker, M. Becker, H. Billiris, B. Bürki, H.-G. Kahle, I. Marson, L. Pezzoli, B. Richter, C. Romagnoli, M. Sztobryn, P. Tomasi, M. Tsimplis, G. Veis, and G. Verrone, 1996: Sea level in the Mediterranean: a first step towards separating crustal movements and absolute sea-level variations. *Global and Planetary Change*, **14(1–2)**, 1–48.

Section III

Case Studies

Section Coordinators:
STEPHEN O. ANDERSEN (USA), AJAY MATHUR (INDIA)

16

Case Studies

Coordinating Lead Authors:
SUKUMAR DEVOTTA (INDIA), MAITHILI IYER (INDIA),
DANIEL M. KAMMEN (USA)

Lead Authors:
*Saroja Asthana (India), James Aston (Samoa), James Boyd (USA),
Thomas Carlson (USA), Alfonso Carrasco (Peru), William Chandler (USA),
Jorge Corona (Mexico), Otto Doering (USA), Richard Duke (USA),
Yuichi Fujimoto (Japan), Yasuo Hosoya (Japan), Hidefumi Imura (Japan),
Arne Jacobson (USA), Suzana K Ribeiro (Brazil), Stefan Kessler (Switzerland),
Xenia Loizidou (Cyprus), Stephen Magezi (Uganda), Eric Martinot (USA), Indu
Murthy (India), Teruo Okazaki (Japan), Gunter Pauli (Belgium), Peter du Pont
(USA), N H Ravindranath (India), Steve Ryder (USA), Teodoro Sanchez (Peru), P.
Sudha (India), Sergey Surnin (Ukraine), Jeanne Townend (USA), Claudio Volonté
(Uruguay), James Williams (USA), Li Yue (China)*

Review Editors:
Rajendra Shende (India), Othmar Schwank (Switzerland)

CONTENTS

16.1 Introduction

This chapter includes thirty case studies illustrating issues discussed in the earlier chapters of this report. The objective of including these case studies is to demonstrate the distinctive problems and special opportunities that managers and implementers are likely to encounter in dealing with technology transfer.

This chapter is the work of 30 coordinating and lead authors from 13 countries (Belgium, Brazil, China, India, Japan, Mexico, Peru, Samoa, Switzerland, Uganda, Uruguay, Ukraine, and the United States). Additional experts from many countries served as peer reviewers. These experts have been drawn from national and regional government agencies, public and private research organisations, multinational and local companies, industry and environmental non-governmental organisations (NGOs) and by the Intergovernmental Panel on Climate Change (IPCC).

The cases included in this chapter encompass both mitigation and adaptation strategies within the context of climate change. Case studies in mitigation include initiatives to foster dematerialisation, de-carbonisation of energy sources, industrial ecology, dissemination and commercialisation of renewable energy technologies, energy efficiency programmes, and household biomass energy usage. Additional case studies are included where institutional reform and market transformation efforts have led to conservation and the protection and use of indigenous resources. Case studies on adaptation focus primarily on technologies/practices in the agriculture and forestry sectors, mitigation of health impacts, and tools and strategies for coastal management. While some information from these cases appear throughout this volume, this chapter provides a consistent methodological approach to the case study research that enables a comparative analysis of approaches, challenges, and lessons learned.

Over the past decade, government, non-governmental, grassroots, and private sector institutions and organisations have worked—with varying degrees of success—to develop, implement, and commercialise a diverse mix of environmentally sound technologies (ESTs) and resource management methods. These efforts provide valuable insights for actions that can be taken in a variety of sectors for enhancing adoption and use of mitigation and adaptation technologies.

Renewable energy technologies (RETs), for example, are increasingly used throughout the world to address energy shortages and to expand the range of services in both rural and urban areas. In Kenya over 80,000 small (20-100 Wp) photovoltaic solar home systems, battery charging stations and other small enterprises have been commercially financed and installed (Hankins, 1993; Acker and Kammen, 1996), while a government programme in Mexico has disseminated over 40,000 systems. In the Inner Mongolian autonomous region of China over 130,000 portable windmills provide electricity to about one-third of the non-grid-connected households in this region (Byrne *et al.*, 1998). In all these projects, the case studies demonstrate that the competitive market can be used to generate interest in renewables, as long as there is a baseline awareness of the technology and early market deployment is targeted at the parties most willing to participate.

Other lessons are learned from programmes that disseminate and encourage the use of improved biomass cookstoves, which are active in more than fifty nations. In India, a government programme has disseminated over eight million improved cookstoves, while nationally coordinated district-level initiatives in China account for over 120 million stoves (Smith *et al.*, 1993; Barnes *et al.*, 1994), and public-private partnerships in Kenya have introduced improved cookstoves to over half of the urban populace (Kammen, 1995a, b). In India, the construction of biogas digesters is supported by a government subsidy while maintenance and educational outreach is available from a mobile 'technology clinic' provided by a non-governmental organisation (NGO). A variety of international networks exist to promote and support the dissemination of improved cookstoves and other improved efficiency biomass technologies. Dozens of successful programmes have used community outreach to promote the new technologies.

Cases in the industrial sector identify information programmes as an important first step for selecting appropriate technology for a specific service. Programmes that specifically aim to increase consumers' awareness, acceptance, and use of particular technologies are designed to assist consumers in understanding and adopting more efficient technologies and practices. While information is seen to be a key factor in some cases, environmental legislation or the regulatory framework for enforcing standards can prove to be a crucial factor in the adoption of advanced technologies, as in the case of India (TERI, 1997).

These examples provide just a few of the insights gained from these case studies. A similar range of programmes and lessons exists in each of the areas of technology transfer covered in this chapter and in this volume as a whole.

The number and magnitude of these programmes marks a qualitative departure from past efforts to develop and transfer ESTs into general use. Increasingly, lessons from these projects encompass not only technical insights, but equally valuable economic and social analysis as well. The current reliance on market-forces and sustainable institutions for climate protection requires the incorporation of local knowledge in research and implementation of incentive programmes, innovative ownership and leasing arrangements, and in some cases subsidies (Cabraal *et al.*, 1995; Inversin, 1996).

16.2 Case Selection Approach

The objective in choosing the case studies in this report is to provide the reader with a broad picture that highlights the diversity among cases, and maintains a reasonable balance in terms of geographic and sectoral representations. The cases presented herein are not a random sample of the projects in existence. Like any research effort of this nature, the approach of soliciting cases that have important lessons to offer is likely to emphasise successful or partially successful examples of technology introduction and transfer. Interested readers can also consult the extensive literature available about the history of failed examples of technology transfer, both from the perspective of technical mismatch or inability of the local context to support the particular machine, industry, or management practice (see, e.g., Feder and Umali, 1993; Mansfield, 1994), and from the broader cultural perspective where the technology used undermines the local economy or society (Gereffi, 1983; Goulet, 1989). Yet another level of analysis concerns the degree to which research and applied projects in areas most appropriate to local needs and skills are neglected because of fundamental biases in the type of technologies that are valued (Kammen and Dove, 1997).

This variety of potential pitfalls in technology transfer projects is formidable and important. This report largely focuses on successful cases that tended to be the longest running and have the most comprehensive information. The fact that successful cases are common, and so diverse, can be used to draw a number of general lessons and hypotheses about the failures.

The challenges of lack of infrastructure that many nations face, notably in rural areas in developing nations, means that care should be taken when evaluating the opportunity to replicate these experiences. In particular, policymakers should analyse the lessons from the Case Studies in the specific local context where projects are planned. Constraints to local adoption often include: *(i)* lack of monetary income to pay for technologies in the formal market sense; *(ii)* the impediments of up-front costs even for technologies that have reasonable pay-back times; *(iii)* the role of gender and informal markets and industries; and *(iv)* the need for training and information resources concurrent or in advance of the availability of appropriate technology.

These local constraints on the intended beneficiaries of new technologies often critically determine if projects, programs, or commercial opportunities that would benefit the intended recipients (as well as the local and global environment) are economically viable. An important component of successful programmes has been the recognition of these constraints *(i – iv)* and the inclusion of specific planning efforts to provide innovative solutions, such as workshop/training efforts, purchase programs following trial/evaluation periods, or micro-loans to both vendors and end-users.

Furthermore, reports from commercial concerns or advocacy groups that may provide cases for self-promotion have been avoided.

16.3   Interaction of the Local and Global Environment for Technology

The relationships between the specific technology and the local, national, and international contexts all strongly affect the outcome of an attempt to transfer a technology. For example, the local context in which the renewable energy technology is implement-

ed is characterised by climate, topography, local actors, financial and technical institutions, educational resources, and especially by the technology itself. International actors, institutions, and economic and political pressures also play an important role. Within these contexts there are often a number of actors at every level, including the end users, businesses, NGOs, local, state and national governments, and national and multinational development agencies. It is useful to examine the implementation processes as the interaction between two systems: the local environment for the technical and financial operation of the technology or management system and the broader political system in which the technology is implemented. Because both systems are unique, there is no single "best outcome" or path, and the process for achieving success must be locally defined. Additionally, both systems are dynamic, so even in a given context the definition of a "best outcome" is liable to change in time.

While there are few hard rules, some patterns do exist. Changes in certain variables - namely those affecting the relationship between local implementation process and the wider political system - generally produce certain results that are fairly predictable and can be described analytically. For example, a subsidy affects the marketability of a product, often in predictable ways. As a second example, government involvement in programme motivation, popularisation, or in standard setting can significantly accelerate and improve the technology transfer process. However, governmental involvement in detailed technology design or implementation can often lead to inefficient and rigid programmes that do not encourage innovation and entrepreneurial activity.

16.4   Cross-Cutting Issues

A number of cross-cutting issues and lessons have been observed that illustrate a variety of successful policy options[1]. In particular, these issues relate to information, capacity to innovate and utilise technologies, market support, and infrastructural support for the uptake of technologies. In several cases, factors that are known to have contributed to a successful working of advanced technologies include:

- Information: information and awareness about the range of technologies available in the market and the quality of service these technologies provide;
- Infrastructure: institutional arrangement to regulate working, monitor performance, and provide service support to ensure efficient performance of the technology;
- Capacity to innovate and use technologies: including training the technicians using the technologies and maintenance capacity (this includes know-how of the technology as well as access to spares for the upkeep of the equipment); and

---

[1] See also Chapter 4 on enabling environments for technology transfer and Chapter 5 for financing and partnerships for technology transfer.

- Market support: ability of the market to sustain the demand for technology, through adequate financial incentives, and an enabling regulatory structure.

The case studies investigate specific strategies to handle these issues. Four important strategies include subsidies, monitoring the regulatory environment, providing information, and enhancing the capability to choose, adopt, and adapt technologies.

16.4.1   Subsidies

Subsidies are often used to facilitate or accelerate the uptake of specific technologies in an investment environment where competing technologies/practices are at a relative economic advantage. As a tool, subsidies play a problematic, but important role in technology transfer. While there are numerous examples where subsidies have either spoiled the market for a new technology, resulted in an inferior product, or blocked market forces from operating properly (Barnes *et al.*, 1994; Kammen, 1995a,b), there are many instances where subsidies have been critical to successful technology transfer. In a number of case studies (e.g., Case studies 1, 4, and 14) subsidies significantly enhanced the transfer and adoption process because they addressed a specific market imperfection and were either limited in extent or duration, or they took the form of logistical support or training.

These programmes did not entail direct cash payoffs. Key factors for success involve the timing and magnitude of subsidies, the degree to which local co-payment or participation are required, and the type of support provided (direct technology subsidy vs. market support). In specific instances, subsidising the 'soft' side of technology transfer, such as infrastructural support or information programmes may prove far more effective in sustaining the uptake of technologies.

Subsidies can be classified into three major areas: (1) development subsidies; (2) technology sales subsidies; (3) and market-support and educational subsidies. Each type of technology/system support has its advantages and drawbacks. The critical policy decision in support of technology transfer, therefore, is under what conditions do one or more of these approaches most directly and cost effectively address market imperfections, social, economic and environmental needs.

Development Subsidies:
Subsidies in the product development phase typically support the classical 'R&D' phase, pre-market design, or possibly diversification from a prototype to models tailored to particular market niches. These subsidies are often in the form of a direct grant or loan to a particular manufacturer, and frequently on the basis of a promising engineering design. A benefit of this approach is that it can be relatively simple to evaluate the proposal and to chart the impact of the subsidy in terms of product development. One drawback is that funding institutions, international, multinational, or national, may often fall into the trap of 'picking winners' before any feedback via the market from end-users is available

(Cohen and Noll, 1991; Margolis and Kammen, 1999). A number of recent technology and environmental policy efforts have illustrated opportunities to move beyond this roadblock, however, by promoting technologies in competitive programmes where subsidies are provided for combinations of technical and managerial innovations. Recent efforts to promote improved cookstoves in China (Barnes et al, 1994), were based on provincial-level competitions to best meet the energy efficiency and economic needs of households.

Technology Sales Subsidies:
Sales subsidies are also a traditional mechanism to support and develop the market for a new technology. In the classic formulation, end-users receive a rebate from a third party (often the government) for the purchase of a technology. The benefits of this approach are that the subsidy can directly reduce up-front capital cost, which is often the critical obstacle for the dissemination of new technologies (Gupta and Ravindranath, 1997; Duke and Kammen, 1999). Conversely, the drawback of this approach is that lump-sum subsidies may not provide an incentive for the performance of the technology, only the initial sales. In Nepal, however, subsidies for biogas digesters have been provided in stages over several years to guarantee that the systems perform well. These subsidies are incremental to provide the most support to the poorest and most remote households. Finally, the biogas digester subsidy is provided to the installer, who also holds the loan to cover end-user purchases. The advantage of this arrangement is that the risk of a novel, and often untested technology does not fall on the end-user.

Market-Support and Educational Subsidies:
There has been a recent explosion of interest in subsidies that avoid direct financial subsidies while still supporting an emerging new technology or clean energy practice. One way to accomplish this is to subsidise the educational, training, or other knowledge-based aspect of the R&D to commercialisation pipeline. For many technologies, particularly in developing nations, there is only a weak link between a promising new technology development and the marketing skills and resources needed to achieve commercial success. Training programmes, efforts to assist with market development and other such 'soft' subsidies can often make a great deal of difference. The benefits are often far greater than would a direct hardware subsidy. An example of this approach has been the development of improved cookstoves in Kenya (Case Study 1) where marketing was subsidised, but the cost of the stoves themselves reflected the actual production and market costs.

An informative example program that integrates pieces from each of the three subsidy categories is that of Greenfreeze refrigerator program. The Greenfreeze (Greenpeace, 1999) program in Europe brought together scientists who had extensively researched the use of propane and butane as refrigerants, with an East German company. The result was a research effort that identified a particularly effective and economically well-suited mixture of refrigerants that were as effective as traditional refrigerants.

The initial announcement of the plan to market this "Greenfreeze" product resulted in Greenpeace being able to gather tens of thousands of pre-orders for the yet-to-be produced new refrigerator from environmentally conscious consumers in Germany. This overwhelming support from the public secured the capital investment needed for the new 'Greenfreeze' product. Initial industry resistance to this switch in refrigerant was overcome by the demonstration of the market demand, and Greenfreeze rapidly became the dominant technology in Europe. There are now over 100 different Greenfreeze models available for purchase.

Long-Term Sustainability

The types of subsidies discussed above each illustrate the importance of targeting windows of opportunity where financial, institutional, educational, or public relations support for promising technologies can have a market-opening or a lasting impact on technology adoption and sustainability. In many cases, however, the concern exists that short-term technology support could lead to long-term economic dependency on the subsidies or other support mechanisms. Many of the Case Studies in this chapter address this issue (e.g. Cases 1-5, 8, 13, 17, 22-24, 27-28). The over-arching lesson is that well-targeted subsidies, often focusing on market support, information and training, and not direct subsidies of the specific technology can make a dramatic difference in an emerging market. Further, providing support to initiate or jump-start a market can then lead to dramatic product and cost improvements as well as large increases in demand (Duke and Kammen, 1999). The transition from novel idea to important product generally obviate the need for subsidies exactly at a stage when competition within the market becomes a more efficient mechanism to spur innovation and cost reduction. This process both argues for the importance of intelligently designed subsidies that support local (individual, community and private-sector) knowledge and control of a new technology and illustrates the opportunities for long-term sustainability of new ESTs.

16.4.2   Regulatory Environment

Environmental regulation and policies often play a significant role in creating conditions for adopting certain technologies. Generally, environmental standards based on "best available technologies" tend to perpetuate existing control technologies at the expense of long-term innovation. This type of standard setting results in regulations focusing on "end-of-pipe" type pollution control technologies, and can create a strong disincentive for going beyond the proven standards dictated by existing technologies (TERI, 1997).

16.4.3   Information

Detailed information on the performance of technologies, processes and equipment, with specific reference to environmental and financial benefits are the necessary first step in technology assessment and selection. Often, lack of information and understanding of skills and infrastructure required for the suc-

cessful operation of systems using transferred/acquired technologies results in sub-optimal performance. Examples of information strategies include providing educational brochures, energy use feedback programmes, energy audits, and labelling programmes. Case studies in India show various efforts to organise technology users and vendors, and to collect and distribute data. These efforts have been successful and have also helped redefine the goals for technology development (TERI, 1997).

16.4.4   Capability to Choose, Adopt, and Adapt Technologies

Capacity is a function of the ability to access, adapt, and manage technologies. Building an adequate technical infrastructure, and training managers and engineers to incorporate environmental issues in the decision making processes have been continuing problems in both economic development and environmental improvement. Several cases (cited in Chapter 9) show that developing indigenous capacity plays an important role in facilitating informed technological choices, and allows for greater capacity to innovate and adapt the technology to suit local conditions.

16.5   Key Findings and Lessons from Technology Cooperation

Just as policies and policy instruments have varying effects on the marketability of a technology, they also tend to affect different aspects of the use and operation of the new technology. The salient lessons emerging from cases listed in this chapter include the following:

1. Technology cooperation can work remarkably well. The global economy has emerged with advanced communication and information access, experts trained internationally, and strong incentives for cooperation. Cooperation is particularly powerful in helping entrepreneurs choose new investments. At the local level, programmes spur local innovation and competition, when supported by broader cooperative networks that pool resources.
2. Technology cooperation for global environmental protection has proved to be successful in a large number of cases. There is a global concern for environmental protection that is shared by citizens, corporations, and governments in every country. This concern has motivated unprecedented actions including global environmental treaties and accords (e.g. the Kyoto and Montreal Protocols, the Law of the Sea, and regional standard-setting measures for emissions such as SO_2 over Europe).
3. Technology cooperation can be win-win. In addition to being environmentally beneficial, technology cooperation can have additional benefits for participants. Companies gain reputation and access to national technology leaders and markets; engineers are challenged, motivated, and empowered; suppliers find capable customers to help

introduce new technology; and end-users save money and improve their quality of life. The financial savings from efficient technologies, reduction of waste, and improved worker productivity often are greater than the costs of cooperative implementation. Nevertheless, win-win scenarios can be complex and multifaceted. In some cases nations or industries can be coerced or induced to adopt ESTs even if they do not initially perceive the benefits. An example is cogeneration of power in India where governmental regulation forced a change that, despite significant industry objections, has proved to have benefits to both the state through added power resources, and to the companies through additional sales and energy security.

4. Funding is critical to technology cooperation. At a minimum, adequate funding is necessary for organising and managing projects, providing expert consultants, communication, and reporting. In some of the most successful examples of technology cooperation, funds are also available for the incremental costs of the environmentally superior technology (e.g. Montreal Protocol Multilateral Fund, see section 3.3.3 in Chapter 3 and 5.5.7 in Chapter 5).

5. Technology cooperation requires clear goals and motivation as a precondition for success.

6. Corporate and government leadership accelerates progress by creating momentum, jointly overcoming market barriers and by promoting market incentives. Corporations are particularly influential in encouraging suppliers to improve environmental performance. Governments are often successful in streamlining regulatory approvals.

7. Funding institutions as opposed to projects supports institutional capacity and local sustainability. Numerous cases point to an inability on the part of technology recipients to pursue locally appropriate technological solutions (e.g. dispersed renewable energy systems), because international support was only available for technology and not human and institutional resources and capacity.

8. Failure to seek participation from users or local people in technological adaptations can lead to sub-optimal utilisation or even a complete failure in the use of the technology.

9. Almost every successful technology cooperation results from the commitment of individual "champions" who motivate, persuade, and manage the technological, political and economic process that leads to eventual dissemination or commercialisation.

10. Failure to recognise issues of intellectual property rights and proprietary R&D produce significant bottlenecks to greater private-sector involvement in many technology transfer efforts that are environmentally and socially beneficial.

11. Restructuring of markets for energy, recycled and re-used materials, and property rights for natural resources create opportunities for EST adoption. However, structural changes also create major challenges that require institutional solutions to make ESTs competitive or less costly than more resource-intensive technologies.

12. National and international action and, in many cases, cooperation and policy initiatives are needed to attach appropriate prices to natural resources. If instituted, many of these policies could significantly reduce the wasteful use and pollution of limited global resources.

References

Acker, R. H., and D.M. Kammen, 1996: The quiet (energy) revolution: analyzing the diffusion of photovoltaic power systems in Kenya. Energy Policy, 24 (1), 81 - 111.

Barnes, D. F., K. Openshaw, K. R. Smith, and R. van der Plas, 1994: *What makes people cook with improved biomass stoves?* World Bank Technical Paper No. 242, World Bank, Washington, DC.

Byrne, J., B. Shen, and W. Wallace, 1998: The economics of sustainable energy for rural development: a study of renewable energy in rural China. Energy Policy, 26(1), 45-54.

Cabraal, A., M. Cosgrove-Davies, and L. Schaeffer, 1995: Best practices for photovoltaic household electrification programs . World Bank Technical Paper Number 324, Asia Technical Department Series, Washington, DC.

Cohen, L.R. and R.G. Noll, 1991: *The Technology Pork Barrel*. The Brookings Institution, Washington, DC.

Duke, R.D., and D.M. Kammen, 1999: *The economics of energy market transformation initiatives. The Energy Journal, 20 (4),* 15 – 64.

Feder, G., and D.L. Umali, 1993: The adoption of agricultural innovations. Technological Forecasting and Social Change, 43, 215 - 239.

Gereffi, G., 1983: *The pharmaceutical industry and dependency in the Third World.* Princeton University Press, Princeton, NJ.

Goulet, D. ,1989: *The Uncertain Promise: Value Conflicts in Technology Transfer.* New Horizons Press, New York, NY.

Greenpeace, 1999: Website: http://www.greenpeace.org/~ozone/unep_ods/8greenfreeze.html

Gupta, S. and N.H. Ravindranath, 1997: Financial analysis of cooking energy options for India. *Energy Conversion and Management*, **38**, 1869-1876.

Hankins, M., 1993: *Solar Rural Electrification in the Developing World; Four Case Studies: Dominican Republic, Kenya, Sri Lanka, and Zimbabwe.* Solar Electric Light Fund, Washington, DC.

Inversin, A. R., 1996: *New Designs for Rural Electrification: Private Sector Experiences in Nepal.* NRECA, International Programs Division, Washington, DC.

Kammen, D. M., 1995a: Cookstoves for the developing world. *Scientific American*, 273, 72 - 75.

Kammen, D. M., 1995b: From energy efficiency to social utility: Improved cookstoves and the Small is Beautiful Model of development. In *Energy as an Instrument for Socio-economic Development*. J. Goldemberg, T.B. Johansson, (eds.), United Nations Development Programme, New York, NY, pp.50-62.

Kammen, D. M., and M.R. Dove, 1997: The virtues of mundane science. *Environment*, 39(6), 10–15; 38 - 41.

Mansfield, E., 1994: Intellectual property protection, foreign direct investment and technology transfer. International Finance Corporation Discussion Paper No. 19, World Bank, Washington, DC.

Margolis, R., and D.M. Kammen, 1999: Underinvestment: The energy technology and R&D policy challenge. *Science*, **285**, 690 - 692.

Smith, K. R., G. Shuhua, H. Kun, and Q. Daxiong, 1993: 100 million biomass stoves in China: How was it done? *World Development*, 18, 941 - 961.

TERI, 1997: *Capacity Building for Technology Transfer in the Context of Climate Change.* TERI, New Delhi.

Case Study 1
Research, Development and Commercialisation of the Kenya Ceramic Jiko (KCJ)

Daniel M. Kammen
Energy and Resources Group
University of California, Berkeley, CA 94720-3050

Keywords: Kenya, energy efficient stoves, informal sector, subsidies, North⇒South, South⇔South, technology transfer

Summary
The Kenya Ceramic Stove, or *Jiko* (KCJ), is a charcoal-burning stove that is roughly 30% efficient, and if used properly can save 20 – 50% in fuel consumption over simple 'unimproved' stoves or a traditional three-stone fire (Walubengo, 1995). The KCJ was developed after study of a Thai 'bucket' stove that was examined partially through a 'South-South' dialog over stove characteristics and design.

The KCJ is a portable improved charcoal burning stove consisting of an hour-glass shaped metal cladding with an interior ceramic liner that is perforated to permit the ash to fall to the collection cavity at the base. A single pot is placed on the top of the stove. There are now more than 200 businesses, artisans, and micro-enterprise or informal sector manufacturers producing over 13,000 stoves each month. There are over 700,000 KCJs in use in Kenya (Walubengo, 1995). The KCJ is found in over 50% of all urban homes, and roughly 16% of rural homes. Stove models adapted from the KCJ are now being disseminated in many countries across Africa, and wood-burning variants are being introduced and promoted in rural areas as well.

The fuel savings of the KCJ have important economic benefits to the users, who in some cases devote a quarter of family income to charcoal purchases (Kammen, 1995). The stoves can also reduce the pollution exposure of families using the stove (see also Box 14.1). The WHO reports that more than two million premature deaths per year can be attributed globally to the indoor air pollution caused by household solid fuels. Reducing the harmful products of incomplete combustion produced by household stoves is an important benefit from the development of cleaner cookstoves.

Approach
The KCJ is the result of research on stove design, efficiency, and patterns of usage initiated in the 1970's and actively continued through the 1980s (Barnes *et al.*, 1994; Kammen, 1995). A single private sector company, Jerri International, served as the initial manufacturer of the KCJ.

Since 1982 the Kenya Energy and Environment Organisation (KENGO) has organised promotion and outreach efforts to encourage the use of the KCJ. A number of NGOs and national development agencies have played important roles in the evolution of the stove and the stove dissemination process, and have worked both within Kenya and across sub-Saharan Africa to promote the manufacture and sales of the KCJ through a network of informal-sector stove entrepreneurs.

A decision was made not to directly subsidise commercial stove production and dissemination. Initially stoves were expensive (~ US$ 15/stove) sales were slow, and quality control has been a significant problem. Continued research and refinement and expanded numbers and types of manufacturers and vendors increased competition, and spurred innovations in materials used and in production methods. The KCJ can now be purchased in a variety of sizes and styles. Prices for KCJ models have decreased to roughly US$ 1 – 3 (Walubengo, 1995). This decrease is consistent with the 'learning curve' theory of price reductions through innovations that result from experience gained in the manufacturing, distribution, marketing and sales process. Two architects of the stove programme received an international award for their work, which is an important recognition for the need for research on often unheralded but important technologies.

The ceramic liner of the KCJ degrades over time, and needs to be replaced. Street vendors of stoves, and many of the larger stove sales outlets take 'used' stoves back, discounting the purchase of a new stove. The liners of the old stove are then removed, the metal cladding is repaired, if needed, and the stove is reassembled, repainted, and resold. This process has also served to foster a wider informal sector stove economy.

Impacts
The KCJ can reduce fuel use by 30 - 50%, although charcoal production itself can have significant environmental impacts, and the attractiveness of the KCJ may have increased this demand. The KCJ also reduces emissions of trace gases and particulate matter, which contributes to acute respiratory infection, the leading cause of illness in developing nations. Reported levels of emissions reductions from KCJ range up to 50%, although this is a subject of ongoing research (Barnes *et al.*, 1994). The KCJ and the dissemination process used in Kenya has now been widely disseminated (and adapted) across sub-Saharan Africa.

Lessons Learned
While avoiding direct subsidies, a number of organisations provide training, outreach services, publicity, and logistical support for the local commercial industry. This 'soft' subsidy can be particularly effective in facilitating the development and acceptance of a new technology without introducing the price distortions that can be associated with some forms of subsidy.

The lessons for international involvement that can be drawn from the KCJ case include:

Support for research both within developing nations and for research collaborations between developing nations can lead to significant innovations in the performance and commercialisation of what had been regarded by many as a simple and mature technology.

Extended, stable, programme support is invaluable while short-lived, episodic funding can lead to waste and inefficiency. There are significant technical, social, cultural and economic questions that must be addressed even for technologies that may appear simple.

Support for stove programmes need not take the form of direct subsidies. Partnerships between institutional groups, including NGOs and international organisations, involved in R&D, promotion, and training can support commercial producers and sellers if the mechanisms for feedback and cooperation are planned and developed.

References

Barnes, D.F., K. Openshaw, K. Smith, and R. van der Plas, 1994: *What makes people cook with improved biomass stoves?* World Bank Technical Paper No. 242, Energy Series.

Kammen, D. M., 1995b: From energy efficiency to social utility: Improved cookstoves and the *Small is Beautiful* Model of development. In *Energy as an Instrument for Socio-economic Development.* J. Goldemberg, T.B. Johansson, (eds.), United Nations Development Programme, New York, NY, pp. 50 - 62.

Walubengo, D., 1995: Commercialization of improved stoves: the case of the Kenya Ceramic Jiko (KCI). In *Stove Images: a Documentation of Improved and Traditional Stoves in Africa.* B. Westhoff, D. Germann, (eds.), Commission of the European Communities, Brussels.

Contact
Stephen Karekezi, Director
Aafrican Policy Research Network
AFREPREN/FWD
P.O. Box 30979
Nairobi, Kenya
Tel: 254-2-566032/571467
Fax: 254-2-561464/566231/740524
Email: StephenK@africaonline.co.ke

Case Study 2
Public Promotion of Private Investment in Efficient Lighting
Richard Duke and Steve Ryder
Woodrow Wilson School
Princeton University, NJ

Keywords: CFLs; Green Lights; PELP

Summary

Governments and multilateral agencies have promoted efficient lighting markets as a means to reduce greenhouse gases (GHGs). Lessons learned from the Poland Efficient Lighting Project (PELP) and Green Lights (GL) have inspired similar efforts in developing countries.

Background

In 1991, the U.S. Environmental Protection Agency (U.S. EPA), established Green Lights (GL) to encourage Partners to systematically consider efficient lighting investments. Furthermore, it empowers facilities managers to internally advocate for lighting efficiency upgrades.

PELP is a US$5 million initiative (1995 to 1998) funded by the Global Environment Facility (GEF) and executed by the IFC to reduce GHGs by increasing sales of compact fluorescent lamps (CFLs). PELP reduces price and consumer awareness barriers through financial incentives and public education.

Approach

GL is a voluntary programme that encourages public and private institutions to enter into Memoranda of Understanding with U.S. EPA. Partners agree to invest in 90% of profitable lighting efficiency upgrades within five years. Partners appoint implementation managers and submit annual progress reports. GL also recruits utilities and lighting companies to become certified Allies.

U.S. EPA agrees to provide both Partners and Allies with efficient lighting technology workshops, objective information about efficient lighting technology and financing, and assistance publicising Partners' successes.

PELP selected a manufacturer wholesale price reduction approach because it promised the largest increase in CFL sales at the lowest cost. When the subsidy is given to the manufacturer, the sales tax and distribution chain markups are also proportionately reduced. Careful monitoring ensured that subsidies passed down the distribution chain to the retail level. Five Poland-based manufacturers competed to deliver the largest reduction in electricity use for the smallest subsidy (World Bank, 1996).

The subsidy programme was accompanied by an extensive public education component using media, school programmes, environmental fairs and energy efficiency competitions. A PELP logo, along with the allowable retail price, was placed on subsidised products.

Through a competitive process, IFC selected the Netherlands Energy Company (NECO), a Dutch utility entity, to administer PELP. Among others, NECO drew upon services from the Polish Foundation for Energy Efficiency and the International Institute for Energy Conservation. The project also included a pilot CFL-based demand-side management (DSM) programme.

Impacts

GL has over 1,600 Partners (85% corporate, 15% government) and 594 Allies (41% manufacturers, 25% distributors, 21% lighting management companies, and 14% utilities). GL upgrades completed by 1997 reduced U.S. GHG emissions by 0.1% and U.S. EPA projects that this will rise to 0.4% by 2000.

GL does not provide direct subsidies to Partners; however, through 1997, the total administrative budget has been US$90 million and Partners received US$184 million in utility DSM rebates. Through 1997, the private internal rate of return (IRR) for Partners averaged 50%. The social IRR (counting GL programme costs as expenses and excluding DSM transfers from benefits) is still robust at approximately 35%.

In 1995, U.S. EPA launched the Energy Star Buildings programme that employs a similar voluntary approach to encourage efficiency in all aspects of building energy use. GL also inspired a related China Green Lights programme. While U.S. EPA supports international replication of the GL model, programmes must be carefully tailored to country-specific barriers. Moreover, multilaterals and NGOs may prove better positioned to propagate the lessons from this programme given political constraints on U.S. EPA investment in international programmes.

With PELP, US$2.6 million in GEF-funded direct subsidies leveraged US$7.5 million to reduce the cost of 1.2 million CFLs. Thus, US$2.10 of GEF wholesale subsidy per CFL yielded a US$5.90 retail price decrease, a 20% discount. These CFLs will save approximately 725 GWh of electricity and eliminate 206,000 tons of carbon. Public education activities raised consumer awareness of CFLs. Roughly 50% of first-time CFL purchasers learned about CFLs through PELP and 80% said they intend to buy more CFLs.

PELP appears to have achieved the desired market transformation. Polish CFL prices have stabilised and even dropped after the programme ended. This can be attributed to the economies of scale possible in the now-expanded Polish CFL market. GEF recently approved a US$15 million Efficient Lighting Initiative to promote similar programmes in other transitional and developing countries.

Lessons Learned

Principle lessons include: engaging the private sector through manufacturer subsidies or voluntary agreements can reduce barriers to efficient lighting markets; governments can also help transform energy efficient lighting markets by providing credible public information; evaluating programmes like PELP and GL poses analytic challenges; for PELP, the use of respected local companies and NGOs was key to navigating legal and economic obstacles, lowering project costs and promoting capacity for implementing similar projects in the future.

The fundamental programme evaluation obstacle is determining what would have happened had the programme never existed. Free riders and drivers underlie this baseline problem. The free rider effect is the percentage of participants who would have purchased efficient lighting had GL or PELP never existed. The free driver effect refers to non-participants that invest in upgrades due to indirect effects such as spillover of information and price reductions. Duke and Kammen (1999) show that GL has decreased the price of electronic ballasts by approximately 1-2%. This induces additional demand for efficient lighting products.

References

Duke, R., and D.M. Kammen, 1999: The economics of energy market transformation programs. *The Energy Journal,* **20** (4), 1 – 50.

World Bank, 1996: Poland Efficient Lighting Project. GEF Project Document, Washington, DC.

Contact
Richard Duke
Science, Technology and Environmental Policy (STEP) Program
Woodrow Wilson School
Princeton University
Five Ivy Lane
Princeton, NJ 08540
duke@princeton.edu

Case Study 3
Inner Mongolian Household Wind Electric Systems

James H. Williams
Energy & Resources Group
University of California, Berkeley, CA 94720-3050

Keywords: China, Energy, N⇒S

Summary

Since the early 1980s, China's Inner Mongolia Autonomous Region (IMAR) has achieved widespread local production and dissemination of stand-alone wind electric systems among its rural herding population. The IMAR government has employed a combination of market and state planning mechanisms to create markets, adapt foreign technologies, and develop a unified system of design, manufacturing, distribution, and service. By 1998, 130,000 household wind systems provided electricity to over 500,000 people, or one-third of the herding population, and displaced substantial GHG emissions that would otherwise have been produced by coal or diesel alternatives.

Background

The Inner Mongolia steppe is inhabited by ethnic Mongolian livestock herders earning US$120 per capita/yr. Because of low population densities (< 3 persons/km^2) and high costs, only about 5% of the herding population is served by the regional electric grid or stand-alone diesel generators. To exploit IMAR's abundant wind resources, in 1980 the regional government appointed a high-level New Energy Leading Group (NELG) to oversee development, issuing four guidelines: (a) serve the pastoral population (b) emphasize basic household needs (c) make products "reliable to use, convenient to maintain, and affordable to herders" (d) "local management is the key, with the state providing appropriate support." (Byrne *et al.,* 1998).

Approach

NELG-sponsored task forces and pilot projects brought together the agricultural, finance, and planning ministries, universities, research institutes, factories, local governments, and herders. Technical R&D and economic research produced four important results. First, products were developed that met herders' needs and were manufacturable by local industry. Second, demonstration projects familiarised herders with the technology. Third, key questions about programme implementation were resolved empirically, such as the decision to emphasize private ownership and individual household systems. Fourth, a unified network combining R&D, manufacturing, distribution, and service was created, under the leadership of the IMAR New Energy Office (IMARNEO).

The resulting manufacturing system is diverse. Six factories produce 20 different models of wind generators, from 50W to 7.5kW. The development of reliable low-cost turbines under 200W is a key innovation. Eleven factories manufacture batteries, inverters, and charge controllers. Water-pumping windmills and renewable energy products such as electric fences and DC lights are also produced. Service centres in 60 of IMAR's 88 counties, financially supported by county governments, employ 300 technicians who handle distribution, customer education, and service.

Most (>90%) household systems in IMAR include a 100W wind generator, 12V battery, and charge controller. They cost ¥2000-2500 (US$250-300) and typically power lights, radio, and black-and-white television. Recently, increases in household income have shifted demand toward larger systems, typically including a 300W wind generator and two batteries (US$500-600). New wind/PV hybrid systems supply power more reliably in the summer when wind resources are lowest.

Sales of wind systems are supported by a government subsidy of ¥200 (US$25) per 100W of capacity, or 10% of typical system cost. Since 1986, the government has provided ¥26.5 million (US$3 million) in subsidies, peaking at ¥4 million (US$500,000) in 1989 (Li, 1998). Subsidies were initially given directly to households, then to local governments; in both cases, abuses occurred. Since 1988, factories have been given the subsidy directly, which is passed on in reduced purchase prices (Lin, 1997).

Technology transfer has centred on local adaptation of foreign products. The design of the Shangdu Livestock Machinery Factory's 100W and 300W turbines, accounting for over 80,000 units in IMAR, grew from a collaboration between SVIAB of Sweden and the Shangdu plant, which produced turbines for SVIAB in return for the technology license. The key development was local design changes that altered the power curve of Shangdu generators, increasing power output at lower wind speeds. This is a crucial adaptation to the steady but low-speed wind resource in IMAR.

Up to 85% of IMAR systems are reported operational. Common problems involve batteries, blades, inverters, and charge controllers. Service centres keep statistics on failure rates. A technical board determines whether frequent failures result from quality defects or design flaws. Manufacturers address quality defects; R&D organisations such as IMAR Polytechnic University address design flaws. Joint ventures with foreign wind turbine and electronics manufacturers are currently being sought to improve blades and inverters.

Impacts

(a) The IMAR wind programme has led to dramatic changes in quality of life, providing lighting, conveniences, and access to the outside world for low-income inhabitants of this remote region. A common wedding gift is now a wind generator or television, illustrating the extent of the technology's integration into the pastoral lifestyle. (b) Wind generators displace potential carbon emissions of about 10-15 ktC/yr based on an equivalent amount of electricity produced with diesel generators.

Lessons Learned

The IMAR programme exemplifies successful government leadership in rural energy using realistic programme guidelines, adequate preparatory research, multiple-stakeholder participation in goal-setting, coordinated use of plan, market, and subsidy, local project control, and adaptive management. Well-targeted subsidies have been effective. Technology transfer has been important, but has served as an adjunct, rather than as a principal driver of the process.

References

Byrne, J., B. Shen, and W. William Wallace, 1998: The economics of sustainable energy for rural development: a study of renewable energy in rural China. *Energy Policy*, 26(1),45-54.

Lin L., 1997: The development and utilisation of new energy sources in the Inner Mongolia Autonomous Region: review and outlook. *Inner Mongolia Science, Technology and Economy*, 4, 27-30.

Li D., 1998: The current situation of the wind power equipment industry in our country. *New Energy*, 20(1), 37-41.

Case Study 4
Hydrocarbon Refrigerator "Ecofrig" in India

Sukumar Devotta and Saroja Asthana
National Chemical Laboratory, Pune 411 008, India

Keywords: India, Switzerland, Germany, ODS phaseout, HC refrigerator technology, N⇔S, S⇔S.

Summary

The basic objective of the Ecofrig project is to adapt the hydrocarbon technology (HT), developed by German manufacturers, for Indian domestic and commercial refrigeration appliances in cooperation with a few Indian industry and research partners. The project involved a voluntary partnership between the Swiss, German and Indian governments and private industries. In the first phase of the project two pilot facilities to use cyclopentane (CP) as an alternative to CFC-11 as the blowing agent in polyurethane foam insulation used in refrigerators have been commissioned. In the second phase of the project, the use of hydrocarbon (HC) alternative refrigerant to CFC-12 is being implemented in some industries.

Background

The worldwide phaseout of ozone depleting substances (ODSs) in the major sectors, including refrigeration, was active in 1992 to 1993 under the Montreal Protocol (MP). The Indian refrigeration industry is presently engaged in phasing out ODSs. There are many choices as alternatives to chlorofluorocarbons (CFCs). In this project, the emphasis was given for both Ozone Depletion Potential (ODP) and Global Warming Potential (GWP) in the choice of alternatives. With these bases, hydrocarbons with zero ODP and negligible GWP were chosen. The key issue in the HC based technology is the safety aspects of using flammable blowing agents, i.e. CP and the flammable HC refrigerant.

The globalisation of the Indian economy and the integration of the Indian market with international markets have created new challenges as well as business opportunities for Indian industries. These changes have facilitated multinational acquisitions, import of CFC-free refrigerators into India at comparatively cheaper rates, and encouraged Indian companies to plan for international markets.

Approach

The legal basis for the project is the Memorandum of Understanding (MOU) the 'Indo-Swiss collaboration in ecological domestic refrigeration', signed between the Swiss government, represented by the Swiss Agency for Development and Cooperation (SDC) and BUWAL (the Ministry of Environment and Forests), and the Indian government, represented by the Ozone Cell of the Ministry of Environment and Forests (MOEF). The German government, represented by BMZ (the Federal Ministry of Cooperation) and acting through Gesellschaft for technische Zusammenarbeit (GTZ)

Gmbh as a co-contributor to the project, had entered into a separate co-financing agreement with the Swiss government. The collaboration between MOEF and the Indian industry partners, namely, Godrej GE Appliances and Voltas Ltd. was formalised through another MOU. The project coordination was entrusted to INFRAS AG, an independent environmental consulting group from Zurich. The Indian Institute of Technology (IIT), New Delhi; National Chemical Laboratory (NCL), Pune; and Tata Energy Research Institute (TERI), New Delhi are the research partners in the project.

Both the Swiss and German governments had made donor contributions towards the cost of consultancy services for planning and project implementation, equipment, materials and other imports, and consultancy services of Indian research institutes, etc. The Indian private industry partners have contributed about 50% of the cost of the project. The Indian government had facilitated the import of plant machinery under this project as Swiss and German contributions, exempted from all customs and excise taxes.

There are two phases of the Ecofrig project, with Phase I from September 1992 to March 1996 and Phase II from March 1997 to 1999/2000. The two components of hydrocarbon technology for India are the blowing agent in foam (cyclopentane) and the refrigerant (HCs: HC-600a or HC-600a/HC-290 blend). During Phase I, two pilot plants based on CP were set up at Godrej and Voltas/Allwyn. Plant machinery was imported from Italy and Germany. The installation was completed as per the German safety norms. Through the operation of these pilot plants, the two Indian industry partners had acquired the know-how needed for the handling of CP blown foam technology on a commercial scale.

The use of HC refrigerants in compressor and appliance manufacture has been achieved only partially with the transfer of know-how for safe designs of domestic refrigerators from Liebherr, Germany to Voltas and Godrej in India. In compressor design for R600a, the additional activities of adaptive research, pilot production, information dissemination, and networking is expected to continue into Phase-II.

The project includes several elements including plant machinery to manufacture refrigerators using cyclopentane foam and hydrocarbon refrigerant. The compressor and other components were adapted to the new foam blowing process and refrigerant.

Impacts

The experiences gained in CP foam technology, particularly in relation to safety, are being made available to other Indian refrigerator manufacturers and, indeed, for similar projects in other developing countries. The two Indian refrigerator manufacturers have opted for CP blown polyurethane foam for their entire range of refrigerators. This will lead to significant reduction in ODS consumption in India.

At Godrej-GE the conversion to cyclopentane has eliminated 568 MT of CFC11 equivalent to roughly 600,000 MT CO2/a if HCFC or HFC path would have been chosen. 147 MT of CFC12 are in process of being phased out by use of isobutane

Lessons Learned

The project had built capacity in the actual setting up of the pilot (or demonstration) production units (including plant machinery) for CP foam and isobutane, and HC blend-based adaptive research in compressor and appliance design. These are considered essential for sustainable indigenous development. The first hand experiences gained in CP foam technology, particularly in relation to safety, are expected to be made available to other Indian refrigerator manufacturers and, indeed, for similar projects in other developing countries.

The documentation of adaptive research results and dissemination through various international publications by the Indian research partners have been found to be very effective methods for replication and sustainable development.

Contacts
Dr. Othmar Schwank,
Infras, Zurich, Switzerland;
Email: oschwank@infras.ch.

Case Study 5
The Commercial Dissemination of Photovoltaic Systems in Kenya

Daniel M. Kammen
Energy and Resources Group
University of California, Berkeley, CA 94720-3050

Keywords: Kenya, photovoltaics, solar home systems, N⇔S, S⇔S.

Summary

The commercial market for solar photovoltaic (PV) home systems has been active in Kenya for over 10 years. Over 80,000 solar home systems (SHSs) have been sold, providing power to over 1% of the rural population of roughly 25 million people. The total installed capacity is over two megawatts (MWp), with typical individual systems from 10 – 40 Wp (Wp = peak watts), and costs per system between US$250 and US$1,000. By contrast, the national Rural Electrification Programme has connected less than 2% of the rural households to grid-connected power supply. The Kenyan PV market evolved without subsidies or significant government or multinational agency support, although the activities of several private and volunteer organisations helped to disseminate information on photovoltaics and provide opportunities. Today dozens of companies are active in the PV industry in east Africa, and annual sales may exceed 20,000 – 30,000 individual systems and over 300 kWp (Msinga *et al.*, 1997). USAID, GTZ, Care and Bellerive foundations were instrumental in the development and dissemination process of cookstoves.

Background

Today nearly two billion people worldwide remain without electricity or the immediate prospect of grid electrification. While over 1.2 billion people have gained access to grid-connected electricity services over the past 20 years, this has not kept pace with the population that has grown by over 1.5 billion during the same time span. In Africa, grid extension has been the slowest of all the major regions of the 'South'; only half of urban residents and a mere eight per cent of the rural population are served by a grid-based power supply. Meeting the growing demand for power over the next decades will be exceedingly difficult in itself, and will be compounded by the challenge of doing so without massive increases in GHG emissions.

Until the late 1980s PV electrification in east Africa was largely confined to the affluent and to donor projects that often powered wells and bore holes, schools, refrigerators in rural health clinics, and missions (Acker and Kammen, 1996). These systems generally arrived as complete packages, often imported directly from overseas via a local agent, distributor, or development group. Falling prices of crystalline and amorphous panels, increased awareness of the services that photovoltaics can provide, and recognition of the potential to commercialise PV systems all increased activity in the local market.

Approach

A key aspect of the dissemination process for photovoltaics in Kenya is the degree that it has been market driven, and thus the 'approach' taken is best described as 'market realism'. Initially, most systems were relatively large, and, as noted above, either donated or purchased by rural elites. A recent survey of over 400 households (Msinga *et al.*, 1997) found that while the mean income was US$108/month, three-fourths of the system owners earned less than US$100/month, which is close to the national average. Recently the average system size had decreased from over 40 Wp to about 20 Wp, including the sales of large numbers of 12-14 Wp packaged systems.

Expanding interest in SHSs in the late 1980s and early 1990s was accompanied by a change in the retail network and the development of a diverse regional network of assembly, sales, installation, and maintenance businesses. While solar panels are still imported, local companies now manufacture batteries for use in PV systems, and the 'balance-of-system' (electronics, charge controllers, lights and outlets, and other components) is assembled or manufactured in Kenya. Local agents now sell over half of all modules, and three-fourths of the PV batteries. Sales of SHSs in Kenya have been increasing at 10 – 18% per year, and this trend is expected to continue.

Impacts

PV systems have had a significant impact in Kenya and East Africa. Surveys indicate that most of the systems (> 60%) were performing well, with the majority of the remaining systems not in use because the battery had no charge (Acker and Kammen, 1996; Msinga *et al.*, 1997). Most users purchased PV for the combined services of lighting, TV, and radio, and were pleased with the systems and would recommend them to others. In many locations the PV systems are simply cheaper than the kerosene, diesel or other alternatives, and in virtually every case the service provided by the SHS is superior in quality and reliability to these alternatives.

The PV experience in Kenya has also proved to be an important model for SHS introduction efforts in other developing nations (Cabraal *et al.*, 1995), and Kenya as well as a number of other nations are now the target for international aid and development funding to expand the markets for rural PV in developing nations. Kenya has also become the focus for a regional PV market that extends into neighbouring nations, as well as to other regions of Africa.

Lessons Learned

A variety of lessons emerge from the Kenya experience. First, a relatively small set of organisations and individuals providing training and support services can provide a critical infrastructure for an emerging technology. Second, subsidies are not necessarily needed to promote technology transfer, although logistical support, training courses, and performance standards all have central roles that require policy attention and commitments of resources.

Critical to the sustainability of the PV industry has been the diversity of commercial interests. The role for government and international policy action is also significant, and includes setting policies that promote independent power producers, limit or remove taxes and tariffs on desirable clean energy alternatives, and provide credit or financing to both companies and end-users.

References

Acker, R., and D.M. Kammen, 1996: The quiet (energy) revolution: the diffusion of photovoltaic power systems in Kenya. *Energy Policy,* **24,** 81 – 111.

Cabraal, A., M. Cosgrove-Davies, and L. Schaeffer, 1995: *Best Practices for Photovoltaic Household Electrification Programs.* World Bank Technical Paper Number 324, Asia Technical Department Series, Washington, DC.

Msinga, M., M. Hankins, D. Hirsch, and J. de Schutter, 1997: *Kenya Photo-Voltaic Rural Energy Project (KENPREP): Results of the 1997 Market Survey.* Energy Alternatives Africa, Nairobi.

Contacts
Mark Hankins
Energy Alternatives Africa
P. O. Box 76406, Nairobi, Kenya
Tel: +254-2-254-714623 or 716284
Fax: +254-2-720909
Email: energyaf@iconnect.co.ke

Case Study 6
Coal Power Plants in China
Ajay Mathur
Energy-Environment Technology Division
Tata Energy Research Institute
New Delhi – 110 003, India

Keywords: CCT, financing, low NO_x burners, China

Summary
A recent example of innovative financing of a clean coal technology (CCT) in a developing country is the Huaneng Power Project in China. In this case, the CCT manufacturer worked proactively together with export credit agencies (ECAs) and commercial banks to arrange financing, for the benefit of all stakeholders. In this example, the ECAs' contributions were a vital catalyst for the sourcing of funds for the utility to develop its power stations. This highlights the importance of ECAs, and the need for manufacturers to be proactive in arranging finance.

Background
In 1997, Huaneng Power International Inc. (HPI), a Chinese utility and a subsidiary of the Ministry of Electric Power (MOEP), financed three new power stations from funds raised by share issues on the international stock markets and by using export credit finance for the procurement of western equipment. From the CCT manufacturer's perspective, the transaction was a typical sale to an 'on balance sheet' purchase by a utility company in a foreign market. Because of the country risk for this market, the CCT manufacturer and the commercial banks required the involvement of ECAs.

Approach
HPI had a charter from MOEP to procure first-class boiler and turbine equipment from western suppliers. HPI financed its power station development from funds raised by share issues on the international stock markets and by using export credit finance for the procurement of western equipment. The share issues resulted in 25% of the company being owned by international investors. Three new power 700MW power stations were built at Dandong, Dalian and Fuzhou. The western equipment was supplied by Mitsui Babcock Energy Limited (MBEL), Westinghouse and Siemens. MBEL supplied the boiler equipment and Westinghouse and Siemens supplied the steam turbine equipment. The boilers are fitted with 24 low-NO_x burners, a CCT at the commercially available stage. These burners are rated at 51.6 MW_{th}, and NO_x is guaranteed at 300ppm (6% O_2 dry) over a range of coals.

To finance the western power station equipment, buyer credit facilities were arranged by the manufacturers. To achieve this, the manufacturers approached commercial banks and ECAs. The buyer credit was in the form of commercial bank loans for HPI, provided by a consortium of banks. Société Generale and Barclays Bank plc led the lending group of five major banks in

these contracts. To enable these commercial banks to make these loans available, export credit backing was required. This was arranged with the ECAs from the manufacturers' countries: ECGD from the UK and EXIM bank from the USA. ECGD acted as loan guarantor to Barclays, and Société Generale and EXIM bank provided loans.

The export credit financing for the western equipment can be summarised as follows:

	ECG	EXIM
Dandong	US$ 81 million	US$ 158 million US
Dalian	US$ 84 million	US$ 184 million US
Fuzhou	US$ 92 million	US$ 47 million US

In addition, for the Fuzhou project, KfW, a German ECA, supported Siemens in turbine financing.

Impacts
The low-NO_x burners are designed to fire a range of high ash, high volatile coals. There are also wide-scale plans to retrofit existing boilers with low-NO_x burners. There is a reduction of over 60% in NO_x emission levels compared to conventional burners. Emissions of NO_x can typically be reduced from 430-450 g/GJ to 170 g/GJ.

Lessons Learned
This example illustrates typical 'on balance sheet' financing for a CCT and demonstrates the importance of ECAs in export credit financing. The need for manufacturers to be proactive in arranging finance is clearly demonstrated.

Bibliography

IEA/OECD, 1999: *Cleaner Coal Technologies-Financing.* Brochure prepared by ETSU for the UK Department of Trade and Industry on behalf of the IEA Committee on Energy Research and Technology.

Contact
Roshan Kamall
Location 1124
Department of Trade and Industry
1 Victoria Street
London SW 1H 0ET
Fax. 0171 8287969
email. roshan.kamal@hend.dti.gov.uk

Case Study 7
Butane Gas Stove in Senegal
Sukumar Devotta and Saroja Asthana
National Chemical Laboratory
Pune 411 008, India

Keywords: Senegal, technology diffusion, energy efficiency, cookstoves, fuelwood, N⇔S

Summary
This is an energy sector case study and it involves Total, Totalgaz, the Senegal government, the European Development Fund, about 50 distribution companies, and local communities. The consumption of fuel wood was leading to deforestation and desertification in Senegal. But the introduction of butane as a household fuel with a suitable energy efficient stove helped to turn the tide. Since 1974, the sales of these stoves had enabled much of the population to benefit from this modern fuel. The programme has helped the government in reducing fuelwood energy consumption. The stoves are made by semi-industrial companies that also contribute to the economic development of the country.

Background
Senegal is a small country in the African Sahel. For many years, Senegal had been trying to curb the logging that was seriously depleting the country's forests and its limited supply of wood fuel. The government had also tried several means of protecting the forest from the charcoal burners that supplied most of the fuel consumed by Senegal's urban population. Government tried reforestation, establishment of plantations, the introduction of better carbonisation techniques, improved household cooking stoves to make more efficient use of wood fuel supplies, and the substitution of wood fuels by peat, paraffin, and butane. However, these efforts met with limited success due to financial constraints and lack of follow-up actions. The only effort that produced a significant result was the promotion of butane as a household fuel with a suitable stove in 1974.

Approach
Totalgaz was a leader in the Liquefied Petroleum Gas (LPG) market. The Senegalese Government subsidised the use of butane gas and waived a levy on bottles or accessories. The subsidy fell from the equivalent of US$ 17 million a year in 1989 to less than US$ 5 million a year in 1992. The European Development Fund financed the three year Regional Gas Programme, from 1989 to 1992, with more than US$ 14 million. EDF also financed the training of workers in the region. As a result of this financing, the retail distribution has become a flourishing business. Butane was selected as the effective fuel because it was cheap and easy to transport. The first stove design made use of a 2.7 kg gas bottle. However, it was soon withdrawn because the stove could be knocked down easily and gas did not last for long. Total designed a new stove, which was simple to use, stable, cheap and met all the cooking requirements. The stove had a long-lasting 6 kg valve type camper bottle, topped with a special burner. The unit cost in 1993 was the equivalent of US$ 16. This new gas stove, 'Nopale', was a success and is now an everyday part of the Senegalese way of life.

Impacts
Technology cooperation had allowed Senegal to diversify energy sources as well as protect the environment. LPG is now widely used as the domestic fuel in place of wood and charcoal, reducing deforestation. Charcoal consumption of 400,000 tonnes a year was reduced to about 100,000 tonnes a year due to the butane programme, saving 20,000 hectares of Senegalese forests. Supply of LPG in the 6 kg bottle became a main activity of Totalgaz in Senegal. Sales of Nopale cooking gas rose from 402 tonnes in 1983 to more than 22,360 tonnes in 1994. The bottles and the burners for the Nopale were imported, but stands were made locally to reduce costs. The butane programme produced productive partnerships with local people and led to more job opportunities.

Lessons Learned
Technology transfer from a developed country partner to developing countries will be successful if the technology is wanted by the recipient countries. The oil company Total, the main stakeholder, had invested in developing a simple and new energy efficient cooking stove for the local market, which had become a part of every household. The close working relationship between the various stakeholders is an important issue, even if the technology being transferred is a simple one.

Bibliography
Mobil in Indonesia. 1995: In The Oil Industry Experience – Technology Cooperation and Capacity Building: Contribution to Agenda 21. UNEP and IPIEECA, London.

Case Study 8
The Brazilian Fuel Alcohol Programme
Suzana Kahn Ribeiro,
COPPE, Federal University of Rio de Janeiro,
Brazil

Keywords: Brazil; Technology transfer within the country with the potential of S⇔N and S⇔S transfer, Fuel alcohol;

Summary
The energy derived from bio-mass in this case is from a renewable "clean" source. The National Alcohol Programme (PROAL-COOL), was launched in Brazil to substitute the motor vehicle fuel (imported) gasoline with locally produced alcohol. This involved cooperation between the Brazilian government, farmers, alcohol producers and the car manufacturers. There is a significant scope for replication in other developing and developed countries.

Background
The programme was launched in November 1975. Its objectives were to guarantee the steady supply of fuel in the country, to substitute a motor vehicle fuel from a renewable energy source for imported gasoline, and to encourage technological development in connection with the production of sugar cane and alcohol.

Until 1979, the first phase of the Programme, alcohol production concentrated on anhydrous alcohol (99.33% ethanol) for blending with gasoline. The second phase, which began with the second oil crisis (1979), focused on hydrated alcohol used in pure form as car fuel.

Approach
The first cars run solely on fuel alcohol were produced in 1979. By December 1984, the number of cars run on pure hydrated alcohol reached 1,800,000, i.e., 17% of the country's car fleet. A protocol between the car manufacturers and the Brazilian Federal Government was then signed. Independent and autonomous distilleries, outfitted exclusively for producing alcohol directly from sugar cane syrup were constructed. Since its inauguration in 1975, the programme has yielded positive results, though it has suffered a prolonged crisis since 1989.

Impacts
Fuel alcohol is a "clean" fuel because alcohol-run vehicles emit less carbon monoxide, hydrocarbons and sulphur. Furthermore, as an additive to gasoline, alcohol replaces the hazardous tetraethyl lead used to increase the gasoline octane level. The Brazilian Fuel Alcohol Programme may prove to be an important alternative that helps stabilise the level of greenhouse gases in the atmosphere.

The energy potential of sugar cane bagasse, if fully used as a substitute for fuel oil, for the use of the production units proper as well as for the larger network, would help bring down CO_2 emissions further. The carbon emitted through the combustion of motor fuel is reabsorbed by the sugar cane, rendering net emissions practically to zero.

A possibility for increasing Brazilian alcohol competitiveness is its export for use as an additive to gasoline in developed countries since its cost in Brazil is competitive with the USA and Europe. The export of alcohol is currently hindered by commercial barriers aimed to protect U.S. and European agriculture. Developed countries could finance alcohol production in developing countries at lower cost to CO_2 abatement than in developed countries proper. Both sides may find the arrangement advantageous, as this entails the implementation of international commercial relationships and increased employment in the developing countries due to the labour intensive production of sugar cane.

The main obstacle to practical agreements on concrete projects with high potential for large-scale abatement such as fuel alcohol is the mistrust resulting from the historical tradition of zero-sum relations between North and South.

The advantages of fuel alcohol as a renewable energy source posing fewer hazards to the environment and reducing local atmospheric pollution are significant. The major and indisputable contribution of fuel alcohol is its potential for reducing CO_2 emissions, considered a key factor in the intensification of the greenhouse effect.

The UN Convention on Climate Change contemplates measures for controlling CO_2 emissions in the mid and long term. Reducing the uses of fossil fuels is an important item on the agenda. This, together with considerations on the possible increase in petroleum prices, has prompted European countries and the United States to look seriously into the question of restricting the use of fossil fuels. In this context, Brazil may have a head start in the employment of a "clean" energy source such as fuel alcohol and sugar cane bagasse. In addition to the country's contribution to the transnational efforts for controlling CO_2 emissions, Brazil will also benefit from reduced local pollution levels, greater employment and the securing of a national energy source.

Lessons Learned
The following are to be avoided: (1) The transfer of old technologies, even if they are more efficient than the host country's prevalent technology. (2) The transfer of heavy, energy intensive industry. (3) The transfer of credits for low cost emissions abatement now, at the price of high cost abatement in the future. (4) The transaction cost of the additionality issue.

The following are desirable: (1) Global emissions will be lower with technology transfer than without it. (2) The technology will slow the rate of growth in GHG emissions of developing countries. (3) Projects will support national development priorities in the host countries.

Bibliography

Ribeiro, K., 1995: Thesis (D.Sc.). COPPE, Federal University of Rio de Janeiro.

Case Study 9
Bamboo Fibre Reinforced Cement Board for Carbon Sequestration
Gunter Pauli
ZERI Foundation

Keywords: Japan, Indonesia, Columbia, construction materials, bamboo, cement, N⇔S, S⇔S.

Summary
Cement board has become a standard construction material in the tropics. The mixing of cement with mineral fibres or synthetic substitutes has evolved into a major industry. Taiheyo Cement, the second largest cement producer in the world, invested in a special research programme in association with the Zero Emissions programme to substitute the mineral and synthetic fibres with natural ones. This has now been successfully implemented.

The cement industry is under tremendous pressure to improve its carbon dioxide balance. The massive production has been partly reduced and marginally offset with reforestation programmes, but the overall balance remains unfavourable. The search for natural fibres, which sequester carbon dioxide, was considered a practical option. Scientists were sceptical, however, since the mixing of organic material (bamboo fibres) with inorganic (cement) has always been difficult. The residual sugars from organic material inhibit the crystallisation of cement, reducing the quality and price performance.

Approach
Researchers identified bamboo-specific fungi that would eliminate all sugars after crushing the bamboo. This process saves water and offers a good quality fibre with no residual sugars. The blending of 50% cement with 50% bamboo fibres reverses the carbon dioxide balance. Since the cement board has an expected life of 30 years, the fast growing species like Bambusa vulgaris offers a unique opportunity for the construction industry to adhere to the Kyoto Protocol.

The research was undertaken in Japan, but the first pilot plant was located in Java, Indonesia, just one hour outside Jakarta. The proximity to a cheap and abundant supply of bamboo is critical in the financial viability of the operation. On the basis of this first experience, an improved version of the production technology has been obtained. A second factory is now being planned in Manizales, Colombia, the centre of bamboo forests in Latin America. This permits the fast and pragmatic transfer of technologies developed in Japan to be fine-tuned in Indonesia and then transferred to Latin America.

Impacts
The use of bamboo fibre has a positive effect on the carbon balance, but it also changes the look of the city. Whereas asbestos or synthetic fibre cement boards were strong, their grey look rendered the horizon of any city unattractive. The wheat-yellow look of the bamboo fibre cement board favourably changes the impression a city offers to the outside world.

Since there are some 1,300 species of bamboo and the material is abundant and easy to grow, the use of bamboo fibre offers numerous benefits. Bamboo forests are known to have a positive influence on the hydrobalance of a region, are effectively used as a measure against soil erosion, and reinstate a natural habitat that dominated the tropical highlands until agricultural and industrial development began one hundred and fifty years ago.

Lessons Learned
This programme demonstrates that there are unexpected opportunities for carbon sequestration. These approaches require innovative biotechnologies such as the effective use of sugar digesting fungi enzymes and the combination of organic and inorganic materials. The programme of Taiheyo Cement has demonstrated that this is not only feasible, it is commercially viable. And even better, the consumer has access to a quality product that leaves a beautiful impression.

Bibliography

ZERI website: www.zeri.org

Contact
Mr. Noriaki Hayama
Director Research and Development
Taiheyo Cement
Fax: 81-43-498 38 09

Case Study 10
Demand Side Management (DSM) in Ukraine

Sergey Surnin, ARENA-ECO, Kiev
Stefan Kessler, INFRAS, CH 8002 Zurich
William U. Chandler, Battelle Washington

Keywords: Ukraine, Switzerland, USA, buildings, energy efficiency, government-driven transfer, bilateral and multilateral assistance, business & dissemination model, N⇔N.

Summary

This case study highlights the transfer mechanisms leading to municipally and cooperatively owned energy efficiency investments at the end user level in public or residential buildings in Kiev. The transfer has been managed by the Agency for Rational Energy Use and Ecology (ARENA-ECO, 1998) Kiev, supported by the bilateral cooperation agencies of the US and Switzerland. The cooperation has strengthened the capacity for implementing larger multilateral assistance projects.

Background

In Kiev, as in many other Central and Eastern European cities, the energy for heating and warm water is supplied to buildings through district heating systems with high transmission/distribution losses and low energy efficiency at the end-user level. The supplied heat is produced on the basis of natural gas imported mostly from the Russian Federation – the consumption is paid by public, commercial and private building owners in local currency. With the goal of economic transformation, energy pricing and related government policies are in a painful process of restructuring to make prices reflect market realities at a politically acceptable social cost. The energy pricing reform is interlinked with investments increasing the efficiency in generation, distribution and use of the energy. A loan agreement to upgrade the district heating supply infrastructure and to strengthen capacity in consumption-based billing was signed between the World Bank and Kiev Energo in 1998. The US Dept. of Energy (DoE) and the Swiss Agency for Cooperation and Development (SDC), as bilateral donors, have since 1996 supported the development of transfer models for demand-side management in selected public and residential buildings in cooperation with the local private agency ARENA-ECO. ARENA-ECO was in charge of project implementation and institutional networking. At the micro level the Swiss sponsored residential building project investigated the social acceptability of DSM investments at soft loan terms. The U.S. sponsored project focused on feasibility of cost-effective efficiency improvements in public buildings and financing arrangements that involved Ukrainian budgetary organisations and international financial institutions.

Approach

Both bilateral projects have demonstrated the feasibility of DSM investments (regulated building substations with energy meters, building energy conservation of short payback below 3 years) and software innovations in demonstration projects at four schools and two cooperatively owned apartment buildings in one district of Kiev.

Prior to investment in the residential buildings (which was pre-financed jointly by the local district authorities and the Swiss agency) the cooperatives had to sign a repayment agreement based on a monthly rate of 75% of the estimated reduction in the energy bill. If energy prices are increased the monthly rate increases in proportion. The repayments establish a fund at the level of the district authorities.

Investments in public buildings are financed jointly from the municipal budget and a World Bank loan, and the loan agreement requires ratification of sovereign guarantees by the Ukrainian Parliament. The investments are repaid by reducing budgetary expenditures for heat supply.

Impacts

By spreading information and raising awareness about the potential for energy saving in the buildings sector, the two projects increase acceptability of the transfer model among end-users and policymakers. The projects also help develop energy service infrastructure, create new job and business opportunities and improve the situation for investing public and private money in energy efficiency, making the effort replicable and self-sustainable. By reducing energy consumption at the end-user level by 15-25%, suggested efficiency measures reduce fuel consumption by generating facilities lowering emissions of GHG and other combustion-related pollutants, and reducing reliance on imported fuel supplies.

Lessons Learned

A key barrier to energy efficiency in residential or public buildings for economies in transition is not the technology itself, but the policy framework for the transfer mechanism needed to catalyse investments. An entirely market-based technology transfer approach is not socially or politically acceptable. There should be financial facilities providing softer financing, as well as mechanisms encouraging investments in energy efficiency and support at the district and municipal levels. Projects will only be self-sustainable if soft public finance remains available.

Bilateral donors have played an important role in enhancing the policy dialogue with stakeholders involved at the local level, in capacity-building and in facilitating institutional learning.

References

ARENA ECO, 1998: Experiences gained under the public buildings project Newsletter 3/1998, Kiev.

Case Study 11
Mitigating Transport Sector GHG Emissions: Options for Uganda

Stephen A.K. Magezi
AFREPREN Climate theme group
P.O. Box 7025, KAMPALA, Uganda

Keywords: NMT, mass transit, fuel efficiency

Summary
This study outlines some options to achieve fuel savings in the mass transit sector in Uganda. It also looks at the likely impacts of introducing non-motorised transport (NMT) to complement motorised transport (MT), especially within Kampala City. Four cluster options were analysed: mass transit (MS), road management and maintenance, NMT, and enhanced telecommunications. Out of the four, mass transit and NMT were further analysed for mitigation options to reduce GHG emissions.

Background
The study addresses land transportation including the motor vehicle, rail transport and NMT. Historical emissions were estimated using both the top-down and bottom-up approaches of IPCC recommended methodologies. The energy demanded by mass transit was estimated using the Long-Range Energy Alternative Planning model (LEAP) and focused on mini-buses, large buses and trains.

Approach
The following assumptions were made:
- The vehicle fleet is ageing and poorly maintained.
- There is low NO_x control in diesel vehicles.
- Petrol motor vehicles conform to unspecified European standards.
- The energy utilisation industry is liable to change for various mitigation scenarios.

The following observations influenced the analysis:
- The growth rate of mini-buses was very high, increasing GHG emissions as well as the gap between the national diesel budget and the petrol budget.
- Mass transit in the form of buses was non-existent in urban areas, while in up country both the coach and train were easily outcompeted by mini-buses.
- Fuel prices in Uganda were the highest within the Great Lakes region and have led to increased fuel smuggling into the country.
- Fiscal considerations rather than energy efficiency or environmental considerations influenced the national energy policy for transport.
- Within urban centres (Kampala) the designed-for average daily traffic (ADT) had been exceeded, leading to severe urban traffic congestion with increased fuel consumption and GHG emissions.

- The study on comparing energy intensity and transportation capacity found that most of the petrol consuming vehicles had a lower capacity (passengers per kilogram of fuel per vehicle - kilometre), and that the bus was the best choice for mass transit in Uganda.
- Pedal bicycle counts showed that between 1988 and 1996, bicycle transport had become a major transport option (growing by 212%) on Kampala roads, leading to some increased traffic congestion.
- Telephone density was the lowest in the region, at 0.17/100 inhabitants.

Impacts
In view of those findings, policies should aim at replacing mini-buses with buses and should estimate different emissions scenarios.

NMT lanes should be provided for 42 selected kilometres of urban roads, with a further option of making the Central Business District (CBD) a traffic free zone.

It was also found that there is a considerable saving in energy demand depending on the scenario:

ENERGY DEMAND IN GIGA JOULES BY 2010		
	Mass Transit	NMT
Base Case	1.4×10^8	1.4×10^8
Alternative Scenario	1.2×10^8	8.0×10^7
Aggressive Scenario	8.2×10^7	7.8×10^7

COSTS PER TON OF CO_2 SAVED AT 1993 PRICES			
MODE	CO_2 SAVED MILLION KG	COST OF MEASURES MILLION US$	COST/TON CO_2 AT 1993 PRICES US$
MT	640	1.7	2.65
NMT + MT	725	2.0+1.7	5.10

The analysis had considerable impact on policy consultation. Key players in the transportation sector were consulted during the survey, and an exchange between the researchers and the policymakers recorded. Also, at a workshop on incremental costs of implementing the UNFCCC in the transport sector, key players were invited and gave their recommendations. At this workshop, all climate theme researchers participated.

Current Status
Several roads within the city have been turned into one way streets to reduce on traffic congestion.

A company to organise parking space within Kampala has been contracted, and had the effect of reducing traffic congestion considerably (yet to be assessed).

Kampala City Council is in agreement with the creation of NMT lanes but has yet to mobilise resources.

Mini-buses continue to dominate mass transit, but the Government of Uganda has plans to privatise motor vehicle inspection to improve both fuel utilisation efficiency and general fitness of the motor vehicle fleet.

The Government of Uganda has recently put in place a Road Agency Formulation Unit (RAFU) with the objective to prepare modalities for a Uganda Roads Authority. The Authority will be autonomous and will advise on transport policy as well as management of roads. To date the Chief Executive of RAFU has been appointed.

Lessons Learned

There is no direct forum through which research could be fed into government policies. Nevertheless, use of senior government and research personnel in their respective fields has led to considerable interaction between policymakers and scientists. In the future, it may be advisable to arrange workshops or seminars for policymakers so research will be more influential and remove some of the barriers against the transfer of technology.

Acknowledgement

This case study has been made possible through the kind support of the Swedish International Development Cooperation Agency (SIDA), the Norwegian Agency for Development Cooperation (NORAD) and the African Energy Policy Research Network (AFREPREN).

Case Study 12
USIJI as a Technology Transfer Process

Ajay Mathur
Energy-Environment Technology Division
Tata Energy Research Institute
New Delhi – 110 003, India

Keywords: AIJ, USA, GHG reduction, N⇒S, N⇒E, Joint Implementation

Summary

The US IJI was launched in 1993 to facilitate private sector AIJ projects between the United States (US) and foreign partners in developing countries and economies in transition. Of the 110 proposals that were submitted to the Initiative until July 1998, only 30 met the additionality and sustainability criteria for IJI approval. Finally, only 13 were fully financed. The experience underscores the need for clear and widely understood goals and procedures; as well as for increased project financing. Transaction costs in the process were often substantial.

Background

The first Conference of the Parties (CoP) to the Climate Change Convention authorised a five year pilot phase of activities implemented jointly (AIJ) to learn more about the operation of inter-country climate mitigation projects in which the carbon emissions reductions were transferred from the host country to the investor country. The AIJ phase sought to build confidence in the approach and to develop a framework for the international implementation of such projects. The US government established the US Initiative on Joint Implementation (USIJI) in 1993 to facilitate US based private investors to finance climate mitigation projects in developing countries and CEITs.

Objective

The USIJI encourages private investment in climate change mitigation projects in developing and transitional economies, and enables the development and promotion of a broad range of projects to test and evaluate methodologies for measuring, tracking and verifying costs and benefits. The USIJI experience will contribute towards the formulation of international criteria for joint implementation projects.

Approach

The USIJI developed a set of criteria for accepting projects. These criteria include the requirements set out in the decision of the CoP (which include additionality in financing and in emissions reductions), as well as other criteria that help document the emissions reductions and other non-GHG benefits. By July 1998, 110 proposals had been submitted. Of these, 30 were approved as having met the established criteria after seven rounds of proposal reviews. Finally, 13 of the accepted proposals were fully financed. These projects are currently at various stages of implementation. The results of project evaluations will be released soon.

Impacts

The coal bed methane recovery project in Poland launched in October 1994 with an investment of US$ 7.5 million for a commercial scale plant is estimated to reduce net GHG emissions by the equivalent of more than 450 kilotons of CO_2. Similarly the fuel switching/cogeneration in the Czech Republic was commissioned in September 1996. USIJI estimates that over the 25 year lifespan of the project, GHG emissions will be reduced by the equivalent of more than 600 kilotons of CO_2. While it is still too early to assess the level of carbon emission reduction and their cost effectiveness, the process has had a major impact on the design of processes to promote inter-country carbon emission reduction projects. Provisions establishing the CDM in the Kyoto Protocol reflect some of the experiences of the Initiative.

Lessons Learned

In the early cases, transaction costs in establishing the projects were substantial. While the nature of transaction costs changed over time, they still remain somewhat considerable. Project proponents regard USIJI acceptance as a major hurdle that illustrates the need for clear and widely understood goals and procedures for investor country approval. Some USIJI projects also became entangled in the kinds of problems that bedevil other transactions in developing and transitional economies, reflecting the importance of the economic and legal context of the host country in project finalisation. The lack of adequate project financing remains the ultimate challenge even after projects meet the approval criteria. This lack of demand for GHG reduction projects needs to be urgently addressed.

Bibliography

Lile R., M. Powell, and M. Toman, 1998: Implementing the Clean Development Mechanism: lessons from US private sector participation in activities implemented jointly. Discussion Paper 99-08.

US Initiative on Joint Implementaion (USIJI), 1996: The USIJI International Partnerships Report, vol.2, no.3 (June).

US Initiative on Joint Implementation (USIJI), 1998: Activities Implemented Jointly. Second report to the Secretariat of the United Nations Framework Convention on Climate Change: accomplishments and descriptions of projects accepted under the US Initiative on Joint Implementation, vol.1 (January).

Contact

Dr Robert K. Dixon, Acting Director, USIJI Secretariat, 1000 Independence Avenue SW, PO-63, Washington DC 20585 Tel: 202 426 0011; Fax: 202 426 1540/1551; E-mail: rdixon@igc.apc.org Office Location: 600 Maryland Avenue SW, Suite 200, Washington DC

Case Study 13
Technology Cooperation in Indonesia for Natural Gas Production

Sukumar Devotta and Saroja Asthana
National Chemical Laboratory
Pune 411 008, India

Keywords: Indonesia, energy sector, capacity-building, N⇔N, N⇔S

Summary

This case study, for the energy sector, highlights the technology cooperation between Arun Natural Gas Field (northern Sumatra), Mobil Oil Corporation, and Pertamina, the Indonesian national oil company. The natural gas is produced, liquefied, and shipped to Japan and Korea for use in power plants, displacing the alternative use of coal or heavy oil. Mobil has utilised the best available technology and upgrades it continuously. The development of Arun Natural Gas Field is an example of both technology transfer and capacity building. The project has the potential for replication under similar operations worldwide.

Background

In 1971, Mobil discovered the Arun Natural Gas Field in rural Indonesia. This is one of the biggest and most important finds both for Indonesia and Mobil. The Operation at the Arun Field was the processing of liquefied natural gas (LNG) for exportation to Japan and Korea as well as for use in Indonesia. The Arun Field was located in a rural area that lacked industrial infrastructure. There was no competent work force available for operation. The project not only had to provide that infrastructure, but it also had to provide a competent indigenous workforce to operate these facilities. The project also faced political and cultural barriers.

Approach

The project is a cooperative effort between Mobil Oil Corporation and Pertamina, the Indonesian national oil company. Mobile Oil Indonesia (MOI) was incorporated as a production-sharing contractor for Pertamina in 1967. Although the development of Arun Gas Field and its associated P.T. Arun LNG plant required a significant financial investment, its operating costs were relatively low and its profitability high. Hence, it was considered a low financial risk for MOI.

Mobil encouraged Indonesian participation through training and capacity building. In 1998, 96% of MOI's more than 1,100 employees were Indonesian. At the P.T. Arun LNG plant, only twelve of the nearly 1,750 employees were non-Indonesian. To achieve this high rate of Indonesian participation in the workforce, Mobil conducted an extensive series of training courses. In 1993, Mobil expanded its training effort to Pertamina to train 100 Pertamina professionals over a two-year period in the latest environmental, health and safety technology.

Impacts

Since the first deliveries of Arun in 1977, the growth in sales from Arun has been quite remarkable. In addition to long LNG sales contracts with Japan and Korea, Pertamina has been able to provide incremental cargoes of LNG from Arun to satisfy additional short-term market demands. Both Indonesia and Mobil continue to share significant economic benefits from this operation. The gas produced at Arun has opened the way for new commercial activities in North Sumatra, thus helping to build a strong local economy.

At Arun, there were five discrete areas and four hydrocarbon producing clusters. This cluster approach had minimised the use of cultivated land, centralised process equipment and reduced the distance that high-pressure gas had to be transported. The field also supplied gas to two fertiliser plants and a local paper mill, creating additional job opportunities. The success at Arun is due to a very competent Indonesian workforce and the application of technology improvements to increase the plant throughput and field production, providing for the continued growth in LNG and gas liquids sales.

The project provided significant economic benefits both to Indonesia and Mobil. The continued operation demonstrates the success of technical cooperation for environmental issues.

In addition, the technology transfer and capacity building have made a significant contribution to the people of Indonesia. Mobil assisted the development of the country's infrastructure and workforce. In other local projects, Mobil had drilled water wells, built and repaired roads and bridges, and assisted in the construction of schools and mosques

Lessons Learned

Gas development at Arun has been a success both technically and economically. This success can be attributed to several factors, the most pertinent of which is the close working partnership between Pertamina, the LNG buyers and the production sharing contractor, Mobil. The mutual commitment assuring the best technology and a competent workforce was the key to the success of this operation.

Bibliography

UNEP & IPIEECA, 1995: Mobil in Indonesia. In *The Oil Industry Experience – Technology Cooperation and Capacity Building: Contribution to Agenda 21*

Contacts

Lenny S. Bernstein
Tel: (703) 846-3530
Fax: (703) 846-2972

Susan J. Sonnenberg
Tel: (703) 846-4752
Fax: (703) 846-2972

Case Study 14
Rural Electrification Using Photovoltaics in Ladakh, India

Arne Jacobson
Energy and Resources Group
University of California, Berkeley, CA 94720-3050

Keywords: Ladakh, India, photovoltaic, rural electrification, S⇔S.

Summary
In the Ladakh region of India, government funded photovoltaic (PV) systems are used for rural electrification. The system capital costs are covered by the government; homeowners must pay for maintenance costs. Systems have been installed by several agencies. One, an NGO called the Social Work and Research Centre (SWRC), has successfully installed long lasting systems. Two contributing factors to their success are the SWRC's development of a maintenance infrastructure for PV systems and Indian government standards on the design and installation of PV systems.

The SWRC's success contributes to a high demand for PV among Ladakhi villagers. Local policymakers have responded by allocating money for PV. However, the focus is on installing new systems, not on building a maintenance infrastructure. There is a danger that despite the SWRC's positive example, a large number of PV systems will be installed without provision for maintenance.

Background
Ladakh is an isolated high desert in northern India. The 100,000 sq. km region is sparsely populated by subsistence farmers and nomadic herders (pop. 170,000).

Approximately 85% of the population has some access to electricity. Plans are in place to electrify the remaining portion in the next few years, mostly using PV.

Barriers for PV electrification include mountainous terrain, bad roads, a harsh winter climate, low population density, and a lack of skilled technicians and managers. To date, low rural per capita income (estimated at $US 45/yr) has not proven problematic, but only because of large subsidies.

Approach
PV electrification is funded by the Indian Government. Funds are distributed to NGOs and private contractors through state and local government agencies.

PV systems cost about $US 450, and consist of a 35 watt solar panel, a 75 Ahr deep cycle battery, a charge controller with low voltage disconnect, and 2 fluorescent lamps (9-11 watt). The winter solar resource in Ladakh allows for three hours of lighting per day.

PV systems are popular among villagers and policymakers. Reasons include low cost per electrified family, short installation

times, and a reputation for reliability. Of 200 surveyed systems, 71% were fully functional, 27% were partially functional, and 2% were non-functional. However, the average age of the surveyed systems was only 2.4 years; more failures are likely to occur in time.

The short-term success of the systems is in part because government standards require the use of quality components.

PV System Maintenance:
There are two arrangements for maintaining PV systems. The most common is a one year warranty on parts and labour. After one year these systems generally are not properly maintained, as family incomes are low and no businesses in Ladakh service PV systems or sell the appropriate components.

Approximately 1,000 of the systems were installed by the SWRC, a local NGO in the Tilonia school network in India. The SWRC has done an excellent job of developing a village level maintenance infrastructure for servicing its PV systems. The programme includes an extensive hands-on training programme for village technicians. Homeowners pay a monthly fee ($US 0.70, covers ~50% of the post-technician training maintenance cost) in exchange for a ten year maintenance contract. Many of the SWRC systems still operate with their original batteries; a few of these systems are ten years old.

Unfortunately, system owners who are not in the SWRC network do not have access to their maintenance programme; this is due in part to the large distances between villages.

Success of PV Leads to More Installations:
The demand for electricity in Ladakh and the success of the SWRC's PV electrification programme have created pressure on local officials to install more systems. The response was to allocate funds for 2,000 systems for 1997 to 1999.

However, most of these installations will not follow the SWRC's methodology. Instead, the focus is on installing new systems; little money or time is allocated to developing a maintenance infrastructure. This approach will likely result in future problems, as systems that are not maintained will fail prematurely.

Impacts
Approximately 5,000 PV systems are installed in Ladakh; 25% of the population is receiving electricity from solar energy. However, 70% of these installations have occurred in the last three years, and lack of maintenance may prove problematic.

Lessons Learned
- The long life of the SWRC installed systems contributes to the positive reputation of PV, but has not lead to the development of a maintenance infrastructure for the larger number of PV systems installed by other agencies.
- Several elements contribute to the success of SWRC installations:
  - Systems are installed using quality components according to government design standards.

      - Systems are maintained carefully by village level tech-
         nicians. Spare parts are available.
Lack of funding is not an issue due to subsidies. Replication of
this work will require access to capital.

Bibliography

Jacobson, A., 1997: *Renewable Energy Resource Data Collection in Ladakh,
    India.*Technical Report Delivered to the Ladakh Ecological Development
    Group.

Maithel, Malhotra, Prasad, and Singh, 1998: *Renewable Energy Plan for Ladakh
    Region.* Tata Energy Research Institute, New Delhi.

MNES, 1996: Solar Energy Group of the Ministry of Non-Conventional Energy
    Sources, Document No. 32/371/96-97/PV/SE, Government of India.

Contacts
Mr. Sonam Dawa
Executive Councillor
Ladakh Autonomous Hill Development Council
Leh
Ladakh 194101
India. Tel: 91-1982-52397 Fax: 91-1982-52212.

Mr. Anchuk Colon
Director Social Work and Research Center (SWRC)
Leh
Ladakh 194101
India

Case Study 15
Blast Furnace Hot Stove Heat Recovery Technology for Chinese Steel Industries

Teruo Okazaki

Nippon Steel Corporation, Japan

Keywords: Japan, China, steel industry, industrial energy conservation, certified emission reduction, N⇔S

Summary

This industrial sector case study is on the technology for energy saving in steel manufacturing, transferred from Nippon Steel Corporation (NSC) to steel industries in the Peoples Republic of China (P.R. China). This project was a part of the Green Aid Plan proposed by the Ministry of International Trade and Industry (MITI), Japan for energy saving and environmental protection in P.R. China. Two examples of actual technology transfer projects illustrate the barriers encountered and countermeasures for overcoming them.

Background

In the Peoples Republic of China, roughly three quarters of the energy consumed comes from coal. The total amount of CO_2 emitted in P.R. China is the second highest in the world. P.R. China produces the largest amount of crude steel in the world. But the specific energy consumption for producing each tonne of crude steel is higher than that of some developed countries such as Japan and the US. Japan had achieved a 20% reduction in energy consumption through technology upgrades as early as 1993. Japan supplies steel products to both international and domestic markets with the lowest energy consumption. As a part of the Green Aid Plan to promote energy savings and environmental protection in China, MITI in Japan launched the "Japan - China model project for blast furnace, hot stove, waste heat recovery."

Approach

The project, located in Laiwu, demonstrates and disseminates the blast furnace hot stove waste heat recovery technology. The operation of the equipment started in October 1995. The waste heat in the exhaust gas, at a temperature from 200 to 300°C, was recovered and used efficiently to preheat the combustion air and the fuel gas by adopting an organic oil as a heating medium. This process technology can be applied to existing hot stoves of blast furnaces, because the heat exchangers are separately installed and connected with the transfer lines where the hot oil is circulated by pumps. As a result, the operating index has been improved significantly, reducing the fuel ratio of blast furnaces by approximately 7%. Further studies in this area were done for diffusing this technology within P.R. China. For successful implementation, the operating skills were developed under a set of training courses for operators using the actual equipment used in Japan. Through this training, the operators could get information related to both operation and improvement techniques. Japanese engineers were also sent to the site for the further transfer of related information.

Impacts

Incentives and benefits applied to both NSC and P.R. China. NSC profited through technology upgrade. For P.R. China, facing increasingly competitive international and domestic markets, being able to reduce the cost of steel produced by saving energy while reducing GHG emissions is a positive development. There were certain barriers associated with technology transfer, such as the import of equipment components. It will be beneficial for them to be locally made to cut down the costs. However, these parts purchased domestically could cause some reliability problems.

Lessons Learned

Maintenance technology is also essential for sound operation. It is important to transfer not only maintenance procedure manuals, but also maintenance skills by providing actual training on-site. For most developing countries, indigenous availability of parts appears to be critical for economic viability and sustainable development.

Contact address:

Global Environmental Affairs Department

Environmental Affairs Division

Nippon Steel Corporation

6-3 Otemachi 2 Chiyoda-ku

Tokyo 100-8071

Japan

Case Study 16
Coastal Zone Management for Cyprus: Trans-national Technology Transfer and Diffusion

Xenia I Loizidou
Ministry of Communications and Works
Cyprus

Keywords: Cyprus, The Netherlands, shoreline management plans, technologies to combat erosion

Summary
In 1993 the Government of Cyprus and Delft Hydraulics jointly started the Study "Coastal Zone Management for Cyprus" under the framework of the European Union programme MEDSPA (MEDiterranean Strategy and Plan Action). The contract involved transfer of technology and know-how on shoreline management and erosion control technologies from the Dutch Consultants to the staff of the Coastal Unit of the Public Works Department of Cyprus.
The major objectives were:
- capacity building and development of local expertise within the Coastal Unit through transfer of technology and on-the-job training
- development of proper methods to protect the coastline and improve the quality of the beach where necessary, without any serious consequences to the environment

The outcome of the study formed the basis for further work on Shoreline Management in Cyprus, which is currently being conducted by the Coastal Unit.

Background
Cyprus is the third largest island in the Mediterranean. The Coastal zone is of vital economic importance for the island and is under increasing pressure mainly due to tourism development. During the late 1970s and 1980s, erosion of the coastline and shortage of fine sandy beaches led to a sectoral approach to managing the receding coastlines. This included construction of coastal defence structures without considering any long-term effects. This approach resulted in a series of problems such as erosion of neighbouring areas, water quality problems, disruption of the ecological balance in the areas with coastal structures, etc. In light of these problems and considering the coasts as a natural resource, the Government of Cyprus recognised an immediate need for an integrated approach for managing the coastal zone.

Approach
The Study was co-financed by the Government of Cyprus (50% of the funding) and the European Union through the MEDSPA programme. The free coastline of Cyprus was divided into 12 littoral cells. A monitoring system was set up in order to generate scientific data concerning morphology and dynamics of the coastal zone. Numerical models were used to analyse and process the data in order to identify and describe the natural system components.

The technical committee appointed to oversee the operations included scientists from eight governmental departments. The committee also sought active participation of the local authorities and private sector though a series of workshops and meetings. The study resulted in the preparation of Shoreline Management plans for three selected coastal areas. Furthermore, environmental impact studies were carried out for the proposed coastal structures.

Impacts
The technologies proposed by the Study to combat erosion included: 1) retreat management in areas with no coastal development and 2) beach nourishment supported by the construction of low-crested offshore breakwaters, and demolition of the existing vertical groynes in highly developed coastal areas. All stakeholders accepted the proposals. However, the implementation process was constrained by specific barriers. For instance, budget allocation took three years of negotiations with the involvement of the Council of Ministers. The recent decision (in 1998) on this matter defines that coastal protection work is going to be co-financed by the Government and the local authorities. However, since some local authorities have financial problems with their contribution, the implementation of the study is still partly pending.

The task pertaining to defining specific set-back lines for each coastal area is still pending due to lack of funding for the study and limited number of scientific personnel.

Lack of borrow material to proceed with beach nourishment. The seabed from the coastline up to 20 m depth is protected by the Barcelona Convention (protection of Posedonia Meadows), thus making dredging a difficult option. Detailed investigation around the coasts of Cyprus revealed that the entire seabed in waters deeper than 15 m consists of extremely fine sediment, which is unsuitable for beach nourishment. The investigation of land borrow pits was also unsuccessful. The option of importing sediment from neighbouring countries was rejected due to the extremely high costs and other ecological considerations. Based on the above, it was decided that beach nourishment was not feasible as a solution for general application in Cyprus. A small-scale pilot project has monitored the coastal area the last three years and the results support this decision (Famagusta Nautical Club, Limassol). The stakeholders accepted this proposal and new plans for combating erosion are now under implementation.

Lessons Learned
It is very important to have Shoreline Management plans for coastal areas under rapid development. Over the last three years, these plans helped Cyprus to control coastal development. However, decisions on implementing appropriate technological solutions can take time. Knowledge of local conditions proved important. Thus, setting up an efficient monitoring system is necessary. This case study also illustrates the importance of using local expertise for decisions on appropriate solutions. The

approach adopted in this case involved all stakeholders in the process of decision making which proved to be very efficient and flexible.

Bibliography

Delft Hydraulics, Coastal Zone Management for Cyprus. Reports I, II, III, 1993, 1994, 1996, The Netherlands.

Coastal Unit, Internal Reports 1993-1996. Public Works Department, Ministry of Communication and Works, Cyprus.

Contact

Xenia I. Loizidou
Executive Coastal Engineer
Coastal Unit, Public Works Department,
Ministry of Communication and Works, Cyprus,
Tel +357 2 806622, Fax +357 2 498934;
e-mail: xenia@logos.cy.net

Case Study 17
CFC (ODS) Solvent Phaseout in Mexican Electronic Industries

Stephen O. Andersen, US Environmental Protection Agency, Washington DC
Jorge Corona, IMAAC, Mexico

Keywords: Mexico, U.S. EPA, ODS phaseout; Technology transfer North-South (N⇔S); Technology diffusion within the country with potential of a South-South (S⇔S) transfer.

Summary

This industrial sector case study exemplifies the Mexican government's leadership and technology cooperation by multinational companies to set in place the technology know-how and information necessary to rapidly eliminate CFC solvents used in the Mexican electronic manufacturing industry.

The CFCs were phased out under the Montreal Protocol by using HFCs, which are regulated under the Kyoto Protocol. HFCs have much lower GWP as compared to CFCs. The lesson learned from this technology transfer may be useful for similar transfer under the Climate Change Convention.

Background

In several impressive ways the Government and industry in Mexico supported the stratospheric ozone layer protection and this had motivated technology cooperation. Mexican leadership attracted the support of private companies and the U.S. EPA. Diplomatically, Mexico supported the Montreal Protocol by advocating its passage in 1987, by being the first country to sign the Protocol and by being the first country to complete ratification. With the support of environmental NGOs, the Mexico aerosol industry association "Instituto Mexicano del Aerosol A.C." (IMAAC) organised the first voluntary sectoral phaseout of CFCs by any developing country, achieving the phaseout in cosmetic and pesticide products at least five years faster than the European Union. This recognition caused the industry association to reconsider conventional wisdom and to propose that Mexico phaseout on the same schedule as developed countries rather than taking the additional ten years allowed by the Protocol. At a series of industry workshops, it was identified that the major barriers to phasing out ODS solvents used in Mexican electronics manufacturing were information and access to alternatives, agreement of foreign business partners and customers, and technical implementation. Jorge Corona had been the President of IMAAC at the time of the Mexican aerosol phaseout and also was an expert member of the Montreal Protocol Solvents Technical Options Committee. This joint experience gave the Mexican industry associations full confidence in the technical alternatives to CFC solvents and the ability to help Mexico skip CFC solvents as they expanded and modernised factories, in collaboration with international government and business partners.

Approach

The partnership strategy was to fully involve Mexican factory managers and solvent cleaning experts through workshops, technology study tours, and training. Nortel assigned a full-time coordinator to identify barriers and work with the Mexican government and industry authorities to speed technology change.

Impacts

The leadership of the government of Mexico in announcing a rapid CFC phaseout goal motivated multinational companies to concentrate on modernising their Mexican production facilities. Within three years of starting the project, AT&T built their first CFC-free factory in the world, demonstrating the technical superiority of aqueous cleaning; Nortel built a new factory using "no clean" soldering that eliminated the need to clean with CFCs. The technology was later duplicated throughout the world. With the help of multinational companies and associations, the new ozone-safe technologies were demonstrated and Mexican experts were trained to implement alternatives. The project was ultimately successful when the United States required products made either with or containing ozone-depleting substances to be labelled. As a result of the technology cooperation, Mexican industry was poised for the change when the labelling law took effect. The workshops had educated Mexican experts and helped prepare the infrastructure to accept new technology.

Lessons Learned

The initial slow progress in implementing technology at Mexican factories has four explanations: 1) local companies initially lacked motivation to change, 2) customers of Mexican products were slow to specify environmental criteria, 3) funds from the Multilateral Fund of the Montreal Protocol were delayed by government procedures, and 4) CFC solvent suppliers discounted prices and undercut market incentives.

Bibliography

Le Prestre, P.G., J.D. Reid, and E. Thomas Morehouse (eds.), 1998: Protecting the Ozone Layer: Lessons, Models and Prospects. Kluwer Publishers, Dordrecht.

National Academy of Engineering, 1992: Cross-Border Technology Transfer To Eliminate Ozone-Depleting Substances. Report on the International Workshop on Technology Transfer to Eliminate Ozone-Depleting Chemicals, National Academy Press, Washington, DC.

Schmidheiny, S., 1992: Changing Course: A Global Business Perspective on Development and the Environment. The Massachusetts Insitute of Technology Press, Cambridge, MA.

UNEP, 1995: Report of the Solvents, Coatings and Adhesives Technical Options Committee; 1995 Assessment. United Nations Environment Programme.

UNEP, 1995: Report of the Technology and Economic Assessment Panel; 1995 Assessment. United Nations Environment Programme.

UNEP, 1999: The Implications to the Montreal Protocol of the Inclusion of HFCs and PFCs in the Kyoto Protocol, HFC and PFC Task Force of the Technology and Economic Assessment Panel (October).

Contacts:

Stephen O. Andersen
U.S. EPA
Director of Strategic Climate Projects
401 M Street SW (6202J)
Washington, DC 20460
Phone: (202) 564-9069
Fax: (202) 565-2135
E-mail: andersen.stephen@epa.gov

Jorge Corona
Environmental Commission Vice-Chair
Camara Nacional de la Industria de Transformacion (CANAC-INTRA)
Cto. Misioneros G-8, Apt. 501
Cd. Satelite, Naucalpan
53100, Edo. de Mex.
Mexico
Phone: (525) 393-3649
Fax: (525) 572-9346
E-mail: jcoronav@supernet.com.mx; jocorona@prodigy.net.mx

Ms. Cintia Mosler
Environmental Information Technical Secretary
Cuauhtemoc 84
Col. Toriello Guerra,Del. Tlalpam
14050, México,D.F.
Phone: (52-5)606-1043 or (52-5)606-0793
Fax: (52-5)606-1785
e-mail: cmosler@servidor.unam.mx

Case Study 18
Swedish Government Programmme for Biomass Boiler conversions in the Baltic States

Eric Martinot, Stockholm Environment Institute--Boston,
11 Arlington St., Boston, MA 02116 ; martinot@seib.org

Keywords: Baltic States, district heating, biomass, capacity building, developed country to countries-in-transition (CEITs)

Summary
Technology for utilising local biomass fuels in the Baltic States was transferred from Swedish firms to local Baltic heating enterprises. The projects were considered highly viable from an economic and financial perspective. In addition to providing financing and risk-reduction mechanisms, Sweden provided capacity building that emphasized new managerial and financial skills that did not exist in the formerly planned economies. Joint ventures in the host countries were established.

Background
In 1993, the Swedish government, through the agency NUTEK (Swedish National Board for Industrial and Technical Development), began a programme to assist district-heat-supply companies in the Baltic region (Northwest Russia, Poland, Estonia, Latvia, and Lithuania) to convert coal-fired and oil-fired heating boilers to burn local biomass fuels. The technology is well-proven, having been used at 4,000 sites in Sweden and in several other countries over 25 years. The technology is compatible with many of the small boilers used in the Baltics, an important factor for the viability of the programme. One programme goal was to reduce CO_2, SO_2, and NO_x emissions on a sustainable basis through cost-effective projects. Another goal was to encourage partnerships between Swedish and Baltic commercial firms and promote technology transfer of boiler conversion technology to the Baltic firms.

Approach
The programme was designed to overcome key barriers to technology transfer: lack of financing, project performance risks, unavailability of the technology in the Baltics, lack of understanding by boiler owners of the feasibility and economic benefits of the technology, missing management skills, and lack of financial analysis and competitive-bidding procurement capabilities.

NUTEK administered financing of the boiler conversion projects with 10-year loans at 7-8% interest and a three-year grace period. NUTEK reduced project performance risk associated with future biomass fuel prices by guaranteeing a 15% minimum fuel-cost reduction independent of relative biomass and oil prices. NUTEK provided technical assistance, in the form of local consultants able to speak the local languages, for understanding the technical feasibility and economic benefits of the boiler conversions. NUTEK staff also provided technical assistance to boiler owners for issuing tenders, evaluating bids, selecting suppliers, negotiating biofuels supply contracts, and operating the converted boilers. Boiler owners were responsible for making decisions and applying for participation in the programme, and awarding equipment bids. They are also responsible for repaying the loan to the Swedish government. Almost SEK 300 million was spent on the programme, mostly as loans.

To encourage joint ventures between Swedish and Baltic companies, the programme encouraged foreign firms to have a local Baltic partner when bidding on the projects, and to include local suppliers and service firms whenever possible.

Impacts
The first conversion, a 6 MW boiler in Estonia, was completed in 1993 in a six-month period. By 1997, about 30 boiler conversion projects had been completed. These projects were considered highly viable from an economic and financial perspective, with simple payback times of about 3 to 6 years at the beginning of the programme. As the prices of biomass fuels increased over time relative to conventional fuels, payback times increased but remained attractive.

The programme has been successful in promoting technology transfer, capacity building, and institutional development in the Baltics. The programme has helped create a self-sustaining biomass-equipment industry, biomass-fuel markets and supply infrastructure, and a positive public attitude toward biomass fuels. A commercial biomass-fuel market did not exist in these countries prior to the programme, but the programme has now provided a market for industrial wood wastes. Two major joint ventures between Baltic and Swedish manufacturers were established, and a third Baltic manufacturer forged technology cooperation links with Swedish and Danish firms.

Lessons Learned
Technology transfer has been greatly facilitated by the capacities built among local companies - especially the managerial and financial skills that were lacking in the formerly planned economies. The success of the programme results from providing not just financing, but also technical assistance for learning how to make the most effective use of capital, estimate costs, conduct financial rate-of-return evaluations, bid and evaluate international tenders, write contracts, and manage the projects. Projects have had low per-ton carbon abatement costs. Sweden has chosen to finance projects primarily by soft loans. Project financing through loans on favourable terms rather than grants has promoted commercial accountability and sustainability of the projects.

Bibliography

Martinot, E. 1999: Renewable Energy in Russia: Markets, Development and Technology Transfer. *Renewable & Sustainable Energy Reviews,* 3, 49-75.

Contact
Swedish National Energy Administration (STEM)
Stockholm, Sweden.

Case Study 19
Dissemination of Biogas Digester Technology
Li Yue
Chinese Academy of Agricultural Sciences (CAAS)

Keywords: China, rural energy, biogas, training, S⇔S.

Summary
Small-scale biogas digesters have been reasonably successful in China and India for providing clean energy and high quality fertilisers in rural areas. The Asia-Pacific Region Biogas Research and Training Centre (BRTC), founded primarily to facilitate diffusion of biogas digester technology, has contributed significantly to promoting the development of biogas digester technology in developing countries. This case is a unique example of a South-South technology transfer.

Background
Global methane emissions from livestock manure were estimated to be 20-30 Tg/yr. Manure management systems that store manure under anaerobic conditions contribute about 60% to this source. Biogas digesters have a proven record as an environmentally sound technology and find considerable acceptability in China and India. These digesters are designed to enhance the anaerobic decomposition of organic material and to maximise methane production and recovery. Moreover, this technology has proven to be suitable for temperate as well as tropical climatic conditions. In addition to reducing methane emissions from livestock manure, anaerobic digesters are very well suited to meeting rural energy requirements. This technology also reduces the demand for commercial fertilisers and, thus, helps protect the environment and improve human health.

Approach
The BRTC was established in 1981, in Chengdu, China, to spread the use of the technology in other countries. BRTC has since been responsible for training technical engineers in countries from Africa and the Asia-Pacific region. The parties involved in setting up this centre were the United Nations Development Programme (UNDP), the Ministry of Agriculture and the Ministry of Foreign Economic Relations and Trade in China. The Chinese government was interested in the project because it provided a good opportunity to publicise biogas digesters in developing countries at a reasonably low cost. However, one of the major barriers to initiating this project was the financial resources needed to support trainees from developing countries. Furthermore, small farmers found it difficult to raise enough financial resources to cover the initial costs of constructing a biogas digester.

During the first six years of its operation, the centre received financial support from UNDP. Subsequently, the Chinese government undertook the financial responsibility for training engineers. Since the time of its inception, the centre has conducted 21 training workshops with over 270 participants from over 71 countries. During the period 1980 to 1990, this centre assisted the construction of over 70 digesters in 22 developing countries.

Impacts
Most participants of the programme acquired the skills to construct, operate and maintain small-scale biogas digesters in their countries. The centre proved to be a valuable tool in demonstrating the usefulness of capacity building in transferring technology. In addition to sequestering methane emissions, the technology provided clean and convenient ways of energy generation.

Lessons Learned
Important lessons from this case are:
- Technology transfer from one developing country to another works remarkably well.
- Technology transfers among developing countries are limited because most advanced technologies are developed and owned by industrialised countries.
- Greater emphasis on joint development of technology is likely to increase developing country participation.
- In a technology transfer case between developing countries, financing can become an issue.

References
Hu Ronglu, 1998: Personal communication.
Safley, L.M., M.E. Casada, J.W. Woodbury, and K.F. Roos, 1992: *Global Methane Emissions from Livestock and Poultry Manure*. U.S. EPA Report 400/1-91/048, Office of Air and Radiation, Washington, DC, 68pp.
Zhao, Y., 1990: International exchange on biogas technology during 1980 to 1990. In *China Biogas 1980-1990*. Environmental Protection and Energy Department of Agricultural Ministry of China, Chinese Scientific and Technological Publishing House. pp. 53-57.

Contacts
Fang Guoyuan
Deputy Director
Biogas Institute of CAAS
13 Renminnan Road
Chengdu 610041
China.

Case Study 20
Caribbean Planning for Adaptation to Global Climate Change (CPACC): Design and Establishment of Sea-Level/Climate Monitoring Network
Claudio R. Volonté
Organisation of American States

Keywords: sea-level, monitoring, project implementation, adaptation

Summary

CPACC's overall objective is to support twelve Caribbean countries to cope with potential impacts of global climate change, particularly sea-level rise, through vulnerability assessment, adaptation planning, and capacity building. The Organisation of American States (OAS) is CPACC's executing agency and recipient of the US$6.7 million grant from the Global Environment Facility (GEF) through the World Bank (as the GEF implementing agency). Funds are transferred to the Regional Project Implementation Unit (RPIU) established at the University of the West Indies Centre for Environment and Development (UWICED), Barbados, to coordinate and manage project activities at the regional level. All participating countries have established National Implementation Coordinating Units (NICUs) that facilitate project implementation at the national level. The project became effective on April 1997 for a four-year implementation period.

Approach

Adaptation to climate change is achieved through a regional approach to strengthen technical and institutional capacity of national and regional institutions. These capacities include: *(i)* monitoring and analysis of climate and sea-level dynamics and trends to determine potential impacts of global climate change; *(ii)* identifying areas vulnerable to climate change and sea-level rise; and *(iii)* developing an integrated management and planning framework for cost-effective responses and adaptations, as well as policy options and instruments to impacts of climate change on coastal areas.

Design and establishment of the Sea-Level/Climate Monitoring Network is one of 9 components of the CPACC programme. The overall objectives of this component are to establish a state-of-the-art telemetry sea-level and meteorological monitoring network in countries participating in CPACC, and to develop the region's capacity to take charge of the network over the course of the project. The network provides data to be used for analysis and prediction of oceanographic and atmospheric phenomena related to sea level change and global warning. These stations comply with the minimum standards of the Global Sea Level Observing System (GLOSS) network. The capacity of data acquisition, analysis, archiving and dissemination is the responsibility of the Caribbean Meteorological Institute (CMI) in Barbados. The Institute of Marine Affairs (IMA) in Trinidad and Tobago acts as a regional support for archiving and disseminating sea-level

data. A training and technology transfer programme has been designed to ensure that participating regional and national institutions acquire the capacity to coordinate the network, support data analysis and dissemination and maintain its operational integrity during and after the project.

Parties to the project

Parties to the project include the Global Environment Facility Trust Fund (GEF), the General Secretariat/Organisation of American States (GS/OAS), RPIU at University of the West Indies Centre for Environment and Development (UWICED), Meteorological and Land Survey Offices of 12 participating countries, the Caribbean Meteorological Institute and Institute of Marine Affairs, and the US NOAA Coastal Survey Department and National Geodetic Survey Laboratory.

Type of agreement

- Legal agreement between GS/OAS and The World Bank.
- Technical cooperation agreement between GS/OAS and the University of the West Indies.
- Technical cooperation agreements between GS/OAS and 12 participating governments, two regional institutions (CMI and IMA) and US NOAA (specific for this component).
- CMI was designated as the lead regional agency for the implementation of this component.

Commercial considerations

Financial considerations (e.g., investment, operating costs, profitability):
The project receives 100% financing from the GEF. The OAS, UWICED, and national and regional institutions provide in kind contribution (staff, office space and equipment, etc.). The budget for Component 1 is US$811,500 for four years, divided in technical assistance/training (US$410,900), equipment (US$350,600) and maintenance/replacement trust fund at CMI (US$50,000). So far, mechanisms for cost recovery have not been developed.

Risk considerations (e.g., investment, country, intellectual property issues):
- Effective and efficient implementation due to the multiplicity of countries and institutions involved.
- Maintenance and operation by national institutions, although the equipment selected is designed to provide high quality continuous data with minimal maintenance.
- Safety of equipment from vandalism and natural hazards.
- Project sustainability after GEF funding ends.

Lessons Learned

It is clear that project ownership is a key element for implementation success. The extensive consultation effort during project preparation has paid off. Another key element of project success is that technology transfer should be accompanied with intensive training of local experts, both formally, through workshops, and informally, hands-on. Furthermore, the CPACC framework appears conducive for replication.

Contacts
Jan C. Vermeiren
USDE/OAS, USA
Fax: +1 202 4583560
E-mail: jvermeiren@oas.org

Neville Trotz
CPACC/RPIU Project Manager
CERMES Building
University of the West Indies
Barbados
Fax: +1 246 4244204
E-mail: utrotz@ndlc.com

Case Study 21
Concrete Armouring for the Coast - Government to Private Sector Technology Transfer
James G. Boyd
Coastal Services Center
National Oceanic and Atmospheric Administration (NOAA)

Keywords: breakwater concrete armouring, coastal protection, licensing,

Summary
In 1993, two researchers at a United States federal laboratory, the U.S. Army Engineers Waterways Experiment Station (WES), developed and patented a new design for breakwater concrete armouring. The transfer of this innovative technology followed the classical pathway by which national government-sponsored research and development may be adopted and applied by the international private sector. The significant features of this case were new government policy, which provided additional incentives for the government laboratory and the inventors to pursue commercialisation of their innovation, and the presence of a global market.

Geographical Focus
The transfer pathway included government-to-government transfer within the U.S. Army Corps of Engineers (COE), but has focused intensively on transfer from government to the international private sector. Five companies are now permanent regional licensees covering North America, South America, Europe, Asia, the Far East, Japan, South Africa, and the Middle East.

Background
The COE research programme established to investigate new design methods for concrete armour units funded the initial development and testing of this technology. This innovation has wide-ranging potential application, as there are sites worldwide in need of coastal protection for land, property, and life. However, for some, the cost of this type of technology is a prohibitive factor. While this technology has been shown to potentially cost less and be more reliable than other armouring technologies, the initial investment is still a consideration, especially for those countries with limited resources. Altogether, this innovative technology seems to hold great promise, while the case study provides an excellent example of a product, market, and policy coming together to provide an opportunity for the transfer of technology.

Approach
Research and development funding was provided exclusively by the Corps. Once testing was completed, technical presentations were made at coastal engineering conferences in Japan, the United Kingdom, South Africa, and Spain, and local seminars were used to introduce the technology to prospective users. A U.S. patent was awarded in 1995, but the international patents were significantly more difficult for the patentees to obtain. By retaining an experienced international patent attorney, this obstacle was overcome. In the United States, the Federal Technology Transfer Act of 1986 (FTTA) made technology transfer a responsibility of all federal laboratory scientists, allowed laboratories to enter into CRADAs and negotiate licensing agreements, and established principles for royalty sharing for federal inventors and other incentives for the laboratories. The developers and WES took advantage of the FTTA and entered into a CRADA with a South African firm interested in using the armouring units for a project. Difficulties in construction and the substantial time required administering the agreements provided WES with the incentive to establish permanent licenses with a few select experienced private companies. The patentees continued to promote the technology during licensing negotiations and provided technical support to another CRADA partner as well as subsequent licensee projects.

Impacts
The application and promotion of the technology resulted in the unit being proposed and used worldwide very early in its patent life, which is particularly uncharacteristic for coastal protection structures. A significant feature in gaining the quick acceptance of this new civil engineering technology was that the researchers addressed their clients' anxiety risk by streamlining the usual "chain-of-command" responses, as provided by the FTTA. At this time, WES involvement in projects is limited, as the licensees have begun to actively promote the technology, conduct their own research, and administer all projects.

Lessons Learned
It is important to note that much of the early technology transfer was done without specific government funding and was accomplished beyond the normal duties of the government researchers. However, the incentives provided by the FTTA have resulted in nearly US$1 million in engineering revenue, license fees, and royalties for the WES. In addition, the FTTA allowed the government inventors to share the royalty revenue. This provided motivation for the researchers to commit a great deal of personal time and energy to develop the product and an associated construction technique, and implement the technology transfer. It also provided incentive for WES to allow the researchers to perform this work. Without these types of incentives, technology transfer of this type would be much more difficult, or might require a completely different type of transfer pathway.

Bibliography
Jeffrey, M. (Research Hydraulic Engineer, WES), Personal communication.
Phillip, S. (Technology Transfer Office, WES), Personal communication.
Melby, J., and G. Turk. U.S. 1997: Army Corps of Engineers Waterways Experiment Station Technical Report CHL-97-4 (March).

Contacts
Jeffrey A. Melby (CEWES-CN-S)
USAE Waterways Experiment Station
Coastal and Hydraulics Laboratory
fax: 601/634-3433; 3909
email: j.melby@cerc.wes.army.mil

Case Study 22
World Bank/GEF India Alternative Energy Project

Eric Martinot

Stockholm Environment Institute--Boston, 11 Arlington St.,
Boston, MA 02116

Keywords: India, wind, solar, tax policy, capacity building,
joint ventures

Summary

A multilateral development project helped catalyse important technology transfers and market changes for wind power and solar photovoltaic systems in India. Tax incentives, capacity building, and new market delivery mechanisms promoted private-sector activity. In just a few years, 968 MW of wind farms were installed and operating in India, almost all commercial and privately operated. The wind industry jumped from three companies to 26, many of them joint ventures. Technology development and exports accelerated and costs declined.

Background

The India Alternate Energy project was started in 1991 by the World Bank and Global Environment Facility (GEF) to promote commercialisation of wind power and solar PV technologies in India. The project was designed to support existing government policies to promote wind power through special tax incentives.

Approach

The project was designed to pioneer financing and market delivery mechanisms based on private-sector intermediaries and suitable incentive schemes and policies for small independent power producers. Markets for these technologies were catalysed through large-scale demonstration, increased consumer confidence, and enhanced willingness to pay. The project strengthened the capabilities of the India Renewable Energy Development Agency (IREDA) to promote and finance private-sector investments, and channelled financing through that agency. One component of the project directly financed wind farm installations by private-sector developers, with a target of 85MW to be financed through the GEF and the International Development Association (IDA), and with co-financing from the Danish government and from other resources mobilised by IREDA. A second component provided a marketing campaign, credit facilities, and subsidies to rural consumers for purchasing solar PV systems. The project targeted 2.5 to 3.0 MWp of solar PV. The project also supported policies that encourage small-scale independent power producers to invest in wind farms and mini-hydro installations.

Impacts

By 1998, over 270 MW of wind power had been financed by IREDA and commissioned, including 41 MW commissioned with GEF and IDA financing and 10 MW commissioned with Danish funds. Only 0.3 MWp of PV had been commissioned, but an additional 1.0 MWp was being prepared. In parallel with these direct project impacts, a total of 968 MW of wind farms were installed and operating, of which 917 MW were commercial and privately operated. Highly favourable investment tax policies strongly influenced these commercial installations. New suppliers entered the wind power and solar PV markets. Before the project there were three major companies involved in the wind industry, but by 1998 as many as 26 companies were engaged in the wind turbine manufacturing industry, many with foreign partners. High-technology wind turbine designs up to 600-kW with variable speed operation were produced by 14 companies. Domestic production of blades began, and exports of blades and synchronous generators to Europe were underway. Wind turbine exports to other countries also began. The installed costs of wind turbines in India declined. Domestic production capacity for solar modules went from 3 MWp in 1991 to 8 MWp by 1996, and the number of companies involved in the PV industry went from 16 in 1991 to more than 70 in 1998.

The World Bank/GEF project indirectly helped catalyse these market changes and technology transfers by helping to raise awareness among investors and banking institutions on the viability of wind power technology and lobbying for lower import tariffs for both wind and solar PV systems. Many more financial institutions decided to offer financing for wind farms, and a wind-power loan portfolio among commercial banking institutions emerged (this was a key project goal). The number of Indian consultants capable of developing wind power investment projects increased dramatically, in part because of GEF-supported training and networking activities for consultants, technicians and private firms (a roster of consultants was available from IREDA for reference by investors). Promotional efforts and numerous business meetings organised by IREDA increased awareness of various PV applications among potential users.

Lessons Learned

Investment tax credits appear to be a powerful stimulus to technology transfer and market development, and have had a huge impact in a very short time. However, the sustainability and viability of the markets and joint ventures is uncertain if the credits are removed.

There have been problems in extending credit to potential rural PV consumers. Evidence suggests that financial institutions perceive rural consumers as unwilling to repay loans and therefore have not extended credit. Also, the lack of infrastructure for after-sales support and service has emerged as an additional difficulty in rural markets.

Bibliography

Martinot, E., 1998: *Monitoring and Evaluation of Market Development in World Bank/GEF Climate-Change Projects: Framework and Guidelines.* World Bank Environment Department Paper, Washington, DC.

Contact

India Renewable Energy Development Agency (IREDA)
New Delhi.

Case Study 23
CFC-free Refrigerators in Thailand
Yuichi Fujimoto, JICOP, Japan
Stephen Andersen, U.S. EPA, Washington DC, USA

Keywords: Japan, USA, Thailand, N⇔S, technology transfer, CFC-free refrigerators, ODS phaseout, HFC-134a

Summary
This case study is for the industrial sector and the stakeholders are Government of Thailand, Hitachi, Matsushita, Mitsubishi Electric, Toshiba of Japan, and the Japan Electrical Manufacturers Association (JEMA). The government of Thailand and the multinational companies decided to speed the phase-out of ozone depleting substances (ODSs) through a voluntary pledge. The technical challenge of achieving this environmental objective for the manufacture of ODS-free domestic refrigerators required an unprecedented degree of technical cooperation between fierce competitors in Thailand and Japan.

The CFCs were phased out under the Montreal Protocol by using HFCs, which are regulated under the Kyoto Protocol. HFCs have much lower GWP as compared to CFCs. The lesson learned from this technology transfer may be useful for similar transfer under Climate Change Convention.

Background
In a 1990 seminar conducted for the Association of South-East Asian Nations (ASEAN), it was reported that electronics and refrigerator enterprises from Japan, the United States and Europe were primarily responsible for dramatic rates of increase in CFC consumption in Thailand. This case study considers the technology cooperation undertaken to halt the use of CFCs in refrigerators. Seven Japanese companies and their joint ventures in Thailand produced household refrigerators in Thailand. Three of these companies imported all the critical parts necessary to complete the phaseout, but the other four companies used compressors supplied by a single Thai company. It was therefore crucial that the Thai compressor manufacturer redesign their products to use CFC alternatives, and that they manufacture reliable compressors.

Approach
The government of Thailand announced its target date for eliminating CFCs from Thai household refrigerators as the end of 1996. This phaseout date for CFCs used as the refrigerant and as the blowing agent for insulating foam in domestic refrigerators was particularly aggressive. The first option was to keep to the phaseout schedule by getting compressors from manufacturers other than the local companies. This option was rejected because Japanese companies had pledged to help Thailand to build local manufacturing capability. The second option was to postpone the phaseout and wait for a supply of reliable compressors from the local companies. This option was also rejected, because the Thai government was eager to keep the phase-out on schedule. The

only other option required additional technology cooperation with the local compressor company. The refrigerator parent companies that were buying compressors in Thailand, i.e. Hitachi, Matsushita, Mitsubishi Electric, Toshiba of Japan, and the Japan Electrical Manufacturers Association (JEMA) launched a voluntary emergency joint project to assist the local companies to improve reliability and to help with its implementation. The U.S. Environmental Protection Agency attended follow-up meetings in Thailand and encouraged the Thai compressor manufacturers' parent companies in the United States to promote their participation in the technical support project.

Impacts
As a result of the corporate leadership by multinational companies, ambitious goals were announced that ultimately required unprecedented technical cooperation to achieve. The seven Japanese refrigerator companies and their joint ventures all achieved the complete CFC phaseout by the end of 1996, the first phaseout of CFCs used in domestic refrigerators by any developing country in the world.

Because the Thai Government realised the importance of phasing out of CFCs, they took measures to prohibit the manufacture or import of foreign CFC-based refrigerators after 1996. This leadership action and support by the Thai government encouraged the Thai refrigerator manufacturers to work hard to achieve the CFC phaseout. It also created a strong motivation to phase out CFCs in refrigerators manufactured in other developing countries. It is worth noting that the Japanese companies, which helped the Thai companies in developing HFC-134a compressors, were in direct competition to supply the other three refrigerator manufacturers in Thailand.

Lessons Learned
There are three crucial aspects to the successful achievement of global environmental objectives: global leadership and commitment to cooperation for global environmental protection, an early action plan from industry, and cooperation between government and industry.

This extraordinary technical cooperation can only be replicated when all of these aspects can be orchestrated and when there is a strong commitment from corporate management and national governments.

Bibliography
Japan Industrial Conference for Ozone Layer Protection (JICOP): Guide (9909-1500).

Le Prestre, P.G., J.D. Reid, and E. Thomas Morehouse (eds.), 1998: Protecting the Ozone Layer: Lessons, Models and Prospects. E Kluwer Publishers, Dordrecht.

UNEP, 1995: Report of the Technology and Economic Assessment Panel; 1995 Assessment, United Nations Environment Programme.

UNEP, 1999: The Implications to the Montreal Protocol of the Inclusion of HFCs and PFCs in the Kyoto Protocol. HFC and PFC Task Force of the Technology and Economic Assessment Panel (October).

U.S. EPA Office of Air and Radiation (6205J), 1997: Champions of the World: Stratospheric Ozone Protection Ozone Protection Awards (August), EPA 430-R-97-023

U.S. EPA Office of Air and Radiation (6205J), 1998: Newest Champions of the World: Winners of the 1997 Stratospheric Ozone Protection Awards (February), EPA 430-K-98-003.

Contacts:

Stephen O. Andersen
U.S. EPA
Director of Strategic Climate Projects
401 M Street SW (6202J)
Washington, DC 20460
Phone: (202) 564-9069
Fax: (202) 565-2135
E-mail: andersen.stephen@epa.gov

Mr. Yuichi Fujimoto
Advisor
Japan Industrial Conference for Ozone Layer Protection (JICOP)
Hongo-Wakai Bldg.
2-40-17, Hongo
Bunkyo-ku
Tokyo 113-0033
Japan
Phone: (813) 5689-7981 or 7982
Fax: (813) 5689-7983
E-mail: jicop@nisiq.net

Mr. Thanavat Junchaya
Regional Network Coordinator of Ozon Action Programme, Bangkok
United Nations Environment Programme
Regional Office for Asia and the Pacific
United Nations Building, Rajadamnern Avenue,
Bangkok 10200, Thailand
Phone: 662-2881870, 2881234
Fax: 662-2803829
e-mail: junchaya.unescap@un.org

Mr. Wiraphon Rajadanuraks
Director
Hazardous Substances Control Bureau
Department of Industrial Works
75/6 Rama 6 Road
Bangkok 10400
Thailand
Phone: (662) 202-4201
Fax: (662) 202-4015
E-mail: hazard@narai.diw.go.th

Mr. Viraj Vithoontien
Montreal Protocol Environmental Specialist
World Bank
1818 H Street, NW
Washington, DC 20433
Phone: (202) 458-1913

Case Study 24
Financing Micro-hydro Energy Dissemination in Perú
Alfonso Carrasco, ITDG Peru
Teodoro Sanchez, ITDG Peru

Keywords: Peru, micro-hydro, financial mechanisms, dissemination, clean energy, replication potential

Summary

This case study describes an experience of dissemination of micro-hydro plants in isolated rural areas in Peru, based on the provision of soft loans plus technical assistance. A total of 15 plants are being installed and a Revolving Fund has been established. The micro-hydro plants replace fossil fuel-based energy generation, and have an expected life of 25 years.

Background

Only about 5% of the rural population in Perú has access to electricity. For the remaining 95%, the main alternative –if available- is the use of diesel stand-alone generators that are usually unreliable and very dirty. There are no financial mechanisms available to facilitate or promote the investment in small renewable energy schemes in the rural areas. Local technology on small hydro energy has also had a poor reputation and considered to be expensive. ITDG, an international technical NGO, has been promoting the dissemination of micro-hydro energy in Peru during the last 20 years. For this purpose an integrated approach has been developed, including technology development, training, pilot projects, research on institutional issues and advocacy work. One of the elements of this approach was the creation, with support of the IDB, of a financial mechanism to facilitate access to funds for small micro-hydro plants to the rural population.

Approach

The IDB provided US$ 520,000 as a loan to ITDG. This loan had two components:

1. US$ 400,000 to be allocated as small credits up to US$ 30,000 at an interest rate of 8% annually and maximum 5 years payback time. Clients should be rural entrepreneurs or local institutions (municipalities, communities, cooperatives)
2. US$ 120,000 as a grant, for the provision of technical support, including promotion, technical and financial assistance, beneficiary organisation and implementation.

A small local consultant group was hired for financial studies, preparation of credit instruments, control of expenditures during the implementation time and recovery of loans. Technical assistance (introducing low-cost technologies), promotion of credits and training of the users were the responsibility of ITDG.

Impacts

15 small hydro plants were installed by late 1997, allocating US$ 465,000. In addition to this sum, US$ 1.4 million co-funding has been raised for the same purpose from various sources, the most important ones being the regional government and the national poverty-focused programme FONCODES. Loans repayment is going well with an average delay of 2.5%. The 15 plants provide access to electricity for about 1,800 people; lead to the creation of small industries such as carpentry, welding, battery charging; and improve basic services such as light for evening education, health posts, communication services, etc. The project provided an alternative to the diesel option, reducing pollution and greenhouse gas emissions. Total installed power of 602 kW will be added to the national power capacity. The expected life of each plant is 25 years. Currently this is the only existing scheme of this type in Perú.

Lessons Learned

It is possible to disseminate the option of clean micro-hydro electricity for isolated rural populations using appropriately designed financial mechanisms. A combination of credit, subsidised promotion and technical assistance seems necessary. Also, very close contact with the population in order to promote and market the scheme is required. Significant cost reduction can be obtained using less expensive but reliable technology.

Bibliography

Alternatives for Rural Electrification. An analysis of the experience gained in the project Credit for Micro-hydro plants ITD-ITDG. 1998: Alfonso Carrasco. Dic.
Energía para el Area Rural del Peru. 1998: ITDG Peru.

Contact

Author: Teodoro SanchezInstitution
ITDG-PeruAv
Jorge Chavez
275 Miraflores
Lima
Peru
Fax : +(511) 4466621
Phone : (511) 4475127, 4467324
Email: teo@itdg.org.pe

Case Study 25
Tree Growers' Cooperatives: A Participatory Approach to Reclaim Degraded Lands

Sudha, P., Indu K. Murthy & N.H. Ravindranath
Centre for Ecological Sciences, Indian Institute of Science,
Bangalore – 560 012, Karnataka, India

Keywords: India, forestry, S⇒S

Summary

National Tree Growers' Cooperative Federation (NTGCF) was established in India in 1988 with the main objective of restoring the ecological security of village communities in eco-fragile and marginalised zones. The NTGCF has promoted the organisation and establishment of village-level Tree Growers' Cooperative Societies (TGCS) in six states of the country, and provided technical and financial support for regenerating degraded forests, conservation of natural resources through community protection, and activities related to Joint Forest Management. Nearly 11 million trees have been planted, resulting in fuelwood, fodder and employment benefits for the villagers, as well as in indirect benefits such as enhanced carbon sequestration and the setting up of processing units for higher value-added products.

Background

The NTGCF was created in 1988 with the objective of restoring and protecting the ecological security of the country through the creation of self-sustaining village TGCS, and supporting the activities of these Cooperatives through training, financing, and institutional development. The concept of the NTGCF was based on the recognition of land degradation as a major concern, and informed by the earlier success of the National Dairy Development Board in the enhancement of milk production through milk cooperatives.

Financial assistance by the Swedish International Development Authority in 1991 and the Canadian International Development Agency in 1993 enabled the NTGCF to establish activities in six states of India: Gujarat, Orissa, Andhra Pradesh, Rajasthan, Karnataka and Uttar Pradesh.

Objective

The main objective of the TGCS has been the restoration of the biological productivity of marginally productive and unproductive degraded lands by establishing sustainable fuelwood and fodder plantations to meet the essential needs of the villagers. Additional produce caters to the urban demand for fuelwood, timber and tree-based products. Concurrent objectives include the strengthening of existing village institutions and empowerment of women.

Approach

NTGCF facilitates the organisation and establishment of TGCSs in villages, and assists the villagers in acquiring village wastelands on long leases. In cases where village-level forest-management institutions already exist, NTGCF helps them frame management rules based on specific village needs. The degraded lands are restored to a productive state through natural regeneration plantations and soil and water conservation measures. In some areas, large scale seed sowing of forest trees and shrubs was done. For the seeds to germinate on difficult sites, alternations such as wildings at the super-abundant stage of germination and replanting on village lands, soil stabilisation and moisture conservation work had been undertaken. The villagers are supplied seeds, saplings, cuttings and biofertilisers. NTGCF promotes community protection of forests by ensuring that forest-management rules link the needs of the villagers to the sustainability of the forests.

NTGCF has had to contend with many issues such as common property rights, sustainable extraction of rare and endangered medicinal and aromatic plant species, appropriate market channels for surplus produce, legislative enactments, etc. This has lead to capacity building in the Cooperatives and in NTGCF to influence policy with regard to forest dwelling communities and conservation.

In order to manage biomass demand, NTGCF and the TGCS encourage and facilitate the adoption of energy conserving techniques and devices such as improved stoves, biogas plants, pressure cookers and solar cookers. They have also developed facilities for the processing and marketing of wood at Cooperative and regional levels, and created opportunities for the value-addition of forest products, such as a Neem biocide plant.

Impacts

As of March 1998, NTGCF had organised TGCS and other village institutions in 518 villages involving 40,237 members. These institutions have revegetated 8,911 hectares of the 15,343 hectares of the common lands made available, and planted 10.74 million trees on their common land as well as private lands. This work has generated 1.22 million workdays of employment in these villages. Indirect benefits to the leased land has included soil fertility improvements and moisture retention, and benefits to nearby land has included wind protection and decreased salinity.

The experience of the older TGCS promotes the organisation of new Cooperatives in surrounding villages, and proposals to organise TGCS on non-revenue wasteland, which are either equally degraded or belong to the village, are being introduced.

The global benefits include carbon sequestration in standing trees, litter and soil. Furthermore, fuelwood supply from the new forests leads to conservation of C sinks in forests and village trees. The NTGCF has not conducted ecological studies of these plantations with regard to the survival of seedlings and the standing biomass, but based on the NTGCF experience, it is estimated that its activities have lead to the sequestration of 52,183 tCO_2. Carbon density of litter and soils also increases. Thus, in addition to restoration of degraded lands, carbon sequestration is an additional benefit of the projects. If the TGCS concept is extended to cover all the degraded forest land, pasture, and cultivatable wasteland of 41.62 million ha, the annual C sequestered could be ~ 20.81 MtC.

Lessons Learned

One of the major reasons for the success of the project is its ability to secure tenure of the land in the form of a lease. This kind of ownership allows communities to take a long-term view, and encourages better management of the common lands. The NTGCF experience indicates that the village institutions that have become self-sustaining are those that the communities developed themselves and are built on local knowledge and needs. A mid-term review concluded that the sustainability of TGCSs in terms of social and economic aspects is closely linked to the sustained collective action of the community. Introduction of alien rules and regulations leads to confusion and non-compliance.

The protection and management of land leased by the cooperative societies has, however, shifted the pressure on surrounding common lands. NTGCF proposes to take a holistic look at the entire village landscape while dealing with the common land resources with secured tenure.

Creating awareness, capacity building and establishment of a formal institution such as TGCS at the village level is critical to revegetation of degraded village commons. An external institution such as NTGCF is necessary for providing technical assistance and training to local communities to protect and manage village commons.

The other effects of the projects are leadership development, spread of collective action and cooperation into other spheres of village life, institutional development, resource enrichment and expansion, and improvement in the quality of life of rural communities.

Bibliography

National Tree Growers' Cooperative Federation: Annual Report 1997-98.
A Decade of Learning. 1996: A participatory mid term review summary of Sida and CIDA Assisted Tree Growers' Cooperative Project (November).
Qualitative/quantitative aspects – mid term review/evaluation. 1996: Data Review (November).
Seebaur, M., 1992: Review of social forestry programmes in India. GWB Gesselschaft Fur Walderhaltung und Waldbewirtschaftung GMBH, Michelstadt, GermanBH, Michelstadt, Germany.

Contact
Coordination Office
National Tree Grower's
Cooperative Federation Ltd.
Anand – 388 001
Phone: 02692-41303
Fax: 02692-42087
Email:theo@ntgcf.irm.ernet.in

Case Study 26
Carbon Sequestration Benefits of Reduced-Impact Logging

Sudha, P & N.H. Ravindranath
Centre for Ecological Sciences, Indian Institute of Science,
Bangalore – 560 012, Karnataka, India

Keywords: Malaysia, USA, Australia, forestry, training, monitoring, N⇒S

Summary
A Reduced-Impact Logging (RIL) project to decrease greenhouse emissions was taken up by Innoprise Corporation in the forest of Sabah, Malaysia, with financial aid from a US utility. Malaysian foresters were trained for RIL operations, and an independent team monitored and verified project activities. The associated reduction in carbon emissions was estimated at 65tC/ha over the logging cycle of the coupe at an estimated cost of US$4/tC.

Background
The carbon-offset project for reducing greenhouse gas emissions through RIL was initiated by Innoprise Corporation (ICSB, Malaysia) in August 1992 with financial aid from the US-based New England Electric System (NEES). RIL, in contrast to conventional logging, maintains the capacity of forests to sequester carbon because of the reduced number of trees logged. This reduces biomass losses and decreases carbon emissions from the decay of logging debris.

Objective
This project aimed at demonstrating reductions in carbon emissions from logging activities. The project also included field studies for quantifying carbon storage and fluxes; development of a model to simulate changes in biomass and carbon pools following logging and a simple projection model to generate an estimate of the carbon benefit.

Approach
Training in RIL techniques: Traditionally, Malaysian forest rangers are trained in mensuration and inventory methods, with limited expertise in harvesting techniques. Sawyers and bulldozer drivers receive no explicit training although they are apprenticed for several years. The RIL project sponsored training at several levels in the forest department. One of the first project activities was a visit by senior Innoprise staff and logging contractors to areas managed by the Queensland Forest Service in Australia. Several of the Australian foresters who trained the logging contractors then visited Sabah as advisors for implementing RIL guidelines. Ten tractor drivers and 15 ICSB field staff worked with experienced Australian foresters for three weeks. During this training period, timber in a logging block of approximately 50 ha was harvested. During two five-day training programmes, sawyers were trained by a Swedish specialist in directional felling.

Monitoring: Compliance with the RIL guidelines and verification of reduction in logging damage was assessed by an independent team comprised of NEES, ICSB and the Center for International Forestry Research. The team conducted site inspections and met the ground staff, and verified the records and calculations of the logging damage studies and of the carbon offsets.

Impacts
The associated reduction in carbon emissions and enhanced sequestration is estimated at 65 tC/ha over the logging cycle of the coupe, at an estimated cost of less than US$4/tC.

Lessons Learned
The pilot project brought together NEES, ICSB and the Queensland Forest Service. NEES gained the capacity to plan and implement international carbon-offset projects in anticipation of US legal requirements to control emissions from its coal-based power plants. NEES financing enabled the transfer of RIL expertise and technology from the Queensland Forest Service to ICSB. ICSB gained funds to train staff; the opportunity to improve harvesting practices in a part of their 1 million-hectare, 99-year concession; and the increase in value of residual forest in the project area. The financing paid the costs of training operators and implementing improved harvesting practices. However, the sustainability of RIL is not guaranteed because of the lack of continuing financial incentives. The expansion of RIL is likely to depend on acceptance of the Clean Development Mechanism by both developed and developing countries. However, RIL is not easy to classify under the three defined activities (reforestation, afforestation and deforestation) currently allowed under the Kyoto Protocol in the Land Use Change and Forestry Sector.

The people most responsible for the RIL success were the ICSB forest rangers. The rangers participated in all the RIL activities. They not only consistently used good judgement in implementing the guidelines, but also suggested a number of innovations that were incorporated into the project, including excellent methods for record keeping. The trained forest rangers may serve as future instructors, a situation that can offer considerable advantage in effectiveness, cost and ease of implementation.

Bibliography

Pinard, M. A., F.E. Putz, J. Tay, and T.E. Sullivan, 1995: Creating Timber Harvest Guidelines for a Reduced-Impact Logging project in Malaysia. Journal of Foestry, 93, 411-45.

Pinard, M.A., and F.E. Putz, 1997: Monitoring carbon sequestration benefits associated with a reduced-impact logging project in, Malaysia. Mitigation and Adaptation Strategies for Global Change, 2, 203 -215.

Tropical Forests in the Kyoto Protocol. 1998: Tropical Forest Update, 8 (4), 5-8.

Contact
Innoprise Corportion Sdn Bhd
88817 Kota Kinabalu
Sabah, Malaysia

Case Study 27
Technology Information Assessment and Dissemination in India

Ajay Mathur
Energy-Environment Technology Division
Tata Energy Research Institute
New Delhi – 110 003, India

Keywords: Institutional framework, technology development, information system, forecasting

Summary
The Technology Information, Forecasting and Assessment Council (TIFAC), established as an information clearinghouse by Government of India, is playing a significant role in technology development of the country. TIFAC can serve as a role model institutional infrastructure in a developing country for technology information flows, analysis and technology development.

Background
Established in 1988, during a period of transition when India was moving from a state-regulated economy to a market driven system, TIFAC is an autonomous organisation under the Department of Science and Technology (DST), Government of India (GoI). Its mission is to undertake technology assessment and forecasting studies in key areas of the national economy, keep track of global trends in technology and formulate preferred options for India, and establish a nationally accessible technology information system. Though it does not explicitly state protecting the global environment as one of its objectives, TIFAC was chosen for this case study because it represents a developing country's effort to promote technology assessment and forecasting that are vital for a successful realisation of the initiatives under the Climate Change Convention.

Approach
Activities at TIFAC include information dissemination, studies on technology-linked business opportunities, technology sourcing, commercialisation of technology projects, and extending patent information service to scientists from academic institutions. TIFAC's mandate includes functioning as a clearinghouse on information, and is facilitated through a technology information system aimed to disseminate techno-commercial information and business opportunities, especially to entrepreneurs and business planners. At the macro level, TIFAC has produced a series of documents on India's efforts in science and technology development in all sectors (government, private and public, etc.). TIFAC disseminates information through newsletters, participation in national and international technology expositions, industry meetings, and one-to-one mail campaigns. TIFAC provides assistance in technology sourcing worldwide. There is a Patent Facilitating Cell, with the support of a panel of attorneys, to assist scientists patent their inventions by providing financial and technical support. In addition, the functions include infor-

mation dissemination to create awareness about intellectual property rights while keeping abreast of global developments and to facilitate the use of patent information as inputs to R&D programmes. In order to be in line with global developments, TIFAC has international linkages with institutions/organisations including the Association of South East Asian Nations, the United Nations Industrial Development Organisation, the World Association of Industrial and Technological Research Organisation, the International Association of Technology Assessment and Forecasting Institutions, the National Science and Technology Board (Singapore), and the National Technical Information Service (US Department of Commerce).

Impacts
About 170 studies have been undertaken encompassing a broad spectrum on materials, chemicals, energy, environment, and information technology. Studies aimed at establishing the technology status in the country vis-à-vis emerging global trends based on 'market-pull' and 'technology-push'. The results have been used for investment risk evaluation by financial institutions, and have been instrumental in planning the technology development efforts of many government departments. To prepare the technology vision for the country, 5000 experts from the industry and government, R&D, and academic institutions were brought together in a span of two years to identify the focal areas of the country. TIFAC's technology sourcing database contains more than 4,000 technology offers and business opportunities. It has also established several national centres of excellence for joint technology promotion and development in areas like leather, aerospace, and chemical technology.

Lessons Learned
There are three lessons from this experience:
- Access to information is the first step in a programme which aims at 'technology development'. This includes creating awareness among the user groups of the existence of an information clearinghouse.
- The importance to evolve the objectives/activities of institutions to suit the needs of users/stakeholders.
- The extent of involvement of the stakeholders determines the success of the venture.

Bibliography

TERI, 1997: Capacity building for technology transfer in the context of climate change.

Contact
The Registrar, TIFAC
Dept. of Science & Technology
Technology Bhawan
New Mehrauli Road
New Delhi 110016
Fax. 011-6857643, 6863866

Case Study 28
Medicinal Plants vs. Pharmaceuticals for Tropical Rural Health Care
Thomas J. Carlson,
Department of Integrative Biology,
University of California, Berkeley, CA 94720-3050

Keywords: Medicinal plants, pharmaceuticals, biodiversity, ethnolinguistic diversity, tropical countries, USA, Europe

Summary

Tropical rural communities may receive treatment from locally available traditional botanical medicines and/or modern pharmaceuticals. Assessment of these two medical systems generates interesting comparisons when evaluating local indigenous versus external control, access, availability, affordability, long term sustainability and ability to safely and effectively use each medical system. Each medical system maintains a characteristic capital flow between North and South and rural and urban. Traditional botanical medicine is based on local indigenous resources and knowledge. Climate change and ecosystem damage diminish the local biological resources available to tropical rural communities to contribute to their health care needs.

Background

While pharmaceutical companies conduct advertising campaigns in tropical countries to increase consumption of pharmaceuticals in urban areas, most people living in rural areas have limited access to or can not afford these drugs. The World Health Organisation (WHO) estimated that 80% of people in the world use medicinal plants as their primary health care medicines. Research on the bioactivity of tropical medicinal plants has demonstrated that most are safe and effective therapies. Unfortunately, tropical public health programmes do not usually recognise the therapeutic value of traditional medicine and instead encourage widespread use of pharmaceuticals to treat diseases already adequately managed by locally available traditional botanical medicines.

Due to the research, development, formulation, packaging, distribution, and refrigeration costs, pharmaceuticals are capital and energy intensive forms of medicine that are under external urban and/or Northern control resulting in capital flow from rural to urban and South to North. These capital and energy inputs make the cost of pharmaceuticals high and reduce access for tropical rural communities. If pharmaceuticals reach these communities there is often not a continuous supply or available refrigeration. If refrigeration is available it requires a high capital input and energy consumption. Many donated pharmaceuticals have exceeded their expiration date and are often for ailments that are rare or not present in the recipient communities. When pharmaceuticals are used by rural populations, they are often given inappropriately (wrong dose and/or for wrong disease) because modern medical professionals are seldom present to correctly administer these medicines.

Approach

Locally available medicinal plants can contribute to health care needs and generate economic benefits for tropical rural communities. The WHO Traditional Medicine Programme and other research programmes have conducted research on tropical medicinal plants that have demonstrated safety and efficacy for the treatment of common tropical diseases including malaria and infections of the skin, lungs, and gastrointestinal tract.

Collaborative agreements may be established that enable tropical rural communities to harvest botanical medicines from their local ecosystems and sell them to northern or tropical urban areas as herbal medicines or for the extraction of pharmaceuticals. Twenty five per cent of modern medical drug prescriptions written in the United States are pharmaceuticals derived from plant species. In compliance with the Convention on Biological Diversity, these collaborative relationships between rural communities and research institutions can include agreements that will entitle communities to receive long term benefits if marketable pharmaceuticals or herbal medicines are derived from their botanical resources.

Impacts (Achieved Benefits)

Tropical rural traditional medicines under local indigenous control are more affordable, available, and sustainable forms of medicine, because they do not require the capital and energy inputs needed for pharmaceuticals. These botanical medicines are typically more therapeutic and safe because the medicine source is locally harvested and knowledge of its medicinal use is known by the local ethnolinguistic group. Use of ethnobotanical knowledge can also generate economic benefits that result in capital flow from urban to rural and North to South enabling local communities to use these resources to establish land demarcation, community-based medicinal plant reserves, traditional medicine hospitals, infrastructure support for a traditional healers' union, supplies for schools, and clean water systems. These health care and economic benefits derived from ethnobotanical knowledge generate incentives for tropical rural communities to conserve their biological and ethnolinguistic diversity.

Lessons Learned

Table 16.1 Tropical Botanical Medicines Versus Pharmaceuticals		
	BOTANICAL MEDICINES	**PHARMACEUTICALS**
Indigenous control	↑↑↑↑↑	
External control	↑	↑↑↑↑↑
Indigenous access/availability	↑↑↑↑↑	↑
Long-term sustainability for indigenous community	↑↑↑↑↑	↑
Ability of indigenous community to use medicine appropriately	↑↑↑↑↑	↑
Cost of medicine	↑	↑↑↑↑↑
Commodity produced & sold by indigenous community	↑↑↑↑↑	
Capital flow South to /North		↑↑↑↑
Capital flow North to South	↑↑↑↑↑	
Capital flow rural to urban		↑↑↑↑
Capital flow urban to rural	↑↑↑↑↑	

First, as demonstrated in Table 16.1, many barriers exist to safe, effective, affordable, and sustainable use of modern medicines in tropical rural communities. Tropical botanical medicines under local rural indigenous control are more affordable, available, therapeutically beneficial, and sustainable for these communities.

Second, modern medical health care programmes should work to complement rather than replace the local traditional botanical medical systems. The use of modern pharmaceuticals at the local rural level should be reserved to only treat those diseases not well managed by the local botanical medicines. When pharmaceuticals are used for specific ailments, there should be careful monitoring of the treatment by modern medical professionals to make sure the correct dose is given and the appropriate disease is being treated.

Third, the local traditional medical systems should be included as integral components of tropical health care programmes.

Fourth, use of ethnobotanical knowledge and harvesting of non-timber medicinal plant products from their local ecosystems can generate economic benefits for rural communities.

Fifth, the health and economic benefits of botanical medicines can establish incentives for rural tropical peoples to conserve their ecosystems, ethnobotanical knowledge, and languages.

Sixth, and perhaps most relevant to the climate change issue, all of these efforts to encourage the use of indigenous, ethnobotanical and local resources can help to preserve these areas, and thereby contribute to the mitigation of climate change.

Bibliography

Carlson, T. J., R. Cooper, S.R. King, and E.J. Rozhon, 1997: Modern Science and Traditional Healing. *Royal Society of Chemistry,* Special Publication 200 (Phytochemical Diversity), 84-95.

Farnsworth, N.R., O. Akerele, A.S.Bingel, D.D. Soejarto, Diaja, and Zhengang Guo, 1985. Medicinal Plants in Therapy. *Bulletin of the World Health Organization,* 63 (6), 965-81.

WHO, 1992: The Use of Essential Drugs: Model List of Essential Drugs: Fifth Report of the WHO Expert Committee, 1992. *World Health Organization Technical Report Series,* 825, 1-75.

Case Study 29
ROK-5 Mangrove Rice Variety in Sierra Leone
Otto Doering
Purdue University, West Lafayette,
Indiana 47907-1077, USA

Keywords: Sierra Leone, Africa, mangrove rice, improved yields, intra-national transfer

Summary
The development of a new mangrove rice variety in Africa is an important case study of technology development and diffusion that relates to the opportunity for agriculture to contribute to increased output of food while at the same time reducing the impact of agriculture on global climate change per unit of food produced. This development involved both international cooperation and a critical commitment of local resources to be successful and has proved itself to be an important contributor to increasing rice production in Sierra Leone with important potential contributions to similar areas in Africa.

Background
This project developed and extended a rice variety for mangrove rice production that increased yield per unit of land and per unit of inputs as well as adapting to changes in climate that had already occurred. The diminishing rain forest resulted in less rainfall and in a reduction of the fresh water available to mangrove rice systems. Thus, a shorter-season variety to capture available seasonal rainfall was a logical potential adaptation. Pest resistance was also added which reduced energy consumption from the use of pesticides. The incentive for this was a national and regional concern over adequate food production. The path used was the development of a new rice variety to improve yields and reduce the level of inputs required. Stakeholders included consumers and farmers in Sierra Leone, Sierra Leone agricultural researchers at Rokupr, and the West African Rice Research Development Association (WARDA). Critical to the choice of this path was the farming systems research which had been carried out years earlier. This identified the problem, provided an understanding of mangrove rice farming systems, and described the parameters under which new technology would be successful and successfully transferred. The barriers to technology transfer included a civil war, a collapsed transportation system, no capital for investment, and the lack of government resources to mount a large-scale technology transfer effort.

Approach
Much of the success of this effort hinged on the accident of a critical mass of researchers at the government rice research station in Rokupr Sierra Leone and the interest of WARDA in this effort. WARDA provided additional resources to the station at Rokupr to carry out the development of a new rice variety to meet the changed climate conditions and improve yields above those previously achieved. The development and diffusion took only years to achieve, not decades. There were no real commercial consid-

erations here. It is critically important to recognise that in agriculture a genetic improvement such as a new variety usually does not require an accompanying change or increase in the capital stock or equipment of a farming operation.

There were no royalties or commercial benefits involved. The Sierra Leone agricultural research establishment was able to demonstrate the value of their rice research effort to the food supply of the nation, and WARDA was able to demonstrate to their financial supporters their value in contributing to this new technology and its transfer. There was a German seed distribution project that helped with some seed distribution, but farmers themselves undertook most of the technology transfer to other farmers once the success of the new variety became apparent.

Impacts
The impact was an increase in rice output from mangrove rice production where climate change was beginning to reduce yields from this agricultural resource. This led to improved conditions for farmers and more rice available beyond farm-family subsistence for distribution and sale to local consumers. This case is replicable, but the key was the existence of local research and development capacity such as the Rokupr rice research station with a critical mass of competent staff. (Due to civil war, this institution is no longer a viable operation.) Its viability was also a result of the willingness of WARDA to give it the marginal resources to do the job that the government of Sierra Leone was unwilling or unable to supply.

Lessons Learned
For successful technology transfer there must be locally-based and supported institutions with a primary stake in the technology and in its successful transfer. There may also be instances where an international organisation can provide marginal resources to get the job done. However, the transfer of the technology by the users themselves depended upon the ease of local adaptability to the technology, the confidence farmers had in their local institution and its products, and effective demonstration of its success. The development of technology also depends upon a degree of civil order and investment by the local government in technology development and transfer. Where there are similar crops and farming systems in different countries there is also a natural opportunity for several countries to combine and pool resources for technology development and transfer.

The key lesson learned is that if there is an essential need and the technology is good enough agriculture technology can be transferred by farmers without new formal technology transfer institutions or efforts. A new effort may assist and speed up the transfer. However, in the case of a variety or practice that does not require new investment and changes in basic skills, farmers will adopt and pass this technology on to their neighbours rapidly.

The international community needs to recognise that parachuting in a new technology from the outside often does not work and faces especially severe problems in technology transfer. What is key is how international resources are spent. In cases like this, international resources must involve and contribute to the national programme to ensure appropriateness and transfer based on local ownership and local institutions that are already trusted in technology transfer.

Bibliography

Adesina, A., and M. Zinnah, 1993: Technology characteristics, farmers' perceptions and adoption decisions: A tobit model application in Sierra Leone. Agricultural Economics, 9, 297-311.

Agyen-Sampong, M., S. Anoop, K. Sandu, M.P. Prakah-Asante, C.A. Jones, S.F. Dixon, W.A.E. Fannah, W.A.E. Cole, and H. M. Bernard, 1986: A Guide to Better Mangrove Swamp Rice Cultivation in the WARDA Region. WARDA Regional Mangrove Swamp Rice Research Station, Rokupr, Sierra Leone.

Contact
Jess Lowenberg-DeBoer
Dept. of Ag. Economics
Purdue University
West Lafayette, IN 47907
Phone: (765) 494-4230
Fax: (765) 494-9176
Email: LOWENBERG-DEBOER@AGECON.PURDUE.EDU

Case Study 30
Use of Indigenous Technologies/Community-driven
Pathways in the South Pacific
James Aston
SPREP

Keywords: Tuvalu, coastal protection, community, national government, N⇒S

Summary
The transfer of conventional coastal protection technology followed the classical international aid route where the recipient developing country applied for assistance and often installed unsuitable coastal protection technologies, which eventually failed. An integrated system using a combination of indigenous and conventional technology has now been initiated through the efforts of the local community in Tuvalu. The significant feature of this case is that the benefits of an integrated approach to coastal protection were apparent to the community.

Background
In the past, protection of coastal assets in Tuvalu was not as high a priority as it is today since many of the structures were made of naturally abundant renewable materials, such as coconut leaf fibres. Traditional coastal protection systems focused on dealing with the effect of coastal erosion as a result of wind and wave action (Seluka, 1997). However, beach mining, land reclamation and imported structures have changed the nature and value of this technology. The increased intensity and frequency of tropical storms and cyclones predicted as a consequence of GHG emissions may also limit the utility of these approaches.

Approach
In the early 1980s, the Public Works Department of the Tuvalu government submitted a request for assistance from the European Union to install a coastal protection system for eroded areas of foreshore within Fongafale, Tuvalu. Conventional systems such as gabion baskets, concrete blocks and concrete walls were installed along the foreshore on the lagoon side. In the case of concrete walls and blocks, the aggregates had to be sourced by dredging in the lagoon, causing environmental problems.

The gabion baskets proved the most effective of these systems but only while the encasing remained intact. Nevertheless, local communities could not afford this type of coastal protection system because of the costs of maintenance, the limited supply of large stones, and difficulty in transporting materials (Seluka, 1997).

The Tuvaluans have detailed knowledge of seasonal, countervailing patterns in lagoon currents and tidal flows. Some of the older residents can allegedly predict the dispersal and concentration of sediments across the lagoon and barrier reef. Traditional responses to coastal erosion include coconut leaf walls, coconut fibre stone units, wooden walls and platforms, coral stone walls, fish traps, trees, relocation of housing and a combination of these methods (Seluka, 1997). These methods are low cost and they are effective in reducing wave energy and stabilising sediments along the foreshore.

In the last few years, through a combination of government assistance and community effort, an integrated system using a combination of indigenous and conventional technology has been initiated. The Agricultural Department assists this process by advising the community on the use of particular plant and tree species with multiple uses (e.g., timber for houses, canoes, and soil erosion control) (Seluka, 1997).

Impacts
The potential of traditional coastal protection systems, such as those practised in Tuvalu, are as yet not known for their effectiveness in addressing problems of coastal erosion. The impacts of the traditional methods are, however, expected to be beneficial as they involve local people in the seawall construction and design. Nevertheless, difficulties of *effective* design and *appropriate* materials compared to those of the popularly perceived and most easily-acquired materials (see Section 15.5.2) can likely prove to be a barrier to the dissemination of this technology. Furthermore, national or government level decision-making may have a limited impact on the more traditional practices. These impacts are significant for integrating indigenous knowledge and experience in the villages.

Lessons Learned
Community has a legitimate role to play in the formulation of project proposals and requests for assistance. However, community cannot be simply defined in advance by outsiders, but broad local participation in planning, design and decision making should be encouraged in line with various ecological, social and economic linkages.

Responses or adaptation strategies to coastal erosion in the South Pacific islands need to consider more than coastal engineering solutions. The nature of the problem and the exiting socio-economic, cultural and environmental situation must be understood.

The critical mass of research and development capacity to develop technology does not exist in most Pacific island nations. The more modern coastal protection technologies are constrained by agreements or high costs that governments willing to pay for the associated benefits. However, neither modern nor traditional technologies necessarily provide a long-term solution to coastal management problems. In other words, traditional coastal protection systems may be just as appropriate as the more modern systems. The integration of traditional and modern systems has the potential to improve the effectiveness of coastal protection systems (Seluka, 1997). This gives an opportunity for the community to focus on a diverse suite of coastal protection mechanisms within an integrated planning scheme.

References

Seluka, S., 1997: Traditional and Historical Responses to Coastal Erosion in
    Tuvalu. A paper prepared for the Australia/SPREP Vulnerability Initiative
    Workshop, Tarawa, Kiribati, 10-13 February, 1997. SPREP, Samoa.
Seluka, S., 1999: Personal communications.
Hviding, E., 1992: Upstream Development and Coastal Zone Conservation.
    Pacific Islands Perspectives on Holistic Planning. A Discussion Paper com-
    missioned by Greenpeace Pacific Campaign.

Contacts
Mr. S Seluka
Ministry of Natural Resources & Environment
Environment Office, Private Mail Bag
Vaiaku, Funafuti
Fax: 688) 20-826

Annex 16 - 1: Case Study by Chapter Cross Reference

The following lists case studies in relation to chapters.

CHAPTER 1: Case (s) 1, 15, 17, 19, 27. These cases reflect transfer processes that cover applications/technologies ranging from small scale (e.g. cookstoves) to large scale (e.g. heat recovery in the Chinese steel industry).

CHAPTER 2: Case studies illustrate inter-country technology transfer (forestry –case 25); South-South transfer (cookstoves – case 1); North-South (many including CFC-free refrigerators case 23); and North-East (biomass boilers in the Baltics – case 18). The cases also illustrate TT processes based on best practices (DSM – case 10, CPACC – Case 20), as well as cases meant for demonstration purposes of stimulating technology transfer (Case 18).

CHAPTER 3: The case studies bring out the experiences from international agreements such as the Montreal Protocol (cases 4, 17 and 23), and the AIJ process (case 12).

CHAPTER 4: Cases that bring out the importance of institutional arrangements as an enabling environment include: DSM in Thailand (case 10), TIFAC (case 27).

CHAPTER 5: A large number of case studies address the use of subsidies to promote market development. These include: wind in Inner Mongolia (case 3), Butane gas stove by TOTAL (case 7), renewables in Ladakh (case 14). Innovative private sector initiatives include: Mobil (case 13), Green Lights (case2), PV in Kenya (case 5), micro-hydro in Peru (case 25), GEF in India (case 22). Public-private partnerships are illustrated by cases 4, 17, 22 and 23

CHAPTER 7: Cases relevant are: cookstoves (case 1), Green Lights (case 2), Inner Mongolia Wind (case 3), PV in Kenya (case 5), Butane in Senegal (case 7), Ladakh renewables (case 14), CFC-free refrigerators in Thailand (case 23).

CHAPTER 8: Cases relevant: Brazilian Ethanol programme (case8), Transport in Uganda (case 11).

CHAPTER 9: Cases relevant fall in the category of technologies, new innovations and corporate leadership. Cases reflecting a technological initiative are Ecofrig in India (case 4), heat recovery in Chinese steel industry (case 15), ODS in Mexico (case 17), Biomass boilers in the Baltics (case 18), CFC-free refrigerators in Thailand (case 23). Mobil (case 13) reflects corporate leadership. The Bamboo reinforced cement case (case 9) represents the new innovations in technologies and an application from North-South transfer to South-South transfer.

CHAPTER 10: Cases cover decentralised and centralised applications. Decentralised applications include: Coal power plant (case 6), Inner Mongolia Wind (case 3), PV in Kenya (case 5). Centralised applications include micro-hydro in Peru (case 24).

CHAPTER 11: Cases available: mangrove rice variety (case 29)

CHAPTER 12: Cases relevant are: tree growers' cooperative (case 25), reduced-impact logging (case 26), and ethnomedicine (case 28).

CHAPTER 13: Cases available: biogas digester (case 19).

CHAPTER 14: Ethnomedicine (case 28)

CHAPTER 15: Cases include indigenous technologies (now case 16, earlier case 30), CPACC (case 20), concrete armouring (21), Coastal zone management in Cyprus (case 16).

A

Section Coordinators, Coordinating Lead Authors, Lead Authors, Contributing Authors, Review Editors and Expert Reviewers

SECTION I: FRAMEWORK FOR ANALYSIS: TECHNOLOGY TRANSFER TO ADDRESS CLIMATE CHANGE

Section Coordinators:

Bert Metz   Co-Chair IPCC Working Group III, The
                               Netherlands
Kilaparti Ramakrishna   Woods Hole Research Center, USA

Chapter 1: Managing Technological Change in Support of the Climate Change Convention: A Framework for Decision Making

Coordinating Lead Author:

Sergio C. Trindade   SE2T International, USA

Lead Authors:

Toufiq Siddiqi   Global Environment and Energy in the
                                21st Century (GEE-21), USA
Eric Martinot   The World Bank, USA

Contributing Authors:

Richard J.T. Klein   Potsdam Institute for Climate Impact
                                Research (PIK), Germany
Mary-Rene Dempsey
-Clifford   Department of Public Enterprise,
                                Government of Ireland, Ireland

Review Editors:

Li Liyan   State Planning Commission , China
Roberto Schaeffer   Federal University of Rio de Janeiro,
                                Brazil

Chapter 2: Trends in Technology Transfer

Coordinating Lead Author:

Mark Radka UNEP, France

Lead Authors:

Jacqueline Aloisi de
Larderel   UNEP, France
J.P. Abeeku Brew-
Hammond   KITE, Ghana
Xu Huaqing   Center for Energy Environment and
                                Climate Change Research, China

Contributing Authors:

Julia Benn   OECD, France
Woodrow W. Clark, Jr.   University, Denmark
Andrew Dearing   World Business Council for Sustainable
                                Development, Switzerland
Kevin Fay   Alliance for Responsible Atmospheric
                                Policy (ARAP), USA

Doug McKay   Shell International Ltd., United
                                Kingdom
Paul Metz   European Business Council for a
                                Sustainable Energy Future, The
                                Netherlands
K.P. Nyati   Confederation of Indian Industry, India
Luiz Pinguelli Rosa   Federal University of Rio de Janeiro,
                                Brazil

Review Editors:

Prodipto Ghosh   Asian Development Bank, The
                                Philippines
Ramon Pichs-Madruga   Centro de Investigaciones de Economía
                                Mundial (CIEM), Cuba

Chapter 3: International Agreements and Legal Structures

Coordinating Lead Authors

Michael Grubb   Imperial College, United Kingdom
Kilaparti Ramakrishna   Woods Hole Research Center, USA

Lead Authors:

Raekwon Chung   Ministry of Foreign Affairs and Trade ,
                                South Korea
Jan Corfee-Morlot   OECD, Paris
Michael Gollin   VENABLE, BAETJER, HOWARD &
                                CIVILETTI, LLP   , USA
Patricia Iturregui   Consejo Nacional de Medio Ambiente,
                                Peru
Jorge Leiva   Ozone Layer Protection Program,Chile
Atiq Rahman   Bangladesh Centre for Advanced
                                Studies (BCAS), Bangladesh
Terence Thorn   Enron Corp., USA
Gerardo Trueba
Gonzalez   Ministerio de Ciencia Tecnología y
                                Medio Ambiente (CITMA), Cuba

Review Editors:

Woodrow W. ClarkJr.   Aalborg University, Denmark
Richard Odingo   University of Nairobi, Kenya

Chapter 4: Enabling Environments for Technology Transfer

Coordinating Lead Authors:

Merylyn McKenzie
Hedger   University of Oxford, United Kingdom
Eric Martinot   The World Bank, USA
Tongroj Onchan   Thailand Environment Institute,
                                Thailand

Lead Authors:

Dilip Ahuja	Global Environment Facility Secretariat, USA
Weerawat Chantanakome	Thailand Research Fund (TRF), Thailand
Michael Grubb	Imperial College, United Kingdom
Joyeeta Gupta	Institute for Environmental Studies, The Netherlands
Thomas C. Heller	Stanford University, USA
Li Junfeng	Energy Research Institute, China
Mark Mansley	Claros Consulting, United Kingdom
Charles Mehl	Mae Fah Luang Foundation/Mekong Environment Institute, Thailand
Bhaskhar Natarajan	India-Canada Environment Facility (ICEF), India
Theodore Panayotou	Harvard University, USA
John Turkson	UNEP Collaborating Centre on Energy and Environment, Denmark
David Wallace	OECD, France

Contributing Author:

Richard J.T. Klein	Potsdam Institute for Climate Impact Research (PIK), Germany

Review Editor:

Karen R. Polenske	Massachusetts Institute for Technology, USA

Chapter 5: Financing and Partnerships for Technology Transfer

Coordinating Lead Authors:

Mark Mansley	Claros Consulting, United Kingdom
Eric Martinot	The World Bank, USA

Lead Authors:

Dilip Ahuja	Global Environment Facility Secretariat, USA
Weerawat Chantanakome	Thailand Research Fund (TRF), Thailand
Stephen Decanio	University of California, USA
Michael Grubb	Imperial College, United Kingdom
Joyeeta Gupta	Institute for Environmental Studies, The Netherlands
Li Junfeng	Energy Research Institute, China
Merylyne McKenzie Hedger	University of Oxford, United Kingdom
Bhaskhar Natarajan	India-Canada Environment Facility (ICEF), India
John Turkson	UNEP Collaborating Centre on Energy and Environment, Denmark
David Wallace	OECD, France

Contributing Authors:

Ron Benioff	National Renewable Energy Laboratory, USA

Ibrahim Abdel Gelil	Egyptian Environmental Affairs Agency (EEAA), Egypt

Review Editors:

Karen R. Polenske	Massachusetts Institute for Technology, USA

Section II: Technology Transfer: A Sectoral Analysis

Section Coordinators:

Jayant Sathaye	Lawrence Berkeley National Laboratory, USA
Youba Sokona	ENDA TM - Programme Energie, Senegal
Ogunlade Davidson	Co-chair of IPCC Working Group III, Sierra Leone
William Chandler	Battelle Pacific Northwest National Laboratory, USA

Chapter 6: Introduction to Section II

Coordinating Lead Author:

Jayant Sathaye	Lawrence Berkeley National Laboratory, USA

Lead Authors:

William Chandler	Battelle Pacific Northwest National Laboratory, USA
John Christensen	UNEP Collaborating Centre on Energy and Environment, Denmark
Ogunlade Davidson	Co-chair of IPCC Working Group III, Sierra Leone
Youba Sokona	ENDA TM - Programme Energie, Senegal

Chapter 7: Residential, Commercial, and Institutional Buildings Sector

Coordinating Lead Author:

John Millhone	U.S. Department of Energy , USA

Lead Authors:

Odon de Buen R.	Comision Nacional Para el Ahorro de Energía, Mexico
Gautam Dutt	Dirección de Uso Racional de Energía, Argentina
Tom Otiti	Makerere University, Uganda
Yuri Tabunschikov	ABOK, Russia
Tu Fengxiang	Energy Efficiency Association, China
Mark Zimmermann	EMPA-KWH, Switzerland

Contributing Authors:

Marilyn Brown	Oak Ridge National Laboratory, USA

Jim Crawford	The Trane Company, USA
Howard Geller	American Council for an Energy Efficient Economy, USA
Joe Huang	Lawrence Berkeley National Laboratory, USA
Martin Liddament	Oscar Faber Group, United Kingdom
Eric Martinot	The World Bank, USA
John Novak	Edison Electric Institute EEI, USA

Review Editor:

Ewaryst Hille	Polish Foundation for Energy Efficiency, Poland

Chapter 8: Transportation

Coordinating Lead Author:

Ogunlade Davidson	Co-chair of IPCC Working Group III, Sierra Leone

Lead Authors:

Oyuko Mbeche	University of Nairobi, Kenya
Laurie Michaelis	Mansfield College, United Kingdom
Lee Schipper	International Energy Agency, France
Suzana Kahn Ribeiro	Federal University of Rio de Janeiro, Brazil
Romeo Pacudan	Asian Institute of Technology, Thailand
Michael P. Walsh	International Consultant , USA
Yang Honghian	Institute of Comprehensive Transportation, China

Review Editor:

Lars Sjöstedt	Chalmers University of Technology, Sweden

Chapter 9: Industry

Coordinating Lead Authors:

Ernst Worrell	Lawrence Berkeley National Laboratory, USA
Mark Levine	Lawrence Berkeley National Laboratory, USA

Lead Authors:

Rene van Berkel	John Curtin International Institute , Australia
Zhou Fengqi	Energy Research Institute, China
Christoph Menke	University of Applied Sciences Trier, Germany
Roberto Schaeffer	Federal University of Rio de Janeiro, Brazil
Robert O. Williams	United Nations Industrial Development Organization, Austria

Contributing Authors:

Sanghoon Joo	Research Institute of Industrial Science & Technology, South Korea
Xiulian Hu	Energy Research Institute, China

Review Editors:

Prosanto Pal	Tata Energy Research Institute , India
Doug McKay	Shell International Ltd., United Kingdom

Chapter 10: Energy Supply

Coordinating Lead Author:

Jose Roberto Moreira	Biomass Users Network (BUN), Brazil

Lead Authors:

Jos Bruggink	Netherlands Energy Research Foundation, The Netherlands
Hisashi Ishitani	University of Tokyo, Japan
P.R. Shukla	Indian Institute of Management, India
Katia J. Simeonova	UNFCCC, Germany
John J. Wise	IPIECA, USA

Contributing Authors:

Youba Sokona	ENDA TM - Programme Energie, Senegal
Helena Li Chum	National Renewable Energy Laboratory, USA
Eric Martinot	The World Bank, USA

Review Editors:

R.S. Agarwal	Indian Institute of Technology, India
Steven Bernow	Tellus Institute , USA

Chapter 11: Agricultural Sector

Coordinating Lead Author:

Lin Erda	Agrometeorology Institute, China

Lead Authors:

Carlos Clemente Cerri	Centro de Energia Nuclear na Agricultura, Brazil
George Frisvold	University of Arizona, USA
Katsuyuki Minami	National Institute of Agro-Environmental Sciences, Japan
Otto Doering	Purdue University, USA
Neil Sampson	The Sampson Group Inc., USA
Paul Waggoner	The Connecticut Agricultural Experiment Station (CAES), USA

Contributing Authors:

Don Plucknet	Agricultural Research and Development International, USA

Heinz Ulrich Neue	UFZ-Centre for Environmental Research, Germany
Karim Makarim	Institut Pertanian Bogor, Indonesia
Kenneth Hubbard	University of Nebraska , USA
Li Jiusheng	Chinese Academy of Agricultural Sciences, China
Li Yu'e	Chinese Academy of Agricultural Sciences, China
Vernon Ruttan	University of Minnesota , USA

Review Editor:

Walter Baethgen	IFDC, Uruguay

Chapter 12: Forestry Sector

Coordinating Lead Author:

N.H. Ravindranath	Indian Institute of Science, India

Lead Authors:

Philip M. Fearnside	Instituto Nacional de Pesquisas da Amazonia, Brazil
Willy Makundi	Lawrence Berkeley National Laboratory, USA
Omar Masera	Departamento de Ecologia de los Recursos Naturales, Mexico
Robert Dixon	US Country Studies Program, USA

Contributing Authors:

Kenneth Andrasko	Environmental Protection Agency, USA
Neil Byron	CIFOR (Centre for International Forestry Research), Indonesia
Antony DiNicola	Counterpart International, USA
Nandita Mongia	Global Environment Facility GEF, USA
P. Sudha	Indian Institute of Science, India

Review Editor:

David Hall	King's College, United Kingdom

Chapter 13: Solid Waste Management and Wastewater Treatment

Coordinating Lead Author:

Dina Kruger	Atmospheric and Pollution Prevention Division, USA

Lead Authors:

Tom Beer	CSIRO Environmental Risk Network, Australia
Ron Wainberg	CRC for Waste Management and Pollution Control, Australia
Xu Huaqing	Center for Energy Environment and Climate Change Research, China

Review Editor:

Carlos Pereyra	ERM Argentina S.A., Argentina

Chapter 14: Human Health

Coordinating Lead Author:

Anthony McMichael	London School of Hygiene & Tropical Medicine, United Kingdom

Lead Authors:

Ulisses Confalonieri	School of Public Health - Flocruz, Brazil
Andrew Githeko	Kenya Medical Research Institute, Kenya
Pim Martens	International Centre for Integrative Studies, The Netherlands
Sari Kovats	London School of Hygiene & Tropical Medicine, United Kingdom
Jonathan Patz	Johns Hopkins School of Public Health, USA
Alistair Woodward	Wellington School of Medicine, New Zealand
Andrew Haines	Royal Free Hospital Medical School, United Kingdom
Akihiko Sasaki	National Institute of Public Health, Japan

Contributing Authors:

Gregg Greenough	Johns Hopkins School of Hygiene & Public Health, USA
Simon Hales	Wellington School of Medicine, New Zealand
Larry Kalkstein	University of Delaware, USA
Pete Kolsky	Associate Director WELL Resource Centre in Water, United Kingdom
Len Lerer	INSEAD, France
Rudi Slooff	retired, France
Kirk Smith	London School of Hygiene and Tropical Medicine , United Kingdom

Review Editor:

Tord Kjellstrom	The University of Auckland, New Zealand

Chapter 15: Coastal Adaptation Technologies

Coordinating Lead Author:

Richard J.T. Klein	Potsdam Institute for Climate Impact Research (PIK), Germany

Lead Authors:

James Aston	South Pacific Regional Environment Programme (SPREP), Western Samoa
Earle N. Buckley	Coastal Technology Services, USA
Michele Capobianco	Tecnomare S.p.A., Italy
Norimi Mizutani	Nagoya University, Japan
Robert J. Nicholls	Middlesex University, United Kingdom
Patrick D. Nunn	The University of the South Pacific, Fiji
Sachooda Ragoonaden	Meteorological Services, Mauritius

Contributing Authors:

Darius J. Bartlett	University College Cork, Ireland
James G. Boyd	Coastal Services Center, USA
Eugene Lecomte	Institute for Business and Home Safety, USA
Xenia I. Loizidou	Ministry of Communications and Works, Cyprus
Claudio R. Volonté	Organization of American States, USA

Review Editor:

Isabelle Niang-Diop	Université Cheikh Anta Diop, Senegal

Section III: Case Studies

Section Coordinators:

Stephen O. Andersen	U.S. Environmental Protection Agency, USA
Ajay Mathur	Tata Energy Research Institute, India

Chapter 16: Case Studies

Coordinating Lead Authors:

Sukumar Devotta	National Chemical Laboratory, India
Maithili Iyer	University of Delaware, USA
Daniel M. Kammen	University of California, USA

Lead Authors:

Saroja Asthana	National Chemical Laboratory, India
James Aston	South Pacific Regional Environment Programme (SPREP), Samoa
James Boyd	Coastal Services Center, USA
Thomas Carlson	University of California, USA
Alfonso Carrasco	ITDG, Peru
William Chandler	Battelle Pacific Northwest National Laboratory, USA
Jorge Corona	IMAAC, Mexico
Otto Doering	Purdue University, USA
Peter Du Pont	IIEC-Asia, Bangkok
Richard Duke	Princeton University, USA
Yuichi Fujimoto	JICOP, Japan
Yasuo Hosoya	Tokyo Electric Power Company, Japan
Hidefumi Imura	Institute of Environmental Systems, Japan

Arne Jacobson	University of California, USA
Suzana Kahn Ribeiro	Federal University of Rio de Janeiro, Brazil
Stefan Kessler	INFRAS, Switzerland
Xenia Loizidou	Ministry of Communications and Works, Cyprus
Stephen Magezi	AFREPEN, Uganda
Eric Martinot	The World Bank, USA
Indu Murthy	Indian Institute of Science, India
Teruo Okazaki	Nippon Steel Corporation, Japan
Gunter Pauli	ZERI Foundation, Colombia
N. H. Ravindranath	Indian Institute of Science, India
Steve Ryder	Princeton University, USA
Teodoro Sanchez	ITDG, Peru
P. Sudha	Indian Institute of Science, India
Sergey Surnin	ARENA-ECO, Ukraine
Jeanne Townend	ICF Kaiser, USA
Claudio Volonté	Organization of American States, USA
James Williams	University of California, USA
Li Yue	Chinese Academy of Agricultural Sciences, China

Review Editors:

Rajendra Shende	UNEP, France
Othmar Schwank	Infras , Switzerland

Additional Lead Authors of Summary for Policymakers and Technical Summary

Renate Christ	IPCC Secretariat, Switzerland
Ritu Kumar	Commonwealth Secretariat, United Kingdom
Jan-Willem Martens	IPCC Working Group III TSU
Sascha N.M. van Rooijen	IPCC Working Group III TSU

List of Expert Reviewers

Albania
Eglantina Demiraj Bruci
Academy of Sciences, Hydrometeorological Institute, IPCC
National Contact

Argentina
Gerardo M. E. Perillo
Instituto Argentino de Oceanografía

Australia
Australian Greenhouse Office, Climate Change International Team

Nick Harvey	The University of Adelaide, South Australia
Robert Kay	Coastal Management Branch, Department of Transport, Australia
Bill Biu-Pui Lim	Queensland University of Technology
David Shearman	Emeritus Professor, University of Adelaide
Helen Tope	Environment Protection Authority, Victoria, Australia
Rene Van Berkel	John Curtin International Institute, Curtin University of Technology
Tony Weir	Department of Industry, Science and Resources

Austria
Hans-Holger Rogner	Department of Nuclear Energy, Vienna, Austria
Gertraud Wollansky	Republic of Austria
Herwig Dürr	Ministry for Economic Affairs
Klaus Radunsky	Federal Environmental Agency, Austria

Belgium
Erwan Cotard	COGEN Europe/ICA, Brussels
Delia Dimitriu	European Environmental Management Association (EEMA), Brussels

Royal Government of Bhutan
Yeshey Penjor	National Environment Commission

Botswana
Aston C. Chipanshi	University of Botswana

Brazil
Ulisses E.C.Confalonieri	Oswaldo Cruz Foundation
Jose Goldemberg	Universidade de Sao Paulo
Suzana Kahn Ribeiro	Federal University of Rio de Janeiro
José Roberto Moreira	Biomass Users Network/CENBIO

Canada
Donald Forbes	Geological Survey of Canada
John Last	University of Ottawa, Ottawa, Canada
Rodney R. White	Institute for Environmental Studies, University of Toronto, Canada

China
Guanri Tan	Zhongshan University
Guo Liping	Agrometeorology Institute, Chinese Academy of Agricultural Sciences
Hu Xiulian	Energy Research Institute, State Planning Commission, China
Jiusheng Li	Agrometeorology Institute, Chinese Academy of Agricultural Sciences
Li Yu'e	Agrometeorology Institute, Chinese Academy of Agricultural Sciences
Lin Erda	Agrometeorology Institute, Chinese Academy of Agricultural Sciences
Zhou Fengqi	Energy Research Institute, State Planning Commission
Tu Fengxiang	China Building Energy Efficiency Association
Wen Kegang	China Meteorological Administration
Deying Xu	Chinese Academy of Forestry
Song Zhao	IPCC WG II TSU

Cuba
Marlena Castellanos	Ministry of Science, Technology and Environment of Cuba (CITMA)
Julio Torres	Ministry of Science, Technology and Environment of Cuba (CITMA)
Gerardo Trueb	Ministry of Science, Technology and Environment of Cuba (CITMA)

The Czech Republic
Jan Pretel	Czech Hydrometeorological Institute IPCC National Focial Point,

France
Shoichi Ando	OECD, Paris
Jean-Yves Caneill	Electricité de France
Wilfrid Legg	OECD, Paris
Gene McGlynn	OECD, Paris
Jan Coffee-Morlot	OECD, Paris
Philippe Crist	OECD, Paris

Hans-J. Neef,	International Energy Agency (IEA), France	
Reny Paris	OECD, Paris	
Grzegorz Peszko	OECD, Paris	

Germany

Michael Ernst	Federal Ministry for the Environment, Germany
Manfred Kleeman	Research Centre Juelich, STE
Christoph Menke	University of Applied Sciences Trier
H. W. Scharpenseel	Universitat Hamburg

Ghana

Abeeku Brew-Hammond UST, Kumasi, Ghana

India

Dr. P. Khanna	National Environmental Engineering Research Institute (NEERI)
Bhaskar Natarajan	India-Canada Environment Facility (ICEF)

Ireland

Darius Bartlett	University College Cork

Italy

Michele Capobianco	Tecnomare S.p.aA.

Japan

Ichiro Higashi	Socio-economic Research Center (SERC), Central Research Institute of Electric Power Industry (CRIEPI)
Yasuo Hosoya	The Tokyo Electric Power Company
Kazu Kato	Nagoya University
Katsuyuki Minami	National Institute of Agro-Environ. Sciences
Koji Nakui	Ministry of International Trade and Industry
Haruki Tsuchiya	Research Institute for Systems Technology
Hironobu Yokota	Environment Agency of Japan

Kenya

Emily Ojoo-Massawa	Ministry of Environmental Conservation
Paul N. Mbuthi	Ministry of Enerrgy
John K. Ng'ang'a	University of Nairobi
Joseph Kagia Njihia	Kenya Met Department
Phanuel Oballa	Kenya Forestry Research Institute
Richard S. Odingo	University of Nairobi
L. J. Ogallo University of Nairobi	

Republic of Korea

Raekwon Chung	Ministry of Foreign Affairs

The Netherlands

Luitzen Bijlsma	National Institute for Coastal and Marine Management (RIKZ)
Frank van der Meulen	National Institute for Coastal and Marine Management (RIKZ)
Paul E. Metz	European Business Council for a Sustainable Energy Future
Leo A. Meyer	Ministry of the Environment
Maarten Scheffers	National Institute for Coastal and Marine Management (RIKZ)
Sible Schone	World Wildlife Foundation-NL
Dr. R. Slooff	Retired
Rob Swart	TSU IP -WG-III/RIVM
Richard Tol	Vrije U ersiteit Amsterdam
Rene van Berkel	IVAM /ironmental Research, Unive / of Amsterdam

New Zealand

Peter L. Read	Mas University

Norway

Uno Abrahamsen	Institute of Technology, Oslo
Georg Borsting	Ministry of Environment
Oyvind Christophersen	Ministry of Environment
Harold Leffertstra	Norwegian State Pollution Control Authority
Petter Neksa	Stiftelsen for industriell og tcknisk forskning ved NTH
Knut H. Sorensen	The Norwegian Research Council

Peru

Alfonso Carrasco	Intermediate Technology (ITDG)
Mateo Casaverde	National Council for Science and Technology (CONCYTEC)
Dora Cortijo	Universidad Peruana de Ciencias
Benjamin Marticorena	
Rosa Morales	National Council for the Environment (CONAM)
Paul Remy	Executive Director, CONAM

The Philippines

Bernarditas Muller	Philippines Government Negotiator for UNFCCC

Poland

Anna Olecka	Polish UNFCCC Executive Bureau
Wojciech Suchorzewski	Warsaw University of Technology

Singapore

Dr. Beng Wah Ang — National University of Singapore

Sweden

Bengt Boström — Swedish National Energy Administration

Sven Jansson — ABB STAL AB (previously ABB Carbon AB)

Bengt Johansson — Swedish Environmental Protection Agency

Marianne Lilliesköld — Swedish Environmental Protection Agency

Lars Nilsson — Lund Institute of Technology

Ulf Silvander — Swedish Environmental Protection Agency

Håkan Staaf — Swedish Environmental Protection Agency

Sune Westermark — Swedish National Energy Administration

Switzerland

John L. Innes — Swiss Federal Institute for Forest, Snow and Landscape Research

Rudi Slooff — Retired from WHO

Mark Zimmermann — EMPA-KWH

Tanzania

B.S. Nyenzi — IPCC Focal Point, Tanzania

Thailand

Amara Pongsapich — Chulalongkorn University Social Research Institute

Uganda

S.A.K. Magezi — Department of Meteorology, Uganda

United Kingdom

David Banister — University College London

Andrew Barnett — Sussex Research Assoc. Ltd./UNDP/Shell

Simon Collings — ETSU

Louise Collins — Department of the Environment

Prof B.E.A. Fisher — University of Greenwich, UK

M. Jefferson — Global Energy & Environmental Consultants

Michael Jefferson — World Energy Council

Archie McCulloch — ICI Chemicals & Polymers Ltd.

Richard A W Shock — ETSU AEA

David Warrilow — Global Atmosphere Division Department of Environment, Transport and the Regions

United States of America

Margot Anderson — US Department of Agriculture

Ken Andrasko, — Environmental Protection Agency

Lee Beck — Environmental Protection Agency Office of Research and Development

Ron Benioff — National Renewable Energy Laboratory (NREL)

Lois E. Boland — Patent and Trademark Office, Office of Legislative and International Affairs

Bob Borgwardt — Environmental Protection Agency

Jean Brennan — US Department of State

Joel Brown — US Department of Agriculture

Laurence Campbell — US Department of Commerce, Office of Policy

Aaron Cohen — Health Effects Institute

Helena L. Chum — National Renewable Energy Laboratory

Anthony DiNicola — Independent

Meredydd Evans — Battelle Pacific Northwest National Laboratory

Cynthia Gage — Environmental Protection Agency, Office of Research and Development

Michael A. Gollin — Spencer & Frank

Vivien Gornitz — Columbia Univ. and Goddard Inst. for Space Studies

Thomas J. Grahame — US Department of Energy, Energy Office of Policy

Abe Haspel — US Department of Energy

Kate Hayes — US Department of Agriculture

Elmer Holt — US Department of Energy, Office of Energy Efficiency

James Hrubovcak — US Department of Agriculture International Petroleum Industry Environmental Conservation Association (IPIECA)

Leonard S. Bernstein

Abyd Karmali — ICF Kaiser International: ICF Consulting Group

Gregory Kats — US Department of Energy

Pat Keegan — National Renewable Energy Laboratory (NREL)

Ann Kinzig — Office of Science and Technology Policy,

Duncan Marsh — US Department of State

Edward J. McInerney — GE Appliances

Jeff Miotke — US Department of State

Goray Mookerjee — US Department of Energy, Office of Science and Technology Policy Analysis

Arvin R. Mosier — US Department of Agriculture/Agricultural Research Service

John Novak — Edison Electric Institute

Timothy Pieper — US Global Change Research Program

Karen R. Polenske — Massachusetts Institute of Technology

Raymond Prince — US Department of Energy

Don Pryor — National Oceanic and Atmospheric Administration (NOAA)

Gene I. Rochlin — University of California-Berkeley

Jim Rubin	Department of Justice/ENRD
Paul Schwengels	Environmental Protection Agency
Turman Semans	US Treasury Department
Toufiq Siddiqi	Global Environment and Energy in the 21st Century (GEE-21)
Susan Thorneloe	Environmental Protection Agency Office of Research and Development
Jeanne C. Townend	ICF Kaiser International
Don Trilling	US Department of Transportation
John Wise	Princeton, NJ
Ernst Worrell	Lawrence Berkeley National Laboratory, LBNL
Dana Younger	International Finance Corp (IFC)
Alice T. Zalik	Patent and Trademark Office, Office of Legislative and International Affairs

Venezuela

Jose I. Pons	Spray Quimica C.A.

United Nations Organizations
UN Economic Commission for Africa (UNECA)
Jacques Hamel

United Nations Environment Programme-HQ - Kenya
Megumi Seki

United Nations Environment Programme
Division of Technology, Industry and Economics, Paris
Jacqueline Aloisi De Larderel

UN Food and Agriculture Organization

Louise Fresco,	Chairperson of the "FAO ad hoc Interdepartmental Working Group on Climate in relation to Agriculture and Food Security"

UN FCCC
Katia Jeleva Simeonova

UN World Health Organization
C. Corvalan

B

Glossary of Terms

Adaptation
Adjustment in natural or human systems in response to actual or expected climatic stimuli or their effects, that moderates harm or exploits beneficial opportunities.

Adjustment
A modification of a technology or practice to reflect actual/local conditions. The term is used to denote one of the five basic stages of technology transfer as defined in this Report.

Adoption
Taking up and practicing or using as one's own.

Afforestation
Artificial establishment of forest stands on lands that previously have not supported forests for more than 50 years.

Agenda 21
Blueprint for sustainable development into the 21st Century. Its basis was agreed during the "Earth Summit" at Rio in 1992, and signed by 179 Heads of State and Government.

Agreement
An arrangement or contract as to a course of action; the language or instrument embodying such a contract. The term is used to denote one of the five basic stages of technology transfer as defined in this Report.

Annex I Countries
The countries listed in the Annex I to the UN Framework Convention on Climate Change that, as Parties, are committed to adopt national policies and take measures to mitigate climate change. Annex I Parties consist of the 24 original countries belonging to the Organization for Economic Cooperation and Development (OECD), the European Union and countries designated as Economies in Transition.

Annex II Countries
The countries listed in Annex II to the UN Framework Convention on Climate Change. These countries have a special obligation to help developing countries with financial and technological resources. They include the 24 original OECD members plus the European Union.

Anthropogenic
Derived from human activities.

Assessment
Determination of the importance, size, or other characteristics of a country or entity's technology
needs, market opportunities for technology transfer. The term is used to denote one of the five basic stages of technology transfer as defined in this Report.

Barriers
Factors that prevent or impede the transfer of technologies or practices.

Capacity Building
Increasing skilled personnel and technical and institutional capacity.

Carbon Dioxide
The greenhouse gas whose concentration is being most affected directly by human activities. CO_2 also serves as the reference to compare all other greenhouse gases (see carbon dioxide equivalents). The major source of CO_2 emissions is fossil fuel combustion. CO_2 emissions are also a product of forest clearing, biomass burning, and non-energy production processes such as cement production.

Carbon Sequestration
The biochemical process through which carbon in the atmosphere is absorbed by biomass such as trees, soils and crops.

Chlorofluorocarbons and Related Compounds
This family of anthropogenic compounds includes chlorofluorocarbons (CFCs), bromofluorcarbons (halons), methyl chloroform, carbon tetrachloride, methyl bromide, and hydrochlorofluorcarbons (HCFCs). These compounds have been shown to deplete stratospheric ozone, and therefore are typically referred to as ozone depleting substances. The most ozone-depleting of these compounds are being phased out under the Montreal Protocol.

Clean Development Mechanism
Possible agreements between Annex I and non-Annex I Parties as defined in Article 12 of the Kyoto Protocol to help reduce greenhouse gas emissions.

Climate Change
The term 'climate change' is sometimes used to refer to all forms of climatic inconsistency, but because the Earth's climate is never static, the term is more properly used to imply a significant change from one climatic condition to another. In some cases, 'climate change' has been used synonymously with the term, 'global warming'; scientists however, tend to use the term in the wider sense to also include natural changes in climate.

Climate Friendly
Actions conducive to mitigating climate change.

Climate Safe
Actions conducive to adapting to climate change.

Community Driven Pathways
Technology transfers initiated and lead by community organizations and entities with a high degree of collective decision-making.

Conference of the Parties
The Conference of the Parties (CoP) is the collection of nations which have ratified the Framework Convention on Climate Change, currently over 170 strong. The primary role

of the CoP is to keep the implementation of the Convention under review and to take the decisions necessary for the effective implementation of the Convention.

Cooperation
Association of persons or institutions for common benefit. In this Report generally used to denote the collaboration for purposes of spreading technology, practices or know-how.

Diffusion
The spread of technological equipment, practices or know-how from one area or group of people to others by contact.

Economic Potential
The portion of the technical potential for adapting to or mitigation climate change that could be achieved cost-effectively in the absence of market barriers. The achievement of the economic potential requires additional policies and measures to break down market barriers.

Economies In Transition
Countries in Central and East Europe and the Former Members of the Soviet Union that are in transition to a market economy.

Emission Standard
A level of emission that under law may not be exceeded.

Environmentally Sound Technologies
Technologies which protect the environment, are less polluting, use all resources in a more sustainable manner, recycle more of their wastes and products, and handle residual wastes in a more acceptable manner than the technologies for which they were substitutes and are compatible with nationally determined socio-economic, cultural and environmental priorities. In this Report environmentally sound technologies imply mitigation and adaptation technologies. The term includes hard and soft technologies.

Evaluation
Determination by careful appraisal of the profitability and quality of a technology transferred, the appropriateness of the technology to a local situation, and other results of a project or program. The term is used to denote one of the five basic stages of technology transfer as defined in this Report.

Externalities
By-products of activities that affect the well-being of people or damage the environment, where those impacts are not reflected in market prices. The costs (or benefits) associated with externalities do normally not enter standard cost accounting schemes.

Fluorocarbons
Carbon-fluorine compounds that often contain other elements such as hydrogen, chlorine, or bromine. Common fluorocarbons include chlorofluorocarbons and related compounds (also known as ozone depleting substances), hydrofluorocarbons (HFCs), and perfluorocarbons (PFCs).

Foreign Direct Investment
Capital invested for the purpose of acquiring a long term interest in an enterprise and of exerting a degree of influence on that enterprise's operations.

Framework Convention on Climate Change
The international treaty signed at the United Nations Conference on Environment and Development (UNCED) at Rio in 1992. The UN Framework Convention on Climate Change (FCCC) commits signatory countries to stabilize anthropogenic (i.e., human-induced) greenhouse gas concentrations to 'levels that would prevent dangerous anthropogenic interference with the climate system'. The FCCC also requires that all signatory Parties develop and update national inventories of anthropogenic emissions of all greenhouse gases not otherwise controlled by the Montreal Protocol.

Full-cost Pricing
The pricing of commercial goods - such as electric power - that would include in the final prices faced by the end user not only the private costs of inputs, but also the costs of the externalities created by their production and use.

General Circulation Model
A global, three-dimensional computer model of the climate system which can be used to simulate human-induced climate change. GCMs are highly complex and they represent the effects of such factors as reflective and absorptive properties of atmospheric water vapor, greenhouse gas concentrations, clouds, annual and daily solar heating, ocean temperatures and ice boundaries. The most recent GCMs include global representations of the atmosphere, oceans, and land surface.

Global Warming
An increase in the near surface temperature of the Earth. Global warming has occurred in the distant past as the result of natural influences, but the term is most often used to refer to the warming predicted to occur as a result of increased emissions of greenhouse gases.

Global Warming Potential
The index used to translate the level of emissions of various gases into a common measure in order to compare the relative radiative forcing of different gases without directly calculating the changes in atmospheric concentrations.

Government Driven Pathways
Technology transfers initiated by governments who play a leading role in the transfer as well.

Greenhouse Effect
The effect produced as greenhouse gases allow incoming solar radiation to pass through the Earth's atmosphere, but prevent most of the outgoing infra-red radiation from the surface and lower atmosphere from escaping into outer space.

Greenhouse Gas
Any gas that absorbs infra-red radiation in the atmosphere.

Greenhouse Gas Reduction Potential
Possible reductions in emissions of greenhouse gases (quantified in terms of absolute reductions or in percentages of baseline emissions) that can be achieved through the use of technologies and measures.

Halocarbons
Chemicals (belonging to the ODSs) consisting of carbon, sometimes hydrogen, and either chlorine, fluorine bromine or iodine.

Halons
Chemical compounds (belonging to the ODSs) developed from hydrocarbons by replacing atoms of hydrogen with atoms of halogens, such as fluorine, chlorine, or bromine. CFCs are halons.

Hydrocarbons
Substances containing only hydrogen and carbon. Fossil fuels are made up of hydrocarbons. Some hydrocarbon compounds are major air pollutants.

Hydrofluorocarbons
Chemicals (along with perfluorocarbons) introduced as alternatives to ozone depleting substances in serving many industrial, commercial, and personal needs. HFCs are emitted as byproducts of industrial processes and are also used in manufacturing. They do not significantly deplete the stratospheric ozone layer, but they are powerful greenhouse gases.

Implementation
Carrying out; giving practical effect to and ensuring of actual fulfillment of technology transfer processes by concrete measures. The term is used to denote one of the five basic stages of technology transfer as defined in this Report.

Infra-red Radiation
The heat energy that is emitted from all solids, liquids, and gases. In the context of the greenhouse issue, the term refers to the heat energy emitted by the Earth's surface and its atmosphere. Greenhouse gases strongly absorb this radiation in the Earth's atmosphere, and reradiate some back towards the surface, creating the greenhouse effect.

Innovation
The introduction of something new; a new idea, method, or device.

Intellectual Property Right
An intangible asset, such as a copyright or patent.

Intergovernmental Panel on Climate Change
The Intergovernmental Panel on Climate Change (IPCC) was established jointly by the United Nations Environment

Programme and the World Meteorological Organization in 1988. The purpose of the IPCC is to assess information in the scientific and technical literature related to all significant components of the issue of climate change. The IPCC draws upon hundreds of the world's expert scientists as authors and thousands as expert reviewers. Leading experts on climate change and environmental, social, and economic sciences from all over the world have helped the IPCC to prepare periodic assessments of the scientific underpinnings for understanding global climate change, its consequences and potential adaptation and mitigation responses.

Invention
A device, contrivance, or process originated after study and experiment.

Joint Implementation
Possible agreements between Annex I Parties as defined in Article 6 of the Kyoto Protocol to help reduce greenhouse gas emissions. Some aspects of this approach are being tested as Activities Implemented Jointly (AIJ).

Joint Venture
An alliance between two or more entities to carry out a single business enterprise by pooling property, money, equipment, and/or know-how.

Kyoto Protocol
Protocol belonging to the UN Framework Convention on Climate Change agreed in Kyoto (Japan) in December 1997.

Market Barriers
Conditions which prevent or impede the transfer of cost-effective technologies or practices which could adapt to or mitigate climate change.

Market Potential
The portion of the economic potential for adapting to or mitigation climate change that could be achieved under existing market conditions, assuming no new policies and measures.

Market-Based Incentives
Measures intended to directly change relative prices and overcome market barriers.

Methane
A hydrocarbon that is a greenhouse gas produced through anaerobic (without oxygen) decomposition of waste in landfills, animal digestion, decomposition of animal wastes, production and distribution of natural gas and oil, coal production, and incomplete fossil fuel combustion.

Mitigation
An anthropogenic intervention to reduce the emissions or enhance the sinks of greenhouse gases.

Montreal Protocol
Protocol to the Vienna Convention on Substances that Deplete the Ozone Layer originally signed in 1987 and amended in 1990 and 1992.

Multilateral Fund
Facility under the Montreal Protocol established in 1990 to assist the developing countries, operating under Article 5 (1), to achieve compliance with the Protocol.

Nitrogen Oxides
Gases consisting of one atom of nitrogen and varying numbers of oxygen atoms. Nitrogen oxides are produced in the emissions of vehicle exhausts and from power stations. In the atmosphere, nitrogen oxides can contribute to formation of photochemical ozone (smog) which is a greenhouse gas.

Nitrous Oxide
A powerful greenhouse gas emitted through soil cultivation practices, especially the use of commercial and organic fertilizers, fossil fuel combustion, nitric acid production, and biomass burning.

No Regret Measures
Measures whose benefits - such as improved performance or reduced emissions of local/regional pollutants, but excluding the benefits of climate change mitigation - equal or exceed their costs. They are sometimes known as "measures worth doing anyway."

Official Aid
Aid meeting the test of Official Development Assistance but directed to countries on Part II of the OECD Development Assistance Committee list of aid recipients.

Official Development Assistance
Flows to developing countries and multilateral institutions provided by official agencies, including state and local governments, each transaction of which 1) is administered with the promotion of the economic development and welfare of developing countries as its main objective and 2) is concessional in character, with a grant element of at least 25 percent.

Ozone
Ozone consists of three atoms of oxygen bonded together in contrast to normal atmospheric oxygen which consists of two atoms of oxygen. Ozone is an important greenhouse gas found in both the stratosphere (about 90% of the total atmospheric loading) and the troposphere (about 10%). Ozone has other effects beyond acting as a greenhouse gas. In the stratosphere, ozone provides a protective layer shielding the Earth from ultraviolet radiation and subsequent harmful health effect on humans and the environment. In the troposphere, oxygen atoms in ozone combine with other chemicals and gases (oxidization) to cause smog.

Particulates
Tiny pieces of solid or liquid matter, such as soot, dust, fumes, or mist.

Partnership
Close cooperation between parties having specified and joint rights and responsibilities.

Party
A state (or regional economic integration organization) that ratifies or accedes to an international agreement. In this Report it mostly refers to the UNFCCC.

Pathway
A route through which technology transfer takes place, composed of a combination of processes and involving different stakeholders.

Perfluorocarbons
A group of human-made chemicals composed of carbon and fluorine only: CF_4 and C_2F_6. These chemicals, specifically CF_4 and C_2F_6, (along with hydrofluorocarbons) were introduced as alternatives to the ozone depleting substances. They are powerful greenhouse gases.

Policies
Procedures developed and implemented by government(s) regarding the goal of adapting to or mitigating climate change through the use of technologies and measures.

Private Sector Driven Pathways
Technology transfers primarily between commercially oriented private-sector entities.

Radiative Forcing
A change in the balance between incoming solar radiation and outgoing infra-red radiation. Without any radiative forcing, solar radiation coming to the Earth would continue to be approximately equal to the infra-red radiation emitted from the Earth. The addition of greenhouse gases traps and increased fraction of the infra-red radiation, reradiating it back toward the surface and creating a warming influence (i.e., positive radiative forcing because incoming solar radiation will exceed outgoing infra-red radiation).

Reforestation
Forest stands established artificially on lands that have supported forests within the last 50 years.

Regulatory Measures
Rules or codes enacted by governments that mandate product specifications or process performance characteristics.

Replication
The final stage of the five basic stages of technology transfer (assesment, agreement, implementation, evaluation and adjustment, replication), defined in this Report as a combination of actions that lead to the deployment of a given technology, once transferred, to meet a new demand elsewhere.

Sinks
A process that removes greenhouse gases from the atmosphere, either by destroying them through chemical processes or storing them in some other form. Carbon dioxide is often stored in ocean water, plants, or soils, from where it can be released at a later time.

Solar Radiation
Energy from the Sun. Also referred to as short-wave radiation. Of importance to the climate system, solar radiation includes ultra-violet radiation, visible radiation, and infra-red radiation.

Stakeholders
Actors such as governments, private sector entities, financial institutions, NGOs and research/education institutions, involved in a technology transfer process.

Standards/Performance Criteria
Set of rules or codes mandating or defining product performance (grades, dimensions, characteristics, test methods, rules for use).

Subsidiary Body for Scientific and Technological Advice
UNFCCC committee serving as the link between the information and assessments provided by expert sources (such as the IPCC) on the one hand, and the policy-oriented needs of the CoP on the other.

Sulfur Dioxide
A compound composed of one sulfur and two oxygen atoms. Sulfur dioxide emitted into the atmosphere through natural and anthropogenic processes is changed in a complex series of chemical reactions in the atmosphere to sulfate aerosols. These aerosols result in negative radiative forcing (i.e., tending to cool the Earth's surface).

Sulfur Hexafluoride
A very powerful greenhouse gas, composed of one sulfur and six fluorine atoms, used primarily in electrical transmission and distribution systems.

Sustainable Development
Development that ensures that the use of resources and the environment today does not damage prospects for their use by future generations.

Technical Potential
The amount by which it is possible to address climate change by using a technology or practice in all applications in which it could technically be adopted, without consideration of its costs or practical feasibility.

Technology
A piece of equipment, technique, practical knowledge or skills for performing a particular activity.

Technology Transfer
The broad set of processes covering the exchange of knowledge, money and goods amongst different stakeholders that lead to the spreading of technology for adapting to or mitigating climate change. In an attempt to use the broadest and most inclusive concept possible, the Report uses the word 'transfer' to encompass both diffusion of technologies and cooperation across and within countries.

Volatile Organic Compound
The term used to describe the organic gases and vapours that are present in the air. They are believed to be involved in ground-level ozone formation. Some VOCs are toxic air pollutants.

Voluntary Measures
Measures to address climate change that are adopted by firms or other actors in the absence of government mandates.

Vulnerability
The degree to which a systems is susceptible to, and unable to cope with, injury damage or harm.

C

Acronyms and Abbreviations

ADB	Asian Development Bank	**ERUs**	Emission Reduction Credits
ADT	Average Daily Traffic	**ESCOs**	Energy Service Companies
AfDB	African Development Bank	**EST**	Environmentally Sound Technologies
AHWGTTC	Ad Hoc Working Group on Technology Transfer and Co-Operation	**EU**	European Union
		FAO	UN Food and Agriculture Organization
AIC	Appreciation Influence Control	**FCCC**	UN Framework Convention on Climate Change
AIJ	Activities Implemented Jointly		
ALGAS	Asia Least Cost Greenhouse Gas Abatement Strategy	**FDI**	Foreign Direct Investment
		FPEI	Foreign Portfolio Equity Investment
APEC	Asian Pacific Economic Council	**FSC**	Forest Stewardship Council
ASEAN	Association of South East Asian Nations	**GA**	Gender Analysis
		GATT	General Agreement on Tariffs and Trade
ASEP	ASEAN Environment Programme		
ATCS	Advanced Traffic Control Systems	**GCM**	General Circulation Model
AWDN	Automated Weather Data Network	**GDP**	Gross Domestic Product
BA	Beneficiary Assessment	**GEF**	Global Environment Facility
BOO	Build-Own-Operate	**GEMS**	Global Environmental Monitoring System
BOOT	Build-Own-Operate-Transfer		
BOT	Build-Operate-Transfer	**GHG**	Greenhouse Gas
CBD	Convention on Biological Diversity	**GIS**	Geographical Information Systems
CDM	Clean Development Mechanism	**GL**	Green Lights
CEC	Commission of the European Communities	**GLOSS**	Global Sea Level Observing System
		GM	General Motors
CEIT	Countries with economies in transition	**GNP**	Gross National Product
		GREENTIE	Global Remedy for the Environment and Energy Use - Information Exchange
CEO	Chief Executive Officer		
CEPA	Commonwealth Environment Protection Agency	**GS/OAS**	General Secretariat/Organization of American States
CER	Certified Emission Reductions		
CFLs	Compact Fluorescent Lamps	**GTZ**	Gesellschaft für Technische Zusammenarbeit
CGIAR	Consultative Group on International Agricultural Research		
		GWh	Gigawatt-hours
CIF	Carbon Investment Fund	**GWP**	Global Warming Potential
CIMEP	Community Involvement in Management of Environmental Pollution	**HT**	Hydrocarbon Technology
		IACCA	Interagency Committee for the Climate Agenda
CNG	Compressed Natural Gas	**IARCs**	International Agricultural Research Centers
CoP	Conference of the Parties		
CPACC	Caribbean Planning for Adaptation to Climate Change	**IBRD**	International Bank for Reconstruction and Development
CRS	Creditor Reporting System	**ICLEI**	International Council for Local Environmental Initiative
CS	Climate Safe		
CTI	Climate Technology Initiative	**ICRAF**	International Centre for Research in Agroforestry
CTIC	Conservation Technology Information Center		
		IDA	International Development Association
DAC	Development Assistance Committee		
DC	Developing Countries	**IDB**	Inter-American Development Bank
DSM	Demand Side Management	**IEA**	International Energy Agency
EBRD	European Bank for Reconstruction and Development	**IET**	International Emissions Trading
		IFC	International Finance Corporation
ECAs	Export Credit Agencies	**ILO**	International Labour Organisation
EIE	Environmental Impact Evaluations	**IPCC**	Intergovernmental Panel on Climate Change
EIT	Economies In Transition		
EMS	Environmental Management System	**IPMVP**	International Performance Measurement and Verification Protocol
ERMA	Environmental Risk Management Authority		

Rs	Intellectual Property Rights	RTD	Research and Technology Development
R	Internal Rate of Return		
EE	Information Service on Energy Efficiency	SADC	Southern African Development Community
O	International Standards Organization	SAR	Second Assessment Report
	Joint Implementation	SAVE	Specific Actions for Vigorous Energy Efficiency
	Joint Ventures		
AP	Long-Range Energy Alternative Planning model	SBSTA	Subsidiary Body for Scientific and Technological Advice
NG	Liquefied Natural Gas	SFM	Sustainable Forest Management
PG	Liquefied Petroleum Gas	SHS	Solar Home System
&A	Mergers & Acquisitions	SIS	Small Island States
DB	Multilateral Development Bank	SME	Small and Medium Sized Enterprise
EA	Multilateral Environmental Agreements	SMPs	Shoreline Management Plans
		SPM	Summary for Policymakers
F	Multilateral Fund	TAP	Technology Assessment Panel
NC	Multinational Corporations	TCAPP	Technology Cooperation Agreement Pilot Project (US)
OP	Meeting of the Parties		
OU	Memorandum of Understanding	TEPs	Tradable Emission Permits
T	Motorized Transport	TNC	Transnational Corporations
VA	Manufacturing Value Added	TP	Technical Paper
W	Mega Watt	TP	Technology Partnership
ARSs	National Agricultural Research Systems	TRIP	Trade Related Aspects of Intellectual Property
FFO	Non Fossil Fuel Obligation (UK)	TT	Technology Transfer
FIP	National Flood Insurance Program	UN	United Nations
GO	Non Governmental Organization	UNCED	United Nations Conference on Environment and Development
C	Newly Industrialising Countries		
CUs	National Implementation Co-ordinating Units	UNCHS-Habitat	United Nations Centre for Human Settlements
MT	Non-Motorized Transport	UNCSD	United Nations Commission for Sustainable Development
OUs	National Ozone Units		
SI	National Systems of Innovation	UNCTAD	United Nations Conference on Trade and Development
A	Official Aid		
AS	Organization of American States	UNDESA	United Nations Department of Social and Economic Affairs
DA	Official Development Assistance		
DP	Ozone Depletion Potential	UNDP	United Nations Development Programme
DS	Ozone Depleting Substances		
ECD	Organization for Economic Cooperation and Development	UNECOSOC	United Nations Economic and Social Council
PEC	Organization of Petroleum Exporting Countries	UNEP	United Nations Environment Programme
AOs	Proportional Abatement Obligations	UNFCCC	United Nations Framework Convention on Climate Change
EI	Portfolio Equity Investment		
LP	Poland Efficient Lighting Project	UNGASS	United Nations General Assembly Special Session
CCAP	Pacific Islands Climate Change Assistance Programme	UNIDO	United Nations Industrial Development Organization
Ms	Policies, Programs and Measures		
RA	Participatory Rural Appraisal	UNRISD	United Nations Research Institute for Social Development
V	Photovoltaic		
&D	Research and Development	VA	Voluntary Agreement
D&D	Research, Development and Demonstration	WBCSD	World Business Council for Sustainable Development
EEF	Renewable Energy and Energy Efficiency Projects	WCHE	World Commission on Health and the Environment
ETs	Renewable Energy Technologies	WCMC	World Conservation and Monitoring Centre
IL	Reduced Impact Logging		

WEC	World Energy Council
WHO	World Health Organization
WIPO	World Intellectual Property Organization
WMO	World Meteorological Organization
Wp	Watt Peak
WRI	World Resources Institute
WTO	World Trade Organization

Chemical Symbols

C_2F_6	Perfluoroethane
CF_4	Perfluoromethane
CFC	Chlorofluorocarbon
CH_4	Methane
CO	Carbon monoxide
CO_2	Carbon dioxide
HCFC	Hydrochlorofluorocarbon
HFC	Hydrofluorocarbon
N_2O	Nitrous oxide
NMVOCs	Nonmethane volatile organic compounds
NO_x	Nitrogen oxides
O_3	Ozone
Pb	Lead
PFC	Perfluorocarbon
PM	Particulate matter
SF_6	Sulfur hexafluoride
SO_2	Sulfur dioxide
SO_x	Sulfur oxides
VOC	Volatile organic compound

D

List of Major IPCC Reports

Climate Change—The IPCC Scientific Assessment
The 1990 Report of the IPCC Scientific Assessment Working Group (also in Chinese, French, Russian, and Spanish)

Climate Change—The IPCC Impacts Assessment
The 1990 Report of the IPCC Impacts Assessment Working Group (also in Chinese, French, Russian, and Spanish)

Climate Change—The IPCC Response Strategies
The 1990 Report of the IPCC Response Strategies Working Group (also in Chinese, French, Russian, and Spanish)

Emissions Scenarios
Prepared for the IPCC Response Strategies Working Group, 1990

Assessment of the Vulnerability of Coastal Areas to Sea Level Rise–A Common Methodology
1991 (also in Arabic and French)

Climate Change 1992—The Supplementary Report to the IPCC Scientific Assessment
The 1992 Report of the IPCC Scientific Assessment Working Group

Climate Change 1992—The Supplementary Report to the IPCC Impacts Assessment
The 1992 Report of the IPCC Impacts Assessment Working Group

Climate Change: The IPCC 1990 and 1992 Assessments
IPCC First Assessment Report Overview and Policymaker Summaries, and 1992 IPCC Supplement

Global Climate Change and the Rising Challenge of the Sea
Coastal Zone Management Subgroup of the IPCC Response Strategies Working Group, 1992

Report of the IPCC Country Studies Workshop
1992

Preliminary Guidelines for Assessing Impacts of Climate Change
1992

IPCC Guidelines for National Greenhouse Gas Inventories
Three volumes, 1994 (also in French, Russian, and Spanish)

IPCC Technical Guidelines for Assessing Climate Change Impacts and Adaptations
1995 (also in Arabic, Chinese, French, Russian, and Spanish)

Climate Change 1994—Radiative Forcing of Climate Change and an Evaluation of the IPCC IS92 Emission Scenarios
1995

Climate Change 1995—The Science of Climate Change – Contribution of Working Group I to the Second Assessment Report
1996

Climate Change 1995—Impacts, Adaptations, and Mitigation of Climate Change: Scientific-Technical Analyses – Contribution of Working Group II to the Second Assessment Report
1996

Climate Change 1995—Economic and Social Dimensions of Climate Change – Contribution of Working Group III to the Second Assessment Report
1996

Climate Change 1995—IPCC Second Assessment Synthesis of Scientific-Technical Information Relevant to Interpreting Article 2 of the UN Framework Convention on Climate Change
1996 (also in Arabic, Chinese, French, Russian, and Spanish)

Technologies, Policies, and Measures for Mitigating Climate Change – IPCC Technical Paper I
1996 (also in French and Spanish)

An Introduction to Simple Climate Models used in the IPCC Second Assessment Report – IPCC Technical Paper II
1997 (also in French and Spanish)

Stabilization of Atmospheric Greenhouse Gases: Physical, Biological and Socio-economic Implications – IPCC Technical Paper III
1997 (also in French and Spanish)

Implications of Proposed CO_2 Emissions Limitations – IPCC Technical Paper IV
1997 (also in French and Spanish)

The Regional Impacts of Climate Change: An Assessment of Vulnerability – IPCC Special Report
1998

Aviation and the Global Atmosphere - IPCC Special Report
1999

Land Use, Land Use Change, and Forestry - IPCC Special Report
2000

Emissions Scenarios - IPCC Special Report
2000

ENQUIRIES: IPCC Secretariat, c/o World Meteorological Organization, 7 bis, Avenue de la Paix, Case Postale 2300, 1211 Geneva 2, Switzerland